English-Chinese Mathematics Encyclopedia

Algebra, Geometry, and Pre-Calculus
Second Edition

英汉数学全书

代数、几何与微积分初步

（第2版）

[美] 梁家睿（Jerry Conrad Leung）
吴晓云 ◎著

清华大学出版社
北京

内 容 简 介

本书以美国数学教材的知识体系为框架，按章节顺序系统梳理数学核心概念，提取数学词条，对相关知识点展开介绍。对于每个数学词条，介绍其相应的英汉词义、音标、定义、数学符号、性质、证明、相关短语与应用、例题与答案及章节小结。全书正文分为3部分：第1部分讲述代数初步、代数1、代数2(第1～11章)；第2部分讲述几何(第12～21章)；第3部分讲述微积分初步(第22～34章)。本书附录部分补充高等数学延伸内容，涵盖微积分、概率与统计、线性代数，以及所涉及的符号、用法及英语表达方式，并附有希腊字母表、英制单位等常用数学内容。

为了鼓励读者的探索精神，帮助读者学透基础知识点，作者在编撰本书的过程中，采用"探究式"演绎方式，通过完整呈现定理证明过程("Why"而非仅"What")，帮助读者理解数学逻辑。本书注重读者双语能力的培养，规范数学英语表达，破除语言障碍，助力国际学术阅读与交流。

图书在版编目（CIP）数据

英汉数学全书：代数、几何与微积分初步/（美）梁家睿（Jerry Conrad Leung），吴晓云著. -- 2版.
北京：清华大学出版社，2025.6（2025.8重印）. -- ISBN 978-7-302-68633-0

Ⅰ. O13

中国国家版本馆CIP数据核字第2025AL3883号

责任编辑： 曾　珊
封面设计： 李召霞
责任校对： 李建庄
责任印制： 宋　林

出版发行： 清华大学出版社

网　　址： https://www.tup.com.cn，https://www.wqxuetang.com
地　　址： 北京清华大学学研大厦A座　　**邮　　编：** 100084
社 总 机： 010-83470000　　**邮　　购：** 010-62786544
投稿与读者服务： 010-62776969，c-service@tup.tsinghua.edu.cn
质量反馈： 010-62772015，zhiliang@tup.tsinghua.edu.cn
课件下载： https://www.tup.com.cn，010-83470236

印 装 者： 三河市铭诚印务有限公司
经　　销： 全国新华书店
开　　本： 203mm×260mm　　**印　张：** 42.5　　**字　　数：** 1195千字
版　　次： 2020年1月第1版　　2025年6月第2版　　**印　　次：** 2025年8月第2次印刷
印　　数： 1001～2500
定　　价： 258.00元

产品编号：105502-01

学习攻略

STUDY GUIDE

亲爱的读者朋友,从您拿到这本书开始,崭新的国际数学学习之旅就开始了。您一定不要把它束之高阁,相信您的努力会带给您意想不到的成长惊喜。如下几点提示有助于您高效使用本书。

(1) 本书是一本覆盖了初、高中六个年级全部知识体系的英汉双语数学工具书,内容极为丰富,像一本小型“辞海”,既可以作为工具书查阅词条,又适合循序渐进地自学研读。书中用英汉双语阐释了从基础到高级的数学概念及它们背后的性质、定理和证明,这是数学学习过程中最为基础、最容易被人忽视的部分。大家往往对一些学习“秘籍”更感兴趣,殊不知对于数学的学习,“深抠”基础概念才是最符合数学学习本质的。所以,在这里提醒读者朋友,务必重视对书中基础数学概念的认真学习。

(2) 为了帮助大家对书中重要章节的深度理解和学习,我们录制了内容丰富的精选真题配套视频,带领读者综合运用不同知识点、运用数学性质解题,以期融会贯通。

(3) 为了更好地服务本书大家庭的每一位成员,加强彼此的沟通与切磋,请您关注“11 次方国际数学”微信公众号并加入读者交流群,您将不定期获得各种免费福利,包括国际数学标准化考试、国际数学考前/赛前串讲、定期答疑、国际数学竞赛/夏校/夏令营资讯及备考攻略等。让我们随时陪伴并见证您的进步与成长。

(4) 为了方便自学者学习,给大家提供一个大致内容分布列表,供大家参考。

《英汉数学全书——代数、几何与微积分初步》(第 2 版)内容分布列表

章节 / 年级	Algebra 代数(第 1～11 章)											Geometry 几何(第 12～21 章)										Pre-Calculus 微积分初步(第 22～34 章)												
	1	2	3	4	5	6	7	8	9	10	11	12	13	14	15	16	17	18	19	20	21	22	23	24	25	26	27	28	29	30	31	32	33	34
5-6 年级	●	●	●	●																														
7 年级	●	●				★			●		★	●		●	●	★	●		●		●													
8 年级	●	★	★	●			●																											
9 年级							●					★	★	★	●	★	★	★	★	★	★													
10 年级		●	●	●	★		●	★	★	★															★	●	★	★			★	★		
11 年级							●															★	★	★	★	★	★	★	★	★	★	★	★	★

●表示需要读者完成该章节的部分内容;★表示需要读者完成该章节的所有内容。

祝大家学习之旅愉快!加油!

第2版序

FOREWORD

数学是现代文明崛起的基石。从计算月球的周期到启动控制日常互联网流量的算法，数学一直是人类进化过程中的密切参与者。它超越了文化和语言界限的领域，提供了一个解决问题和创新的通用语言。在这个全球化浪潮澎湃、科技进步飞速的时代，我们站在了一个历史上前所未有的交叉路口。这个时代不仅要求我们拥有扎实的科学和技术知识，还要求我们理解这些领域是如何在全球范围内交织互动的。《英汉数学全书——代数、几何与微积分初步》(第 2 版)的出现标志着国际化中学数学教育中的一个关键时刻，它搭建了跨越语言隔阂的桥梁，并展示了数学作为全球教育连接器的力量。

在我们这个日益紧密相连的世界里，每一个学生，无论起点如何，都应该有机会获取在全球舞台竞争与合作所需的工具和知识。通过采用开创性的双语教学方法，本书不仅涵盖了基础和高级的数学概念，还为中国学生搭建了跨越语言和文化障碍的桥梁。不仅为他们的数学技能和英语能力带来双重提升，还为他们从初中到研究生阶段的国际教育学习习惯和职业生涯奠定坚实的基础。

本书强调自学的重要性，每一章都严格按照国际教育体系和教科书标准进行编排。它超越了简单的知识性讨论，倡导积极、主动地与材料互动，鼓励学生独立探索概念。这对于培养能够适应多样化教育环境并自信面对未来挑战的学习者至关重要。详细的双语解释和全面的示例不仅能够加深读者对数学概念的理解，还能够培养读者用英语学术表达这些概念的能力。此外，作者在国际数学竞赛教育方面多年的经验也使得书中对各种知识点的应用和拓展处理得恰到好处。

在人工智能技术迅猛发展的今天，本书特别强调其结构和学习模式的前瞻性设计，以适应教育改革方面不断演进的需求。通过整合传统教学方法与现代科技工具，旨在为数学教育带来根本性的变革。引入探索式学习、主动思考和个性化学习计划，极大地丰富了教学和学习体验。这种方法不仅关乎效率和深度，更符合当代学生的需求，帮助他们为面对未来社会的挑战做好准备。

本书由梁家睿[美]和吴晓云撰写，他们多年以来致力于深耕国际数学素质教育研究，孜孜以求的学术精神令我钦佩。他们的教学成果，国际数学素质教育理论——“探究式国际数学素质教育与专业数学英语教育的交叉融合”，即将“CLIL”教学法(一种科学地把“学科素养”和“语言能力”相结合的教学法)和“探究式”教学法融合应用，为本书的高质量和实用性提供了坚实的学术理论保障。希望这本书不仅能传授数学知识，更能激发读者对于数学的热爱，引导他们在充满挑战和机遇的新时代中不断探索、学习和成长。

鉴于本书读者都是数学爱好者，借由这个机会带您了解数学的前世与今生，分享一些我对数学的理解与感悟，希望可以激发大家学习数学的兴趣，加深大家对数学的理解与探索。

古代计算的回声

数学的起源如同其发展所涉及的文明一样多样和复杂。从美索不达米亚的肥沃新月地带到古印度繁忙的市场,每一种文化都带来了独特的贡献,共同塑造数学科学的基本性质。这种探索不仅仅是列举成就,更重要的是理解这些古代洞见如何编织进现代数学的结构中。

在古代美索不达米亚,苏美尔人和巴比伦人创造了一套实用且精细的数学体系,足以探索星辰和测量季节。他们发展了楔形文字(这是最早的书写系统之一,最初用于记录贸易账目),不久这种文字便被用来记录天文观察和数学计算。人们引入的六十进制系统——基于数字 60——是他们智慧的证明,这一系统影响了我们今天对时、分和秒的计量方式。

这些发展的重要性是深远的,为管理者管理城市和建筑师建造如通天塔那样的巨大结构提供了工具。同时,尼罗河畔的埃及数学专注于几何学,用于建造金字塔和测量尼罗河泛滥后的土地。他们对几何学的实际应用确保了土地能在每个泛滥季节后重新准确地分配给其合法主人,维持了社会秩序和正义。

在印度次大陆,数学家思考空虚的哲学含义,导致了零的数字表示。十进制的引入进一步简化了计算,使数学更加容易获取。这些工具不仅是抽象概念,它们促进了贸易、发展了天文学,并且是管理庞大帝国的不可或缺的一部分。

在中国,数学的方法深深植根于国家的需要——从计算收入、税收到预测被视为皇帝神圣统治征兆的天体事件。中国人开发了一种早期的代数形式(算盘的先驱)以提高计算效率。在中国漫长历史中出现了浩如烟海的辉煌作品,如《九章算术》,囊括了解决问题的传统方法,这不仅影响了我们自己的文明,也为与其他文化的数学对话提供了基础。

数学的故事也是连接和交换的故事。随着丝绸之路等贸易路线的开放,知识交流的渠道也随之开放。数学思想从印度和中国流向伊斯兰世界,再从那里流向欧洲,激发了文艺复兴时期的科学觉醒。西亚、中亚的学者们翻译并发展了前人的工作,增加了如代数和三角学等创新,这些都深刻影响了欧洲的数学思想。他们的工作为现代数学科学奠定了基础。尽管我们有着不同的起源,但对于知识的追求和应用是人类共同的目标。

丝绸之路与学术交流

丝绸之路以其在丝绸、香料和黄金等珍贵货物贸易中的作用而闻名,它也是知识流动的关键动脉。这一贸易路线网络连接了不同的文明,促进了跨文化交流,这对塑造这些社会的知识风貌起到关键作用。数学概念通过这个网络的传播不仅丰富了当地知识,还催化了数学学科的发展。

在巴格达、撒马尔罕和君士坦丁堡的繁忙市场中,商人和学者不仅交换商品,还交换观点。阿拔斯哈里发推动的智力活动使伊斯兰世界成为从希腊、印度和中国收集来的数学知识的大熔炉。在这里,印度的数字系统转化为我们现在认识的阿拉伯数字。

印度数学家的作品,特别是如零和十进制的概念,被翻译成阿拉伯语。这些想法具有革命性,它们减少了算术的复杂性,并为代数和微积分开辟了新的可能性。Al-Khwarizmi 等伊斯兰学者将这些观念发扬光大,发展出后来成为代数基础的算法,这个词本身就源自阿拉伯语 al-jabr,意为“破碎部分的重聚”。伊斯兰学者的智力努力并未局限于中东。通过十字军东征、西班牙的再征服,以及学者们在穆斯林和基督教土地上的旅行,这些数学概念传入欧洲。在巴格达的智慧之屋以及后来的托莱多和西西里岛等地的翻译运动,见证了将阿拉伯文本翻译成拉丁文,将复杂的数学概念带入欧洲学术界的过程。

这些知识的转移在文艺复兴期间是至关重要的,这发生在欧洲一个深刻的智力觉醒时期。如斐波那契(Fibonacci)等数学家研究了阿拉伯数字系统,并通过其作品将这些方法引入欧洲。他对印度的阿拉伯数字系统的推广,取代了罗马数字,彻底革新了会计和银行业,显著影响了欧洲的商业和经济。

后来的欧洲数学家,如笛卡儿发展的笛卡儿坐标系统,受到了东方基础文本的影响。几何和代数概念的综合构成了现代科学的大部分基础,从物理学到经济学,都展示了这些早期交流的持久影响。

丝绸之路促进了前所未有的文明对话，这种对话在范围和影响上是空前的，为塑造人类未来的科学进步奠定了基础。人类的历史反复向我们强调了智力交流和观念综合在推进人类理解方面的重要性，并强调了丝绸之路在创造跨越地理和文化界限的共享智慧遗产中的作用。

语言是数学思维的透镜

语言和数学都是基本的沟通形式，但它们的相互作用经常被忽视。数学概念的理解、教学和学习方式在不同的语言环境中有显著的差异。我们需要考查语言的内在属性如何影响数学认知和教学法。每种语言都有其独特的结构、语法和词汇，这些可以促进或阻碍人们对数学概念的理解。例如，一些语言有特定的数字系统，使计数和算术运算更直观。以中文为例，其简单、规则的数字命名系统使得一些儿童比具有更不规则系统的语言（如英语或法语）的儿童能够更早地掌握基础算术。此外，数学话语所需的精确性和清晰度往往反映了语言特性。德语和俄语所具有的"格"和"性"的变化，提供了表达数学关系的细微方式，这可能影响学习者处理变量和函数问题的方式。

数学术语不仅仅是定义的集合，而是反映了它起源的文化。诸如"算法"和"代数"等术语源自阿拉伯语，指向伊斯兰学者对数学的历史贡献。这些术语如何融入不同的语言可能会影响这些概念的教学和理解方式。例如，"零"的概念通过阿拉伯文本的翻译引入西方世界，在欧洲的数学发展上产生了深远的影响，从根本上改变了数学景观。

在双语环境中教授数学具有独特的挑战和机会。它需要在语言能力和数学理解之间谨慎平衡。对于在非英语母语环境中学习数学的学生，认知资源被分割为处理语言和解决数学问题，这可能会阻碍学习。然而，这种双语挑战也可以增强认知灵活性，让学生能从多个角度解决数学问题。有效的双语数学教育策略必须考虑这些语言复杂性。整合语言学习与数学训练的项目可以为学生提供更深入、更细致的数学概念理解，为他们在全球化的学术和职业环境中做好准备。

语言塑造了我们对世界的理解，数学也不例外。通过认识和利用数学教育中的语言多样性，教育者可以开发更有效的教学策略，不仅传递数学知识，还可以跨越文化和语言的鸿沟。这种方法丰富了学习体验，促进了更深入、更包容的数学理解，这对于在我们日益相互联系的世界中导航至关重要。

通过双语教育搭建心灵之桥

数学的双语教育既带来挑战，也提供了非凡的机会。通过多种语言教授和学习数学，可以帮助读者加深理解，鼓励认知灵活性，并为学生准备多元文化的世界。探讨双语数学教育的策略和益处，突出成功的模式，并讨论用多种语言教授数学的教学含义，是有非常深远影响的好事。

双语者通常表现出更大的认知灵活性、更好的解决问题能力和增强的执行功能。在数学的语境中，这些技能转化为更容易操纵抽象概念的能力，以及从多样化的视角来解决问题的能力。双语学生在识别模式和在复杂的问题解决场景中协商意义方面通常表现得更好，这些技能在高级数学思维中至关重要。

尽管益处显著，双语数学教育的挑战也不可小觑。主要问题包括以下几点。

① 语言能力障碍：学生可能在第二语言的数学词汇上挣扎，这可能阻碍理解并限制他们清晰表达数学推理的能力。

② 数学习惯的文化差异：数字格式、符号和解题方法在不同文化中可能有所不同，对于不熟悉不同语言环境中使用习惯的学生可能会感到困惑。

③ 教师准备情况：有效的双语数学教育要求教师不仅精通两种语言，还要擅长将语言学习与数学指导相结合。

为应对这些挑战，可以采用以下几种有效的策略。

① 整合语言和数学学习：同时教授数学词汇、概念与语言指导有助于确保学生同时发展语言技能和数学理解。

② 使用视觉和操作材料：视觉辅助、图表和实践材料可以帮助弥合语言差距，并提供抽象概念的具体示例。

③ 文化响应教学：理解并将文化背景融入数学课程可以使来自不同背景的学生的学习更具相关性和吸引力。

④ 教师专业发展：为教师提供双语教育最佳实践的持续培训可以提高教学质量和学生成果。

世界各地有许多的例子展示了双语数学教育的潜力和多样性。在加拿大，法英双语沉浸式项目成功地展示了学生如何在两种语言中都达到高水平的熟练度，同时在数学和其他学术科目上表现出色。新加坡的双语教育政策要求学生用英语学习数学，同时在母语中也非常熟练，这有助于提升该国在国际数学评估中的表现。在欧洲，边境地区的学校经常使用多种语言教授数学，为学生准备跨境的经济和学术机会。数学的双语教育不仅提供了改善教育成果的途径，还增进了跨文化理解和合作。通过有效地应对挑战并利用双语的认知和文化益处，受教育者可以在数学学习和全球交流中开发新的潜力。本书强调采用创新的教育实践的重要性，这些实践把语言多样性作为丰富学习体验和为连接世界的学生做准备的有力工具。

通过人工智能革新教育

人工智能(Artificial Intelligence，AI)正在以戏剧性的方式重塑教育领域，特别是在数学教育方面。随着 AI 技术的不断成熟，它们为个性化教育、自动化行政任务和实时反馈评估提供了前所未有的机会。教育者和学习者正在积极探索如何将 AI 技术融入数学教育，并充分利用其优势，应对随之而来的挑战和潜在的发展前景。

在个性化教育方面，AI 系统可以分析大量来自学生互动的数据，从而创建定制化的学习路径。自适应学习平台利用算法评估学生的现有知识水平，预测难点，并相应地调整教学材料。这种个性化的方法能够满足不同学习风格和速度的需求，确保所有学生都能发挥出他们的全部潜力。例如，MATHia 等程序使用复杂的 AI 技术来适应每个学生的学习进程，提供个性化的提示、反馈和解释，类似于一对一的辅导。而 ASSISTments 平台则将 AI 与传统家庭作业结合，提供即时反馈和详细分析，帮助教师识别学生的难点，并进行有针对性的干预。

AI 还可以改变数学概念的呈现和理解方式。虚拟现实(VR)和增强现实(AR)应用允许学生将复杂的数学理论可视化，并以更具体和互动的方式探索抽象概念。虚拟数学实验室利用 VR 创建沉浸式学习体验，帮助学生通过与虚拟环境中的对象互动来理解空间和几何概念。而 AR 工具可以将数据和图形表示覆盖在现实世界的对象上，帮助学生实时可视化数学问题和解决方案。

尽管 AI 在教育中拥有巨大的潜力，但也带来了数据隐私和安全、偏见和公正性、对技术的依赖等重大挑战。在教育中使用 AI 需要收集大量个人数据，这引发了隐私和数据保护的担忧。如果设计不当，AI 系统可能会持续放大偏见。确保 AI 工具的公平性和无偏性在多元化的教育环境中至关重要。此外，对 AI 工具的过度依赖可能会导致基本数学技能和批判性思维能力的下降。教育者必须平衡技术使用与传统教学方法，以培养全面的理解能力。

展望未来，AI 有望通过促进更为复杂的个性化学习形式，并通过数据分析提供对学习过程的更深入见解，进一步革新数学教育。未来的发展可能包括 AI 驱动的课程开发、全球课堂连接等。在这样的趋势中，AI 可以协助设计适应不断发展的教育标准和学生需求的最佳课程。AI 还可以实现更互联的全球教育系统，使来自全世界不同地区的学生通过智能平台共同合作和学习。

总体来说，人工智能作为一种强大的工具，代表了教育实践(尤其是在数学领域)的进化。它提供了使学习更加可获得、可参与而有效的潜力。然而,利用这一潜力需要仔细考虑伦理问题,承诺维护教育公平,并专注于发展教育中的技术和人力资源。随着我们继续将 AI 整合到教育环境中,保持这些因素的重要性至关重要,以确保技术成为学习者追求知识的桥梁而非障碍。

在中学数学教育中强调数学建模

数学建模是现代数学教育工具箱中的一项重要技能,它是理论数学和实际应用之间的桥梁。在中学教育中引入数学建模可以揭示复杂概念,展示数学在日常生活和专业领域中的相关性。数学建模涉及使用数学概念创建现实世界现象的抽象表示。这一过程帮助学生将数学知识应用于解决实际问题,理解复杂系统,并基于数据分析作出明智的决策。通过数学建模开发的技能对于在许多 STEM 领域及重视批判性思维和分析技能的学科中取得成功至关重要。

通过在数学建模实践中的互动,学生学会了有方法地处理问题、识别变量、构建假设,并根据现实世界数据测试他们的模型。这种实践经验鼓励学生更深入地理解数学原理,并发展一套在学术和专业设置中都极具价值的多功能解决问题的技能。将数学建模纳入中学课程需要周密的计划和资源,包括开发介绍数学建模基础的模块,如问题的制定、算法的使用和结果的解释;创建机会让学生将数学建模应用于科学、技术、经济和社会研究中,增强他们数学技能的相关性和实用性;利用计算工具和软件来促进数学模型的构建和分析,使复杂的计算和模拟对学生可接触。

为了有效教授数学建模,教育者自身必须精通建模的理论和实践。专业发展项目应提供数学建模技术培训,增强教师构建和使用模型的理解,使教师具有将建模任务整合到教学中、有效评估学生作业和鼓励批判性思维的技能。

尽管益处显著,但数学建模在中学教育中的整合面临几个挑战,包括学校可能缺乏进行复杂建模所需的必要技术和软件,传统的测试方法可能无法有效捕捉数学建模理解和技能的深度,以及引入新教学方法需要克服教育者和管理者之间的惯性和怀疑。数学建模不仅是一个教育工具——它是现代数学教育的关键组成部分,为学生准备应对现实世界的复杂性。通过在中学教育中强调数学建模,我们可以为年轻人提供理解世界、创新和塑造世界的技能。我们强调培养一种动态且适用的数学教育的重要性,这种教育能使学生有信心和创造力地应对现实世界问题。

尾声:超越数字——呼吁普遍理解

随着我们从古代起源到现代应用的数学领域旅程的结束,我们得出一个基本的认识:数学不仅仅是数字和方程式的集合。它是一种超越文化和地理界限的语言,一种在不同民族之间搭建桥梁的工具,一种解决世界上一些最紧迫挑战的手段。这个尾声呼吁更广泛地欣赏数学——它不仅作为一门学术学科,更作为促进全球理解和合作的重要工具。

在人类文明史各篇章中,我们看到了数学思想如何在大陆之间无缝流动、影响各种文化和被各种文化所影响。这种跨文化交流突出了数学作为一种能够促进不同文明之间对话和理解的普遍语言。面对需要国际合作的全球挑战时,数学的普遍语言可以作为沟通和解决问题的共同基础。

当今相互联系的世界中,全球公民意识比以往任何时候都更为重要。这需要理解地方行动如何产生全球影响,以及我们如何共同努力实现共同目标。数学教育在这方面发挥着关键作用,通过培养批判性思维、解决问题的能力和对逻辑推理的欣赏,培养这些技能对于负责任的公民身份和明智的决策至关重要,使个体能够深思熟虑和建设性地参与全球复杂问题。可持续性的挑战——环境、经济和社会——本质上是数学问题。无论是通过模拟气候变化、优化资源分配,还是分析经济稳定性,数学都是理解和解决这些问题的核心。通过加强数学素养,我们为未来一代提供了创新和实施造福全人类的可持续解决方案所需的工具。这个尾声是对全世界的教育者、政策制定者、学生和公民的一次行动呼吁,要求他们认识并利用数学的力量。我们必须倡导为所有人提供坚实的数学教育,鼓励整合数学与其他领域的跨学科方法,并培养一个大力欣赏并利用数学思维的环境,跨越社会各个部门。

数学的旅程是对知识和理解永无止境的追求,这不仅丰富了我们的思维,还有潜力加深我们的共同人性。在未来的探索中,让我们将数学不仅视为一门学科,更看作一种思考方式和生活的策略。让我们利用它来解开宇宙的奥秘,解决现实世界的问题,构建一个更加连接且和谐的世界。数学的影响宽广而深远,通过继续探索和重视其普遍语言,我们为一个更明亮、更包容的未来铺平了道路。

段凯耀
芝加哥大学"杰出教育者"奖得主
2024 AP 全球年会发言人
国际化中学数学教育者
资深数学建模导师
2025 年 2 月

第1版序

FOREWORD

亲爱的各位中国朋友以及数学爱好者，大家好！我是 Daniel Flegler，曾是一名从 1965 年开始在美国新泽西州沃尔德威克高中执教且教龄达 26 年的数学教师兼教学督导，同时也是成立于 1977 年的美国数学大联盟(Math League)的联合创始人之一。怀着对数学教育的深深热爱和责任，我在数学教育领域持续耕耘了几十年，今天很荣幸有机会把一些数学教学理念和心得分享给中国的朋友们。数学不分国界，热爱学生的心也没有国界，希望我的分享可以让大家更加了解数学以及美国基础数学教育现状，并带来数学教育方面的思考和启迪，希望大家都能爱上数学。

大家都知道学习数学很重要，相信几乎每所学校都开设了数学课程并非常重视这一学科。可是有谁能够准确地回答：我们为什么要学习数学？在数学课上，学生经常好奇地问："我们为什么要学习这些知识点？毕业后还会用到吗？"一些教师则会幽默地回答："如果你不学习这些知识点，我就失业了。"当然，真正的答案远比这个复杂得多。

我们知道，对于将来从事科学工作的学生，学习数学至关重要，因为数学是一切科学的基础。虽然成为科学家的学生只占一小部分，但是他们却是揭开宇宙奥秘，开启新发明、新创造，引领世界前进最重要的群体。如果没有数学知识作为基础，那么近 200 年的伟大科学发明也就不复存在了。现代生活中我们享受到的一切技术成果，包括互联网、飞机、抗生素、核电、新型武器、汽车、手机、电灯泡、计算机、电视、冰箱、空调、相机等不胜枚举。这些科学发明背后的基础都是数学。现在我们还无法估计 200 年后会出现怎样的新技术，但是有一点可以肯定的是：没有深厚的数学功底，科学家是不可能发明出新技术的。社会进步取决于学生从小就开始培养的数学基础。

现实中大部分学生却不会从事科学工作，那么是不是数学对于那些不打算从事科学研究的学生就不重要了呢？对每个人来说，具备分析和推理思维(数学思维的一部分)能力都是极其重要的，因为可以帮助我们找到解决问题的路径。分析思维能力指的是对世界的批判性思考的能力。推理思维能力指的是我们应对周遭复杂事物时逻辑思考的能力。尽管其他学科也可以训练思维能力，但是没有哪一门学科比数学对思维的训练更有效。所以"我们为什么要学习数学"的答案就是：训练我们的思维能力。只有具备了一定的思维能力，我们才能真正理解和吸收报纸、杂志以及互联网等各种社交媒体的几乎覆盖科学、政治、社会、心理等所有领域的信息。虽然看似把完成一道有关火车速度的数学问题与帮助我们解决生活中的难题联系在一起比较牵强附会，但是找到问题所在、厘清已知和未知条件，按步骤解决数学问题的逻辑思考方式对于解决我们在生活中遇到的问题就非常重要了。这种几乎由方程、数字和字母构成的数学语言和数学规则不需要任何翻译就可以被全世界不同国家、不同种族、不同信仰的人所理解。数学知识就是如此神奇且对每个人都意义重大。

很多中国朋友对美国基础数学教育很好奇,在这里我也简单谈一谈这个话题。美国传统基础数学教育的教学目标是通过州治考试和标准化考试,让大部分初、高中生顺利毕业。在教学生涯中,我观察到一个非常普遍的教学场景:教师讲一些新的知识点,在黑板上做 1~2 道例题,然后布置课堂练习。多数情况下,教师仅仅向学生讲授一种机械的解题套路,而不是带领学生探究解题方法背后的数学依据。对很多重要的数学定理只讲结果,不讲证明过程,这种教学方法导致学生把所学知识点孤立起来理解,无法融会贯通,在遇到较为复杂的数学问题时束手无策。这样的课堂教学模式对于那些有天赋的学生和数学爱好者来说是远远不够的,根本无法满足他们对数学领域深入探究的好奇心。某种意义上讲,也会造成智力资源的巨大浪费。我一直认为,发现问题和解决问题才是提升教学水平的方式。

为了弥补以上基础数学教育的不足,我和我的搭档 Steven Conrad 先生(我们都是总统里根颁发的"杰出数学教学奖"的获得者),于 1977 年一起创立了美国数学大联盟,至今已经走过了 40 多个年头。作为美国最悠久的数学竞赛之一,我们一贯秉承的宗旨是通过让学生解答精挑细选的数学问题,帮助他们建立对数学浓厚的兴趣和信心。我们的这一教学理念点燃了数以万计的美国青少年学习数学的兴趣和热情,使成千上万名优秀学生脱颖而出。我直接教过的学生中至少有 15 人在美国著名科学赛事——英特尔科学人才选拔赛中获得数学论文荣誉奖。鼎盛时期的美国数学大联盟每年参赛者高达一百多万人,极大地促进了美国数学教育的发展。无论这些孩子是否最终沿着数学之路继续走下去,他们都得益于青少年时代参加过的数学竞赛和夏令营的经历,他们取得的数学成就将影响他们的一生,我对此深感欣慰和自豪。

为了帮助学生上更高的台阶,我经常鼓励他们参加一些类似英特尔科学人才选拔赛的数学课外活动并完成数学论文;鼓励他们申请一些含金量高但挑战极大的数学夏令营,如俄亥俄州立大学的 Ross 数学夏令营、波士顿大学的 PROMYS 数学夏令营等。这些数学夏令营可以帮助学生真正领略到数学的精髓,让他们深入接触到重要的数学定理证明;鼓励学生把学过的基础知识运用到极致;鼓励他们不厌其烦地研究一题多解,让他们找到学习数学的快乐;极大地激发学生的创新思维,坚定他们在数学领域深入探究下去的信念。这就是美国数学竞赛和夏校在美国乃至全世界非常受欢迎的原因所在。美国高中生非常重视利用假期参加各种夏令营或课外活动提升自己的学术兴趣和竞争力。竞赛的成绩和参加夏校的履历在申请美国一流大学时也是非常重要的参考指标。想被知名数学夏令营录取必须通过严格的筛选:学生必须用创新思维在一定时间内完成所有试题,要求用流利的英文,逻辑清晰地阐述试题的每一个步骤;同时还需要独立完成有深刻见解的数学论文。

除此之外,我还鼓励学生积极参加 AMC、AIME 等系列数学竞赛,并建议他们在备赛的过程中多练习一些非常规的、挑战性强的历届竞赛真题,这样才能真正提升解题能力。数学竞赛或夏令营中学到的解题技巧对大学期间的高等数学学习也会有很大的帮助。希望更多优秀的中国初中、高中学生积极尝试一些有挑战性的国际数学竞赛或夏校,开阔视野、提升思维创新能力和英文表达能力,激发巨大潜能,客观评价自己在同龄数学佼佼者中的数学水平。相信与优秀的同龄人同台竞技收获的丰富阅历在人生中更为可贵。

知道《英汉数学全书——代数、几何与微积分初步》即将面世,我感到兴奋无比!《英汉数学全书——代数、几何与微积分初步》是一本值得珍藏的经典数学工具书,是我的得意门生 Jerry C. Leung 和他的中国主创研发团队共同努力的丰硕成果,是敬献给中国年轻学子的珍贵礼物。Jerry C. Leung 作为一名我教过的最优秀的学生,多年以来他对数学研究孜孜以求,对数学竞赛命题的研发也一直秉承精益求精的治学态度,创作了很多优秀的作品。我发现大多数中国学生在参加国际数学竞赛或夏令营时都面临诸多的语言障碍,影响着他们的理解和表达,影响着他们的竞赛成绩。我一贯允许参加数学大联盟竞赛的中国学生携带双语数学词典,可是他们很难在中国书店找到一本专业、系统而全面的英汉双语数学词典。如今《英汉数学全书——代数、几何与微积分初步》的问世终于填补了中国英汉双语数学工具书的空白，相信中国学生在这本工具书的帮助下，在未来的数学学习过程中会如虎添翼。《英汉数学全书——代数、几何与微积分初步》吸取美国探究性数学课堂的精髓，无论从内容、结构设计还是知识体系的完整性上，都是一本品质精良的匠心佳作。它能够帮助同学们扫清数学学习过程中遇到的种种障碍，提升数学学习效率和数学专业语言综合水平。我在此衷心祝福广大的中国莘莘学子能够更加热爱数学和探索数学，不仅可以在激烈的数学竞赛和夏校申请中脱颖而出，还能如愿申请到自己的梦想大学。奇妙的数学王国值得意气风发的你们去探索，一旦奠定了深厚的数学基础，你们会得到超乎想象的收获。我这个美国老教师在美国期待着你们每一年都可以超越自我，期待你们的成功！我也会如期在 Math League 夏令营决赛中等着你们，Math League 欢迎你们的到来！

Daniel Flegler

2019 年 8 月

Daniel Flegler 简介

里根总统颁发的"杰出数学教学奖"获得者

普林斯顿大学"杰出中学教学奖项"获得者

美国数学大联盟(Math League)创始人

美国数学竞赛奠基人

美国新泽西州沃尔德威克高中执教并任数学教学督导 26 年

24 本畅销数学竞赛书的撰写者

布朗大学数学系本科和哥伦比亚大学数学教育系研究生院毕业

FOREWORD

Dear Chinese Friends and Math Lovers,

Greetings! My name is Daniel Flegler, a former mathematics teacher and department chairman at Waldwick High School, New Jersey, for 26 years beginning in 1965. In addition, I am the cofounder of Math League in the USA which was begun in 1977. Filled with tremendous enthusiasm and responsibilities of mathematics education, I devoted tens of years in this field. I am very fortunate to share some mathematics education principles and experiences with you. There is no national boundary for mathematics, and there is no national boundary for the affection in students. I hope this discussion can help you learn more about mathematics and know more about America's current mathematics education system. As a result, I hope you all will love math and gain more thoughts and enlightenments in mathematics education.

Everyone knows that it is important to study mathematics. I believe that every school already has developed its mathematics curricula and value these studies significantly. However, who can accurately answer the question "Why should we study mathematics?" In math classes, curious students often ask "Why are we learning these materials? Will I ever use them in my life after school?" The humorous answer that some teachers give is "If you didn't have to learn these materials, I would not have a job." Of course, the real answer is much more complex than this.

For students who will continue in scientific careers, learning mathematics is essential. The reason is that mathematics is the foundation of all sciences. While only a minority of students will become scientists, it is those students who will lead the world in new discoveries about the universe as well as developing and improving new technologies. Without mathematical knowledge, none of the great scientific discoveries of the last 200 years would exist. Many of these inventions, including but not limited to internet, airplanes, antibiotics, nuclear power, new weapons, automobiles, telephones, cell phones, light bulbs, computers, televisions, refrigerators, air conditioners, and cameras, greatly facilitate our lives. The fundamental principle behind all of these discoveries is mathematics. We can never accurately predict all the new technologies that will be developed in the next 200 years, but we can say with certainty that without a strong mathematical background, scientists would not be able to develop these future technologies. The advancement of the society depends on the mathematical background that students develop at their young ages.

However, most students will not pursue scientific careers. So, is mathematics unimportant to these students? In fact, for all individuals, analytical thinking and reasoning skills (part of mathematical thinking) are pivotal. They allow us to search for solutions to problems. Analytical thinking refers to the skill of thinking critically about the world. Reasoning skills refer to our abilities to think logically in complex situations. While there may be other subjects that develop these masteries, no subject does it better than mathematics. So, the real answer to the question "Why should we study mathematics" is that the subject trains our thinking skills. Unless we learn how to think logically, we will never be able to absorb and truly understand all the information that is presented daily in newspapers, magazines, and on the internet via social media. The materials we need to understand include but are not limited to science, politics, society, and psychology. While it may seem farfetched to think that solving a problem about the train's speed can help us in solving a problem in our lives, the skills that we use in framing the problem, identifying the knowns and unknowns, and taking steps to solve the word problems in mathematics are very important strategies that can be applied to the other problems in life. Mathematics, a language and a set of rules that mostly consist of equations, numbers, and Greek letters, can be understood by people with different nationalities, ethnicities, and beliefs, without any difficulties. Mathematical knowledge is so magical, and it is an important life skill for each of us.

Many of you are curious about the current state of mathematics education in America. Here we will discuss this topic. The ultimate goal of traditional American math classes is to succeed at state exams and standardized tests, which are the basic requirements of the secondary school completion. From what I have observed throughout my teaching career, most math classes in the USA are taught in the following manner: The teacher presents a new topic, does one or two sample problems on the board, and then assigns work to the class. In many cases, the teacher only shows the students one algorithmic approach to the problems rather than teaching for understanding. For many important theorems in the subject, the teacher only covers the results without discussing their proofs. As a result, instead of seeing the continuity of mathematics, students view each mathematics topic as a discrete unit, unrelated to any other topics. So, students easily get overwhelmed by more complicated problems. Evidently, this traditional classroom setting is far from sufficient for the best math students and math lovers—it never satisfies their curiosity for the in-depth exploration in the subject. This scenario, in some senses, is a huge waste in intellectual resources. I always believe that to improve the quality of education, we must discover the issues and then resolve them.

To remedy the shortcomings of mathematics education in the USA, my business partner Mr. Steven Conrad and I (both of us recipients of President Ronald Reagan's Presidential awards for "Excellence in Mathematics Teaching") started Math League in the USA in 1977, more than 40 years ago. As one of the earliest math contest companies, our purpose was and remains to build student interest and confidence in mathematics through solving worthwhile problems. This teaching philosophy has triggered the interest and enthusiasm in mathematics studies for thousands of teenagers in the USA. Also, it has helped many excellent math students to stand out from the crowd. More than 15 of my students have been named to the honors group of the Intel National Science Talent Search for the mathematics papers they have written. At the peak period, the number of participants of the Math League contests annually exceeded one million, which greatly promoted the development of mathematics education in the United States. Whether these students ended up in the mathematics career, they were all greatly benefited from the experiences of participating in math contests and math camps during their youths. I am very proud that their accomplishments in mathematics influenced their lives in positive ways.

To shape my students from good to excellent, I often encourage them to do research papers for competitions such as the Intel National Science Talent Search and also have them applied to prestigious summer programs such as the Ross Program at Ohio State University or the PROMYS Program at Boston University. These programs will help students develop a real understanding of mathematics and proofs. These summer camps not only allow students to apply the basic knowledge in extensive directions, but also train their willingness and determination to spend hours on one problem to explore the alternate solutions and generalizations. Through recreational and exploratory problems, math competitions and math camps inspire students to think creatively and encourage them to go extra miles in their studies. This is the core reason that the American math competitions and math camps are so popular in the USA and even in the world. Many high school students in the USA spend their vacations joining academic camps or extra-curricular activities to promote their academic interests and competitive strengths. The scores on math competitions and the history of joining math camps are valued significantly in the applications to the best colleges in the USA. These math camps are highly selective about the applicants: Students must think innovatively to solve the math problems on the application forms within a given period of time. They must clearly and logically explain their detailed solution in English, and they also need to complete some essays related to their profound understandings of mathematics.

Furthermore, I highly recommend that my students participate in math competitions such as AMC and AIME. As a preparation for these competitions, students should engage in solving non-routine questions that have appeared on mathematics competitions in the past. It is only through attempting to solve challenging problems that students extend and build their mathematical strength. Many of the same skills that are required to solve competition mathematics or taught in summer camps are useful in the studies of higher mathematics in college. I hope that many excellent secondary school students in China should proactively get involved in some challenging international math competitions and math summer camps. In this way, they can expand their vision, improve their innovative thinking abilities and fluent English skills, stimulate their potential, and objectively evaluate their math abilities relative to the top students of the same age. I believe that it is an enriching life experience to compete and meet with other students of exceptional mathematics ability and interest.

I am thrilled that the book ***English-Chinese Mathematics Encyclopedia: Algebra, Geometry, and Pre-Calculus*** will be published soon! This is a collectible mathematics reference book authored by my protégé Jerry C. Leung and his development team in China, through diligent work. This encyclopedia will be a valuable gift for the Chinese students. As one of my best former students, Jerry devotes his time in doing math research and authoring math contest questions. He has written some wonderful contest questions through his pursuits in rigor and perfection. I find out that when participating in international math competitions or math camps, many Chinese students face language barriers, which affect their understanding and communication, and eventually their contest results. Therefore, I always allow the Chinese participants of the Math League contests to use bilingual math dictionaries. However, it is very hard to find such professional and comprehensive bilingual dictionaries in China's bookstores that really resolve the issue. Now the publication of the ***English-Chinese Mathematics Encyclopedia: Algebra, Geometry, and Pre-Calculus*** will fill in the gap for this area. I believe that with the help of such a reference book, the Chinese students can excel even more in their future mathematics studies. From the content, organization, and completeness, the ***English-Chinese Mathematics Encyclopedia: Algebra, Geometry, and Pre-Calculus*** is a masterpiece that adapts the advantages of the exploratory mathematics education in the USA. This book can help students conquer their obstacles in studying math, improve their learning efficiency, and enhance their mathematics and English skills. My ultimate and sincere wish is that the Chinese students can ardently love mathematics and explore the subject. Not only can they succeed in these intense math contests and camps, but they also can get admitted to their dream colleges. The kingdom of mathematics is marvelous, waiting for high-spirited individuals like you to explore. Once you grasp its fundamentals, you can gain more than you expect. I, your old teacher from the USA, await for you surpassing yourself every year. I truly wish you succeed! I will also await meeting you in the Math League summer tournaments. Welcome to Math League!

Daniel Flegler

August 2019

About Daniel Flegler

Recipient of President Ronald Reagan's Presidential Awards for "Excellence in Mathematics Teaching"

Recipient of Princeton University's Award for "Distinguished Secondary School Teaching"

Cofounder of the Math League in the USA

Founder of the Mathematics Contests in the USA

Former Mathematics Teacher and Department Chairman at Waldwick High School, New Jersey, USA, for 26 years

Coauthor of 24 Bestseller Mathematics Contest Books

BA in Mathematics from Brown University, MA in Mathematics Education from Columbia University

第2版前言

PREFACE

引子

几年前播下发愿的种子，在时光的催化下，时至今日竟然长成了一棵树。《英汉数学全书——代数、几何与微积分初步》(后文均为《英汉数学全书》)已经与读者见面五载，今天迎来了第2版修订。

《英汉数学全书》的诞生是偶然，也是冥冥中的必然。偶然源自8年前我在北京王府井大街为即将参加国际数学竞赛的女儿寻遍双语数学工具书而不得的一次发愿，必然源自与生俱来的执拗、果决、责任与担当，源自心中多年来一直想要做成点事儿的熊熊火焰。

智能时代，未来已来

无论如何，谁都无法否认也无法阻挡，滚滚向前的历史车轮已经把我们裹挟到了智能时代，以5G、人工智能、机器人技术、基因生物技术等为特征的第四次工业革命(2010—至今)已然到来。在这场革命中，教育行业作为培养人才后备军的行业，紧跟时代步伐，不断创新和突破的需求更为迫切。作为一个教育领域的参与者，为新新人类铺就一条适合他们成长的"天路"是责任、义务，更是使命。

多年来，我们关注最多的是对国际数学教育依赖程度比较大、有留学计划的学生群体和国际学校教师群体。观察发现，如果学生没有过硬的实力，仅靠粗放式刷题和过度包装营造的优秀"人设"，会随着进入真实而陌生的留学后学习环境而崩塌。靠刷题得来的高分并不代表真正具备了留学所需的各项能力，即便凭借优秀"人设"获得了录取资格，也有很大概率无法很好地适应国外课堂而面临种种挑战。如，听不懂教授的讲述，无法融入课堂小组讨论，不能大方自如地表达自我或学期演讲，不能完成论文写作等问题；部分学生入学后跟不上国外课程，还需要依靠线上补课的方式完成GPA的要求；每年因学术成绩不及格而被"退学"或无法如期毕业的学生数目也非常触目惊心。而上述能力不是一蹴而就能获得的，学生需要在系统教学的支撑下，经过长期的学习过程才能实现。

站在一个高的维度来看，这一代"新人类"的生活、学习背景是一个越来越融合的世界大舞台，他们需要不断地打破认知藩篱，拓宽认知边界，他们需要具备活力和创新力，这就要求国际数学素质教育基本面不断升级，成为一个更具"穿透力"的学科。一方面助力莘莘学子升入理想梦校并顺利毕业，为他们的职业、事业生涯奠定深远的基础。另一方面，希望他们通过对丰富多元数学理念的系统学习，让他们拥有更聪明、更强劲的大脑、更卓尔不群的思维力，更是为这些新新人类深谋远虑、为计深远。

我们常年与国内外知名数学教育家保持密切联系，不断切磋交流与探索新知。美国著名数学教育家Mr. Daniel Flegler、美国著名数学教育企业AoPS创始人Mr. Richard Rusczyk，以及斯坦福终身教授Mr. Brian Conrad都是我们成长路上的指路人。我们的国际数学素质教育理论——"探究式国际数学素质教育与专业数学英语教育的交叉融合"，即将CLIL教学法(注：即Content and Language

Integrated Learning，是一种科学地把“学科素养”和“语言能力”相结合的教学法）和“探究式”教学法融合应用于国际数学素质教育中的教学理论。“探究式教学法”是一种以解决问题为导向的主动学习方法，旨在激发学生的好奇心和自主学习的兴趣。它鼓励学生提出问题、进行实验和探索新思路，培养批判性思维、创新能力和解决问题的能力。整个教学过程以学生为中心，教师作为引导者，设计开放性问题情境，引导学生通过观察、实验、讨论和推理，深入理解和掌握数学概念及其背后的逻辑。这样，学生不仅能够牢固掌握知识点，还能灵活运用所学知识，解决复杂的数学问题，培养终身学习和自主探索的精神。“探究式”国际数学素质教育与专业数学英语交叉融合是我们目前阶段探索出的系统性解决方案的最优解。我们已经将这一教学理念深深扎根于研发的每一部作品中，希望给大家带来良好的学习体验。

因《英汉数学全书》结缘，我们结识了许多工作在教学一线的国际数学精英教师，段凯耀老师便是其中一员。他是一位名副其实的智能时代教育改革的领跑者和杰出贡献者，他精通各种国际课程体系，2022 年被芝加哥大学授予“杰出教育者”称号，也是迄今为止受邀参加 AP 全球年会的中国大陆地区唯一发言人。多年以来，他积极探索并应用人工智能技术和数据分析教学方法，大幅提升了学生学习效率并优化了学习体验。他带领着他的“学生军”左突右击，斩获国际数学竞赛大奖无数。非常巧合的是，我们的开创性双语国际数学素质教学法——“CLIL”教学法和“探究式”教学法的交叉融合理念与段凯耀老师的某些教学思路不谋而合。段老师说他一直在找一本既可以为学生搭建一座跨越语言和文化障碍的桥梁的好书，让他们在数学素养和英语能力方面双重提升，又能促进他们初、高中阶段国际教育学习习惯养成，从而为未来职业生涯奠定坚实而长远的基础。他一直求而不得，直到某一天偶遇了《英汉数学全书》。《英汉数学全书》覆盖了初、高中六个年级的所有词条，用英汉双语比较专业地阐释了从基础到高级的数学概念以及它们背后的性质、定理和证明，对学生学习 IG 等国际课程起到多层次的承接作用。经过几届学生的用书实践之后，段凯耀老师把《英汉数学全书》推荐为“学生人手一册的双语数学案头工具书”。

每一次破茧的努力，只为您攀登的路上多一块垫脚石

这本书既要帮助读者掌握地道的英语数学词汇和标准的惯用表达方式，又要激发他们在数学世界的探索精神、汲取多元的数学理念，大大缩短他们与世界接轨的路径，它更像一本小型双语国际数学基础教育的“辞海”，而非一本仅仅罗列数学词汇的词典。尽管当时从各个环节、从结构到内容细节都力求做到最好，但是后来发现不足的地方依然很多。面世以来，听到很多来自一线数学教师和家长的褒奖与肯定，但更有价值的是那些宝贵的改进意见，激发了我们很多内容优化灵感，促成了此次修订版的实现。读者反馈主要集中在以下几方面：

- 需要修正的错误之处；
- 题量不够多、不够深，希望能有更多经典或拔高试题供学生练习以巩固所学；
- 本书篇幅较大，作为工具书查阅时没有问题，但如果让学生高效独自学完这个“大部头”，不仅挑战很大，而且边界不够清晰。有的家长希望在专业教师陪伴下帮助孩子“啃完”这本内容丰富的好书，希望有专业教师的答疑。

此次修订包括勘误、增加附录内容、推荐书目，以及为了提升读者学习效率和增大试题量而研发的配套双语伴读视频系列。采用“探究式教学”，除了“扫清”书中要点和难点，还加入了更丰富的内容，逐步引领学生发现规律、归纳总结及探究数学性质背后的证明，着重强调对学生数学英语综合水平和创新数学思维的培养，陪伴大家高效学习。

在内容方面，根据不同年级对知识体系精准切割，优化了知识体系的系统性和逻辑性。在紧扣对应年级知识点和考点的基础上，适度扩展知识宽度和深度，有梯度性地精选配套练习真题，弥补了第 1 版习题量不足和试题难度较浅的缺憾，满足学生从课堂到标准化考试的平滑过渡，再到国际数学竞赛、国际知名数学夏校/夏令营申请的梯度性需求。设计的一些非常规经典难题可以满足有余力的学生巩固所学，培养创新思维，进一步提高数学水平。

在数学英语学习方面，为了帮助学生快速识记并掌握专业数学英语，载明词条，并对拓展词汇、数学符号、单位换算、数学固定短语等进行系统串讲，精心设计的英汉双语伴读教学过程，营造出一个培养学生听、说、读、写全方位立体“浸泡式”语言环境，让有语言障碍的学生快速高效提升数学英语综合应用能力，以扫清升学考试语言障碍，同时缩短留学后适应期。

致敬每位亲爱的读者——感谢与您的不期而遇

非常感谢您选择我们、信赖并支持我们，感谢您对我们诸多不足给予的包容与海涵。这本书的主要服务对象是成千上万努力成长的学子，他们背后是一个个努力托举他们成才的家庭，父母不易，孩子们也非常辛苦，他们不仅面临着沉重的学业压力，还承载着自己或家庭的梦想和未来。这份深深的懂得，是我们深耕国际数学素质教育、不断创新的原动力，期待我们的努力可以为大家交付更多、更优质的作品。让我们一起努力，期待下一版更精彩。

感恩默默付出的每一位幕后英雄

(以下顺序不分主次)承蒙美国数学竞赛先河的开创者、美国数学大联盟的创始人、美国里根总统颁发的“杰出数学教学奖”获得者、11次方国际数学终身首席顾问——Mr. Daniel Flegler先生的信赖，他将凝结毕生心血、覆盖从小学到高中全套共24本国际数学素质教育优秀数学作品独家授权给我们11次方国际数学。他的命题极富特色：他设计的题干生动有趣，思考题的设计更是可以让学生把学过的基础知识点运用发挥到极致。他是可以将学生融会贯通能力培养到极致的“数学艺术大师”。本书伴读课程的很多精彩试题均选自他的作品。

在此对赵青老师的辛勤付出表示深深的感谢！赵青老师对国际数学教育事业的敬业和热爱之情令人叹服！她常年进取，多次国内外访学，对国际数学先进教学理念和模式孜孜以求，培养出无数优秀学生，其中不乏世界名校的佼佼者，可谓桃李满天下。为了让《英汉数学全书》更适合中国学生的学情和习惯，《英汉数学全书》研发团队诚挚邀请赵青老师担纲了最为艰巨又乏味的校对工作。她不辞辛苦通读了整本书，结合多年教学经验，提出了近百处修订意见，相信我们的《英汉数学全书》修订版会因为赵青老师的辛苦付出而拥有更精彩的呈现和用户体验。

段凯耀老师作为一名坚守在国际数学教育一线的成绩斐然的名师，被芝加哥大学授予“杰出教育者”称号，也是迄今为止受邀参加AP全球年会的中国大陆唯一发言人。他超前的国际数学教学理念对我们的研发工作影响至深，感谢他成了我们的“义务编外顾问”，为我们打开更宽广的学术视野和科技前瞻。此次修订版附录的补充内容和推荐书目为段老师倾情推荐，希望读者朋友们好好加以利用。

感谢数学名师金莹莹老师多年以来对我们教研团队的教学指导，她作为《英汉数学全书》内容结构的奠基人，一直非常谦逊和低调，她说“在这里一句话带过我就行了”。那么这一句话应该是：她是中国千万个默默奉献的优秀教育工作者的缩影和代表，他们是一群平凡而伟大的“灵魂雕刻师”。衷心感谢您，尊敬的金老师！

吴晓云于北京

2025年2月

第1版前言

PREFACE

时光飞逝！4 年前为了帮女儿找到一本参加美国数学竞赛需要的英汉双语数学工具书，跑遍了京城书店而不得的沮丧情形依然历历在目。后来我们决定不再等待，自己着手组建团队开始编写《英汉数学全书——代数、几何与微积分初步》，直至成书。这 4 年付出了太多，现在终于看到了这新诞生的“孩子”，心中百感交集。

《英汉数学全书——代数、几何与微积分初步》的诞生顺应了智能化时代背景下中国与世界日益增强的国际文化交流过程中更为专业化、精细化的需求，填补了中国英汉数学工具书的空白。虽然这本工具书还有很多需要改进的地方，但不影响本书成为双语专业工具书领域的里程碑。愿它的面世可以起到抛砖引玉的作用，带动中国更多其他学科领域的英汉双语工具书尽早问世，百花齐放，以飨读者。愿这本双语工具书成为中国学子们的“加油站”，为科技强国贡献一分力量。

数学水平一直以来是我们国人心头的骄傲。可是事实果真如此吗？2002 年 8 月，北京成功地举办了第 24 届国际数学家大会。这是第一次在中国召开，也是第一次在发展中国家召开的国际数学家大会。此次大会成为衡量我国数学水平在国际数学界地位的象征。当时许多媒体发表了一篇名为《中国数学已经达到世界一流水平》的文章。记者曾就此观点采访第一位获得菲尔兹奖的华人数学家丘成桐：“中国数学真的像一些人所说的那样已经接近世界一流水平了吗？”丘成桐先生沉吟片刻，吐出 5 个字：“差得还很远。”2018 年 8 月 1 日在里约热内卢举办了第 28 届国际数学家大会，颁布了四年一度的菲尔兹奖后，中国数学强与弱的问题再一次被热议，其中有一句话特别刺耳，“从 1936 年颁发第一次菲尔兹奖以来，人口如此庞大的中国，却诞生不了几位获奖得主，谁来救救中国数学，它真的还很年轻”。中国数学强吗？强，曾经我们真的很强。在 14 世纪以前，中国作为世界上最发达的国家之一，取得过许多辉煌的成就，出现过很多杰出的数学家。在 14 世纪，世界上重大的数学成就有 15 项，中国占了 9 项。而从 14 世纪到 19 世纪，世界重大数学成就有 100 项，中国 0 项。①

中国的数学教育以大量刷题、重记忆、重解题技巧而轻逻辑和抽象思维能力为特点，学生的基本功非常扎实，计算速度、解题技巧可以“秒杀”外国人。中国学生带着这个优势出国留学后，都会觉得外国的数学比较简单。他们在班上的数学考试和在数学竞赛中都能表现优异。但是我们发现一个有趣的现象：在高中阶段微积分学得非常好的中国学生，学习抽象代数、拓扑学时会比较吃力，外国的学生反而学得比较轻松，到了大学阶段面对更高端的数学课，外国学生在思维方面表现得更加敏捷。中国的基础数学教得很难、很扎实，但是却只有极少数中国学生选择数学专业继续深造。反观那些计算能力略逊一

---

① 数据来源：谁来救救中国数学，它才 82 岁. http://www.jinciwei.cn/b348586.html。

筹的国外学生,往往成长为数学科研奖项得主。这些现象值得我们深思。数学学习旨在培养人的计算、逻辑和抽象思维三项能力,三者不可偏废。

现在国家已经深刻认识到数学教育改革的重要性,越来越在国家层面重视数学在科技发展中的战略性地位。2018年,国务院发布《关于全面加强基础科学研究的若干意见》(国发〔2018〕4号),提出"潜心加强基础科学研究,对数学、物理等重点基础学科给予更多倾斜"。为切实加强我国数学科学研究,科技部、教育部、中科院、自然科学基金委于2019年7月联合制定了《关于加强数学科学研究工作方案》,值得我们期待。华裔数学大师陈省身曾在《怎样把中国建为数学大国》中说:我们要创办世界水平的、一流的数学研究院,开办一些基本的数学先进课程,将科学的种子播下并在中国本土生根发芽。我们既要避免盲目的夜郎自大,又要鼓足勇气在挑战中砥砺前行。相信我们可以在中西合璧、取长补短的数学教育理念中找到答案。本书不仅仅是一本全面系统的工具书,还是一个了解国际数学,尤其是美国数学教育的窗口,希望借此机会激发大家展开一场关于数学教育的讨论,找到我们的不足之处,然后弥补不足而进步。

数学是科学的"皇后",其重要性不言而喻。数学是自然科学的基础,也是重大技术创新发展的基础。数学实力往往影响着国家实力,几乎所有的重大发现都与数学的发展与进步相关,数学已成为航空航天、国防安全、生物医药、信息、能源、海洋、人工智能、先进制造等领域不可或缺的重要支撑。最近读到华为舵手任正非先生的一篇精彩演讲文稿,里面提及中国和美国的差距就两个字——软件。今天的信息技术时代,互联网、大数据、云计算、人工智能靠什么来支撑呢?就是靠软件。手机、计算机都离不开软件,它是最基础的科技。毋庸置疑,人类社会最终要走向人工智能,这是一个软件至上的时代!软件的背后又是什么呢?软件的背后是算法。算法的背后就是数学。2019年中美贸易摩擦不断升级的背景下,任正非于2019年5月21日接受《面对面》节目记者董倩独家专访时提到:"我们国家修桥、修路、修房子……已经习惯了只要砸钱就行,但是芯片砸钱不行,得砸数学家、物理学家、化学家……中国要踏踏实实在数学、化学、神经学、脑科学各方面努力地去改变,我们才可能在这个世界上站得起来。"2万字的采访实录,任老先生27处提及"数学",多次举例说明数学对华为科技竞争力的重要性,我们深深地被他采访中体现的爱国情怀和高瞻远瞩所激励和感动。中国著名自然语言处理和搜索专家吴军博士于2017年11月在北京作了一篇精彩的主题演讲,题目为《教育改变命运》。其中提到了"我们为什么学习数学"这个话题,他对数学学习的真知灼见非常有必要在此分享。他提出:数学的重要性在于它培养科学的常识和基本的素养,讲究逻辑和方法。每一种解决数学题的方法都可以用于工作和创业。当你遇到一个不会解决的问题时,以前那些解数学题的逻辑全都用得上。学好数学需要三个要素:读题要读对;基础知识要牢固;逻辑能力要具备。把题读懂了,从已知到未知,读懂题就是知道自己现在的位置在哪儿,未来要去哪儿。分析和逻辑就是找到这条路,基础知识是你的工具,你的逻辑自己学会把这条路走通了。学好数学就是这三条,遗憾的是,大部分人学数学只知道第二条——基础知识牢固。老师把第二条全教给了你,关键是你题也没读懂,也没找到逻辑。在今天这样一个竞争激烈的时代背景下,科技强国梦需要更先进的数学理念和成果作为支撑,需要培养出更多、更杰出的数学人才。当然,世界是多元的,我们也不必期待所有的数学尖子生都成为数学家,但是如果他们能够领悟到数学之美并从中得到很好的逻辑思维训练,日后进入他们感兴趣的领域发展,很好地生活和工作,也是我们备感欣慰的事情。

上文提及读懂题至关重要,在这里有一个不能回避的现实问题:中国留学生在熟练运用专业数学语言能力方面还比较有限。学术英文能力的欠缺在方方面面严重制约了学生的学业、学术进步和长远的职业发展。探讨、沟通、写论文、读专业文章以及获取优质的各种课程资源都需要坚实的语言基础。本书除引入美国数学教学元素供读者体会之外,还有一个重要的目的就是提升中国学生的数学英文水平，帮助他们扫清语言障碍。

本书的大部分读者应该是一群志在四方的年轻人，他们背负着“以梦为马，不负韶华”的梦想负笈海外，让我的思绪不禁穿越到147年前。回首当年，在第一个毕业于耶鲁大学的中国人——爱国者容闳先生18年不懈的努力和积极推动下，清政府终于同意向美国派驻中国第一批官派赴美留学生。这些留学生是一群平均年龄只有12岁的孩子。1872年8月11日，一艘即将从上海出发远涉重洋的轮船在码头冒着白烟，送别的亲人泪眼婆娑……因为在那个时代，这次分别意味着长久的离别。这些留学生要在波涛汹涌的大海中航行25天才能抵达美国旧金山，然后还要乘坐7天的蒸汽火车才能抵达留学目的地——美国东北部的新英格兰地区。清政府先后分四批派出120名留美幼童，他们聪敏、勤奋，取得了令美国人都惊叹不已的优异成绩。他们当中除了在美早亡和在中法战争、甲午战争中捐躯的23名外，剩下的97名都成了建设国家的栋梁，完成了他们在那个时代被赋予的历史使命。生活在今天的我们是如此幸运，科技高度发达，飞机可以让我们日行万里，不必再历经数月跋山涉水旅途劳顿；我们拥有各种现代化通信工具，可以随时随地跟亲人“天涯若比邻”。在此提及这个尘封已久的故事用以勉励即将留学或已经在海外留学的每一个年轻人，一定要好好学习，追求梦想，不要辜负青春好年华。

我自己的女儿也是一名留美高中生，一路陪伴她成长的历程让我深知孩子们留学前准备过程的艰辛和留学期间长期的高负荷学习压力，他们在外面留学非常不容易。带着对这群孩子深深的责任和爱，我和我的中美研发团队一起完成此书。尽管我们已经尽力发挥匠人精神，对本书精雕细琢，但难免还有疏漏和不足之处，还望广大读者批评指正，以待进一步修改和完善。在本书的编写期间得到了北京资深数学名师金莹莹女士和耿喜良先生的大力指导和帮助，他们都是中国数学教育领域顶级的数学导师，他们为本书倾情贡献了积累多年的宝贵数学经验和智慧，在此表示深深谢意；同时非常感谢清华大学出版社首席策划编辑盛东亮先生的慧眼识珠以及他的整个编辑团队的倾情奉献，使本书得以尽早面世，以飨读者。感谢我的女儿李佳蓉(Li Jiarong)，是她参加美国数学竞赛时在国内基本找不到英汉双语数学工具书激发了我们编写本书的灵感，是她积极促成了本书。她留学之前就屡屡在数学大赛中获奖，良好的数学基础使她有能力胜任本书某些章节的创作和编辑工作，这使她的数学实践能力再次得到了巩固和提升。

这本《英汉数学全书——代数、几何与微积分初步》终于完稿，终于可以为中国学子提供一本好用的案头双语数学工具书了。我们对本书的编写也可以暂告一段落，可以静下心来喝杯茶，然后继续酝酿下一本敬献给中国学子们的好书。

“中国是数学大国，大不一定强，想成为数学强国，要靠你们！”
——陈省身先生于92岁时对中国少年的寄语

吴晓云于北京

2019年7月

视频目录

VIDEO CONTENTS

视频名称	位　　置	内容简介	视频时长
视频 1　分解质因数精讲精练	第 14 页 1.2.2.5 小节	与分解质因数性质配套的精选试题解析	11min24s
视频 2　运算种类精讲精练	第 48 页 1.6.2 小节	与减法性质拓展及循环规律分析配套的精选试题解析	13min57s
视频 3　指数定律精讲精练	第 55 页 1.6.3 小节	与指数定律综合运用配套的精选试题解析	8min11s
视频 4　整式的运算精讲精练	第 70 页 2.2.4 小节	与四舍五入及添/去括号性质配套的精选试题解析	14min47s
视频 5　四项分配法精讲精练	第 74 页 2.2.5 小节	与四项分配法及平方差公式性质配套的精选试题解析	7min49s
视频 6　因式分解精讲精练	第 77 页 2.2.6 小节	与因式分解技巧综合运用配套的精选试题解析	11min40s
视频 7　比例精讲精练	第 82 页 2.3.2 小节	与比例日常生活应用配套的精选试题解析	12min43s
视频 8　根式精讲精练	第 86 页 2.4 小节	与简化根式技巧综合运用配套的精选试题解析	9min14s
视频 9　一元一次方程精讲精练	第 96 页 3.2.2 小节	与一元一次方程分析及应用配套的精选试题解析	11min59s
视频 10　绝对值方程精讲精练	第 98 页 3.3.1 小节	与含有多重绝对值方程配套的精选试题解析	5min49s
视频 11　二元一次方程组精讲精练	第 105 页 3.4.2 小节	与二元一次方程巧解及应用配套的精选试题解析	9min21s
视频 12　一元二次方程精讲精练	第 110 页 3.5.2 小节	与一元二次方程分析及应用配套的精选试题解析	18min43s
视频 13　行程问题精讲精练	第 115 页 3.6 小节	与速度/时间/路程配套的精选试题解析	11min57s
视频 14　一元一次不等式精讲精练	第 118 页 4.2 小节	与一元一次不等式分析及应用配套的精选试题解析	10min17s
视频 15　一次函数精讲精练	第 155 页 7.3.5 小节	与一次函数分析及应用配套的精选试题解析	13min38s
视频 16　二次函数精讲精练	第 171 页 7.3.6 小节	与二次函数分析及应用配套的精选试题解析	10min55s
视频 17　分段函数精讲精练	第 185 页 7.4.3 小节	与分段函数性质配套的精选试题解析	13min21s
视频 18　数列精讲精练	第 193 页第 8 章	与等差数列、等比数列和一般数列配套的精选试题解析	9min22s
视频 19　统计数据精讲精练	第 201 页 9.2.1 小节	与统计数据性质配套的精选试题解析	8min51s
视频 20　计数精讲精练	第 253 页 11.1 小节	与计数技巧综合运用配套的精选试题解析	9min12s
视频 21　概率精讲精练	第 258 页 11.2 小节	与概率技巧综合运用配套的精选试题解析	10min13s

续表

视频名称	位置	内容简介	视频时长
视频 22　命题精讲精练	第 299 页 13.1.5 小节	与命题推理配套的精选试题解析	10min32s
视频 23　平行线精讲精练	第 316 页 14.3.2 小节	与平行线和截线所形成角的关系配套的精选试题解析	5min46s
视频 24　全等三角形精讲精练	第 333 页 15.5.3 小节	与等腰三角形性质配套的精选试题解析	10min54s
视频 25　三角形的特殊线段精讲精练	第 336 页 15.6.2 小节	与三角形特殊线段性质综合运用配套的精选试题解析	15min50s
视频 26　勾股定理精讲精练	第 355 页 15.9.1 小节	与勾股定理应用及分析配套的精选试题解析	22min44s
视频 27　相似三角形精讲精练	第 363 页 15.10.3 小节	与相似三角形和角平分线定理配套的精选试题解析	13min06s
视频 28　平行四边形精讲精练	第 374 页 16.2 小节	与三角形特殊线段和平行四边形性质配套的精选试题解析	11min19s
视频 29　多边形精讲精练	第 386 页第 17 章	与多边形的角度和周长计算配套的精选试题解析	15min54s
视频 30　合同变换精讲精练	第 395 页 18.2 小节	与平移和反射性质综合运用配套的精选试题解析	13min43s
视频 31　圆的基础知识精讲精练	第 403 页 19.1 小节	与圆的基础知识综合运用配套的精选试题解析	15min09s
视频 32　圆的切线精讲精练	第 415 页 19.4 小节	与圆的切线性质综合运用配套的精选试题解析	16min02s
视频 33　圆的定理精讲精练	第 419 页 19.5 小节	与圆的定理综合运用配套的精选试题解析	18min13s
视频 34　多面体精讲精练	第 430 页 20.2 小节	与棱锥和立方体性质综合运用配套的精选试题解析	13min32s
视频 35　非多面体精讲精练	第 436 页 20.3 小节	与圆锥和平截头体性质综合运用配套的精选试题解析	11min52s
视频 36　周长/面积/表面积/体积精讲精练	第 442 页第 21 章	与立方体和棱锥体积公式综合运用配套的精选试题解析	8min48s
视频 37　三元一次方程组精讲精练	第 456 页 23.4 小节	与三元一次方程分析及应用配套的精选试题解析	12min17s
视频 38　复数精讲精练	第 471 页第 25 章	与复数性质配套的精选试题解析	17min24s
视频 39　多项式除法精讲精练	第 479 页 26.2 小节	与多项式除法性质配套的精选试题解析	11min52s
视频 40　完全分解精讲精练	第 483 页 26.3 小节	与因式分解性质配套的精选试题解析	11min04s
视频 41　对数精讲精练	第 495 页 28.2 小节	与对数性质配套的精选试题解析	12min11s
视频 42　部分分式分解精讲精练	第 511 页 29.4 小节	与部分分式分解性质配套的精选试题解析	17min21s
视频 43　数列与级数精讲精练	第 529 页第 31 章	与等差数列和等比数列配套的精选试题解析	18min27s
视频 44　二项式定理精讲精练	第 536 页 32.2 小节	与二项式定理配套的精选试题解析	8min41s
视频 45　排列与组合精讲精练	第 539 页 32.3 小节	与计数和概率技巧综合运用配套的精选试题解析	12min46s
视频 46　三角函数精讲精练	第 544 页 33.2 小节	与三角恒等式综合运用配套的精选试题解析	15min28s
视频 47　正弦定理与余弦定理精讲精练	第 557 页 33.3 小节	与正弦定理和余弦定理配套的精选试题解析	11min06s
视频 48　导数精讲精练	第 573 页 34.2 小节	与导数和反导数方程及图像综合分析配套的精选试题解析	20min14s

目录

CONTENTS

Part 1: Pre-Algebra, Algebra 1, Algebra 2　第1部分：代数初步，代数1，代数2

Part 1: Pre-Algebra, Algebra 1, Algebra 2[①]

第1部分：

代数初步，代数1，代数2[①]

代数(Algebra)是数学的一个分支。代数就是将真实多变的数用一般性的符号和变量(通常是 x 和 y)表示，并探讨数之间的关系。代数学使我们用数学语言表达"文字问题"成为可能。代数是出生于波斯帝国的伊斯兰学者、数学家阿尔·花剌子模留给后人的遗产之一。他是代数与算术的整理者，被誉为"代数之父"。公元 825 年左右，他写了一本影响深远的书 *Kitab al-Jabr wa-l-Mugabala*，首次引入了 al-jabr 这个词，意思是"恢复平衡"，在这里指的是一项代数运算——移项完成后，等式两端又恢复平衡的意思。这就是 Algebra 的由来。中国对代数早有研究。到了清朝咸丰九年(1859 年)，数学家李善兰首次把 Algebra 译成"代数学"，从此我国正式使用"代数"一词。

本书第 1 部分包括代数初步(对应美国 6～7 年级学生)、代数 1(对应美国 8～9 年级学生)、代数 2(对应美国 10～11 年级学生)。通过对这部分的学习，学生会熟悉实数及表达式的种类、一元一次方程(组)、二元一次方程、因式分解、一次函数、二次函数、不规则函数、不等式方程(组)、数列、计数与概率和基本统计学的整个知识体系以及惯用的数学专业英文表达方式。本书第 9～10 章为统计学，也是 AP(Advanced Placement，大学预修课程)统计学考试的基本知识点之一。

---

① Tom Jackson. Mathematics: An Illustrated History of Numbers[M]. Shelter Harbor Press, 2012.

Algebra　代数

n. [ˈældʒəbrə]

Definition: The branch of mathematics that deals with general statements of relations. It uses symbols and variables to represent certain numbers, especially unknown numbers, along with their relations.　代数是数学的一个分支，专门处理关系命题。它用符号和变量表示一些数(特别是未知数)，并且探讨其相互关系。

1. Real Numbers　实数

1.1 Introduction　介绍

Real Number　实数

n. [ˈriəl ˈnʌmbər]

Definition: A number that can be found on the number line.　实数是存在于数轴上的数。

Notations: The set of real numbers is denoted by $\mathbb{R}$.　实数集用$\mathbb{R}$表示。

Properties: Closure Property, Commutative Property, Associative Property, Distributive Property, Identity Property, Inverse Property of addition and multiplication.　实数的加法与乘法具有封闭性、交换律、结合律、分配律、单位元性质、逆元性质。

Examples: $1, 0.5, -233, 2/3, -\pi, \sqrt{3}, \sqrt[3]{5}, \cdots$

Note: In this encyclopedia, all numbers mentioned are real numbers, unless specified.　本书中提到的数均为实数，除非特别指定。

1.1.1 Positive, Negative, Zero　正数、负数、零

Positive　正(数)

adj. [ˈpɑzətɪv]

Definition: Greater than zero.　正数是大于 0 的数。

Notation: The set of positive numbers is denoted by $\mathbb{R}^+$.　正数集用$\mathbb{R}^+$表示。

The symbol "+" is in front of a positive number, but usually we omit it. e. g. "+3" is expressed as "3"; "+6" is expressed as "6", etc.　正数前面"+"可省略。

"3" is read as "positive three", or simply "three".

Properties: 1. Positive numbers are always at the right side of 0 on the number line.　正数在数轴上 0 的右边。

2. Closure Property, Commutative Property, Associative Property, Distributive

Property, of addition and multiplication, Identity Property and Inverse Property of multiplication. 正数的加法与乘法具有封闭性,满足交换律、结合律、分配律。乘法具有单位元性质与逆元性质。

3. The union of the set of positive numbers and zero is also known as the set of nonnegative numbers. 正数集与零的并集是非负数集。

Examples: 2, 3.5, $\sqrt{5}$, 3/4, 71/3, π, …

Phrases: ~real number (short for ~number) 正实数, ~integer 正整数, ~fraction 正分数, ~rational number 正有理数, ~irrational number 正无理数, ~direction 正方向

Questions: 1. What are all positive integers less than 6? 小于6的正整数有哪些?

2. What is the only positive solution for $x^2=9$? 方程 $x^2=9$ 的唯一正数解是什么?

Answers: 1. 1, 2, 3, 4, 5.

2. 3.

Negative 负(数)

adj. ['negətɪv]

Definition: Less than zero. 负数是小于0的数。

Notation: The set of negative numbers is denoted by $\mathbb{R}^-$. 负数集用$\mathbb{R}^-$表示。

The symbol "−" must be in front of a negative number. 负号必须在负数前面。

"−3" is read as "negative three".

Properties: 1. Negative numbers are always at the left side of 0 on the number line. 负数在数轴上0的左边。

2. Commutative Property, Associative Property, Distributive Property, of addition and multiplication, Closure Property of addition. 负数的加法与乘法满足交换律、结合律、分配律。加法具有封闭性。

3. The union of the set of negative numbers and zero is also known as the set of nonpositive numbers. 负数集与零的并集是非正数集。

Examples: −2, −3.5, $-\sqrt{5}$, −3/4, −71/3, $-\pi$, …

Phrases: ~real number 负实数, ~integer 负整数, ~fraction 负分数, ~rational number 负有理数, ~irrational number 负无理数, ~direction 负方向

Questions: 1. What are all negative integers greater than −6? 大于−6的负整数有哪些?

2. What is the only negative solution for $x^2=9$? 方程 $x^2=9$ 的唯一负数解是什么?

3. If John has \$3 and lunch costs \$5, how much will John have after buying lunch?

Answers: 1. −5, −4, −3, −2, −1.

2. −3.

3. After buying lunch, John will have \$3 − \$5 = −\$2 left. In the context of the problem, −\$2 means that he owes \$2. 在这个题目里,负数表示欠款的意思。

Zero 零

number ['zɪəroʊ]

Definition: Representation of a "nothing-quantity", also is a placeholder for rounding numbers. 零

代表一个"什么都没有"的数量,亦是四舍五入后的占位符。

Notation: 0

Properties: 1. The number 0 is neither positive nor negative.　0 既不是正数,也不是负数。

2. Identity Property of Addition: $a+0=a$. 0 is the identity element.　加法单位元性质: $a+0=a$。0 是单位元。

3. Inverse Property of Addition: $a+(-a)=0$. The numbers a and $-a$ are additive inverses of each other.　加法逆元性质: $a+(-a)=0$。a 和 $-a$ 互为加法逆元。

4. The product of 0 and any number is 0: $a \cdot 0=0$.　任何数和 0 的积均为 0。

5. The quotient of 0 and any nonzero number is 0: $0/a=0$, for which $a \neq 0$.　0 和任何非零数的商均为 0。

6. The number 0 serves as a placeholder for rounding numbers.　0 是四舍五入的占位符。

Questions: 1. How many even numbers greater than 2 are primes?　有多少个大于 2 的偶数为质数?

2. What is the value of $8+3-3+0-8+123$?

3. Round 4.999 to the nearest ones, tenths, and hundredths.　分别四舍五入 4.999 到个位、十分位、百分位。

Answers: 1. 0.

2. 123.

3. 4.999 rounded to the nearest ones is 5.　4.999 四舍五入到个位是 5。

4.999 rounded to the nearest tenths is 5.0.　4.999 四舍五入到十分位是 5.0。

4.999 rounded to the nearest hundredths is 5.00.　4.999 四舍五入到百分位是 5.00。

Note that the 0's at the end are the placeholders that represent the precisions of rounding. Therefore they cannot be omitted.　小数最后的 0 是占位符,代表的是四舍五入的精确性。它们不能被忽略。

1.1.2 One 一

One 一

number [wʌn]

Definition: The smallest positive integer.　一是最小的正整数。

Notation: 1

Properties: 1. Identity Property of Multiplication: $a \cdot 1=a$. The number 1 is the identity element.　乘法单位元性质: $a \cdot 1=a$。1 是单位元。

2. Inverse Property of Multiplication: $a \cdot 1/a=1$, for which $a \neq 0$. The numbers a and $1/a$ are multiplicative inverses of each other.　乘法逆元性质: $a \cdot 1/a=1(a \neq 0)$。a 和 $1/a$ 互为乘法逆元。

3. All positive integers are divisible by 1.　所有正整数都能被 1 整除。

4. The number 1 is neither a prime nor a composite.　1 既不是质数,也不是合数。

Questions: What is the smallest integer greater than 0?　大于 0 的最小整数是什么?

Answers：1.

1.1.3 Summary 总结

1. Figure 1-1 shows the overview of the number line. 图 1-1 展示了数轴的概貌。
   The number 0 is neither positive nor negative. 0 既不是正数，也不是负数。

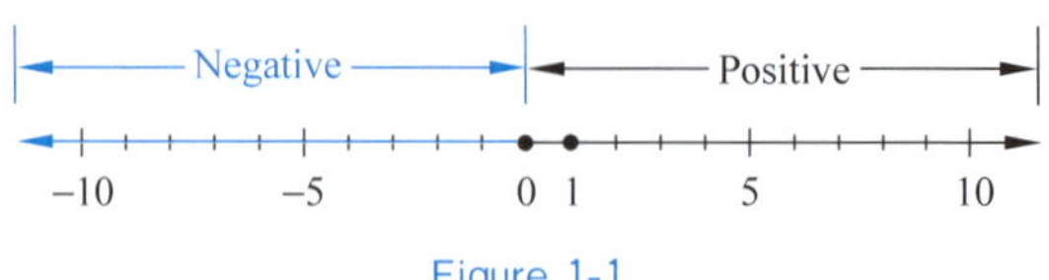

Figure 1-1

2. (1) Identity Property of Addition：$a+0=a$. The number 0 is the identity element. 加法单位元性质：$a+0=a$。0 是单位元。
   (2) Inverse Property of Addition：$a+(-a)=0$. The numbers a and $-a$ are additive inverses of each other. 加法逆元性质：$a+(-a)=0$。a 和 $-a$ 互为加法逆元。
3. (1) Identity Property of Multiplication：$a\cdot 1=a$. The number 1 is the identity element. 乘法单位元性质：$a\cdot 1=a$。1 是单位元。
   (2) Inverse Property of Multiplication：$a\cdot 1/a=1$, for which $a\neq 0$. The numbers a and $1/a$ are multiplicative inverses of each other. 乘法逆元性质：$a\cdot 1/a=1$。a 和 $1/a$ 互为乘法逆元 $(a\neq 0)$。
4. See zero (Section 1.1.1) and one (Section 1.1.2) for more properties. 更多性质见零(1.1.1 节)和一(1.1.2 节)的介绍。

1.2 Rational Numbers 有理数

1.2.1 Introduction 介绍

Rational Number 有理数

n. [ˈræʃənəl ˈnʌmbər]

Definition：A number that can be written as the quotient p/q, where p and q are integers such that $q\neq 0$. 有理数是能被商 p/q 表示的数，其中 p 与 q 均为整数，且 $q\neq 0$。

Notation：The set of rational numbers is denoted by $\mathbb{Q}$. 有理数集用$\mathbb{Q}$表示。

Properties：Closure Property, Commutative Property, Associative Property, Distributive Property, Identity Property, Inverse Properties of addition and multiplication. 有理数的加法与乘法具有封闭性、交换律、结合律、分配律、单位元性质、逆元性质。

Examples：$2/5, 4/7, 6, -982, 0.\overline{23}, 123.45, -0.\overline{3}, \cdots$ （数字上的线表示循环节）

Questions：Identify all the rational numbers below：

$23, 1.2, \sqrt{3}, -\sqrt[3]{5}, -123, 5.\overline{8}, 123/456, -\sqrt{3}/5$

Answers：$23, 1.2, -123, 5.\overline{8}, 123/456$.

Integer 整数

n. [ˈɪntədʒər]

Definition：A number with no fractional or decimal part：$\cdots, -3, -2, -1, 0, 1, 2, 3, \cdots$
不含分数或小数部分的数：$\cdots, -3, -2, -1, 0, 1, 2, 3, \cdots$

Notation: The set of integers is denoted by $\mathbb{Z}$.　整数集用$\mathbb{Z}$表示。

Properties: (1) The set of integers consists of:

- Positive Integers: 1,2,3,4,5,…
- Zero: 0
- Negative Integers: −1, −2, −3, −4, −5,…

整数集包含:正整数、零、负整数。

(2) Closure Property, Commutative Property, Associative Property, Distributive Property, Identity Property, Inverse Property of addition and multiplication.　整数的加法与乘法具有封闭性、交换律、结合律、分配律、单位元性质、逆元性质。

(3) The union of the set of positive integers and zero is known as the set of natural numbers, which is denoted by $\mathbb{N}$.　正整数集与0的并集为自然数集,用$\mathbb{N}$表示。

(4) Integers are rational because every integer N can be written as $N/1$. It follows that N and 1 are relatively prime.　根据有理数的定义,整数均为有理数。

Phrases: positive～　正整数, negative～　负整数, nonpositive～　非正整数, nonnegative～　非负整数

Questions: How many integers are there from −5 to 6, inclusive?　从−5至6有多少个整数?

Answers: 12.

1.2.2　Integers　整数

1.2.2.1　Parity　奇偶性

Even 偶(数)

adj. [ˈivən]

Definition: Integers that are divisible by 2.　偶数是能被2整除的整数。

Properties: 1. The nth nonnegative even number is $2(n-1)$.　第n个非负偶数是$2(n-1)$。

The 1st (smallest) nonnegative even number is 0.　第一个(最小的)非负偶数是0。

2. The ones digit of an even number must be one of 0,2,4,6, or 8.　偶数的个位数是0、2、4、6、8中任一个。

Examples: 0,2,4,6,8,…

Phrases: ～number

Questions: 1. What are all the even numbers greater than 2 and less than 10?　比2大且比10小的偶数有哪些?

2. What is the 2017th nonnegative even number?　第2017个非负偶数是多少?

3. How many even numbers are there from 8 to 888, inclusive?　从8至888有多少个偶数?

Answers: 1. 4,6,8.

2. $2(2017-1)=4032$.

3. 8 is the 5th nonnegative even number, and 888 is the 445th nonnegative even number. The question is the same as "how many integers are there from 5 to 445, inclusive?". The answer is $445-5+1=441$.

8 是第 5 个非负偶数,888 是第 445 个非负偶数。现在的问题是,从 5 至 445 有多少个整数？答案是 $445-5+1=441$。

Odd 奇(数)

adj. [ɑd]

Definition：Integers that are not divisible by 2. 奇数是不能被 2 整除的整数。

Properties：1. The nth positive odd number is $2n-1$. 第 n 个正奇数是 $2n-1$。
The 1st (smallest) positive odd number is 1. 第一个(最小的)正奇数是 1。
2. The ones digit of an odd number must be one of 1,3,5,7,or 9. 奇数的个位数是 1、3、5、7、9 中的任一个。

Examples：1,3,5,7,9,…

Phrases：～number

Questions：1. What are all the odd numbers greater than 2009 and less than 2017? 比 2009 大且比 2017 小的奇数有哪些?
2. What is the 2017th positive odd number? 第 2017 个正奇数是多少?
3. How many odd numbers are there from 111 to 999,inclusive? 从 111 至 999 有多少个奇数?

Answers：1. 2011,2013,2015.
2. $2(2017)-1=4033$.
3. 111 is the 56th positive odd number,and 999 is the 500th positive odd number. The question is the same as "how many integers are there from 56 to 500,inclusive?". The answer is $500-56+1=445$.
111 是第 56 个正奇数,999 是第 500 个正奇数。现在的问题是,从 56 至 500 有多少个整数？答案是 $500-56+1=445$。

1.2.2.2 Primes and Composites 质数与合数

Prime/Prime Number 质数/素数

adj./n. [praɪm/praɪm 'nʌmbər]

Definition：A positive integer whose only positive divisors are 1 and itself (exactly two positive divisors). 质数是正约数只有 1 和它本身的正整数(质数只有两个正约数)。

Properties：1. There are infinitely many prime numbers. 质数是无穷多的。
2. The number 1 is not a prime. 1 不是质数。
3. The number 2 is the smallest prime,and it is the only even prime. 2 是最小的质数,也是唯一一个偶质数。
4. The number 3 is the smallest odd prime. 3 是最小的奇质数。
5. All primes greater than 3 can be written as either $6n+1$ or $6n+5$,where n is a nonnegative integer. 所有大于 3 的质数可表示为 $6n+1$ 或 $6n+5$ 的形式,其中 n 为非负整数。

Examples：2,3,5,7,11,13,17,19,…,2017,…

Phrases：odd～ 奇质数,even～ 偶质数,～number 质数,～factor 质因数,～factorization 分解质因数

Questions：List all the prime numbers from the first 100 positive integers.　从前 100 个正整数中列出所有质数。

Answers：2,3,5,7,11,13,17,19,23,29,31,37,41,43,47,53,59,61,67,71,73,79,83,89,97.

Sieve of Eratosthenes　埃拉托色尼筛选法

n. [sɪv ʌv ˌɛrə'tɑsθəˌniz]

Definition：A simple ancient algorithm that finds all prime numbers up to a certain limit n, inclusive.　埃拉托色尼筛选法是能找出不超过 n 的所有质数的古老算法。

Algorithm：Since 2 is the smallest prime, the Sieve of Eratosthenes finds all prime numbers in the list L containing integers from 2 to n, inclusive.　因为 2 是最小的质数,用埃拉托色尼筛选法能找出列表 L(从 2 至 n 的整数列表)的质数：

L = 2,3,4,5,6,7,8,9,10,11,12,13,14,15,16,17,18,19,20,21,22,23,24,25,26,27,28,29,30,…

1. Circle the first uncrossed integer 2, then cross out every other multiples of this circled integer 2 in L. i.e. Cross out 4,6,8,10,12,14,…　圈起第一个没被划去的整数 2,然后划去这个整数 2 的所有倍数：

   ②,3,~~4~~,5,~~6~~,7,~~8~~,9,~~10~~,11,~~12~~,13,~~14~~,15,~~16~~,17,~~18~~,19,~~20~~,21,~~22~~,23,~~24~~,25,~~26~~,27,~~28~~,29,~~30~~,…

2. Circle the second uncrossed integer 3, then cross out every other uncrossed multiples of this circled integer 3 in L. i.e. Cross out 9,15,21,…　圈起第二个没被划去的整数 3,然后划去这个整数 3 没被划去的所有倍数：

   ②,③,~~4~~,5,~~6~~,7,~~8~~,~~9~~,~~10~~,11,~~12~~,13,~~14~~,~~15~~,~~16~~,17,~~18~~,19,~~20~~,~~21~~,~~22~~,23,~~24~~,25,~~26~~,~~27~~,~~28~~,29,~~30~~,…

3. Repeat Step 2 until every integer from 2 through n is either circled or crossed out. For example, Step 3 would be: Circle the third uncrossed integer 5, then cross out every other uncrossed multiples of this circled integer 5 in L. i.e. Cross out 25,35,…　重复第 2 步,直到 2～n 的每个数被圈起或划去。例如,第 3 步是：圈起第三个没被划去的整数 5,然后划去这个整数 5 没被划去的所有倍数：

   ②,③,~~4~~,⑤,~~6~~,7,~~8~~,~~9~~,~~10~~,11,~~12~~,13,~~14~~,~~15~~,~~16~~,17,~~18~~,19,~~20~~,~~21~~,~~22~~,23,~~24~~,~~25~~,~~26~~,~~27~~,~~28~~,29,~~30~~,…

   At the end, the list of circled integers is the list of all prime numbers up to n.　最后,所有被圈起来的整数就是所有不超过 n 的质数：

   ②,③,~~4~~,⑤,~~6~~,⑦,~~8~~,~~9~~,~~10~~,⑪,~~12~~,⑬,~~14~~,~~15~~,~~16~~,⑰,~~18~~,⑲,~~20~~,~~21~~,~~22~~,㉓,~~24~~,~~25~~,~~26~~,~~27~~,~~28~~,㉙,~~30~~,…

Notes：When an integer is crossed out, it is a composite because it has (prime) factors other than 1 and itself, namely the last circled integer so far.　当一个整数被划去,可得知它是合数。原因是它有 1 和它本身以外的(质)因数。这个质因数正是目前为止最后一个被圈起来的整数。

Below is the list of primes up to 1000：

2,3,5,7,11,13,17,19,23,29,31,37,41,43,47,53,59,61,67,71,73,79,83,89,97,101,103,107,109,113,127,131,137,139,149,151,157,163,167,173,179,181,191,193,197,

199,211,223,227,229,233,239,241,251,257,263,269,271,277,281,283,293,307,311,313,317,331,337,347,349,353,359,367,373,379,383,389,397,401,409,419,421,431,433,439,443,449,457,461,463,467,479,487,491,499,503,509,521,523,541,547,557,563,569,571,577,587,593,599,601,607,613,617,619,631,641,643,647,653,659,661,673,677,683,691,701,709,719,727,733,739,743,751,757,761,769,773,787,797,809,811,821,823,827,829,839,853,857,859,863,877,881,883,887,907,911,919,929,937,941,947,953,967,971,977,983,991,997.

Composite/Composite Number　合数

adj./n. [kɑm'pɑzɪt/kɑm'pɑzɪt 'nʌmbər]

Definition：A positive integer that has positive divisors other than 1 and itself (at least three positive divisors).　合数是正约数不止 1 和它本身的正整数(至少有 3 个正约数)。

Properties：1. There are infinitely many composites.　合数是无穷多的。

2. The number 1 is not a composite.　1 不是合数。

3. The number 4 is the smallest composite, and is the smallest even composite.　4 是最小的(偶)合数。

4. The number 9 is the smallest odd composite.　9 是最小的奇合数。

Examples：4,6,8,9,10,12,14,15,16,18,20,…

Phrases：odd～　奇合数，even～　偶合数，～number　合数，～factor　合数因数

Questions：List all the composite numbers from the first 40 positive integers.　从前 40 个正整数中列出所有合数。

Answers：4,6,8,9,10,12,14,15,16,18,20,21,22,24,25,26,27,28,30,32,33,34,35,36,38,39,40.

1.2.2.3　Factors and Multiples　约数与倍数

Factor/Divisor　乘数、因子(因数、约数)

n. ['fæktər/dɪ'vaɪ.zɚ]

Definition：1. Definition of Factor：The positive integers to be multiplied in a multiplication.　乘数是乘法算式中相乘的数。

2. Definition of Factor/Divisor：For positive integers a and b, a is a factor/divisor of b if b is divisible by a.　因子(因数、约数)的定义：a 和 b 为正整数。若 b 能被 a 整除，则 a 是 b 的因子/因数/约数。

Properties：1. The number a is always a factor/divisor of a.　a 是它本身的约数。

2. If a is a factor/divisor of b and b is a factor/divisor of a, then $a=b$.　若 a 是 b 的约数，且 b 是 a 的约数，则 $a=b$。

3. All positive integers other than 1 have at least 2 factors/divisors：1 and itself.　所有除 1 以外的正整数至少有两个约数：1 和它本身。

4. The number 1 only has one factor/divisor：itself.　1 只有一个约数：它本身。

5. Factors/divisors of a positive integer n smaller than the n itself are known as the proper divisors of n.　对于正整数 n，小于 n 的约数叫作 n 的真约数。

Examples：1. (Definition 1) In the multiplication $3\cdot5=15$, 3 and 5 are known as the factors.　在乘法算式 $3\times5=15$ 中，3 和 5 均为乘数。

2.（Definition 2）45 is a factor/divisor of 90.　45 是 90 的约数。

3.（Definition 2）1 is a factor/divisor of 2017.　1 是 2017 的约数。

4.（Definition 2）2017 is a factor/divisor of 2017.　2017 是 2017 的约数。

Phrases：common～　公约数，greatest common～（GCF or GCD）　最大公约数，prime～　质因数，composite～　合数因数，odd～　奇约数，even～　偶约数

Questions：1. List all the factors of 60.　列出 60 的所有约数。

2. List all the proper divisors of 60.　列出 60 的所有真约数。

Answers：1. 1,2,3,4,5,6,10,12,15,20,30,60.

2. 1,2,3,4,5,6,10,12,15,20,30.

Common Factor/Common Divisor　公约数

n. [ˌkɒm.ən ˈfæk.tər / ˌkɒm.ən dɪˈvaɪ.zɚ]

Definition：A common factor/common divisor of two or more positive integers is a positive integer that divides all of these positive integers.　两个或更多的正整数的公约数是能整除它们的正整数。

Properties：1. The number 1 is always a common factor/divisor of all groups of positive integers, and is their least common factor/divisor.　1 是所有正整数组的公约数，是它们的最小公约数。

2. If 1 is the only common factor of two positive integers, then these two positive integers are known as relatively prime.　若 1 是两个正整数的唯一公约数，则这两个正整数互质。

Examples：50 is a common factor of 100 and 150.

Phrases：greatest～（GCF/GCD）　最大公约数

Questions：1. What are all the common factors of 20 and 30?

2. What are all the common factors of 40,50,and 75?

Answers：1. Factors of 20：1,2,4,5,10,20.

Factors of 30：1,2,3,5,6,10,15,30

Common factors：$\boxed{1,2,5,10}$.

2. Factors of 40：1,2,4,5,8,10,20,40

Factors of 50：1,2,5,10,25,50

Factors of 75：1,3,5,15,25,75

Common factors：$\boxed{1,5}$.

Greatest Common Factor（GCF）/Greatest Common Divisor（GCD）　最大公约数

n. [ˈgreɪtəst …]

Definition：See definition for common factor. GCF is the largest of those common factors.　见公约数的定义，最大公约数是公约数里最大的数。

Notation：GCD(#,#,…) or GCF(#,#,…)

Properties：1. When the GCF of two positive integers is 1, the positive integers are said to be relatively prime.　若两个正整数的最大公约数为 1，则它们互质。

2. If a is a multiple of b, then $\text{GCF}(a,b)=b$.　若 a 是 b 的倍数，则 $\text{GCF}(a,b)=b$。

3. GCF(1,…,…)=1.

4. The GCF can be found in the following ways:
最大公约数有下面几种算法:
(1) Euclidean Algorithm 欧几里得算法;
(2) Short Division 短除法;
(3) Prime Factorization 分解质因数。

Examples: 1. GCF(8,12)=4.
2. GCF(10,23)=1 (Property 1).
3. GCF(80,20)=20 (Property 2).

Questions: Jazmin is completing an art project. She has two pieces of construction paper. The first piece is 56 centimeters wide and the second piece is 70 centimeters wide. Jazmin wants to cut the paper into strips that are equal in width and are as wide as possible. How wide should Jazmin cut for each strip?

Answers: Since GCF(56,70)=14, the answer is 14 centimeters.

Multiple (正)倍数

n. ['mʌltəpəl]

Definition: For positive integers a and b, a is a multiple of b if a is divisible by b. a 和 b 均为正整数。若 a 能被 b 整除,则 a 是 b 的倍数。

Properties: 1. The number a is always a multiple of a. a 是它本身的倍数。
2. If a is a multiple of b and b is a multiple of a, then $a=b$. 若 a 是 b 的倍数,且 b 是 a 的倍数,则 $a=b$。
3. Each positive integer has itself as a multiple. 每个正整数都是它本身的倍数。
4. Every positive integer is a multiple of 1. 所有正整数都是 1 的倍数。

Examples: 15 is a multiple of 1, 3, 5, and 15 because 15/1=15, 15/3=5, 15/5=3, 15/15=1.

Phrases: common～ 公倍数, least common～ 最小公倍数

Questions: List five smallest multiples of 40.

Answers: 40, 80, 120, 160, 200.

Common Multiple 公倍数

n. [ˌkɒm.ən 'mʌltəpəl]

Definition: A common multiple of two or more positive integers is a positive integer that is divisible by all of these positive integers. 两个或更多正整数的公倍数是能被它们整除的正整数。

Properties: In a group of positive integers, the product of all these positive integers is one of their common multiples. 在任何正整数组中,这些正整数的积是它们的其中一个公倍数。

Examples: 900 is a common multiple of 100 and 150.

Phrases: least～(LCM)

Questions: 1. What are some common multiples of 20 and 30?
2. What are all the common multiples of 3, 4, and 6?

Answers: 1. Multiples of 20: 20, 40, 60, 80, 100, 120, 140, 160, 180, …

Multiples of 30：30,60,90,120,150,180,⋯

Common multiples：60,120,180,⋯(any multiple of 60)

2. Multiples of 3：3,6,9,12,15,18,21,24,27,30,33,36,⋯

Multiples of 4：4,8,12,16,20,24,28,32,36,⋯

Multiples of 6：6,12,18,24,30,36,⋯

Common multiples：12,24,36,⋯(any multiple of 12)

Least Common Multiple (LCM)　最小公倍数

n. [list ⋯]

Definition：See definition for common multiple. The LCM is the smallest of these common multiples.　见公倍数的定义。最小公倍数是公倍数中最小的数。

Notation：LCM(#,#,⋯)

Properties：1. If two positive integers are relatively prime, then their LCM is their product. Phrasing differently, if $\text{GCF}(a,b)=1$, then $\text{LCM}(a,b)=ab$.　若两个正整数互质，则它们的最小公倍数是它们的积。

2. If a is a multiple of b, then $\text{LCM}(a,b)=a$.　若 a 是 b 的倍数，则 $\text{LCM}(a,b)=a$。

3. $\text{LCM}(1,x,y,\cdots)=\text{LCM}(x,y,\cdots)$.

4. The LCM can be found in the following ways:

最小公倍数有下面几种算法：

(1) Short Division　短除法；

(2) Prime factorization and observation　分解质因数。

5. For positive integers a and b, $\text{GCF}(a,b)\cdot\text{LCM}(a,b)=a\cdot b$.　若 a 和 b 为正整数，则 $\text{GCF}(a,b)\cdot\text{LCM}(a,b)=a\cdot b$。

Property 1 comes from Property 5.　性质 1 是从性质 5 推导出来的。

Proofs：To prove Property 5, see prime factorization in Section 1.2.2.6.　性质 5 的证明见 1.2.2.6 节分解质因数(方法)。

Examples：1. $\text{LCM}(8,12)=24$.

2. $\text{LCM}(10,23)=230$ (Property 1).

3. $\text{LCM}(80,20)=80$ (Property 2).

Questions：In a bakery shop, every 12th customer wins a cookie, and every 15th customer wins a cupcake. What is the value of n if the nth customer is the first one who wins both a cookie and a cupcake?

Answers：Since $\text{LCM}(12,15)=60$, the 60th customer is the first who wins both a cookie and a cupcake. Thus $n=\boxed{60}$.

1.2.2.4 Relatively Prime/Coprime　互质

Relatively Prime/Coprime　互质

adj. [ˈrel.ə.tɪv.li praɪm/ˈkəʊprʌɪm]

Definition：Two positive integers are said to be relatively prime/coprime if their only common factor is 1, or their greatest common factor is 1.　若两个正整数的公约数只有1(最大公约数为 1)，则它们互质。

Positive integers a and b are said to be relatively prime/coprime if GCF$(a,b)=1$. 若正整数 a 和 b 满足 GCF$(a,b)=1$,则它们互质。

Properties: 1. If a and b are relatively prime/coprime, then GCF$(a,b)=1$ and LCM$(a,b)=ab$. 若 a 和 b 互质,则 GCF$(a,b)=1$ 和 LCM$(a,b)=ab$。

2. If n is a positive integer, n and $n+1$ are relatively prime/coprime. 若 n 为正整数,则 n 和 $n+1$ 互质。

3. 1 and any positive integers are relatively prime/coprime. 1 和所有正整数互质。

4. Two even numbers can never be relatively prime/coprime. 两个偶数永远不互质。

Proofs: 1. To prove Property 1, use the definition of relatively prime/coprime. and Property 5 of least common multiple. 要证明性质 1,可用互质的定义以及最小公倍数的性质 5。

2. To prove Property 2, use Euclidean Algorithm. 要证明性质 2,可用欧几里得算法。

3. To prove Property 3, note that GCF$(1,\cdots,\cdots,\cdots)=1$ (Property 3 of greatest common factor (GCF)/greatest common divisor (GCD)). 要证明性质 3,可参见最大公约数的性质 3。

4. To prove Property 4, note that all even numbers have 2 as a common factor. 要证明性质 4,注意 2 是所有偶数的公约数。

Examples: 1. 10 and 17 are relatively prime.

2. 7 and 8 are relative prime (Property 2).

3. 4 and 9 are relatively prime.

Questions: 1. Are 15 and 24 relatively prime?

Are 20 and 27 relatively prime?

2. What are the GCF and LCM of 20 and 27?

Answers: 1. No, since GCF$(15,24)=3$.

Yes, since GCF$(20,27)=1$.

2. Since 20 and 27 are relatively prime, GCF$(20,27)=1$ and LCM$(20,27)=540$.

1.2.2.5 Factorization 分解

视频 1

Factorization 分解

n. [fæktərʌɪ'zeɪʃən]

Definition: Rewriting a positive integer as a product of two or more integers greater than 1. 分解是指把一个正整数改写成两个或更多个大于 1 的整数的积。

Properties: Only composites are factorable, by its definition. 根据定义,只有合数能被分解。

Examples: Some possible ways to factor 24:

4×6

2×12

$2\times 2\times 6$

$2\times 2\times 2\times 3$

Phrases: prime factorization 分解质因数

Questions: Evaluate each of the following efficiently. Hint: Factoring. 速算下面各题。提示:用分解方法。

(1) 25×24.

(2) 125×32.

Answer: Recognize that $25\times4=100$ and $125\times8=1000$.

(1) $25\times24=25\times(4\times6)=(25\times4)\times6=100\times6=600$.

(2) $125\times32=125\times(8\times4)=(125\times8)\times4=1000\times4=4000$.

Factor Tree　因子树

n. [ˈfæktər tri]

Definition: A visual view of factorization, which is used to help doing prime factorization. Only used in composite.　因子树是指分解方法的图解,对分解质因数有帮助。只适用于分解合数。

Properties: The numbers at the end of the branches must be primes.　因子树最底层的数字均为质数。

By convention, in a branch, the number on the left branch does not exceed that on the right branch.　根据惯例,在一个分支里,左边的数不超过右边的数。

Examples: Figure 1-2 and Figure 1-3 show two possible factor trees for 24 and their analogues.

图 1-2 与图 1-3 展示了两种对 24 进行分解的方法以及因子树的对照。

$24=3\times8$　　　　$24=2\times12$

$\quad=3\times2\times4$　　　　$\quad=2\times3\times4$

$\quad=3\times2\times2\times2$　　　　$\quad=2\times3\times2\times2$

Can you come up with another one?

Questions: What is one possible factor tree for 60?

Answers: Figure 1-4 shows the answer.

$60=6\times10$

$60=2\times3\times2\times5$

$60=2\times2\times3\times5$

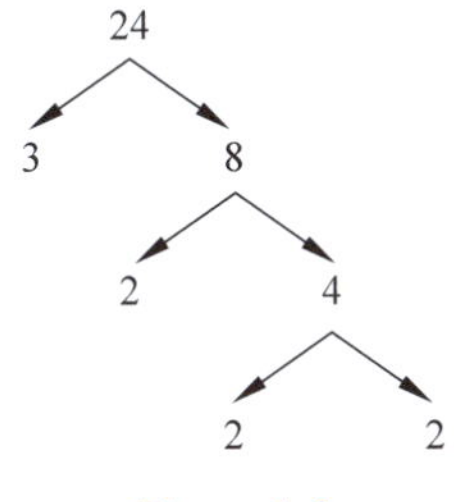

Figure 1-2

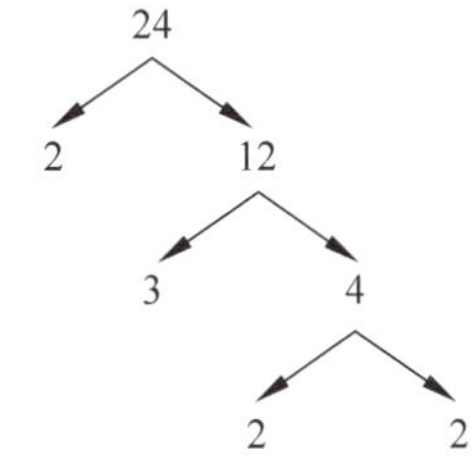

Figure 1-3

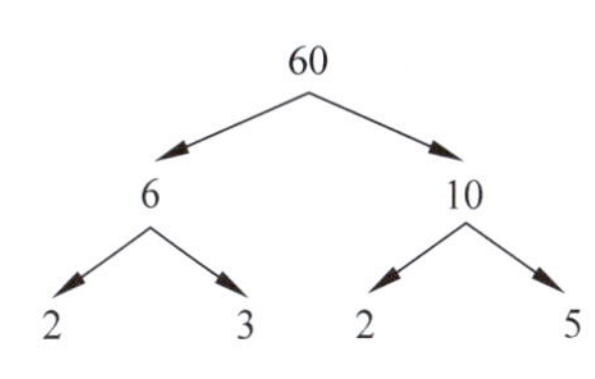

Figure 1-4

Prime Factorization　分解质因数

n. [prɑɪm fæktərʌɪˈzeɪʃən]

Definition: Rewrite a composite as a product of primes, for which are not necessarily distinct.　分解质因数是指把一个合数改写成若干质数的积,这些质数可以是相同的。

The prime factorization of a prime is just the prime itself. (trivial case)　一般质数的分解质因数是它本身(特殊情况)。

Notation: The prime factorization of a positive integer m can be written uniquely as $m=p_1^{n_1}p_2^{n_2}p_3^{n_3}\cdots$, for which each p is a prime, and $p_1<p_2<p_3<\cdots$, and each n is a nonnegative integer.　正整数 m 的分解质因数是唯一的: $m=p_1^{n_1}p_2^{n_2}p_3^{n_3}\cdots$, p 均是质

数,其中 $p_1<p_2<p_3<\cdots$。n 是一个非负整数。

If a certain n (say, n_i) equals to 0, then we omit that factor $p_i^{n_i}$ since it is equal to 1, according to Property 6 of exponent. 若某个 n(设 n_i)等于 0,则可以不写出该因子 $p_i^{n_i}$,原因是它等于 1。见指数的性质 6。

Properties: 1. There is only one prime factorization for each composite. The proof will use some techniques from number theory, which is out of the scope of middle school algebra. 每个合数的分解质因数都是唯一的。这个证明会用到数论的某些知识点。这些知识点超出了中学代数的范围。

2. Prime factorization can be done based on a factor tree. 分解质因数可以根据因子树写出来。

Examples: 1. The prime factorization of 24 is $2^3\times3$(Refer to the example of factor tree).

2. The prime factorization of 60 is $2^2\times3\times5$(Refer to the example of factor tree).

3. The prime factorization of 150 is $2\times3\times5^2$.

Questions: What is the prime factorization of 90?

Answers: Every time, we will pull out a prime factor from the remaining factors.

The remaining factor in each step is written in blue:

$$\begin{aligned}90&=2\times45\\&=2\times3\times15\\&=2\times3\times3\times5\\&=\boxed{2\times3^2\times5}.\end{aligned}$$

1.2.2.6 Techniques of Finding GCF and LCM 求最大公约数和最小公倍数的方法

Euclidean Algorithm 欧几里得算法

n. [ju'klɪd.i.ən 'ælgəˌrɪðəm]

Purpose: A method to find the GCF of two positive integers. 欧几里得算法是求两个正整数的最大公约数的方法。

Method: To find GCF of positive integers a and b, if $a>b$, then $\mathrm{GCF}(a,b)=\mathrm{GCF}(b,a \bmod b)=\cdots=\mathrm{GCF}(d,0)=d$.

In this method, note that:

1. "a mod b" represents the remainder of a divides by b. It is guaranteed that (a mod b)$<b$ because the remainder in a division is always smaller than the divisor. "a mod b" 代表的是"a 除以 b 的余数"。它比 b 小,因为除法中的余数总是比除数小。

2. $\mathrm{GCF}(d,0)=d$ for all positive integers d. This is a degenerated case for GCF (involving a nonpositive integer, 0). Note that 0 is divisible by every nonzero integer. 对于所有的正整数 d,$\mathrm{GCF}(d,0)=d$。这是最大公约数的简化情况(涉及非正数 0)。注意:0 能被任何非零的整数整除。

Lemmas: If positive integers a and b, for which $a>b$, are multiples of a positive integer, k, then both $a+b$ and $a-b$ are divisible by k. 若正整数 a 和 b(满足 $a>b$)都是正整数 k 的倍数,则 $a+b$ 和 $a-b$ 均能被 k 整除。

We can show this lemma by letting $a=mk$ and $b=nk$, for some positive integers m and

n. Therefore, $a+b=mk+nk=(m+n)k$, and $a-b=mk-nk=(m-n)k$. By the Closure Property of integers over addition, both $m+n$ and $m-n$ are integers, so both $a+b$ and $a-b$ are divisible by k. 要证明这个引理,设 $a=mk$ 和 $b=nk$。m 和 n 均为正整数。所以 $a+b=mk+nk=(m+n)k$,且 $a-b=mk-nk=(m-n)k$。因为整数加法的封闭性,$m+n$ 和 $m-n$ 均为整数。所以 $a+b$ 和 $a-b$ 均能被 k 整除。

Proof: Suppose d is a common factor of a and b (such that $a>b$). We can skip-count to a by d's, landing on b along the way. So, we can skip-count from 0 to b, by d's. In addition, we can use d to skip-count the remaining $a-b$ units from b to a. 设 d 是 a 和 b 的公约数($a>b$)。可以以 d 为基数从 0 跳数到 a,期间在 b 上停留。也就是说,可以以 d 为基数从 0 跳数到 b。此外,还可以以 d 为基数从 b 跳数 $a-b$ 个单位到 a。

So, if d is a common factor of a and b, then it is also a factor of $a-b$. Therefore, d is a common factor of b and $a-b$. This is true for all common factors of a and b: they are factors of $a-b$. 若 d 是 a 和 b 的公约数,则它也是 $a-b$ 的约数,即 d 是 b 和 $a-b$ 的公约数。故所有 a 和 b 的公约数也是 $a-b$ 的约数。

Additionally, any common factor of b and $a-b$ can be used to skip-count from 0 to a, stopping at b. 另外,可以以 b 和 $a-b$ 的任何公约数为基数,从 0 跳数到 a,期间在 b 上停留。So, any common factor of b and $a-b$ is also a factor of a. 也就是说,b 和 $a-b$ 的公约数也是 a 的约数。

We have shown that all common factors of a and b are factors of $a-b$, and all common factors of b and $a-b$ are factors of a. Therefore, we know that the GCF maintains in the following statement:

已证明所有 a 和 b 的公约数也是 $a-b$ 的约数,且所有 b 和 $a-b$ 的公约数也是 a 的约数。由此可知以下求最大公约数的公式成立.

(1) $\mathrm{GCF}(a,b)=\mathrm{GCF}(a-b,b)$.

Another version is:

另一种表示:

(2) $\mathrm{GCF}(a+b,b)=\mathrm{GCF}(a,b)$.

As we can use this algorithm repeatedly, which gives $\mathrm{GCF}(a,b)=\mathrm{GCF}(b,a \bmod b)$ as a shortcut. 当需要重复地使用这个算法时,可以使用得到的简便公式: $\mathrm{GCF}(a,b)=\mathrm{GCF}(b,a \bmod b)$。

Notes: Once the GCF is obvious, we can stop the algorithm and give the answer. 当最大公约数很明显的时候,可以不用算法,直接写出答案。

Questions: Using the Euclidean Algorithm, determine the GCF for each pair of positive integers below.

(1) 8,12

(2) 40,401

(3) 899,928

(4) 12,1208

Answers: (1) $\mathrm{GCF}(12,8)=\mathrm{GCF}(8,12 \bmod 8)=\mathrm{GCF}(8,4)=\mathrm{GCF}(4,8 \bmod 4)=\mathrm{GCF}(4,0)=4$.

(2) $\mathrm{GCF}(401,40)=\mathrm{GCF}(40,401 \bmod 40)=\mathrm{GCF}(40,1)=1$.

(3) $\mathrm{GCF}(928,899)=\mathrm{GCF}(899,928 \bmod 899)=\mathrm{GCF}(899,29)=29$.

(4) GCF(12,1208) = GCF(12,1208 mod 12) = GCF(12,8) = 4, as shown in Question 1.

Short Division 短除法

n. [ʃɔrt dɪ'vɪʒən]

Purpose: A method to find the GCF and LCM of two or more positive integers. 短除法是求两个或更多正整数的最大公约数和最小公倍数的方法。

Method: To find the GCF and LCM of positive integers a and b, the procedure using short division would be:

要求正整数 a 和 b 的最大公约数和最小公倍数,使用短除法的步骤如下:

1. Divide each of a and b by a common factor greater than 1, if possible. 可以将 a 和 b 分别除以它们大于 1 的公约数。
2. If the quotients are not relatively prime, divide each of the quotients by a common factor greater than 1. 若商不互质,继续将商分别除以它们大于 1 的公约数。
3. Repeat Step 2, until the quotients are relatively prime. 继续步骤 2,直到商互质。

GCF(a, b) = product of those common factors. 最大公约数为所有那些公约数的积。

LCM(a, b) = product of those common factors and quotients. 最小公倍数为所有那些公约数和商的积。

The common factors mentioned in steps 1~3 are prime factors by convention. However, regardless of its primality, we will arrive at the same GCF and LCM. 在步骤 1~3 中的公约数一般都是质因数,但是不论它们是否为质数,最后得到的最大公约数和最小公倍数都是相同的。

Generalization: The same procedure also works in finding GCF and LCM of three or more positive integers. 上述步骤可应用到求 3 个或更多个正整数的最大公约数和最小公倍数上。

In steps 1, 2, and 3:

To find GCF, if the positive integers/quotients do not have a common factor greater than 1, stop the short division and multiply the common factors pulled out so far—this is the GCF. See questions (5) and (6) for examples. 求最大公约数:若正整数/商没有大于 1 的公约数,停止短除法并把分解出来的公约数相乘,得出最大公约数。

To find LCM, if two or more positive integers/quotients have a common factor greater than 1, then pull out that common factor. For positive integers/quotients that are not divisible by that common factor, leave them alone. The product of the pulled out common factors and the quotients is the LCM. See questions (5) and (6) for examples. 要找出最小公倍数,若两个或更多的正整数/商有大于 1 的公约数,则把这个公约数分解出来。若正整数/商不能被这个公约数整除,则保持不变。最小公倍数为分解出来的公约数与商的积。

Questions: Using short division, determine the GCF and LCM for each group of positive integers below.

(1) 8, 12

(2) 15, 21

(3) 30, 40

(4) 12, 25

(5) 30,35,48

(6) 22,28,36,40

Answers: (1) Figure 1-5 gives the short division.

GCF(8,12) = 2×2 = 4.

LCM(8,12) = 2×2×2×3 = 24.

(2) Figure 1-6 gives the short division.

GCF(15,21) = 3.

LCM(15,21) = 3×5×7 = 105.

(3) Figure 1-7 gives the short division.

GCF(30,40) = 2×5 = 10.

LCM(30,40) = 2×5×3×4 = 120.

Figure 1-5

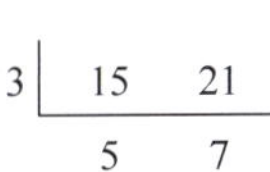

Figure 1-6

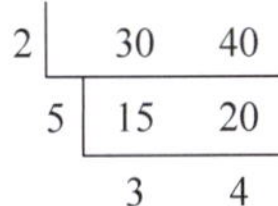

Figure 1-7

(4) Since 12 and 25 do not have any common prime factor, or, they are relatively prime, we have

GCF(12,25) = 1.

LCM(12,25) = 12×25 = 300.

(5) Since 30,35, and 48 do not have any common prime factor, we have

GCF(30,35,48) = 1.

Since at least two of 30,35, and 48 have a common divisor greater than 1, we can use short division from here, as shown in Figure 1-8:

LCM(30,35,48) = 2×3×5×1×7×8 = 1680.

(6) Figure 1-9 shows the short division to find GCF.

GCF(22,28,36,40) = 2.

Figure 1-10 shows the short division to find LCM.

LCM(22,28,36,40) = 2×2×11×7×9×10 = 27 720.

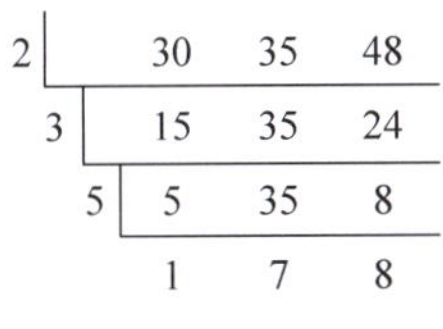

Figure 1-8

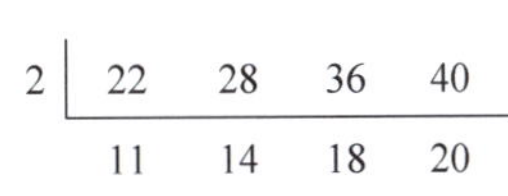

Figure 1-9

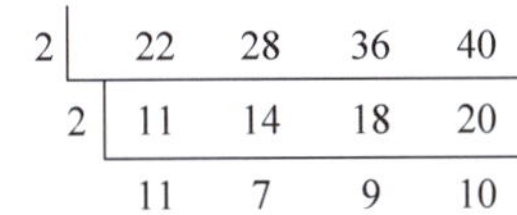

Figure 1-10

Prime Factorization　分解质因数(方法)

n. [praɪm fæktərʌɪ'zeɪʃən]

Purpose: A method to find the GCF and LCM of two or more positive integers.　分解质因数是求两个或更多个正整数的最大公约数和最小公倍数的方法。

Method: For positive integers a and b such that $a=p_1^{m_1}p_2^{m_2}p_3^{m_3}\cdots$, and $b=p_1^{n_1}p_2^{n_2}p_3^{n_3}\cdots$, for which each $p_i(i=1,2,\cdots)$ is a prime with $p_1<p_2<p_3<\cdots$, and each $m_i(i=1,2,\cdots)$ and $n_i(i=1,2,\cdots)$ is a nonnegative integer, we have:

若 $a=p_1^{m_1}p_2^{m_2}p_3^{m_3}\cdots$，且 $b=p_1^{n_1}p_2^{n_2}p_3^{n_3}\cdots$，$p_i(i=1,2,\cdots)$均为质数，$m_i(i=1,2,\cdots)$和 $n_i(i=1,2,\cdots)$均为非负整数，则得到

$\text{GCF}(a,b)=p_1^{\min(m_1,n_1)}p_2^{\min(m_2,n_2)}p_3^{\min(m_3,n_3)}\cdots$,

$\text{LCM}(a,b)=p_1^{\max(m_1,n_1)}p_2^{\max(m_2,n_2)}p_3^{\max(m_3,n_3)}\cdots$.

This is an intuitive way to find GCF and LCM. 这是一个求最大公约数和最小公倍数的直观方法。

Results: We can use this method to prove Property 5 of LCM: For two positive integers a and b, $\text{GCF}(a,b)\times\text{LCM}(a,b)=a\cdot b$.

Using Property 1 of exponent in Section 1.6.3, we have:

用 1.6.3 节指数的性质 1 能得到下面这个结果：

$$\text{GCF}(a,b)\times\text{LCM}(a,b)=(p_1^{\min(m_1,n_1)}p_2^{\min(m_2,n_2)}p_3^{\min(m_3,n_3)}\cdots)\cdot(p_1^{\max(m_1,n_1)}p_2^{\max(m_2,n_2)}p_3^{\max(m_3,n_3)}\cdots)$$
$$=p_1^{\min(m_1,n_1)+\max(m_1,n_1)}p_2^{\min(m_2,n_2)+\max(m_2,n_2)}p_3^{\min(m_3,n_3)+\max(m_3,n_3)}\cdots.$$
$$a\times b=p_1^{\min(m_1,n_1)+\max(m_1,n_1)}p_2^{\min(m_2,n_2)+\max(m_2,n_2)}p_3^{\min(m_3,n_3)+\max(m_3,n_3)}\cdots.$$

We have shown the property.

Generalization: For three or more positive integers, each has the prime factorization of $p_1^{n_1}p_2^{n_2}p_3^{n_3}\cdots$ for which each $p_i(i=1,2,\cdots)$ is a prime, and $p_1<p_2<p_3<\cdots$, and each exponent is a nonnegative integer, we have:

对于 3 个或更多的正整数，满足每个正整数的分解质因数是 $p_1^{n_1}p_2^{n_2}p_3^{n_3}\cdots$，$p_i(i=1,2,\cdots)$均是质数，使得 $p_1<p_2<p_3<\cdots$，且指数都是非负数，可得到

$\text{GCF(all)}=p_1^{\min(\text{all exponents of } p_1)}p_2^{\min(\text{all exponents of } p_2)}p_3^{\min(\text{all exponents of } p_3)}\cdots$,

$\text{LCM(all)}=p_1^{\max(\text{all exponents of } p_1)}p_2^{\max(\text{all exponents of } p_2)}p_3^{\max(\text{all exponents of } p_3)}\cdots$.

Questions: Using prime factorizations, determine the GCF and LCM for each group of positive integers below.

(1) 8,12

(2) 15,21

(3) 30,40

(4) 12,25

(5) 30,35,48

(6) 22,28,36,40

Answers: (1) $8=2^3$

$12=2^2\times3$

$\text{GCF}(8,12)=2^2=4$.

$\text{LCM}(8,12)=2^3\times3=24$.

(2) $15=3\times5$

$21=3\times7$

$\text{GCF}(15,21)=3$.

$LCM(15,21)=3\times5\times7=105$.

(3) $30=2\times3\times5$

$40=2^3\times5$

$GCF(30,40)=2\times5=10$.

$LCM(30,40)=2^3\times3\times5=120$.

(4) $12=2^2\times3$

$25=5^2$

$GCF(30,40)=1$.

$LCM(30,40)=2^2\times3\times5^2=300$.

(5) $30=2\times3\times5$

$35=5\times7$

$48=2^4\times3$

$GCF(30,35,48)=1$.

$LCM(30,35,48)=2^4\times3\times5\times7=1680$.

(6) $22=2\times11$

$28=2^2\times7$

$36=2^2\times3^2$

$40=2^3\times5$

$GCF(22,28,36,40)=2$.

$LCM(22,28,36,40)=2^3\times3^2\times5\times7\times11=27\,720$.

1.2.2.7 Perfect *n*th Powers　完全 *n* 次方

Perfect Square　完全平方

n. [ˈpɜrfɪkt skweər]

Definition：The product of an integer and itself.　完全平方是指整数和它本身的乘积。

A perfect square is sometimes called a "square" for short.

Properties：The ones digit of a perfect square cannot be 2,3,7,or 8.　完全平方的个位数不能是2、3、7、8。

Note that for integers,the ones digit cycles：0,1,2,3,4,5,6,7,8,9. The ones digit of perfect squares cycles correspondingly：0,1,4,9,6,5,6,9,4,1.　整数的个位数以0,1,2,3,4,5,6,7,8,9的顺序循环,对应的完全平方的个位数以0,1,4,9,6,5,6,9,4,1的顺序循环。

Examples：$0^2=0,1^2=1,2^2=4,3^2=9,4^2=16,5^2=25,6^2=36,7^2=49,8^2=64,9^2=81,10^2=100,\cdots$

Phrases：The word "square" has two meanings in algebra：

在代数里,square一词有两种含义：

1. (noun) A perfect square.　完全平方。
2. (verb) To raise a number to the second power.　取一个数的二次幂/平方。

   Squaring 5,we get 25.　取5的平方得25。

   7 squared is 49.　7的平方是49。

Questions：1. What is the next perfect square of 144?

2. List all perfect squares between 101 and 200.

Answers：1. Since $144 = 12^2$，the next perfect square is $13^2 = 169$.

2. Since $(10^2 = 100) < 101 < (11^2 = 121)$ and $(14^2 = 196) < 200 < (15^2 = 225)$，the list of perfect squares is $11^2, 12^2, 13^2, 14^2$，or $\boxed{121,144,169,196}$.

Perfect Cube 完全立方

n. [ˈpɜrfɪkt kjub]

Definition：The product of three equal integers，or，the product of an integer and its square. 完全立方是指三个相等的整数的积，或者说是一个整数与它平方的积。

A perfect cube is sometimes called a "cube" for short.

Properties：Note that for integers，the ones digit cycles：0，1，2，3，4，5，6，7，8，9. The ones digit of perfect cubes cycles correspondingly：0，1，8，7，4，5，6，3，2，9. 整数的个位数以 0，1，2，3，4，5，6，7，8，9 的顺序循环，对应的完全立方的个位数以 0，1，8，7，4，5，6，3，2，9 的顺序循环。

Examples：$0^3 = 0, 1^3 = 1, 2^3 = 8, 3^3 = 27, 4^3 = 64, 5^3 = 125, 6^3 = 216, 7^3 = 343, 8^3 = 512, 9^3 = 729, 10^3 = 1000, \cdots$

Also，note that perfect cubes can be negative：完全立方可以为负数。

$(-1)^3 = -1, (-2)^3 = -8, (-3)^3 = -27, (-4)^3 = -64, (-5)^3 = -125, (-6)^3 = -216, (-7)^3 = -343, (-8)^3 = -512, (-9)^3 = -729, (-10)^3 = -1000, \cdots$

Phrases：The word "cube" has two meanings：

在代数中，cube 一词有两种含义：

1. (noun) A perfect cube. 完全立方。
2. (verb) To raise a number to the third power. 取一个数的三次幂/立方。

   Cubing 5，we get 125. 取 5 的立方得 125。

   7 cubed is 343. 7 的立方是 343。

Questions：1. What is the next perfect cube of 343?

2. List all perfect cubes between 1001 and 2000.

Answers：1. Since $343 = 7^3$，the next perfect cube is $8^3 = 512$.

2. Since $(10^3 = 1000) < 1001 < (11^3 = 1331)$ and $(12^3 = 1728) < 2000 < (13^3 = 2197)$，the list of perfect cubes is $11^3, 12^3$，or $\boxed{1331,1728}$.

Perfect *n*th Power 完全 *n* 次幂(方)

n. [ˈpɜrfɪkt enθ ˈpɑʊər]

Definition：The product of n equal integers，where n is an integer greater than or equal to 2. 完全 n 次幂是指 n 个相等整数的积，其中 n 为大于或等于 2 的整数。

A perfect nth power is sometimes called an nth power for short. 完全 n 次幂简称 n 次幂。

Properties：We refer a perfect 2nd power $(n = 2)$ as a perfect square. 完全二次幂称为完全平方。

We refer a perfect 3rd power $(n = 3)$ as a perfect cube. 完全三次幂称为完全立方。

Examples：1. Examples of perfect squares：

完全平方的例子：

$0^2=0, 1^2=1, 2^2=4, 3^2=9, 4^2=16, 5^2=25, 6^2=36, 7^2=49, 8^2=64, 9^2=81, 10^2=100, \cdots$

2. Examples of perfect cubes:
   完全立方的例子:
   $0^3=0, 1^3=1, 2^3=8, 3^3=27, 4^3=64, 5^3=125, 6^3=216, 7^3=343, 8^3=512, 9^3=729, 10^3=1000, \cdots$
3. Examples of perfect 4th powers:
   完全四次幂的例子:
   $0^4=0, 1^4=1, 2^4=16, 3^4=81, 4^4=256, 5^4=625, 6^4=1296, 7^4=2401, 8^4=4096, 9^4=6561, 10^4=10\,000, \cdots$
4. The numbers 0 and 1 are always perfect nth powers because $0^k=0$ (for all real numbers k such that $k>0$) and $1^k=1$ (for all real numbers). 0 和 1 必是完全 n 次幂,原因是对于所有正数 k,$0^k=0$ 成立;对于所有实数 k,$1^k=1$ 成立。

Phrases: We read a^n as "raising a to the nth power". a^n 读作求 a 的 n 次幂。

Questions: 1. What is the next perfect 4th power of 14 641? (You may need a calculator for this question.)

2. List all perfect 5th powers between 100 001 and 200 000. (You may need a calculator for this question.)

Answers: 1. Since $14\,641=11^4$, the next perfect 4th power is $12^4=20\,736$.

2. Since $(10^5=100\,000)<100\,001<(11^5=161\,051)<200\,000<(12^5=248\,832)$, the list of perfect 4th powers is 11^5, or $\boxed{161\,051}$.

1.2.3 Fractions 分数

1.2.3.1 Introduction 介绍

Fraction 分数

n. [ˈfrækʃən]

Definition: The representation of parts of a whole (1). 分数是指对一个整体中的部分的表示法。

Figure 1-11 shows the parts of a fraction.

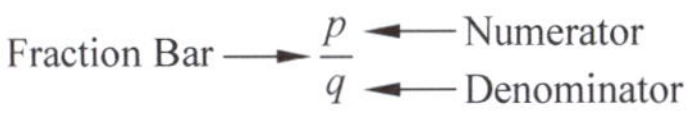

Fraction Bar → $\frac{p}{q}$ ← Numerator ← Denominator

Figure 1-11

Notation: p/q or $\frac{p}{q}$, in which p and q are numbers, in which p is known as the numerator and q is known as the denominator. $q\neq0$.

在 p/q 中,p 和 q 分别是分数的分子和分母,分母不为 0。

p/q is read as "p divided by q", or "p over q", or "p out of q". In the context of many problems, p/q means to divide a whole (1) into q equal parts, and take p of the parts.

分数亦可读作"分子除以分母"或"分母分之分子"。在很多问题中,可根据上下文解释分数 p/q:把一个整体平均分成 q 份,占其中的 p 份。

Properties: 1. Fractions can be represented by divisions, which gives the decimal representation of a fraction. For example, $2/5=2\div5=0.4$. In application, we can say John eats 2/5 of a pancake (which means to divide the pancake into 5 equal parts and eat 2 of the parts), or John eats 0.4 pancake. 分数可以用除法表示——除法给出了与分数等价的小数值。

2. A common fraction is a fraction whose numerator and denominator are both integers. 简分数是分子与分母均为整数的分数。

3. A common fraction (regardless of signs) is called a simplest fraction if its numerator and denominator are coprime. 不考虑正负符号,最简分数是分子与分母互质的简分数。

4. The statements below are equivalent in determining whether two or more fractions are equal fractions:

   判断两个或更多的分数是否相等,以下命题等价:

   (1) They have the same simplest form. 它们的最简形式相同。

   (2) They have the same decimal value. 它们的小数值相同。

   (3) They have the same exact location on the number line. 它们在数轴上的位置完全相同。

5. Multiplying or dividing the numerator and the denominator by the same nonzero number does not change the value of the fraction. 分子和分母同时乘以或除以同一个非零数,分数的值不变。

   In short, if $b, c \neq 0$, then $\frac{a}{b} = \frac{ac}{bc} = \frac{a \div c}{b \div c}$.

6. Regardless of signs, two fractions are like fractions if they have the same denominator. Otherwise they are unlike fractions. 不考虑正负符号,若两个分数的分母相同,则它们被称为同分母分数。否则它们被称为异分母分数。

7. To add/subtract two like fractions, the numerator of the sum/difference is the sum/difference of the numerators. The denominator of the sum/difference is the denominator of one of the fractions. 加/减两个同分母分数,则所得分子是两个分子的和/差,分母不变。

   In short, $\frac{a}{b} \pm \frac{c}{b} = \frac{a \pm c}{b}$.

8. To add/subtract two unlike fractions, find the least common denominator (LCD) and rewrite each fraction by the LCD. Then follow Property 7. 加/减两个异分母分数,先求出最小公分母,再根据最小公分母改写每个分数。最后参照性质 7 进行加/减。

9. To multiply two fractions, the numerator of the product is the product of the numerators. The denominator of the product is the product of the denominators. Simplify using Properties 3 and 5. 两个分数相乘,则所得分子是两个分数分子的积,所得分母是两个分数分母的积。根据性质 3 和 5 简化。

   In short, $\frac{a}{b} \cdot \frac{c}{d} = \frac{ac}{bd}$.

10. To divide by a number, multiply by its reciprocal. 除以一个数,相当于乘它的倒数。

In short, $a \div \frac{b}{c} = a \cdot \frac{c}{b}$.

11. A fraction is a proper fraction if its absolute value is less than 1. It is an improper fraction otherwise. Improper fraction can be written as a mixed number or an integer. 若一个分数的绝对值小于1,则它是真分数。否则它是假分数。假分数可以用带分数或整数表示。
12. Any integer n can be written as improper fractions: $n/1$. 整数 n 可以被写成假分数 $n/1$。
13. Fractions are useful in representing ratios. 分数能代表"比"的概念。
14. $-(p/q) = -p/q = p/(-q)$, $q \neq 0$ (Properties 1 and 2 of division). 见除法的性质1和2。
15. In a fraction, the numerator or the denominator, or both, can be irrational numbers. 在分数中,分子或分母或两者均可为无理数。
16. Fractions of the form $1/n$ are called unit fractions, for which n is an integer greater than 1. $1/n$ 是单位分数,其中 n 是大于1的整数。

Examples: $-3/5$, 2017/2016, 8/9, 4/7, 1.5/3.8, 48/90.6, …

Phrases: common～ 简分数, proper～ 真分数, improper～ 假分数, like～ 同分母分数, unlike～ 异分母分数

Questions: 1. (Property 1) What is the decimal representation of 8/40? 用小数表达8/40。

2. (Properties 2,3,11) Of the fractions below:

Which of the following are common fractions? 下面哪些是简分数?

Which of the following are simplest fractions? 下面哪些是最简分数?

Which of the following are proper fractions? 下面哪些是真分数?

Which of the following are improper fractions? 下面哪些是假分数?

32/13, $-50/14$, 20.5/75, 10/40, 9/9, 10/3,

$-7/9$, $-5/12$, 4.5/6.8, $-1.2/3.3$, $-7/6.3$

3. (Property 4,5) Which of the two fractions below are equivalent fractions? 下列分数中哪两个是等分数?

13/20, 4/9, 25/40, 18/80, 65/100

4. (Property 6) Identify one pair of like fraction and unlike fractions. 在下列分数中认出同分母分数和异分母分数。

13/17, 4/13, $-7/13$, 9/20

5. (Property 7,8) Evaluate each of the following.

(1) $2/19 + 5/19$

(2) $3/4 + 1/8$

(3) $3/10 + 1/6$

6. (Property 9,10)

(1) What is the product of 4/15 and 3/14? 求两个分数的积。

(2) What is the quotient of 9/20 and 3/40? 求两个分数的商。

Answers: 1. $8/40 = 8 \div 40 = \boxed{0.2}$.

2. Common fractions: 32/13, −50/14, 10/40, 9/9, 10/3, −7/9, −5/12
   Simplest fractions: 32/13, 10/3, −7/9, −5/12
   Proper fractions: 20.5/75, 10/40, −7/9, −5/12, 4.5/6.8, −1.2/3.3
   Improper fractions: 32/13, −50/14, 9/9, 10/3, −7/6.3
3. $\boxed{13/20 \text{ and } 65/100}$. Their decimal values are 0.65. Multiplying both the numerator and the denominator of 13/20 by 5, we get 65/100.
4. Like fractions: 4/13, −7/13.
   Unlike fractions: 13/17, 4/13.
5. (1) $2/19+5/19=(2+5)/19=\boxed{7/19}$.
   (2) LCD = 8.
   So, $3/4+1/8=(3/4)(2/2)+1/8=6/8+1/8=(6+1)/8=\boxed{7/8}$.
   (3) LCD = 30.
   So, $3/10+1/6=(3/10)(3/3)+(1/6)(5/5)=9/30+5/30=(9+5)/30=14/30=\boxed{7/15}$.
6. (1) $4/15\times3/14=(4\times3)/(15\times14)=12/210=\boxed{2/35}$.
   (2) $9/20\div3/40=9/20\times40/3=(9\times40)/(20\times3)=360/60=\boxed{6}$.

1.2.3.2 Numerator and Denominator 分子与分母

Numerator and Denominator 分子与分母

n. [ˈnuməˌreɪtər] & n. [dɪˈnɑməˌneɪtər]

Definition: In the fraction $\frac{p}{q}$, q is the denominator, representing how many equal parts the whole (1) is divided into ($q\neq0$). p is the numerator, representing the number of equal parts out of q equal parts that the fraction takes. 在分数$\frac{p}{q}$中,q 是分母,代表的是一个整体被分成的等份份数($q\neq0$);p 是分子,代表的是占这 q 等份中的份数。

Notation: p/q is read as "p divided by q", or "p over q", or "p out of q". p/q 亦读作 p 除以 q,或 q 分之 p。

Examples: 2/7—The numerator is 2, and the denominator is 7.
5/14—The numerator is 5, and the denominator is 14.

Phrases: least common denominator 最小公分母

Questions: 1. John ate 2/5 of a cake. What does that mean?
2. John ate 5/3 of a cake. What does that mean?

Answers: 1. Divide the cake into 5 equal parts, and John ate 2 out of the 5 parts.
2. Divide the cake into 3 equal parts, and John ate 5 out of the 3 parts. Clearly, he ate 1 cake and 2 of the 3 parts of the second cake.

Least Common Denominator (LCD) 最小公分母

n. [list ˈkɑmən dɪˈnɑməˌneɪtər]

Definition: The LCD for two or more common fractions is the LCM of their denominators (Assume that the denominators are positive.). 两个或更多的简分数的最小公分母是它们分母的最小公倍数(假设它们的分母是正的)。

Notation: LCD(Fraction 1, Fraction 2, …)

Examples: LCD(1/8, 2/7) = LCM(8, 7) = 56.

Questions: 1. What is the LCD of 2/9 and 7/12?

2. What is the LCD of −5/4, 6/5, and 7/6?

Answers: 1. LCD(2/9, 7/12) = LCM(9, 12) = $\boxed{36}$.

2. LCD(−5/4, 6/5, 7/6) = LCM(4, 5, 6) = $\boxed{60}$.

1.2.3.3 Mixed Numbers　带分数

Mixed Number　带分数

n. [ˌmɪkst ˈnʌmbər]

Definition: The sum/difference of a nonzero integer and a proper fraction. 带分数是非零整数与真分数的和/差。

A mixed number is one of: positive mixed number and negative mixed number. 带分数有正带分数和负带分数。

Notation: $a\,\frac{b}{c}$, for which a is a nonzero integer and $\frac{b}{c}$ is a proper fraction.

写作 $a\,\frac{b}{c}$,其中 a 为非零整数,$\frac{b}{c}$ 为真分数。

Positive Mixed Number　正带分数

n. [ˈpɑzətɪv ˌmɪkst ˈnʌmbər]

Notation: $a\,\frac{b}{c}$, for which a is a positive integer and $\frac{b}{c}$ is a proper fraction.

写作 $a\,\frac{b}{c}$,其中 a 为正整数,$\frac{b}{c}$ 为真分数。

Properties: 1. All positive mixed numbers are greater than 1. 所有正带分数大于1。

2. $a\,\frac{b}{c} = a + \frac{b}{c}$.

3. Converting a positive mixed number to an improper fraction:
转换正带分数为假分数:

$$a\,\frac{b}{c} = a + \frac{b}{c} = \frac{ac}{c} + \frac{b}{c} = \boxed{\frac{ac+b}{c}}.$$

Examples: $1\,\frac{2}{3}$, $3\,\frac{2}{25}$.

Questions: 1. Find the sum: $1\,\frac{2}{3} + 3\,\frac{1}{3}$. Simplify your answer.

2. Find the sum: $1\,\frac{2}{3} + 4\,\frac{2}{5}$. Simplify your answer.

3. Find the difference: $5\,\frac{2}{3} - 1\,\frac{1}{3}$. Simplify your answer.

4. Find the difference: $5\dfrac{2}{3}-1\dfrac{3}{7}$. Simplify your answer.

5. Find the difference: $5\dfrac{3}{7}-1\dfrac{2}{3}$. Simplify your answer.

Answers: 1. $1\dfrac{2}{3}+3\dfrac{1}{3}=4\dfrac{3}{3}=\boxed{5}$.

2. The LCD is 15. Thus $1\dfrac{2}{3}+4\dfrac{2}{5}=1\dfrac{10}{15}+4\dfrac{6}{15}=5\dfrac{16}{15}=\boxed{6\dfrac{1}{15}}$.

3. $5\dfrac{3}{5}-1\dfrac{1}{5}=\boxed{4\dfrac{2}{5}}$.

4. The LCD is 21. Thus $5\dfrac{2}{3}-1\dfrac{3}{7}=5\dfrac{14}{21}-1\dfrac{9}{21}=\boxed{4\dfrac{5}{21}}$.

5. The LCD is 21. Thus $5\dfrac{3}{7}-1\dfrac{2}{3}=5\dfrac{9}{21}-1\dfrac{14}{21}$. Since $\dfrac{9}{21}<\dfrac{14}{21}$, we will borrow $1=\dfrac{21}{21}$ from 5:

$$5\frac{9}{21}-1\frac{14}{21}=4\frac{30}{21}-1\frac{14}{21}=\boxed{3\frac{16}{21}}.$$

Negative Mixed Number 负带分数

n. [ˈnegətɪv ˌmɪkst ˈnʌmbər]

Notation: $a\dfrac{b}{c}$, for which a is a negative integer and $\dfrac{b}{c}$ is a proper fraction.

写作 $a\dfrac{b}{c}$,其中 a 为负整数,$\dfrac{b}{c}$ 为真分数。

Properties: 1. All negative mixed numbers are less than -1. 所有负带分数小于 -1。

2. $a\dfrac{b}{c}=a-\dfrac{b}{c}$.

3. Converting a negative mixed number to an improper fraction:
转换负带分数为假分数:

$$a\frac{b}{c}=a-\frac{b}{c}=\frac{ac}{c}-\frac{b}{c}=\boxed{\frac{ac-b}{c}}.$$

Examples: $-2\dfrac{4}{5}$, $-5\dfrac{10}{31}$.

Questions: 1. Find the sum: $-1\dfrac{2}{3}+3\dfrac{1}{3}$. Simplify your answer.

2. Find the sum: $1\dfrac{2}{3}+\left(-4\dfrac{2}{5}\right)$. Simplify your answer.

3. Find the difference: $-5\dfrac{2}{3}-1\dfrac{3}{7}$. Simplify your answer.

4. Find the difference: $-5\dfrac{3}{7}-1\dfrac{2}{3}$. Simplify your answer.

Answers：1. $-1\frac{2}{3}+3\frac{1}{3}=-1-\frac{2}{3}+3+\frac{1}{3}=2-\frac{1}{3}=\boxed{1\frac{2}{3}}$.

2. The LCD is 15. Thus $1\frac{2}{3}+\left(-4\frac{2}{5}\right)=1\frac{10}{15}+\left(-4\frac{6}{15}\right)=1+\frac{10}{15}+\left(-4-\frac{6}{15}\right)=1+\frac{10}{15}-4-\frac{6}{15}=-3+\frac{4}{15}=\boxed{-2\frac{11}{15}}$.

3. The LCD is 21. Thus $-5\frac{2}{3}-1\frac{3}{7}=-5\frac{14}{21}-1\frac{9}{21}=-5-\frac{14}{21}-\left(1+\frac{9}{21}\right)=-5-\frac{14}{21}-1-\frac{9}{21}=-6-\frac{23}{21}=-7-\frac{2}{21}=\boxed{-7\frac{2}{21}}$.

4. The LCD is 21. Thus $-5\frac{3}{7}-1\frac{2}{3}=-5\frac{9}{21}-1\frac{14}{21}=-5-\frac{9}{21}-\left(1+\frac{14}{21}\right)=-5-\frac{9}{21}-1-\frac{14}{21}=-6-\frac{23}{21}=-7-\frac{2}{21}=\boxed{-7\frac{2}{21}}$. This has the same answer as part (d) after expanding the mixed numbers using Property 2.

In the future we will not use mixed numbers often, as they can create confusions between positive and negative mixed numbers.

Instead, for positive mixed number $a\frac{b}{c}$ and negative mixed number $d\frac{e}{f}$, we will either rewrite them as improper fractions, or rewrite them as $a+\frac{b}{c}$ and $d-\frac{e}{f}$, respectively, which are what mixed numbers mean anyway.

以后会少用带分数,因为读者容易对正带分数和负带分数产生混淆。

要表达正带分数 $a\frac{b}{c}$ 或负带分数 $d\frac{e}{f}$,应把它们都改写为假分数,或者分别改写为 $a+\frac{b}{c}$ 和 $d-\frac{e}{f}$,这本来也是带分数的含义。

1.2.4 Percents 百分数

1.2.4.1 Introduction 介绍

Percent 百分数

n. [pər'sent]

Definition：The number out of 100. i.e. $p\%=p/100$.

在 100 等份里占有的份数。$p\%$ 也就是 $p/100$。

Notation：Number followed by %.

Properties：1. $p\%=p/100$.

2. $100\%=1$, representing the whole. 100%代表的是一个整体,等于 1。

3. (1) To convert from a decimal to a percent, move the decimal point two places to the right and add the sign "%" at the end of the number. 从小数转换到百分数,把小数点向右移两位,并且在数字最后加上百分号。

(2) To convert from a percent to a decimal, move the decimal point two places to the

left and remove the sign "%" at the end of the number. 从百分数转换到小数，把小数点向左移两位，并且移除数字后面的百分号。

4. (1) To convert from a fraction to a percent, convert fraction to decimal first (numerator is divided by the denominator), then convert from decimal to percent by Property 3(1). 从分数转换到百分数，先把分数转换到小数(分子除以分母)，然后根据性质3(1)从小数转换到百分数。

Alternatively, rewrite the fraction so that its denominator is 100 (see Property 1 of proportion—cross product in Section 2.3.2), then use the definition of percent to rewrite in percent. 或者，把这个分数改写成一个分母为100的分数(见2.3.2节比例的性质1交叉积)，然后用百分数的定义改写成百分数。

(2) To convert from percent to fraction, use the definition of percent then simplify. 从百分数转换到分数，用百分数的定义，然后化简。

Examples: −1%, −0.5%, 4%, 12%, 100.50%, 300%, …

Phrases: ~change, ~increase, ~decrease

Questions: 1. Convert each of the decimals to a percent:
把下面的小数转换为百分数：

0.45, 0.0123, −0.888, 1, 1.234

2. Convert each of the percents to a decimal:
把下面的百分数转换为小数：

100%, 32%, 12.3%, 0.001%, 0.023%

3. Convert each of the fractions to a percent:
把下面的分数转换为百分数：

3/4, 2/5, 17/20, 3/3

4. Convert each of the percents to a common fraction:
把下面的百分数转换为简分数：

2%, 2.5%, 0.40%, 50%

Answers: 1. Using Property 3(1),

0.45 = 45%.

0.0123 = 1.23%.

−0.888 = −88.8%.

1 = 100%.

1.234 = 123.4%.

2. Using Property 3(2),

100% = 1.

32% = 0.32.

12.3% = 0.123.

0.001% = 0.00001.

0.023% = 0.00023.

3. Using Property 4(1),

3/4 = 0.75 = 75%. Or, 3/4 = 75/100 = 75%.

$2/5=0.4=40\%$. Or, $2/5=40/100=40\%$.

$17/20=0.85=85\%$. Or, $17/20=85/100=85\%$.

$3/3=1=100\%$. Or, $3/3=100/100=100\%$.

4. Using Property 4(2),

$2\%=2/100=1/50$.

$2.5\%=2.5/100=25/1000=1/40$.

$0.40\%=0.40/100=40/10\,000=1/250$.

$50\%=50/100=1/2$.

1.2.4.2 Percent Problems and Applications　百分数常见问题和应用

Common Percent Problems　百分数常见问题

n. [ˈkɑmən pərˈsent ˈprɑbləmz]

Properties: Let a, b, and p be known positive numbers.　设 a、b、p 为正数。

1. "Percent of" Problems:
   关于"百分之几"的问题:
   (1) What number is $p\%$ of a?　a 的 $p\%$ 等于多少?
   (2) a is what percent of b? /What percent of b is a?　a 是 b 的百分之几? /b 的百分之几等于 a?
2. Increase/Decrease by a Percent Problems:
   关于增加/减少百分数的问题:
   (1) What is the result when a is increased by $p\%$?　a 增加它的 $p\%$ 后的结果是多少?
   (2) What is the result when a is decreased by $p\%$?　a 减少它的 $p\%$ 后的结果是多少?
3. Percent Change Problems:
   关于百分数变化问题:
   What is the percent change from a to b? (See Percent Change)　从 a 到 b,百分数的变化是多少?(见百分数变化)

Answers: 1. (1) $a\cdot p\%=\boxed{\dfrac{ap}{100}}$. "Of" means multiplication.

(2) $\boxed{\dfrac{a}{b}}\cdot 100\%$.

$\dfrac{a}{b}$ represents "what fraction of b is a", whose conversion to percent is the answer.

$\dfrac{a}{b}$表示的是"a 是 b 的几分之几"。转换到百分数就是答案。

2. (1) The numerical increase is $a\cdot p\%=\dfrac{ap}{100}$.　数值的增加量为$\dfrac{ap}{100}$。

The result is $a+\dfrac{ap}{100}=\boxed{a\left(1+\dfrac{p}{100}\right)}$.　结果为 $a\left(1+\dfrac{p}{100}\right)$。

(2) The numerical decrease is $a \cdot p\% = \frac{ap}{100}$. 数值的减少量为$\frac{ap}{100}$。

The result is $a - \frac{ap}{100} = \boxed{a\left(1 - \frac{p}{100}\right)}$. 结果为 $a\left(1 - \frac{p}{100}\right)$。

3. $\boxed{\frac{|b-a|}{a} \cdot 100\%}$.

If $a<b$, then it is a percent increase. 若 $a<b$,则这是百分数增加。

If $a>b$, then it is a percent decrease. 若 $a>b$,则这是百分数减少。

If $a=b$, then the percent change is 0 (There is no change.). 若 $a=b$,则百分数变化为 0(没有变化)。

Questions: 1. (Property 1)

(1) 20 is what percent of 50? 20 是 50 的百分之几?

(2) 25 is what percent of 5? 25 是 5 的百分之几?

(3) 100 is 25% of what number? 100 是什么数的 25%?

(4) What number is 20% of 80? 什么数是 80 的 20%?

2. (Property 2)

(1) What is the result when 50 is increased by 20%? 当 50 增加 20%后的结果是多少?

(2) What is the result when 60 is decreased by 25%? 当 60 减少 25%后的结果是多少?

(3) When 50 is increased by 20% then the new number decreased by 20%, what is the result? 当 50 增加 20%,然后新数减少 20%,结果是多少?

(4) What number results 80 when increased by 25%? 什么数增加 25%后的结果是 80?

3. (Property 3)

(1) What is the percent change from 40 to 50? 从 40 到 50,百分数的变化是多少?

(2) What is the percent change from 50 to 40? 从 50 到 40,百分数的变化是多少?

(3) The percent increase from what number to 100 is 400%? 什么数的百分数增加 400%为 100?

Answers: 1. (1) $\frac{20}{50} \times 100\% = \boxed{40\%}$.

(2) $\frac{25}{5} \times 100\% = \boxed{500\%}$.

(3) Let the number be x.

$x \times 25\% = 100$, from which $x = \boxed{400}$.

(4) $80 \times 20\% = \boxed{16}$.

2. (1) $50\left(1 + \frac{20}{100}\right) = \boxed{60}$.

(2) $60\left(1 - \frac{25}{100}\right) = \boxed{45}$.

(3) When 50 is increased by 20%, the result is 60 as shown in part (1). When 60 is

decreased by 20%, the result is $60\left(1-\frac{20}{100}\right)=\boxed{48}$.

(4) Let that number be x.

$x\left(1+\frac{25}{100}\right)=80$, from which $x=\boxed{64}$.

3. (1) new = 50; original = 40. Since new > original, this is a percent increase. Use the formula for percent increase:

$\frac{|50-40|}{40}\times 100\%=\boxed{25\%\text{ increase}}$.

(2) new = 40; original = 50. Since new < original, this is a percent decrease. Use the formula for percent decrease:

$\frac{|40-50|}{50}\times 100\%=\boxed{20\%\text{ decrease}}$.

(3) Method 1:

new = 100; original = x; new > original. Use the formula for percent increase.

$$\frac{100-x}{x}\times 100\%=400\%$$

$$\frac{100-x}{x}=4$$

$$100-x=4x$$

$$100=5x$$

$$x=\boxed{20}.$$

Method 2 (Use Property 2):

Rephrasing the question: When x is increased by 400%, the result is 100. What is the value of x? Using the formula in Property 2(1):

$$x\left(1+\frac{400}{100}\right)=100$$

$$x=\boxed{20}.$$

Percent Change 百分数变化

n. [pər'sent tʃeɪndʒ]

Definition: The extent to which a variable gains value or loses value. The figures are arrived at by comparing the initial (or before) and final (or after) quantities according to a specific formula. It is assumed that both the initial and the final quantities are positive. 百分数变化是指变量增值或贬值的情况。可用特定的公式对比它的初始值和最终值。可以假设初始值和最终值为正数。

Percent change is one of percent increase or percent decrease or no change. 百分数变化可以是百分数增长、百分数减少、没有变化中的一种。

See Property 3 of common percent problems for a short introduction of percent change. 百分数常见问题的性质 3 是百分数变化的简要介绍。

Properties: 1. Percent Change $=\frac{|\text{new}-\text{original}|}{\text{original}}\times 100\%$, for which new > 0 and original > 0. 百分

数变化 $=\dfrac{|新数-原数|}{原数}\times 100\%$,其中新数和原数均大于 0。

- $|new-original|$ represents the absolute numerical change. $|新数-原数|$表示的是绝对数值变化。
- $\dfrac{|new-original|}{original}$ represents the absolute ratio of numerical change to the original number. $\dfrac{|new-original|}{original}\times 100\%$ converts this ratio to a percent, representing the percent change. $\dfrac{|新数-原数|}{原数}$表示的是绝对数值变化和原数之比。$\dfrac{|新数-原数|}{原数}\times 100\%$表示把比值转换成百分数,代表的是百分数变化。

2. Let $p\%$ = Percent Change:
   设 $p\%$ = 百分数变化:
   - If new $>$ original, then we say that the original is increased by $p\%$. This is a percent increase. We can simplify the formula for percent change:
     若新数>原数,则原数增加了 $p\%$。这是百分数增长。可以简化百分数变化的公式如下:
     $$\text{Percent Change}=\frac{|\text{new}-\text{original}|}{\text{original}}\times 100\%=\frac{\text{new}-\text{original}}{\text{original}}\times 100\%$$
   - If new $<$ original, then we say that the original is decreased by $p\%$. This is a percent decrease. We can simplify the formula for percent change:
     若新数<原数,则原数减少了 $p\%$。这是百分数减少。可以简化百分数变化的公式如下:
     $$\text{Percent Change}=\frac{|\text{new}-\text{original}|}{\text{original}}\times 100\%=\frac{\text{original}-\text{new}}{\text{original}}\times 100\%$$
   - If new = original, then we say that the original has percent change is 0, or no change.
     若新数 = 原数,则原数的百分数变化为 0,也就是没有变化。
3. To summarize, percent change is deciding which of original or new is larger. Then use the formula $\dfrac{\text{large}-\text{small}}{\text{original}}\times 100\%$, and finally specify whether it is a percent increase or percent decrease. 总体来说,百分数变化是先判断原数和新数哪个大,然后运用公式:$\dfrac{大-小}{原}\times 100\%$决定是百分数增加还是百分数减少。

Questions: See Question 3 on Common Percent Problem. 见百分数常见问题的问题 3。

Percent Change Application 百分数变化的应用

n. [pər'sent tʃeɪndʒ ˌæplɪ'keɪʃən]

Examples: 1. "Mark up the price" 涨价

The following statements are equivalent:

以下表述是等价的:

- Mark up the price by $p\%$. 涨价 $p\%$。

- The price is increased by $p\%$ (See Property 2(1) of Common Percent Problems). 价格增加 $p\%$(见百分数的常见问题性质 2(1))。

2. "Mark down the price/Discount"　降价/折扣

The following statements are equivalent:

以下表述是等价的:

- Mark down the price by $p\%$. 降价 $p\%$。
- The price is decreased by $p\%$ (See Property 2(2) of Common Percent Problems). 价格减少 $p\%$ (见百分数的常见问题性质 2(2))。
- The price has a $p\%$ discount. 价格有 $p\%$的折扣。
- $p\%$ off the price. 降价 $p\%$。

3. Interest and Interest Rate　利息与利率

The following statements are equivalent:

以下表述是等价的:

- The interest if the interest rate is $p\%$. 利率为 $p\%$的利息。
- $p\%$ of the money put in the bank. 存款 $p\%$的钱。

1.2.5 Summary 总结

1. Integers are rational because every integer N can be written as $N/1$. It follows that N and 1 are relatively prime. 根据有理数的定义,整数均为有理数。

2. (1) The nth nonnegative even number is $2(n-1)$. 第 n 个非负偶数是 $2(n-1)$。

   0 is the smallest nonnegative even number. 0 是最小的非负偶数。

   The ones digit of an even number is one of 0,2,4,6,or 8. 偶数的个位数是 0、2、4、6、8 中任一个。

   (2) The nth positive odd number is $2n-1$. 第 n 个正奇数是 $2n-1$。

   1 is the smallest positive odd number. 1 是最小的正奇数。

   The ones digit of an odd number is one of 1,3,5,7,or 9. 奇数的个位数是 1、3、5、7、9 中任一个。

3. (1) A prime is a positive integer whose only positive divisors are 1 and itself (exactly two positive divisors). 质数是正约数只有 1 和它本身的正整数(只有两个正约数)。

   2 is the smallest prime, and it is the only even prime. 3 is the smallest odd prime. 2 是最小的质数,且是唯一的偶质数。3 是最小的奇质数。

   All primes up to a given limit can be found by the Sieve of Eratosthenes. 可用埃拉托色尼筛选法找出指定上限内的所有质数。

   (2) A composite is a positive integer that has positive divisors other than 1 and itself (at least three positive divisors). 合数是正约数不止 1 和它本身的正整数(至少有 3 个正约数)。

   4 is the smallest composite, and 9 is the smallest odd composite. 4 是最小的合数。9 是最小的奇合数。

   1 is neither prime nor composite. 1 既不是质数,也不是合数。

4. (1) The GCF of two or more positive integers can be found by:

   两个或更多数的最大公约数可用以下方法求出:

- Euclidean Algorithm (for 2 numbers only) 欧几里得算法(只限于两个数)
- Short Division 短除法
- Prime Factorization 分解质因数

(2) The LCM of two or more positive integers can be found by:
两个或更多数的最小公倍数可用以下方法求出:

- Short Division 短除法
- Prime Factorization 分解质因数

(3) For any two positive integers a and b: $\text{GCF}(a,b)\cdot\text{LCM}(a,b)=a\cdot b$.
When a and b are relatively prime:
a 和 b 互质的情况:

- $\text{GCF}(a,b)=1$.
- $\text{LCM}(a,b)=ab$.

5. The expression a^n represents the product of n number of a's. It is the perfect nth power of a. a^n 代表的是 n 个 a 的积。它是 a 的 n 次幂。

6. In the fraction p/q or $\frac{p}{q}$, in which p and q are numbers: p is known as the numerator and q is known as the denominator with $q\neq 0$. It is read as "p divided by q", or "p over q", or "p out of q". In the context of many problems, p/q means to divide a whole (1) into q equal parts, and take p of the parts. $\frac{p}{q}$中 p 和 q 分别是分数的分子和分母,分母不为 0。分数亦可读作"分子除以分母"或"分母分之分子"。在很多问题中,可根据上下文解释分数 p/q:把一个整体平均分成 q 份,占其中的 p 份。

(1) Fractions whose absolute values are less than 1 are proper fractions. 绝对值小于 1 的分数是真分数。

(2) Fractions whose absolute values are greater or equal to 1 are improper fractions, which can be written as integers or mixed numbers. 绝对值大于或等于 1 的分数是假分数,可被写成整数或带分数。

Review properties of fractions for different types of fractions, as well as their operations. 复习关于不同的分数种类以及分数之间的运算见分数的性质。

Also, review mixed numbers for the conversion of improper fractions to mixed numbers, and vice versa. 复习关于假分数和带分数的相互转换见带分数。

7. A percent is a number out of 100. i.e. $p\%=p/100$. 百分数是指在 100 等份里占的份数。$p\%$也就是 $p/100$。

See percent for the following conversion techniques:
以下转换的方法见百分数:

(1) Percent ↔ Decimal. 百分数和小数的相互转换。

(2) Percent ↔ Fraction. 百分数和分数的相互转换。

We know that (fraction → decimal) is by a simple division. 分数到小数的转换是相对简单的除法。

We will cover (decimal → fraction) in decimal in Section 1.4. 小数到分数的转换见 1.4 节的小数。

8. See common percent problems and percent change application in Section 1.2.4.2 for practical

uses of percents. 百分数实用的例子见 1.2.4.2 节百分数的常见问题和百分数变化的应用。

1.3 Irrational Numbers 无理数

Irrational Number 无理数

n. [ɪ'ræʃənəl 'nʌmbər]

Definition: A number that cannot be written as the quotient p/q, where p and q are integers such that $q\neq 0$. 无理数是不能被商 p/q 表示的数,其中 p 与 q 均为整数,且 $q\neq 0$。

Notation: The set of irrational numbers is denoted by $\mathbb{R}\setminus\mathbb{Q}$, which is the result when the set of rational numbers is subtracted from the set of real numbers. "\" is the set subtraction symbol. 无理数集用$\mathbb{R}\setminus\mathbb{Q}$表示,也就是实数集减去有理数集的结果。反斜线是集合的减法符号。

Properties:
1. Commutative Property, Associative Property, Distributive Property of addition and multiplication. 无理数的加法和乘法满足交换律、结合律、分配律。
2. The decimal form of an irrational number is neither terminating nor repeating. 无理数的小数形式既不是有限(小数),也不循环。
3. If a and b are rational, but $\sqrt{b}$ is irrational, then $(a+\sqrt{b})(a-\sqrt{b})$ is rational. 若 a 和 b 均为有理数,但$\sqrt{b}$是无理数,则$(a+\sqrt{b})(a-\sqrt{b})$是有理数。

Proofs:
1. To prove Property 2, use proof by contradiction. 用反证法证明性质 2。
   Suppose the decimal form of an irrational number D is terminating. We can write D as $D/1$ and multiply both numerator and denominator of $D/1$ by the same power of 10 (so that both the numerator and denominator of the fraction are integers), and simplify from there. We can conclude that D is rational, which contradicts the assumption. 设无理数 D 的小数形式有限。先把 D 改写成$D/1$,再在分子和分母乘上相同的 10 的次幂(然后分子和分母都是整数了),最后从这里简化。可得知 D 是有理数,从而产生矛盾。
   Suppose the decimal form of an irrational number D is repeating. We can write D as a fraction using the property of a geometric series(see Chapter 32). We can conclude that D is rational, which contradicts the assumption. 设无理数 D 的小数形式循环。可以用等比级数(见第 31 章)的性质把 D 写成一个简分数。最后得出 D 是有理数的结论,从而产生矛盾。
2. To prove Property 3, by the Difference of Squares formula, $(a+\sqrt{b})(a-\sqrt{b})=a^2-\sqrt{b}^2=a^2-b$. By the Closure Property of rational numbers over addition and multiplication, this is rational. 要证明性质 3,用平方差公式:$(a+\sqrt{b})(a-\sqrt{b})=a^2-(\sqrt{b})^2=a^2-b$。根据有理数中加法和乘法的封闭性,可知$(a+\sqrt{b})(a-\sqrt{b})$是有理数。

Examples:
1. Special irrational numbers: π, e, …
2. If positive integer a is not a perfect nth power, then $\sqrt[n]{a}$ is irrational: 若正整数 a 不

是完全 n 次幂,则$\sqrt[n]{a}$是无理数。

$\sqrt{2}$,$\sqrt{6}$,$\sqrt[3]{2}$,$\sqrt[3]{40}$,$\sqrt[4]{55}$,…

Phrases: positive~ 正无理数,negative~ 负无理数

Questions: 1. Determine whether each number below is irrational. 判断下面各数是否为无理数。

$$-\pi,\sqrt{3},\sqrt[3]{\frac{27}{125}},\frac{\sqrt{26}}{6},\sqrt{100}$$

2. Must the sum of any two irrational numbers an irrational number? (Are irrational numbers closed under addition?) 两个无理数的和是否必为无理数?(无理数有加法的封闭性吗?)

3. Must the product of any two irrational numbers an irrational number? (Are irrational numbers closed under multiplication?) 两个无理数的积是否必为无理数?(无理数有乘法的封闭性吗?)

4. Prove that $\sqrt{2}$ is irrational. 证明$\sqrt{2}$是无理数。

Answers: 1. The number $-\pi$ is irrational by definition.

The number $\sqrt{3}$ is irrational by Example 2.

The number $\sqrt[3]{\frac{27}{125}}$ is rational because it is $\frac{3}{5}$.

The number $\frac{\sqrt{26}}{6}$ is irrational.

The number $\sqrt{100}$ is rational because it is $10=10/1$.

2. No. For example, $\pi+(-\pi)=0$, but 0 is a rational number.

3. No. For example, $(\sqrt{2})^2=2$, but 2 is a rational number.

4. See prove by contradiction in Section 13.3.2.3. 见 13.3.2.3 节的反证法。

1.4 Decimals 小数

1.4.1 Introduction 介绍

Decimal/Decimal Number 小数

n. [ˈdesəməl/ˈdesəməl ˈnʌmbər]

Definition: A number that is written in decimal notation. 小数是指用十进制表示法表示的数。

It is one of terminating decimal and non-terminating decimal. 小数包括有限小数和无限小数。

Properties: 1. All decimals consist of two parts: whole (integer) part and fractional part. The two parts are separated by a decimal point ".". The whole part is on the left of the decimal point. The fractional part is on the right of the decimal point. For positive decimals, the sum of whole part and decimal part make up this decimal value. The negative decimals, the difference of whole part and decimal part make up this decimal value. 小数包括两部分:整数部分和小数部分。这两部分被小数点"."隔开。整数部分在小数点的左边。小数部分在小数点的右边。对于正小数而言,整数

部分和小数部分的和为小数的值。对于负小数而言,整数部分和小数部分的差为小数的值。

2. The fractional part is always less than 1, and usually nonempty, otherwise the decimal is an integer.　小数部分总是比 1 小,而且不是空的,否则就是整数了。
3. Each digit in a decimal has its corresponding position, called its digit. Each digit has a name.　小数里的每个数字都有对应的位置,叫作数位。每个数位都有名称。
4. For a decimal whose absolute value is less than 1, we can omit the "0" on the ones digit.
写绝对值小于 1 的小数时,可省略个位上的"0"。
For example, $0.6 = .6$; $-0.87 = -.87$.
5. To compare two positive decimal numbers, whoever has the greater whole part is greater. If their whole parts are the same, whoever has the greater tenths digit is greater. If their tenths digits are the same, whoever has the greater hundredths digit is greater, and so on.　比较两个正小数的大小时,整数部分大的数较大；若整数部分相同,十分位大的数较大；若十分位相同,百分位大的数较大……
To compare two negative decimal numbers, disregard the negative sign, and whoever has the greater whole part is less. If their whole parts are the same, whoever has the greater tenths digit is less. If their tenths digits are the same, whoever has the greater hundredths digit is less, and so on.　要比较两个负小数的大小,先忽略负号,整数部分大的数较小；若整数部分相同,十分位大的数较小；若十分位相同,百分位大的数较小……
6. To convert a fraction to decimal, treat the fraction bar as the division sign: numerator divided by the denominator.　分数转换成小数,把分数线当作除号,分子除以分母即可。
7. To convert a rational decimal D to a common fraction:
把有理小数 D 转换成简分数:
If D is terminating, then write D as $D/1$ and multiply both numerator and denominator of $D/1$ by the same power of 10 (so that both the numerator and denominator of the fraction are integers), and simplify from there.　若 D 为有限小数,则把 D 改写成 $D/1$,并把分子和分母乘上一个相同的 10 的次幂(使得分数的分子和分母都为整数)进行简化。
If D is a repeating, then write D as a fraction using the property of a geometric series.　若 D 为循环小数,可用等比级数改写成分数。

Examples: $0.2, -5.67, 3.141592653589\cdots, 1.23456, 0.8999\cdots, -5.201, \cdots$

Phrases: terminating～　有限小数, non-terminating　～无限小数, repeating～　循环小数, pure recurring　～纯循环小数, mixed recurring　～混循环小数

Questions: 1. What are the whole part and fractional part of 12345.6789? (Property 1)求整数部分和小数部分。
2. What is the fractional part of 12345? (Property 2)求小数部分。
3. Order this list of decimals from the least to the greatest:
从小到大排列下面的小数:

$$5.6, -4.8, 4.89, 4.79, 5.87, 6$$ (Property 5)

4. Convert each of the following fractions to a decimal.
将下面的分数转换为小数。

4/5,20/60,5/18 (Property 6)

Answers: 1. Whole part: 12345; Fractional part: 0.6789. Whole + Fractional = 12345 + 0.6789 = 12345.6789.

2. 0,and 12345 is known as an integer.

3. −4.8,4.79,4.89,5.6,5.87,6.

4. 4/5 = 4 ÷ 5 = 0.8. This is known as a terminating decimal. 这是一个有限小数。

20/60 = 0.333… This is known as a non-terminating repeating decimal and a pure recurring decimal,with "3" as its repetend. 这是一个无限循环小数——纯循环小数。循环节为3。

5/18 = 0.2777… This is known as a non-terminating repeating decimal and a mixed recurring decimal,with "7" as its repetend. 这是一个无限循环小数——混循环小数。循环节为7。

Under the words terminating decimal,pure recurring decimal,and mixed recurring decimal,we will show how to convert from those decimals to fractions. 在有限小数、纯循环小数、混循环小数的词条里会介绍如何转换小数为分数。

Digit 数字/数位/位数

n. ['dɪdʒɪt]

Definition: 1. Any integer from 0 to 9,inclusive. 数字:从0至9的整数。

2. The position that a digit. Each position has a name. Its usage is in the "Phrase" section. 数位:数字的位置。每个数位都有名字,见短语部分。

3. The number of digits of an integer(or the "length"). 位数:数字的个数,也是整数的长度。

Properties: Adopts the decimal system in addition and multiplication. 在加法和乘法中都使用十进制。

Phrases: 1. Different digits have different names,as shown in Table 1-1.
不同数位的名称如表1-1所示。

Table 1-1

ones~个位	
tens~十位	tenths~十分位
hundreds~百位	hundredths~百分位
thousands~千位	thousandths~千分位
ten-thousands~万位	ten-thousandths~万分位
hundred-thousands~十万位	hundred-thousandths~十万分位
millions~百万位	millionths~百万分位
ten-millions~千万位	ten-millionths~千万分位
hundred-millions~亿位	hundred-millionths~亿分位
billions~十亿位	billionths 十亿分位

The left column consists of digits of the whole part.　左列的是整数部分的数位。
The right column consists of digits of the fractional part.　右列的是小数部分的数位。

2. Number of digits:
位数的意思:
12345 is a five-digit number.　12345 是五位数。
4567 is a four-digit number.　4567 是四位数。
89 is a two-digit number.　89 是两位数。

Questions: 1. What does the number 123456789 read as?　写出题中数的读法。
2. What does the number 12345.6789 read as?　写出题中数的读法。
3. What is the sum of the thousands digit and tenths digit in the number 12345.6789?　求 12345.6789 中千位和十分位的和。
4. What does the number 100023.045 read as?　写出题中数的读法。

Answers: 1. One hundred twenty-three million four hundred fifty-six thousand seven hundred eighty-nine.
2. "Twelve thousand three hundred forty-five point six seven eight nine" or "twelve thousand three hundred forty-five and six thousand seven hundred eighty-nine ten-thousandths".
3. $2+6=\boxed{8}$.
4. "One hundred-thousand **and** twenty-three point zero four five" or "one hundred-thousand **and** twenty-three **and** forty-five thousandths".

Note: The word "and" is used when:
单词 and 用于:
(1) Reading zeros according to method of reading zeros.　与中文读零法的"零"用法相同。
(2) Reading the decimal point without saying the word "point".　读小数不读"点"字的方法。

1.4.2 Types of Decimals　小数的种类

Terminating Decimal　有限小数

n. [ˈtɜrməˌneɪtɪŋ ˈdesəməl]

Definition: A decimal number with digits that do not go on forever.　数字有限的小数。

Properties: 1. To convert a terminating decimal to a common fraction, divide it by 1 and rewrite it as a common fraction using Property 5 of fraction, then simplify.　要把有限小数转换成分数,用它本身除以 1,再用分数的性质 5 写成最简分数。
2. Since all terminating decimals can be converted to common fractions, and common fractions can be written in the simplest forms, they must be rational.　因为有限小数可被写成简分数,且简分数又能被写成最简分数,所以有限小数是有理数。

Examples: $-0.123, 0.4, 1.89, -20.16, 4.58786165, \cdots$

Questions: Convert each of the following terminating decimals into fractions.　把下列有限小数写成分数。

1.125,0.35,−6.42,−3.025

Answers: $1.125=\frac{1.125}{1}=\frac{1125}{1000}=\boxed{\frac{9}{8}\text{(improper fraction)}=1\frac{1}{8}\text{(mixed number)}}$.

$0.35=\frac{0.35}{1}=\frac{35}{100}=\boxed{\frac{7}{20}}$.

$-6.42=-\frac{6.42}{1}=-\frac{642}{100}=\boxed{-\frac{321}{50}\text{(improper fraction)}=-6\frac{21}{50}\text{(mixed number)}}$.

$-3.025=-\frac{3.025}{1}=-\frac{3025}{1000}=\boxed{-\frac{121}{40}\text{(improper fraction)}=-3\frac{1}{40}\text{(mixed number)}}$.

Non-Terminating Decimal 无限小数

n. [ˌnɑːn'tɜrməˌneɪtɪŋ ˌdesəməl]

Definition: A decimal number with digits that go on forever. 数字无限的小数。

Properties: 1. There are two types of non-terminating decimals:
无限小数有两种:
(1) irrational decimals 无理小数
(2) repeating decimals 循环小数
① pure recurring decimals 纯循环小数
② mixed recurring decimals 混循环小数

2. In repeating decimals, the fractional part appears in cycles. If the cycle begins with the tenths digit such as $2/9=0.222\cdots=0.\overline{2}$, it is a pure recurring decimals. If the cycle does not begin with the tenths digit such as $5/18=0.2777\cdots=0.2\overline{7}$, it is a mixed repeating decimal. 在循环小数中,小数部分循环出现。若十分位是循环的开始,则这个小数是纯循环小数,否则这个小数是混循环小数。

1.4.3 Repeating Decimals 循环小数

Repetend 循环节

n. [ˌrɛpɪ'tɛnd]

Definition: The part of the fractional component of a repeating decimal that appears in cycles. 小数中循环的部分。

Notation: In repeating decimals, the repetend is denoted by an upper bar. i.e. $2/9=0.222\cdots=0.\overline{2}$; $5/18=0.2777\cdots=0.2\overline{7}$. 在循环小数中,循环节用上画线表示。

Properties: The repetend goes on forever. 循环节无休止地循环下去。

Examples: $1/9=0.111\cdots$(repetend is 1), $2/9=0.222\cdots$(repetend is 2), …, $8/9=0.888\cdots$(repetend is 8).

Questions: Identify the repetend and rewrite using the repetend notation for each of the following. 在下面各题确认循环节,并用循环节写法写出小数。

1. $4/3=1.333\cdots$
2. $5/36=0.138\,88\cdots$
3. $457/999=0.457\,457\,457\cdots$

Answers: 1. The repetend is 3, and $4/3=1.333\cdots=\boxed{1.\overline{3}}$.

2. The repetend is 8, and $5/36 = 0.138\,88\cdots = \boxed{0.13\overline{8}}$.

3. The repetend is 457, and $457/999 = 0.457\,457\,457\cdots = \boxed{0.\overline{457}}$.

Pure Recurring Decimal　纯循环小数

n. [pjʊr rɪˈkɜː.ɪŋ ˈdesəməl]

Definition: A decimal in which all the digits in the fractional part are repeated (appear according to a cycle).

小数部分所有的数字都是循环的。

Properties: To convert a pure recurring decimal to a simplest fraction, one possible way is to use some algebra, as shown in the Questions section.

把纯循环小数转换成最简分数,一种方法是用一些代数知识,见问题(Questions)部分。

We will learn other methods of converting pure recurring decimals to simplest fractions in geometric series in Chapter 31.

在第31章的等比级数中,将介绍从纯循环小数转换成最简分数的其他方法。

Examples: $0.\overline{5}$, $0.\overline{567}$, $3.\overline{125}$, $-5.\overline{123\,45}$.

Questions: Convert each of the following to a common fraction:

把下列每个小数转换成最简分数:

1. $0.\overline{7}$
2. $0.\overline{12}$
3. $5.\overline{271}$

Answers: 1. Let $x = 0.\overline{7}$. We know that $10x = 7.\overline{7}$.

Subtracting, we get $9x = 7$, for which $x = \boxed{\dfrac{7}{9}}$.

2. Let $x = 0.\overline{12}$. We know that $100x = 12.\overline{12}$.

Subtracting, we get $99x = 12$, for which $x = \dfrac{12}{99} = \boxed{\dfrac{4}{33}}$.

3. Let $x = 0.\overline{271}$. We know that $1000x = 271.\overline{271}$.

Subtracting, we get $999x = 271$, for which $x = \dfrac{271}{999}$.

The final answer is $\boxed{5 + \dfrac{271}{999}}$, or $\boxed{\dfrac{5266}{999}}$.

Mixed Recurring Decimal　混循环小数

n. [mɪkst rɪˈkɜː.ɪŋ ˈdesəməl]

Definition: A decimal in which not all the digits in the fractional part are repeated (appears according to a cycle). i.e. The first or the first several digits after the decimal point are standing by itself, different from the repetend.

小数部分不是所有的数字都是循环的。小数点后最前面的一位或几位是独立的,不同于循环节。

Properties: To convert a mixed recurring decimal to a simplest fraction, one possible way is to use some algebra, as shown in the Questions section.

把混循环小数转换成最简分数，一种方法是用代数知识，见问题(Questions)部分。

We will learn other methods of converting mixed recurring decimals to simplest fractions in geometric series in Chapter 31.

在第 31 章的等比级数中，将介绍从混循环小数转换成最简分数的其他方法。

Examples：$0.2\overline{3}, 1.2\overline{5}, 3.5\overline{67}, \cdots$

Questions：Convert each of the following to a common fraction：

把下列每个小数转换成最简分数：

1. $0.3\overline{2}$
2. $0.3\overline{45}$
3. $6.25\overline{13}$

Answers：1. Let $x = 0.0\overline{2}$. We know that $10x = 0.2\overline{2}$.

Subtracting，we get $9x = 0.2$，for which $x = \frac{2}{90}$.

The final answer is $0.3 + \frac{2}{90} = \frac{3}{10} + \frac{2}{90} = \boxed{\frac{29}{90}}$.

2. Let $x = 0.0\overline{45}$. We know that $100x = 4.5\overline{45}$.

Subtracting，we get $99x = 4.5$，for which $x = \frac{45}{990} = \frac{1}{22}$.

The final answer is $0.3 + \frac{1}{22} = \frac{3}{10} + \frac{1}{22} = \boxed{\frac{19}{55}}$.

3. Let $x = 0.00\overline{13}$. We know that $100x = 0.13\overline{13}$.

Subtracting，we get $99x = 0.13$，for which $x = \frac{13}{9900}$.

The final answer is $6.25 + \frac{13}{9900} = \frac{625}{100} + \frac{13}{9900} = \frac{61\,888}{9900} = \boxed{\frac{15\,472}{2475}} = \boxed{6 + \frac{622}{2475}}$.

1.4.4 Summary 总结

1. Figure 1-12 shows the relationship among different types of decimals. 图 1-12 展示了不同种类的小数。

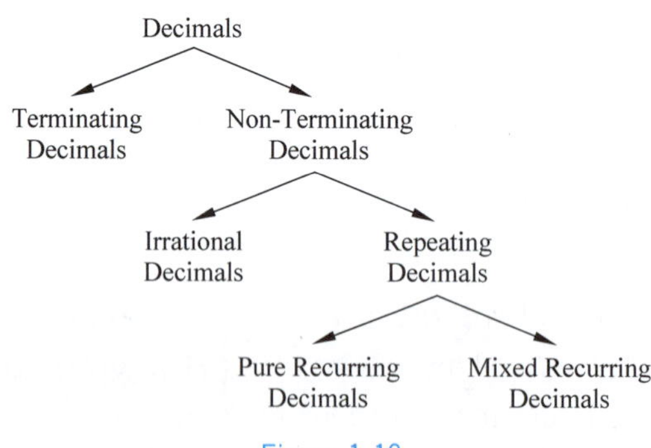

Figure 1-12

2. Review digits for numbers on different places of a decimal. 复习数位名称。

3. Review terminating decimal, pure recurring decimal, and mixed recurring decimal for rules

converting decimals to fractions.　复习有限小数、纯循环小数、混循环小数，掌握小数到分数的转换方法。

1.5 Number Line 数轴

1.5.1 Introduction 介绍

Number Line 数轴

n. [ˈnʌmbər ˌlaɪn]

Definition：A straight line with a "zero" point in the middle, with positive and negative numbers listed on right and left of zero, respectively, both going on indefinitely. It is used in illustrating simple numerical operations.　数轴是指一条以 0 为中心值、正数和负数分别在 0 的右边和左边的轴。轴的两端无限延长。数轴能图解简单的数字运算。

Notation：The illustration of a number line is shown in Figure 1-13.　数轴的图解如图 1-13 所示。

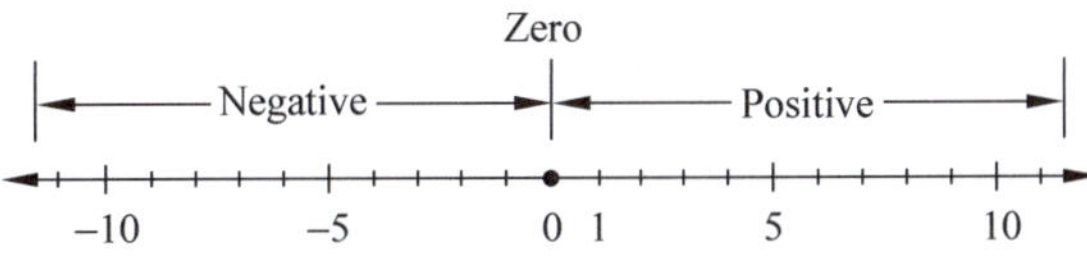

Figure 1-13

Properties：1. See "Real Number" in Section 1.1.　见 1.1 节实数。

2. We define the unit of the number line 1, which is the increment from one integer to the next. For example:
数轴的单位为 1，是一个整数到下一个整数的增量。例如：
From 4 to 5, there is 1 unit.
From 7 to 20, there are 13 units.
From −6 to −1, there are 5 units.

Questions：How do you represent each of the following on the number line?

1. $3+5$
2. $10-6$
3. $5\cdot 7$
4. $-5\cdot 7$

Answers：1. Starting with 3 and moving 5 units to the right, we end up with 8.
Or, starting with 5 and moving 3 units to the right, we end up with 8.
See addition.　见加法。

2. Starting with 10 and moving 6 units to the left, we end up with 4.
See subtraction, which is a special form of addition.　见减法，它是特殊的加法。
Or, $10-6=10+(-6)$.
Starting with 10 and moving −6 units to the right, we end up with 4.
Or, starting with −6 and moving 10 units to the right, we end up with 4.
See addition.　见加法。

3. Starting with 0 and adding 5 seven times, we end up with 35.
Or, starting with 0 and adding 7 five times, we end up with 35.
See multiplication, which is a convenient way of addition.　见乘法，它是简便的加法。

4. Starting with 0 and adding -5 seven times, we end up with -35.
   Or, starting with 0 and subtracting 5 seven times, we end up with -35.

Distance (on number line)　距离(数轴上)

n. [ˈdɪstəns]

Definition: The numerical length between two points (numbers) on the number line.　数轴上两点(数字)的数值长度。

Properties: 1. The distance between a and b on the number line is $|b-a|$, as shown in Figure 1-14. 数轴上 a 和 b 两点的距离是两点之差的绝对值,如图 1-14 所示。

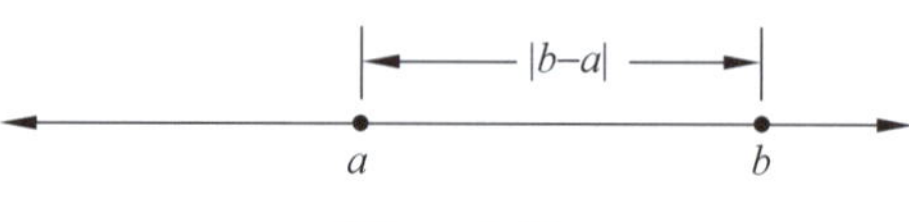

Figure 1-14

2. The distance between a and b on the number line is the same as the distance between b and a. a 和 b 的距离与 b 和 a 的距离是相同的。
3. The distance between 0 and a on the number line is $|a|$. This can be concluded from Property 1 and the properties of absolute value. 0 和 a 的距离是 $|a|$。见绝对值的性质 1 和其他性质。

Examples: 1. The distance between 2 and 5 is $|5-2|=3$.
2. The distance between -10 and 5 is $|5-(-10)|=15$.
3. The distance between -10 and -2 is $|-2-(-10)|=8$.
4. The distance between 0 and 8 is 8.
5. The distance between -10 and 0 is 10.

1.5.2 The Relationship between Two Numbers　两数之间的关系

Opposite　相反数

n. [ˈɑpəsɪt]

Definition: Two numbers are opposites if they each has the same distance away from zero, but are on the opposite sides of the number line.　若两数与 0 等距,且在 0 的两边,则它们是相反数。

Notation: The opposite of a is $-a$ for all real numbers of a. Note that $-a=-1\cdot a$. See multiplication. The illustration of opposites is shown in Figure 1-15.　对于所有实数 a, a 的相反数是 $-a$。注意 $-a=-1\cdot a$。见乘法。相反数的图解见图 1-15。

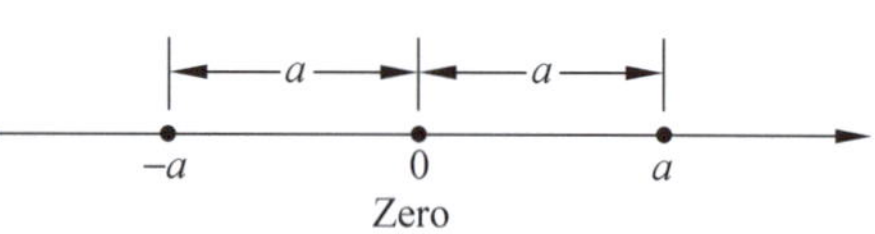

Figure 1-15

Properties: 1. (1) a and $-a$ are opposites of each other. a 和 $-a$ 互为相反数。
(2) From 1(1), it can be deduced that 0 is its own opposite.　从性质 1(1)里可推导出 0 是它自己的相反数。
2. Another name for “opposite” is additive inverse.　相反数的另一个名称是加法逆元。
3. The sum of a number and its opposite is 0.　一个数与它相反数的和是 0。

Questions: Find the opposite for each of the following: 40, -5, $x+6$, $10-x$.　求相反数。

Answers: The opposite of 40 is -40.

The opposite of -5 is $-(-5)=5$. See multiplication.　见乘法。

The opposite of $x+6$ is $-(x+6)=-x-6$. See distributive property.　见分配律。

The opposite of $10-x$ is $-(10-x)=-10+x$. See distributive property.　见分配律。

Reciprocal　倒数

n. [rɪ'sɪprəkəl]

Definition：The reciprocal of a number is the quotient of 1 and that number.　一个数的倒数为 1 除以它本身。

Notation：The reciprocal of a is $1/a$ for all nonzero numbers of a.　对于所有非零数 a，a 的倒数是 $1/a$。

Properties：1. Nonzero numbers a and $1/a$ are reciprocals of each other.　非零数 a 和 $1/a$ 互为倒数。

2. The reciprocal of a number has the same sign as that number.　一个数和它的倒数正负符号相同。

3. 0 does not have a reciprocal because $1/0$ is undefined.　0 没有倒数，因为 $1/0$ 无意义。

4. The product of a nonzero number and its reciprocal is always 1：$a\cdot\frac{1}{a}=1$，for which $a\neq 0$.　非零数和它的倒数之积必然为 1，因为对于非零数 a，$a\cdot\frac{1}{a}=1$。

Questions：What are the reciprocals of each of the following：

求下列数的倒数：

$3,-4,1/5,-1/8$。

Answers：$1/3$，

$-1/4$，

$1\div(1/5)=5$，

$1\div(-1/8)=-8$.

1.5.3　Absolute Values　绝对值

Absolute Value　绝对值

n. ['æbsəˌlut 'vælju]

Definition：The absolute value of a number is the distance on the number line from that number to 0.　一个数的绝对值是它离 0 的距离。

Notation：The absolute value of a is denoted by $|a|$. The illustration of absolute values of two numbers，a and b，is shown in Figure 1-16. a 的绝对值写作 $|a|$。图 1-16 展示了两个数（a 和 b）的绝对值。

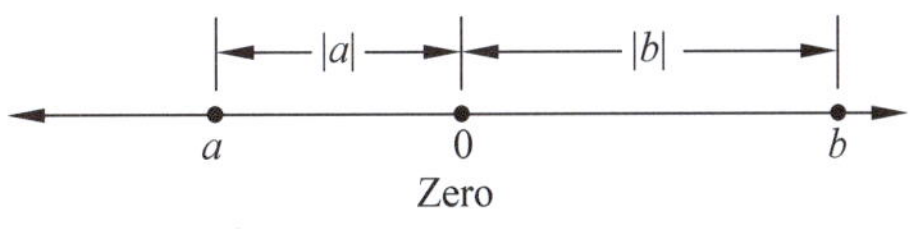

Figure 1-16

Properties：1. The absolute value of a number is always nonnegative.　一个数的绝对值是非负的。

2. The absolute value of 0 is 0，and the absolute value of any other number is positive.　0 的绝对值为 0。其他数的绝对值均大于 0。

3. The absolute value of a number is equal to that of its opposite：$|a|=|-a|$.　一个数

的绝对值与它相反数的绝对值相等。

4. If a is nonnegative, then $|ab| = a \cdot |b|$. 若 a 为非负数,则 $|ab| = a \cdot |b|$。

Examples: $|2| = 2, |15| = 15, |-14| = 14, |-6| = 6, |0| = 0$.

1.5.4 Summary 总结

1. A number line is a straight line with a "zero" point in the middle, with positive and negative numbers listed on right and left of zero, respectively, both going on indefinitely. It is used in illustrating simple numerical operations. 数轴是一条以 0 为中心值,正数和负数分别在 0 的右边和左边的轴。轴的两端无限延长。数轴能图解简单的数字运算。
2. For all real numbers a, a and $-a$ are opposites. Their sum is 0. Opposites have the same absolute value, and are on different sides of 0. 对于所有实数 a,a 和 $-a$ 互为相反数。它们的和为 0。相反数的绝对值相同。它们在 0 的对边。
3. For all nonzero numbers a, a and $1/a$ are reciprocals. Their product is 1. Reciprocals have the same sign, and 0 does not have a reciprocal. 对于所有非零数 a,a 和 $1/a$ 互为倒数。它们的积为 1。倒数的正负符号相同。0 没有倒数。
4. The absolute value of a number a represents the distance between 0 and a on the number line. It is written as $|a|$, which is nonnegative. Note that $|a| = 0$ if $a = 0$, and $|a| > 0$ otherwise. 实数 a 的绝对值是数轴中 0 到 a 的距离。写作 $|a|$,是非负数。若 $a = 0$,则 $|a| = 0$,否则 $|a| > 0$。

1.6 Operations 运算

1.6.1 Introduction 介绍

Operation 运算

n. [ˌɑpəˈreɪʃən]

Definition: The interaction of two or more numbers in order to form a new number, including addition, subtraction, multiplication, division, power, absolute value, and many more. The new number is known as the result. 运算是指两个或更多数的"互动",从而组成新的数。运算包括加、减、乘、除、幂运算、绝对值等。新的数也称为得数。

1.6.2 Types of Operations 运算种类

视频 2

Addition 加法

n. [əˈdɪʃən]

Definition: The process of uniting two numbers into one sum. 加法是合并两个数至一个和的过程。

Notation: The numbers involved in an addition, namely the addends, are connected by the addition symbol "+". 加数用加号"+"连接。

Properties: 1. The numbers involved in an addition are known as the terms. More specifically, the addends.

terms 和 addends 均有加数的意思。通常用后者,因为前者亦有"项"的意思。

2. The result of an addition is known as the sum. 加法的得数叫作和。

3. The Commutative Property of Addition: When adding two numbers, the order does not matter: $a + b = b + a$.　加法交换律。
4. The Associative Property of Addition: When adding three numbers, the grouping does not matter: $(a + b) + c = a + (b + c)$.　加法结合律。
5. The Identity Property of Addition: When adding any number to 0, the result is that number: $a + 0 = a$. The number 0 is known as the identity element of addition.　加法的单位元性质：0 为单位元。
6. The Inverse Property of Addition: When adding any number with its opposite, the result is 0: $a + (-a) = 0$. The numbers a and $-a$ are additive inverses of each other.　加法的逆元性质：a 和 $-a$ 互为加法逆元。
7. When adding two positive numbers, it is straight forward (use the decimal system as learned in elementary school).　两个正数相加时，相对来说简单(用小学学的十进制)。
8. When adding two negative numbers, add their absolute values, then put the negative sign in front of the result.　两个负数相加时，先把它们的绝对值相加，然后在结果前面添上负号。
9. When adding a positive and a negative numbers, first take the positive difference of their absolute values. The sum takes the sign of the number with greater absolute value.　正数和负数相加时，先取它们绝对值的正差。和的正负符号取决于绝对值较大的数的正负符号。

Examples: $3 + 4 = 7$.

$6 + 8 = 14$.

$3 + (-2) = 1$.

$-6 + (-7) = -13$.

Questions: 1. (Property 5) Find the sum: (1) $5 + 0$; (2) $0 + 9$.

2. (Property 6) Find the sum: (1) $6 + (-6)$; (2) $10 + (-10)$.
3. (Property 7) Find the sum: (1) $5 + 6$; (2) $8 + 9$.
4. (Property 8) Find the sum: (1) $-6 + (-9)$; (2) $-50 + (-45)$.
5. (Property 9) Find the sum: (1) $-6 + 9$; (2) $5 + (-20)$.
6. John spent \$1 and earned \$2 on Day 1, spent \$3 and earned \$4 on Day 2, and spent \$5 and earned \$6 on Day 3. How much money did he earn in these three days?
7. John earned \$1 and spent \$2 on Day 1, earned \$3 and spent \$4 on Day 2, and earned \$5 and spent \$6 on Day 3. How much money did he earn in these three days?

Answers: 1. (1) 5; (2) 9.

2. (1) 0; (2) 0.
3. (1) 11; (2) 17.
4. (1) $-6 + (-9) = -(|-6| + |-9|) = -(6 + 9) = -15$.

   (2) $-50 + (-45) = -(|-50| + |-45|) = -(50 + 45) = -95$.

5. (1) The addends are -6 and 9. Since $|-6|<|9|$, the sum is positive.
$-6+9=|9|-|-6|=3$.
(2) The addends are 5 and -20. Since $|5|<|-20|$, the sum is negative.
$5+(-20)=-(|20|-|5|)=-15$.
注意确认加数,比较绝对值大小并取绝对值的正差。和的正负符号取决于绝对值较大的数的正负符号。

6. On Day 1, John earned $-1+2=1$ dollar.
On Day 2, he earned $-3+4=1$ dollar.
On Day 3, he earned $-5+6=1$ dollar.
He earned $1+1+1=\boxed{3\text{ dollars}}$ these three days.

7. On Day 1, John earned $1+(-2)=-1$ dollar.
On Day 2, he earned $3+(-4)=-1$ dollar.
On Day 3, he earned $5+(-6)=-1$ dollar.
He earned $-1+(-1)+(-1)=\boxed{-3\text{ dollars}}$ these three days. This is the same as spending/losing 3 dollars these three days.

Subtraction 减法

n. [səbˈtrækʃən]

Definition: The process of taking one number or amount away from another. 减法是从一个数里拿走另一个数的过程。

Notation: The numbers involved in a subtraction are connected by the subtraction symbol "$-$". 跟减法有关系的数用减号"$-$"连接。

Properties: 1. To subtract a number, add its opposite. 减去一个数,相当于加上它的相反数。
2. The result of a subtraction is known as the difference. 减法的得数叫作差。
3. Subtracting 0 does not change anything: $a-0=a$. 减去0后,被减数不变。
4. Any number subtracting itself results 0: $a-a=0$. 任何数减去它本身得0。
5. $(a-b)$ and $(b-a)$ are opposites of each other. $(a-b)$和$(b-a)$互为相反数。

Examples: $8-6=2$.
$6-8=-2$.
$-5-3=-8$.
$-7-(-9)=2$.

Questions: Find the difference in each of the following:
(1) $80-45$.
(2) $30-50$.
(3) $40-(-20)$.
(4) $50-(-10)$.
(5) $-10-8$.
(6) $-5-9$.
(7) $-10-(-30)$.
(8) $-12-(-20)$.

Answers：Following Property 1：

(1) $80-45=80+(-45)=\boxed{35}$.

(2) $30-50=30+(-50)=\boxed{-20}$.

(3) $40-(-20)=40+20=\boxed{60}$.

(4) $50-(-10)=50+10=\boxed{60}$.

(5) $-10-8=-10+(-8)=\boxed{-18}$.

(6) $-5-9=-5+(-9)=\boxed{-14}$.

(7) $-10-(-30)=-10+30=\boxed{20}$.

(8) $-12-(-20)=-12+20=\boxed{8}$.

Multiplication　乘法

n. [ˌmʌltəplɪˈkeɪʃən]

Definition：A shorthand notation for addition. The expression $a \cdot b$ means the sum of b number of a's, or the sum of a number of b's.　乘法是加法的简便写法。表达式 $a \cdot b$ 代表的是“b 个 a 的和”或“a 个 b 的和”。

Notation：The numbers involved in a multiplication, namely the factors, are connected by the multiplication symbol “·” or “×”.　因子用乘号“·”或“×”连接。

When we multiply two variables a and b, we usually drop the multiplication sign：$a \cdot b = ab$.　两个变量相乘时，通常省略掉乘号。

Properties：
1. The numbers involved in a multiplication are known as factors.　乘数也称为因子。
2. The result of a multiplication is known as the product.　乘法的得数叫作积。
3. Definition of the negative sign：$-a=-1 \cdot a$.　负号的定义。
4. Multiplying 0 by any number is 0：$0 \cdot a = 0$.　任何数乘 0 都得 0。
5. When multiplying two numbers of the same sign, the result is positive.　同号相乘得正。
6. When multiplying two numbers of different signs, the result is negative.　异号相乘得负。
7. The Commutative Property of Multiplication：When multiplying two numbers, the order does not matter：$a \cdot b = b \cdot a$.　乘法的交换律。
8. The Associative Property of Multiplication：When multiplying three numbers, the grouping does not matter：$(a \cdot b) \cdot c = a \cdot (b \cdot c)$.　乘法的结合律。
9. The Identity Property of Multiplication：When multiplying any number by 1, the result is that number：$a \cdot 1 = a$. The number 1 is known as the identity element of multiplication.　乘法的单位元性质：1 为单位元。
10. The Inverse Property of Multiplication：When multiplying any nonzero number by its reciprocal, the result is 1：$a \cdot (1/a) = 1$. The numbers a and $1/a$ are multiplicative inverses of each other. Note that $a \neq 0$ because 0 does not have a reciprocal.　乘法的逆元性质：a 和 $1/a$ 互为乘法逆元。注意 a 不能等于 0，因为 0

没有倒数。

11. The Distribute Property of Multiplication: $a(b+c)=ab+ac$. More generally, $a(b+c+d+\cdots)=ab+ac+ad+\cdots$. 乘法的分配律。

12. The product of a number and 2 is sometimes called the double of that number. 一个数字和 2 的积有时候被称为这个数的两倍。

    The product of a number and 3 is sometimes called the triple of that number. 一个数字和 3 的积有时候被称为这个数的三倍。

    The product of a number and 4 is sometimes called the quadruple of that number. 一个数字和 4 的积有时候被称为这个数的四倍。

Questions: 1. Find the product on each of the following:

(1) 5×7; (2) -5×7; (3) $3\times(-8)$; (4) $-3\times(-8)$; (5) $5\times(1/5)$; (6) $6\times(-1/6)$; (7) 0×7; (8) -9×0; (9) 1×10; (10) $-1\times(-12)$.

2. Write and evaluate an expression in multiplication in calculating the amount earned for each of the following:

在下列各种情况中,写出代表赚取的钱的表达式:

(1) Earning \$5 three times.

(2) Spending \$5 three times.

(3) Before earning \$5 three times.

(4) Before spending \$5 three times.

Answers: 1. (1) 35; (2) −35; (3) −24; (4) 24; (5) 1; (6) −1; (7) 0; (8) 0; (9) 10; (10) 12。

2. (1) \$5 · 3 = \$15. Gained \$15.

(2) − \$5 · 3 = − \$15. Lost \$15.

(3) \$5 · (−3) = − \$15. Lost \$15.

(4) − \$5 · (−3) = \$15. Gained \$15.

Division 除法

n. [dɪ'vɪʒən]

Definition: "a is divided by b" means the number of items in each portion if we divide a items into b equal portions. a 除以 b 的意思是若把 a 个物品分成 b 等份,求每份的数量。

Notation: "a is divided by b" is denoted by a/b, or $a\div b$, or $\frac{a}{b}$. 除法可用除号或者分数线表示。

Properties: 1. If a and b have the same sign, then a/b is positive. 同号相除得正。

2. If a and b have different signs, then a/b is negative. 异号相除得负。

From Properties 3-7, let a, q, and r be nonnegative integers and b be a positive integer. 对于下面的性质 3~7,设 a、q、r 为非负整数,b 是正数。

3. Remainder Notation: If $a\div b=q\cdots r$, a is called the dividend; b is called the divisor; q is called the quotient; r is called the remainder. $b>r$ must be true. For example, $16\div3=5\cdots1$. 余数表示法:在 $a\div b=q\cdots r$ 中,a 为被除数,b 为除数,q 为商,r 为余数。除数必须大于余数。

4. In Property 3, $a\div b=q\cdots r$. If $a<b$, then $q=0$ and $r=a$. For example, $5\div13=$

0 ⋯ 5.　若商为 0,则余数等于被除数。

5. In Property 3, $a \div b = q \cdots r$. If $r = 0$, then b and q are factors of a. For example, $18 \div 3 = 6 \cdots 0$. Clearly, 3 and 6 are factors of 18.　若余数为 0,则 b 和 q 均为 a 的约数。

6. In Property 3, if $a \div b = q \cdots r$, and $b > r$, then $a = bq + r$. For example, in $16 \div 3 = 5 \cdots 1$, $16 = 3 \cdot 5 + 1$.　根据性质 3, $a \div b = q \cdots r$ 的等价表达式为 $a = bq + r$。

7. Fraction Notation: To convert the Remainder Notation to a fraction, from $a = bq + r$ of Property 6, dividing both sides by b, we get $\frac{a}{b} = q + \frac{r}{b}$.　分数表示法:在 $a = bq + r$ 的两边除以 b 可得 $\frac{a}{b} = q + \frac{r}{b}$。

8. The divisor can never be 0. Otherwise the result is undefined.　除数不能为 0,否则得数无意义。

9. To divide by a number, multiply by its reciprocal:
除以一个数,相当于乘上它的倒数:

$$a \div b = a \cdot \frac{1}{b}, \quad b \neq 0.$$

10. When 0 is divided by any nonzero number, the result is 0:
0 除以任何非零的数得 0:

$$0 \div a = 0, \quad a \neq 0.$$

Questions: 1. Using the Remainder Notation, evaluate each of the following:
用余数表示法求下面各式的值:
(1) 18/4　　(2) 104/20　　(3) 10/81

2. Using the Fraction Notation, evaluate each of the following:
用分数表示法求下面各式的值:
(1) 18/4　　(2) 104/20　　(3) 10/81

3. Using the decimal notation, evaluate each of the following:
用小数表示法求下面各式的值:
(1) 18/(−4)　　(2) 104/20　　(3) (−10)/(−81)

4. Evaluate:
(1) −15/(1/3)　　(2) 12/(−1/4)　　(3) 0/5

Answers: 1. (1) $18/4 = 4 \cdots 2$.
(2) $104/20 = 5 \cdots 4$.
(3) $10/81 = 0 \cdots 10$.

2. (1) $18/4 = 4 + 2/4 = 4 + 1/2$.
(2) $104/20 = 5 + 4/20 = 5 + 1/5$.
(3) $10/81 = 0 + 10/81 = 10/81$, as it is.

3. (1) $18/(-4) = -4.5$.
(2) $104/20 = 50.2$.
(3) $(-10)/(-81) = 0.\overline{123456790}$.

4. Using Property 9 and 10：

(1) $-15/(1/3) = -15 \times 3 = -45$.

(2) $12/(-1/4) = 12 \times (-4) = -48$.

(3) $0/5 = 0$.

Remainder 余数

n. [rɪˈmeɪndər]

Definition：The amount "left over" after a division. 余数是指除法中余下的数量。

Notation：For positive integers a and b：

"a mod b" represents the remainder when a is divided by b.

"a mod b" is the abbreviation of "a modulo b". mod 的含义是"模"。是数学中余数的运算符号。

Properties：1. In a division, the remainder is always less than the divisor. 在除法中，余数必须比除数小。

2. If the dividend is a multiple of the divisor, then the remainder is 0. Or, if a is divisible by b, then a mod $b = 0$. 若被除数是除数的倍数，则余数为 0。

3. If the dividend is smaller than the divisor, then the remainder is the dividend. Or, if $a < b$, then a mod $b = a$. 若被除数小于除数，则余数等于被除数。

4. If a mod $b = b$ mod a, then $a = b$.

5. Special remainder shortcut：For a positive integer n：
对于正整数 n，求余数的简便方法：

(1) When n is divided by 2, the remainder is 1 if the ones digit of n is odd, 0 if the ones digit of n is even. 当 n 除以 2 时，若 n 是奇数，余数为 1；若 n 是偶数，余数为 0。

(2) When n is divided by 3, the remainder is the same as that of when the sum of digits of n is divided by 3. n 除以 3 的余数与 n 的数位和除以 3 的余数相同。

(3) When n is divided by 5, the remainder is the same as that of when the ones digit of n is divided by 5. n 除以 5 的余数与 n 的个位除以 5 的余数相同。

(4) When n is divided by 9, the remainder is the same as that of when the sum of digits of n is divided by 9. n 除以 9 的余数与 n 的数位和除以 9 的余数相同。

(5) When n is divided by 10, the remainder is the ones digit of n. n 除以 10 的余数是 n 的个位数。

Questions：1. What is the value for each of the following?

(1) 14 mod 4　(2) 2016 mod 5　(3) 166 mod 6

(4) 50 mod 25　(5) 123456 mod 3　(6) 524 mod 9

(7) 1001 mod 2　(8) 123456 mod 10　(9) 3 mod 80

2. Joshua wrote his name 100 times. What is the 100th letter he wrote?

Answers：1. (1) 2　(2) 1　(3) 4

(4) 0　(5) 0　(6) 2

(7) 1　(8) 6　(9) 3

2. 100 mod 6 = 4, and the 4th letter is [h].

Power　幂

n. [ˈpaʊər]

Figure 1-17 labels the parts of a power.

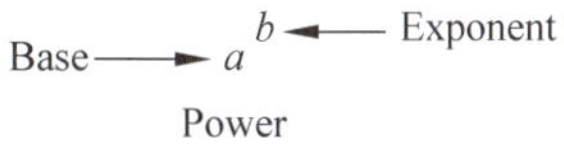

Figure 1-17

Definition: A shorthand notation of multiplying the same number by itself many times. Its representation is a combination of a base and an exponent. The base represents the number being multiplied, and the exponent represents how many bases are being multiplied.　幂是指一个数乘上它本身若干遍的简便写法。乘上若干遍的数称为底数。在乘法里,底数的个数称为指数。

Notation: a^b is read as "a to the bth power", or "the bth power of a". It represents the product of b number of a's.

a^b 读作 a 的 b 次幂,代表的是 b 个 a 相乘的积。

a is known as the base, and b is known as the exponent.　a 是底数,b 是指数。

Properties: 1. $a^b=b^a$ is not necessarily true for all a and b.

2. See the properties of exponents.　见指数性质。

3. $a^{b^c}=a^{(b^c)}$ according to PEMDAS.　此性质根据运算顺序(详见 1.7.2 节)。

4. $1^a=1$ for all a.

5. $0^a=0$ for all $a>0$.

Examples: $3^4=3\times3\times3\times3=81$.

$2^8=2\times2\times2\times2\times2\times2\times2\times2=256$.

Phrases: a is known as the first～of a, but we always say "a" instead of saying the first～of a.　a 是它本身的一次幂,但通常省略不说。

a^2 is known as the second～(square) of a.　二次幂也叫平方。

a^3 is known as the third～(cube) of a, and so on.　三次幂也叫立方。

Questions: 1. What is the 10th power of 2?　求 2 的十次方。

2. 625 is the ________ power of 5.　625 是 5 的多少次方?

3. What is the value of 4^{2^3}?

Answers: 1. $2^{10}=2\times2\times2\times2\times2\times2\times2\times2\times2\times2=\boxed{1024}$.

2. Since $5^4=625$, 625 is the $\boxed{\text{fourth}}$ power of 5.

3. $4^{2^3}=4^8=\boxed{65\,536}$.

1.6.3　Parts of a Power　幂的部分

视频 3

Base　底数

n. [beɪs]

Definition: Part of the power that represents what is being multiplied.　乘上若干遍的数是幂的底数。

Notation: In a^b, a is the base.

Exponent　指数

n. [ɪkˈspoʊnənt]

Definition: Part of a power that represents how many bases are being multiplied.　相乘中,底数的个

数叫作指数。

Notation: In a^b, b is the exponent.

Properties: 1. $a^m \cdot a^n = a^{m+n}$.

a^m is the product of m number of a's, and a^n is the product of n number of a's. Thus their product is $(m+n)$ number of a's.

a^m 是 m 个 a 的积。a^n 是 n 个 a 的积。因此,两者的积有 $(m+n)$ 个 a。

2. $\dfrac{a^m}{a^n} = a^{m-n}$, if $a \neq 0$.

a^m is the product of m number of a's, and a^n is the product of n number of a's. Factors cancel in both numerator and denominator.

理由同性质1。注意分数中分子和分母的相同因子可消去。

3. $(a^m)^n = a^{mn}$.

a^m is the product of m number of a's. $(a^m)^n$ is the product of n number of a^m's. The exponent is the result of adding "m" n times, which is mn.

4. $(ab)^m = a^m b^m$.

$(ab)^m$ is the product of m number of (ab)'s. The numbers a and b each appears m times in the product.

5. $\left(\dfrac{a}{b}\right)^m = \dfrac{a^m}{b^m}$. Same reasoning of Property 4.

6. $a^0 = 1$, if $a \neq 0$.

From Property 2, $\dfrac{a^m}{a^m} = a^{m-m} = a^0$. Also, $\dfrac{a^m}{a^m} = 1$, if $a \neq 0$. Therefore, $a^0 = 1$.

7. $a^{-m} = \dfrac{1}{a^m}$, if $a \neq 0$.

From Property 2, $\dfrac{a^0}{a^m} = a^{-m}$. Also $\dfrac{a^0}{a^m} = \dfrac{1}{a^m}$ from Property 6. Therefore, by transitive, $a^{-m} = \dfrac{1}{a^m}$. 根据传递性。

Phrases: positive～ 正指数, negative～ 负指数, zero～ 零指数

Questions: 1. (1) If $5^3 \cdot 5^{10} = 5^k$, what is the value of k?

(2) If $7^5 \cdot 7^k = 7^{30}$, what is the value of k?

2. (1) If $\dfrac{6^{20}}{6^3} = 6^k$, what is the value of k?

(2) If $\dfrac{8^{15}}{8^k} = 8^2$, what is the value of k?

3. (1) If $(5^6)^8 = 5^k$, what is the value of k?

(2) If $(9^7)^k = 9^{63}$, what is the value of k?

4. (1) If $(5 \cdot 7)^{50} = 5^j \cdot 7^k$, what are the values of integers j and k?

(2) If $(6^3 \cdot 17^5)^{20} = 6^j \cdot 17^k$, what are the values of integers j and k?

5. (1) If $\left(\dfrac{5}{8}\right)^{30} = \dfrac{5^j}{8^k}$, what are the values of integers j and k?

(2) If $\left(\frac{2^3}{7^5}\right)^{30}=\frac{2^j}{7^k}$, what are the values of integers j and k?

6. What is the value for each of the following?

(a) 9^0　(b) 14^0　(c) 10^{-2}　(d) 5^{-5}

7. If $2^{22}+2^{22}=2^k$, what is the value of k?

8. If $5^3\cdot 25^{10}=5^k$, what is the value of k?

Answers: 1. (1) $k=13$.

(2) $k=25$.

2. (1) $k=17$.

(2) $k=13$.

3. (1) $k=48$.

(2) $k=9$.

4. (1) $j=k=50$.

(2) $(6^3\cdot 17^5)^{20}=(6^3)^{20}\cdot(17^5)^{20}=6^{60}\cdot 17^{100}$.

Therefore, $j=60$ and $k=100$.

5. (1) $j=k=30$.

(2) $\left(\frac{2^3}{7^5}\right)^{30}=\frac{(2^3)^{30}}{(7^5)^{30}}=\frac{2^{90}}{7^{150}}$. Therefore, $j=90$ and $k=150$.

6. (1) 1.

(2) 1.

(3) $\frac{1}{10^2}$.

(4) $\frac{1}{5^5}$.

7. $2^{22}+2^{22}=2\cdot 2^{22}=2^{23}$. Thus $k=23$.

8. $5^3\cdot 25^{10}=5^3\cdot(5^2)^{10}=5^3\cdot 5^{20}=5^{23}$. Thus $k=23$.

1.6.4 Translating Words to Mathematical Expressions　从文字到表达式

Questions of Writing Expressions from Words　文字题

Questions: Write an expression for each of the following.

1. The sum of a and b minus the product of c and d.

a 与 b 的和减去 c 与 d 的积。

2. The quotient of a and b plus the product of c and d.

a 与 b 的商加上 c 与 d 的积。

3. The sum of a and b multiplies by the kth power of 3.

a 与 b 的和乘上 3 的 k 次幂。

4. 3 more than the difference of a and b.

比 a 与 b 的差多 3。

5. 8 less than the sum of 2 to the mth power and 6 to the nth power.

比 2 的 m 次幂与 6 的 n 次幂的和少 8。

Answers: 1. $(a+b)-cd$.

2. $\frac{a}{b}+cd$.

3. $(a+b)3^k$.

4. $3+(a-b)$.

5. 2^m+6^n-8.

1.6.5 Application of Powers of 10: Scientific Notation 10 的次幂应用：科学记数法

Scientific Notation 科学记数法

n. [ˌsaɪənˈtɪfɪk noʊˈteɪʃən]

Definition: The notation of a number in a way that it is the product of a number whose absolute value is greater than or equal to 1 and less than 10 and a power of 10. See notation below. 科学记数法是指把一个数改写成一个“绝对值大于或等于 1 且小于 10 的数”与“一个 10 的次幂”的积。见以下记数。

Notation: $c \cdot 10^n$, for which $1 \leqslant |c| < 10$ and n is an integer.

Properties: 1. To convert a number to scientific notation, first determine what c is. Then observe how many places the decimal point of c moved to the right. This number is known as n (Moving to the left makes n negative.). 要把一个数改写成科学记数法，先求出 c 的值，然后观察小数点要向右移多少位从而决定 n。假如是向左移，则 n 为负数。

2. To compare scientific notations of two numbers:
要比较科学记数法的两个数的大小：

(1) If they have different signs, then the positive number is always greater. 若正、负符号不同，则正数比负数大。

(2) If they have the same sign:
若符号相同：

- If they are positive, whoever has the larger value of n is greater. If their values of n are the same, whoever has the larger value of c is greater. 若都是正数，带有较大的 n 的数较大。若 n 的值相同，带有较大的 c 的数较大。
- If they are negative, whoever has the larger value of n is less. If their values of n are the same, whoever has the larger value of $|c|$ is less. 若都是负数，带有较大的 n 的数较小。若 n 的值相同，带有较大的 $|c|$ 的数较小。

Examples: 1. The age of Earth, in years, is 4 600 000 000. In scientific notation, this is 4.6×10^9.
地球的年龄用科学记数法表示为 4.6×10^9 年。

2. In chemistry, an atom mass unit is 0.000 000 000 000 000 000 000 000 001 66 kilograms, or 1.66×10^{-27} kg. 原子质量单位表示为 1.66×10^{-27} kg。

Scientific notation is a really convenient way to express long numbers. Making them easy to write and compare. 科学记数法便于表示长数字，使它们更容易表达和比较。

Questions: 1. Convert each of the following to scientific notation.

(1) $N = 12\ 345\ 000\ 000\ 000$

(2) $N = -9\ 870\ 000\ 000$

(3) $N = 0.000\ 000\ 000\ 012\ 6$

(4) $N = -0.000\ 008$

2. Convert each of the following to standard form.

(1) 1.42×10^{10}

(2) 5.2×10^{-8}

Answers: 1. (1) Note that $c = 1.2345$. The decimal point of c moves 13 places to the right to form N. Thus $n = 13$. The answer is $\boxed{1.2345 \times 10^{13}}$.

(2) Note that $c = -9.87$. The decimal point of c moves 9 places to the right to form N. Thus $n = 9$. The answer is $\boxed{-9.87 \times 10^{9}}$.

(3) Note that $c = 1.26$. The decimal point of c moves 11 places to the left to form N. Thus $n = -11$. The answer is $\boxed{1.26 \times 10^{-11}}$.

(4) Note that $c = -8$. The decimal point of c moves 6 places to the left to form N. Thus $n = -6$. The answer is $\boxed{-8 \times 10^{-6}}$.

2. (1) 14 200 000 000

(2) 0.000 000 052

1.6.6 Summary 总结

1. The result of an addition is called the sum. 加法的结果称为和。
   The result of a subtraction is called the difference. 减法的结果称为差。
   The result of a multiplication is called the product. 乘法的结果称为积。
   The result of a division is called the quotient. 除法的结果称为商。
2. Subtractions are special additions—subtracting a number is the same as adding its opposite. 减法是特殊的加法——减去一个数相当于加上它的相反数。
   Divisions are special multiplications—dividing by a nonzero number is the same as multiplying by its reciprocal. 除法是特殊的乘法——除以一个非零数相当于乘上它的倒数。
3. Addition and multiplication have Commutative Property, Associative Property, Identity Property, and Inverse Property. Their combination has the Distributive Property. 加法和乘法均有交换律、结合律、单位元性质、逆元性质。它们的组合有分配律。
   The identity element of addition is 0: $a + 0 = a$, and $a + (-a) = 0$. 加法的单位元是 0。
   The identity element of multiplication is 1: $a \cdot 1 = a$ (for all real numbers), and $a \cdot 1/a = 1$ ($a \neq 0$). 乘法的单位元是 1。
4. Review the concept of remainder in division. 复习除法中余数的概念。
5. When we raise a base to an exponent, we get power. 底数和指数结合成为幂。
   Review the rules of exponent, special powers: of 0 and of 1. 复习次幂的法则以及特殊的次幂：0 的次幂和 1 的次幂。
6. Review scientific notation—a notation that helps writing and comparing long numbers. 复习科学记数法——一种方便书写较长数字和比较它们大小的记数法。

1.7 Rules of Operations 运算法则

1.7.1 Introduction 介绍

Parentheses 括号

n. [pə'ren.θə.siːz]

Definition: The grouping operation for which the operations inside must be done first. 括号用于说明组合运算的顺序。括号里的运算优先。

Notation: (…),[…]

Properties: 1. In a numerical expression, parentheses is the priority. 数字表达式中,括号运算优先。

2. When there are parentheses within parentheses, the innermost one(s) has the priority. 若括号中包含括号,则最里面的括号优先。

Questions: 1. Evaluate $20-10-(10+5-8)$.

2. Evaluate $(10+20)\div(17-7)$.

3. Evaluate $5\cdot[10+(20-10)-(50-10)]$.

Answers: 1.

$$\begin{aligned}&20-10-(10+5-8)\\=&20-10-7\\=&\boxed{3}.\end{aligned}$$

2.

$$\begin{aligned}&(10+20)\div(17-7)\\=&30\div 10\\=&\boxed{3}.\end{aligned}$$

3.

$$\begin{aligned}&5\cdot[10+(20-10)-(50-10)]\\=&5\cdot[10+10-40]\\=&5\cdot[-20]\\=&\boxed{-100}.\end{aligned}$$

1.7.2 Order of Operations 运算顺序

PEMDAS 运算顺序

Definition: Abbreviation for "Parentheses→Exponents→[Multiplication, Division]→[Addition, Subtraction]". "括号→指数→【乘除】→【加减】"的缩写。

Properties: To explain the phrase in words: in a numerical expression, perform the steps in this order: 在数字表达式中,用文字概括步骤如下:

1. Remove the parentheses first. If there are nested parentheses, remove the parentheses from the innermost layer. 先运算括号中的项,去除括号。若有多重括号,先去除最里面的括号。
2. Compute the exponents and their associated powers. 再计算指数以及它们相应的幂。
3. Compute multiplications and divisions, whoever come first (from left to right). 乘法和除法从左到右依次计算。
4. Compute additions and subtractions, whoever come first (from left to right). 加法

和减法从左到右依次计算。

Questions：1. Evaluate $10+4\cdot(2^3-5)-8\div 4$.

2. -10^4

Answers：1.
$$\begin{aligned}&10+4\cdot(2^3-5)-8\div 4\\=&10+4\cdot(8-5)-8\div 4\\=&10+4\cdot 3-8\div 4\\=&10+12-8\div 4\\=&10+12-2\\=&22-2\\=&\boxed{20}.\end{aligned}$$

2.
$$\begin{aligned}&-10^4\\=&-1\cdot 10^4\\=&-1\cdot 10\,000\\=&\boxed{-10\,000}.\end{aligned}$$

1.7.3　Properties of Operations　运算性质

Commutative Property　交换律

n. [kə'mjuːtəˌtɪv 'prɑpərti]

Definition：The property in addition and multiplication such that the order of the addends/factors does not matter.　交换律是指在加法和乘法里,加数/乘数的顺序不重要。

Properties：1. Commutative Property of Addition：$a+b=b+a$.

加法交换律：$a+b=b+a$。

2. Commutative Property of Multiplication：$a\cdot b=b\cdot a$.

乘法交换律：$a\cdot b=b\cdot a$。

Associative Property　结合律

n. [ə'soʊʃiətɪv 'prɑpərti]

Definition：The property in addition and multiplication such that the grouping of addends/factors does not matter.　结合律是指在加法和乘法里,加数/乘数的组合不重要。

Properties：1. Associative Property of Addition：$(a+b)+c=a+(b+c)$.

加法结合律：$(a+b)+c=a+(b+c)$。

2. Associative Property of Multiplication：$(a\cdot b)\cdot c=a\cdot(b\cdot c)$.

乘法结合律：$(a\cdot b)\cdot c=a\cdot(b\cdot c)$。

Distributive Property　分配律

n. [dɪ'strɪbjətɪv 'prɑpərti]

Definition：To multiply a sum by a number, multiply each addend in the sum separately by that number, then add up the products.　把若干数的和与另一个数相乘,先把这个和的每一个加数与那个数分别相乘,然后把这些积加起来。

Properties：1. $(a+b)\cdot c=ac+bc$ (Main Property)

2. $(x+y+z+\cdots)\cdot c=xc+yc+zc+\cdots$ (Generalized Property 1)

3. $-(a+b-c)=-a-b+c$ (All signs in the parentheses alter if the parentheses is multiplied by a negative number. 若乘上一个负数,则括号里的所有符号翻转。

To show this Property, note that (1) $-a=-1\cdot a$ for all real numbers a, and (2) subtracting a number is the same as adding its opposite.

要证明此性质,注意:(1) 负号的定义;(2) 减去一个数相当于加上它的相反数。

$$\begin{aligned}&-(a+b-c)\\&=-(a+b+(-c))\\&=-1(a+b+(-c))\\&=(-1\cdot a)+(-1\cdot b)+((-1)\cdot(-c))\\&=-a+(-b)+c\\&=-a-b+c\end{aligned}$$

Closure Property 封闭性

n. [ˈkloʊʒər ˈprɑpərti]

Definition: A set of numbers is said to be closed under a specific mathematical operation if any two numbers in the set (not necessarily distinct) produce a number that is also from the same set under that operation. If a set of numbers is closed for a particular operation then it is said to possess the Closure Property for that operation. 在一个数集中,若对数集中的任意两个数(可以相同)进行某种运算,得数也在这个数集中,则这个数集对这个运算封闭。若一个数集对一个运算封闭,则这个数集对这个运算有封闭性。

For any $a, b\in S$, if $a\oplus b\in S$, then S is said to be closed under the operation $\oplus$.

Examples: 1. (1) The set of real numbers is closed under addition because the sum of any two real numbers is always a real number. 实数集对加法封闭,因为两个实数的和必为实数。

(2) The set of real numbers is closed under multiplication because the product of any two real numbers is always a real number. 实数集对乘法封闭,因为两个实数的积必为实数。

2. (1) The set of positive integers is closed under multiplication because the product of any two positive integers is a positive integers. 正整数集对乘法封闭,因为两个正整数的积必为正整数。

(2) The set of positive integers is not closed under division because 4 and 5 are positive integers, but 4/5 or 5/4 is not. 正整数集对除法不封闭,因为 4 和 5 是正整数,但 4/5 和 5/4 不是。

Identity Property 单位元性质

n. [aɪˈdentɪti ˈprɑpərti]

Definition: A set of numbers is said to have the Identity Property of an operation, when there exists a number a in the set such that when that operation is performed on a and any other number (can be a as well) in the set, the result is that other number. a is known as the identity element of set in that operation. 若一个数集存在数 a,使得 a 和这个数集中的另外一个数(也可以是 a 本身)进行运算且得数是另外那个数,则这个数集对于这个运算有单位元性质,a 为这个运算的单位元。

For a number $a \in S$, if $a \oplus b = b$ for all $b \in S$, then S is said to have the Identity Property in the $\oplus$ operation, and a is the identity element of $\oplus$ in S.

Properties: 1. Identity Property of Addition: $a + 0 = a$. The number 0 is the identity element of addition.　加法的单位元性质：$a + 0 = a$。0 是单位元。

2. Identity Property of Multiplication: $a \cdot 1 = a$. The number 1 is the identity element of multiplication.　乘法的单位元性质：$a \cdot 1 = a$。1 是单位元。

Inverse Property　逆元性质

n. [ɪn'vɜrs 'prɑpərti]

Definition: A set of numbers is said to have the Inverse Property of an operation, when for each number in the set, there exists a number (can be that number itself as well) so that when the operation is performed on these two numbers, the result is the identity element. These two numbers are known as inverses.　在一个数集里，若对于里面的每个数都存在一个数，使得这两个数运算的得数为单位元，则这个数集对于这个运算有逆元性质。这两个数互为逆元。

If for every number $a \in S$, there exists $b \in S$ for which $a \oplus b = e$ (e is the identity element), then S is said to have the Inverse Property in the $\oplus$ operation. The numbers a and b are known as the inverses of $\oplus$ in S.

Properties: 1. Inverse Property of Addition: $a + (-a) = 0$. The numbers a and $-a$ are inverses of each other. The number 0 is the identity element of addition.　加法的逆元性质：a 与 $-a$ 互为逆元。0 是加法的单位元。

2. Inverse Property of Multiplication: $a \cdot \frac{1}{a} = 1$ for all real numbers of a except 0. The numbers a and $\frac{1}{a}$ are inverses of each other. The number 1 is the identity element of multiplication.　乘法的逆元性质：a 与 $1/a$ 互为逆元，其中 a 不等于 0。1 是乘法的单位元。

1.7.4　Exercises of Properties of Operations　运算性质的练习

Exercises of Properties of Operations　运算性质的练习

Questions: 1. A new operation, $\oplus$, is defined over {0, 1, 2, 3}, as shown in Table 1-2.

Table 1-2

$\oplus$	0	1	2	3
0	0	1	2	3
1	1	2	3	0
2	2	3	0	1
3	3	0	1	2

Key: $1 \oplus 2 = 3$, as highlighted in blue.

(1) Does $\oplus$ have the Closure Property?　$\oplus$有封闭性吗？

(2) Does $\oplus$ have the Commutative Property?　$\oplus$有交换律吗？

(3) Does $\oplus$ have the Associative Property?　$\oplus$有结合律吗？

(4) List, if any, the identity elements of $\oplus$. 如果$\oplus$有单位元，试列出所有单位元。

(5) List, if any, the inverse elements of 0, 1, 2, and 3, in order. 如果$\oplus$有逆元，试依次列出 0、1、2、3 的逆元。

2. Evaluate each of the following:

(1) $1+2+3+4+5+6+9994+9995+9996+9997+9998+9999$.

(2) $37\times46+37\times35+37\times19$.

(3) $2\times3\times4\times5\times\frac{1}{2}\times\frac{1}{3}\times\frac{1}{4}\times\frac{1}{5}$.

Answers: 1. (1) Yes. For any two elements, a and b (not necessarily distinct), of $\{0,1,2,3\}$, it is true that $a\oplus b\in\{0,1,2,3\}$.

(2) Yes. For any two elements, a and b (not necessarily distinct), of $\{0,1,2,3\}$, it is true that $a\oplus b=b\oplus a$. Note the table's symmetry along the gray part of entries, as shown in Table 1-3. 注意表 1-3 沿阴影对称。

Table 1-3

$\oplus$	0	1	2	3
0	0	1	2	3
1	1	2	3	0
2	2	3	0	1
3	3	0	1	2

(3) Yes. For any three elements, a, b, and c (not necessarily distinct), of $\{0,1,2,3\}$, it is true that $(a\oplus b)\oplus c=a\oplus(b\oplus c)$.

(4) The number 0 is the identity elements. For any element $a\in\{0,1,2,3\}$, it is true that $a\oplus0=0\oplus a=a$.

(5) The inverse element of 0 is 0 because $0\oplus0=0$.

The inverse element of 1 is 3 because $1\oplus3=3\oplus1=0$.

The inverse element of 2 is 2 because $2\oplus2=0$.

The inverse element of 3 is 1 because $3\oplus1=1\oplus3=0$.

2. (1) Use Associative Property several times to group the numbers:

用结合律计算如下：

$$\begin{aligned}&1+2+3+4+5+6+9994+9995+9996+9997+9998+9999\\&=(1+9999)+(2+9998)+(3+9997)+(4+9996)+(5+9995)+(6+9994)\\&=10000+10000+10000+10000+10000+10000\\&=60000.\end{aligned}$$

(2) Use the Distributive Property:

用分配律计算如下：

$$\begin{aligned}&37\times46+37\times35+37\times19\\&=37(46+35+19)\\&=37(100)\\&=3700.\end{aligned}$$

(3) Use the Inverse Property:

用逆元性质计算如下:

$$2\times3\times4\times5\times\frac{1}{2}\times\frac{1}{3}\times\frac{1}{4}\times\frac{1}{5}$$
$$=\left(2\times\frac{1}{2}\right)\times\left(3\times\frac{1}{3}\right)\times\left(4\times\frac{1}{4}\right)\times\left(5\times\frac{1}{5}\right)$$
$$=1\times1\times1\times1$$
$$=1.$$

1.7.5 Summary　总结

1. Review PEMDAS.　复习运算顺序:括号→指数→乘除→加减。
2. For a set S and a binary operation $\oplus$:对于集 S 与二元运算$\oplus$:
   - For all elements $a,b\in S$, if $a\oplus b=b\oplus a$, then S has the Commutative Property under $\oplus$.　交换律。
   - For all elements $a,b,c\in S$, if $(a\oplus b)\oplus c=a\oplus(b\oplus c)$, then S has the Associative Property under $\oplus$.　结合律。
   - For all elements $a,b\in S$, if $a\oplus b\in S$, then S has the Closure Property under $\oplus$.　封闭性。
   - For an element $a\in S$ and for all elements $b\in S$, if $a\oplus b=b$, then S has the Identity Property under $\oplus$, and a is the identity element.　单位元性质。
   - For every element $a\in S$, if there exists an element $b\in S$ such that $a\oplus b=e$, where e is the identity element, then has the Inverse Property under $\oplus$, and a and b are inverses of each other.　逆元性质。
3. Review the Distributive Property.　复习分配律。

2. Expressions　表达式

2.1 Introduction　介绍

Expression　表达式

n. [ɪk'spreʃən]

Definition: A mathematical phrase that can contain numbers, variables (like x or y), operators (like signs for addition, subtraction, multiplication, division, power, and many other advanced operations), and grouping symbols (parentheses).　表达式是指由数字、变量、运算符、组合符号(括号)组成的运算组合。

Properties: 1. An expression that only consists of numbers, operators, and parentheses is known as a numerical expression. 只包括数字、运算符、括号的表达式叫作数字表达式。

2. An expression that consists of variables is known as a variable expression. 包含变量的表达式叫作变量表达式。

Examples: 1. Examples of numerical expressions:

数字表达式的例子:

(1) $10+3-8\cdot(14-5)$

(2) $1+4\cdot(2^3-5)-8\div4$

(3) 5

2. Examples of variable expressions:

变量表达式的例子:

(1) $5xy+2x-3y-9$

(2) $6xyz+4xy-3yz-8xyz+2-3xy+5x$

(3) $2x+\dfrac{x+y+z}{x+3y-5z}$

2.2 Integral Expressions 整式

Integral Expression 整式

n. [ˈɪntəgrəl ɪkˈspreʃən]

Definition: An expression with following restrictions:

整式是指有以下限制的表达式:

1. In division, the divisor cannot contain variables. 在除法里,除式不能包含变量。
2. In power, the exponent cannot contain variables and must be nonnegative integers. 在幂里,指数不能包含变量,且必须为非负整数。

An integral expression is one of monomial and polynomial, according to the number of terms. 整式是单项式和多项式的一种,取决于项数。

Examples: 1. Examples of integral expressions that are numerical expressions:

下面是数字表达式的整式例子:

(1) $10+3-8\cdot(14-5)$

(2) $1+4\cdot(2^3-5)-8\div4$

(3) 5

2. Examples of integral expressions that are variable expressions:

下面是变量表达式的整式例子:

(1) $5xy+2x-3y-9$

(2) $6xyz+4xy-3yz-8xyz+2-3xy+5x$

(3) $2x$

2.2.1 Parts of an Integral Expression 整式的组成部分

Variable 变量/未知数

n. [ˈveəriəbəl]

Definition: A symbol for a number that we do not know its value yet. The most popular variable names are x and y.　变量是指一个未知数(其值暂时未知)。最常见的是 x 和 y。“变量”和“未知数”可以互用。

Properties：When multiplying something with a variable, we omit the multiplication sign. We usually write “$6x$” instead of “$6 \cdot x$”.　当一个数/变量与一个变量相乘,通常会省略乘号。

Examples：Letters (usually lowercase) in the alphabet.　变量一般用小写字母表示。

Term　加数/项

n. [tɜrm]

Definition：1. In addition, the addends are also known as terms.　加数。

2. In variable expressions, a term either is a constant, or the product of a coefficient and variable(s). A term is a monomial.　在变量表达式中,每个项可以是常数或系数与变量的积。每个项均为单项式。

3. Refers to each number in a sequence. See sequence in Section 8.1.　指的是数列中的一个数。见 8.1 节的数列。

Questions：What are the terms on each of the following?

(1) $1+4 \cdot (2^3-5)-8 \div 4$

(2) $2xy-7x-y$

Answers：(1) Terms(加数)：$1, 4 \cdot (2^3-5), -8 \div 4$.

(2) Terms(项)：$2xy, -7x, -y$.

Like Terms　同类项

n. [laɪk tɜrmz]

Definition：Two or more terms are like terms if they have the same variable parts.　若两个项的变量部分相同,则它们是同类项。

Properties：Constants are always like terms since their variable parts are the same—no variable part.　常数总是同类项,因为它们的变量部分都是相同的——都没有变量部分。

Examples：1. Terms $4x$ and $-7x$ are like terms because their variable parts are x.

2. Terms 40 and 999 are like terms because they do not have variables, which means their variable parts are the same.

3. Terms $7x^2y^3$ and $5x^2y^3$ are like terms because their variable parts are x^2y^3.　以上例子根据同类项定义来判断两个项是否为同类项。

Phrases：combine～

Questions：Identify the like terms for the following polynomial:

在以下多项式里分出同类项：

$3xy+8x-10y+20+6xy-4x-5y+10$.

Answers：$3xy$ and $6xy$, $8x$ and $-4x$, $-10y$ and $-5y$, 20 and 10.

Coefficient　系数

n. [ˌkoʊɪ'fɪʃənt]

Definition：The coefficients of an integral expression are the numbers that are multiplied by the variables.　在整式中,系数是与变量相乘的数。

Questions: What are the coefficients for the following expression? 下式的系数是什么?

$3xy - 6x + 7y - 9$

Answers: 3, −6, 7.

Constant 常数/常值/常量

n. [ˈkɑnstənt]

Definition: 1. The constants of an integral expression are the numbers that are not multiplied by the variables. 常数:在整式中,常数是不与变量相乘的数。

2. The constants of an equation are numbers that are given/known. 常值:与已知数的定义相同。

Constant terms' values do not change regardless of the values of variables. 不论变量取什么值,常数项的值永远不变。

Questions: What are the constants for the following expression? 下式的常数是什么?

$3xy - 6x + 7y - 9$

Answers: −9.

Degree 次数

n. [dɪˈgri]

Definition: 1. The degree of a monomial/term is the sum of exponents of its variables. 单项式/项的次数是变量的指数之和。

2. The degree of a polynomial is the degree of the term with the highest degree. 多项式的次数是最高次项的次数。

Questions: On the following polynomial, what is the degree of each of the terms? What is the degree of the polynomial? 在下列的多项式里,每个项的次数是多少?多项式的次数是多少?

$3xy^2z + 7x^{40}y^3 - 5x + 10$

Answers: Degree of $3xy^2z$: $1+2+1=4$.

Degree of $7x^{40}y^3$: $40+3=43$.

Degree of $-5x$: 1

Degree of 10: 0

Degree of the polynomial: $\max(4, 43, 1, 0) = 43$.

Leading Coefficient 最高次项系数

n. [ˈlidɪŋ ˌkoʊɪˈfɪʃənt]

Definition: In an integral expression, the leading coefficient is the coefficient of the term with the highest degree. 在整式里,最高次项系数是有最大的次数项的系数。

Questions: What is the leading coefficient of $3xy^2z + 7x^{40}y^3 - 5x + 10$?

Answers: 7.

2.2.2 Types of Integral Expressions 整式的种类

Monomial 单项式

n. [mɑˈnoʊmiəl]

Definition: An integral expression that consists of only one term. 单项式是指只有一个项的整式。

Examples：$x, 2y, -5x, 7xy, 1$.

Polynomial　多项式

n. [ˌpɑləˈnoʊmiəl]

Definition：An integral expression that consists of more than one term.　多项式是指有多于一个项的整式。

Properties：1. Integral expressions that contain exactly two terms are known as binomials specifically.　有两个项的整式称为二项式。

2. Integral expressions that contain exactly three terms are known as trinomials specifically.　有 3 个项的整式称为三项式。

3. Two or more polynomials are equivalent polynomials if the values of these polynomials are the same, regardless of the values of the variables.　无论变量取什么值,若两个或多个多项式在代入变量值后的得数相同,则它们被称为相等的多项式。In other words, two or more polynomials are equivalent if they have the same terms.　若两个或多个多项式的所有项相同,则它们相等。

Examples：$x+y-5, 2xy-7x-y+5, \cdots$

Questions：Identify the variables, terms, coefficients, constants, and degree of the following polynomial: $3xy-4x+5y-6$.　在下面的多项式中分辨出变量、项、系数、常数、次数。

Answers：Variables：x, y.

Terms：$3xy, -4x, 5y, -6$.

Coefficients：$3, -4, 5$.

Constants：-6.

Degree：2.

2.2.3　Forms of a Polynomial　多项式的形式

Descending Order of Powers　降幂排列

n. [dɪˈsen.dɪŋ ˈɔrdər əv ˈpaʊərs]

Definition：The arrangement of terms in a polynomial such that the degrees of the terms are decreasing from left to right.　降幂排列是指根据项的次数从大到小从左到右排列多项式的各项。

This is known as the standard form of a polynomial, or the standard form of an integral expression. Like terms must be combined (see combine like terms) in the standard form.　这是多项式的一般形式,或整式的一般形式。在一般形式里必须合并同类项。

Questions：Rewrite $7xy+8x^2y^4-5x-10+5xy^2$ in descending order of power.

Answers：$8x^2y^4+5xy^2+7xy-5x-10$.

Ascending Order of Powers　升幂排列

n. [əˈsen.dɪŋ ˈɔrdər əv ˈpaʊərs]

Definition：The arrangement of terms in a polynomial such that the degrees of the terms are increasing from left to right.　升幂排列是指根据项的次数从小到大从左到右排列多项式的各项。

Questions: Rewrite $7xy+8x^2y^4-5x-10+5xy^2$ in ascending order of power.

Answers: $-10-5x+7xy+5xy^2+8x^2y^4$.

2.2.4 Operations of Integral Expressions 整式的运算

视频 4

Value 值

n. ['vælju]

Definition: 1. The value of a numerical expression is the resulting number. 数字表达式的值是其得数。

2. The value of a variable is the number that the variable represents. 变量的值是指变量所代表的数字。

Examples: 1. The value of $6+7$ is 13.

2. If the value of x is 3, then the value of $x+7$ is 10.

Evaluate 求值

v. [ɪ'væljuˌeɪt]

Definition: To calculate the value of a numerical expression.

Questions: Evaluate each of the following:

(1) $5+3-7\cdot 2$

(2) $8-2^4+10\cdot 3$

Answers: Using PEMDAS,

用运算顺序,得

(1) -6

(2) 22

Round 四舍五入

v. [raʊnd]

Definition: To keep a number simpler (has fewer digits), but keep the estimated value close to the number itself. 约算一个数(使它的数位更少),但让约算后的结果与数字本身相近,四舍五入是一种约算方法。

Properties: 1. To round a number to a nearest digit (call it X), look at the digit immediately after it (call it Y). 要四舍五入一个数到一个数位(假设为 X),先看 X 后面的那位数(假设为 Y)。

If $Y=0,1,2,3,$ or 4:

- If Y is before the decimal point, erase all digits behind the decimal point, and, from Y to the ones digit, change all digits to 0.
  若 Y 在小数点之前,则去掉小数点后的所有数。从 Y 到个位数,所有数位改为 0。
- If Y is after the decimal point, erase Y and the digits after it.
  若 Y 在小数点之后,则去掉 Y 以及它后面的数。

If $Y=5,6,7,8,$ or 9:

- If Y is before the decimal point, erase all digits behind the decimal point, and, from Y to the ones digit, change all digits to 0. Finally, increase X by 1.

若 Y 在小数点之前,则去掉小数点后的所有数。从 Y 到个位数,所有数位改为 0,最后 X 加 1。

- If Y is after the decimal point, erase Y and the digits after it. Finally increase X by 1. 若 Y 在小数点之后,则去掉 Y 以及它后面的数,最后 X 加 1。

(1) For example, to round 13579.2468 to the nearest hundreds, we have $X=5$ and $Y=7$. Look at the digits 7,9,2,4,6, and 8. Erase 2,4,6, and 8 because they are behind the decimal point. Replace the 7 and 9 with 0s since they are in front of the decimal point.

Since $Y=7\geqslant5$ we need to increase X by 1, our answer is 13 600.

(2) For example, to round 13 579.2468 to the nearest tenths, we have $X=2$ and $Y=4$. Look at the digits 4,6, and 8. Erase 4,6, and 8 since they are behind the decimal point.

Since $Y=4<5$, our answer is 13 579.2.

2. When we round a number to a digit that is after the decimal point, we must write out the number at that digit in our answer, even though it is a 0 because it indicates precision. 若精确到小数后的一个数位,则必须写出那个数位。即使是 0 也须写出,因为应表明其精确性。

We sometimes call 0 a placeholder for this. For example, to round 19.998 to the nearest hundredths, the answer is 20.00 instead of 20. Even though $20.00=20$, their meaning in scientific report differs quite a lot. 0 是占位符,见下面的实例。

Suppose a student measures the length of a pencil. Reporting that the length of the pencil is 20.00cm means that the student's measurement precision is at the nearest hundredths. The actual length of the pencil must be somewhere between 19.995cm and 20.004cm. 若一个学生测量一支铅笔的长度为 20.00cm,则这个长度精确到百分位。可以知道它的实际长度在 19.995～20.004cm。

On the other hand, reporting that the length of the pencil is 20cm means that the student's measurement precision is at the nearest ones. The actual length of the pencil must be somewhere between 19.5cm and 20.4cm. 若报告的长度为 20cm,则这个长度精确到个位。可以知道它的实际长度在 19.5～20.4cm。

Questions: Round 135.79 to the nearest

(1) tenths.

(2) ones.

(3) tens.

(4) hundreds.

Answers: (1) The tenth digit is 7, and the digit immediately to its right is 9. The 9 should be erased since it is after the decimal point. Since $9\geqslant5$, 7 should increase by 1. The answer is 135.8.

(2) 136.

(3) 140.

(4) 100.

Substitute 代入

v. [ˈsʌbstɪˌtut]

Definition: To replace the variable with the given (numerical) value. 代入是指用给出的数值替换变量。

Phrases: Substitution (noun) 代入法

Questions: 1. Evaluate xyz when $x=3, y=4$, and $z=5$.

2. Evaluate $a^b - cd$ if $a=4, b=3, c=-2$, and $d=1$.

Answers: 1. $xyz = (3)(4)(5)$. (This step is known as substitution.)

$= \boxed{60}$.

2. $a^b - cd = 4^3 - (-2)(1)$

$= 64 + 2$

$= \boxed{66}$.

Add/Remove Parentheses 添/去括号

v. [æd/rɪˈmuv pəˈren. θə. siːz]

Definition: Grouping (adding parentheses) or ungrouping (removing parentheses) techniques without changing the value of the expression. 添/去括号是指在不改变表达式值的条件下的组合(添括号)或取消组合(去括号)的方法。

Properties: 1. To add/remove parentheses after a "+" sign, simply add/remove the parentheses as wished. Adding/removing a parentheses after a "+" sign is the same as distributing a 1. 要在加号后面添/去括号,直接添/去即可。在加号后面添/去括号相当于括号中各项乘以1(运用乘法分配律)。

2. To add/remove parentheses after a "−" sign, negate all the +/− signs in the parentheses after adding/removing. Adding/removing a parentheses after a "−" sign is the same as distributing a −1. 要在减号后面添/去括号,添/去括号后须改变括号里的所有加减符号(加号改为减号,减号改为加号)。在减号后面添/去括号相当于括号中各项乘以−1(运用乘法分配律)。

Questions: 1. Remove the parentheses on each of the following:

(1) $a+(b+c)$

(2) $a+(b-c)$

(3) $a-(b+c)$

(4) $a-(b-c)$

2. Add parentheses on each of the following to group b and c:

(1) $a+b+c$

(2) $a+b-c$

(3) $a-b+c$

(4) $a-b-c$

3. Evaluate each of the following efficiently:

(1) $4100-99-1$

(2) $195+64+36$

(3) $375+188-88$

(4) $4100-199+99$

(5) $101+(99+58)$

(6) $777-(77+99)$

Answers: 1. (1) $a+(b+c)=a+1(b+c)=a+1b+1c=a+b+c$ (Property 1).

(2) $a+b-c$ (Property 1), same reasoning as Answer 1(1).

(3) $a-(b+c)=a+(-1)(b+c)=a+(-1)b+(-1)c=a-b-c$ (Property 2).

(4) $a-b+c$ (Property 2), same reasoning as Answer 1(3).

2. (1) $a+(b+c)$ (Property 1).

(2) $a+(b-c)$ (Property 1).

(3) $a-(b-c)$ (Property 2).

(4) $a-(b+c)$ (Property 2).

3. (1) $4100-99-1=4100-(99+1)=4100-100=4000$.

(2) $195+64+36=195+(64+36)=195+100=295$.

(3) $375+188-88=375+(188-88)=375+100=475$.

(4) $4100-199+99=4100-(199-99)=4100-100=4000$.

(5) $101+(99+58)=101+99+58=200+58=258$.

(6) $777-(77+99)=777-77-99=700-99=601$.

Combine Like Terms　合并同类项

v. [kəm'baɪn laɪk tɜrmz]

Definition: To combine like terms by addition or subtraction, add/subtract the like terms' coefficients; the variable part of the like terms stays the same.　要通过加/减法合并同类项,把同类项的系数相加/相减;同类项的变量部分不变。

Properties: This is based on the Distributive Property: If ac and bc are like terms, for which a and b are the coefficients and c is the variable part, then $ac \pm bc = (a \pm b)c$. Exactly matching the definition.　合并同类项的步骤原理参照乘法分配律的定义。

Questions: Combine like terms on each of the following:

(1) $5x+4x$

(2) $8xy-10xy$

(3) $40y-10+2y+16$

(4) $3xy+8x-10y+20+6xy-4x-5y+10$

Answers: (1) $9x$.

(2) $-2xy$.

(3) $$\begin{aligned}&40y-10+2y+16\\&=(40y+2y)+(-10+16)\\&=42y+6.\end{aligned}$$

(4) $$\begin{aligned}&3xy+8x-10y+20+6xy-4x-5y+10\\&=(3xy+6xy)+(8x-4x)+(-10y-5y)+(20+10)\\&=9xy+4x-15y+30.\end{aligned}$$

2.2.5 Applications of the Distributive Property 分配律的应用

视频 5

FOIL 四项分配法

v. [fɔɪl]

Definition: "First Outer Inner Last" when taking product of two binomials. FOIL 法则是求两个二项式积的方法。

Properties: In $(a+b)(c+d)$, "First, Outer, Inner, Last" refers to the products ac, ad, bc, and bd, respectively. The expanded form of $(a+b)(c+d)$ is $ac+ad+bc+bd$. FOIL 是"首、外、内、末"的缩写。

Figure 2-1 shows the FOIL rule using a square.

图 2-1 用正方形展示了四项分配法。

Figure 2-2 provides another illustration of FOIL.

图 2-2 用另一种方法图解了四项分配法。

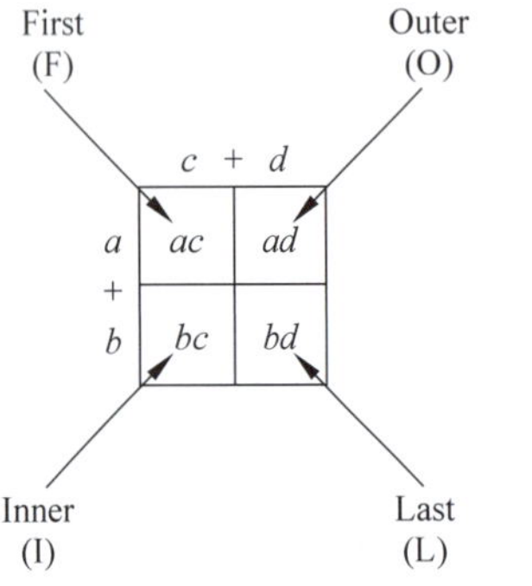

Figure 2-1

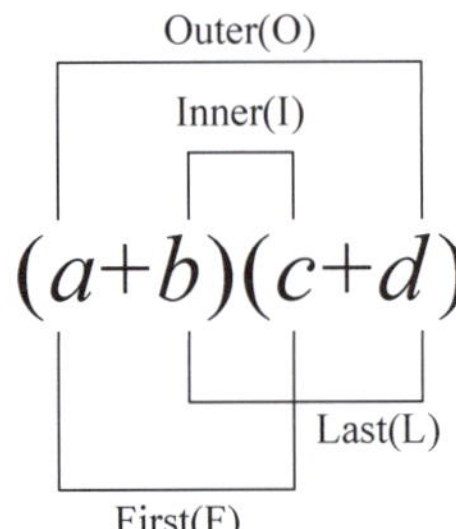

Figure 2-2

Proofs: By Distributive Property, $(a+b)(c+d)=a(c+d)+b(c+d)=ac+ad+bc+bd$. 根据分配律证明四项分配法。

Questions: 1. Expand $(4x+5)(3y+7)$.

2. Expand $(3x-10)(5y-4)$.

Answers: 1. The "First" is $(4x)(3y)=12xy$.

The "Outer" is $(4x)(7)=28x$.

The "Inner" is $(5)(3y)=15y$.

The "Last" is $(5)(7)=35$.

The answer is $\boxed{12xy+28x+15y+35}$.

2. To make the binomials in the form of $(a+b)(c+d)$, rewrite it as $[3x+(-10)][5y+(-4)]$.

The "First" is $(3x)(5y)=15xy$.

The "Outer" is $(3x)(-4)=-12x$.

The "Inner" is $(-10)(5y)=-50y$.

The "Last" is $(-10)(-4)=40$.

The answer is $\boxed{15xy-12x-50y+40}$.

Formulas for Special Products　乘法公式

Notes：These formulas are well-known and can be derived from FOIL. They are special cases of FOIL.　乘法公式是从四项分配法派生的常见公式，是四项分配法的特殊情况。

Formulas：For each of the formulas below, we can use FOIL to show it. Alternatively, we provide a pictorial view for each of the formulas below.　对于下列的公式，既可以用四项分配法证明，又可以参照公式下面的图解证明。

1. Square of A Sum：$(a+b)^2=a^2+2ab+b^2$.　和的平方。

   Figure 2-3 illustrates this property.

   The area of the large square is $(a+b)^2$.

   The sum of areas of these four pieces is $a^2+2ab+b^2$.

   Hence the two expressions are equivalent.

   用两种不同的方法表示大正方形的面积。第一种方法是用大正方形边长的平方表示；另一种是用 4 个小矩形面积之和表示，两者相等。

2. Square of A Difference：$(a-b)^2=a^2-2ab+b^2$.　差的平方。

   Figure 2-4 illustrates this property.

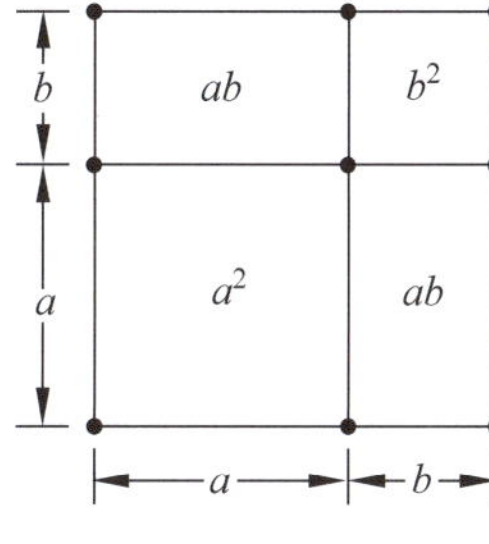

Figure 2-3

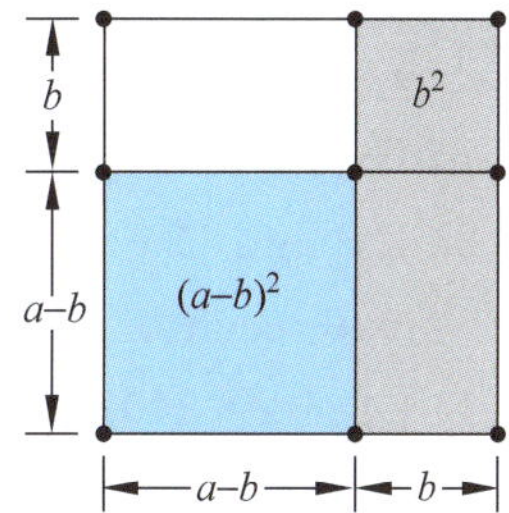

Figure 2-4

The area of the large square is a^2, and the area of the blue square is $(a-b)^2$.

The areas of each "long rectangle" (one is shaded) is ab.

Let the brackets denote the area.

[Blue Square] = [Large Square] − 2[Long Rectangle] + [Small Square]

$(a-b)^2 \quad = \quad a^2 \quad - \quad 2ab \quad + \quad b^2$.

Hence the two expressions are equivalent.

用两种不同的方法表示蓝色正方形的面积。一种方法是用蓝色正方形的边长平方表示；另一种方法是用大正方形的面积进行面积加减法。

Combining this with Formula 1 we will have the Complete Square Formula：结合公式 1 可得以下完全平方公式：

$$(a\pm b)^2=a^2\pm 2ab+b^2.$$

3. Difference of Squares：$a^2-b^2=(a+b)(a-b)$.　平方差公式。

   Figure 2-5 illustrates this property.

   The area of the large square is a^2, and the area of the gray region is a^2-b^2.

   We can cut along the dotted lines and piece the gray rectangles as shown in Figure 2-6.

   The areas of these gray rectangles is $(a+b)(a-b)$.

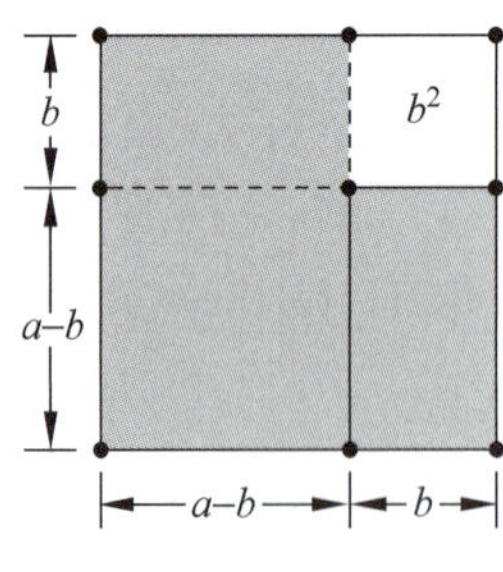

Figure 2-5

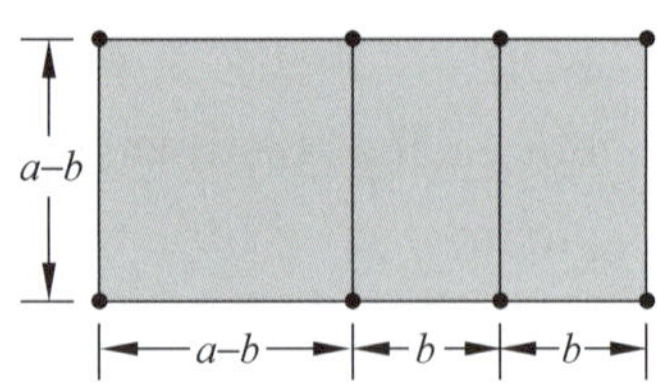

Figure 2-6

Therefore, $a^2 - b^2 = (a+b)(a-b)$.

Hence the two expressions are equivalent.

用两种不同的方法表示灰色区域的面积。一种是用大正方形的面积减去小正方形的面积;另一种是把灰色区域拼凑成矩形以后用矩形的面积公式算面积,两者相等。

Questions: 1. (1) Expand $(x+6)^2$.

(2) Expand $(4x+3)^2$.

2. (1) Expand $(x-5)^2$.

(2) Expand $(2x-3)^2$.

3. (1) Expand $(x-10)(x+10)$.

(2) Expand $(2x+3y)(2x-3y)$.

4. Evaluate each of the following efficiently:

(1) $24^2 + 2\times 24\times 16 + 16^2$

(2) 201^2

(3) $144^2 - 2\times 144\times 44 + 44^2$

(4) $9.99^2 - 0.01^2$

(5) 399×401

Answers: 1. Using the Square of a Sum formula (Formula 1), we have:

(1) $(x+6)^2 = (x)^2 + 2(x)(6) + (6)^2 = x^2 + 12x + 36$.

(2) $(4x+3)^2 = (4x)^2 + 2(4x)(3) + (3)^2 = 16x^2 + 24x + 9$.

2. Using the Square of a Difference formula (Formula 2), we have:

(1) $(x-5)^2 = (x)^2 - 2(x)(5) + (5)^2 = x^2 - 10x + 25$.

(2) $(2x-3)^2 = (2x)^2 - 2(2x)(3) + (3)^2 = 4x^2 - 12x + 9$.

3. Using the Difference of Squares formula (Formula 3), we have:

(1) $(x-10)(x+10) = (x)^2 - (10)^2 = x^2 - 100$.

(2) $(2x+3y)(2x-3y) = (2x)^2 - (3y)^2 = 4x^2 - 9y^2$.

4. (1) $24^2 + 2\times 24\times 16 + 16^2 = (24+16)^2 = 40^2 = 1600$.

(2) $201^2 = (200+1)^2 = 200^2 + 2\times 200\times 1 + 1^2 = 40\ 000 + 400 + 1 = 40\ 401$.

(3) $144^2 - 2\times 144\times 44 + 44^2 = (144-44)^2 = 100^2 = 10\ 000$.

(4) $9.99^2 - 0.01^2 = (9.99+0.01)\times(9.99-0.01) = 10\times 9.98 = 99.8$.

(5) $399\times 401 = (400-1)\times(400+1) = 400^2 - 1^2 = 160\ 000 - 1 = 159\ 999$.

2.2.6 Factoring 因式分解

视频 6

2.2.6.1 Introduction 介绍

Factor 因式

n. [ˈfæktər]

Definition: A factor of an integral expression P is an integral expression F for which the quotient of P and F is another integral expression. 若整式 P 与整式 F 的商为整式,则 F 是 P 的因式。

Examples: 1. In the monomial $4x^4$, all possible factors are $\pm 1, \pm 2, \pm 4, \pm x, \pm 2x, \pm 4x, \pm x^2, \pm 2x^2, \pm 4x^2, \pm x^3, \pm 2x^3, \pm 4x^3, \pm x^4, \pm 2x^4, \pm 4x^4$. For now, let's focus those monomials that have positive coefficients or are positive constants. 单项式 $4x^4$ 的所有因式列举如上。目前重点关注那些为正系数或正常数的单项式。

2. In the polynomial $x^2 - y^2$, some possible factors are: $1, x+y, x-y, x^2-y^2$.

3. In the polynomial $x^3 - y^3$, some possible factors are: $1, x-y, x+xy+y, x^3-y^3$ (We are not expected to know this for now.).

Notes: Refer to the definition, in the future, if F is a factor of P, we wish P and P/F each have degree at least 1, and each have positive coefficients whenever possible.

Common Factor 公因式

n. [ˌkɒm.ən ˈfæk.tər/dɪˈvaɪ.zɚ]

Definition: The common factor of two or more integral expressions is an integral expression F so that F is a factor of all of these integral expressions. By convention, the degree of F is at least 1. 两个或更多个整式的公因式是整式 F,因此 F 为所有这些整式的因式。根据惯例,F 的次数至少为 1。

This is an analogue of common factor/divisor. 公因式与公约数类似。

Properties: We are usually interested in the highest common factors, or the factors whose degrees are as large as possible. 我们通常对最高公因式(公因式的次数尽可能大)感兴趣。

The highest common factor is an analogue of the greatest common factor. 最高公因式与最大公约数类似。

Conventionally, we want the resulting integral expression (after dividing by the highest common factor) to have integer coefficients whenever possible. If so, keep pulling out constant factors until the coefficients' absolute values' GCF is 1. 习惯上,我们希望整式的得数(原式除以最大公因式后)的系数为整数。若系数为整数,则继续提取常数因数,直到这些系数绝对值的最大公约数为 1。

Phrases: Pull out～

Examples: For integral expressions $16x^8, 24x^6$, and $32x^4$, some possible highest common factors are $2x^4, 4x^4, 8x^4, 9x^4, \cdots$ Any monomial whose variable part is x^4 is a highest common factor. 变量部分为 x^4 的单项式均为 $16x^8$、$24x^6$、$32x^4$ 的最大公因式。

Conventionally, we will consider $8x^4$ to be the highest common factor because $\frac{16x^8}{8x^4} =$

$2x^4$, $\frac{24x^6}{8x^4}=3x^2$, and $\frac{32x^4}{8x^4}=4$. Note that 2, 3, and 4 are integers, and GCF(2,3,4) = 1. 习惯上，会认为 $8x^4$ 为最大公因式，原因是 $\frac{16x^8}{8x^4}=2x^4$、$\frac{24x^6}{8x^4}=3x^2$、$\frac{32x^4}{8x^4}=4$ 后，2、3、4 的最大公约数为 1。

For integral expressions $16x^8$, $24x^6$, and $32x^4$, some common factors are $2x$, $6x^2$, x^3.

2.2.6.2 Methods of Factoring Polynomials 因式分解多项式的方法

Factor 因式分解

v. [ˈfæktər]

Definition: To factor a polynomial P, rewrite it as a product of two or more factors. The degree of each factor must be at least 1 and less than the degree of P. 因式分解是指把多项式 P 改写成两个或更多个因式的积。每个因式的次数必须至少为 1 且小于 P 的次数。

Some possible ways to factor a polynomial are:

因式分解的方法有：

1. Factor by Pulling Out the Highest Common Factor. 提取最大公因式。
2. Factor by Grouping. 分组分解。

Examples: 1. Some possible ways to factor $4x^4$:

$$4x^4 \\ =2x(2x^3) \\ =x^2(4x^2).$$

2. To factor x^2-y^2:

$x^2-y^2=(x+y)(x-y)$.

3. To factor x^3-y^3:

$x^3-y^3=(x-y)(x+xy+y)$. We are not expected to know this yet.

4. Some possible ways to factor $18m^5-24m^4+12m^3$:

$$18m^5-24m^4+12m^3 \\ =2m^3(9m^2-12m+6) \\ =6m^3(3m^2-4m+2) \\ =3m(6m^4-8m^3+4m^2).$$

Factor by Pulling Out the Highest Common Factor 提取最大公因式

Definition: One method of factoring by pulling out the highest common factor (with degree at least 1). 因式分解的一种方法——提取最大公因式(次数至少为 1)。

Properties: This is based on the Distributive Property: Suppose c is the highest common factor of ac and bc, we have $ac+bc=c(a+b)$. 根据乘法分配律可知其原理。

Questions: Pull out the highest common factor for each of the following polynomials:

对下面各多项式提取最大公因式：

(1) $12x-16$

(2) $2xy+8xz$

(3) x^2-x

(4) $16x^8+24x^6-32x^4$

Answers：(1) The highest common factor of $12x$ and -16 is 4.

$$12x-16=4\left(\frac{12x}{4}-\frac{16}{4}\right)=4(3x-4).$$

The degree of the common factor 4 is 0. In this case we are not really factoring the binomial, but we rewrite the coefficients with smaller numbers.

(2) The highest common factor of $2xy$ and $8xz$ is $2x$.

$$2xy+8xz=2x\left(\frac{2xy}{2x}-\frac{8xz}{2x}\right)=2x(y-4z).$$

(3) The highest common factor of x^2 and x is x.

$$x^2-x=x\left(\frac{x^2}{x}-\frac{x}{x}\right)=x(x-1).$$

(4) The highest common factor of $16x^8$, $24x^6$, and $32x^4$ is $8x^4$.

$$16x^8+24x^6-32x^4=8x^4\left(\frac{16x^8}{8x^4}+\frac{24x^6}{8x^4}-\frac{32x^4}{8x^4}\right)=8x^4(2x^4+3x^2-4).$$

Factor by Grouping　分组分解

Definition：When we have a polynomial in the form of $ac+bc+ad+bd$, we can factor using the Distributive Property：

$ac+bc+ad+bd=c(a+b)+d(a+b)=(c+d)(a+b)$. This is undoing the FOIL!　这是四项分配法的复原步骤。

This method usually involves factor by pulling out the common factor, as shown above.　这个步骤经常与提取公因式相结合，如上。

Questions：Factor the following by grouping：

(1) $2x+4x^2+6x^3+12x^4$

(2) $5x+15x^2-8x^3-24x^4$

(3) $6xy-9y-20x+30$

Answers：(1) $2x+4x^2+6x^3+12x^4=2x(1+2x)+6x^3(1+2x)=(2x+6x^3)(1+2x)=\boxed{2x(1+3x^2)(1+2x)}$. The last step involves factoring by pulling out a common factor.　最后一步就是提取公因式。

(2) $5x+15x^2-8x^3-24x^4=5x+15x^2-(8x^3+24x^4)=5x(1+3x)-8x^3(1+3x)=(5x-8x^3)(1+3x)=\boxed{x(5-8x^2)(1+3x)}$. The last step involves factoring by pulling out a common factor.　最后一步就是提取公因式。

(3) $6xy-9y-20x+30=6xy-9y-(20x-30)=3y(2x-3)-10(2x-3)=\boxed{(3y-10)(2x-3)}$.

2.2.7 More on Operations of Polynomials　其他多项式运算

Factor Completely　完全分解

Definition：Factor in a way that it is impossible to factor further by pulling out a common factor or by grouping, for any of the factors in the factored form.　完全分解是指分解以后，在任

何因式中不能再提取公因式或分组分解。

Examples: See the questions from factor by grouping. The last step is to factor completely. 见分组分解的问题。最后一步就是完全分解。

Simplify Polynomials 简化多项式

Definition: The result of simplifying polynomials involves as few terms as possible. 改写多项式，项数越少越好。

Usually we want to rewrite the polynomial in the standard form, which is the descending order of power. The process is a combination of removing parentheses, performing FOIL, and combining like terms. 一般都会把多项式写成标准形式(降幂排列)。简化多项式是去括号、四项分配法、合并同类项的组合。

Questions: 1. Simplify each of the following:

(1) $5xyz-6xy+3xyz$

(2) $(a-5)(a+6)+10$

(3) $5x+7(x+2y-3z)$

(4) $-x+2y-6z-3(4x-10y+2z)+2(x+1)$

2. What is the value of $-x+2y-6z-3(4x-10y+2z)+2(x+1)$ if $x=2, y=1$, and $z=3$?

Answers: 1. (1) $8xyz-6xy$.

(2) $(a-5)(a+6)+10=a^2+a-30+10=a^2+a-20$.

(3) $5x+7(x+2y-3z)=5x+7x+14y-21z=12x+14y-21z$.

(4)
$$\begin{aligned}&-x+2y-6z-3(4x-10y+2z)+2(x+1)\\&=-x+2y-6z-12x+30y-6z+2x+2\\&=-11x+32y-12z+2.\end{aligned}$$

2. Working with the simplified version is much easier (see 1(4)):

$$\begin{aligned}&-x+2y-6z-3(4x-10y+2z)+2(x+1)\\&=-11x+32y-12z+2\\&=-11(2)+32(1)-12(3)+2\\&=-22+32-36+2\\&=-24.\end{aligned}$$

2.2.8 Summary 总结

1. An expression is a mathematical phrase that can contain numbers, variables (like x or y), operators (like signs for addition, subtraction, multiplication, division, power, and many other advanced operations), and grouping symbols (parentheses). 表达式是由数字、变量、运算符以及组合符号(括号)组成的运算组合。

An integral expression with following restrictions:

整式是指有以下限制的表达式：

- In division, the divisor cannot contain variables. 在除法里，除式不能有变量。
- In power, the exponent cannot contain variables and must be nonnegative integers. 在幂里，指数不能包含变量，且必须为非负整数。

2. Parts of an integral expression: variables, terms, like terms, coefficients, constants. 整式的组成部分：变量、项、同类项、系数、常数。

   Characteristics of an integral expression: degree of a term, degree of the integral expression, leading coefficient. 整式的描述：单项的次数、整式的次数、最高次项系数。

   Describing integral expressions by the number of terms: monomial (1), polynomial (2 or more), binomial (2), trinomial (3),… 根据项数描述整式，有单项式、多项式(二项式、三项式)……

   Forms of an integral expression: ascending order of power, descending order of power (standard form). Terms are connected by additions. Subtracting a term is the same as adding its opposite. 整式的形式有升幂排列和降幂排列(一般形式)。项以加法连接。减去一个项相当于加上它的相反数。

3. The most common actions: evaluate, round, substitute, add/remove parentheses, combine like terms. Most of them are useful when simplifying polynomials. 最常见的操作有：求值、四舍五入、代入、添/去括号、合并同类项。很多操作对于简化多项式有用。

4. FOIL is based on the Distributive Property to distribute everything out. Special cases of FOIL are the formulas of special products: 根据乘法分配律可知四项分配法的原理——把一切都分配出来。四项分配法的特殊情况是乘法公式。
   - Complete Square Formula: $(a \pm b)^2 = a^2 \pm 2ab + b^2$. 完全平方公式。
   - Difference of Squares Formula: $a^2 - b^2 = (a+b)(a-b)$. 平方差公式。

5. Factoring is the process of undoing the FOIL—it is based on the Distributive Property. The two methods are: 因式分解是四项分配法的复原步骤——也是基于乘法分配律。它的两种方法是：
   - Factor by Pull Out The Highest Common Factor. 提取最大公因式。
   - Factor by Grouping. 分组分解。

6. Factoring completely is to factor in a way that it is impossible to factor further by pulling out a common factor or by grouping, for any of the factors in the factored form. 完全分解是分解到以下程度：分解以后，在任何因式中不能再提取公因式或分组分解。

2.3 Fractions 分式

2.3.1 Ratios 比

Ratio 比

n. [ˈreɪʃɪˌoʊ]

Definition: The quantitative relation between two amounts showing the number of times one contains or is being contained by the other. 比是指两个量的关系，表示其中一个量包含另一个量或者被另一个量包含的次数。

Notation: The ratio "p to q" can be written as $p:q$, p/q, or $\frac{p}{q}$.

p 比 q 可写作 $p:q$, p/q 或者$\frac{p}{q}$。

Properties: 1. In $p:q$, p is known as the antecedent and q is known as the consequent. The consequent cannot be 0. 在 $p:q$ 中，p 是前项，q 是后项。后项不能为 0。

2. The antecedent and consequent are expected to have the same unit. If not, it must be

specified. 前项和后项的单位必须相同,否则必须指定。

3. Ratios can be interpreted as fractions. See properties of fraction in Section 1.2.3.1. 比可以理解成为分数,见 1.2.3.1 节分数的性质。

Questions: Jack's height is 160cm and Jill's height is 1.4m. What is the ratio of Jack's height to Jill's height?

Answers: The ratio of Jack's height to Jill's height is "160cm to 1.4m", but we want to express this as a fraction for which the numerator and the denominator have the same units. Let's convert everything to cm: Jill's height is 140cm. Thus the ratio is 160cm/140cm = 16/14 = $\boxed{8/7}$.

2.3.2 Two or More Ratios 两个或多个比

视频 7

Proportion 比例

n. [prə'pɔrʃən]

Definition: Two or more equivalent ratios. 两个或多个比相等的式子。

Notation: One example of proportion is $\frac{a}{b}=\frac{c}{d}$, for which $b\neq 0$ and $d\neq 0$. The quantities a, b, c, and d are known as the proportionals. Furthermore, a and d are the extremes and b and c are the means. 在比例$\frac{a}{b}=\frac{c}{d}$中(b 和d 均不为 0),a、b、c、d 均为比例项。其中 a 和 d 为外项,b 和c 为内项。

Properties: If $\frac{a}{b}=\frac{c}{d}$, for which $b\neq 0$ and $d\neq 0$, then the following are true.

1. $ad=bc$. Cross Product 交叉积。
2. $\frac{a+c}{b+d}=\frac{a}{b}$.
3. $\frac{b}{a}=\frac{d}{c}$. Invertendo Property 反比性质。
4. $\frac{a}{c}=\frac{b}{d}$. Alternendo Property 更比性质。
5. $\frac{a+b}{b}=\frac{c+d}{d}$. Componendo Property 合比性质。
6. $\frac{a-b}{b}=\frac{c-d}{d}$. Dividendo Property 分比性质。
7. $\frac{a+b}{a-b}=\frac{c+d}{c-d}$. Componendo and Dividendo Property 合分比性质。

Property 1 is very well known, while the rest is useful and can be derived fairly easily. See the proofs below. 性质 1 很常用。其他的性质比较有用,证明起来也比较容易。证明如下。

The original names for Property 3-7 are from Latin. 性质 3~7 的原名来自于拉丁语。

Proofs: Property 1: From $\frac{a}{b}=\frac{c}{d}$, multiply both sides by bd:

$$\frac{a}{b}(bd)=\frac{c}{d}(bd)$$

$$ad=bc.$$

Property 2: Let $\frac{a}{b}=\frac{c}{d}=k$. We have $a=bk$ and $c=dk$, for which

$$\frac{a+c}{b+d}=\frac{bk+dk}{b+d}=\frac{k(b+d)}{b+d}=k=\frac{a}{b}.$$

Property 3: Since $\frac{a}{b}=\frac{c}{d}$,

$$1\div\frac{a}{b}=1\div\frac{c}{d}$$

$$1\cdot\frac{b}{a}=1\cdot\frac{d}{c}$$

$$\frac{b}{a}=\frac{d}{c}.$$

Property 4: From $\frac{a}{b}=\frac{c}{d}$, multiply both sides by $\frac{b}{c}$:

$$\frac{a}{b}\left(\frac{b}{c}\right)=\frac{c}{d}\left(\frac{b}{c}\right)$$

$$\frac{a}{c}=\frac{b}{d}.$$

Property 5: From $\frac{a}{b}=\frac{c}{d}$, add 1 to both sides:

$$\frac{a}{b}+1=\frac{c}{d}+1$$

$$\frac{a}{b}+\frac{b}{b}=\frac{c}{d}+\frac{d}{d}$$

$$\frac{a+b}{b}=\frac{c+d}{d}.$$

Property 6: From $\frac{a}{b}=\frac{c}{d}$, subtract 1 from both sides:

$$\frac{a}{b}-1=\frac{c}{d}-1$$

$$\frac{a}{b}-\frac{b}{b}=\frac{c}{d}-\frac{d}{d}$$

$$\frac{a-b}{b}=\frac{c-d}{d}.$$

Property 7: We have

$$\frac{a+b}{b}=\frac{c+d}{d} \qquad \text{(Property 5)}$$

$$\frac{a-b}{b}=\frac{c-d}{d}. \qquad \text{(Property 6)}$$

Dividing the first equation by the second, we get $\frac{a+b}{a-b}=\frac{c+d}{c-d}$. 两等式相除。

Questions：1. What is the value of x if $\frac{x}{6}=\frac{10}{12}$?

2. To bake a cake, for every 5 cups of flour, we need 3 cups of sugar. If the cake needs 20 cups of flour, how many cups of sugar do we need?

Answers：1. Method 1：Note that doubling the denominator on the left gets the denominator on the right ($6\times 2=12$). Therefore, doubling the numerator on the left gets the numerator on the right ($2x=10$), from which $x=\boxed{5}$. 注意在左边的分式中可以同时把分子和分母乘上 2，原值不变。

Method 2：Using Property 1, we have $12x=6\times 10$, from which $12x=60$, and $x=\boxed{5}$. 用性质 1(交叉积性质)。

2. Suppose we need x cups of sugar. Setting up the proportion (each ratio is [number of cups of flour] over [number of cups of sugar]) gives $\frac{5}{3}=\frac{20}{x}$. Cross multiplying (by Property 1), we have $5x=3\times 20$, from which $x=\boxed{12}$.

The proportion can be also set in these ways：$\frac{3}{5}=\frac{x}{20}$ ([number of cups of sugar] over [number of cups of flour], from which it works by Property 3), or $\frac{5}{20}=\frac{3}{x}$ (scale over actual, from which it works by Property 4).
此题用交叉积求解。注意，其中有多种方法列出式子。例如，[糖的杯数]/[面粉的杯数]，或者分子和分母对换。

2.3.3 More on Fractions 更多与分式有关的知识

Simplify Fractions 约分

v. ['sɪmpləˌfɑɪ 'frækʃənz]

Definition：To simplify a fraction, we multiply or divide both its numerator and the denominator by the same nonzero number for which the resulting fraction's numerator and denominator are relatively prime.
要约分/简化一个分数，把分子和分母同时乘上或除以一个非零数，使得得数的分子和分母互质。

Questions：Simplify each of the following：

(1) $\frac{40}{50}$ (2) $\frac{50}{58}$ (3) $\frac{72}{99}$ (4) $\frac{1.8}{12}$

Answers：(1) $\frac{40}{50}=\frac{40\div 10}{50\div 10}=\frac{4}{5}$.

(2) $\frac{50}{58}=\frac{50\div 2}{58\div 2}=\frac{25}{29}$.

(3) $\frac{72}{99}=\frac{72\div 9}{99\div 9}=\frac{8}{11}$.

(4) $\frac{1.8}{12}=\frac{1.8\cdot 10}{12\cdot 10}=\frac{18}{120}=\frac{18\div 6}{120\div 6}=\frac{3}{20}$.

Simplest Form of a Fraction　最简分数

Definition：A fraction (regardless of sign) is in the simplest form if the numerator and denominator are integers that are relatively prime.　忽略正负符号,若一个分数的分子和分母互质,则它是最简分数。

Questions：Which of the following fractions are in simplest form?　找出最简分数。

1/8,2/9,5/25,7/28,9/21,－8/7,－5/9,－10/85

Answers：1/8,2/9,－8/7,－5/9.

Reduction of Fractions to a Common Denominator　通分

Definition：See Property 8 of fraction. 见分数的性质 8(见 1.2.3 节)。

Questions：Evaluate each of the following:

(1) $\frac{3}{5}+\frac{1}{3}$

(2) $\frac{3}{10}-\frac{11}{30}$

(3) $\frac{0.1}{0.6}+\frac{3}{4}$

Answers：(1) LCD＝15.

$$\frac{3}{5}+\frac{1}{3}=\frac{9}{15}+\frac{5}{15}=\boxed{\frac{14}{15}}.$$

(2) LCD＝30.

$$\frac{3}{10}-\frac{11}{30}=\frac{9}{30}-\frac{11}{30}=-\frac{2}{30}=\boxed{-\frac{1}{15}}.$$

(3) Note that $\frac{0.1}{0.6}=\frac{1}{6}$, we have $\frac{0.1}{0.6}+\frac{3}{4}=\frac{1}{6}+\frac{3}{4}$.

LCD＝12.

$$\frac{1}{6}+\frac{3}{4}=\frac{2}{12}+\frac{9}{12}=\boxed{\frac{11}{12}}.$$

2.3.4　Summary　总结

1. A ratio can be interpreted as a fraction—it is the quantitative relation between two amounts showing the number of times one contains or is being contained by the other.　比可以理解成为分数——它表示两个量的关系,表示其中一个量包含另一个量或者被另一个量包含的次数。
2. A proportion is two or more equivalent ratios. Review properties of proportion.　比例是两个量或更多量的比相等。复习比例的性质。
3. Review the definition of simplest fractions and the procedure of simplifying fractions and adding/subtracting unlike fractions.　复习最简分数的定义、约分的步骤、加减异分母分数(需要通分)。

2.4 Radicals 根式

视频 8

2.4.1 Introduction 介绍

Radical 根式

n. [ˈrædɪkəl]

Definition: The radical of a number is sometimes called the number's "nth root". Saying b is an nth root of a means that $b^n = a$. 一个数的根式也称为该数的 n 次方根。b 是 a 的 n 次方根表示的是 $b^n = a$。

Notation: Figure 2-7 shows the parts of a radical.

Radical Sign——根号

Radicand——被开方数

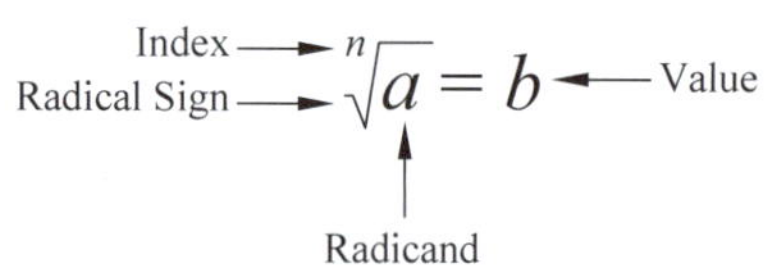

Figure 2-7

A radical has a value. We will use the words "radical" and "value of the radical" interchangeably. 根式的值为 b。"根式"与"根式的值"两者可以互用。

The nth power of the radical is equivalent to the radicand: $b^n = a$. 根式的 n 次幂等于它的被开方数: $b^n = a$。

The nth root of the radicand is equivalent to the radical: $\sqrt[n]{a} = b$. 被开方数的 n 次方根等于根式。

Properties: Properties 1-2 refer to $\sqrt[n]{a} = b$:

1. (1) If a is an integer but not a perfect nth power, then b is irrational. 若 a 是整数但不是完全 n 次幂,则 b 是无理数。

   (2) When $n = 2$, we call this the second root of a, or the square root of a. We do not write out the index 2 in the radical. Any radical without an index is known as a square root. 当 $n = 2$,称为 a 的二次方根或 a 的平方根。一般不把指数写出来。没有指数的根式都是指平方根。

   (3) When $n = 3$, we call this the third root of a, or the cube root of a. 当 $n = 3$,称为 a 的三次方根或 a 的立方根。

2. If n is even, then $a \geqslant 0$, otherwise the radical is not real. 若 n 为偶数,则 $a \geqslant 0$,否则根式不为实数。

3. To solve for b in $b^n = a$:

   在 $b^n = a$ 下解出 b,以 n 的奇偶性讨论。

   (1) If n is odd, then $b = \sqrt[n]{a}$.

   (2) If n is even, then $b = \pm\sqrt[n]{a}$ because $b^n = (-b)^n$.

   For n is even, when we calculate b in $\sqrt[n]{a} = b$, we are calculating the arithmetic nth root, which must be nonnegative. When we calculate b in $b^n = a$, we are calculating all values of b for which the nth power of b is a. This can be negative.

   For instance, $\sqrt[4]{81} = 3$ but not -3.

   On the other hand, for $b^4 = 81$, we have $b = \pm 3$ because $3^4 = 81$ and $(-3)^4 = 81$.

4. To convert from a radical to a power: $\sqrt[n]{a^m}=a^{m/n}$. 从根式到幂的转换。

5. $\sqrt[n]{a}\sqrt[n]{b}=\sqrt[n]{ab}$. If n is even, then both a and b must be nonnegative. Note that the (even)th root of a negative number is not real.

   We can show this using Property 4:

   $\sqrt[n]{a}\sqrt[n]{b}=a^{1/n}\cdot b^{1/n}=(ab)^{1/n}=\sqrt[n]{ab}$.

6. If n is odd, then $\sqrt[n]{a^n}=a$.

   If n is even, then $\sqrt[n]{a^n}=|a|$. Recall that in Property 3, we know that if n is even, then the value of the radical must be nonnegative. 注意,如性质3所述,若知道 n 为偶数,则根式必须为非负数。

7. $(\sqrt[n]{a})^n=a$, assuming that $\sqrt[n]{a}$ is real.

8. If n is odd, then $\sqrt[n]{-a}=-\sqrt[n]{a}$ because $(-1)^n=-1$.

Examples: $\sqrt{144}=12$; $\sqrt[3]{-216}=-6$; $\sqrt[4]{256}=4$.

Phrases: positive～, negative～

Questions: 1. Evaluate each of the following:

(1) $-\sqrt{100}$　(2) $\sqrt[3]{125}$　(3) $\sqrt[5]{1024}$

2. Convert the following radicals to powers:

从根式转换成幂:

(1) $-\sqrt{100}$　(2) $\sqrt[3]{125}$　(3) $\sqrt[5]{1024}$

3. Convert the following powers to radicals:

从幂转换成根式:

(1) $20^{-3/5}$　(2) $30^{8/9}$　(3) $40^{19/12}$

4. (1) What is a side-length of a square whose area is 81? 面积为81的正方形的边长是多少?

   (2) What is an edge-length of a cube whose volume is 343? 体积为343的立方体的棱长是多少?

5. (1) What is the value of $\sqrt{49}$?

   (2) What are all possible values for x for which $x^2=49$?

   (3) What are all possible values for x for which $x^9=-512$?

6. Is it always true that $(\sqrt[n]{a})^n=\sqrt[n]{a^n}$ if n is an integer greater than 1?

Answers: 1. (1) Since $10^2=100$, $\sqrt{100}=10$, and $-\sqrt{100}=\boxed{-10}$.

(2) Since $5^3=125$, $\sqrt[3]{125}=\boxed{5}$.

(3) Since $4^5=1024$, $\sqrt[5]{1024}=\boxed{4}$.

2. (1) $-100^{1/2}$.

   (2) $125^{1/3}$.

   (3) $1024^{1/5}$.

3. (1) $\sqrt[5]{20^{-3}}$.

(2) $\sqrt[9]{30^8}$.

(3) $\sqrt[12]{40^{19}}$.

4. (1) Let the side-length be s, for which $s>0$. Since area is s^2, we have $81=s^2$. Since $9^2=81$, we have $s=\boxed{9}$.

(2) Let the edge-length be s, for which $s>0$. Since volume is s^3, we have $343=s^3$. Since $7^3=343$, we have $s=\boxed{7}$.

5. (1) $\boxed{7}$. We should ignore -7 because this is the arithmetic square root.

(2) We have $x^2=49$. Thus $x=\pm\sqrt{49}=\boxed{\pm 7}$ (Property 3b).

(3) We have $x^9=-512$. Thus $x=\sqrt[9]{-512}=\boxed{-2}$ (Property 3a).

6. No. Suppose $a=-5$ and $n=4$, we have:

LHS $=(\sqrt[n]{a})^n=(\sqrt[4]{-5})^4$. Since $\sqrt[4]{-5}$ is undefined by Property 2, this is undefined.

RHS $=\sqrt[n]{a^n}=\sqrt[4]{(-5)^4}=|-5|=5$.

Note that $\sqrt[4]{(-5)^4}=\sqrt[4]{625}=5$, by the way.

Therefore, LHS $\neq$ RHS.

2.4.2 Simplifying Radicals 简化根式

Simplest Radical Form 最简根式

n. [ˈsɪmplɪst ˈrædɪkəl fɔrm]

Definition: Let $\sqrt[n]{a}=b\sqrt[n]{c}$. If c is not a multiple of the nth power of any prime, then $b\sqrt[n]{c}$ is in the simplest form. 若 c 不是任何质数的 n 次幂的倍数,则 $b\sqrt[n]{c}$ 为最简根式。

Properties: To rewrite a radical in the simplest form, rewrite the radicand as its prime factorization, then follow Properties 5, 6, and 8 of radical. 要把根式改写成最简根式,先写出被开方数的分解质因数,然后运用根式的性质5、6、8。

Questions: 1. Simplify $\sqrt{20}$.

2. Simplify $\sqrt{250}$.

3. Simplify $\sqrt[3]{-16}$.

4. Simplify $\sqrt[5]{320}$.

Answers: 1. $\sqrt{20}=\sqrt{2^2\cdot 5}=\sqrt{2^2}\cdot\sqrt{5}=\boxed{2\sqrt{5}}$.

2. $\sqrt{250}=\sqrt{2\cdot 5^3}=\sqrt{2}\cdot\sqrt{5^3}=\sqrt{2}\cdot\sqrt{5\cdot 5^2}=\sqrt{2}\cdot\sqrt{5}\cdot\sqrt{5^2}=\sqrt{2}\cdot\sqrt{5}\cdot 5=\sqrt{10}\cdot 5=\boxed{5\sqrt{10}}$.

3. $\sqrt[3]{-16}=-\sqrt[3]{16}=-\sqrt[3]{2^4}=-\sqrt[3]{2\cdot 2^3}=-\sqrt[3]{2}\cdot\sqrt[3]{2^3}=-\sqrt[3]{2}\cdot 2=\boxed{-2\sqrt[3]{2}}$.

4. $\sqrt[5]{320}=\sqrt[5]{2^6\cdot 5}=\sqrt[5]{2^6}\cdot\sqrt[5]{5}=\sqrt[5]{2^5\cdot 2}\cdot\sqrt[5]{5}=\sqrt[5]{2^5}\cdot\sqrt[5]{2}\cdot\sqrt[5]{5}=2\cdot\sqrt[5]{2}\cdot\sqrt[5]{5}=\boxed{2\sqrt[5]{10}}$.

Irrational Conjugate 共轭无理数

n. [ɪˈræʃənəl ˈkɑndʒəˌgeɪt]

Definition: If a, b, and c are rational numbers and $\sqrt{c}$ is irrational, then $a+b\sqrt{c}$ and $a-b\sqrt{c}$ are irrational conjugates of each other. 若 a、b、c 为有理数,且$\sqrt{c}$为无理数,则 $a+b\sqrt{c}$ 与 $a-b\sqrt{c}$ 互为共轭无理数。

Properties: The sum and product of irrational conjugates are rational. 共轭无理数的和与积均为有理数。

Proofs: To prove the property, we have

$(a+b\sqrt{c})+(a-b\sqrt{c})=2a$,

and

$(a+b\sqrt{c})(a-b\sqrt{c})$

$=a^2-(b\sqrt{c})^2$ Difference of Squares formula 平方差公式

$=a^2-b^2c$.

Since rational numbers are closed under addition and multiplication (The sum and product of two rational numbers are rational.), the product of irrational conjugates is rational. 因为有理数封闭于加法和乘法(两个有理数的和与积均为有理数),共轭无理数的积为有理数。

Questions: What is the irrational conjugate for each of the following?

(1) $2+\sqrt{5}$

(2) $-5+\sqrt{6}$

(3) $8-\sqrt{7}$

(4) $\sqrt{10}$

Answers: (1) $2-\sqrt{5}$.

(2) $-5-\sqrt{6}$.

(3) $8+\sqrt{7}$.

(4) $-\sqrt{10}$.

Rationalize the Denominator 分母有理化

v. [ˈræʃənəlˌaɪz ðə dɪˈnɑməˌneɪtər]

Definition: To rationalize the denominator of a fraction, we need to rewrite the fraction in a way that the denominator is a rational number. 分母有理化是指把分数改写成一个分母为有理数的分数。

Properties: Depending on the scenario, there are different ways to rationalize the denominator: 分母有理化的方法有多种:

1. If the denominator is in the form $\sqrt[n]{a^k}$, for which n and k are positive integers, and a is rational: 若分母的形式为$\sqrt[n]{a^k}$,其中 n 和 k 为正整数,a 为有理数,则根据 k 和 n 的大小讨论:

   If $k<n$, then multiply the fraction by $\frac{\sqrt[n]{a^{n-k}}}{\sqrt[n]{a^{n-k}}}$ so that the new denominator becomes $\sqrt[n]{a^k}\cdot\sqrt[n]{a^{n-k}}=\sqrt[n]{(a^k)(a^{n-k})}=\sqrt[n]{a^n}=a$.

If $k=n$, then the denominator, which is equal to a, is already rationalized.

If $k>n$, then write $\sqrt[n]{a^k}$ in the simplest radical form and proceed to the case of $k<n$ or $k=n$.

2. If the denominator is in the form $a-b\sqrt{c}$, for which a, b, and c are rational but $\sqrt{c}$ is irrational, multiply the fraction by $\frac{a+b\sqrt{c}}{a+b\sqrt{c}}$. See irrational conjugate. 若分母的形式为 $a-b\sqrt{c}$,其中 a、b、c 为有理数,但 $\sqrt{c}$ 为无理数,则分数乘上 $\frac{a+b\sqrt{c}}{a+b\sqrt{c}}$。见共轭无理数。

Questions: Rationalize the denominator for each of the following:

(1) $\frac{1}{\sqrt{2}}$

(2) $\frac{5}{\sqrt[3]{6}}$

(3) $\frac{6}{\sqrt[4]{49}}$

(4) $\frac{1}{\sqrt[4]{64}}$

(5) $\frac{8}{5+\sqrt{6}}$

(6) $\frac{10}{7-3\sqrt{5}}$

Answers: (1) $\frac{1}{\sqrt{2}}=\frac{1}{\sqrt{2}}\cdot\frac{\sqrt{2}}{\sqrt{2}}=\frac{\sqrt{2}}{\sqrt{2^2}}=\boxed{\frac{\sqrt{2}}{2}}$.

(2) $\frac{5}{\sqrt[3]{6}}=\frac{5}{\sqrt[3]{6}}\cdot\frac{\sqrt[3]{6^2}}{\sqrt[3]{6^2}}=\frac{5\sqrt[3]{6^2}}{\sqrt[3]{6^3}}=\boxed{\frac{5\sqrt[3]{36}}{6}}$.

(3) $\frac{6}{\sqrt[4]{49}}=\frac{6}{\sqrt[4]{7^2}}=\frac{6}{\sqrt[4]{7^2}}\cdot\frac{\sqrt[4]{7^2}}{\sqrt[4]{7^2}}=\frac{6\sqrt[4]{7^2}}{\sqrt[4]{7^4}}=\boxed{\frac{6\sqrt[4]{49}}{7}}$.

We can further simplify the numerator: $\frac{6\sqrt[4]{49}}{7}=\frac{6\sqrt[4]{7^2}}{7}=\frac{6\cdot 7^{2/4}}{7}=\frac{6\cdot 7^{1/2}}{7}=\boxed{\frac{6\sqrt{7}}{7}}$.

(4) $\frac{1}{\sqrt[4]{64}}=\frac{1}{\sqrt[4]{2^6}}=\frac{1}{2\sqrt[4]{2^2}}=\frac{1}{2\sqrt[4]{2^2}}\cdot\frac{\sqrt[4]{2^2}}{\sqrt[4]{2^2}}=\frac{\sqrt[4]{2^2}}{2^2}=\boxed{\frac{\sqrt[4]{4}}{4}}$.

We can further simplify the numerator and get $\boxed{\frac{\sqrt{2}}{4}}$.

(5) $\frac{8}{5+\sqrt{6}}=\frac{8}{5+\sqrt{6}}\cdot\frac{5-\sqrt{6}}{5-\sqrt{6}}=\frac{8(5-\sqrt{6})}{5^2-\sqrt{6}^2}=\boxed{\frac{8(5-\sqrt{6})}{19}}$.

(6) $\frac{10}{7-3\sqrt{5}}=\frac{10}{7-3\sqrt{5}}\cdot\frac{7+3\sqrt{5}}{7+3\sqrt{5}}=\frac{10(7+3\sqrt{5})}{7^2-(3\sqrt{5})^2}=\frac{10(7+3\sqrt{5})}{4}=\boxed{\frac{5(7+3\sqrt{5})}{2}}$.

2.4.3 Summary 总结

1. Review the properties of radicals. Pay particular attention to:
   - $\sqrt[n]{a}$ must be nonnegative when n is even. 当 n 为偶数时,$\sqrt[n]{a}$ 必为非负数。
   - To solve for b in $b^n=a$:
     要在 $b^n=a$ 下解出 b,以 n 的奇偶性讨论:
     (1) If n is odd, then $b=\sqrt[n]{a}$.
     (2) If n is even, then $b=\pm\sqrt[n]{a}$ because $b^n=(-b)^n$.
   - Properties 4-8 of radical.
2. Review the method of putting $\sqrt[n]{a}$ into the simplest form. 复习简化根式的方法。
3. Review the method of finding the irrational conjugate of an irrational number. 复习求共轭无理数的方法。
4. Review the method of rationalizing the denominator:
   复习分母有理化的方法:
   - Multiplying both the numerator and the denominator by a radical so that the denominator is rationalized afterward. 分子和分母同时乘上一个根式,使得分母被有理化(为有理数)。
   - Multiplying both the numerator and the denominator by the conjugate of the denominator so that the denominator is rationalized afterward. 分子和分母同时乘上分母的共轭无理数,使得分母被有理化(为有理数)。

3. Equations 等式/方程(式)

3.1 Introduction 介绍

Golden Rule: Whatever operation we do on one side of the equation, we must do exactly the same thing to the other side.

黄金法则:在等式一边做任何运算,在另一边也要做同样的运算。

Equation 等式/方程(式)

n. [ɪ'kweɪʒən]

Definition: 1. A statement saying that the two mathematical expressions are equivalent. 等式:展示两个表达式相等。

2. An equation that contains variables. 方程(式)：含有变量的等式。

方程(式)是特殊的等式，但讲到 equation 可以把方程和等式的用法互换。

Notation：(Expression 1) = (Expression 2).

Properties：1. $a = a$ Reflexive Property 反身性

2. If $a = b$, then $b = a$. Symmetric Property 对称性

3. If $a = b$ and $b = c$, then $a = c$. Transitive Property 传递性

4. If $a = b$, then $a + k = b + k$.

5. If $a = b$, then $a - k = b - k$.

6. If $a = b$, then $ak = bk$.

7. If $a = b$, then $a/k = b/k$, provided that $k \neq 0$.

Properties 1-3 are useful in mathematical proofs. Properties 4-7 are useful in solving mathematical equations. Properties 4-7 can be rephrase as "whatever operation we do on one side of the equation, we must do exactly the same thing to the other side (Golden Rule)".

性质 1～3 常用于数学证明。性质 4～7 常用于解方程。性质 4～7 可用黄金法则概括：在等式一边做任何运算，在另一边也要做同样的运算。

Examples：1. $2+3=5$. 等式

2. $x+2x=3x$. 方程式

3. $2x+1=5$. 方程式

4. $x+y+z=w$. 方程式

Phrases：conditional～ 条件等式，paradox～ 矛盾等式，linear～with one unknown 一元一次方程，linear～with two unknowns 二元一次方程，system of linear～ 一元方程组，absolute value～ 绝对值方程

3.1.1 Types of Equations 等式的种类

Identity 恒等式

n. [ɑɪ'dentɪti]

Definition：An equation that is always true regardless of the values that the variables take. 恒等式是指无论变量取什么值都成立的等式/方程式。

Equations without variables are always identities. 没有变量的等式是恒等式。

Examples：1. $2x+x=3x$. This is always true regardless of the value of x. 对于 $2x+x=3x$，无论 x 取什么值，等式均成立。

2. $2x+1=2x-2+5-2$. This is always true regardless of the value of x. 对于 $2x-1=2x-2+5-2$，无论 x 取什么值，等式均成立。

3. $2+3=5$.

Conditional Equation 条件等式

n. [kən'dɪʃənəl ɪ'kweɪʒən]

Definition：An equation that is true only if the variable(s) takes specific value(s). Conditional equations always have variables. 条件等式是指只有变量取某个/某些值才会成立的等式/方程式。

Examples：1. $2x+1=3$ is only true when $x=1$.

2. $x-8=18$ is only true when $x=26$.

3. $|x-3|+(y+2)^2=0$ is only true when $x=3$ and $y=-2$. Note that both $|x-3|$ and $(y+2)^2$ are nonnegative. In order for the equation to be true, we need LHS $=0+0=0=$ RHS.

注意 $|x-3|$ 与 $(y+2)^2$ 均为非负数,因此左边必须等于右边,即 $0+0=0$。

Paradox Equation　矛盾等式

n. [ˈpærəˌdɑks ɪˈkweɪʒən]

Definition：A false equation (The expressions on both sides of the equal sign are never equal). 矛盾等式即假等式(等号两边的表达式永远不等)。

Examples：1. $2x=2x+1$ is a paradox equation because subtracting $2x$ from both sides results $0=1$, which is always false.

2. $5=6$ is a paradox equation because it is always false.

3.1.2 Solutions　方程的解

Solution　解/方程的解

n. [səˈluʃən]

Definition：A value we can put in place of a variable that makes the equation true. Can use interchangeably with the word root. 方程的解是指使得方程成立的变量值。方程的解可与方程的根互用。

In an equation with many variables, a solution refers to a combination of values of the variables. 对于有多个变量的方程,解指的是这些变量值的组合。

Examples：1. In the equation $3x+1=7$, $x=2$ is a solution because $3(2)+1=7$ results $7=7$, which is always true. However, $x=3$ is not a solution because $3(3)+1=7$ results $10=7$, which is false.

$x=2$ 是方程 $3x+1=7$ 的解,但 $x=3$ 不是。

2. In the equation $(x-7)(x-9)=0$, $x=7$ is a solution because $(7-7)(7-9)=0$ results $0(-2)=0$, which is always true. $x=9$ is also a solution because $(9-7)(9-9)=0$ results $2(0)=0$, which is always true. However, $x=11$ is not a solution because $(11-7)(11-9)=0$ results $8=0$, which is false.

$x=7,9$ 是方程 $(x-7)(x-9)=0$ 的解,但 $x=11$ 不是。

Root　根/方程的根

n. [rut]

Definition：See solution. The word "roots" is usually used to describe the solutions of quadratic equations or equations with higher degrees. It can be used interchangeably with solution. 见"解/方程的解"词条。根通常描述二次方程或更高次方程的解。"根"可与"解"互用。

3.1.3 Summary　总结

1. Review all properties of equation, as well as the Golden Rule. 复习等式/方程的性质,以及黄金法则。

2. Types of equations:
   等式的种类:
   - Identities—equations that are always true, regardless of the values of the variables. 恒等式——不论变量取什么值,永远成立的等式。
   - Conditional Equations—equations that are only true when variable(s) takes specific value(s). 条件等式——变量只有取某些值才成立的等式。
   - Paradox Equations—equations that are always false. They have no solutions. 矛盾等式——永远不成立的等式。它们无解。
3. Review the definition of solution. 复习解的定义。

3.2 Linear Equations with One Unknown 一元一次方程

3.2.1 Introduction 介绍

Linear Equation with One Unknown 一元一次方程

n. [ˈlɪniər ɪˈkweɪʃən wɪθ wʌn ʌnˈnoʊn]

Definition: An equation with only one variable, and the variable term(s) has degree 1. 只有一个变量的方程,且变量项的次数为1。

Properties: There are three cases to linear equations with one unknown: 一元一次方程有3种情况:

1. Conditional Equation: There is only *one* solution. 条件等式:只有一个解。
2. Paradox Equation: There is no solution. 矛盾等式:无解。
3. Identity: All real numbers are solutions. 恒等式:所有实数都是解(无数个解)。

Examples: 1. $x+1=3$ is a conditional equation because it is only true when $x=2$. $x+1=3$ 是条件等式的例子。

2. $x=x+1$ is a paradox equation because the right side is always greater than the left side (LHS never equals to RHS.). $x=x+1$ 是矛盾等式的例子。

3. $x+1=x+1$ is an identity because it is always true regardless of the value of x. $x+1=x+1$ 是恒等式的例子。

Inverse Operation 逆运算

n. [ɪnˈvɜrs ˌɑpəˈreɪʃən]

Definition: Inverse operations are opposite operations that undo each other. Addition and subtraction are inverse operations. Multiplication and division are inverse operations. 逆运算是指反方向的运算,能彼此抵消。加法和减法是逆运算。乘法和除法是逆运算。

Properties: 1. Inverse operations are effective in solving one-step linear equations. 逆运算对解单步一元一次方程很有用。

2. One-step linear equations are of the form $x+a=b$, or $ax=b$, for which a and b are constants. 单步一元一次方程的形式是 $x+a=b$ 或 $ax=b$,其中 a 和 b 为常值。

To solve $x+a=b$, perform the following inverse operation: subtract a from both sides. This results $x=b-a$. 要解 $x+a=b$,用逆运算,即两边减去 a。

To solve $ax=b$, perform the following inverse operation: both sides divided by a.

This results $x = b/a$, for which $a \neq 0$. 要解 $ax = b$,用逆运算,即两边除以 a,前提条件是 a 不为 0。

3. Inverse operations are based on the Golden Rule of Equations (Properties 4-7 of equation). 逆运算是根据等式的黄金法则(可参照等式/方程的性质 4~7)。
4. Solving multi-step linear equations involves inverse operations quite often. See the properties of solving linear equation with one unknown in Section 3.2.2. 要解多步一元一次方程,常运用逆运算。参见 3.2.2 节"解一元一次方程"的性质。

Questions: Solve each of the following:

(1) $9 + x = 28$

(2) $-15 + x = -6$

(3) $50x = 250$

(4) $-6x = -84$

Answers: (1) Subtracting 9 from both sides gives $x = 19$.

(2) Adding 15 to both sides gives $x = 9$.

(3) Dividing both sides by 50 gives $x = 5$.

(4) Dividing both sides by -6 gives $x = 14$.

Isolate 解出/隔离

v. [ˈaɪsəˌleɪt]

Definition: To isolate a term means to use inverse operations to separate that term from one side of the equal sign. In other words, we express that term in terms of the other terms. See also inverse operation. 解出一个项,就是用逆运算把那个项隔离到等号的一边,即把那个项通过其他的项表示出来,见逆运算。

Questions: Isolate a on each of the following:

(1) $a + b = c$.

(2) $5a = b$.

(3) $ab + c = d$, for which $b \neq 0$.

(4) $a/b = c$, for which $b \neq 0$.

(5) $\frac{a}{b} - c = d$, for which $b \neq 0$.

(6) $ab + ac = d$, for which $b \neq 0$ and $c \neq 0$.

Answers: (1) Subtracting b from both sides, we get $\boxed{a = c - b}$.

(2) Dividing both sides by 5, we get $\boxed{a = b/5}$.

(3) Subtracting c from both sides, we get $ab = d - c$. Dividing both sides by b, we get $\boxed{a = (d - c)/b}$.

(4) Multiplying both sides by b, we get $\boxed{a = bc}$.

(5) Adding c to both sides, $\frac{a}{b} = c + d$. Multiplying both sides by b, we get $\boxed{a = b(c + d)}$.

(6) From $ab+ac=d$, we factor the left side by pulling out a common factor: $a(b+c)=d$. 注意提取公因式的步骤。

Dividing both sides by $(b+c)$, we get $\boxed{a=d/(b+c)}$.

3.2.2 Solving Linear Equations with One Unknown 解一元一次方程

视频 9

Solving Linear Equations with One Unknown 解一元一次方程

Definition: Isolate the variable through manipulating the terms in order to get the value of the variable. 解一元一次方程是指通过移项,解出变量从而求出变量的值。

Properties: 1. Whatever operation we do on one side of the equation, we must do the exact same thing to the other side (Golden Rule). 参照等式/方程的黄金法则。

2. Remove all parentheses using the rules of remove parentheses and the Distributive Property, then combine like terms whenever possible. 用相应的法则和分配律去括号,然后尽可能合并同类项。

3. If both sides of the equation have variables, subtract one term with variables from both sides so that the variable only appears on one side of the equation afterward. 若方程两边都有变量,则两边同时减去一个带有变量的项,使得新方程只有一边有变量。

For example, if a, b, c, and d are constants $(a, c \neq 0)$, and
举例如下: a、b、c、d 均为常值$(a, c \neq 0)$,且

$$ax+b=cx+d,$$

subtracting cx from both sides gives $ax-cx+b=d$, or
两边减去 cx 可得出以下结果:

$$(a-c)x+b=d.$$

Now the variable appears on one side of the equation only. 现在的方程只有一边含有变量。

4. If only one side of the equation has variables, isolate the variable term by adding/subtracting a constant to/from both sides. 若方程的一边有变量,通过加减一个常数,隔离含有变量的项。

For example, if a, b, and c are constants $(a \neq 0)$, and
若 a、b、c 为常值$(a \neq 0)$,且

$$ax+b=c,$$

subtracting b from both sides gives
两边减去 b 得

$$ax=c-b.$$

Now the variable appears on one side of the equation only. 现在的方程只有一边含有变量。

5. To isolate the variable completely, use the inverse operations. 要解出变量,用逆运算。

For example, if a and b are constants $(a \neq 0)$, and
若 a、b 为常数$(a \neq 0)$,且

$$ax=b,$$

Dividing both sides by a gives:
两边除以 a 可得:

$$x = b/a.$$

Now the variable is isolated completely. 现在变量被完全解出了。

Notes: The procedures in the properties are based on the Golden Rule (Property 1). 这些步骤都是根据等式的黄金法则(性质 1)。

Examples: 1. Solve the equation $2x+1=11$.

$2x+1=11$	Original equation.
$2x=10$	Isolate the variable term $2x$ by subtracting 1 from both sides.
$x=5$	Isolate the variable x by dividing both sides by 2.

2. Solve the equation $4(x-5)=24$.

$4(x-5)=24$	Original equation.
$4x-20=24$	Remove the parentheses.
$4x=44$	Isolate the variable term $4x$ by adding 20 to both sides.
$x=11$	Isolate the variable x by dividing both sides by 4.

Alternatively, this equation can be solved the following way:

$4(x-5)=24$	Original equation.
$x-5=6$	Divide both sides by 4.
$x=11$	Isolate the variable x by adding 5 to both sides.

3. Solve the equation $3(x-1)=-4(x+6)$.

$3(x-1)=-4(x+6)$	Original equation.
$3x-3=-4x-24$	Remove the parentheses.
$7x-3=-24$	Add $4x$ to both sides so that only one side of the equation has the variable now.
$7x=-21$	Isolate the variable term $7x$ by adding 3 on both sides.
$x=-3$	Isolate the variable x by dividing both sides by 7.

4. Solve the equation $2(x-5)=4(x-6)-2x$.

$2(x-5)=4(x-6)-2x$	Original equation.
$2x-10=4x-24-2x$	Remove the parentheses.
$2x-10=2x-24$	Combine like terms.
$-10=-24$	Subtract $2x$ from both sides to get rid of the variable on one of the sides.

It happens that variables are gone from both sides, and we are left with $-10=-24$, which is quite unsound. Thus this is a paradox equation. There is no solution.

5. Solve the equation $-5(x-8)=-5(x-6)+10$.

$-5(x-8)=-5(x-6)+10$	Original equation.
$-5x+40=-5x+30+10$	Remove the parentheses.
$-5x+40=-5x+40$	Combine like terms on both sides.
$40=40$	Add $5x$ to both sides so that one of the sides does not contain variable.

It happens that variables are gone from both sides, and we are left with $40 = 40$, which is an identity. All real numbers are solutions.

3.2.3 Summary 总结

1. Inverse operations are operations that undo each other in order to cancel terms and constants in equations. 逆运算是相互抵消的运算,使得在等式里可以消除一些项以及常数。
   Addition and subtraction are inverse operations. 加法和减法互为逆运算。
   Multiplication and division are inverse operations. 乘法和除法互为逆运算。
   Inverse operations must follow the Golden Rule of Equations: "Whatever operation we do on one side of the equation, we must do the exact same thing to the other side." 逆运算必须按照等式/方程的黄金法则:在等式两边的运算要一致。
2. Isolation of a term is the process of expressing that term in terms of the others. This process is based on inverse operations. 解出一个项是把这个项通过其他的项表示出来的过程。这个过程是通过逆运算进行的。
3. Solving linear equations with one unknown is the process of isolating the unknown and get the answer—the value of the unknown. 解一元一次方程就是解出变量的一个过程,从而求出变量的值。
4. A linear equation with one unknown has exactly 0, 1, or infinitely many solutions. 一元一次方程有 0、1 或无穷个解。

3.3 Absolute Value Equations 绝对值方程

视频 10

3.3.1 Introduction 介绍

Absolute Value Equation 绝对值方程

n. [ˈæbsəˌlut ˈvælju ɪˈkweɪʒən]

Definition: An equation in which the variable is inside the absolute value bars. See both absolute value equations with variable on one side and absolute value equations with variable on both sides. 绝对值方程是指在绝对值符号里带有变量的方程,包括在方程一边有未知数的绝对值方程和方程两边都有未知数的绝对值方程。

Absolute Value Equation with Variable on One Side 在方程一边有未知数的绝对值方程

Notation: $|(\text{variable expression})| = k$, for which $k \geqslant 0$ (The absolute value of any real number must be nonnegative, otherwise there is no solution.). 任何实数的绝对值都必须为非负数,否则无解。

Properties: Let k be a nonnegative number. 设 k 为非负数。

1. If $|x| = k$, then $x = k$ or $-k$.
2. In general, if $|(\text{variable expression})| = k$, then set up two equations:
   (1) $(\text{variable expression}) = k$
   (2) $(\text{variable expression}) = -k$
   Solve each separately.
   绝对值为 k 的量有两种情况:这个量的值为 $-k$ 或 k,需要分情况讨论。

3. In |(variable expression)| = k, there are no solutions if $k<0$, one solution if $k=0$, and two solutions if $k>0$.
在|变量表达式| = k 中,
若 $k<0$,则无解;
若 $k=0$,则有一个解;
若 $k>0$,则有两个解。

Questions: Solve each of the following:

(1) $|x|=6$

(2) $|2x-5|=35$

Answers: (1) $|x|=6$

Case 1: $x=6$

Case 2: $x=-6$

$\boxed{x=6,-6}$.

(2) $|2x-5|=35$

Case 1: $2x-5=35$

$2x=40$　Add 5 to both sides.

$x=20$　Divide both sides by 2.

Case 2: $2x-5=-35$

$2x=-30$　Add 5 to both sides.

$x=-15$　Divide both sides by 2.

$\boxed{x=20,-15}$.

Absolute Value Equation with Variable on Both Sides　在方程两边都有未知数的绝对值方程

Notation: |(variable expression 1)| = (variable expression 2)

Properties: If |(variable expression 1)| = (variable expression 2), then we can set up two equations:

这是在方程一边有未知数的绝对值方程的应用,重点还是分情况讨论绝对值里面的量。

(1) (variable expression 1) = (variable expression 2).

(2) (variable expression 1) = −(variable expression 2).

Solve each separately.

IMPORTANT!!! After getting the solutions, be sure to substitute in the **original** equation and see if they check. We need to ensure that after the substitution, (variable expression 2) ≥ 0.　得出解后,必须代入原方程验根。需要验证等式的右边是否为非负数。

Checking the solution is important. See check in Section 3.3.2.　验根很重要,见3.3.2节"验根"。

Notes: The property above is based on the fact that if the absolute values of two numbers are equal, then the two numbers must be the same or opposites.　上面的性质是根据事实:若两个数的绝对值相等,则这两个数相等或互为相反数。

Examples: 1. $|4x+36|=2x+120$

Case 1: $4x+36=2x+120$

$2x+36=120$ Subtract $2x$ from both sides.

$2x=84$ Subtract 36 from both sides.

$x=42$ Divide both sides by 2.

Case 2: $4x+36=-(2x+120)$

$4x+36=-2x-120$ RHS is the same as $-1\cdot(2x+120)$

$6x+36=-120$ Add $2x$ to both sides.

$6x=-156$ Subtract 36 from both sides.

$x=-26$ Divide both sides by 6.

Now it is time to check the solutions.

Check whether $x=42$ is a solution:

$|4(42)+36|\overset{?}{=}2(42)+120$ Arithmetics.

$204=204$ It worked!!!

Note: The symbol "$\overset{?}{=}$" serves as a question "whether the two quantities are equal".

Check whether $x=-26$ is a solution:

$|4(-26)+36|\overset{?}{=}2(-26)+120$ Arithmetics.

$68=68$ It also worked. Nice!

Both solutions check.

Final answer: $\boxed{x=42,-26}$.

2. $|3x+11|=4x-4$

Case 1: $3x+11=4x-4$

$11=x-4$ Subtract $3x$ from both sides.

$15=x$ Add 4 to both sides.

$x=15$ Switch LHS and RHS.

Case 2: $3x+11=-(4x-4)$

$3x+11=-4x+4$ RHS is the same as $-1\cdot(4x-4)$.

$7x+11=4$ Add $4x$ to both sides.

$7x=-7$ Subtract 11 from both sides.

$x=-1$ Divide by 7 on both sides.

Now it is time to check the solutions.

Check whether $x=15$ is a solution:

$|3(15)+11|\overset{?}{=}4(15)-4$ Arithmetics.

$56=56$ $x=15$ is indeed a solution.

Check whether $x=-1$ is a solution:

$|3(-1)+11|\overset{?}{=}4(-1)-4$ Arithmetics.

$8=-8$ Uh-oh!

By the way, in $|3(-1)+11|=4(-1)-4$, the RHS is clearly negative. The absolute value of any real number must be 0 or greater. So we know that this won't work right away.

Only one solution checks.

Final Answer: $\boxed{x=15}$.

3.3.2 Checking the Solutions　验根

Check　验根

v. [tʃek]

Definition: To check a solution, plug the solution back into the **original** equation and see if the result is an identity. The solutions that satisfy the original equation are valid solutions. The solutions that do not satisfy the original equation are from careless mistakes or extraneous solutions.　把解代入原方程,检验是否成立。满足原方程的解是有效的解。不满足原方程的解是粗心造成的错误或者增根。

Properties: It is always a good habit to check the solution(s).　验根永远是好习惯。

Questions: 1. Check if $x=5$ is a solution for $3x+2=17$.

2. Check if $x=4$ is a solution for $(x-3)(x-8)=0$.

Answers: 1. Since $3(5)+2=17$, $x=5$ is a solution.

2. $\text{LHS}=(4-3)(4-8)=-4$, which is not equal to 0. Thus $x=4$ is not a solution.

Extraneous Solution　增根

n. [ekˈstreɪniəs səˈluʃən]

Definition: A solution that is found by following all the rules of mathematics from the original equation, but does not satisfy the original equation.　根据数学法则解方程得出的但不能满足原方程的根。

Properties: The following types of equations are very likely to involve extraneous solutions:
下面这些方程很可能有增根:

1. Radical Equations: $\sqrt{\text{variable expression}}=\text{expression}$.　根式方程。
2. Absolute value equations with variable on both sides:　在方程两边都有未知数的绝对值方程。
   $|(\text{variable expression})|=\text{variable expression}$.
3. Equations that involve domain restrictions.　有定义域限制的方程。

Examples: 1. See absolute value equations with variable on both sides (Section 3.3.1) Question 2: $x=-1$ is an extraneous solution.　见 3.3.1 节两边都有未知数的绝对值方程词条的题 2: $x=-1$ 是增根。

2. Solve $5\cdot\dfrac{1}{\dfrac{1}{x}}=x$ and check the solution(s).

Simplifying fractions, this is $5x=x$. Subtracting x from both sides, we get $4x=0$, from which $x=0$.

Check if $x=0$ is a solution:

$$5\cdot\frac{1}{\frac{1}{0}}\overset{?}{=}0\cdots$$

The LHS is undefined! Thus $x=0$ does not satisfy the equation and thus is an extraneous solution. 根据运算法则得出 $x=0$,但 $x=0$ 不满足原方程。

3. Solve $\sqrt{x}=-1$ and check the solution(s).

Right away, we know that the arithmetic square root of a number cannot be negative, thus we have no solution. Let's suppose we do not know this fact and solve it formally:

Squaring both sides, we get $x=1$.

Check if $x=1$ is a solution: $\sqrt{1}=-1$ is same as $1=-1$, which is false. Thus $x=1$ is extraneous.

根据运算法则得出 $x=1$,但 $x=1$ 不满足原方程。

3.3.3 Summary 总结

1. When we have |variable expression| $=k$, we have to consider two cases and solve each separately:

   当变量表达式的绝对值等于 k 时,应考虑下面两种情况并分类讨论:

   - (variable expression) $=k$.
   - (variable expression) $=-k$.

   We have to check that $k\geqslant 0$ is true in the original equation, otherwise there will be no valid solutions and we will get extraneous solutions. 必须确定另一个量在原方程里是非负数,否则会无解且有增根。

2. Checking the solutions is necessary—to avoid extraneous solutions and careless mistakes. 验根非常必要——避免增根和粗心造成的错误。

3.4 Linear Equations with Two Unknowns 二元一次方程

3.4.1 Introduction 介绍

Linear Equation with Two Unknowns 二元一次方程

Definition: An equation with two variables. The degree of each term is at most 1. 含有两个变量的方程,每个项的次数最多为 1。

Properties: 1. There are either 0 or infinitely many solutions in a linear equation with two unknowns. 二元一次方程有 0 个或无数个解。

2. We usually wish to isolate one of the variables in a linear equation with two unknowns. 通常解出其中一个变量。

Examples: 1. The equation $x+y=x+y+1$ has no solutions because the RHS is always 1 greater than the LHS. 无解的例子。

2. The equation $x+2y=100$ has infinitely many solutions because we can pick any value for x, and for each value of x, there is one value for y that satisfies the equation.

有无数个解的例子。

3. The equation $x + y = x + 1$ has infinitely many solutions. Subtracting x from both sides, we have $y = 1$. We know that y is always 1, and x can be any real number.　有无数个解的例子。

Questions: In each of the following, (1) Isolate x. (2) Isolate y.

1. $3x + 4y = 12$
2. $5y = 10x - 25$
3. $x + 2y - 3 = 4x - 5y + 6$

Answers: 1. (1) $3x + 4y = 12$　Original equation.

$3x = -4y + 12$　Subtract $4y$ from both sides.

$x = -\frac{4}{3}y + 4$　Divide both sides by 3.

(2) $3x + 4y = 12$　Original equation.

$4y = -3x + 12$　Subtract $3x$ from both sides.

$y = -\frac{3}{4}x + 3$　Divide both sides by 4.

2. (1) $5y = 10x - 25$　Original equation.

$-10x + 5y = -25$　Subtract $10x$ from both sides.

$-10x = -5y - 25$　Subtract $5y$ from both sides.

$x = \frac{1}{2}y + \frac{5}{2}$　Divide both sides by -10.

(2) $5y = 10x - 25$　Original equation.

$y = 2x - 5$　Divide both sides by 5.

3. (1) $x + 2y - 3 = 4x - 5y + 6$　Original equation.

$-3x + 2y - 3 = -5y + 6$　Subtract $4x$ from both sides.

$-3x - 3 = -7y + 6$　Subtract $2y$ from both sides.

$-3x = -7y + 9$　Add 3 to both sides.

$x = \frac{7}{3}y - 3$　Divide both sides by -3.

(2) $x + 2y - 3 = 4x - 5y + 6$　Original equation.

$x + 7y - 3 = 4x + 6$　Add $5y$ to both sides.

$7y - 3 = 3x + 6$　Subtract x from both sides.

$7y = 3x + 9$　Add 3 to both sides.

$y = \frac{3}{7}x + \frac{9}{7}$　Divide both sides by 7.

System of Linear Equations　一次方程组

n. [ˈsɪstəm ʌv ˈlɪniər ɪˈkweɪʃənz]

Definition: A collection of two or more linear equations with the same set of unknowns. The solution of a system of linear equations must satisfy every equation in the system.　由两个或更多个一次方程构成的组合。它们的未知数相同。一次方程组的解必须满足方程组

里的每个方程。

See system of linear equations with two unknowns and system of linear equations with three unknowns, which is in Section 23.4. They are the two most common and practical systems. 见二元一次方程组和23.4节的三元一次方程组。

Properties: Every system of linear equations satisfies exactly one of the following:
每个一次方程组满足下列条件之一:

1. has no solution. 无解。
2. has exactly one solution. 有一个解。
3. has infinitely many solutions. 有无数个解。

Examples: 1. $\begin{cases} x+2y=5 \\ x-5y=-10 \end{cases}$ system of linear equations with two unknowns 二元一次方程组.

2. $\begin{cases} x+y+z=10 \\ x+2y+3z=24 \\ 5x+5y+z=30 \end{cases}$ system of linear equations with three unknowns 三元一次方程组.

System of Linear Equations with Two Unknowns 二元一次方程组

Definition: A collection of two or more (usually two) linear equations with the same set of two unknowns. The solution of a system of linear equations with two unknowns is a combination of the values of the two unknowns that must satisfy every equation in the system. The solution is usually denoted by an ordered pair. 由两个或更多个二元一次方程构成的组(通常是两个)。它们的未知数相同。二元一次方程组的解是两个未知数的值的组合,使其满足组里的每一个方程。解通常用有序对表示。

Properties: 1. For a system of two linear equations with two unknowns, there are three main ways of solving it:
二元一次方程组的解法有以下3种:
(1) Graphing. 画图法。
(2) Substitution. 代入消元法。
(3) Elimination. 加减消元法。
Each of these methods works with two equations, and has its own advantages and disadvantages. 每一种方法应用到两个方程中,各有优点和缺点。
We will proceed with (2) and (3) in the next sections and get to (1) in Section 7.3.5.4.

2. For a system of three or more linear equations with two unknowns, pick any two equations and perform the methods mentioned in Property 1. Check the solution against all the equations in the system. 对于有3个或更多个二元一次方程,选择两个方程并运用性质1,然后对其他的方程进行验根。

3. We usually write our solution in the ordered pair (x, y), suppose the variable names are x and y. 假设未知数用x,y表示,通常以有序对(x, y)代表解。

Examples: Set up a system of linear equations with two unknowns for each of the following. 列出二元一次方程组。

(1) In a cage, there are only two types of animals: chicken and rabbits. Together, there

are 50 heads and 160 legs. How many chicken are there in the cage?
这是“鸡兔同笼问题”。

(2) A calculator store only sells scientific and graphing calculators. A scientific calculator costs \$ 80 and a graphing calculator costs \$ 100. I want to spend exactly \$ 1000 in the store and buy 11 calculators. How many graphing calculators should I buy?

Answers: (1) Let x be the number of chicken and y be the number of rabbits.

A chicken has 2 legs and a rabbit has 4 legs.

To set up the equation regarding to the total number of animals, we have $x+y=50$.

To set up the equation regarding to the total number of legs, we have $2x+4y=160$.

Thus the answer is $\boxed{\begin{cases} x+y=50 \\ 2x+4y=160 \end{cases}}$.

(2) Let x be the number of scientific calculators and y be the number of graphing calculators.

To set up the equation regarding to the total number of calculators, we have $x+y=11$.

To set up the equation regarding to the total price of the calculators, we have $80x+100y=1000$.

Thus the answer is $\boxed{\begin{cases} x+y=11 \\ 80x+100y=1000 \end{cases}}$.

3.4.2 Methods of Solving Systems of Linear Equations with Two Unknowns 二元一次方程组的解法

Solving Systems of Linear Equations by Substitution　代入消元法

视频 11

Properties: 1. In a system of linear equations with two unknowns (call the unknowns x and y):

在二元一次方程组里,设未知数为 x 和 y:

(1) Isolate one of the unknowns (say x) in one equation.　在其中一个方程里解出其中一个变量(设这个变量为 x,解出 x 的意思是把 x 改写成带有 y 的表达式)。

(2) In the other equation, replace all occurrences of x in terms of y according to the isolation of x.　在另一个方程里根据所得的 x 的等式,把所有的 x 用 y 表示。

(3) Solve for y in the equation from step (2).　根据步骤(2)解出 y。

(4) Solve for x according to the isolation.　解出 x。

Isolating y will lead to the same result.　亦可把 y 用 x 表示,殊途同归。

2. In a system of linear equations with three or more unknowns, substitution usually is not the most efficient way. We either need to do it multiple times (which is cumbersome) or use elimination to solve the system efficiently.　在三元一次方程组里,代入消元法通常不是最有效的方法,要代入若干次或者直接用加减消元法解方程组。

3. (1) Advantage: Solving the system this way is efficient if isolating one of the unknowns in one equation is obvious and easy.　若解出一个变量容易,代入消元

法更为高效。

(2) Disadvantage: If isolating one of the unknowns in one equation involves fractions, the chance of making mistakes is higher. 若解出一个变量后,一个方程出现分数,出错的概率较大。

Questions: 1. Solve the equation from Example (1) of system of linear equations with two unknowns (Section 3.4.1), $\begin{cases} x+y=50 \\ 2x+4y=160 \end{cases}$, by substitution. 用代入消元法解 3.4.1 节"二元一次方程组"词条的例题(1)。

2. Solve the equation from Example (2) of system of linear equations with two unknowns (Section 3.4.1), $\begin{cases} x+y=11 \\ 80x+100y=1000 \end{cases}$, by substitution. 用代入消元法解 3.4.1 节"二元一次方程组"词条的例题(2)。

Answers: 1. From equation 1, we get $x=50-y$. Substitute this into equation 2, we have

$$\begin{aligned} 2x+4y&=160 \\ 2(50-y)+4y&=160 \\ 100-2y+4y&=160 \\ 100+2y&=160 \\ 2y&=60 \\ y&=30. \end{aligned}$$

Substitute $y=30$ into $x=50-y$, we get $x=20$. We have $(x, y)=(20,30)$. To answer the original question, we have 20 chicken and 30 rabbits.

2. This is similar from the previous example. This time let's isolate y.

From equation 1, we get $y=11-x$. Substitute this into equation 2, we get

$$\begin{aligned} 80x+100y&=1000 \\ 80x+100(11-x)&=1000 \\ 80x+1100-100x&=1000 \\ 1100-20x&=1000 \\ -20x&=-100 \\ x&=5. \end{aligned}$$

Substitute $x=5$ into $y=11-x$, we get $y=6$. We have $(x, y)=(5,6)$. To answer the original question, I bought 5 scientific calculators and 6 graphing calculators.

Solving Systems of Linear Equations by Elimination 加减消元法

Properties: 1. In a system of linear equations with two unknowns (call the unknowns x and y):

在未知数为 x 和 y 的二元一次方程组中:

(1) If the coefficients of the same variable are opposites in two equations of the system, we add these equations to cancel out that variable. Then we will solve for the other variable and finally solve for the variable we cancelled out, by substitution. 若在方程组的两个方程中,同一个变量的系数互为相反数,则把这两个方程相加,从而消除这个变量。然后解出另一个变量,最后用代入消元法解出

被消除的变量。

(2) If the coefficients of the same variable are the same in two equations of the system, we subtract one equation from the other to cancel out that variable. Then we will solve for the other variable and finally solve for the variable we cancelled out, by substitution. 若在方程组的两个方程中，同一个变量的系数相等，则把这两个方程相减，从而消除这个变量。然后解出另一个变量，最后用代入消元法解出被消除的变量。

(3) If the coefficients of the same variable are neither same nor opposite in two equations of the system, we multiply one of the equations by a constant, then perform either Property (1) or (2). 若在方程组的两个方程中，同一个变量的系数既不互为相反数，也不相等，则把其中一个方程乘上一个常数，然后用上面的性质(1)或(2)解出方程组。

2. This method is also very efficient in systems of linear equations with three or more unknowns, but it needs to be applied several times. 加减消元法对于三元一次方程组也很有效，但要做若干次。

3. Advantage: Generally and fastest and the most common method of solving systems. 优点：加减消元法通常是解方程组的最快方法。

Disadvantage: Sometimes substitution can be faster if isolation is easy. 缺点：若一个变量容易被解出，不如代入消元法快。

Proofs: 1. Why adding and subtracting equations work? 为什么等式能应用加减消元法？

If $a = b$ and $c = d$, then $a + b = c + d$, and $a - b = c - d$ (Following the Golden Rule of equations, if we add/subtract a number on one side, we must add/subtract the same quantity on the other side.). 根据等式的黄金法则：若在等式一边加减一个量，则在等式的另一边必须加减相同的量。

Therefore, in the system of equations $\begin{cases} \text{variable side 1} = \text{constant 1} \\ \text{variable side 2} = \text{constant 2}, \\ \cdots \end{cases}$

(variable side 1) ± (variable side 2) = (constant 1) ± (constant 2).

2. Why multiplying and dividing an equation by a constant work?

If $a = b$, then $ka = kb$ and $a/k = b/k$, for which $k \neq 0$ (Following the Golden Rule of equations, if we multiply/divide a number on one side, we must multiply/divide the same quantity on the other side.). 根据等式的黄金法则：若在等式一边乘上或除以一个量，则在等式的另一边必须乘上或除以相同的量。

Therefore, in the equation (variable side) = (constant), it is true that $k \cdot$ (variable side) = $k \cdot$ (constant), for which k is a constant.

Questions: 1. Solve the system $\begin{cases} 2x + 3y = 31 \\ 2x - 3y = -11 \end{cases}$ by elimination.

2. Solve the system $\begin{cases} x + y = 11 \\ 80x + 100y = 1000 \end{cases}$ by elimination. This is from Example (b) of

system of linear equations with two unknowns.

3. Solve the system $\begin{cases} 2x+5y=55 \\ 4x-7y=59 \end{cases}$ by elimination.

Answers: 1. Method 1: Eliminate x.

Subtracting the second equation from the first, we get $2x+3y-(2x-3y)=31-(-11)$, which is $6y=42$. We get $y=7$.

Substituting $y=7$ in either equation (say, equation 1) in the system and solve for x, we get $2x+3(7)=31$, from which $x=5$. Thus the solution of the system is $(x,y)=(5,7)$.

Method 2: Eliminate y.

Adding the equations, we get $2x+3y+(2x-3y)=31+(-11)$, which is $4x=20$. We get $x=5$.

Substituting $x=5$ in either equation (say, equation 2), we get $2(5)+3y=31$, from which $y=7$. Thus the solution of the system is $(x,y)=(5,7)$.

2. Method 1: Eliminate x.

Multiplying the first equation by 80, we get the new system $\begin{cases} 80x+80y=880 \\ 80x+100y=1000 \end{cases}$. Subtracting the first equation from the second, we get $80x+100y-(80x+80y)=1000-880$, from which $20y=120$, and $y=6$.

Substituting $y=6$ into any equation (say, equation 1 from the original system), we get $x+6=11$, from which $x=5$. Thus the solution of the system is $(x,y)=(5,6)$.

Method 2: Eliminate y.

Multiplying the first equation by 100, we get the new system $\begin{cases} 100x+100y=1100 \\ 80x+100y=1000 \end{cases}$. Subtracting the second equation from the first, we get $100x+100y-(80x+100y)=1100-1000$, from which $20x=100$, and $x=5$.

Substituting $x=5$ into any equation (say, equation 1 from the original system), we get $5+y=11$, from which $y=6$. Thus the solution of the system is $(x,y)=(5,6)$.

To answer the original question, I bought 5 scientific calculators and 6 graphing calculators.

3. Method 1: Eliminate x (easier).

Multiplying the first equation by 2, we get the new system $\begin{cases} 4x+10y=110 \\ 4x-7y=59 \end{cases}$.

From here, subtracting the second equation from the first equation gives $17y=51$, and $y=3$.

Substituting $y = 3$ into any equation (say, equation 1 from the original system), we get $2x + 5(3) = 55$, from which $x = 20$. Thus the solution of the system is $(x, y) = (20, 3)$.

Method 2: Eliminate y.

Multiplying the first equation by 7 and the second equation by 5, we get the new system $\begin{cases} 14x + 35y = 385 \\ 20x - 35y = 295 \end{cases}$.

From here, adding the equations gives $34x = 680$, from which $x = 20$.

Substituting $x = 20$ into any equation (say, equation 1 from the original system), we get $2(20) + 5y = 55$, from which $y = 3$. Thus the solution of the system is $(x, y) = (20, 3)$.

3.4.3 Summary　总结

1. A linear equation with two unknowns is an equation with two variables. The degree of each term is at most 1. Such equation can have 0 or infinitely many solutions. We usually want to isolate one variable.　二元一次方程是指含有两个变量的方程,每个项的次数最多为 1。这种方程有 0 或无数个解。通常通过解出一个变量来解方程。
2. A system of linear equations is a collection of two or more linear equations with the same set of unknowns. The solution of a system of linear equations must satisfy every equation in the system. The solution is usually denoted by an ordered n-tuple.　一次方程组是两个或更多个一次方程构成的组。有多少个未知数就有多少个方程。一次方程组的解必须满足方程组里的每个方程。解通常用有序 n 元组表示。
3. There are three ways to solve a system of linear equations with two unknowns:　有以下 3 种方法解二元一次方程组:
   - Graphing—will cover in Section 7.3.5.4.　画图法,见 7.3.5.4 节。
   - Substitution—easy to use when one unknown is easy to isolate.　代入消元法。
   - Elimination—easy to use when the coefficients of one unknown are equal or opposites of each other.　加减消元法。

3.5 Quadratic Equations　一元二次方程

3.5.1 Introduction　介绍

Quadratic Equation　一元二次方程

n. [kwɑ'drætɪk ɪ'kweɪʒən]

Definition: The standard form of a quadratic equation is

一元二次方程的一般形式为

$$ax^2 + bx + c = 0,$$

for which a, b, and c are constants, and $a \neq 0$.

其中 a、b、c 为常数,$a \neq 0$。

Properties: 1. Quadratic equations can be solved in one of the three ways:

有下面 3 种方法解一元二次方程:

(1) Factoring. 因式分解法。

(2) Completing the square. 配方法。

(3) Quadratic Formula. 一元二次方程求根公式。

Each of these methods has its advantages and disadvantages. 每种方法都有它的优点和缺点。

2. Every quadratic equation satisfies one of the following:

每个一元二次方程都满足下面中的任一项:

(1) has exactly 0 real roots. 有 0 个实数根。

(2) has exactly 1 real root. 有 1 个实数根。

(3) has exactly 2 real roots. 有 2 个实数根。

3.5.2 Methods of Solving Quadratic Equations 一元二次方程的解法

视频 12

Solving Quadratic Equations by Factoring 因式分解法

Properties: 1. Zero-Product Property: If $ab=0$, then $a=0$ or $b=0$. 零积性质。

2. In $x^2+bx+c=0$, for which b and c are constants (usually integers), think of two real numbers (usually integers, and call them p and q) such that $p+q=b$ and $pq=c$. We can factor as $x^2+bx+c=x^2+(p+q)x+pq=(x+p)(x+q)=0$. The roots are $-p$ and $-q$.

方程 $x^2+bx+c=0$,其中 b 和 c 为常数(通常是整数),假设两个实数 p 和 q(通常是整数),使得 $p+q=b$ 和 $pq=c$。则可以进行因式分解: $x^2+bx+c=x^2+(p+q)x+pq=(x+p)(x+q)=0$。根为 $-p$ 和 $-q$。

3. In $ax^2+bx+c=0$, for which a, b, and c are constants (usually integers), and $a\neq 0$. Use educated trial and error to factor: $ax^2+bx+c=C_1C_2x^2+(C_1q+C_2p)x+pq=(C_1x+p)(C_2x+q)=0$. We know that $C_1C_2=a$ and $pq=c$. The roots are $-p/C_1$ and $-q/C_2$.

方程 $ax^2+bx+c=0$,其中 a、b、c 为常数(通常是整数),且 $a\neq 0$。用有根据的试错进行因式分解: $ax^2+bx+c=C_1C_2x^2+(C_1q+C_2p)x+pq=(C_1x+p)(C_2x+q)=0$。可以知道 $C_1C_2=a$ 和 $pq=c$。根为 $-p/C_1$ 和 $-q/C_2$。

Alternatively, from $ax^2+bx+c=0$, we can divide both sides by a and get $x^2+\frac{b}{a}x+\frac{c}{a}=0$. Then use Property 2 in a similar manner. However, this is very difficult because b/a and c/a can be fractions (so can the roots). 或者可以在 $ax^2+bx+c=0$ 两边除以 a,再运用性质 2。但是带有分数的系数就不那么好计算了。

4. Advantage: If the roots are rational (integers, in particular), this method provides a shortcut and finds them efficiently, which is much easier and faster than the Quadratic Formula. 优点:若根是有理数(特别是整数),则这个方法是解一元二次方程的捷径,比一元二次方程求根公式快。

Disadvantage: Not every quadratic equation has rational roots. Irrational roots need

Quadratic Formula or completing the square to find them. Also, this method cannot detect whether the quadratic equation has no real root. 缺点：不是所有的二元一次方程的根都是有理数。若是无理根要用一元二次方程求根公式或配方法来解。而且，这个方法不能检验二元一次方程是否没有实数根。

Questions: Solve each of the following by factoring:

1. $x^2+11x+30=0$
2. $x^2-10x+24=0$
3. $x^2+4x-16=5$
4. $x^2-10x-57=-1$
5. $2x^2-x-6=0$
6. $2x^2+17x+30=0$
7. $6x^2-23x+21=0$
8. $12x^2+19x-10=0$

Answers:

1. Since $5+6=11$ and $5\cdot6=30$, we can factor as $(x+5)(x+6)=0$, from which $\boxed{x=-5,-6}$.
2. Since $-4+(-6)=-10$ and $-4\cdot(-6)=24$, we can factor as $(x-4)(x-6)=0$, from which $\boxed{x=4,6}$.
3. Subtracting 5 from both sides, we get $x^2+4x-21=0$.
   Since $7+(-3)=4$ and $7\cdot(-3)=-21$, we can factor as $(x+7)(x-3)=0$, from which $\boxed{x=-7,3}$.
4. Adding 1 to both sides, we get $x^2-10x-56=0$.
   Since $4+(-14)=-10$ and $4\cdot(-14)=-56$, we can factor as $(x+4)(x-14)=0$, from which $\boxed{x=-4,14}$.
5. We have $(2x\cdots)(x\cdots)=0$. The product of the constant terms in the factors is -6. By educated trial and error, we get $(2x+3)(x-2)=0$, from which $\boxed{x=-3/2,2}$.
6. We have $(2x\cdots)(x\cdots)=0$. The product of the constant terms in the factors is 30. By educated trial and error, we get $(2x+5)(x+6)=0$, from which $\boxed{x=-5/2,-6}$.
7. We have either $(x\cdots)(6x\cdots)=0$ or $(2x\cdots)(3x\cdots)=0$. The product of the constant terms in the factors is 21. By educated trial and error, we get $(2x-3)(3x-7)=0$, from which $\boxed{x=3/2,7/3}$.
8. We have one of $(x\cdots)(12x\ \cdots)=0$, $(2x\cdots)(6x\cdots)$, or $(3x\cdots)(4x\cdots)$. The product of the constant terms in the factors is -10. By educated trial and error, we get $(12x-5)(x+2)=0$, from which $\boxed{x=5/12,-2}$.

Solving Quadratic Equations by Completing the Square　配方法

Definition: To solve $x^2+bx+c=0$, we wish to rewrite the LHS as the [square of a sum of x and a constant] plus/minus a constant. We can rewrite the LHS as follows:
要解 $x^2+bx+c=0$,可以把方程左边改写成 x 与一个常数和的平方加减一个常数的形

式，即

$$x^2+bx+c=0$$

$$x^2+bx+\left(\frac{b}{2}\right)^2+c-\left(\frac{b}{2}\right)^2=0$$

$$\left(x+\frac{b}{2}\right)^2+\left(c-\left(\frac{b}{2}\right)^2\right)=0$$

The last step is the result of completing the square—the square of a sum involving the variable $\left(x+\frac{b}{2}\right)^2$ plus a constant $\left(c-\left(\frac{b}{2}\right)^2\right)$. 上面的步骤最后一步叫作配方——配成变量与一个常数和的平方$\left(x+\frac{b}{2}\right)^2$加上一个常数$\left(c-\left(\frac{b}{2}\right)^2\right)$的形式。

Properties：1. From $\left(x+\frac{b}{2}\right)^2+\left(c-\left(\frac{b}{2}\right)^2\right)=0$, we will continue solving the equation this way：配方以后继续求解：

$$\left(x+\frac{b}{2}\right)^2=-\left(c-\left(\frac{b}{2}\right)^2\right)$$ Subtract $\left(c-\left(\frac{b}{2}\right)^2\right)$ from both sides.

$$\left(x+\frac{b}{2}\right)^2=\left(\frac{b}{2}\right)^2-c$$ Manipulate RHS.

$$\left(x+\frac{b}{2}\right)^2=\frac{b^2}{4}-c$$

$$\left(x+\frac{b}{2}\right)^2=\frac{b^2-4c}{4}$$

$$x+\frac{b}{2}=\pm\frac{\sqrt{b^2-4c}}{2}$$ Solve equations that involve squares.

$$x=-\frac{b}{2}\pm\frac{\sqrt{b^2-4c}}{2}$$

$$x=\frac{-b\pm\sqrt{b^2-4c}}{2}$$

This produces identical result as using the Quadratic Formula in solving $ax^2+bx+c=0$, when $a=1$. 在 $ax^2+bx+c=0$ 里，若 $a=1$，以上就是一元二次方程求根公式的结果。

2. Advantage：Easy to use in solving $ax^2+bx+c=0$ if b is even and $a=1$. It is easier than the Quadratic Formula in these cases. It is another way that is as quick as factoring in when roots are integers. 优点：当 b 是偶数且a 为 1 的时候，用配方法解 $ax^2+bx+c=0$ 比用一元二次方程求根公式简单，在根为整数的情况下配方法与因式分解的效率差不多。

Disadvantage：If the roots are rational and b is not even, then completing the square involves many fractions in the calculation, which is more cumbersome than factoring. 缺点：若根为有理数，且 b 不为偶数，则配方法会含有分数，比因式分解法麻烦。

Notes：This property is based on the Complete Square Formula $(a\pm b)^2=a^2\pm 2ab+b^2$. Thus we need $x^2\pm bx+c=(x\pm\cdots)^2+\text{constant}=x^2+2(\cdots)x+\text{constant}$. Thus number in the ⋯ spot should be $b/2$. In short, ⋯ should be half of the x-term coefficient. 这个性质是

根据完全平方公式$(a \pm b)^2 = a^2 \pm 2ab + b^2$得出的。

Questions：Solve each of the following by completing the square：

1. $x^2 + 4x - 16 = 5$ This is the same as Question 3 in solving quadratic equations by factoring. 题目同因式分解法的题3。
2. $x^2 - 10x + 24 = 0$ This is the same as Question 2 in solving quadratic equations by factoring. 题目同因式分解法的题2。
3. $x^2 + 11x + 30 = 0$ This is the same as Question 1 in solving quadratic equations by factoring. 题目同因式分解法的题1。
4. $2x^2 - x - 6 = 0$ This is the same as Question 5 in solving quadratic equations by factoring. 题目同因式分解法的题5。

Answers：1. Subtracting 5 from both sides, we get $x^2 + 4x - 21 = 0$.

$x^2 + 4x + 4 - 21 - 4 = 0$ Complete the square.

$(x + 2)^2 - 25 = 0$ Simplify.

$(x + 2)^2 = 25$ Isolate the square of a sum.

$x + 2 = \pm 5$ Solve square equation.

$x = -2 \pm 5$ Isolate x.

$\boxed{x = -7, 3}$ Because $x = -2 - 5$ or $-2 + 5$.

2. We will skip some steps from Question 1：

LHS $= x^2 - 10x + 24 = x^2 - 10x + 25 + 24 - 25 = (x - 5)^2 - 1 = 0$.

Therefore, $(x - 5)^2 = 1$, and $x - 5 = \pm 1$, from which $x = 5 \pm 1$, $\boxed{x = 4, 6}$.

3. LHS $= x^2 + 11x + 30 = x^2 + 11x + \left(\frac{11}{2}\right)^2 + 30 - \left(\frac{11}{2}\right)^2 = \left(x + \frac{11}{2}\right)^2 - \frac{1}{4} = 0$.

Therefore, $\left(x + \frac{11}{2}\right)^2 = \frac{1}{4}$, and $x + \frac{11}{2} = \pm \frac{1}{2}$, from which $x = -\frac{11}{2} \pm \frac{1}{2}$.

Thus $\boxed{x = -5, -6}$.

Since 11 is not even, this involves a lot of calculations in fractions, which is not very convenient. 因为11不是偶数,所以含有分数,计算量大,相对来说没那么方便。

4. Dividing both sides by 2 leads to $x^2 - \frac{1}{2}x - 3 = 0$.

LHS $= x^2 - \frac{1}{2}x - 3 = x^2 - \frac{1}{2}x + \left(\frac{1}{4}\right)^2 - 3 - \left(\frac{1}{4}\right)^2 = \left(x - \frac{1}{4}\right)^2 - \frac{49}{16} = 0$.

Therefore, $\left(x - \frac{1}{4}\right)^2 = \frac{49}{16}$, and $x - \frac{1}{4} = \pm \frac{7}{4}$, from which $x = \frac{1}{4} \pm \frac{7}{4}$.

Hence, $\boxed{x = -3/2, 2}$.

Again, this approach involves a lot of calculations in fractions. Solving it by factoring is easier. Also, the Quadratic Formula is a generalized way of solving it. 该题用配方法有比较多的分数出现。用因式分解会稍微简单些。一元二次方程求根公式是配方法的通用方式。

Quadratic Formula 一元二次方程求根公式

n. [kwɑˈdrætɪk ˈfɔrmjələ]

Definition: To solve $ax^2+bx+c=0$, the solutions are $x=\dfrac{-b\pm\sqrt{b^2-4ac}}{2a}$. This is a generalized way of completing the square. 用一元二次方程求根公式求解 $ax^2+bx+c=0$ 是配方法的通用方式。

Properties: 1. b^2-4ac is the discriminant and has the following characteristics:

b^2-4ac 是判别式，有以下性质：

Assume that $a\neq 0$:

- If $b^2-4ac>0$, then there are 2 distinct real roots: $\dfrac{-b+\sqrt{b^2-4ac}}{2a}$ and $\dfrac{-b-\sqrt{b^2-4ac}}{2a}$. Moreover, if b^2-4ac is a perfect square, and a, b, and c are rational, then the roots are rational.
- If $b^2-4ac=0$, then there is 1 real root: $\dfrac{-b}{2a}$.
- If $b^2-4ac<0$, then there are 0 real roots.

根据判别式的大小判断根的个数。若判别式为完全平方，且 a、b、c 为有理数，则根为有理数。

2. Advantage: Works for all quadratic equations in standard form $ax^2+bx+c=0$. Once a, b, and c are identified, solving it by the Quadratic Formula is only a matter of substitution. In addition, the Quadratic Formula can check the number of real roots the quadratic equation has. 优点：一元二次方程求根公式对所有标准的一元二次方程 $ax^2+bx+c=0$ 有效。a、b、c 一旦确定，用一元二次方程求根公式解方程只是代入 a、b、c 的值。而且，一元二次方程求根公式能检验一元二次方程根的个数。

Disadvantage: For quadratic equations with rational roots, solving by factoring is much faster. Sometimes, solving by completing the square is much faster too, depending on preferences. 缺点：对于有有理根的一元二次方程，用因式分解或配方法比较快。

Proofs: Why does the Quadratic Formula work?

To solve $ax^2+bx+c=0$, for which a, b, and c are constants, and $a\neq 0$:

$$x^2+\frac{b}{a}x+\frac{c}{a}=0 \quad \text{Divide } a \text{ on both sides.}$$

$$\left[x^2+\frac{b}{a}x+\left(\frac{b}{2a}\right)^2\right]+\left[\frac{c}{a}-\left(\frac{b}{2a}\right)^2\right]=0 \quad \text{Complete the square.}$$

$$\left(x+\frac{b}{2a}\right)^2+\left[\frac{c}{a}-\left(\frac{b}{2a}\right)^2\right]=0$$

$$\left(x+\frac{b}{2a}\right)^2=-\left[\frac{c}{a}-\left(\frac{b}{2a}\right)^2\right]$$

$$\left(x+\frac{b}{2a}\right)^2=\left(\frac{b}{2a}\right)^2-\frac{c}{a}$$

$$\left(x+\frac{b}{2a}\right)^2=\frac{b^2}{4a^2}-\frac{c}{a}$$

$$\left(x+\frac{b}{2a}\right)^2=\frac{b^2-4ac}{4a^2}$$

$$x+\frac{b}{2a}=\pm\frac{\sqrt{b^2-4ac}}{2a}$$

$$\boxed{x=\frac{-b\pm\sqrt{b^2-4ac}}{2a}}.$$

Questions：Solve each of the following by the Quadratic Formula：

1. $x^2+4x-16=5$ 　This is the same as Question 3 in solving quadratic equations by factoring.　题目同因式分解的题 3。
2. $x^2-10x+24=0$ 　This is the same as Question 2 in solving quadratic equations by factoring.　题目同因式分解的题 2。
3. $2x^2-x-6=0$ 　This is the same as Question 5 in solving quadratic equations by factoring.　题目同因式分解的题 5。

Answers：1. Subtracting 5 from both sides leads to $x^2+4x-21=0$. $x=\frac{-4\pm\sqrt{4^2-4(1)(-21)}}{2(1)}=$ $\frac{-4\pm\sqrt{100}}{2(1)}=\frac{-4\pm10}{2}$. Therefore, $\boxed{x=-7,3}$.

2. We will skip some steps from Question 1：

$$x=\frac{10\pm\sqrt{(-10)^2-4(1)(24)}}{2(1)}=\frac{10\pm\sqrt{4}}{2(1)}=\frac{10\pm2}{2}.$$ Therefore, $\boxed{x=4,6}$.

3. $x=\frac{1\pm\sqrt{(-1)^2-4(2)(-6)}}{2(2)}=\frac{1\pm\sqrt{49}}{2(2)}=\frac{1\pm7}{4}$. Therefore, $\boxed{x=-3/2,2}$.

3.6　Applications of Equations　方程的应用

视频 13

Verbal Model　文字模型/文字表达式

n. [ˈvɜrbəl ˈmɑdəl]

Definition：A relational expression that involves quantities that are described using words. It is useful for setting up the equation.　文字模型/文字表达式是指用语言表达定量的关系式。对于列方程是非常有用的。

Questions：For each of the following, write down the verbal model that can be used to set up an equation.

1. I travel at 10km/hr. If my trip is 180km, how long must I travel?

2. The cost of one-time membership fee is \$50. The cost of each visit to the gym is \$3. If John paid \$140 to the gym this year, including both membership fee and visit fee, how many times did John go to gym?
3. I bought 100 boxes of chocolate and cookies and spent \$1180. If a box of chocolate costs \$10 and a box of cookies costs \$12, how many boxes of chocolate did I buy?

Answers: 1. (The speed of travel) · (The time I travel) = (Total distance).
行驶速度×行驶时间=总距离。
2. (Membership fee) + (Fee of each visit) · (Number of visits) = (Total paid).
会员费+每次访问费用×访问次数=总费用。
3. (Price of 1 box of chocolate) · (Number of boxes of chocolate) + (Price of 1 box of cookies) · (Total number of boxes − Number of boxes of chocolate) = (Total money spent).
每盒巧克力的价格×巧克力的盒数+每盒曲奇饼的价格×(总盒数−巧克力的盒数)=总价。

Common Application Vocabulary 常见应用词汇

Definition: Here we introduce some common relations of quantities in different application problems. It helps us to come up with verbal models. 这里介绍应用题里常见量的关系式,有助于列出文字模型。

Examples: 1. Work 工作, Efficiency 效率, Time 时间
Work = Efficiency · Time
2. Distance 路程, Speed 速度
Distance = Speed · Time
3. Total Price 总价, Unit Price 单价, Quantity 数量
Total Price = Unit Price · Quantity
4. For personal:
New Balance 结余, Original Balance 本金, Income 收入, Expense 支出
New Balance = Original Balance + Income − Expense
5. For company:
Cost 成本, Revenue 收入, Profit 利润
Profit = Revenue − Cost
6. Actual 实际, Expected 期望, Tolerance 容错度
$|\text{Actual} - \text{Expected}| \leqslant \text{Tolerance}$
7. Interest Rate 利率
$\text{New Amount} = (\text{Original Amount})(1 + \text{Interest Rate})^{\text{Time}}$
8. Ways to express real life problems:
应用题解决方法:
(1) Verbal Model 用文字描述模型
(2) Table 绘制表格
(3) Equation and Inequality 列等式或不等式
(4) Graph 在坐标上画图

4. Inequalities　不等式

4.1 Introduction　介绍

Inequality　不等式

n. [ɹɪnˈkwɑlɪti]

Definition：A mathematical statement that involves using one or more of the following signs：＜,＞,⩽,⩾.　带有＜,＞,⩽,⩾符号的数学表达式。

Notation：The definitions of the signs are as follows：

＜(less than)：$a<b$ is read as　小于

- "a is less than b"
- "a is smaller than b".

＞(greater than)：$a>b$ is read as　大于

- "a is greater than b"
- "a is larger than b".

⩽(less than or equal to)：$a \leqslant b$ is read as　小于或等于/不大于

- "a is less than or equal to b"
- "a is at most b"　a 的值最多为 b
- "a is b or less"　a 的值为 b 或更少
- "a does not exceed b"　a 不超过 b
- "a is not greater than b".　a 不大于 b

⩾(greater than or equal to)：$a \geqslant b$ is read as　大于或等于/不小于

- "a is greater than or equal to b"
- "a is at least b"　a 的值至少为 b
- "a is b or more"　a 的值为 b 或更多
- "a is not less than b".　a 不小于 b

Properties：1. To compare two real numbers, a and b, exactly one of the following is true：
若 a 和 b 均为实数，下面 3 项中只有一项成立：

- $a=b$
- $a>b$
- $a<b$

2. Flipping ＜, we get ＞; flipping ＞, we get ＜.
Flipping ⩽, we get ⩾; flipping ⩾, we get ⩽.
＜,＞,⩽,⩾不等式符号书写翻转后的结果分别为＞,＜,⩾,⩽。

Examples: Note that statement such as $2 \geqslant 2$ is true since it is saying that "2 is greater than 2 or 2 is equal to 2". Satisfying one statement connected by the "or" satisfies the whole statement.

$2 \geqslant 2$ 读作"2 大于或等于 2"。此命题成立,原因是在有"或"的命题中,只要满足其中一项,整个命题就正确。

Phrases: linear~ 线性不等式, absolute value~ 绝对值不等式, compound~ 混合不等式, and~ 和不等式, double~ 双不等式, or~ 或不等式

Questions: 1. Decide whether each number below satisfies the inequality: $x>233$.

(1) 2333

(2) 0

(3) -666

2. Translate each of the following to inequality.

(1) The speed on the highway is at most 60 mph (miles per hour).

(2) The cost of a calculator is at least $100.

(3) It takes me more than 3 hours to do the homework.

(4) John goes to work no more than 40 hours per week.

(5) Andy went to a stationery shop and bought a pencil box for $20 and some pencils for $0.50 each. If he spent no more than $40 in the shop, how many pencils can he buy?

Answers: 1. (1) yes (2) no (3) no.

2. (1) Let s be the speed on the highway in mph: $s \leqslant 60$.

(2) Let c be the cost of the calculator in dollars: $c \geqslant 100$.

(3) Let t be the time for me to do the homework, in hours: $t>3$.

(4) Let n be the number of hours John works per week: $n \leqslant 40$.

(5) Let n be the number of pencils Andy bought: $20+0.5n \leqslant 40$.

4.2 Linear Inequalities with One Unknown 一元一次不等式

视频 14

Linear Inequality with One Unknown 一元一次不等式

n. [ˈlɪniər ˌɪnɪˈkwɑlɪti]

Definition: Same as a linear equation with one unknown, EXCEPT that the "=" sign is replaced with one of the four inequality sign: $<, >, \leqslant, \geqslant$. 一元一次不等式除了等号改成不等式符号外,其余与一元一次方程相同。

Properties: To solve linear inequalities with one unknown, we use the same procedure as solving linear equations with one unknown, and the inequality sign maintains, EXCEPT we have to flip it when both sides multiply/divide by a negative number. 除了两边乘上/除以一个负数的情况要翻转不等式符号外,一元一次不等式的解法与一元一次方程的解法相同。

To flip the inequality sign, see Property 2 of inequality in Section 4.1. 要翻转不等式符号,见 4.1 节不等式的性质 2。

Proof：Why should we flip the inequality sign when multiplying/dividing by a negative number? 为什么在两边乘上/除以一个负数的情况下要翻转不等式符号?

For all real numbers a and b, if $a>b$, then $-a<-b$, as shown in Figure 4-1. 对于实数 a 和 b，若 $a>b$，则 $-a<-b$，如图 4-1 所示。

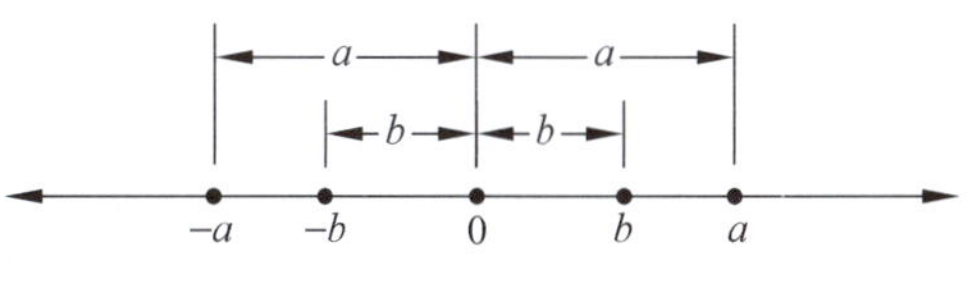

Figure 4-1

For a concrete example, we know that $5>3$ is true, so $-5<-3$ must be true. 例如，已知 $5>3$，则 $-5<-3$。

Questions：1. Solve $4x+5>25$.

2. Solve $233x-1\leqslant-467$.

3. Solve $-6x+1>25+2x$.

4. Solve $-266x-1\leqslant-33x-700$.

Answers：

1. $4x+5>25$ — Original inequality.

$4x>20$ — Subtract 5 from both sides.

$\boxed{x>5}$ — Divide both sides by 4.

2. $233x-1\leqslant-467$ — Original inequality.

$233x\leqslant-466$ — Add 1 to both sides.

$\boxed{x\leqslant-2}$ — Divide both sides by 233.

3. $-6x+1>25+2x$ — Original inequality.

$-8x+1>25$ — Subtract $2x$ from both sides.

$-8x>24$ — Subtract 1 from both sides.

$\boxed{x<-3}$ — Divide both sides by -8… FLIP THE SIGN!!

4. $-266x-1\leqslant-33x-700$ — Original inequality.

$-233x-1\leqslant-700$ — Add $33x$ to both sides.

$-233x\leqslant-699$ — Add 1 to both sides.

$\boxed{x\geqslant3}$ — Divide both sides by -233… FLIP THE SIGN!!

4.3 Methods of Expressing the Solutions of Inequalities 不等式解集表示法

4.3.1 Introduction 介绍

Methods of Expressing the Solutions of Inequalities　不等式解集表示法

Definition：Here are the most common methods：

1. Inequality Notation　不等式符号表示法
2. Set Notation　集合表示法
3. Interval Notation　区间表示法
4. Graphing Inequalities　图解不等式法

Questions：Express each of the following in (1) inequality notation, (2) set notation, (3) interval notation, and (4) graphing inequalities.

分别用不等式符号表示法、集合表示法、区间表示法和图解不等式法表示以下各题。

1. $x<3$
2. $\{x \mid x>3\}$
3. $(-\infty,3]$
4. Refer to Figure 4-2.

Answers: 1. (1) As it is. 如题目本身的表示。

(2) $\{x \mid x<3\}$

(3) $(-\infty,3)$

(4) Refer to Figure 4-3.

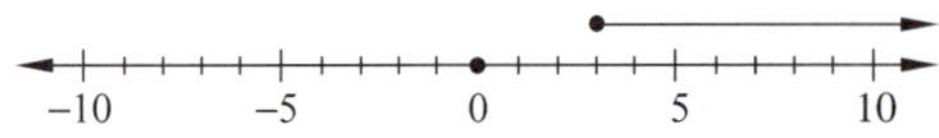

Figure 4-2

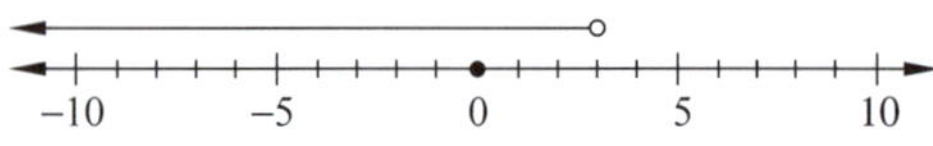

Figure 4-3

2. (1) $x>3$

(2) As it is. 如题目本身的表示。

(3) $(3,+\infty)$

(4) Refer to Figure 4-4.

3. (1) $x\leqslant 3$

(2) $\{x \mid x\leqslant 3\}$

(3) As it is. 如题目本身的表示。

(4) Refer to Figure 4-5.

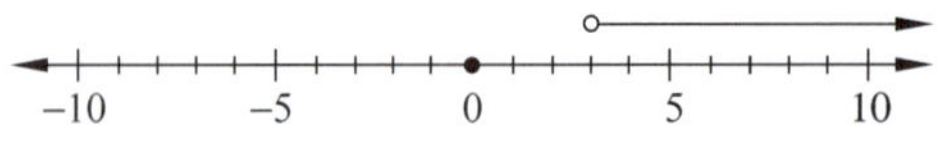

Figure 4-4

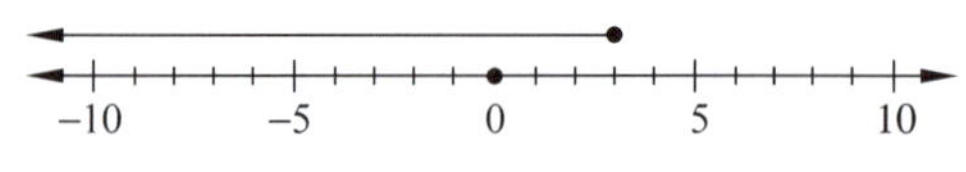

Figure 4-5

4. (1) $x\geqslant 3$

(2) $\{x \mid x\geqslant 3\}$

(3) $[3,+\infty)$

(4) As it is. 如题目本身的表示。

4.3.2 Methods 方法

Inequality Notation 不等式符号表示法

n. [ˌɪnɪˈkwɑlɪti noʊˈteɪʃən]

Notation: (quantity 1) (inequality sign) (quantity 2) (inequality sign) (quantity 3) ⋯

Examples: 1. See the finals answers for the questions in linear inequalities with one unknown in Section 4.3.1.

2. Some examples are shown below:

(1) $x\geqslant 3$.

(2) $x<10$.

(3) $1\leqslant a\leqslant b\leqslant c\leqslant d\leqslant 6$.

Set Notation　集合表示法

n. [set noʊ'teɪʃən]

Notation：{variable(s) | inequality notation}

Properties：{variable(s) | (condition)} is read as "variable(s) such that the condition is true".

{变量|条件}读作"满足条件的变量"。

For example，$\{x \mid x>5\}$ is read as "x such that x is greater than 5".

Examples：See the answers to the questions in methods of expressing the solutions of inequalities in Section 4.3.1.　见4.3.1节不等式解集表示法习题的答案。

Interval Notation　区间表示法

n. ['ɪntərvəl noʊ'teɪʃən]

Notation：One of (a, b)，$[a, b)$，$(a, b]$，$[a, b]$，for which a is the lower limit and b is the upper limit.　a 为下限，b 为上限。

"(" indicates that the lower limit does not include a.

"[" indicates that the lower limit includes a.

")" indicates that the upper limit does not include b.

"]" indicates that the upper limit includes b.

圆括号代表不包含上限/下限；方括号代表包含上限/下限。

Properties：Let x be a variable and a be a constant，

To express $x<a$，we use $(-\infty, a)$.

To express $x>a$，we use $(a, +\infty)$.

To express $x\leqslant a$，we use $(-\infty, a]$.

To express $x\geqslant a$，we use $[a, +\infty)$.

$-\infty$ always is preceded by "("，and ∞ is always followed by ")".

负无穷的前面必须为"("，正无穷的后面必须为")"。

Examples：See the answers to the questions in methods of expressing the solutions of inequalities in Section 4.3.1.　见4.3.1节不等式解集表示法习题的答案。

Graphing Inequalities　图解不等式法

n. ['græfɪŋ ˌɪnɪ'kwɑlɪtis]

Notation：Let x be a variable and a be a constant，　设 x 为变量，a 为常数。

If it is $x<a$，then we have an empty/open dot at a and the arrow points to the left，as shown in Figure 4-6.　若 $x<a$，则在 a 上面画个空心点以及左箭头，如图4-6所示。

If it is $x>a$，then we have an empty/open dot at a and the arrow points to the right，as shown in Figure 4-7.　若 $x>a$，则在 a 上面画个空心点以及右箭头，如图4-7所示。

Figure 4-6　　Figure 4-7

If it is $x\leqslant a$，then we have a filled/closed dot at a and the arrow points to the left，as shown in Figure 4-8.　若 $x\leqslant a$，则在 a 上面画个实心点以及左箭头，如图4-8所示。

If it is $x \geqslant a$, then we have a filled/closed dot at a and the arrow points to the right, as shown in Figure 4-9. 若 $x \geqslant a$,则在 a 上面画个实心点以及右箭头,如图 4-9 所示。

Figure 4-8　　Figure 4-9

Examples: See the answers to the questions in methods of expressing the solutions of inequalities in Section 4.3.1. 见 4.3.1 节不等式解集表示法习题的答案。

4.4 Compound Inequalities 复合不等式

4.4.1 Introduction 介绍

Compound Inequality 复合不等式

n. [ˈkɑmpaʊnd ˌɪnɪˈkwɑlɪti]

Definition: The combination of two or more inequalities, indicated by the words "and" or "or". See "and-inequality" in Section 4.4.2 and "or-inequality" in Section 4.4.3. 复合不等式是指两个或更多个不等式的组合,用 and(且)或 or(或)连接。见 4.4.2 节"同时成立的不等式组"与 4.4.3 节"逻辑或不等式"。

Notation: (inequality 1) (and/or) (inequality 2), for which inequality 1 and inequality 2 can be simple or compound inequalities. 写作"不等式 1 且/或不等式 2"。不等式 1、不等式 2 也可以是复合不等式。

4.4.2 And-Inequalities 同时成立的不等式组

And-Inequality 同时成立的不等式组

n. [ænd ˌɪnɪˈkwɑlɪti]

Definition: A compound inequality for which in order to be true. A solution must satisfy **all** the inequalities connected by the word "and". "同时成立的不等式组"是复合不等式的一种。解必须满足所有被 and(且)连接的不等式。

Notation: Two or more inequalities that are connected by the word "and". 表示成用 and(且)连接的两个或更多的不等式。

Properties: 1. To express an and-inequality using the inequality notation, simply do what the notation above suggests. 用不等式符号表示法表示"同时成立的不等式组",见上面的表示方式。

2. To express an and-inequality using set notation, see set notation. 用集合表示法表示"同时成立的不等式组",见集合表示法。

3. (1) To express an and-inequality using interval notation, use the interval notation to represent individual inequality first, then connect these intervals using the intersection symbol $\cap$. 用区间表示法表示"同时成立的不等式组",先用区间表示法逐一表示以 and(且)连接的不等式,然后用交集符号 $\cap$ 连接。

(2) To express an and-inequality "(inequality 1) and (inequality 2) and (inequality 3) …" using interval notation without using the intersection symbol $\cap$, let a be the maximum of the lower limits of all the inequalities involved, and b be the

minimum of the upper limits of all the inequalities involved, our answer is … a, b …. If all inequalities include a, put "[" before a, else put "(" before a. If all inequalities include b, put "]" after b, else put ")" after b. **In short, this interval is contained by all the inequalities that are connected with the words "and".** Property 4 (graphing) illustrates this property clearer. 不使用交集符号∩,若要用区间表示法表示"同时成立的不等式组",则以 a 为所有不等式中下限的最大值,b 为所有不等式中上限的最小值。答案是…a,b…。若最后的答案包含 a,则在 a 的左边写上"[",否则写上"("。若最后的答案包含 b,则在 b 的右边写上"]",否则写上")"。**简而言之,所有的不等式都包含这个区间**。性质 4(图解不等式法)能更明晰地表示这个概念。

4. As shown in Figure 4-10, to express an and-inequality by graphing, graph each individual inequality first (as shown in black), then take the intersection of the inequalities (as shown in blue). This helps understanding the interval notation. 如图 4-10 所示,要用图解不等式法表示"同时成立的不等式组",先图解各个不等式(用黑色表示),再取它们的交集(用蓝色表示)。图解不等式法能更好地理解区间表示法。

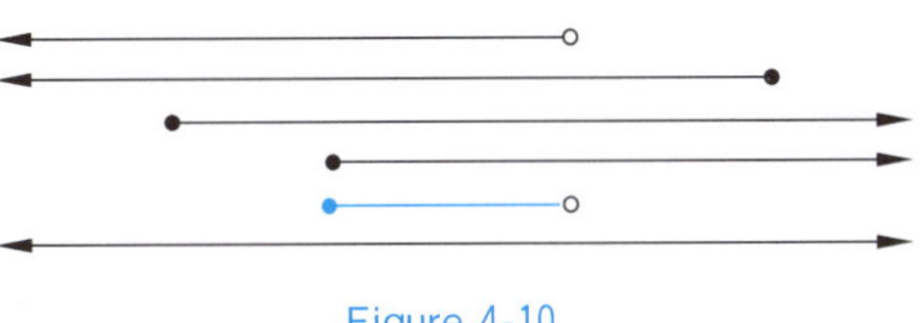

Figure 4-10

Or, first write the and-inequality using the interval notation without the intersection symbol ∩ by Property 3(2), follow the properties in graphing inequalities. 或者,不使用交集符号∩,先根据"同时成立的不等式组"的性质 3(2)把不等式用区间表示法表达出来,再根据图解不等式法的性质画图。

See the "Questions" section for specific examples.

5. To solve an and-inequality, solve each inequality separately. 要解"同时成立的不等式组",可采取逐一击破的方法。
6. "And-inequalities" can be rewritten as double inequalities. "同时成立的不等式组"能被改写成"双重不等式"。

Examples: Here's a real life example:

On the highway, our speed cannot be slower than 45 mph but cannot be faster than 55 mph. Let s be the speed in mph, and we have "$s \geqslant 45$ and $s \leqslant 55$".

And-inequalities are useful when there is a lower limit and an upper limit. 有上限和下限的时候,"同时成立的不等式组"比较有用。

Questions: 1. Decide whether each number below satisfies the and-inequality:

$x \leqslant 233$ and $x > 1$.

(1) -8

(2) 1

(3) 166

(4) 233

(5) 666

2. Rewrite the following and-inequality using (1) inequality notation, (2) set notation,

(3) interval notation with and without the symbol $\cap$, and (4) graphing: $x>-5$ and $x\leqslant 7$.

用不等式表示法、集合表示法、区间表示法、图解不等式法表示“同时成立的不等式组”: $x>-5$ and $x\leqslant 7$。

3. Rewrite the following and-inequality using (1) inequality notation, (2) set notation, (3) interval notation with and without the symbol $\cap$, and (4) graphing: $x>-8$ and $x\geqslant 0$ and $x<10$.

用不等式表示法、集合表示法、区间表示法、图解不等式法表示“同时成立的不等式组”: $x>-8$ and $x\geqslant 0$ and $x<10$。

4. Solve each of the and-inequalities.

(1) $2x+1>5$ and $2x+1<11$

(2) $-2x+1>5$ and $-2x+1<11$

Answers: 1. (1) No, because $x=-8$ does not satisfy $x>1$.

(2) No, because $x=1$ does not satisfy $x>1$.

(3) Yes, because $x=166$ satisfies both $x\leqslant 233$ and $x>1$.

(4) Yes, because $x=233$ satisfies both $x\leqslant 233$ and $x>1$.

(5) No, because $x=666$ does not satisfy $x\leqslant 233$.

2. (1) As it is. 如题目本身的表示。

(2) $\{x \mid x>-5 \text{ and } x\leqslant 7\}$.

(3) With the $\cap$ operator: $(-5,+\infty)\cap(-\infty,7]$.

Without the $\cap$ operator: (Use Property 3(2)).

The lower limit of $x>-5$ is -5.

The lower limit of $x\leqslant 7$ is $-\infty$.

The maximum of the lower limits is -5.

The upper limit of $x>-5$ is $+\infty$.

The upper limit of $x\leqslant 7$ is 7.

The minimum of the upper limit is 7.

The inequality does not include -5 but includes 7, thus our answer is $\boxed{(-5,7]}$.

(4) From $(-5,7]$, graphing should be easy, as shown in Figure 4-11.

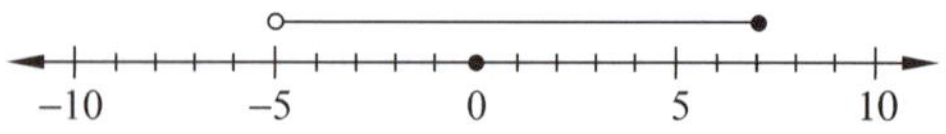

Figure 4-11

3. (1) As it is. 如题目本身的表示。

(2) $\{x \mid x>-8 \text{ and } x\geqslant 0 \text{ and } x<10\}$.

(3) With the $\cap$ operator: $(-8,\infty)\cap[0,\infty)\cap(-\infty,10)$.

Without the $\cap$ operator: (Use Property 3(2)).

The maximum of the lower bounds is 0.

The minimum of the upper bounds is 10.

The inequality includes 0 but does not include 10, thus our answer is $\boxed{[0,10)}$.

(4) From [0, 10), graphing should be easy, as shown in Figure 4-12.

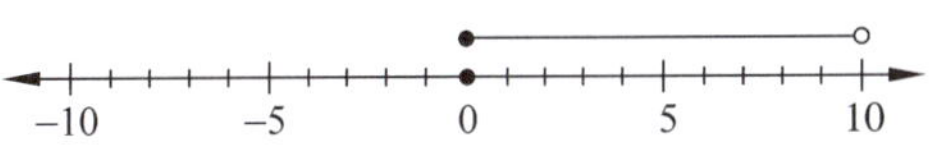

Figure 4-12

4. (1) $x>2$ and $x<5$.

(2) $x<-2$ and $x>-5$. Be sure to flip the inequality signs when multiplying or dividing both sides by a negative number. 当两边都乘上或除以一个负数的时候,必须翻转不等式符号。

Double Inequality 双重不等式

n. [ˈdʌbəl ˌɪnɪˈkwɑlɪti]

Definition: A shorthand notation of the and-inequality—the quantities are arranged in either ascending order (connected by $<$ or $\leqslant$) or descending order (connected by $>$ or $\geqslant$). 双重不等式是同时成立的不等式组的简写——量既可以按照升序排列(用$<$或$\leqslant$连接),也可以按照降序排列(用$>$或$\geqslant$连接)。

Notation: (Quantity 1) ⋯ (Quantity 2) ⋯ (Quantity 3) ⋯

Properties: Let x be a variable, 设 x 为一个变量,

1. If "x ($>$ or $\geqslant$) a and x ($<$ or $\leqslant$) b", then our double inequality will be $\boxed{a\ (<\text{or}\leqslant)\ x\ (<\text{or}\leqslant)\ b}$. If $a>b$, then the inequality is unsound and has no solution.

   Note that in an and-inequality, a solution must satisfy all inequalities connected by the word "and". 同时成立的不等式组的解必须满足被 and(且)连接的每一个不等式。

   It is conventional to arrange quantities in ascending order. 根据惯例,双重不等式一般都用升序排列。

2. If "x ($>$or$\geqslant$) a_1 and x ($>$or$\geqslant$) a_2 and ⋯ and x ($<$or$\leqslant$) b_1 and x ($<$or$\leqslant$) b_2⋯", then our double inequality will be

$$(\max a_1, a_2, \cdots)\ (<\text{or}\leqslant)\ x\ (<\text{or}\leqslant)\ (\min b_1, b_2, \cdots),$$

   depending on whether the inequality includes$(\max a_1, a_2, \cdots)$ or $(\min b_1, b_2, \cdots)$.

   This is an analogue of the interval notation of and-inequality without the intersection symbol $\cap$. 这与同时成立的不等式组的区间表示法(不用交集符号$\cap$)类似。

3. To solve a double inequality, what you do on one quantity, do the same thing on each of the other quantities. **Be sure to flip the signs when multiplying or dividing by a negative number!!!** 解双重不等式,量的运算必须同步。**当乘上或除以负数的时候,不等式符号必须翻转!!!**

4. Let x be a variable and a and b be constants. To graph a ($<$or$\leqslant$) x ($<$or$\leqslant$) b, draw a line segment whose endpoints are a and b, then decide whether each endpoint should be open or closed depending on the inequality signs. 设 x 为一个变量,a 和 b 为常数。要图解 a ($<$or$\leqslant$) x ($<$or$\leqslant$) b,画一条端点为 a 和 b 的线段。线段两头各是实心点或空心点取决于解集是否包含 a 和 b。

Examples: Here's a real life example (same as the example from and-inequality):

On the highway, our speed cannot be slower than 45 mph but cannot be faster than 55 mph. Let s be the speed in mph, and we have "$s \geqslant 45$ and $s \leqslant 55$". Using double inequality, this is $\boxed{45 \leqslant s \leqslant 55}$.

Questions: 1. Rewrite the following and-inequalities as double inequalities.

(1) $x>10$ and $x<233$

(2) $x<30$ and $3 \leqslant x$

(3) $80>x$ and $x \geqslant 4$

(4) $50<x$ and $233 \geqslant x$

(5) $x>10$ and $x \geqslant 20$ and $x \leqslant 80$ and $x<100$ and $x \leqslant 200$

2. Solve each of the following inequalities:

(1) $-7<2x+1<19$

(2) $2<-3x+8<11$

Answers: 1. (1) $10<x<233$ (Property 1)

(2) $3 \leqslant x<30$ (Property 1)

(3) $4 \leqslant x<80$ (Property 1)

(4) $50<x \leqslant 233$ (Property 1)

(5) $20 \leqslant x \leqslant 80$ (Property 2)

2. (1) $-7<2x+1<19$ Original inequality.

$-8<2x<18$ Subtract 1 from everything.

$-4<x<9$ Divide by 2 from everything.

The answer is $\boxed{-4<x<9}$.

(2) $2<-3x+8<11$ Original inequality.

$-6<-3x<3$ Subtract 8 from everything.

$2>x>-1$ Divide by -3 from everything, FLIP THE SIGNS!!!

$-1<x<2$ The conventional way.

The answer is $\boxed{-1<x<2}$.

4.4.3 Or-Inequalities 逻辑或不等式

Or-Inequality 逻辑或不等式

n. [ɔr ˌɪnɪˈkwɑlɪti]

Definition: A compound inequality for which in order to be true. A solution must satisfy **at least one** of the inequalities connected by the word "or". "逻辑或不等式"是复合不等式的一种。解必须满足至少一个被 or(或)字连接的不等式。

Notation: Two or more inequalities that are connected by the word "or". 表示成两个或更多被 or(或)连接的不等式。

Properties: 1. To express an or-inequality using the inequality notation, simply do what the notation above suggests. 用不等式符号表示法表示"逻辑或不等式",见上面的表示方式。

2. To express an or-inequality using set notation, see set notation. 用集合表示法表示

"逻辑或不等式",见集合表示法。

3. To express an or-inequality using interval notation, use the interval notation to represent individual inequality first, then instead of writing the word "or", use the union symbol $\cup$.　用区间表示法表示"逻辑或不等式",先用区间表示法逐一表示以 or(或)连接的不等式,然后用并集符号$\cup$连接。

4. As shown in Figure 4-13, to express an or-inequality by graphing, graph each individual inequality first (as shown in black), then take the union of the inequalities (as shown in blue).　如图 4-13 所示,用图解不等式法表示"逻辑或不等式",先图解每个不等式(用黑色表示),再取它们的并集(用蓝色表示)。

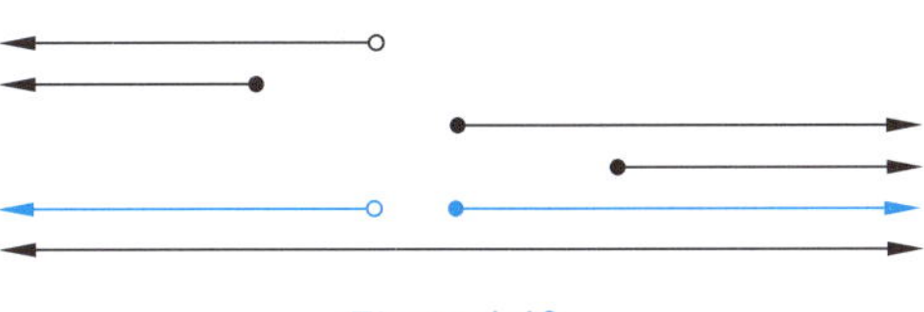

Figure 4-13

5. To solve an or-inequality, solve each inequality separately and maintain the word "or".　要解"逻辑或不等式",保留 or(或),逐一求解每个不等式。

Examples: See absolute value inequalities in Section 4.4.4.　见 4.4.4 节绝对值不等式。

Questions: 1. Decide whether each number below satisfies the or-inequality: $x<233$ or $x>555$.

(1) -8

(2) 1

(3) 233

(4) 456

(5) 666

2. Rewrite the following or-inequality using (1) inequality notation, (2) set notation, (3) interval notation (4) graphing: $x<-5$ or $x\geqslant 10$.　用不等式表示法、集合表示法、区间表示法、图解不等式法表示逻辑或不等式: $x<-5$ or $x\geqslant 10$。

3. Solve this or-inequality: $2x+1<-11$ or $2x+1>11$.

4. Solve this or-inequality: $3x+2>92$ or $-10x-9>91$.

Answers: 1. (1) Yes, because $x=-8$ satisfies $x<233$.

(2) Yes, because $x=1$ satisfies $x<233$.

(3) No, because $x=233$ satisfies neither of $x<233$ nor $x>555$.

(4) No, because $x=456$ satisfies neither of $x<233$ nor $x>555$.

(5) Yes, because $x=666$ satisfies $x>555$.

2. (1) As it is.

(2) $\{x \mid x<-5 \text{ or } x\geqslant 10\}$

(3) $(-\infty,-5)$ or $[10,+\infty)$

(4) As shown in Figure 4-14.

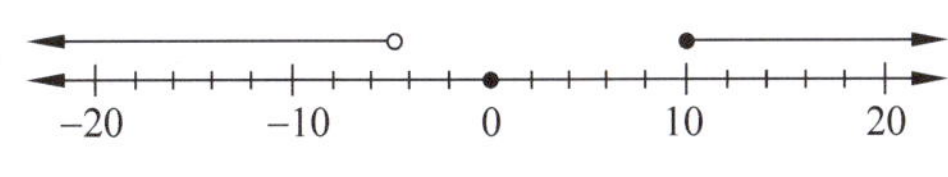

Figure 4-14

3. $2x+1<-11$ or $2x+1>11$　Original inequality.

$2x<-12$ or $2x>10$　Subtract 1 from everything.

$x<-6$ or $x>5$ Divide by 2 from everything.

4. $3x+2>92$ or $-10x-9>91$ Original inequality.

$3x>90$ or $-10x>100$

$x>3$ or $x<-10$ FLIP THE SIGNS!!

4.4.4 Absolute Value Inequalities 绝对值不等式

Absolute Value Inequality 绝对值不等式

n. [ˈæbsəˌlut ˈvælju ˌɪnɪˈkwɑlɪti]

Definition: An inequality that involves absolute values and the variable is inside the absolute value bars. 绝对值不等式是指包含绝对值的不等式(变量在绝对值符号里面)。

Notation: |(variable expression)| (inequality sign) constant.

Properties: Let x be a variable and a be a nonnegative constant, 设 x 为变量,a 为非负的常数。绝对值不等式可以改写成"同时成立的不等式组"以及"逻辑或不等式"。

1. If $|x|<a$, then $-a<x<a$ (and-inequality), as shown in Figure 4-15.
   If $|x|\leqslant a$, then $-a\leqslant x\leqslant a$ (and-inequality), as shown in Figure 4-16.

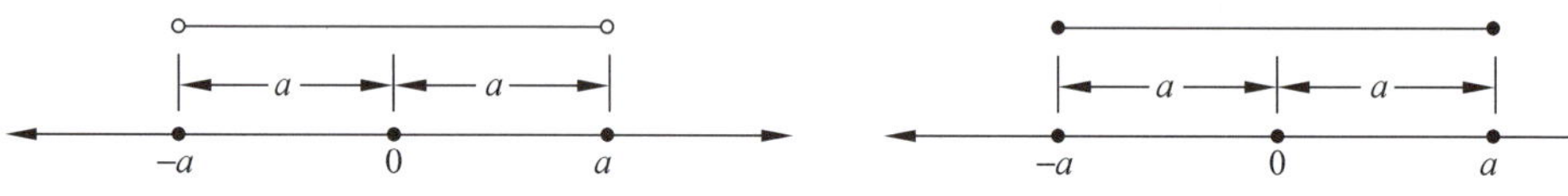

Figure 4-15 Figure 4-16

2. If $|x|>a$, then $x<-a$ or $x>a$ (or-inequality), as shown in Figure 4-17.
   If $|x|\geqslant a$, then $x\leqslant -a$ or $x\geqslant a$ (or-inequality), as shown in Figure 4-18.

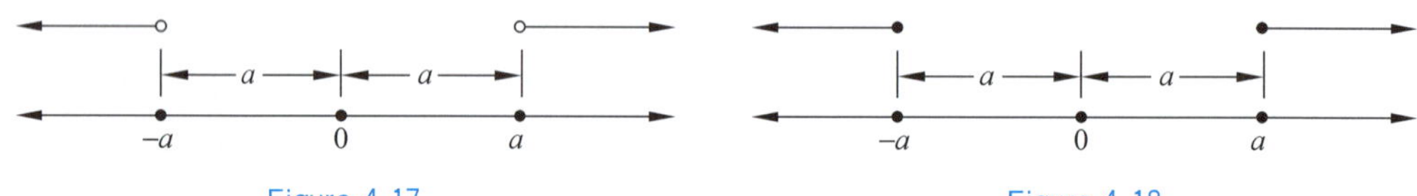

Figure 4-17 Figure 4-18

In general:

3. If |variable expression|$<a$,
   then $-a<$variable expression$<a$ (and-inequality).
   If |variable expression|$\leqslant a$,
   then $-a\leqslant$variable expression$\leqslant a$ (and-inequality).
   绝对值不等式改写成"同时成立的不等式组"后,可再次改写为"双重不等式"。
4. If |variable expression|$>a$,
   then [variable expression$<-a$ or variable expression$>a$] (or-inequality).
   If |variable expression|$\geqslant a$,
   then [variable expression$\leqslant -a$] or [variable expression$\geqslant a$] (or-inequality).
   绝对值不等式改写成"逻辑或不等式"的结果。

Questions: 1. Solve each of the following:

(1) $|2x+8|>20$

(2) $|9x-18|<81$

(3) $|-5x+20|>200$

(4) $|-2x-81|\leqslant 99$

2. In a gym, the registration fee is \$30 and the fee for each visit is \$7. Joe wants to spend \$200 on gym this month, with no more than \$50 difference in the actual expense. How many times can he go to gym this month?

Answers: 1. (1) $2x+8<-20$ or $2x+8>20$

$x<-14$ or $x>6$

(2) $-81<9x-18<81$

$-7<x<11$

(3) $-5x+20<-200$ or $-5x+20>200$

$x>44$ or $x<-36$

(4) $-99\leqslant -2x-81\leqslant 99$

$9\geqslant x\geqslant -90$

$-90\leqslant x\leqslant 9$

2. Refer to Common Application Vocabulary in Section 3.6.　见3.6节常见应用词汇。

Note that $|\text{actual cost}-\text{expected cost}|\leqslant \text{tolerance}$.

$|$实际费用－期望费用$|\leqslant$容错度。

Let x be the number of times Joe visits the gym this month.

$|(30+7x)-200|\leqslant 50$

$|7x-170|\leqslant 50$

$-50\leqslant 7x-170\leqslant 50$

$120\leqslant 7x\leqslant 220$

$17\frac{1}{7}\leqslant x\leqslant 31\frac{3}{7}$

Joe must visit gym more than 17 times and fewer than 32 times.

4.4.5 Summary　总结

1. In an and-inequality, a solution must satisfy every inequality connected by the word "and".　在"同时成立的不等式组"中,解必须满足所有被 and(且)连接的不等式。

   And-inequalities can be rewritten as double inequalities, which are easier to read and graph.　"同时成立的不等式组"可被改写为"双重不等式",这样可以更容易地理解并进行图解。

2. In an or-inequality, a solution must satisfy at least one inequality connected by the word "or".　在"逻辑或不等式"中,解必须满足至少一个被 or(或)连接的不等式。

3. An absolute value inequality can be rewritten as a compound inequality (and-inequality or or-inequality).　绝对值不等式可以被改写成复合不等式("同时成立的不等式组"或"逻辑或不等式")。

   For nonnegative constant k,　对于非负常数 k,

   - If $|(\text{variable expression})|(<\text{or}\leqslant)k$, then this is an and-inequality.　同时成立的不等式组的情况。
   - If $|(\text{variable expression})|(>\text{or}\geqslant)k$, then this is an or-inequality.　逻辑或不等式的情况。

5. Sets 集合

5.1 Introduction 介绍

Set 集合

n. [set]

Definition: A group of objects represented as a unit. The objects are called elements. 集合代表一个单元的一组对象。每个对象称为元素。

Notation: (1) Explicit: {element 1, element 2, …, element n}, this is a set of n elements. 集合用列举法表示为:{元素 1,元素 2,…,元素 n},该集合是一个含有 n 个元素的集合。

For example, {1,3,5,7,9} is a set with 5 elements.

例如,{1,3,5,7,9}集合有 5 个元素。

(2) Implicit: $\{x \mid \cdots(\text{condition})\cdots\}$, read as "the set of all x such that (the condition is true)". For example $\{x \mid x \text{ is odd and less than } 10\}$ is read as "the set of all x such that x is odd and less than 10". To be represented explicitly, this is {1,3,5,7,9}. 集合用描述法表示为:$\{x \mid \text{条件}\}$,读作"包含所有满足条件的元素 x 的集合"。例如 $\{x \mid x$ 是小于 10 的奇数$\}$,表示的是"包含所有小于 10 的奇数的集合"。用列举法表示,为{1,3,5,7,9}。

Properties: 1. Duplication does not affect the uniqueness of a set. 重复的元素不改变集合的独特性。

For example, {1,2,2,3,3,3} = {1,2,3}.

2. Ordering does not affect the uniqueness of a set. 排序不改变集合的独特性。

For example, {1,2,3} = {3,2,1}.

3. A set can have both finite and infinite number of elements. In the examples below, A has a finite number of elements, and B has an infinite number of elements. 集合的元素个数可以为有限的,亦可以为无限的。在下面的例子中,集合 A 的元素个数为有限的,集合 B 的元素个数为无限的。

$A = \{0\} = \{x \mid x \text{ is neither positive nor negative}\}$ (There are many other ways to define it implicitly, but the explicit definition, {0}, will always be the same.) 本例用列举法比用描述法更方便。

$B = \{x \mid x > 0\}$ (There are infinitely many positive numbers.) 正数的个数是无限的。

Element 元素

n. ['eləmənt]

Definition: Refers to a distinct member that makes up the set. 元素是指集合中与众不同的每个对象。

Notation：If a is an element of set A, then we write $a \in A$.

Properties：Elements do not have to be numbers.　元素不一定是数字。

Examples：Consider the following sets：

$A = \{1,2,3,3,4,5\}$

$B = \{1,2,3,a,b,c,\text{apple},\text{pear}\}$

$C = \{x \mid x \text{ is even}\}$

1. The elements of A are 1,2,3,4,and 5. Set A equals to $\{1,2,3,4,5\}$. See Property 1 of set.
2. The elements of B are $1,2,3,a,b,c$,apple,and pear.
3. Some elements of C are $0,2,4,6,8,\cdots$.

Cardinality　基数

n. [ˈkɑː. dɪnˈæl. ə. ti]

Definition：Refers to the size, or the number of distinct elements, of a set.　基数指的是集合的大小，即元素的个数。

Notation：The cardinality of set A is denoted as $|A|$.

Properties：1. For sets that have infinitely many elements, the cardinality is infinity.　对于有无穷多元素的集合，基数为无穷。

2. For the empty set, the cardinality is 0. 空集的基数为 0。

Examples：Consider the following sets：

$A = \{1,2,3,4,5\}$

$B = \{1,2,2,3,3,3,4,4,4,4,5,5,5,5,5\}$

$C = \{x \mid x \text{ is even}\}$

1. $|A| = 5$.
2. $|B| = 5$.
3. $|C| = \infty$.

Empty Set　空集

n. [emp. tiˈset]

Definition：The only set that has 0 elements. i.e. The only set whose cardinality is 0.　空集是唯一的只有 0 个元素的集合，是唯一的基数为 0 的集合。

Notation：$\varnothing$ or $\{\}$.

Properties：$|\varnothing| = 0$, and $\varnothing \neq \{0\}$.

5.2　Relationship between Two Sets　两个集合之间的关系

5.2.1　Subsets and Supersets　子集与超集

Subset and Superset　子集与超集

n. [ˈsʌbˌset] & n. [ˈsuːpəset]

Definition：If all distinct elements in set A appear in set B, then we say A is a subset of B, or B is a superset of A.　若所有在集合 A 里的不同元素都在集合 B 中出现，则 A 是 B 的子集，或者说 B 是 A 的超集。

If all distinct elements in set A appear in set B, and $|A|<|B|$, then we say A is a proper subset of B, or B is a proper superset of A. 若所有在集合 A 里的不同元素都在集合 B 中出现,且$|A|<|B|$,则 A 是 B 的真子集,或者说 B 是 A 的真超集。

We won't use the words "proper subset" or "proper superset" often. 一般很少会用到"真超集"或"真子集"。

Notation: We denote the statement "A is a subset of B" as "$A \subseteq B$". 若 A 是 B 的子集,则可写为 $A\subseteq B$。

We denote the statement "B is a superset of A" as "$B \supseteq A$". 若 B 是 A 的超集,则可写为 $B\supseteq A$。

We denote the statement "A is a proper subset of B" as "$A \subset B$". 若 A 是 B 的真子集,则可写为 $A\subset B$。

We denote the statement "B is a proper superset of A" as "$B \supset A$". 若 B 是 A 的真超集,则可写为 $B\supset A$。

Properties: 1. For sets A and B, if $A \subseteq B$ and $B \subseteq A$, then $A = B$.

2. The empty set is a subset of any given set. 空集是所有集合的子集。
Any given set is a superset of the empty set. 所有集合都是空集的超集。

3. Any given set is a subset of itself. 任何集合都是它本身的子集。
Any given set is a superset of itself. 任何集合都是它本身的超集。

Examples: 1. If $A=\{1,2,3\}$ and $B=\{1,2,3,4,5\}$, then $A \subseteq B$ because all distinct elements of A (1, 2, and 3) appear in B. Also, $B \supseteq A$.

2. If $A=\{1,2,3\}$ and $B=\{1,2,2,3,3,3\}$, then $A \subseteq B$ because all distinct elements of A (1, 2, and 3). Similarly, $B \subseteq A$. Therefore, $A = B$.

5.2.2 Mathematical Conjunctions 数学连接符

Table 5-1 is a truth table for the conjunctions "and" and "or".

Table 5-1

p	q	p and q	p or q
T	T	T	T
T	F	F	T
F	T	F	T
F	F	F	F

Table 5-2 is an extension of Table 5-1 with an additional clause.

Table 5-2

p	q	r	p and q and r	p or q or r
T	T	T	T	T
T	T	F	F	T
T	F	T	F	T
T	F	F	F	T
F	T	T	F	T
F	T	F	F	T
F	F	T	F	T
F	F	F	F	F

And　且

n. [ænd]

Definition：The mathematical conjunction such that a statement involving "and" is true if all clauses are true.　"且"是数学连接符的一种，表示当所有被"且"连接的分句都成立，由分句组成的命题才成立(是真的)。

Notation："(clause 1) and (clause 2)"，or "(clause 1)∩(clause 2)". The symbol "∩" in the second notation is the same as the intersection symbol.　"且"连接符亦可用交集符号∩表示。

Properties：In order to show that an "and" statement is false，simply find a clause that is false. Once the false clause is found，the statement is false regardless of the truth values of other clauses.　要证明一个有"且"连接符的命题是假的，找出一个假分句即可。只要找出了假分句，不论其他分句的真假值如何，该命题都是假的。

Examples：1. "$1+1=2$ and $2+2=3$" is false because $2+2=3$ is false.

2. "$1+1=3$ and $2+2=3$" is false because $1+1=3$ is false.

3. "$1+1=2$ and $2+2=4$" is true because both clauses are true.

Or　或

n. [ɔr]

Definition：The mathematical conjunction such that a statement involving "or" is true if at least one clause is true.　"或"是数学连接符的一种，表示当至少一个被"或"连接的分句成立，由分句组成的命题就是真的。

Notation："(clause 1) or (clause 2)"，or "(clause 1)∪(clause 2)". The symbol "∪" in the second notation is the same as the union symbol.　"或"连接亦可用并集符号∪表示。

Properties：1. In order to show that an "or" statement is true，simply find a clause that is true. Once the true clause is found，the statement is true regardless of the truth values of the other clauses.　要证明一个带有"或"连接符的命题是真的，找出一个真分句即可。只要找出了真分句，不论其他分句的真假值如何，该命题都是真的。

2. In life，the definition of "or" differs from that of in math. Suppose the father asks the son "Which dessert do you want? Ice cream or cheesecake?" Presumably the son will answer one of "ice cream" or "cheesecake" in life. However，in mathematics，saying "I want both" is acceptable because in a true statement that involves "or"，it is possible that all clauses are true，provided that at least one clause is true. In life，the word "or" is exclusive，but in math it is inclusive.　在生活中，常说的"A 或 B"是指在 A 与 B 两者之间只取其一，但在数学中表述的"A 或 B"是指在 A 和 B 两者之间至少取其一。

Examples：1. "$1+1=2$ or $2+2=3$" is true because $1+1=2$ is true.

2. "$1+1=3$ or $2+2=3$" is false because neither of the clauses is true.

3. "$1+1=2$ or $2+2=4$" is true because at least one clause is true. In this case，both clauses are true.

5.2.3　Intersections and Unions　交集与并集

Figure 5-1 shows the relationship of two sets，A and B. The parts are labeled using the intersection and union symbols. 图 5-1 用交集和并集符号展示了 A 和 B 两个集合的关系。

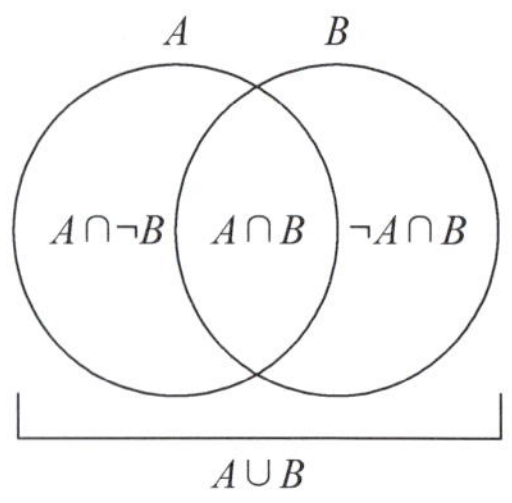

"¬" denotes the symbol for negation: $\neg A$ means the set of elements that are not in set A.

"¬"是否定符号。$\neg A$ 表示包含所有不在集合 A 中的元素的集合。

A B

$A\cap\neg B$ $A\cap B$ $\neg A\cap B$

$A\cup B$

Figure 5-1

Intersection 交集

n. [ˈɪntərˌsekʃən]

Definition: The intersection of two sets is the set that only contains all elements that appear in both sets. 两个集合的交集是指仅包含均在两个集合中出现过的所有元素的集合。

Notation: The intersection of sets A and B is denoted by $A\cap B$.

Properties: For sets A and B:

1. If A and B have no common element, then $A\cap B=\varnothing$.
2. If $A\subseteq B$, then $A\cap B=A$.

Examples: Consider the following sets:

$A=\{1,2,3,4,5\}$

$B=\{1,3,5,7,9\}$

$C=\{2,4,6,8,10\}$

$D=\{1,2,3,4,5,6,7,8,9,10\}$

We have:

$A\cap B=\{1,3,5\}$; $A\cap C=\{2,4\}$;

$B\cap C=\varnothing$; $A\cap D=\{1,2,3,4,5\}=A$;

$B\cap D=\{1,3,5,7,9\}=B$; $C\cap D=\{2,4,6,8,10\}=C$;

Note that A, B, and C are subsets of D.

Union 并集

n. [ˈjunjən]

Definition: The union of two sets is the set that contains all (distinct) elements from at least one set. 两个集合的并集是指包含两个集合中任意一个集合所有不同元素的集合。

Notation: The union of sets A and B are denoted by $A\cup B$.

Properties: For sets A and B:

1. $|A\cup B|=|A|+|B|-|A\cap B|$.
2. If $A\subseteq B$, then $A\cup B=B$.
3. If A and B have no common element, then $|A\cap B|=0$, and $|A\cup B|=|A|+|B|$.

Examples: Consider the following sets:

$A=\{1,2,3,4,5\}$

$B=\{1,3,5,7,9\}$

$C=\{2,4,6,8,10\}$

$D=\{1,2,3,4,5,6,7,8,9,10\}$

We have:

$A \cup B = \{1,2,3,4,5,7,9\}$; $A \cup C = \{1,2,3,4,5,6,8,10\}$;

$B \cup C = \{1,2,3,4,5,6,7,8,9,10\} = D$;

$A \cup D = \{1,2,3,4,5,6,7,8,9,10\} = D$;

$B \cup D = \{1,2,3,4,5,6,7,8,9,10\} = D$;

$C \cup D = \{1,2,3,4,5,6,7,8,9,10\} = D$;

Note that A, B, and C are subsets of D.

From this example, it is clear that Properties 1-3 hold. See intersection for its computation on the same sets A, B, C, and D.

5.2.4 Summary 总结

1. See subsets and supersets. 见子集和超集。
2. To satisfy a statement with the conjunction "and", all clauses connected by "and" must be satisfied. "and" is an analogue of set intersection. 要满足一个有连接符 and(和)的命题,该命题中所有分句都必须被满足。and 与交集符号类似。
3. To satisfy a statement with the conjunction "or", at least one clause connected by "or" must be satisfied. "or" is an analogue of set union. 要满足一个有连接符 or(或)的命题,该命题中至少一个分句要被满足。or 与并集符号类似。
4. For any sets A and B, $|A \cup B| = |A| + |B| - |A \cap B|$.

   $|A \cup B|$: How many distinct elements are there in the union of the two sets?

   $|A|$: How many distinct elements are there in set A?

   $|B|$: How many distinct elements are there in set B?

   $|A \cap B|$: How many distinct elements are there in the intersection of the two sets? Or, how many elements have been counted twice?

   If A and B do not have any element in common, then $|A \cap B| = 0$ and $|A \cup B| = |A| + |B|$.

6. Coordinate Plane/Cartesian Plane 坐标平面

6.1 Introduction 介绍

Coordinate Plane/Cartesian Plane 坐标平面

n. [koʊˈɔrdənət ˌpleɪn/kɑːrˈti.ʒən ˌpleɪn]

Definition: The plane determined by a horizontal number line, called the x-axis, and a vertical number line, called the y-axis. The x-axis and y-axis are perpendicular to each other and intersecting at a point called the origin. 坐标平面是在同一平面上由横数轴(x 轴)与纵数轴(y 轴)确定的平面。x 轴与 y 轴相互垂直,且交于坐标原点。

Notation：Figure 6-1 shows the axes, quadrants, and the origin of the coordinate plane.
图 6-1 展示了坐标轴、象限和坐标原点。

Properties：1. The x-axis and y-axis divide the coordinate plane into four parts, called the quadrants. x 轴和 y 轴把坐标平面平分成 4 部分，每部分叫作象限。

2. Each point on the coordinate plane can be specified by an ordered pair of numbers, as shown in Figure 6-2. 坐标平面上的每个点都能用有序对表示，如图 6-2 所示。

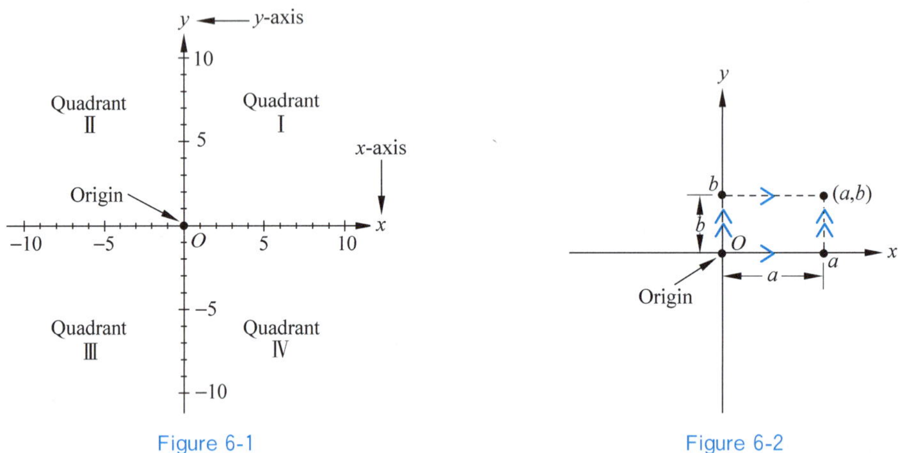

Figure 6-1　　Figure 6-2

6.1.1 Coordinate Axes 坐标轴

Coordinate Axis 坐标轴

n. [koʊ'ɔrdənˌeɪt 'æksəs]

Definition：Refer to the x-axis or the y-axis. 坐标轴一般指 x 轴或 y 轴。

x-Axis x 轴

n. [eks 'æksəs]

Definition：The horizontal number line of the coordinate plane, as shown in Figure 6-3 in blue. x 轴是指坐标平面的横轴，如图 6-3 蓝色所示。

Phrases：positive～ 正 x 轴，negative～ 负 x 轴

y-Axis y 轴

n. [waɪ 'æksəs]

Definition：The vertical number line of the coordinate plane, as shown in Figure 6-4 in blue. y 轴是指坐标平面的纵轴，如图 6-4 蓝色所示。

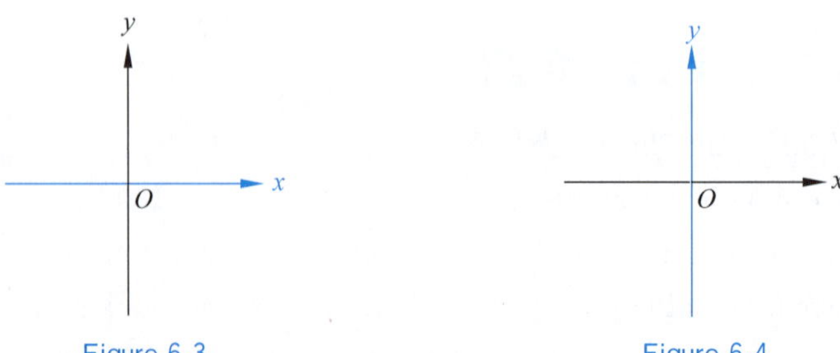

Figure 6-3　　Figure 6-4

Phrases: positive～ 正 y 轴, negative～ 负 y 轴

6.1.2 Coordinates 坐标

Coordinate 坐标

n. [koʊ'ɔrdənˌeɪt]

Definition: Refer to the x-coordinate or the y-coordinate. 坐标是指 x 坐标或 y 坐标。

The coordinates of a point is represented by an ordered pair. 每个点的坐标都可以用有序对表示。

x-Coordinate x 坐标

n. [eks koʊ'ɔrdənˌeɪt]

Definition: Indication of a point's horizontal distance to the origin. x 坐标表示一个点离坐标原点的横向距离。

As shown in Figure 6-5, the x-coordinate of point P is a.

Phrases: positive～ 正 x 坐标, negative～ 负 x 坐标

y-Coordinate y 坐标

n. [waɪ koʊ'ɔrdənˌeɪt]

Definition: Indication of a point's vertical distance to the origin. y 坐标表示一个点离坐标原点的纵向距离。

As shown in Figure 6-6, the y-coordinate of point P is b.

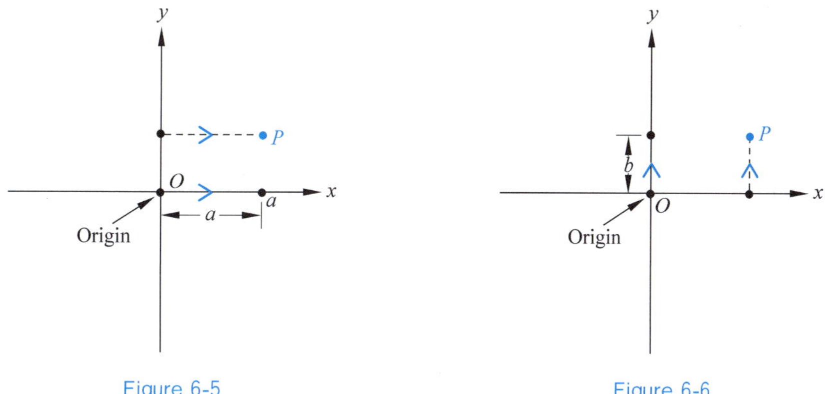

Figure 6-5 Figure 6-6

Phrases: positive～ 正 y 坐标, negative～ 负 y 坐标

6.1.3 Ordered Pairs 有序对

Ordered Pair 有序对

n. ['ɔːr.dɚd peər]

Definition: The representation of the location (as a point) on the coordinate plane. 有序对用于表示点在坐标平面上的位置。

Notation: (x, y), in which x represents the x-coordinate of the point and y represents the y-coordinate of the point. 在(x, y)中, x 代表的是点的 x 坐标, y 代表的是点的 y 坐标。

If point A is located at (a,b), then we also say A has coordinates of (a,b). 若 A 的位置为(a,b)，也就是说 A 的坐标为(a,b)。

Properties: 1. Each ordered pair (a,b) on the coordinate plane is located in the intersection of the vertical line $x=a$ and horizontal line $y=b$. 平面坐标上的有序对(a,b)位于垂直线 $x=a$ 的水平线 $y=b$ 的交点处。

2. Points along the x-axis have y-coordinates equal to 0. 在 x 轴上的点的 y 坐标均为 0。

3. Points along the y-axis have x-coordinates equal to 0. 在 y 轴上的点的 x 坐标均为 0。

4. The origin has coordinates of $(0,0)$. 坐标原点的坐标为$(0,0)$。

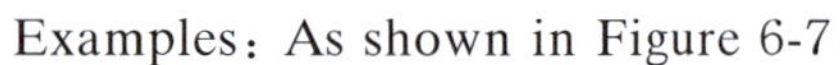

Examples: As shown in Figure 6-7:

The coordinates of point A are $(2,3)$.

The coordinates of point B are $(4,-3)$.

The coordinates of point C are $(-2.5,-1)$.

The coordinates of point D are $(-1,1)$.

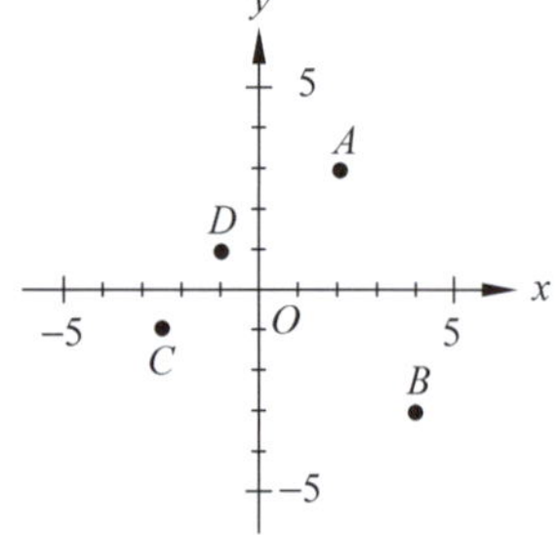

Figure 6-7

6.1.4 Points 点

Point 点

n. [pɔɪnt]

Definition: The visual representation on the coordinate plane of an ordered pair. 点是有序对在坐标平面中的表示。

Origin 坐标原点

n. ['ɔrədʒɪn]

Definition: The intersection of the coordinate axes, as shown in Figure 6-8 in blue. The origin has $(0,0)$ as its coordinates. 坐标原点是指坐标轴的交点，见图 6-8 中蓝色圆点。坐标原点的坐标为$(0,0)$。

Notation: The origin is denoted by the letter O on the coordinate plane. 坐标原点以字母 O 表示。

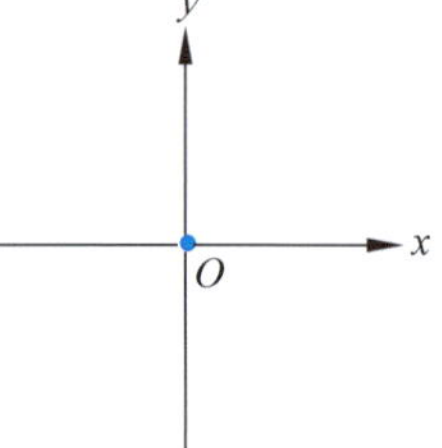

Figure 6-8

6.1.5 Quadrants 象限

Quadrant 象限

n. ['kwɑdrənt]

Definition: One of the four parts in the coordinate plane that are divided by the coordinate axes. 坐标平面上被坐标轴划分的 4 部分其中之一称为象限。

Properties: Each quadrant has its number, as shown in Figure 6-9. 每个象限都有编号，见图 6-9。

Points' coordinates in each quadrant have properties of signs, as described below:

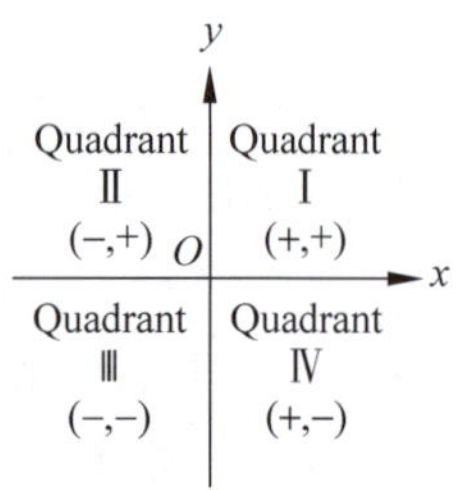

Figure 6-9

每个象限里点的坐标都有正负，描述如下：

- In Quadrant Ⅰ, both x and y coordinates are positive.　在象限Ⅰ，x 坐标和 y 坐标均是正的。
- In Quadrant Ⅱ, the x-coordinate is negative and the y-coordinate is positive.　在象限Ⅱ，x 坐标是负的，y 坐标是正的。
- In Quadrant Ⅲ, both x and y coordinates are negative.　在象限Ⅲ，x 坐标和 y 坐标均是负的。
- In Quadrant Ⅳ, the x-coordinate is positive and the y-coordinate is negative.　在象限Ⅳ，x 坐标是正的，y 坐标是负的。

Coordinate axes do not belong to any quadrant.　坐标轴不属于任何象限。

6.1.6　Summary　总结

1. The coordinate plane is the plane determined by a horizontal number line, called the x-axis, and a vertical number line, called the y-axis. The x-axis and y-axis are perpendicular to each other and intersecting at a point called the origin.　坐标平面是在同一平面上由横数轴（x 轴）与纵数轴（y 轴）确定的平面。x 轴和 y 轴相互垂直，且交于坐标原点。
2. The location/coordinates of point in the coordinate plane can be represented by an ordered pair (x, y): x is the x-coordinate (horizontal distance from that point to the origin), and y is the y-coordinate (vertical distance from that point to the origin). The origin has the coordinates $(0, 0)$.　坐标平面上点的位置/坐标可用有序对（x，y）表示：x 代表 x 坐标（点到坐标原点的横向距离），y 代表 y 坐标（点到坐标原点的纵向距离）。坐标原点的坐标为（0，0）。
3. The coordinate axes divide the coordinate planes into four quadrants. Points in the quadrants have the following sign property:

   坐标轴把坐标平面分成 4 个象限。在象限中的点有以下正负符号的性质：

   Quadrant Ⅰ: $(+, +)$

   Quadrant Ⅱ: $(-, +)$

   Quadrant Ⅲ: $(-, -)$

   Quadrant Ⅳ: $(+, -)$.

   Points along the coordinate axes do not belong to any quadrant.　在坐标轴上的点不属于任何象限。

6.2　Formulas　公式

Midpoint Formula　中点公式

n. [ˈmɪdˌpɔɪnt ˈfɔrmjələ]

Definition: Given two points A and B, the Midpoint Formula gives the midpoint of AB.　中点公式是指给出两个点连成的线段中点的式子。

Notation: Suppose point A has the coordinates (x_1, y_1) and point B has the coordinates (x_2, y_2), the Midpoint Formula is $M = \left(\frac{x_1+x_2}{2}, \frac{y_1+y_2}{2}\right)$, for which $AM = BM$, as shown in Figure 6-10.

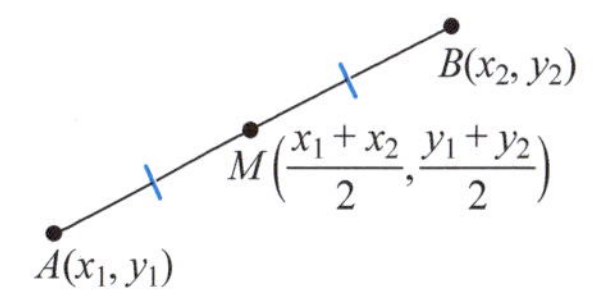

Figure 6-10

Properties：1. Note that the x and y coordinates of M are independent of each other. 计算中点时，横坐标与纵坐标相互独立。

2. Switching points (x_1, y_1) and (x_2, y_2) leads to the same result. This is due to the commutative property of addition. 根据加法交换律，把两个点互换后代入公式，结果相同。

Questions：What is the coordinates of the midpoint of AB for each of the following if：

(1) $A=(4,7)$ and $B=(14,15)$

(2) $A=(-5,-8)$ and $B=(-11,-20)$

(3) $A=(-11,-10)$ and $B=(51,60)$

Answers：(1) $\left(\frac{14+4}{2},\frac{15+7}{2}\right)=(9,11)$.

(2) $\left(\frac{-5+(-11)}{2},\frac{-8+(-20)}{2}\right)=(-8,-14)$.

(3) $\left(\frac{-11+51}{2},\frac{-10+60}{2}\right)=(20,25)$.

Distance Formula 距离公式

n. [ˈdɪstəns ˈfɔrmjələ]

Definition：Given two points A and B, the Distance Formula gives the length of AB. 距离公式是指给出两个点连成的线段长度的式子。

Notation：Suppose point A has the coordinates (x_1, y_1) and point B has the coordinates (x_2, y_2), the Distance Formula is $\sqrt{(x_2-x_1)^2+(y_2-y_1)^2}$. Sometimes it is also written in $\sqrt{(\Delta x)^2+(\Delta y)^2}$, for which Δx denotes the change in x and Δy denotes the change in y. Both Δx and Δy are nonnegative.

The illustration is shown in Figure 6-11.

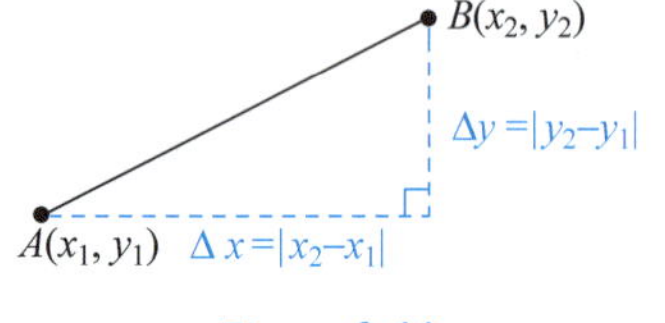

Figure 6-11

Properties：1. This formula is based on the Pythagorean Theorem. Given a line segment on the coordinate plane, we can always build a right triangle and use that line segment as the hypotenuse, such that the legs are parallel to the coordinate axes. Then, we use Pythagorean Theorem to find the length of that line segment, as shown above. 距离公式源于勾股定理。在坐标平面给出一条线段，可以以那条线段为斜边画一个直角三角形(直角边与坐标轴平行)，并用勾股定理求出那条线段长度，如上所述。

If AB is parallel to a coordinate axis, then finding its length is trivial, as one of Δx or Δy is 0. 若 AB 平行于一条坐标轴，则为求它的长度的特殊情况：Δx 或 Δy 为 0。

2. Switching points (x_1, y_1) and (x_2, y_2) leads to the same result. This is due to the fact that the squares of a number and its opposite are equal: $a^2=(-a)^2$. 把两个点互换后代入公式，结果相同，原因是一个数的平方与它相反数的平方相等。

Note that the following formulas are equivalent.

下面的公式是等价的：

$$AB=\sqrt{(x_2-x_1)^2+(y_2-y_1)^2}.$$

$$AB=\sqrt{|x_2-x_1|^2+|y_2-y_1|^2}.$$

$$AB=\sqrt{(\Delta x)^2+(\Delta y)^2}.$$

Questions：What is the distance of $\overline{AB}$ for each of the following if：

(1) (4,1) and (7,5)

(2) (−20，−16) and (−15，−4)

(3) (−2，−8) and (6,7)

Answers：(1) $\sqrt{(7-4)^2+(5-1)^2}=\sqrt{3^2+4^2}=5$.

(2) $\sqrt{(-15-(-20))^2+(-4-(-16))^2}=\sqrt{5^2+12^2}=13$.

(3) $\sqrt{(6-(-2))^2+(7-(-8))^2}=\sqrt{8^2+15^2}=17$.

7. Relations and Functions　多值函数与单值函数

7.1 Introduction　介绍

Figure 7-1 shows the relationship between relations and functions.
图 7-1 展示了多值函数与单值函数的关系。

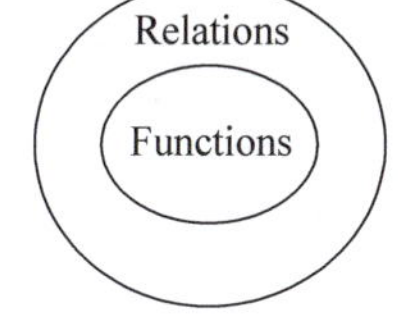

Figure 7-1

7.1.1 Relations　多值函数

Relation　多值函数

n. [rɪˈleɪʃən]

Definition：A relation from set X to set Y is a collection of ordered pairs. Each ordered pair has the form of (x,y) for which x is an element of X and y is an element of Y. In the ordered pair (x,y), x and y are said to be related：x maps to y in the relation.　从集合 X 到集合 Y 的多值函数是一个有序对组。每个有序对以(x,y)表示，其中 x 是 X 的元素，y 是 Y 的元素。在有序对(x,y)中，x 和 y 有关联，即在多值函数中 x 映射到 y。

Properties：1. Not all the elements in set X are involved in the relation.　多值函数不一定会涉及所有在集合 X 中的元素。

2. Not all the elements in set Y are involved in the relation.　多值函数不一定会涉及所有在集合 Y 中的元素。

3. One element from set X can relate to two or more elements from set Y.　在集合 X 中的一个元素可以同时映射两个或更多个在集合 Y 中的元素。

4. One element from set Y can be mapped from two or more elements from set X.　在

集合 Y 中的一个元素可以同时被两个或更多个在集合 X 中的元素映射。

5. Functions are special examples of relations. 单值函数是特殊的多值函数。

Examples: 1. In the relation from set X to set Y, as shown in Figure 7-2, we know that:

a (from X) maps to v (from Y).

b (from X) maps to w (from Y).

c (from X) maps to x (from Y).

d (from X) maps to y (from Y).

e (from X) maps to z (from Y).

2. In the relation from set X to set Y, as shown in Figure 7-3, we know that:

a (from X) maps to v (from Y).

Both b and d (from X) map to w (from Y).

c (from X) maps to both x and y (from Y).

e (from X) maps to z (from Y).

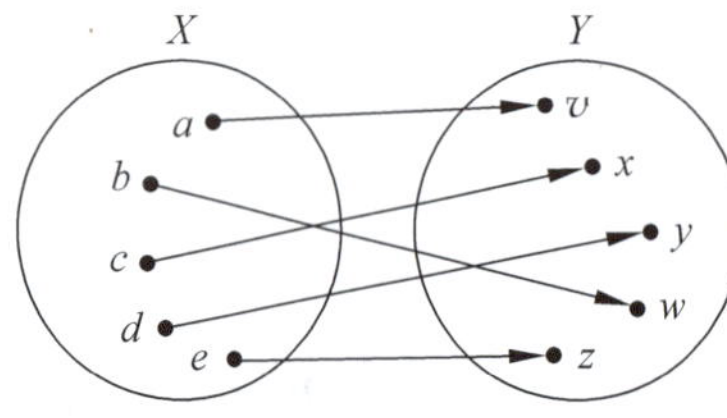

Figure 7-2

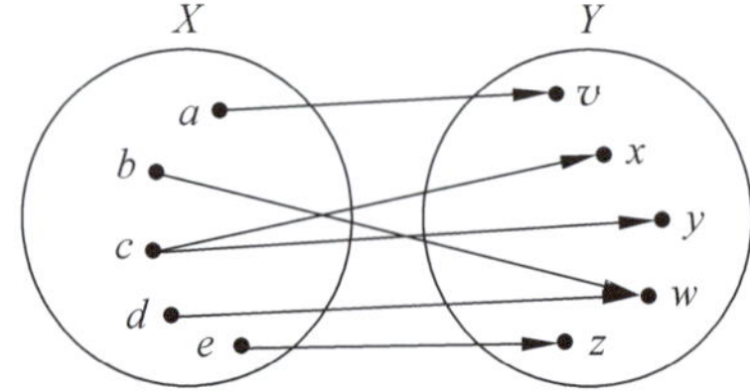

Figure 7-3

7.1.2 Functions 单值函数

Function 单值函数

n. [ˈfʌŋkʃən]

Definition: A function from set X to set Y is a relation with the additional characteristics in the Properties section below. X is called the domain and Y is called the range. 从集合 X 到集合 Y 的单值函数是一个具有附加性质的多值函数(见下面单值函数的性质)。X 是定义域,Y 是值域。

A function is a special relation:

单值函数是特殊的多值函数:

- A function must be a relation. 单值函数必为多值函数。
- A relation is not necessarily a function. 多值函数不一定为单值函数。

Notation: A function from set X to set Y is sometimes called "Y is a function of X". We use the notation f to denote a function, for which $f(x)$ is the element from Y for which x is mapping to. Expressions $f(x)$ and y can be used interchangeably, but $f(x)$ represents the function notation. 从集合 X 到集合 Y 的单值函数有时候被称为"Y 是 X 的单值函数",一般记作 f。$f(x)$代表 x 映射到集合 Y 中的元素。$f(x)$和 y 可以互用,但 $f(x)$是函数表示法。

Notes: "f" is the name of the function. f 是单值函数的名称。

"$f(x)$" is the element from the range, for which x (from the domain) is mapping to. $f(x)$是值域的元素，是 x(在定义域中)映射的元素。

"$f(x)$" is the output of the function f at the input of x. $f(x)$是单值函数 f 输入 x 的输出。

Properties: 1. Each element from X must be mapped to one element in Y. 每个在 X 中的元素必须映射 Y 中的一个元素。

2. Some elements in Y may not be mapped from with any element in X. 某些在 Y 中的元素不一定被 X 中的元素映射。

3. Two or more elements in X may be mapped to the same element in Y. 两个或更多个在 X 中的元素可以同时映射 Y 中的同一个元素。

4. An element in X cannot be mapped to two or more different elements in Y. 在 X 中的一个元素不能同时映射 Y 中的两个或更多个元素。

Examples: 1. Example 1 of relation is a function from set X to set Y because it satisfies all properties of a function. 多值函数词条中的例 1 是单值函数，因为它满足单值函数的所有性质。

2. Example 2 of relation is not a function from set X to set Y because the element c from X matches with two different elements in Y (x and y), which violates Property 4. 多值函数词条中的例 2 不是单值函数，因为集合 X 中的 c 同时映射集合 Y 中的 x 和 y。

Phrases: one-to-one～　一一对应函数，onto～　映成函数，linear～　一次函数，quadratic～　二次函数，cubic～　三次函数，quartic～　四次函数，polynomial～　多项式函数，exponential～　指数函数，logarithmic～　对数函数，trigonometric～　三角函数

Questions: 1. Determine whether each of the following is a function from set X to set Y in Tables 7-1 to 7-3.

(1)

Table 7-1

X	Y
1	80
2	70
3	60
4	50

(2)

Table 7-2

X	Y
1	80
2	70
2	60
4	50

(3)

Table 7-3

X	Y
1	80
2	80
3	80
4	80

2. In the function shown in Table 7-4, what are the values of each of the following?

(1) $f(1)$

(2) $f(2)$

(3) $f(3)$

Table 7-4

x	$f(x)$
1	5
2	7
3	9

Answers: 1. (1) Yes.

(2) No, because 2 from X maps to both 70 and 60 from Y.

(3) Yes.

2. (1) $f(1)=5$.

(2) $f(2)=7$.

(3) $f(3)=9$.

Vertical Line Test 垂线测试

n. [ˈvɜrtɪkəl lɑɪn test]

Definition: A visual way to determine whether a relation is a function by looking at its graph. 垂线测试是从视觉上判断多值函数是否为单值函数的方法：观察图像。

Properties: If two or more points of the same graph lies in the same vertical line, then the relation is not a function. 若图像中两个或更多的点在同一条垂直线上，则这个多值函数不是单值函数。

Notes: If two or more points of the same graph lies in the same vertical line, then the graph violates Property 4 of a function: any element from X (the domain) cannot be mapped to two or more different elements from Y (the range). 若图像中两个或更多的点在同一条垂直线上，则这个多值函数违反单值函数的性质 4：在 X(定义域)里的一个元素不能同时映射 Y(值域)的两个或更多个元素。

Examples: 1. The relation in Figure 7-4 is a function because no vertical line passes through two or more points of its graph. 图 7-4 的多值函数为单值函数，因为没有垂直线通过图像中两个或更多的点。

2. The relation in Figure 7-5 is not a function because there is a vertical line (shown in

blue) passes through two points of its graph. 图 7-5 的多值函数不是单值函数,因为有垂直线通过图像中两个点,如蓝色所示。

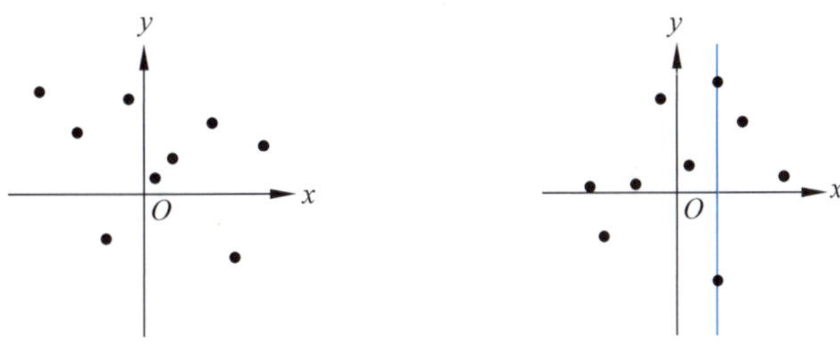

Figure 7-4　　Figure 7-5

Questions: Use the vertical line test to determine whether each of the following relations in Figures 7-6 and 7-7 is a function:

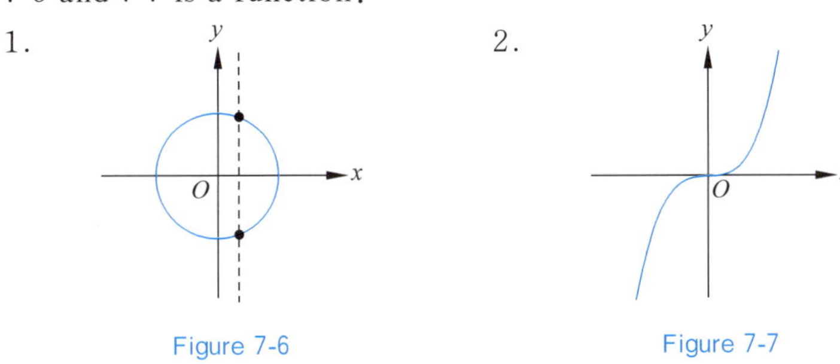

Figure 7-6　　Figure 7-7

Answers: 1. The relation is not a function because there is a vertical line (dashed) passes through two points, as shown in Figure 7-8.

2. The relation is a function because no vertical line passes through two or more points.

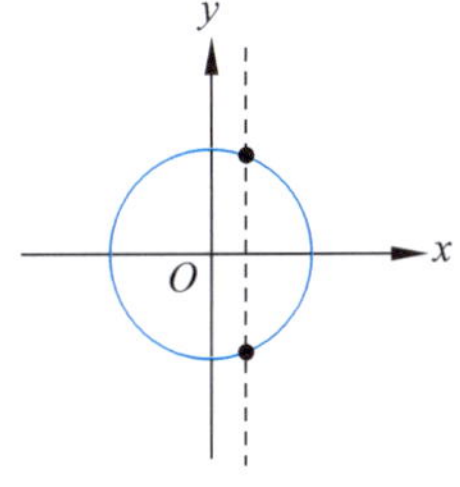

Figure 7-8

7.1.3 Summary 总结

1. A relation from set X to set Y is a collection of ordered pairs. Each ordered pair has the form of (x, y) for which x is an element of X and y is an element of Y. In the ordered pair (x, y), x and y are said to be related: x maps to y in the relation. 从集合 X 到集合 Y 的多值函数是一个有序对组。每个有序对以(x, y)表示,其中 x 是X 的元素,y 是 Y 的元素。在有序对(x, y)中,x 和y 有关联,即在多值函数中 x 映射到y。

   Review properties of relation in Section 7.1.1. 见 7.1.1 节多值函数的性质。

2. A function is a special relation with additional properties. 单值函数是含有附加性质的特殊多值函数。

   Review properties of function in Section 7.1.2 for these additional properties. 附加性质见 7.1.2 节单值函数的性质。

3. A function has to be a relation, but a relation does not have to be a function. 单值函数必为多值函数,但多值函数不一定是单值函数。

4. To test whether a relation is a function, use the additional properties of functions. The vertical line test is an efficient way to determine whether a relation is a function by looking at its graph. 要判断一个多值函数是否为单值函数,可参见单值函数的附加性质。垂线测试是通过观察多值函数的图像,有效地判断多值函数是否为单值函数的方法。

7.2 Basics about Functions 单值函数的基础

7.2.1 Input and Output 输入与输出

Input and Output 输入与输出

n. [ˈɪnˌpʊt] & n. [ˈaʊtˌpʊt]

Definition: In a function f from set X to set Y, the input x is an element from X, and the output y is the element from Y such that $f(x)=y$. 从集合 X 到集合 Y 的单值函数 f 中, X 中的元素 x 为输入,使得 $f(x)=y$ 的 Y 中的元素 y 为输出。

The input is sometimes called the independent variable. 输入也叫作自变量。

The output is sometimes called the dependent variable. 输出也叫作因变量。

The value of the output depends on the value of the input. 输出的值取决于输入的值。

Properties: In $f(a)=b$, a is known as the input, and $f(a)$, or b, is known as the output of the function f at input a. 在 $f(a)=b$ 中, a 为输入, b 或 $f(a)$ 为单值函数 f 在输入为 a 时的输出。

Examples: 1. Same as example 1 from relation. 同多值函数的例 1。

As shown in Figure 7-9 (same as Figure 7-2):

When the input is a, the output is v.

When the input is b, the output is w.

When the input is c, the output is x.

When the input is d, the output is y.

When the input is e, the output is z.

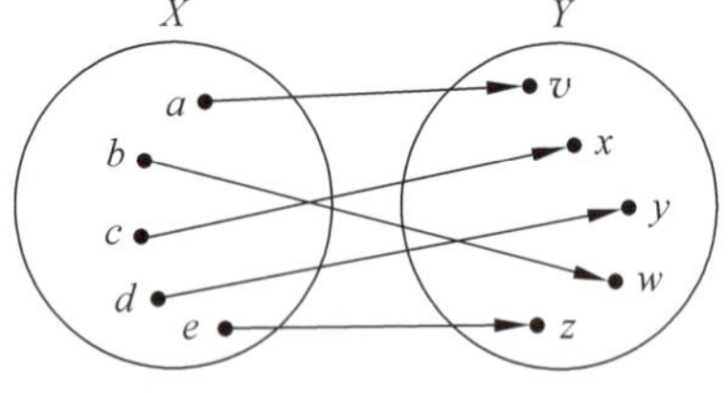

Figure 7-9

2. Same as Question 1(3) of function in Section 7.1.2. 同 7.1.2 节单值函数的习题 1(3)。

As shown in Table 7-5:

When the input is 1, the output is 80.

When the input is 2, the output is 80.

When the input is 3, the output is 80.

When the input is 4, the output is 80.

Table 7-5

X	Y
1	80
2	80
3	80
4	80

Independent Variable and Dependent Variable　自变量与因变量

n. [ˈɪndɪˈpendənt ˈveəriəbəl]

Definition：See input and output.　见输入与输出。

Domain and Range　定义域与值域

n. [doʊˈmeɪn] & n. [reɪndʒ]

Definition：The domain is the set of all input values，and the range is the set of all output values. 定义域是包含所有输入的集。值域是包含所有输出的集。

Examples：1. Same as example 1 from relation.　同多值函数的例1。

As shown in Figure 7-10 (same as Figures 7-2 and 7-9)：

Domain：$X=\{a,b,c,d,e\}$.

Range：$Y=\{v,x,w,y,z\}$.

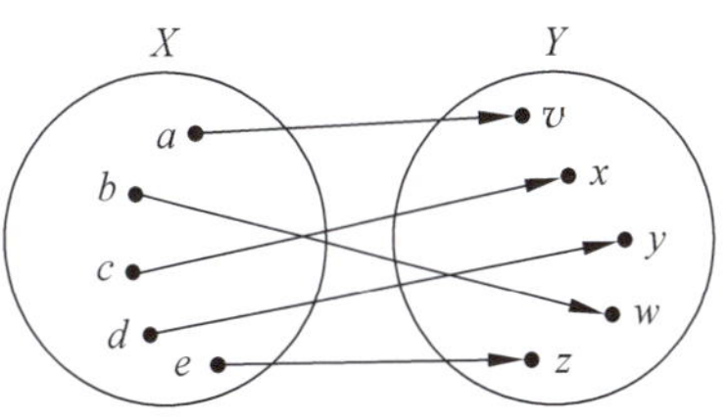

Figure 7-10

2. Same as Question 1(3) of function in Section 7.1.2. 同7.1.2节单值函数的习题1(3)。

As shown in Table 7-6 (same as Table 7-5)：

Domain：{1,2,3,4}.

Range：{80}.

Table 7-6

X	Y
1	80
2	80
3	80
4	80

7.2.2　Formulas　公式

Formula　公式

n. [ˈfɔrmjələ]

Definition：When the input and output of a function behave in a predictable pattern，a formula is a statement that illustrates this pattern. Some well-known patterns include polynomial functions，logarithmic functions，and exponential functions.　当单值函数的输入和输出呈现出规律时，函数公式能表达这种规律。一些常见的包括多项式函数、对数函数、指数函数。

Notation：Formula can be a statement in words or in an equation that contains variables.　函数公式带有变量，既可以用文字表达，也可以用等式表达。

Examples：In each of the following，y is a function of x.　下面各例中，y 是有关 x 的函数。

1. As shown in Table 7-7，the formula is $y=x+5$. Or，"y is 5 more than x".

In function notation，this is $f(x)=x+5$.

Table 7-7

x	1	2	3	4	5	6	7	8
y	6	7	8	9	10	11	12	13

2. As shown in Table 7-8, the formula is $y = x^2$. Or, "y is the square of x", or "y is x multiplied by itself".

   In function notation, this is $f(x) = x^2$.

Table 7-8

x	1	2	3	4	5	6	7	8
y	1	4	9	16	25	36	49	64

3. As shown in Table 7-9, the formula is $y = x^3 + 1$.

   In function notation, this is $f(x) = x^3 + 1$.

Table 7-9

x	1	2	3	4	5	6	7	8
y	2	9	28	65	126	217	344	513

4. As shown in Table 7-10, the formula is $y = 2^x$.

   In function notation, this is $f(x) = 2^x$.

Table 7-10

x	1	2	3	4	5	6	7	8
y	2	4	8	16	32	64	128	256

7.2.3 Intercepts 截距

Intercept 截距

n. [ˌɪntər'sept]

Definition: The number on a coordinate axis in which the graph (of the function) intersects that axis. 截距是指在坐标平面上单值函数的曲线与坐标轴相交的坐标值。

Non-function relations have intercepts as well. Same definition applies. 非单值函数的多值函数也有截距。

See also x-intercept and y-intercept. 见 x 截距和 y 截距。

Examples: As shown in Figure 7-11:

The x-intercept of the graph of f is 2. 函数 f 曲线的 x 截距为 2。

The y-intercept of the graph of f is -2. 函数 f 曲线的 y 截距为 -2。

x-Intercept x 截距

n. [eks ˌɪntər'sept]

Definition: 1. The x-coordinate of the point of intersection of a graph and the x-axis. 曲线与 x 轴

的交点的 x 坐标。

2. A value for x for which $f(x)=0$. 当 $f(x)=0$ 时,x 的值。

Both definitions are equivalent, but they are phrased significantly differently. Definition 1 is based on geometry, and Definition 2 is based on algebra. 上面两个定义等价,但它们表达的方式不同。定义 1 根据几何方法得出,定义 2 根据代数方法得出。

As shown in Figure 7-12, the x-intercept of the graph of f is k. 如图 7-12 所示,函数 f 图像的 x 截距为 k。

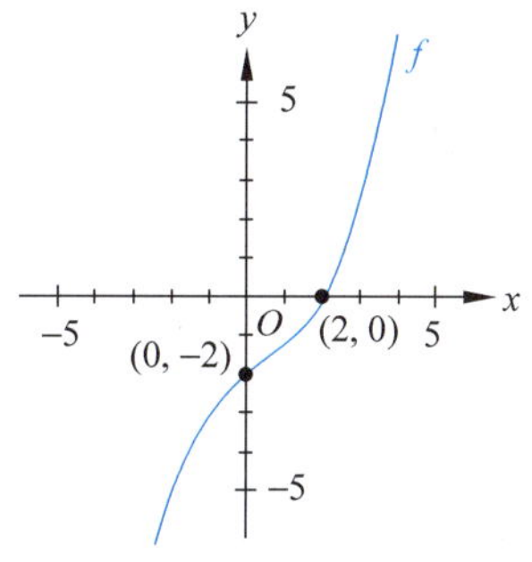

Figure 7-11

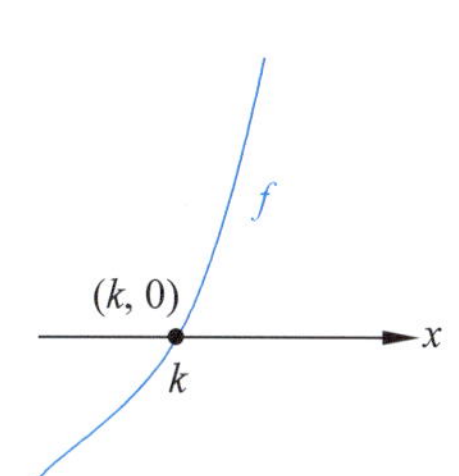

Figure 7-12

Non-function relations have x-intercepts as well. Same definition applies. 非单值函数的多值函数也有 x 截距。定义同上。

Properties: 1. The graph of a function can have 0, 1, or more x-intercepts because two or more elements in the domain may be mapped to the same element (in this case, 0) in the range. 单值函数可以有 0、1 或更多个 x 截距,因为在定义域里的两个或更多的元素可以映射到值域的同一个元素(值域的 0)。

2. If a function f has k as an x-intercept, then:
若函数 f 的一个 x 截距为 k,则

- The graph of f passes through $(k, 0)$. (Definition 1)
  f 的曲线经过$(k, 0)$。
- $f(k)=0$. (Definition 2)

3. Apparently, the x-intercept(s) of a function f is the same as the root(s) of the equation $f(x)=0$. 显然,单值函数 f 的 x 截距与 $f(x)=0$ 的根相同。

4. Note that the x-intercept is a number, not a point. x 截距是一个数值,不是一个点。

Examples: 1. A function can have 0, 1, or more x-intercepts. 单值函数可以有 0、1 或更多个 x 截距。

(1) Figure 7-13 shows the graph of a function with zero x-intercepts.

(2) Figure 7-14 shows the graph of a function with one x-intercept.

(3) Figure 7-15 shows the graph of a function with three x-intercepts.

etc.

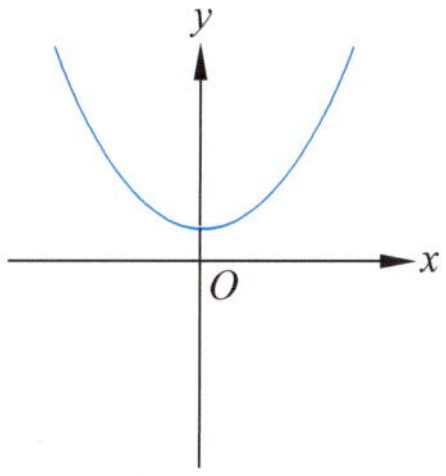

Figure 7-13

2. The x-intercept of $f(x)=3x+18$ is -6 because $f(-6)=0$.

3. The x-intercepts of $f(x)=x^2+7x+12$ are -3 and -4 because $f(-3)=f(-4)=0$.

Questions: What are the x-intercepts for each of the following?

(1) Figure 7-16 shows the graph of a cubic function.

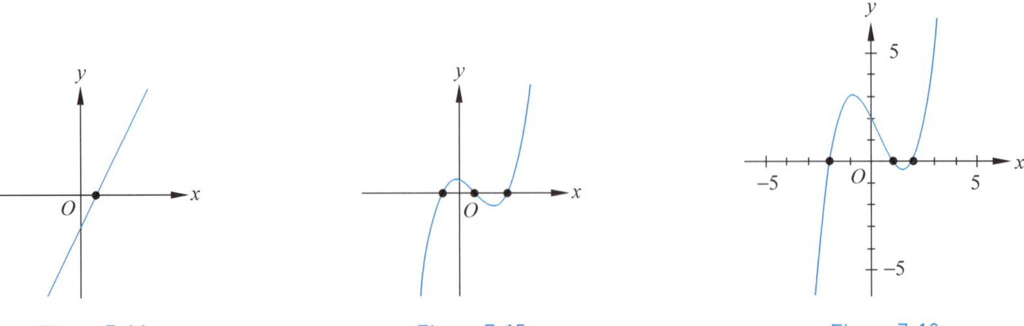

Figure 7-14　Figure 7-15　Figure 7-16

(2) Table 7-11 shows a function with scattered points.

Table 7-11

x	0	1	2	3	4	5	6	7	8
$f(x)$	2	8	6	0	−8	−6	0	13	7

(3) $g(x)=\frac{1}{3}x-20$.

(4) $h(x)=x^2-9x+20$.

Answers: (1) $-2, 1, 2$.

(2) 3 and 6 because $f(3)=f(6)=0$.

(3) 60. Solving the equation $0=\frac{1}{3}x-20$, we get $x=60$, and $g(60)=0$.

(4) 4 and 5. Solving the equation $0=x^2-9x+20$, we get $x=4$ or 5, and $h(4)=h(5)=0$.

y-Intercept　y 截距

n. [waɪ ˌɪntər'sept]

Definition: 1. The y-coordinate of the point of intersection of a graph and the y-axis.　y 截距是指图像与 y 轴的交点的 y 坐标。

2. The value of $f(0)$.

Both definitions are equivalent, but they are phrased significantly differently. Definition 1 is based on geometry, and Definition 2 is based on algebra.　上面两个定义等价,但它们表达的方式不同。定义 1 根据几何方法得出,定义 2 根据代数方法得出。

As shown in Figure 7-17, the y-intercept of the graph of f is k.　如图 7-17 所示,函数 f 图像的 y 截距为 k。

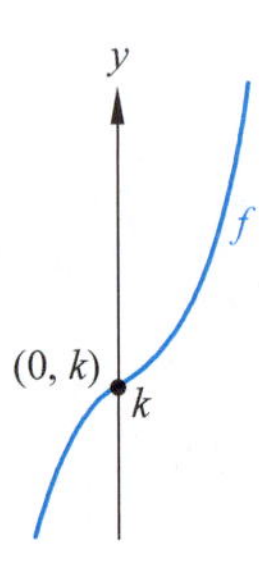

Figure 7-17

Non-function relations have y-intercepts as well. Same definition applies.　非单值函数的多值函数也有 y 截距,定义同上。

Properties: 1. The graph of a function can have no more than one y-intercept, otherwise it fails the vertical line test and will not be a function.　单值函数图像的 y 截距个数不能大于1,否则不满足垂线测试,就不是单值函数。

2. If a function f has k as a y-intercept, then:
若单值函数 f 的 y 截距为 k,则:
   - The graph of f passes through $(0, k)$. (Definition 1)
   f 的图像过$(0, k)$。
   - $f(0) = k$. (Definition 2)

3. Note that the y-intercept is a number, not a point.　y 截距是一个数值,不是一个点。

Examples: 1. (1) Figure 7-18 shows the graph of a function with zero y-intercepts.
(2) Figure 7-19 shows the graph of a function with one y-intercept.

2. The y-intercept of $f(x) = 3x + 18$ is 18 because $f(0) = 18$.

3. The y-intercept of $f(x) = x^2 + 7x + 12$ is 12 because $f(0) = 12$.

Questions: What is the y-intercept for each of the following?

(1) Figure 7-20 shows the graph of a cubic function.

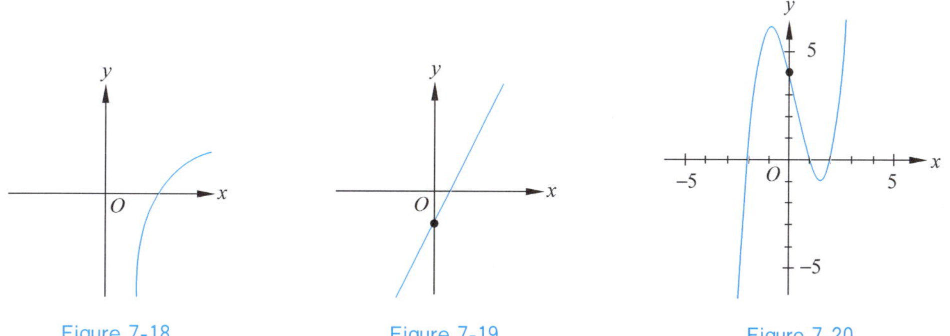

Figure 7-18　Figure 7-19　Figure 7-20

(2) Table 7-12 shows a function with scattered points.

Table 7-12

x	0	1	2	3	4	5	6	7	8
$f(x)$	2	8	6	0	−8	−6	0	13	7

(3) $g(x) = \frac{1}{3}x - 20$.

(4) $h(x) = x^2 - 9x + 20$.

Answers: (1) 4.

(2) 2 because $f(0) = 2$.

(3) −20 because $g(0) = -20$.

(4) 20 because $h(0) = 20$.

7.2.4 Summary 总结

1. In a function f from set X to set Y, X is known as the domain and Y is known as the range. X is the set of all possible input values, and Y is the set of all possible output values. 若 f 是从集合 X 到集合 Y 的函数, X 是定义域, Y 是值域。X 是包含所有输入值的集合, Y 是包含所有输出值的集合。
   If $f(x)=y$, for which $x\in X$ and $y\in Y$, then x is the input and y is the output. 若 $f(x)=y$, 且 $x\in X$ 及 $y\in Y$, 则 x 为输入, y 为输出。
2. We can use words or equations to describe formula—the pattern that the domain and range share. 可用文字或等式描述函数公式——定义域与值域的规律。
3. An intercept is the number on a coordinate axis in which the graph (of the function) intersects that axis. 截距是在坐标平面上单值函数的图像与坐标轴相交的坐标值。
   Review the all definitions of x-intercept and y-intercept (in both algebra and geometry). 复习 x 截距和 y 截距的所有定义(代数方法和几何方法)。

从这里开始,把"单值函数"缩写成"函数"。

7.3 Polynomial Functions 多项式函数

7.3.1 Introduction 介绍

Polynomial Function 多项式函数

n. [ˌpɑləˈnoʊmiəl ˈfʌŋkʃən]

Definition: A function of the form $f(x)=a_nx^n+a_{n-1}x^{n-1}+a_{n-2}x^{n-2}+\cdots+a_1x+a_0$, for which $a_n, a_{n-1}, a_{n-2}, \cdots, a_1$, and a_0 are constants. This is the standard form of polynomial functions. 以 $f(x)=a_nx^n+a_{n-1}x^{n-1}+a_{n-2}x^{n-2}+\cdots+a_1x+a_0$ 的形式表示的单值函数,其中 $a_n, a_{n-1}, a_{n-2}, \cdots, a_1, a_0$ 为常值。这是所有多项式函数的一般形式。

Properties: Here are some specials names for polynomial functions of a certain degrees:
下面是指定次数的多项式函数:
If the degree of f is 0, then f is a constant function. 若 f 的次数为 0, 则 f 为常值函数。
If the degree of f is 1, then f is a linear function. 若 f 的次数为 1, 则 f 为线性函数/一次函数。
If the degree of f is 2, then f is a quadratic function. 若 f 的次数为 2, 则 f 为二次函数。
If the degree of f is 3, then f is a cubic function. 若 f 的次数为 3, 则 f 为三次函数。
If the degree of f is 4, then f is a quartic function. 若 f 的次数为 4, 则 f 为四次函数。

Examples: 1. Examples of constant functions:
常值函数的例子:
$f(x)=1$; $g(x)=-5$; $h(x)=2.89$
2. Examples of linear functions:
线性函数/一次函数的例子:
$f(x)=x$; $g(x)=2x+1$; $h(x)=-5x-8$
3. Examples of quadratic functions:

二次函数的例子：

$f(x)=x^2$；　　$g(x)=3x^2+2$；　　$h(x)=-8x^2-3x+15$

4. Examples of cubic functions：

三次函数的例子：

$f(x)=x^3$；　　$g(x)=x^3-2x^2+3$；　$h(x)=3x^3-5x^2+7x-9$

7.3.2　Graphs of Functions　函数的图像

Line　直线

n. [laɪn]

Definition：A union of vertical lines, graphs of constant functions (horizontal line), and graphs of linear functions (neither horizontal nor vertical lines). 直线是指垂直线、常值函数的图像(水平线)、一次函数的图像(既不是水平线也不是垂直线)的并集。

Curve　曲线

n. [kɜrv]

Definition：Same word for the graph of a function. Lines are special curves. 曲线是函数的图像。直线是特殊的曲线。

It also refers to the graph of a non-function relation. 曲线也包括非单值函数的多值函数图像。

7.3.3　Constant Functions　常值函数

Constant Function　常值函数

n. [ˈkɑnstənt ˈfʌŋkʃən]

Definition：A function of the form $f(x)=k$, for which k is a constant. 常值函数是指形式为 $f(x)=k$ 的函数,其中 k 为常数。

Figure 7-21 shows the graph of a constant function.

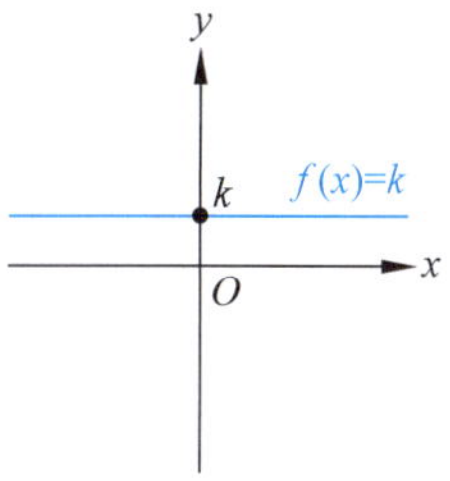

Properties：1. The degree of all constant functions is 0. 常值函数的次数为 0。

2. The output value is fixed regardless of what the input value is. 不论输入是什么,输出固定。

3. The graph of a constant function is a horizontal line, which is either the x-axis or a line that is parallel to the x-axis. To graph $f(x)=k$, first locate a point whose y-coordinate is k, then sketch a horizontal line passing through this point. 常值函数的图像为水平线,它是 x 轴或与其平行的直线。要画出 $f(x)=k$,先找出一个 y 坐标为 k 的点,然后画一条经过这个点的水平线。

y
k　f(x)=k
O　x

Figure 7-21

4. One point (call it (a,b)), determines the equation of a constant function $f(x)=b$. 一个点(a,b)能确定一个常值函数的方程式 $f(x)=b$。

Questions：1. Write the equation of the constant function, as shown in Figure 7-22.

2. Write the equation of a constant function for each of the following：

(1) Passes through (4,9).

(2) Passes through (−5,−8).

(3) Passes through (0,0).

Answers: 1. $y=4$ or $f(x)=4$.

2. (1) $y=9$ or $f(x)=9$.

(2) $y=-8$ or $f(x)=-8$.

(3) $y=0$ or $f(x)=0$.

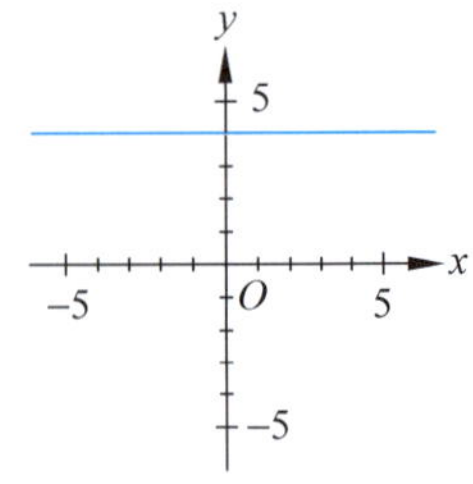

Figure 7-22

7.3.4 Special Lines 特殊的直线

Horizontal Line 水平线

n. [ˌhɔrəˈzɑntəl laɪn]

Definition: Represents the x-axis or a line that is parallel to the x-axis. See constant function. 水平线代表 x 轴或与其平行的直线。见常值函数。

The graph of a constant function is a horizontal line. 常值函数的图像为水平线。

Vertical Line 垂直线

n. [ˈvɜrtɪkəl laɪn]

Definition: A line of the form $x=k$, for which k is a constant. 垂直线是指形式为 $x=k$ 的直线,其中 k 为常数。

The graph of a vertical line is shown in Figure 7-23.

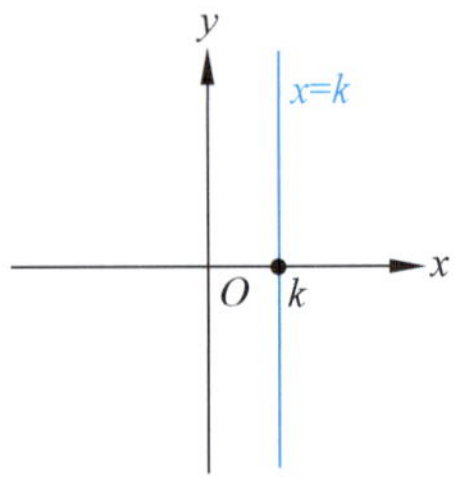

Figure 7-23

Properties: 1. A vertical line is not the graph of any function because it fails the Vertical Line Test. The same input goes with infinitely many different outputs. 垂直线不是单值函数的图像,因为它没有通过垂线测试。同一个输入映射到无数个不同的输出。

2. The input value is fixed regardless of what the output value is. 不论输出是什么,输入固定。

3. The graph of a vertical line is either the y-axis or a line that is parallel to the y-axis. To graph $x=k$, first locate a point whose x-coordinate is k, then sketch a vertical line passing through this point. 垂直线是 y 轴或与其平行的直线。要画出 $x=k$,先找出一个 x 坐标为 k 的点,然后画一条经过这个点的垂直线。

4. One point (call it (a,b)), determines the equation of a vertical line $x=a$. 一个点 (a,b) 能确定一条垂直线的方程式 $x=a$。

Questions: 1. Write the equation of the vertical line, as shown in Figure 7-24.

2. Write the equation of a vertical line for each of the following:

(1) Passes through (4,9).

(2) Passes through (−5,−8).

(3) Passes through (0,0).

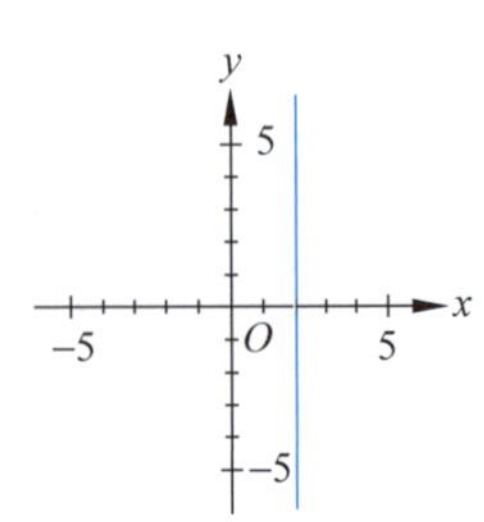

Figure 7-24

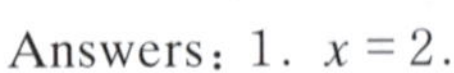

Answers: 1. $x=2$.

2. (1) $x=4$.

(2) $x=-5$.

(3) $x=0$.

7.3.5　Linear Functions　一次函数

视频 15

7.3.5.1　Basics　基础知识

Linear Function　一次函数/线性函数

n. [ˈlɪniər ˈfʌŋkʃən]

Definition：A polynomial function with degree 1.　一次函数是指次数为 1 的多项式函数。

Notation：1. Function Notation/Slope Intercept Form：$f(x) = mx + b$，or $y = mx + b$，for which m is the slope ($m \neq 0$) and b is the y-intercept.　函数表示法/斜截式：$f(x) = mx + b$ 或 $y = mx + b$，其中 m 为斜率($m \neq 0$)，b 为 y 截距。

2. Point-Slope Form：$y - y_0 = m(x - x_0)$，for which m is the slope ($m \neq 0$)，and x_0 and y_0 are constants，and the graph passes through (x_0, y_0).　点斜式：$y - y_0 = m(x - x_0)$，其中 m 为斜率($m \neq 0$)，x_0 和 y_0 为常值，图像经过点(x_0, y_0)。

3. Standard Form：$Ax + By = C$，for which A，B，and C are constants (A and B are not both 0).

一般式：$Ax + By = C$，其中 A、B、C 为常值。A 和 B 不同时为 0。

Note that y and $f(x)$ can be used interchangeably. While y represents the equation of a line，$f(x)$ represents the function notation.　y 和 $f(x)$ 可互用。y 代表的是直线的等式，$f(x)$ 代表的是函数表示法。

See slope-intercept form，point-slope form，and standard form in Section 7.3.5.2.　见 7.3.5.2 节斜截式、点斜式、一般式。

Properties：A line (either the graph of a linear function or a horizontal line or a vertical line) can be determined if any of the following is given：

可以从以下条件之一确定一条直线(一次函数的图像或水平线或垂直线)。

(1) two different points (best to use point-slope form). 两个不同的点(用点斜式)。

(2) a point and a slope (best to use point-slope form or slope-intercept form，especially when the point is located along the y-axis).　点和斜率(可用点斜式或斜截式，特别是那个点在 y 轴上的情况)。

Slope　斜率

n. [sloʊp]

Definition：The slope of a line is the ratio of rise to run，as shown in Figure 7-25.　直线的斜率是上升运行增量和向右运行增量的比值，如图 7-25 所示。

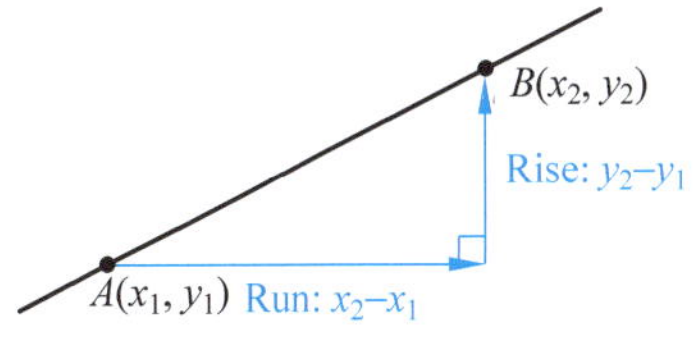

Figure 7-25

For any two different points (call them A and B) on a line，the rise is the difference between B's y-coordinate and A's y-coordinate；the run is the difference between B's x-coordinate and A's x-coordinate.　对于直线上的两点 A 和 B。上升运行增量是 B 的 y 坐标与 A 的 y 坐标的差。向右运行增量是 B 的 x 坐标与 A 的 x 坐标的差。

Notation：The slope is generally represented by the letter "m".　斜率通常用 m 表示。

Properties：1. The slope of a line is always consistent. No matter which two distinct points we pick，our calculation of the slope will not change.　直线的斜率永远不变。不论选哪两个

点,计算出来的斜率不变。

2. Formula for the slope:
斜率公式:
Let point $A=(x_1,y_1)$ and $B=(x_2,y_2)$.
$$m=\frac{\text{rise}}{\text{run}}=\frac{y_2-y_1}{x_2-x_1}.$$

3. It does not matter which point's x-coordinate is x_1 or x_2, and which point's y-coordinate is y_1 or y_2, as long as they are used consistently. 哪个点的 x 坐标为 x_1 或 x_2,和哪个点的 y 坐标为 y_1 或 y_2 都不重要,但计算的时候必须连贯。

Correct: $m=\dfrac{y_2-y_1}{x_2-x_1}$.

Correct: $m=\dfrac{y_1-y_2}{x_1-x_2}$.

Incorrect: $m=\dfrac{y_1-y_2}{x_2-x_1}$.

Incorrect: $m=\dfrac{y_2-y_1}{x_1-x_2}$.

4. From Property 2, we can define slope as the rise when the run is 1. 从性质 2,可以定义斜率为当向右运行增量为 1 个单位时的上升运行增量。

5. There are four types of slopes: positive, negative, zero, and undefined. 斜率可为正数、负数、零、无意义 4 种。
Figure 7-26 shows all these types.

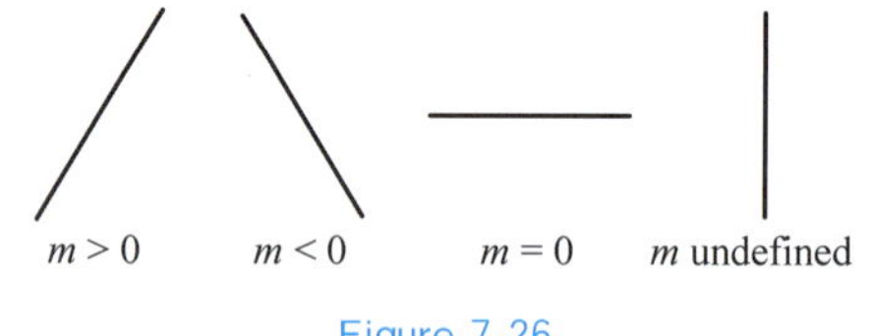

Figure 7-26

6. (1) If the slope of a line is 0, then the line is the graph of a constant function, or simply a horizontal line. 若斜率为 0,则直线为常值函数的图像,也就是水平线。
(2) If the slope of a line is undefined, then the line is a vertical line. 若斜率为无意义,则直线为垂直线。
(3) Otherwise the line is the graph of a linear function. 斜率为其他情况时,直线是一次函数的图像。

Proofs: Property 1 is based on similar right triangles. The ratio of lengths of the longer leg to shorter leg (or the ratio of lengths of shorter leg to longer leg) of similar right triangles is always the same. 上面斜率的性质 1 根据相似直角三角形得出的。因为相似直角三角形中两直角边的长度之比相等。

As shown in Figure 7-27, the two right triangles are similar by AA. 如图 7-27 所示,两个直角三角形因为角角相等而相似。

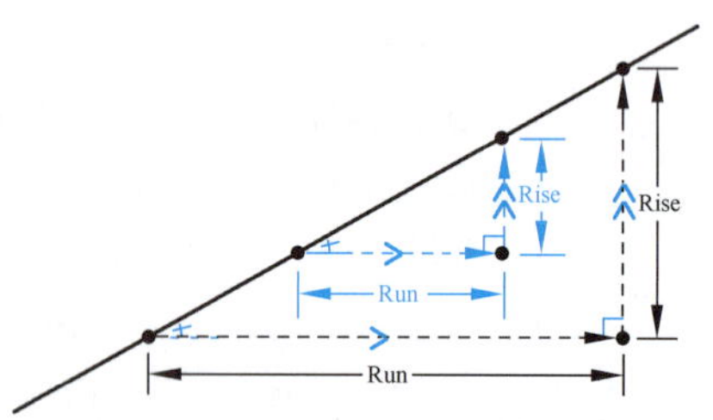

Figure 7-27

Phrases: positive~ 正斜率, negative~ 负斜率, zero~ 零斜率, undefined~ 无意义斜率

Questions: 1. What is the slope of the line that passes

through (2,7) and (7,37)?

2. What is the slope of the line that passes through (6,13) and (3,25)?
3. What is the slope of the line that passes through (2,20) and (10,20)?
4. What is the slope of the line that passes through (−2,10) and (−2,20)?

Answers: 1. $m=\frac{\text{rise}}{\text{run}}=\frac{37-7}{7-2}=\frac{30}{5}=6$.

Alternatively, $m=\frac{\text{rise}}{\text{run}}=\frac{7-37}{2-7}=\frac{-30}{-5}=6$.

2. $m=\frac{\text{rise}}{\text{run}}=\frac{25-13}{3-6}=\frac{12}{-3}=-4$.

Alternatively, $m=\frac{\text{rise}}{\text{run}}=\frac{13-25}{6-3}=\frac{-12}{3}=-4$.

3. $m=\frac{\text{rise}}{\text{run}}=\frac{20-20}{10-2}=\frac{0}{8}=0$.

Alternatively, $m=\frac{\text{rise}}{\text{run}}=\frac{20-20}{2-10}=\frac{0}{-8}=0$.

4. $m=\frac{\text{rise}}{\text{run}}=\frac{20-10}{-2-(-2)}=\frac{10}{0}$. The slope is undefined.

Alternatively, $m=\frac{\text{rise}}{\text{run}}=\frac{10-20}{-2-(-2)}=\frac{-10}{0}$. The slope is undefined.

Rate of Change　变动率

n. [reɪt ʌv tʃeɪndʒ]

Definition: Slope in application problem. 应用题中的斜率。

Properties: The only difference between rate of change and slope is:

变动率与斜率的区别是:

When we say the rate of change A is greater than the rate of change B, we mean $|\text{rate of change } A|>|\text{rate of change } B|$.

When we say the rate of change A is less than the rate of change B, we mean $|\text{rate of change } A|<|\text{rate of change } B|$.　比较变动率大小时,比较的是绝对值。

When we say the slope A is greater than the slope B, we mean slope $A>$slope B.

When we say the slope A is less than the slope B, we mean slope $A<$slope B.　比较斜率大小时,比较的是数值本身。

Questions: Order the following:

(1) from the least slope to the greatest slope.　从小到大排列斜率。

(2) from the least rate of change to the greatest rate of change.　从小到大排列变动率。

A: 8m/4s　B: −9m/3s　C: −25m/5s　D: 20m/5s

E: 14m/2s　F: −100m/10s

Answers: Simplifying each velocity, in m/s, we get A = 2, B = −3, C = −5, D = 4, E = 7, and F = 10.

Thus our answer is:

(1) C, B, A, D, E, F

(2) A, B, D, C, E, F

7.3.5.2 Methods of Expressing Linear Functions 一次函数的表示法

Slope-Intercept Form 斜截式

n. [sloʊp ˌɪntər'sept fɔrm]

Definition: The equation of a line that clearly shows the slope and the y-intercept. 斜截式是指展现斜率和 y 截距的直线方程式。

Notation: $y=mx+b$, for which m is the slope and b is the y-intercept. In function notation, it is $f(x)=mx+b$.

写作 $y=mx+b$,函数表示为 $f(x)=mx+b$。m 为斜率,b 为 y 截距。

Properties: 1. To write the equation of a line in slope-intercept form:

用斜截式写出一条直线的方程式的方法如下:

(1) If two different points are given, calculate the slope first, then substitute x and y with the one point's x-coordinate and y-coordinate respectively, and finally find b. 若已知两个点,先计算出斜率,然后分别代入一点的 x 和 y 坐标值到公式中,最后求出 b。

(2) If the slope and one point are given, substitute x and y with the x-coordinate and y-coordinate of that point, respectively, then find b. 若已知斜率和一个点,分别代入点的 x 和 y 坐标值到公式中,再求出 b。

2. (1) Advantage: ① Given the slope and y-intercept, the equation can be found effortlessly. 已知斜率和 y 截距后,能轻松写出方程式。

② The equation $y=mx+b$ isolates the dependent variable already, and can easily be switched to the function notation $f(x)=mx+b$.

在方程式 $y=mx+b$ 中,因变量已被分隔开,很容易换成函数表示,即 $f(x)=mx+b$。

③ The equation of a line in slope-intercept form is unique, and is in the standard form of a polynomial. 斜截式是独特地表示直线的方程式。同时,斜截式也是多项式函数的一般形式。

(2) Disadvantage: When given the slope and one point, or given two different points, slope-intercept forms require substitutions. It is easier to use the point-slope form to find the equation. 已知斜率和一个点或已知两点,斜截式需要代入坐标值。点斜式比斜截式方便。

3. To graph a line in the form $y=mx+b$, first plot the point on the y-axis: $(0,b)$, then plot the points $(1,b+m)$, $(-1,b-m)$, $(2,b+2m)$, $(-2,b-2m)$, and so on, as shown in Figure 7-28. Recall that we can define slope as the value of the rise when the value of the run is 1. 要画出 $y=mx+b$,先画出在 y 轴上的点$(0,b)$,然后根据斜率画出$(1,b+m)$、$(-1,b-m)$、$(2,b+2m)$、$(-2,b-2m)$,以此类推,如图 7-28 所示。一般可以定义斜率为当向右运行 1 个单位时的上升运行增量。

Finally, connect the points. Note that the slope between any of these two points is consistent：m. 最后连接起来这些点。留意在直线的任意两点中,斜率不变。

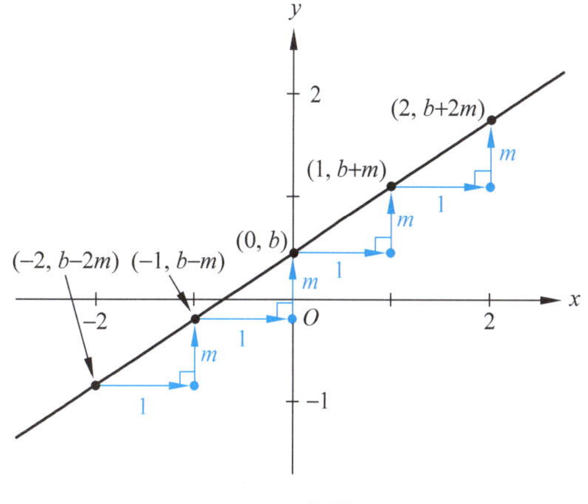

Figure 7-28

4. If $m = 0$, then $y = b$ is a constant function. 常值函数的情况。
   If $b = 0$, then $y = mx$ is a direct variation. 正比例函数的情况。
5. To model real life situations using linear functions in the form of $y = mx + b$, b represents the initial quantity, or one-time quantity, while m represents the rate of change. 要用一次函数斜截式 $y = mx + b$ 表达应用题,b 代表的是起始量或一次性的量,m 代表的是变动率。

Notes: It is clear that b is the y-intercept for the graph of $f(x) = mx + b$ because $f(0) = b$, which means its graph passes through $(0, b)$.
b 为 $f(x) = mx + b$ 图像的 y 截距,因为 $f(0) = b$,图像经过$(0, b)$。

Questions: 1. Write the equation of a line for each of the following in slope-intercept form:
(1) A line that passes through (3,8) and (9,32).
(2) A line with slope of −3 and passes (5, −13).
(3) A line with slope of 2016 and passes through (0,99999).
(4) A line that passes through (4,9) and (15,9).
(5) Aline that passes through (8, −10) and (8,25).

2. Graph the equation of the line: $y = 5x - 3$.

3. Write a linear function for each of the real life situations:
(1) To register for SAT II subject tests, the base fee is \$22, and the cost of each test is \$25. Write the equation of a linear function for the total registration fee VS the number of tests registered.
(2) John picks 70 apples per hour, but he owes Jill 80 apples. Write the equation of a linear function for the number of apples that John has VS the number of hours.
(3) Ben can paint 2 walls in one hour, and he had painted 4 walls today. Write the equation of a linear function that represents the number of walls that he painted VS the number of hours.

4. The number of miles Robot John ran today equals to the number of miles he ran before noon plus the number of miles he ran after noon. **He ran at a constant speed in the afternoon starting from 12 PM.** At 3 PM he has run 25 miles; at 8 PM he has run 55 miles. (1) Write the equation of the function $f(t)$ that represents the number of miles he has run in t hours, where t is the number of hours after 12PM. (2) What is the speed John runs? (3) How many miles did he run before noon?

Answers: 1. (1) $m = \frac{32 - 8}{9 - 3} = 4$.

We have $y=4x+b$ so far. Let's find b.

Substituting (3,8) for (x,y), we get $8=4(3)+b$, for which $b=-4$.

Thus the answer is $y=4x-4$.

(2) We know that $m=-3$.

We have $y=-3x+b$ so far. Let's find b.

Substituting $(5,-13)$ for (x,y), we get $-13=-3(5)+b$, for which $b=2$.

Thus the answer is $y=-3x+2$.

(3) We can find the equation in the same fashion as (1) and (2), but here's a shortcut: with slope of 2016 and y-intercept of 99999, the answer is $y=2016x+99999$.

(4) Clearly $m=0$ (by observation, or use the formula of slope). Thus the answer is $y=9$.

(5) Clearly the slope is undefined (by observation, or use the formula of slope). Thus the answer is $x=8$.

2. Figure 7-29 graphs the line $y=5x-3$.

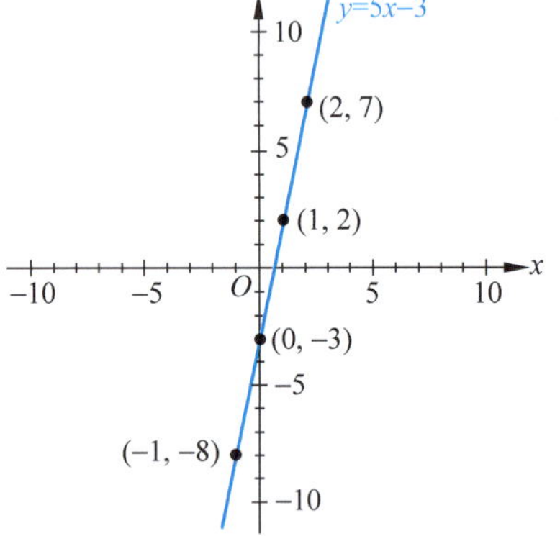

Figure 7-29

3. (1) Let y be the total registration fee, and x be the number of tests taking. The answer is $y=25x+22$.

(2) Let y be the number of apples that John has, and x be the number of hours. The answer is $y=70x-80$. Owing 80 apples is the same as having -80 apples.

(3) Let y be the number of walls Ben painted, and x be the number of hours. The answer is $y=2x+4$.

4. We have two ordered pairs (hour, distance) implicitly: (3,25) and (8,55). The slope is $30/5=6$, representing the speed in miles per hour (mph). We have $f(t)=6t+b$. Substituting (3,25) into the equation, we get $25=6(3)+b$, for which $b=7$. We have $f(t)=6t+7$ and can answer our questions below:

(1) $f(t)=6t+7$, which gives the distance John has run so far in t hours, for which t is the number of hours after 12PM.

(2) 6 miles per hour (mph). The speed is represented by the slope of $f(t)$.

(3) 7 miles. The number of miles he ran before noon (initially) is represented by the y-intercept of $f(t)$, or, $f(0)$.

Point-Slope Form 点斜式

n. [pɔɪnt sloʊp fɔrm]

Definition: The equation of a line that clearly shows the slope and a point. 点斜式是指展现斜率和一个点的直线方程式。

Notation：$y-y_0=m(x-x_0)$，for which m is the slope，and x_0 and y_0 are constants，and the graph passes through (x_0,y_0)，which is called the initial point.

写作 $y-y_0=m(x-x_0)$。m 为斜率，x_0 和 y_0 为常值。图像经过(x_0,y_0)，这个点为初始点。

Properties：1. To write the equation of a line in point-slope form：

用点斜式写出一条直线的方程式的方法如下：

(1) If two different points are given，calculate the slope first. Then substitute (x_0, y_0) with one of the points given.　若已知两个点，则先算出斜率，然后再把一个点代入(x_0,y_0)。

(2) If the slope and one point are given，m is known. All we need is to substitute (x_0,y_0) with one of the points given.　若已知斜率(m 值)和一点，只需要把点代入(x_0,y_0)即可。

2. (1) Advantage：Given the slope and any one point，the point-slope form can be found effortlessly；given any two different points，finding the equation in point-slope form is easier than finding it in the slope-intercept form.　给出斜率和一个点，能立刻得出点斜式；给出两个点，用点斜式比斜截式快捷。

(2) Disadvantage：① The dependent variable is not isolated.　因变量没被分隔开。

② The equation in point-slope form is not unique because it depends on the choice (x_0,y_0).　点斜式不唯一，取决于初始点(x_0,y_0)的选择。

However，different equations for the same line in point-slope form can be rewritten as a unique slope-intercept form.　但是，不同的点斜式最终都能被改写成同样的斜截式。

3. To graph a line in the form of $y-y_0=m(x-x_0)$，first plot the point (x_0,y_0)，then plot the points (x_0+1, y_0+m)，(x_0-1, y_0-m)，(x_0+2, y_0+2m)，(x_0-2, y_0-2m)，and so on. Recall that we can define slope as the value of the rise when the value of the run is 1.　要画出形式为 $y-y_0=m(x-x_0)$的直线，先找出点(x_0,y_0)，然后再画出(x_0+1,y_0+m)、(x_0-1,y_0-m)、(x_0+2,y_0+2m)、(x_0-2,y_0-2m)。以此类推，可以定义斜率为当向右运行 1 个单位时的上升运行增量。

Finally，connect the points. Note that the slope between any of these two points is consistent：m.　最后连接这些点。注意在直线的任意两点中，斜率不变。

The illustration is shown in Figure 7-30.

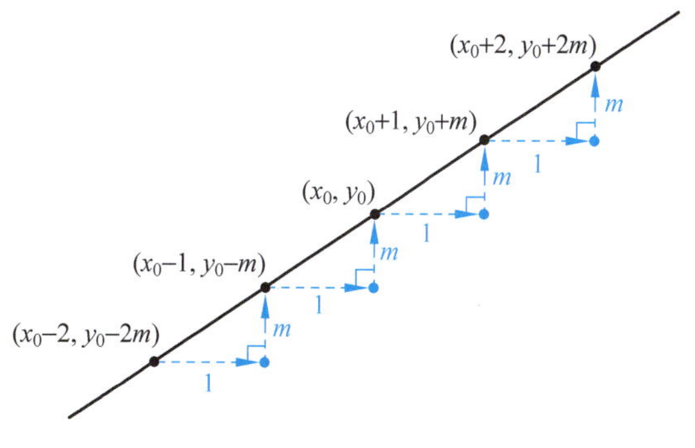

Figure 7-30

4. If $m=0$，then $y-y_0=0$，or $y=y_0$ is a constant function.　若 $m=0$，则 $y-y_0=0$，即 $y=y_0$，这是常值函数的情况。

If $x_0 = y_0 = 0$, then $y = mx$ is a direct variation. 若 $x_0 = y_0 = 0$，则 $y = mx$，这是正比例函数的情况。

5. To model real life situations using linear functions in the form of $y - y_0 = m(x - x_0)$, m represents the rate of change, and (x_0, y_0) represents the initial ordered pair. 要用一次函数点斜式 $y - y_0 = m(x - x_0)$表达应用题，m 代表变动率，(x_0, y_0)代表初始的有序对。

Notes：Point-slope form can derive to slope-intercept form $y = mx + b$：

从点斜式到斜截式：

$y - y_0 = m(x - x_0)$	Original equation.
$y - y_0 = mx - mx_0$	Distribute the m on the RHS.
$y = mx - mx_0 + y_0$	Add y_0 to both sides.
$y = mx + (-mx_0 + y_0)$	Rearrange.

Note that $-mx_0 + y_0$ is a constant, and it is equivalent to b, the y-intercept in the slope-intercept form $y = mx + b$.

注意 $-mx_0 + y_0$ 是常值，相当于斜截式的 b，也就是 y 截距。

Why $-mx_0 + y_0$ is the y-intercept?

Isolating the y-intercept from the slope-intercept form $y = mx + b$, we get $b = -mx + y$. For any point (x_0, y_0) that satisfies the equation, $b = -mx_0 + y_0$ is an identity. Thus $-mx_0 + y_0$ is the y-intercept, depending on our choices of (x_0, y_0).

Questions：1. Write the equation of a line for each of the following in point-slope form (This is the same as Question 1 from slope-intercept form.)：

(1) A line that passes through (3,8) and (9,32).

(2) A line with slope of −3 and passes (5, −13).

(3) A line with slope of 2016 and passes through (0,99999).

(4) A line that passes through (4,9) and (15,9).

(5) A line that passes through (8, −10) and (8,25).

2. Graph the equation of the line：$y - 2 = 4(x - 1)$.

3. To register for SAT II subject tests, the cost of each test is \$25. There is a base fee involved, and taking 4 tests cost \$122. Write the equation of a linear function for the total registration fee VS the number of tests registered.

4. The number of miles Robot John ran today equals to the number of miles he ran before noon plus the number of miles he ran after noon. **He ran at a constant speed in the afternoon starting from 12PM**. At 3 PM he has run 25miles; at 8PM he has run 55miles. Write the equation of the function $f(t)$ that represents the number of miles he has run in t hours, where t is the number of hours after 12PM (This is the same as Question 4 from slope-intercept form.).

Answers：1. (1) $m = \dfrac{32 - 8}{9 - 3} = 4$.

Using the point (3,8), the answer is $y - 8 = 4(x - 3)$.

Using the point (9,32), the answer is $y - 32 = 4(x - 9)$.

(2) $y+13=-3(x-5)$.

(3) $y-99999=2016x$.

(4) Clearly $m=0$ (by observation, or use the formula of slope). Thus the answer is $y=9$.

(5) Clearly the slope is undefined (by observation, or use the formula of slope). Thus the answer is $x=8$.

2. Figure 7-31 graphs the line $y-2=4(x-1)$.
3. Let y be the total registration fee, and x be the number of tests taking. The answer is $y-122=25(x-4)$.
4. We have two ordered pairs (hour, distance) implicitly: (3,25) and (8,55). The slope is $30/5=6$, representing the speed.

Using the point (3,25), we have $f(t)-25=6(t-3)$.

Using the point (8,55), we have $f(t)-55=6(t-8)$.

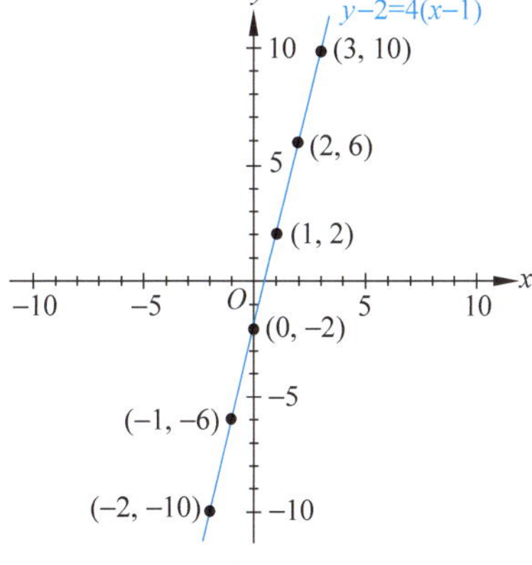

Figure 7-31

Standard Form　一般式

n. [ˈstændərd fɔrm]

Definition: The equation of a line whose LHS is the sum of two terms: one term with variable x and the other with y. The RHS is a constant.　直线方程的一般式为：左边是两个项的和(一个项的变量为 x,另一个项的变量为 y),右边是一个常数。

Notation: $Ax+By=C$, for which A, B, and C are constants, and A and B are not both 0.

Properties: 1. To write the equation of a line in standard form, first write out the line's slope-intercept form, then manipulate the terms: from $y=mx+b=0$ to $-mx+y=b$.　要用一般式写出一条直线的方程式,先用斜截式写出,然后移项。

Note that the standard form of a line is not unique because we can always multiply both sides by a nonzero constant and maintain all the solutions (x, y): $-mx+y=b$ and $-2mx+2y=2b$ are both standard forms of the same line.　注意直线的一般式不唯一,可以在两边都乘上一个非零的常数,使得所有解(x, y)保持不变。

2. (1) Advantage: $Ax+By=C$ gives the relationship between two variables.　一般式给出两个变量的关系。

When the two variables have the same importance in a problem, the standard form represents the problem's context better than the slope-intercept form and the point-slope form.　在问题中,当两个变量的重要性相等的时候,一般式比斜截式和点斜式能更好地体现问题的意思。

(2) Disadvantage: The equations in standard form are not unique.　直线方程式的一般式不唯一。

3. To graph a line in the form of $Ax+By=C$ (where A and B are not both 0), first

find the x-intercept (solve the equation $Ax_0 = C$, and x_0 is the x-intercept) and y-intercept (solve the equation $By_0 = C$, and y_0 is the y-intercept). 要画出一般式 $Ax + By = C$(A 和 B 不同时为 0)的直线，先求出 x 截距(解 $Ax_0 = C$)，再求出 y 截距(解 $By_0 = C$)。

Plot the points $(x_0, 0)$ and $(0, y_0)$ and connect the points. Note that two different points determines a line. 画出$(x_0, 0)$ 和$(0, y_0)$且连接两个点。注意两个点确定一条直线。

4. If $A = 0$ and B and C are nonzero, then we have $By = C$, from which $y = C/B$ is a constant function. 若 $A = 0, B \neq 0, C \neq 0$，则是常值函数的情况。

   If $B = 0$ and A and C are nonzero, then we have $Ax = C$, from which $x = C/A$ is the graph of a vertical line. 若 $B = 0, A \neq 0, C \neq 0$，这是垂直线的情况。

   If $C = 0$ and A and B are nonzero, then we have $Ax + By = 0$, from which $y = -\frac{A}{B}x$ is a direct variation. 若 $C = 0, A \neq 0, B \neq 0$，这是正比例函数。

5. To model real life situations using linear functions of the form $Ax + By = C$, constants A and B usually represent unit quantities (unit price, speed, efficiency …); variables x and y usually represent the number of units (time, number of items …). 用一次函数一般式 $Ax + By = C$ 表达应用题，常值 A 和 B 通常代表单位的量(单价、速度、效率……)；变量 x 和 y 通常代表单位的数量(时间、数量……)。

Questions: 1. Write the equation of a line for each of the following in standard form (This is the same as Question 1 from slope-intercept form.):

(1) A line that passes through (3,8) and (9,32).

(2) A line with slope of -3 and passes $(5, -13)$.

(3) A line with slope of 2016 and passes through (0,99999).

(4) A line that passes through (4,9) and (15,9).

(5) A line that passes through $(8, -10)$ and (8,25).

2. Graph the equation of the line: $3x + 4y = 12$.

3. In a restaurant, a burger meal costs \$7 and a hot dog meal costs \$6. Selling only these two types of meals, the restaurant earned \$700 today. Use the equation of a line in standard form to model this situation.

Answers: 1. (1) The slope-intercept form is $y = 4x - 4$. Subtracting $4x$ from both sides produces the standard form $-4x + y = -4$.

(2) The slope-intercept form is $y = -3x + 2$. Adding $3x$ to both sides produces $3x + y = 2$.

(3) The slope-intercept form is $y = 2016x + 99999$. Subtracting $2016x$ from both sides produces $-2016x + y = 99999$.

(4) Clearly $m = 0$ (by observation, or use the formula of slope). Thus the answer is $y = 9$.

(5) Clearly the slope is undefined (by observation, or use the formula of slope). Thus the answer is $x = 8$.

2. We can convert $3x + 4y = 12$ into slope-intercept form, then graph, but here we will graph by the intercepts.

   Clearly, the x-intercept is 4 and the y-intercept is 3. Therefore we will plot the points (4, 0) and (0, 3) and connect them with a line, as shown in Figure 7-32.

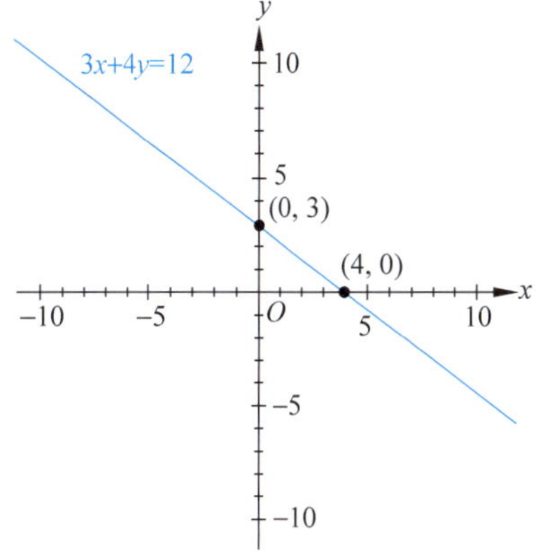

Figure 7-32

3. Let x be the number of burger meals sold, and y be the number of hot dog meals sold. The equation is $7x + 6y = 700$.

   The slope-intercept form of the same equation is $y = -\frac{7}{6}x + \frac{700}{6}$ (Isolate the y to the LHS), which is correct, but does not model the situation as clearly as the standard form does.

   The slope-intercept form gives the number of hot dog meals in terms of the number of burger meals. In this problem, neither of the number of burger meals or the number of hot dog meals really "dominates". Thus the standard form communicates the information better.

7.3.5.3 Special Linear Functions　特殊的一次函数

Direct Variation　正比例函数

n. [də'rekt ˌveəri'eɪʃən]

Definition: 1. A linear function whose y-intercept is 0, or refer to the constant function $y = 0$ (trivial case).　y 截距为 0 的一次函数,或常值函数 $y = 0$(特例)。

2. y is the product of x and a constant: $y = kx$, where $k = 0$ is a trivial case.　y 是 x 和一个常值的积,即 $y = kx$。其中 $k = 0$ 是特例。

3. A linear function for which y/x is a constant, or refer to the constant function $y = 0$ (trivial case).

   y/x 为常值的一次函数,或常值函数 $y = 0$(特例)。

The three definitions above are equivalent.　以上 3 个定义等价。

Figure 7-33 graphs the line $y = kx$.

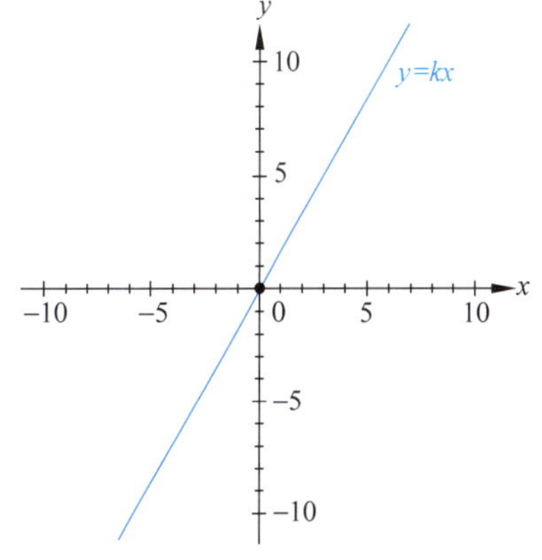

Figure 7-33

Notation: $y = kx$ or $y/x = k$.

A direct variation is similar to the slope-intercept form $y = mx + b$, except that $b = 0$ and m is changed to k (constant slope), for a conventional reason.

跟斜截式相同,$b = 0$,且根据惯例,常值斜率 m 改为 k。

We also say x and y are directly proportional.　x 和 y 成正比。

Properties: 1. If $y = kx$, for which k is a constant, then it is true that $\frac{y}{x}=k$ such that $x\neq 0$. We say y varies directly with x. 若 $y=kx$,其中 k 为常值,则$\frac{y}{x}=k(x\neq 0)$,即 y 和 x 成正比。

2. The graph of a direct variation $y=kx$ is a nonvertical line that always passes through (0,0) by definition. 正比例函数的图像是非垂直线的直线,根据定义,该直线过(0,0)。
3. If $k>0$, y increases proportionally as x increases, and y decreases proportionally as x decreases (Golden Rule). 若 $k>0$,随着 x 增长,y 成比例增长。随着 x 衰减,y 成比例衰减。

Examples: Some real-life examples of direct variations $y=kx$ (k is a positive constant), follow Property 4.

1. x = unit price of items 单价; y = total price of items 总价; k = number of items 数量。

   k is fixed. y increases as x increases, and y decreases as x decreases.
2. x = time used 时间; y = distance traveled 路程; k = speed traveled 速度。

   k is fixed. y increases as x increases, and y decreases as x decreases.
3. x = efficiency of work 工作效率; y = amount of work done 工作总量; k = time 时间。

   k is fixed. y increases as x increases, and y decreases as x decreases.

Questions: 1. What is the equation of a direct variation whose graph passes through (7,3)?

2. In a buffet restaurant, the entry fee per person is \$20. What is the equation of the total price, in dollars, as a function to the number of eaters?

Answers: 1. Method 1: Doing it directly by definition, we have $y=\frac{3}{7}x$.

Method 2: Writing the equation of a line that passes through (7,3) and (0,0), we have $y=\frac{3}{7}x$.

2. Let x be the number of eaters, and let y be the total price, in dollars. We have $y=20x$.

7.3.5.4 More on Graphing Linear Functions 更多一次函数图像的知识点

Solving Systems of Linear Equations by Graphing 画图法

Properties: 1. In a system of linear equations with two unknowns (call the unknowns x and y), graph all the lines. The point of concurrency is the solution. 在二元一次方程组中(设未知数为 x 和 y),画出所有直线。所有直线的交点为解。

2. (1) Advantages: Provides a geometric view of algebra. 以几何的角度解题。

   (2) Disadvantages: Solving systems of linear equations by graphing is much more cumbersome than solving systems of linear equations by elimination. 画图法比加减消元法麻烦很多。

Questions: Solve the following system of linear equations by graphing:

$$\begin{cases} y=6x-10 \\ y=2x-6 \end{cases}.$$

Answers: Graphing the lines, we have the solution $(1,-4)$ as shown in Figure 7-34:

Note that solving this system by substitution is much faster. 注意用代入法解方程组更快。

7.3.5.5 Relationship between Two Lines 两直线之间的关系

Relationship between Two Lines 两直线之间的关系

Definition: The relationship between two lines on the coordinate plane is one of parallel, intersecting, and coincident, as shown in Figure 7-35. 在坐标平面上,两直线的关系可以是平行、相交、重合 3 种之一,如图 7-35 所示。

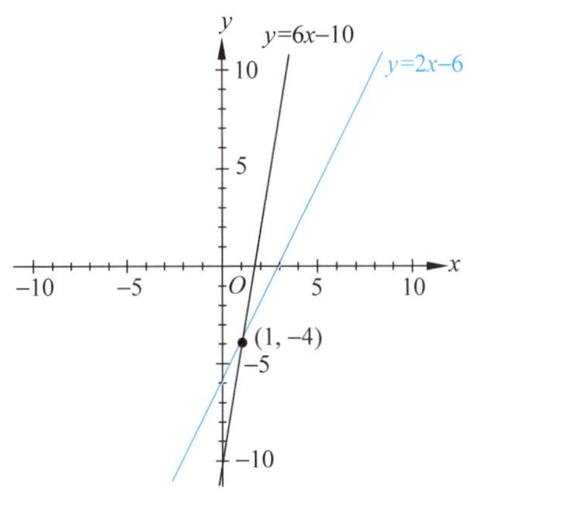

Figure 7-34

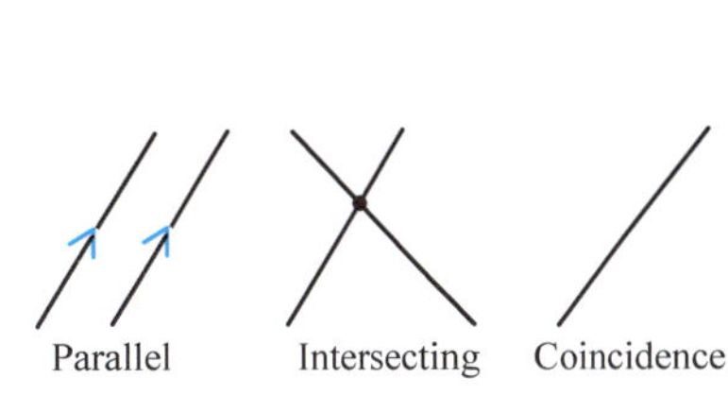

Figure 7-35

Parallel 平行

adj. [ˈpærəˌlel]

Definition: Two (or more) lines are parallel if they have the same slope and no point in common, as shown in Figure 7-36. 若两线(或更多条线)斜率相同且没有共点,则它们平行,如图 7-36 所示。

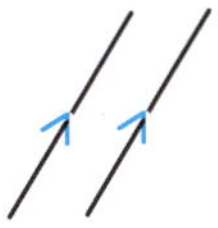

Figure 7-36

Notation: If line l is parallel to line m, we denote their relationship as $l \parallel m$. 若直线 l 平行于直线 m,则写作 $l \parallel m$。

Refer to Table A-5 in Appendix A for the notation of parallel. 平行的标记方式参照附录 A 的表 A-5。

Properties: Parallel lines have no point of intersection. 平行线没有交点。

Examples: The lines $y=3x+8$ and $y=3x-7$ are parallel because they have the same slope and are different lines on the coordinate plane.

Phrases: ~lines 平行线

Questions: 1. Determine which of the lines are parallel:

l: $2x+3y=7$; m: $4x+6y=21$; n: $3x+4y=60$.

2. Write an equation of a line passing $(4,7)$ that is parallel to the given line shown in Figure 7-37.

3. Write an equation of a line passing $(8,10)$ and is parallel to the line $y=7x-5$.

4. Write an equation of a line passing $(7,3)$ and is parallel to the line $8x+4y=999$.

5. Write an equation of a line passing (6,7) and is parallel to the line $y=10$.
6. Write an equation of a line passing (6,7) and is parallel to the line $x=10$.

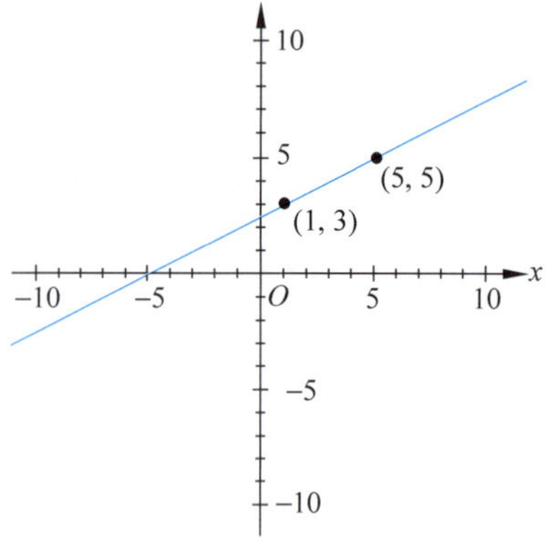

Figure 7-37

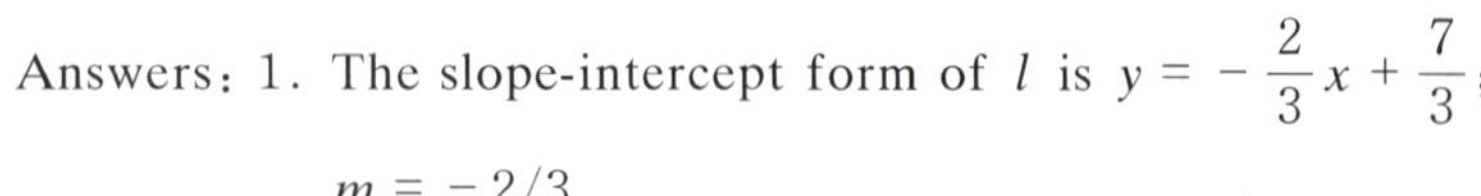

Answers: 1. The slope-intercept form of l is $y=-\frac{2}{3}x+\frac{7}{3}$; $m=-2/3$.

The slope-intercept form of m is $y=-\frac{2}{3}x+\frac{7}{2}$; $m=-2/3$.

The slope-intercept form of n is $y=-\frac{3}{4}x+15$; $m=-3/4$.

Since l and m have the same slope, they are parallel.

2. The slope of the given line is 1/2, which is the same for the slope of our line. Given a slope 1/2 and a point (4,7), we can write our equation in slope-intercept form.

Substituting values for $y=mx+b$, we get $7=\frac{1}{2}(4)+b$, from which $b=5$. Our final answer is $y=\frac{1}{2}x+5$.

3. Because parallel lines have the same slope, the slope of our line is 7.
Given a slope 7 and a point (8,10), we can write our equation in slope-intercept form.
Substituting values for $y=mx+b$, we get $10=7(8)+b$, from which $b=-46$. Our final answer is $y=7x-46$.

4. The slope of our line is the same as the slope of $8x+4y=999$. To find the slope of $8x+4y=999$, we need to rewrite it in slope-intercept form first.

$8x+4y=999$	Original equation.
$4y=-8x+999$	Subtract $8x$ from both sides.
$y=-2x+\frac{999}{4}$	Divide both sides by 4.

The slope of our line is -2. Substituting values into the equation $y=mx+b$, we get $3=-2(7)+b$, for which $b=17$. Thus our answer is $y=-2x+17$.

5. The slope of our line is 0 (horizontal line). Thus our answer is $y=7$.
6. The slope of our line is undefined (vertical line). Thus our answer is $x=6$.

Intersecting 相交

adj. [ˈɪntərˌsektɪŋ]

Definition: Two lines are intersecting if they have exactly one point in common, as shown in Figure 7-38. 若两直线只有一个交点,则它们相交,如图 7-38 所示。

Figure 7-38

Properties：1. The point of intersection satisfies both equations of the lines. In other words, it is the solution for the system of two linear equations.　交点满足两直线的方程式。它是两直线方程组的解。

2. To find the point of intersection of two lines, we can solve it by：
找出两直线的交点,可以采用下面的方法：
(1) Graphing.　画图解方程组。
(2) Substitution.　代入消元法。
(3) Elimination.　加减消元法。
If it turns out that：
(1) the system has no solution, then the lines are parallel.　若方程组无解,则两线平行。
(2) the system has exactly one solution, then the lines are intersecting.　若方程组只有一解,则两直线相交。
(3) the system has infinitely many solutions, then the lines are coincident.　若方程组有无穷多个解,则两直线重合。

3. Perpendicular lines are special cases for intersecting lines. 垂直线是相交线的特殊情况。

Phrases：～lines　相交线

Questions：1. What is the point of intersection between the lines $y=2x+9$ and $y=7x-6$?
2. What is the point of intersection between the lines $5x+7y=19$ and $7x-2y=3$?
3. What is the point of intersection between the lines $3x+7y=10$ and $9x+21y=43$?
4. What is the point of intersection between the lines $2x-6y=10$ and $3x-9y=15$?

Answers：1. The point of intersection is the solution for the system $\begin{cases} y=2x+9 \\ y=7x-6 \end{cases}$.
Subtracting the first equation from the second produces $0=5x-15$, for which $x=3$. Substituting $x=3$ into either equation gives $y=15$. Thus the point of intersection is $(3,15)$.

2. The point of intersection is the solution for the system $\begin{cases} 5x+7y=19 \\ 7x-2y=3 \end{cases}$.
Multiplying the first equation by 7 and the second equation by 5, we get $\begin{cases} 35x+49y=133 \\ 35x-10y=15 \end{cases}$.
Subtracting the second equation from the first produces $59y=118$, for which $y=2$. Substituting $y=2$ into any equation gives $x=1$. Thus the point of intersection is $(1,2)$.

3. The point of intersection is the solution for the system $\begin{cases} 3x+7y=10 \\ 9x+21y=43 \end{cases}$.
Multiplying the first equation by 3, we get $\begin{cases} 9x+21y=30 \\ 9x+21y=43 \end{cases}$.

Subtracting the first equation from the second produces $0 = 13$, which is false. Thus the two lines are parallel and have no point of intersection.

4. The point of intersection is the solution for the system $\begin{cases} 2x - 6y = 10 \\ 3x - 9y = 15 \end{cases}$.

Multiplying the first equation by 1.5, we get $\begin{cases} 3x - 9y = 15 \\ 3x - 9y = 15 \end{cases}$. The equations are identical. Thus the two lines are coincident (infinitely many points of intersection).

Perpendicular 垂直

adj. [ˌpərpənˈdɪkjələr]

Definition: Two lines are perpendicular if they intersect at right angles, as shown in Figure 7-39. 若两直线相交,且相交的角为直角,则它们垂直,如图 7-39 所示。

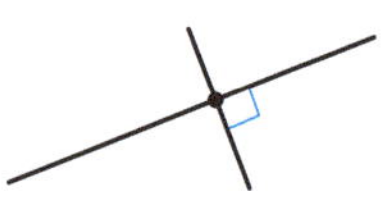

Figure 7-39

Notation: We denote "lines l and m are perpendicular" as "$l \perp m$".
Refer to Table A-5 in Appendix A for the notation of perpendicular.
垂直的标记方式参照附录 A 的表 A-5。

Properties: The slopes of perpendicular lines are negative reciprocals. 垂直线的斜率互为负倒数。

Proofs: We will prove the property in the applications of congruent triangles in Section 15.5.3.
在 15.5.3 节全等三角形的应用可证明以上性质。

Phrases: ~lines 垂直线

Questions: 1. Determine which of the following lines are perpendicular.

$l: 3x + 4y = 12$; $m: 8x + 6y = 45$; $n: 12x - 9y = 40$.

2. Write an equation of a line passing (4,7) that is perpendicular to the given line shown in Figure 7-40.

3. Write an equation of a line passing (5,18) and is perpendicular to the line $y = \frac{1}{2}x$.

4. Write an equation of a line passing (7,3) and is perpendicular to the line $4x + 12y = 999$.

5. Write an equation of a line passing (6,7) and is perpendicular to the line $y = 10$.

6. Write an equation of a line passing (6,7) and is perpendicular to the line $x = 10$.

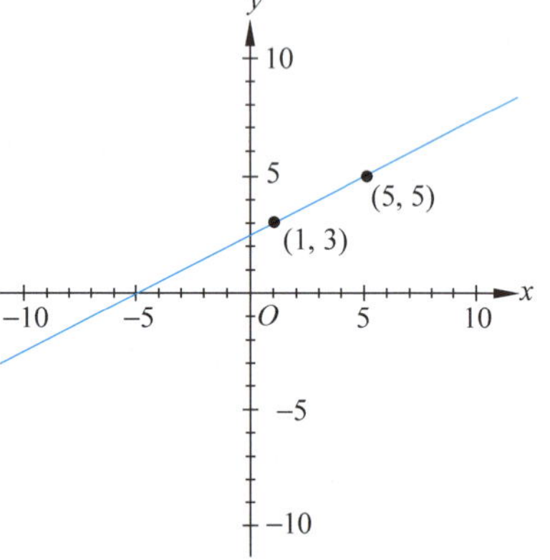

Figure 7-40

Answers: 1. The slope-intercept form of l is $y = -\frac{3}{4}x + 3$; $m = -3/4$.

The slope-intercept form of m is $y = -\frac{3}{4}x + \frac{15}{2}$; $m = -3/4$.

The slope-intercept form of n is $y=\frac{4}{3}x-\frac{40}{9}$; $m=4/3$.

Since the slopes of lines (l and n) and (m and n) are opposite reciprocals, we have $l \perp n$ and $m \perp n$.

2. The slope of the given line is 1/2. The slope of our line is the opposite reciprocal of that, which is -2.
   Given a slope -2 and a point $(4,7)$, we can write our equation in slope-intercept form.
   Substituting values for $y=mx+b$, we get $7=-2(4)+b$, from which $b=15$. Our final answer is $y=-2x+15$.
3. We know that the slope of $y=\frac{1}{2}x$ is 1/2, whose opposite reciprocal is -2. The slope of our line is -2. Substituting $(5,18)$ in gives $18=-2(5)+b$, for which $b=28$. Thus our answer is $y=-2x+28$.
4. The slope-intercept form of $4x+12y=999$ is $y=-\frac{1}{3}x+\frac{333}{4}$. The slope is $-1/3$, whose opposite reciprocal is 3. The slope of our line is 3. Substituting $(7,3)$ in gives $3=3(7)+b$, for which $b=-18$. Thus our answer is $y=3x-18$.
5. The slope of $y=10$ is 0 (horizontal line), whose opposite reciprocal is undefined. Thus our answer is $x=6$, a vertical line.
6. The slope of $x=10$ is undefined (vertical line). By the definition of perpendicular, our line must be horizontal, and it is $y=7$.

Coincident　重合

n. [koʊ'ɪn.sɪ.dənt]

Definition: Two lines that have the same slope and have infinitely many points of intersection, as shown in Figure 7-41.　若两直线斜率相等,且有无数个交点,则它们重合,如图 7-41 所示。

Questions: See Question 4 of intersecting.　见相交词条问题 4。

Figure 7-41

7.3.6 Quadratic Functions　二次函数

视频 16

7.3.6.1 Introduction　介绍

Quadratic Function　二次函数

n. [kwɑ'drætɪk 'fʌŋkʃən]

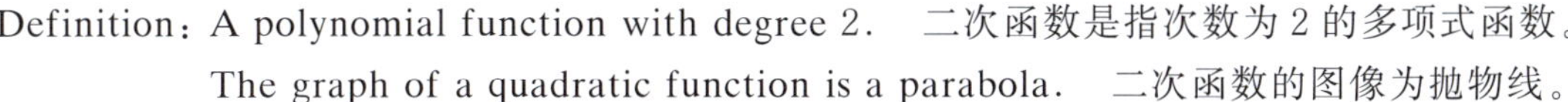

Definition: A polynomial function with degree 2.　二次函数是指次数为 2 的多项式函数。

The graph of a quadratic function is a parabola.　二次函数的图像为抛物线。

Notation: 1. General Form (sometimes called the standard form): $f(x)=ax^2+bx+c$, for which a, b, and c are constants, and $a\neq 0$.　二次函数的一般式为 $f(x)=ax^2+bx+c$,其中 a、b、c 为常值,且 $a\neq 0$。

2. Vertex Form (sometimes called the standard form too, depending on the convention): $f(x)=a(x-h)^2+k$, for which h and k are constants, and $a\neq 0$. The vertex is at $(h,$

k). 二次函数的顶点式(取决于惯例,有时也称为一般式)为 $f(x)=a(x-h)^2+k$,其中 h 和 k 为常值,且 $a\neq0$。

3. Factored Form: $f(x)=a(x-p)(x-q)$, for which p and q are constants, and $a\neq0$. The roots are p and q. 二次函数的交点式为 $f(x)=a(x-p)(x-q)$,其中 p 和 q 为常值,$a\neq0$。

In this section, a always stands for the leading coefficient. 在本节中,a 总是最高次项的系数。

Properties: 1. If $a<0$, then we say the graph is concave down, and the graph has a point of maximum, as shown in Figure 7-42. 若 $a<0$,则二次函数的图像下凹,且有最高值点,如图 7-42 所示。

If $a>0$, then we say the graph is concave up, and the graph has a point of minimum, as shown in Figure 7-43. 若 $a>0$,则二次函数的图像上凹,且有最低值点,如图 7-43 所示。

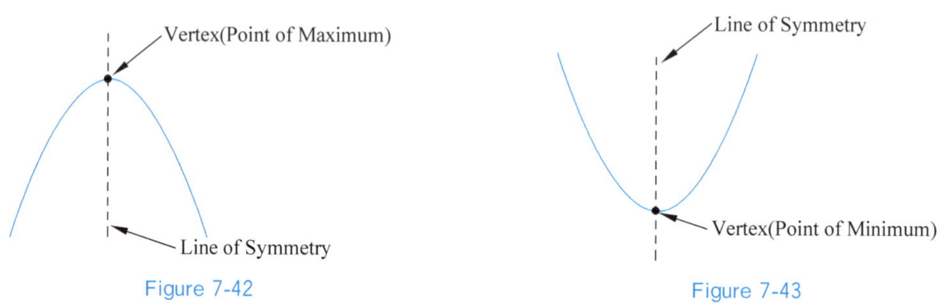

Figure 7-42　　Figure 7-43

The point of maximum or the point of minimum is known as the vertex. 最大值点或最小值点统称顶点。

2. The graph of a quadratic function is symmetric about the vertical line that contains the vertex (see the notes 3 on vertex form in Section 7.3.6.3). 二次函数的图像对称于包含顶点的垂直线(见 7.3.6.3 节顶点式的注意 3)。

Proof: To prove Property 1 above, let's work with the vertex form ($f(x)=a(x-h)^2+k$): 要证明性质 1,观察顶点式 $f(x)=a(x-h)^2+k$:

Note that $(x-h)^2$ is always nonnegative, since the square of any number is always nonnegative. 注意 $(x-h)^2$ 是非负数——任何数的平方都是非负数。

If $a<0$, as x goes farther from h (either left or right), $a(x-h)^2$ is decreasing. The inequality $k>a(x-h)^2+k$ is true for all $x\neq h$. Thus the point of maximum of the graph is (h,k). 若 $a<0$, 则当 x 离 h 越来越远(两边)时,$a(x-h)^2$ 递减。对于所有不等于 h 的 x 值,$k>a(x-h)^2+k$ 成立。

Figure 7-44 shows this case.

If $a>0$, as x goes farther from h (either left or right), $a(x-h)^2$ is increasing. The inequality $k<a(x-h)^2+k$ is true for all $x\neq h$. Thus the point of minimum of the graph is (h,k). 若 $a>0$,则当 x 离 h 越来越远(两边)时,$a(x-h)^2$ 递增。对于所有不等于 h 的 x 值,$k<a(x-h)^2+k$ 成立。

Figure 7-45 shows this case.

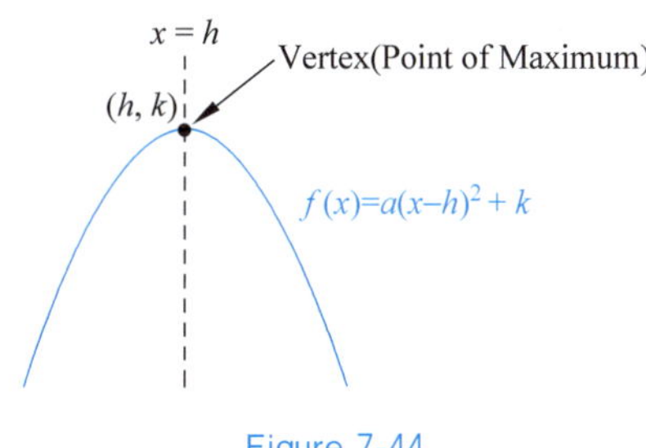

Figure 7-44

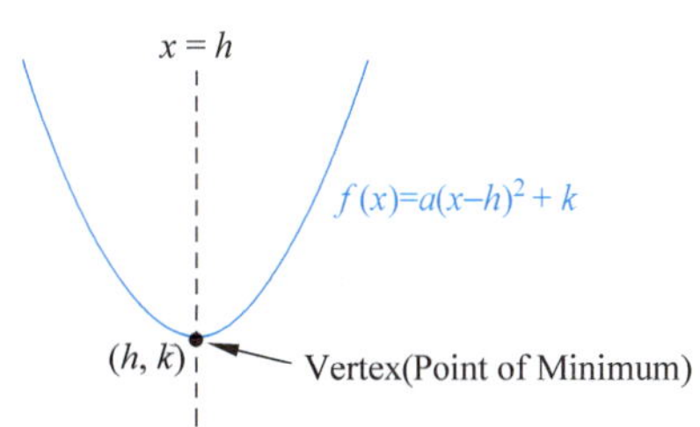

Figure 7-45

All quadratic functions in the vertex form can be converted to standard form, and vice versa. See standard form and vertex form. 顶点式的二次函数能被转换成一般式,反之亦然。见一般式和顶点式。

7.3.6.2 Vertex 顶点

Vertex 顶点

n. ['vɜrteks]

Definition: Refers to the point of maximum or point of minimum of the graph of a quadratic function. 顶点指二次函数图像的最大值点或最小值点。

Point of Maximum 最大值点

n. ['pɔɪnt əv 'mæksəməm]

Definition: A point whose y-coordinate is the greatest. 最大值点是 y 坐标值最大的点。

Refer to the graph of quadratic function, or the graph of any function in general. 指的是二次函数的图像。泛指所有函数的图像。

Examples: If the point of maximum of the graph of a quadratic function f is $(4,10)$, below are several ways of expressing it. Notice how we use the words point of maximum, maximum, and maximum value:

若二次函数 f 的图像的最大值点为$(4,10)$,下面的描述是等价的。注意最大值点和最大值的用法。

The point of maximum of f is $(4,10)$. f 的最大值点是$(4,10)$。

The maximum of f occurs at $x=4$. f 的最大值在 $x=4$。

The maximum value of f is 10. f 的最大值是 10。

Point of Minimum 最小值点

n. ['pɔɪnt əv 'mɪnəməm]

Definition: A point whose y-coordinate is the least. 最小值点是 y 坐标值最小的点。

Refer to the graph of quadratic function, or the graph of any function in general. 指的是二次函数的图像。泛指所有函数的图像。

Examples: If the point of minimum of the graph of a quadratic function f is $(4,10)$, below are several ways of expressing it. Notice how we use the words point of minimum, minimum, and minimum value:

若二次函数 f 的图像的最小值点为$(4,10)$,下面的描述是等价的。注意最小值点和最小值的用法。

The point of minimum of f is (4,10). f 的最小值点是(4,10)。

The minimum of f occurs at $x=4$. f 的最小值在 $x=4$。

The minimum value of f is 10. f 的最小值是 10。

7.3.6.3 Methods of Expressing Quadratic Functions 二次函数的表示法

General Form 一般式

n. [ˈdʒenrəl fɔrm]

Notation: $f(x)=ax^2+bx+c$, for which a, b, and c are constants, and $a\neq 0$.

Figure 7-46 shows the case of $a<0$.

Figure 7-47 shows the case of $a>0$.

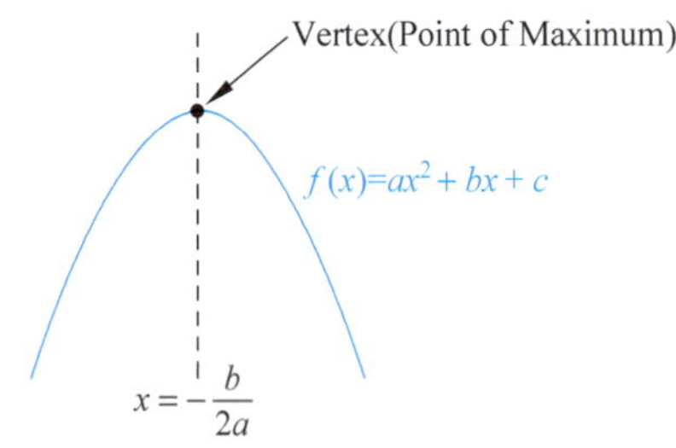

Figure 7-46

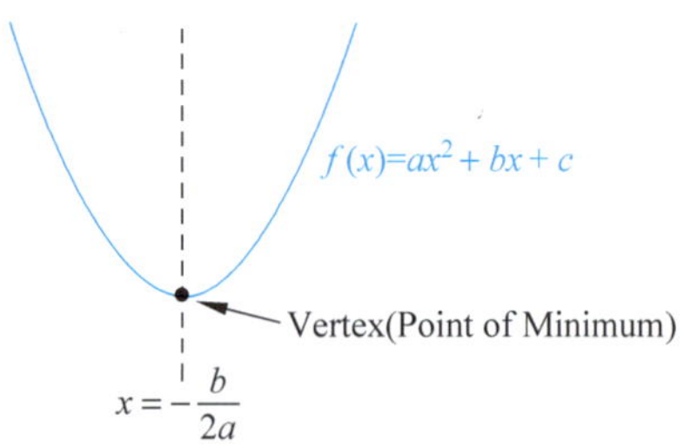

Figure 7-47

Properties: For $f(x)=ax^2+bx+c$:

1. The vertex occurs at $x=-\dfrac{b}{2a}$. 顶点在 $x=-\dfrac{b}{2a}$ 上。
2. The y-intercept is $f(0)=c$. y 截距是 $f(0)=c$。
3. The graph is symmetric about $x=-\dfrac{b}{2a}$. 图像对称于 $x=-\dfrac{b}{2a}$。
4. (1) Advantages: ① Easy for factoring (if possible) or applying the Quadratic Formula, or completing the square when solving $f(x)=0$. 当解方程 $f(x)=0$,更容易运用因式分解法、一元二次方程求根公式、配方法。

   ② Easy to find the y-intercept. 容易求出 y 截距。

   ③ This is the standard form of polynomials. 符合多项式的一般形式。

   (2) Disadvantages: The vertex is not "obvious" enough, which may affect the speed of graphing. 顶点没有明确显示,影响画图的速度。

Proofs: 1. The proof for Property 1 below not only shows the property, but also illustrates how we can rewrite a general form as a vertex form: 以下是性质 1 的证明,而且展示了如何从一般式转换为顶点式。

$$f(x)=ax^2+bx+c$$

$$=a\left(x^2+\frac{b}{a}x\right)+c \qquad \text{Factor out the } a.$$

$$=a\left(x^2+\frac{b}{a}x\right)+\frac{b^2}{4a}+c-\frac{b^2}{4a} \qquad \text{Manipulation.}$$

$$=a\left(x^2+\frac{b}{a}x+\frac{b^2}{4a^2}\right)+\left(c-\frac{b^2}{4a}\right) \qquad \text{Complete the square.}$$

$$= a\left(x+\frac{b}{2a}\right)^2+\left(c-\frac{b^2}{4a}\right) \quad \text{Vertex Form.}$$

Therefore, the vertex of f is located at $x=-\frac{b}{2a}$. 所以,f 的顶点在 $x=-\frac{b}{2a}$ 上。

2. To prove Property 3, see notes 3 of the vertex form. 要证明性质 3,见顶点式词条的注意 3(Notes 3)。

Questions: 1. A projectile is launched from the ground. The height, in feet, of the projectile is given by the function $h(t)=-16t^2+96t+15$, where t is the time in seconds.

(1) What is the initial height of the projectile?

(2) When will the projectile reach the maximum height?

(3) What is the maximum height of the projectile?

2. Rewrite each of the following to the vertex form:

(1) $f(x)=x^2-10x+16$

(2) $g(x)=3x^2+24x+99$

3. If the sum of two real numbers is 100, what is their greatest possible product?

Answers: 1. (1) $h(0)=15$ (feet).

(2) $t=-\frac{96}{2(-16)}=3$ (sec).

(3) $h(3)=159$ (feet).

2. (1) $f(x)=x^2-10x+16$ Original equation.

$=(x^2-10x+25)+16-25$ Complete the square.

$=(x-5)^2-9$ Vertex form.

(2) $g(x)=3x^2+24x+99$ Original equation.

$=3(x^2+8x)+99$ Factor out the 3.

$=3(x^2+8x)+48+99-48$ Manipulation.

$=3(x^2+8x+16)+99-48$ Complete the square.

$=3(x+4)^2+51$ Vertex form.

3. Let one of the numbers be x. It follows that the other is $100-x$. We want to find the maximum value of $x(100-x)=100x-x^2=-x^2+100x$. The maximum occurs at $x=-100/(-2)=50$. Thus the maximum value of $x(100-x)$ is $50(100-50)=2500$.

Vertex Form 顶点式

n. ['vɜrteks fɔrm]

Notation: $f(x)=a(x-h)^2+k$, for which a, h and k are constants, and $a\neq 0$.

Figure 7-48 shows the case of $a<0$.

Figure 7-49 shows the case of $a>0$.

Properties: For $f(x)=a(x-h)^2+k$:

1. The vertex is (h,k). 顶点是(h,k)。

2. The y-intercept is at ah^2+k. y 截距是 ah^2+k。

3. The graph is symmetric about $x=h$. 图像对称于 $x=h$。

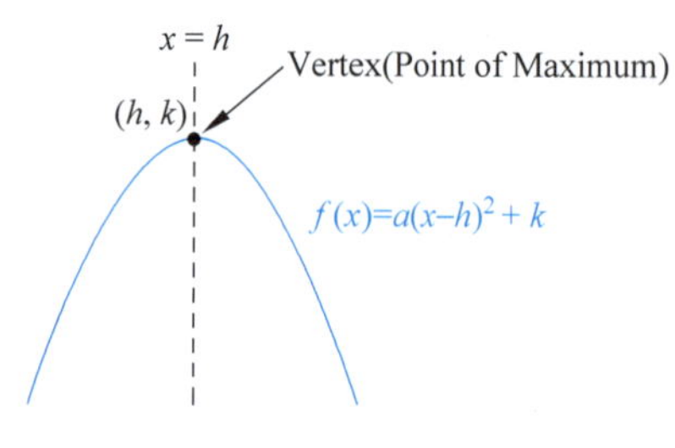

Figure 7-48

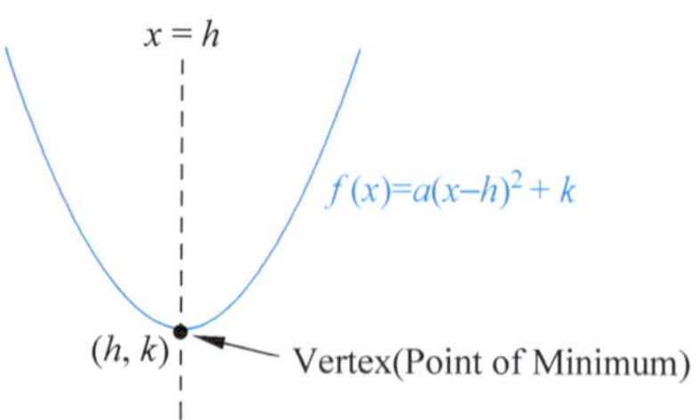

Figure 7-49

4. (1) Advantages: ① The vertex is easy to identify, and graphing is relatively easy. 顶点很明显,画图相对来说比较容易。

② Solving $f(x)=0$ is easy too because the vertex form resembles the result of completing the square. 解方程 $f(x)=0$ 比较简单,用配方法即可。

(2) Disadvantages: The y-intercept takes longer to find. 求解 y 截距的时间长了。

Notes: 1. We will rewrite the vertex form as the general form:

可以把顶点式转换为一般式:

$f(x)=a(x-h)^2+k$

$=a(x^2-2xh+h^2)+k$ Square of a difference.

$=ax^2-2axh+ah^2+k$ Distribute.

$=ax^2-(2ah)x+(ah^2+k)$ General Form.

2. Here is how we will solve $a(x-h)^2+k=0$.

$a(x-h)^2+k=0$

$a(x-h)^2=-k$ Subtract k from both sides.

$(x-h)^2=-\dfrac{k}{a}$ Divide both sides by a.

$x-h=\pm\sqrt{-\dfrac{k}{a}}$ Square root both sides, provided that $-\dfrac{k}{a}\geqslant 0$.

$x=h\pm\sqrt{-\dfrac{k}{a}}$ Add h to both sides.

3. Why does $f(x)=a(x-h)^2+k$ is symmetric about $x=h$?

As shown in Figure 7-50, let Δx be a positive amount. We need to show that $f(h-\Delta x)=f(h+\Delta x)$ for all Δx to prove what we want. 如图 7-50 所示,设 Δx 是一个正的量。需要证明不论 Δx 的值是什么,$f(h-\Delta x)=f(h+\Delta x)$ 成立。

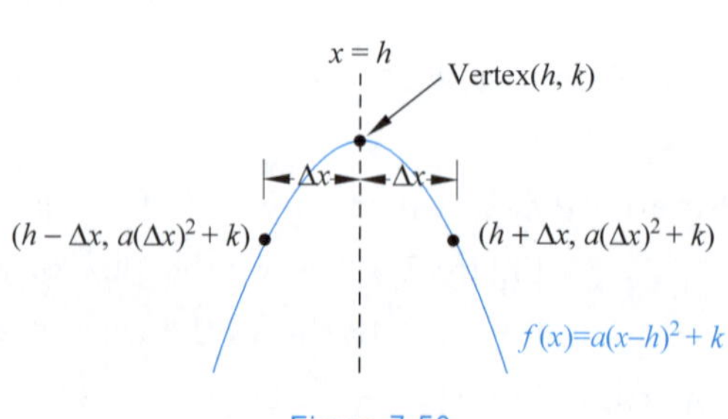

Figure 7-50

$f(h-\Delta x)=a((h-\Delta x)-h)^2+k=a(-\Delta x)^2+k=a(\Delta x)^2+k$.

$f(h+\Delta x)=a((h+\Delta x)-h)^2+k=a(\Delta x)^2+k$.

Since $f(h-\Delta x)=f(h+\Delta x)$, $f(x)$ is symmetric about $x=h$.

Questions: For the function below, (1) Identify the vertex and classify whether it is a point of maximum or a point of minimum. (2) Identify the y-intercept. (3) Find the roots. (4) Rewrite it as the general form.

$$f(x)=6(x-8)^2-54$$

Answers: (1) Vertex at $(8,-54)$. Since $6>0$, it is a point of minimum.

(2) $f(0)=6(0-8)^2-54=330$.

(3)

$6(x-8)^2-54=0$	Original equation.
$6(x-8)^2=54$	Add 54 to both sides.
$(x-8)^2=9$	Divide both sides by 6.
$x-8=\pm 3$	Square root both sides.
$x=8\pm 3$	Add 8 to both sides.
$x=5,11$.	

(4)
$$\begin{aligned} f(x)&=6(x-8)^2-54\\ &=6(x^2-16x+64)-54\\ &=6x^2-96x+384-54\\ &=6x^2-96x+330. \end{aligned}$$

Factored Form　交点式

n. [ˈfæktərd fɔrm]

Notation: $f(x)=a(x-p)(x-q)$, for which p and q are constants. The roots are p and q.

Figure 7-51 shows the case of $a<0$.

Figure 7-52 shows the case of $a>0$.

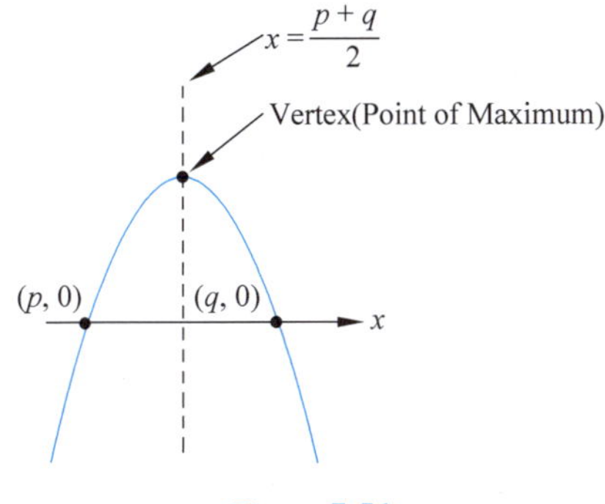

Figure 7-51

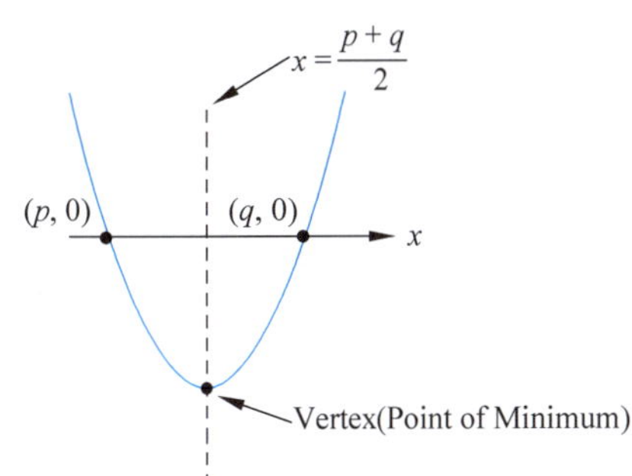

Figure 7-52

Properties: 1. The vertex is at $x=\dfrac{p+q}{2}$.　顶点在 $x=\dfrac{p+q}{2}$ 上。

2. The y-intercept is at apq.　y 截距是 apq。

3. The graph is symmetric about $x=\dfrac{p+q}{2}$.　图像对称于 $x=\dfrac{p+q}{2}$。

4. (1) Advantages: The roots are easy to find. 求根很容易。

(2) Disadvantages: ① Not all quadratic functions can be written in factored form.

Some quadratic functions do not have real roots. 不是所有的二次函数都可以用交点式表示。有些二次函数没有实数根。

② Some quadratic functions have real but irrational roots, which is tough to be written in factored form. Usually the general form or the vertex form is better to represent all these types of quadratic functions. 有的二次函数的根是实数,而且是无理数。用交点式表示比较困难。这种二次函数用一般式或顶点式表示更好。

Proof: We have proved that the graphs of quadratic functions are symmetric. The x-coordinate of the vertex is in the middle of the x-coordinates of the roots. Thus Properties 1 and 3 hold. 已经证出二次函数的图像是对称的。顶点的 x 坐标在两个解的 x 坐标的中间。所以性质 1 和 3 成立。

Questions: For the function below, (1) Identify the vertex and classify whether it is a point of maximum or point of minimum. (2) Identify the y-intercept. (3) Find the roots. (4) Rewrite it as the general form. (5) Rewrite it in vertex form.

$$f(x)=4(x-5)(x+7)$$

Answers: (1) The vertex is at $x=\dfrac{5+(-7)}{2}=-1$. The point is $(-1,-144)$. Since $4>0$, it is a point of minimum.

(2) $f(0)=-140$.

(3) $5, -7$.

(4) $f(x)=4(x-5)(x+7)$
$=4(x^2+2x-35)$
$=4x^2+8x-140$.

(5) From Part (a), we know that the leading coefficient is 4 and $h=-1$, and $k=-144$. Thus, we have $f(x)=4(x+1)^2-144$.

7.3.7 Summary 总结

1. A polynomial function is a function of the form $f(x)=a_nx^n+a_{n-1}x^{n-1}+a_{n-2}x^{n-2}+\cdots+a_1x+a_0$, for which $a_n, a_{n-1}, a_{n-2}, \cdots, a_1$, and a_0 are constants. This is the standard form of polynomial functions. 多项式函数是以 $f(x)=a_nx^n+a_{n-1}x^{n-1}+a_{n-2}x^{n-2}+\cdots+a_1x+a_0$ 的形式表示的单值函数,其中 $a_n, a_{n-1}, a_{n-2}, \cdots, a_1, a_0$ 为常值。这是所有多项式函数的一般形式。
2. When the degree of a polynomial function is 0, we call this a constant function: $f(x)=k$. Its graph is a horizontal line. 当多项式函数的次数为 0,称为常值函数,即 $f(x)=k$,其图像为一条水平线。
3. When the degree of a polynomial function is 1, we call this a linear function: $f(x)=mx+b$. Its graph is a line that is neither horizontal nor vertical. 当多项式函数的次数为 1,称为一次函数/线性函数,即 $f(x)=mx+b$,其图像为倾斜的直线(非水平线或垂直线)。

   Methods of expressing equations of lines include:
   直线方程表示方法如下:

- Slope-Intercept Form：　$y = mx + b, m \neq 0$.　斜截式。
- Point-Slope Form：　$y - y_0 = m(x - x_0), m \neq 0$.　点斜式。
- Standard Form：　$Ax + By = C$, A and B are not both 0.　一般式。

4. Slope $= m = \frac{\text{rise}}{\text{run}} = \frac{y_2 - y_1}{x_2 - x_1}$. The slope of every line is consistent.　斜率 $= \frac{\text{上升运行增量}}{\text{向右运行增量}}$。每条直线只有一个斜率。
   - The slope of a horizontal line is 0.　水平线的斜率为 0。
   - The slope of a vertical line is undefined.　垂直线的斜率无意义。
   - The slope of every other line is either positive or negative.　其他线的斜率是正数或负数。
5. When the y-intercept of a linear function f is 0, then f is a direct variation: $f(x) = kx$, for some constant k.　当线性函数 f 的 y 截距为 0,则 f 为正比例函数,即 $f(x) = kx$,其中 k 为常值。
6. Two lines in the coordinate plane satisfy exactly one of the following:
   在坐标平面的两直线满足以下情况之一：
   - Parallel—Slopes are the same and lines have no points of intersection.　平行。
   - Intersecting—The lines have exactly one point of intersection.　相交。
     Special case: perpendicular—Slopes are opposite reciprocals.　相交的特殊情况：垂直。
   - Coincidence—Slopes are the same, and lines have infinitely many points of intersection.　重合。
7. When the degree of a polynomial function is 2, we call this a quadratic function: $f(x) = ax^2 + bx + c$. Its graph is a parabola.　当多项式函数的次数为 2 时,称为二次函数,即 $f(x) = ax^2 + bx + c$,其图像为抛物线。
   Methods of expressing quadratic functions include:
   二次函数的表示方法如下：
   - Standard Form：　$f(x) = ax^2 + bx + c, a \neq 0$.　一般式。
   - Vertex Form：　$f(x) = a(x - h)^2 + k, a \neq 0$.　顶点式。
   - Factored Form：　$f(x) = a(x - p)(x - q), a \neq 0$.　交点式。

   If $a > 0$, then the graph is concave up and the vertex is a point of minimum.　若 $a > 0$,则二次函数的图像上凹,且顶点是最小值点。
   If $a < 0$, then the graph is concave down and the vertex is a point of maximum.　若 $a < 0$,则二次函数的图像下凹,且顶点是最大值点。
   The graph of every quadratic function is symmetric about the vertical line containing the vertex.　每个二次函数的图像都对称于包含顶点的垂直线。

7.4 Non-Polynomial Functions　非多项式函数

7.4.1 Inverse Variations　反比例函数

Inverse Variation　反比例函数

n. [ɪn'vɜrs ˌveəri'eɪʃən]

Definition: 1. y is a quotient of a nonzero constant and x.
y 为非零常值与 x 的商。
2. A function for which xy is a nonzero constant.

使 xy 为非零常值的函数。

The two definitions above are equivalent. 以上两个定义等价。

Notation: $xy = k$ or $y = k/x$, for which $k \neq 0$.

We also say that x and y are inversely proportional.

$xy = k$ 或 $y = k/x$, $k \neq 0$, 则称 x 和 y 成反比。

Figure 7-53 shows the case of $k<0$.

Figure 7-54 shows the case of $k>0$.

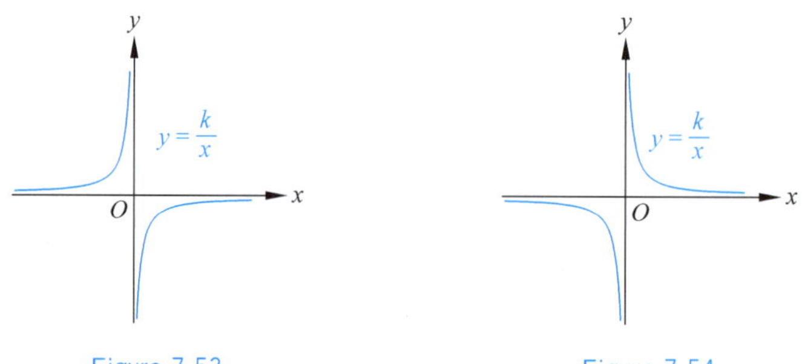

Figure 7-53　　Figure 7-54

Properties: 1. If $y = k/x$, for which k is a nonzero constant, then it is true that the y-intercept does not exist (x cannot be 0, otherwise k/x is undefined.).

若 $y = k/x$, 其中 k 为非零常值,则反比例函数的图像没有 y 截距(x 不能为 0,否则分数无意义)。

2. When $k>0$, y increases proportionally as x decreases, and y decreases proportionally as x increases (Golden Rule). 若 $k>0$,随着 x 减少, y 成比例增加。随着 x 增加, y 成比例减少。

Examples: Some real-life examples of inverse variations $y = k/x$ (k is a positive constant):

生活中反比例函数的例子:

1. x = unit price of items 单价; y = number of items 数量; k = total price of items 总价.

   k is fixed. y increases as x decreases, and y decreases as x increases.

2. x = time 时间; y = speed traveled 速度; k = distance traveled 路程.

   k is fixed. y increases as x decreases, and y decreases as x increases.

3. x = efficiency of work 效率; y = time used 时间; k = amount of work done 工作.

   k is fixed. y increases as x decreases, and y decreases as x increases.

Questions: 1. What is the equation of an inverse variation whose graph passes through (7,3)?

2. The speed John traveled is inversely proportional to the time he used. If he travels at 40mph for 3 hours, how much time, in hours, does he need to travel the same distance at 60mph?

Answers: 1. $xy = 21$.

2. Assume that he needs t hours. We have $40 \times 3 = 60t$, for which $t = 2$.

7.4.2　Exponential Functions　指数函数

7.4.2.1　Introduction　介绍

Exponential Function　指数函数

n. [ˌekspəˈnenʃəl ˈfʌŋkʃən]

Definition: A function that is a constant multiplied by a power with a constant base and a variable exponent.　指数函数是指常数乘幂的函数。在幂中，底为常数，指数为变量。

Notation: $f(x)=ab^x$, for which a and b are constants with $a\neq 0, b>0$ and $b\neq 1$ (If $b=1$, then $f(x)=a$, which is a constant function). We call b the constant base, and x the variable exponent.

$f(x)=ab^x$，其中 a 和 b 为常数，使得 $a\neq 0$、$b>0$、$b\neq 1$(若 $b=1$，则 $f(x)=a$ 为常值函数)。b 为常数底，x 为变量指数。

Properties: For the exponential function $f(x)=ab^x$, for which $a\neq 0, b>0$ and $b\neq 1$,

1. If $0<b<1$, then we say f is an exponential decay, as shown below.　若 $0<b<1$，则 f 为指数衰减的情况。

   Figure 7-55 shows the case of $a<0$.

   Figure 7-56 shows the case of $a>0$.

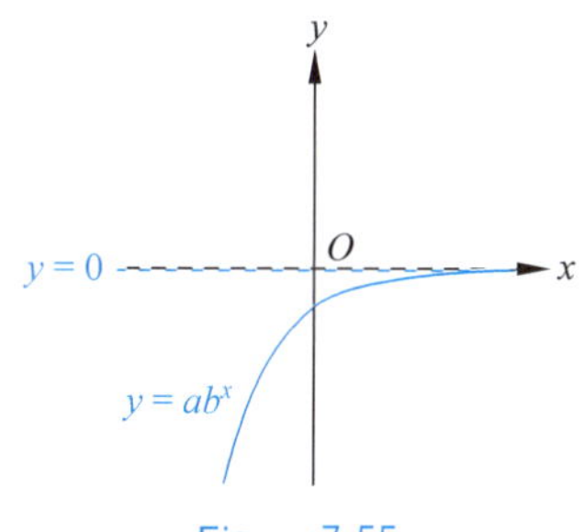

Figure 7-55

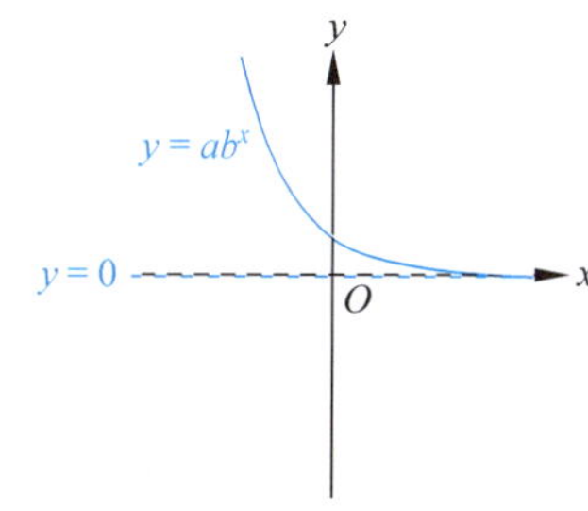

Figure 7-56

2. If $b>1$, then we say f is an exponential growth, as shown below.　若 $b>1$，则 f 为指数增长的情况。

   Figure 7-57 shows the case of $a<0$.

   Figure 7-58 shows the case of $a>0$.

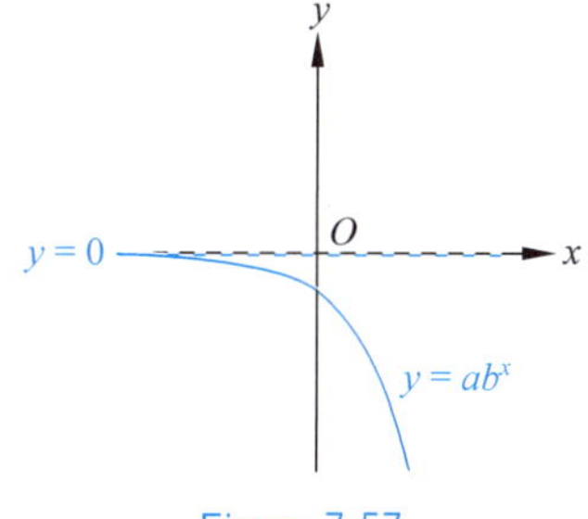

Figure 7-57

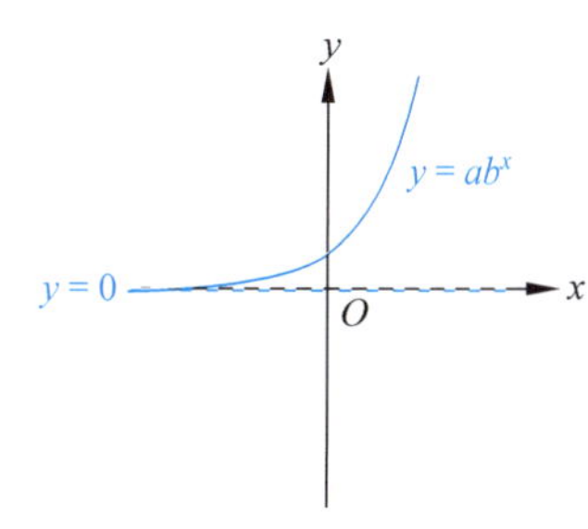

Figure 7-58

3. The horizontal asymptote of f is $y=0$.

   $y=0$ 为 f 的水平渐近线。

Questions: Rewrite each of the following functions in the form of ab^x, for which a and b are constants (Hint: Use the properties of exponents). Then classify whether it is an exponential growth or exponential decay.

(1) $g(x) = -5^{x+2}$.

(2) $h(x) = 6(2^{2x+4})$.

(3) $j(x) = -4\left(\frac{1}{3}\right)^{x-2}$.

(4) $k(x) = \frac{1}{3}\left(\frac{1}{4}\right)^{3x+2}$.

(5) $l(x) = -8^{-2x+3}$.

(6) $m(x) = 6\left(\frac{1}{7}\right)^{-3x-2}$.

Answers: (1) $g(x) = -5^{x+2} = -1 \cdot 5^x \cdot 5^2 = -25(5^x)$. Since $5>1$, $g(x)$ is an exponential growth.

(2) $h(x) = 6(2^{2x+4}) = 6(2^{2x})(2^4) = 6(4^x)(16) = 96(4^x)$. Since $4>1$, $h(x)$ is an exponential growth.

(3) $j(x) = -4\left(\frac{1}{3}\right)^{x-2} = -4\left(\frac{1}{3}\right)^x\left(\frac{1}{3}\right)^{-2} = -4\left(\frac{1}{3}\right)^x(9) = -36\left(\frac{1}{3}\right)^x$. Since $1/3<1$, $j(x)$ is an exponential decay.

(4) $k(x) = \frac{1}{3}\left(\frac{1}{4}\right)^{3x+2} = \frac{1}{3}\left(\frac{1}{4}\right)^{3x}\left(\frac{1}{4}\right)^2 = \frac{1}{3}\left(\frac{1}{64}\right)^x\left(\frac{1}{16}\right) = \frac{1}{48}\left(\frac{1}{64}\right)^x$. Since $1/64<1$, $k(x)$ is an exponential decay.

(5) $l(x) = -8^{-2x+3} = -1 \cdot (8^{-2x})(8^3) = -1 \cdot \left(\frac{1}{64}\right)^x(8^3) = -512\left(\frac{1}{64}\right)^x$. Since $1/64<1$, $l(x)$ is an exponential decay.

(6) $m(x) = 6\left(\frac{1}{7}\right)^{-3x-2} = 6\left(\frac{1}{7}\right)^{-3x}\left(\frac{1}{7}\right)^{-2} = 6(343)^x(49) = 294(343)^x$. Since $343>1$, $m(x)$ is an exponential growth.

7.4.2.2 Exponential Growth and Decay 指数增长与衰减

Exponential Growth 指数增长

n. [ˌekspəˈnenʃəl ɡroʊθ]

Definition: For the exponential function $f(x) = ab^x$, for which $a \neq 0$ and $b>1$. 符合 $a \neq 0$ 和 $b>1$ 的指数函数 $f(x) = ab^x$。

Figure 7-59 shows the case of $a<0$.

Figure 7-60 shows the case of $a>0$.

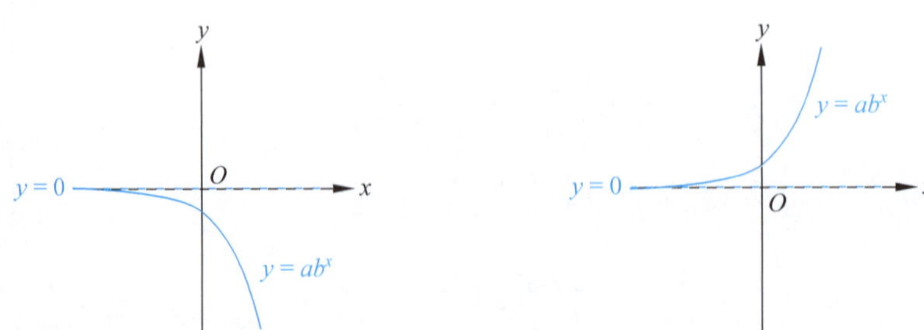

Figure 7-59　　Figure 7-60

Notation：$f(x)=a(1+r)^x$. We rewrite b as $1+r$, for which $r>0$ (Since $b>1$, r must be positive.). 这种表示法与 $f(x)=ab^x$ 等价，同时显示了增长率。

a is said to be the initial quantity since $f(0)=a$, and r is said to be the (exponential) growth rate.

a 为初始的量(因为 $f(0)=a$)。r 为指数增长率。

Questions：The University of Intelligence charges \$50000 for tuition this year. Each year after, it raises the tuition by 10% from the previous year. Find (1) a table that shows the tuition (in dollars) for 0, 1, 2, 3, and 4 years from now, (2) the initial tuition (in dollars) and the growth rate, (3) an exponential function $f(t)$ that represents the tuition (in dollars) t years from now, and (4) the tuition, to the nearest cent, after 7 years.

Answers：(1) Increasing by 10% is the same as multiplying by 1.10 (see percent increase in the common percent problems in Section 1.2.4.2).

We construct a table as shown in Table 7-13.

Table 7-13

Number of years from now	0	1	2	3	4
tuition	50 000	50 000(1.10) = 55 000	55 000(1.10) = 60 500	60 500(1.10) = 66 550	66 550(1.10) = 73 205

(2) Initial quantity：\$50 000. Growth rate：10% = 0.1.

(3) $f(t)=50\,000(1+0.1)^t=50\,000(1.1)^t$.

(4) $f(7)=50\,000(1.1)^7=97\,435.86$.

Exponential Decay　指数衰减

n. [ˌekspəˈnenʃəl dɪˈkeɪ]

Definition：For the exponential function $f(x)=ab^x$, for which $0<b<1$. 符合 $0<b<1$ 的指数函数 $f(x)=ab^x$。

Figure 7-61 shows the case of $a<0$.

Figure 7-62 shows the case of $a>0$.

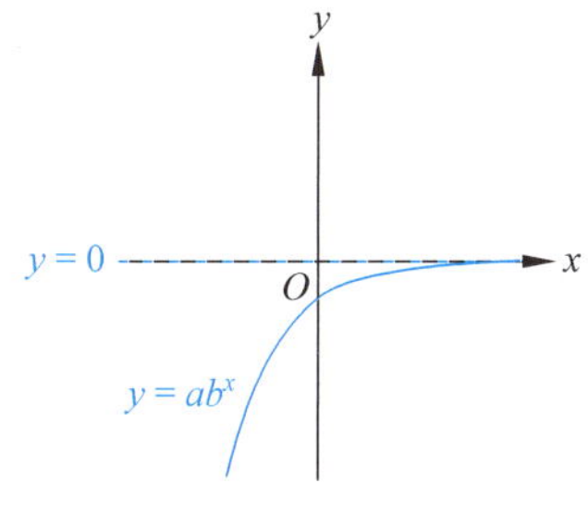

Figure 7-61

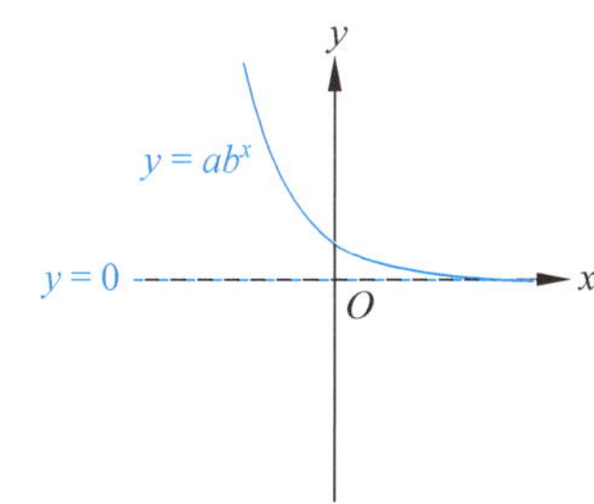

Figure 7-62

Notation：$f(x)=a(1-r)^x$. We rewrite b as $1-r$, for which $0<r<1$ (Since $0<b<1$, r must be positive.). 这种表示法与 $f(x)=ab^x$ 等价，同时显示了衰减率。

a is said to be the initial quantity since $f(0)=a$, and r is said to be the (exponential) decay rate.

a 为初始的量(因为 $f(0)=a$)。r 为指数衰减率。

Questions: In the year 2000, a car is bought in for \$ 40 000. It depreciates 5% each year after. Find (1) a table that shows the value of the car (in dollars) in 2000, 2001, 2002, 2003, and 2004, (2) the initial value of the car (in dollars) and the decay rate, (3) an exponential function $f(t)$ that represents the value of the car (in dollars) t years from the year 2000, and (4) the value of the car, to the nearest cent, in the year 2017.

Answers: (1) Decreasing by 5% is the same as multiplying by 0.95 (see percent decrease in the common percent problems in Section 1.2.4.2).

We construct a table as shown in Table 7-14.

Table 7-14

Number of years from 2000	0	1	2	3	4
tuition	40 000	40 000(0.95) = 38 000	38 000(0.95) = 36 100	36 100(0.95) = 34 295	34 295(0.95) = 32 580.25

(2) Initial quantity: \$ 40 000. Decay rate: 5% = 0.05.

(3) $f(t)=40\,000(1-0.05)^t=40\,000(0.95)^t$.

(4) $f(7)=40\,000(0.95)^7=27\,933.49$.

7.4.2.3 Horizontal Asymptote 水平渐近线

Horizontal Asymptote 水平渐近线

n. [ˌhɔrəˈzɑntəl ˈæsəmpˌtoʊt]

Definition: The line $y=b$ is a horizontal asymptote of the graph of f if $f(x)\to b$ as $x\to+\infty$ or $x\to-\infty$. 若随着 $x\to\infty$ 或 $x\to-\infty$, $f(x)\to b$, 则 $y=b$ 为函数 f 图像的水平渐近线。

The symbol "→" is read as "approaches to, **but never reaches**".

右箭头表示"渐进,但永不到达"。

The horizontal asymptote for the graph of function f is shown in Figure 7-63.

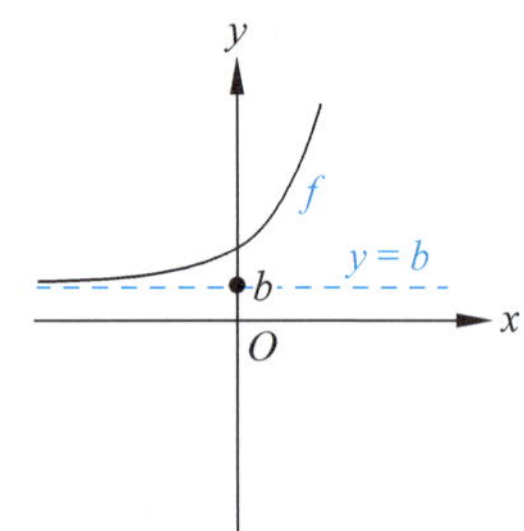

Figure 7-63

Notation: $y=b$, for which b is a constant.

Properties: Exponential function $f(x)=ab^x$ (for which a and b are constants, $b>0$, and $b\neq 1$) has $y=0$ as a horizontal asymptote. 对于所有 $f(x)=ab^x$ 形式的指数函数, $y=0$ 是水平渐近线。

Proof: To prove the property above, let's divide into cases. In $f(x)=ab^x$,

- If $b>1$, then $f(x)\to 0$ as $x\to-\infty$. The reason is that as x becomes very negative, $-x$ becomes very positive. Thus $b^x=\frac{1}{b^{-x}}=\frac{1}{b \text{ to an indefinitely large positive number}}$, which is approaching to 0.
- If $0<b<1$, then $f(x)\to 0$ as $x\to\infty$. The reason is that as x becomes very positive,

b^x approaches to 0.

Therefore, in both cases, we have the horizontal asymptote $y = 0$, by definition.

根据指数增长和指数衰减分情况讨论。

若指数增长,则根据指数性质,证明当 $x \to -\infty$ 时, $f(x) \to 0$。

若指数衰减,则根据指数性质,证明当 $x \to \infty$ 时, $f(x) \to 0$。

Examples: We have a piece of paper.

On the first step, we will cut the paper into two halves and keep one half.

On the second step, we will take the paper we keep from the previous step and cut it in half and keep one half.

On the third step, we will take the paper we keep from the previous step and cut it in half and keep one half.

…

Table 7-15 shows the amount of the paper being kept as a function of the number of steps.

Table 7-15

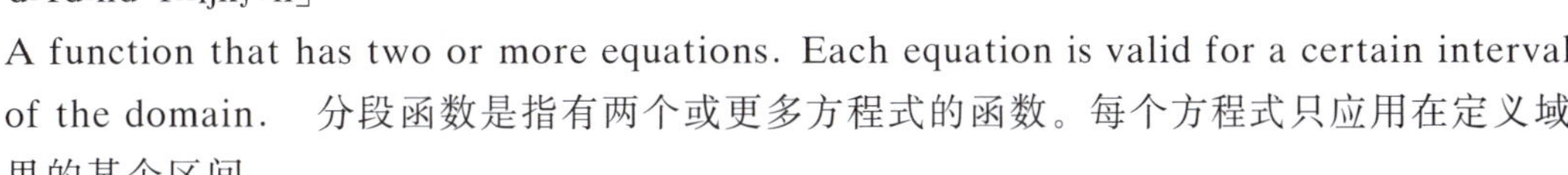

x: Number of Steps	0	1	2	3	4	5	6	7	8
$f(x)$: Amount of Original Paper Left	1	1/2	1/4	1/8	1/16	1/32	1/64	1/128	1/256

We have $f(x) = \left(\frac{1}{2}\right)^x$, an exponential decay. The horizontal asymptote is at $y = 0$. In the context of the problem, as the number of steps increases, the piece of the paper we have becomes infinitesimally small (approaching to 0, but does not reach 0 because no matter how many steps are run, we still have *something* remaining).

7.4.3 Piecewise-Defined Functions 分段函数

视频 17

Piecewise-Defined Function 分段函数

n. [piswaɪz dɪˈfaɪnd ˈfʌŋkʃən]

Definition: A function that has two or more equations. Each equation is valid for a certain interval of the domain. 分段函数是指有两个或更多方程式的函数。每个方程式只应用在定义域里的某个区间。

The graph of a piecewise-defined function is shown in Figure 7-64.

Notation: $f(x) = \begin{cases} \text{(equation 1), interval 1} \\ \text{(equation 2), interval 2.} \\ \cdots \end{cases}$

Each equation is known as a sub-function, and each interval is known as a sub-domain. 每个方程式称为分方程式,每个区间称为分定义域。

Piecewise-defined functions are sometimes called "piecewise functions" in short.

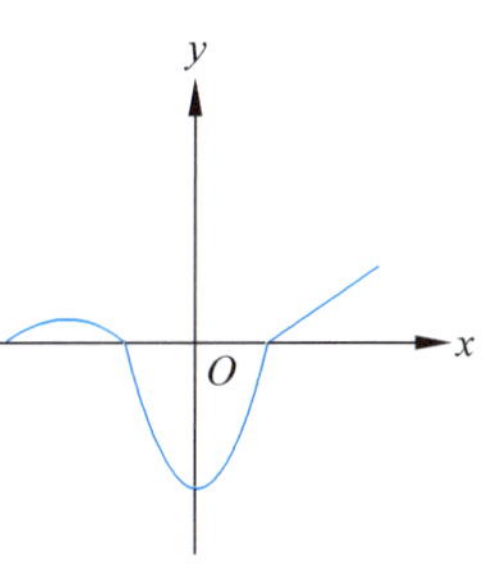

Figure 7-64

Properties: 1. To graph a piecewise-defined function, we graph all the sub-functions in their corresponding subdomain. We use an open circle at the endpoint of each subdomain to indicate that the subdomain does not include that endpoint, or a closed circle at the endpoint of each subdomain to indicate that the sub-function includes that endpoint (See Example 2). 画分段函数图,先在对应的分定义域画出每个分函数的图像。对于端点,若分定义域不包含这个点,则用空心点表示;若分定义域包含这个点,则用实心点表示,见下面例 2。

2. Many aspects of life have more than one rule. Many times, a piecewise-defined functions are the best cases to model these situations (See Questions). 在生活中有很多多于一条规则的情况,用分段函数最能表达这些情况(见下面的问题)。

Examples: 1. The absolute value function is a typical piecewise-defined function: 下面的绝对值函数是一个典型的分段函数:

$$f(x)=|x|=\begin{cases}-x, & x<0\\ x, & x\geqslant 0\end{cases}.$$

Figure 7-65 shows its graph.

2. To graph $f(x)=\begin{cases}x, & x<0\\ 2x-1, & 0\leqslant x<5\\ 10, & x\geqslant 5\end{cases}$, we have:

We graph $y=x$, $y=2x-1$, and $y=10$ on the intervals $x<0$, $0\leqslant x<5$, and $x\geqslant 5$, respectively.

Figure 7-66 shows its graph.

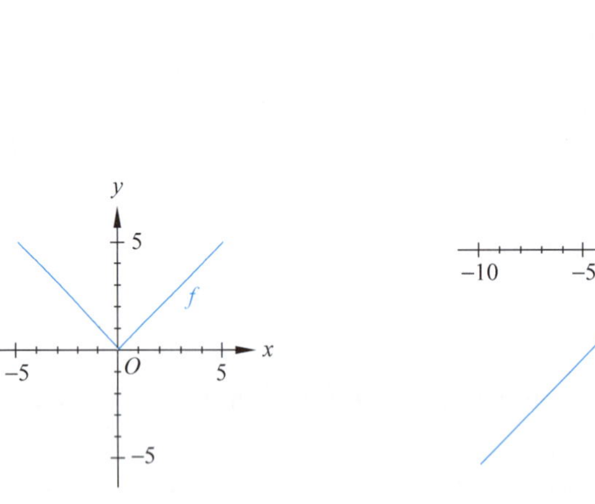

Figure 7-65 Figure 7-66

The subdomain of the first sub-function does not include 0, thus we use an open circle at $x=0$ on the graph of the first sub-function.

The subdomain of the second sub-function includes 0 but excludes 5, thus we use a closed circle at $x=0$ and an open circle at $x=5$ on the graph of the second sub-function.

The subdomain of the third sub-function includes 5, thus we use a closed circle at $x = 5$ on the graph of the third sub-function.

Questions: Superstar Math Museum charges admission to groups according to the policy below:

For a group with fewer than 50 people, the entry fee is \$40/person.

For a group with 50-100 people (inclusive), the entry fee is \$35/person.

For a group with more than 100 people, the entry fee is \$30/person.

Write a piecewise-defined function $C(n)$ representing the total entry fee for a group of n people.

Answers: $C(n)=\begin{cases}40n, & n<50\\ 35n, & 50\leqslant n\leqslant 100.\\ 30n, & n>100\end{cases}$

7.4.4 Summary 总结

1. Inverse variations: The product of x and y is a nonzero constant: $xy = k$ or $y = k/x$, for which $k \neq 0$. 反比例函数：x 和 y 的积为非零常值，即 $xy = k$ 或 $y = k/x$，其中 $k \neq 0$。
2. Exponential functions: $f(x) = ab^x$, for which a and b are constants with $a \neq 0$, $b > 0$ and $b \neq 1$. We call b the constant base, and x the variable exponent. 指数函数：$f(x) = ab^x$，其中 a 和 b 为常数，使得 $a \neq 0$、$b > 0$、$b \neq 1$。b 为常数底，x 为变量指数。
   - If $b > 1$, then f is an exponential growth, and we can rewrite $f(x) = ab^x$ as $f(x) = a(1 + r)^x$, for some $r > 0$. We say that a is the initial quantity and r is the exponential growth rate. 若 $b > 1$，则函数 f 为指数增长。可以把 $f(x) = ab^x$ 改写成 $f(x) = a(1 + r)^x$，其中 $r > 0$。a 为初始量，r 为指数增长率。
   - If $0 < b < 1$, then f is an exponential decay, and we can rewrite $f(x) = ab^x$ as $f(x) = a(1 - r)^x$, for some $0 < r < 1$. We say that a is the initial quantity and r is the exponential decay rate. 若 $0 < b < 1$，则函数 f 为指数衰减。可以把 $f(x) = ab^x$ 改写成 $f(x) = a(1 - r)^x$，其中 $0 < r < 1$。a 为初始量，r 为指数衰减率。
3. The line $y = b$ is a horizontal asymptote of the graph of f if $f(x) \to b$ as $x \to \infty$ or $x \to -\infty$.
   若随着 $x \to \infty$ 或 $x \to -\infty$，$f(x) \to b$，则 $y = b$ 为函数 f 图像的水平渐近线。
   The horizontal asymptote of the exponential function $f(x) = ab^x$ is $y = 0$.
   指数函数的水平渐近线为 $y = 0$。
4. A piecewise-defined function is a function that has two or more equations. Each equation is valid for a certain interval of the domain. 分段函数为有两个或更多方程式的函数。每个方程式只应用在定义域里的某个区间。
   An example of a piecewise-defined function is the absolute value function. 绝对值函数是分段函数的一个例子。
   Review piecewise-defined function in Section 7.4.3. 复习 7.4.3 节分段函数。

7.5 Miscellaneous 其他

7.5.1 Equivalent Statements for $f(x)=0$ $f(x)=0$ 的等价命题

Equivalent Statements for $f(x)=0$ $f(x)=0$ 的等价命题

Properties: For a function f, when $f(a)=0$, the following statements are equivalent. 对于任何函数 f,当 $f(a)=0$ 时,下面的命题等价。

- $x=a$ is a zero of f.
  $x=a$ 是 f 的零值。
- $x=a$ is a solution for the equation $f(x)=0$.
  $x=a$ 是方程 $f(x)=0$ 的解。
- If f is a polynomial function, then $(x-a)$ is a factor of f.
  若 f 为多项式函数,则 $(x-a)$ 是 f 的因式。
- a is an x-intercept for the graph of f.
  a 是 f 的图像的 x 截距。

7.5.2 Descriptions of Functions 函数的描述

Increasing 递增

adj. [ɪn'krisɪŋ]

Definition: A function f is increasing on an open interval if, for any x_1 and x_2 in the interval, $x_1<x_2$ implies $f(x_1)<f(x_2)$.

Decreasing 递减

adj. [di'krisɪŋ]

Definition: A function f is decreasing on an open interval if, for any x_1 and x_2 in the interval, $x_1<x_2$ implies $f(x_1)>f(x_2)$.

Constant 常值

adj. ['kɑnstənt]

Definition: A function f is constant on an open interval if, for any x_1 and x_2 in the interval, $f(x_1)=f(x_2)$.

Refer to Figure 7-67 for the examples of increasing, decreasing, and constant intervals.

The function f is a piecewise-defined function whose domain is $(-10,10)$, as shown in Figure 7-67.

The function f is increasing on the intervals $(-10,-5)$ and $(3,6)$, decreasing on the intervals $(0,3)$ and $(6,10)$, and constant on the interval $(-5,0)$.

Note that at $x=-10,-5,0,3,6,$ or 10, although $f(x)$ is defined, f is none of increasing, decreasing, or constant. Recall that the terms "increasing", "decreasing" and "constant" only describe open intervals.

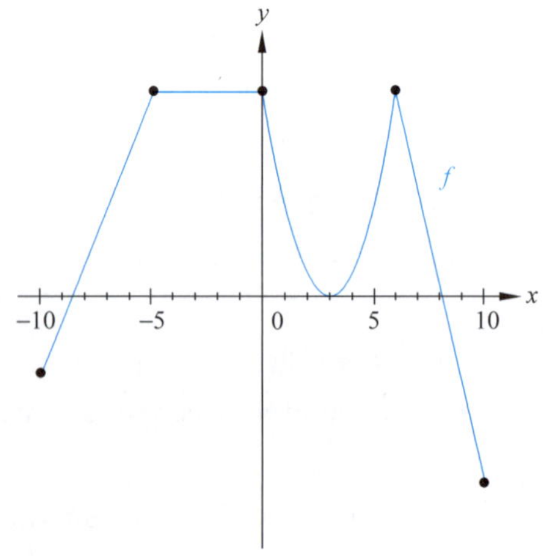

Figure 7-67

Even　偶(函数)

adj. [ˈivən]

Definition：A function f is even if $f(-x)=f(x)$ for all x.

In words, opposite inputs produce the same output.

Figure 7-68 shows an even function.

Properties：Even functions have line symmetry about the y-axis.　偶函数沿 y 轴对称。

Odd　奇(函数)

adj. [ɑd]

Definition：A function f is odd if $f(-x)=-f(x)$ for all x.

In words, opposite inputs produce the opposite output.

Figure 7-69 shows an odd function.

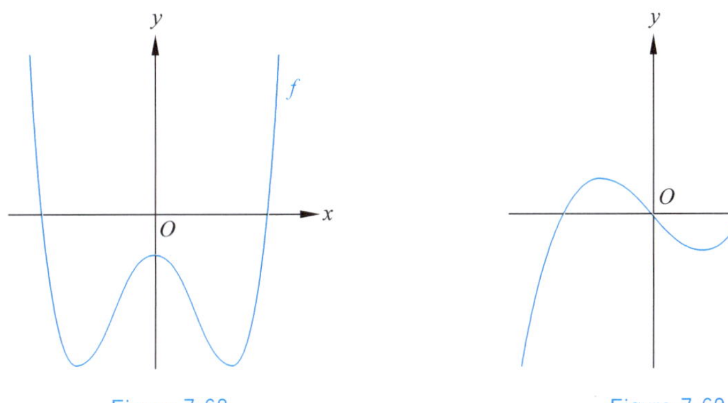

Figure 7-68　Figure 7-69

Properties：Odd functions have rotational symmetry about the origin.　奇函数存在关于坐标原点的中心对称。

Classifying Functions as Even, Odd, or Neither　对函数进行奇偶分类

Notes：Given a function $f(x)$, to determine it is even, odd, or either, we find $f(-x)$ first.

Even functions satisfy that $f(-x)=f(x)$.

Odd functions satisfy that $f(-x)=-f(x)$.

Questions：Classify each function as even, odd, or neither.

(1) $f(x)=x^4+x^2+8$

(2) $g(x)=x^3+x-9$

(3) $h(x)=5x^5-3x^3+7x$

(4) $m(x)=x^3+x^2-x-1$

(5) $n(x)=x^4-x^2+10$

Answers：(1) $f(-x)=(-x)^4+(-x)^2+8=x^4+x^2+8=f(x)$.

It is an even function.

(2) $g(-x)=(-x)^3+(-x)-9=-x^3-x-9$.

It is neither an even nor odd function.

(3) $h(-x)=5(-x)^5-3(-x)^3+7(-x)=-5x^5+3x^3-7x=-h(x)$.

It is an odd function.

(4) $m(-x)=(-x)^3+(-x)^2-(-x)-1=-x^3+x^2+x-1$.

It is neither an even nor odd function.

(5) $n(-x)=(-x)^4-(-x)^2+10=x^4-x^2+10=n(x)$.

It is an even function.

7.6 Functional Inequalities 函数不等式

Functional Inequality 函数不等式

n. [ˈfʌŋkʃənəl ˌɪnɪˈkwɑlɪti]

Notation: Same as expressing a functional equation $f(x)=\cdots$ (equation) $\cdots$ **except** that the "$=$" changes to one of $<$, $>$, $\leqslant$, or $\geqslant$.

This is an analogue to inequalities. 函数不等式与不等式方程类似。

Properties: 1. The graph of a functional inequality is 2-dimensional (because it involves a function, and there are two variables, x and y) as opposed to the graph of an inequality, which is 1-dimensional because it only has one variable. 函数不等式的图像是二维的(因为函数有两个变量 x 和 y),这与不等式的图解有所不同。不等式只有一个变量,其图解是一维图像。

2. To graph a functional inequality, we graph the function like we normally do, then follow the rules below: 要图解函数不等式,先照例画出函数的图像,然后根据以下定律进行图解:

- If we have $f(x)>\cdots$ (equation) $\cdots$, then the curve of $f(x)$ should be dashed/dotted, and we shade the area above the curve.
- If we have $f(x)<\cdots$ (equation) $\cdots$, then the curve of $f(x)$ should be dashed/dotted, and we shade the area below the curve.
- If we have $f(x)\geqslant\cdots$ (equation) $\cdots$, then the curve of $f(x)$ should be solid, and we shade the area above the curve.
- If we have $f(x)\leqslant\cdots$ (equation) $\cdots$, then the curve of $f(x)$ should be solid, and we shade the area below the curve.

Questions: 1. Graph each of the following:

(1) $f(x)>4x-3$

(2) $g(x)<8$

(3) $h(x)\geqslant x^2$

(4) $j(x)\leqslant 2^x$

2. John earns at least \$50 per hour. Write a functional inequality $f(t)$ showing the amount John earns in t hours.

Answers: 1. (1) Figure 7-70 shows the graph of $f(x)>4x-3$.

(2) Figure 7-71 shows the graph of $g(x)<8$.

(3) Figure 7-72 shows the graph of $h(x)\geqslant x^2$.

(4) Figure 7-73 shows the graph of $j(x)\leqslant 2^x$.

2. $f(t)\geqslant 50t$.

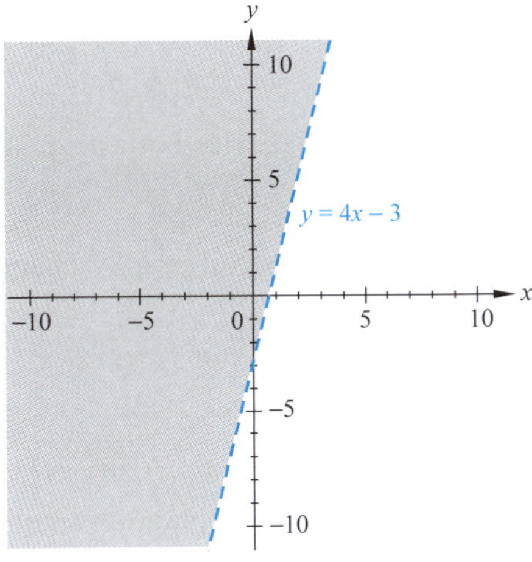

Figure 7-70

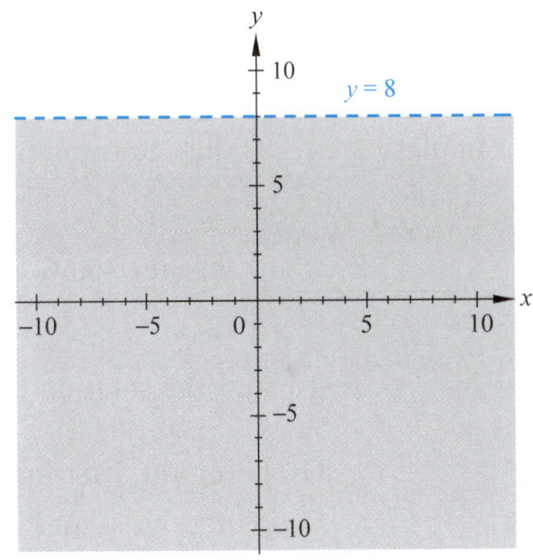

Figure 7-71

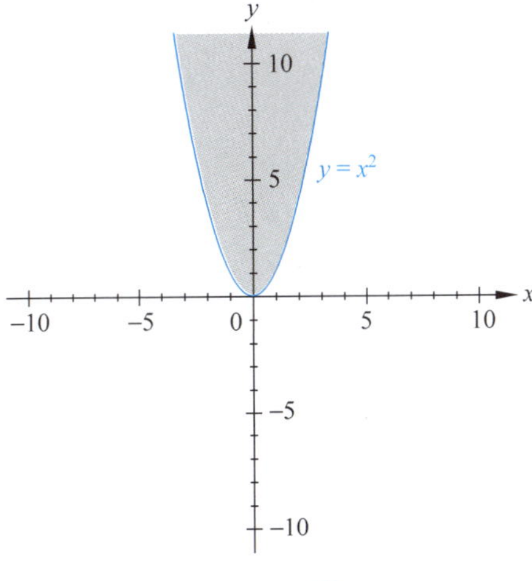

Figure 7-72

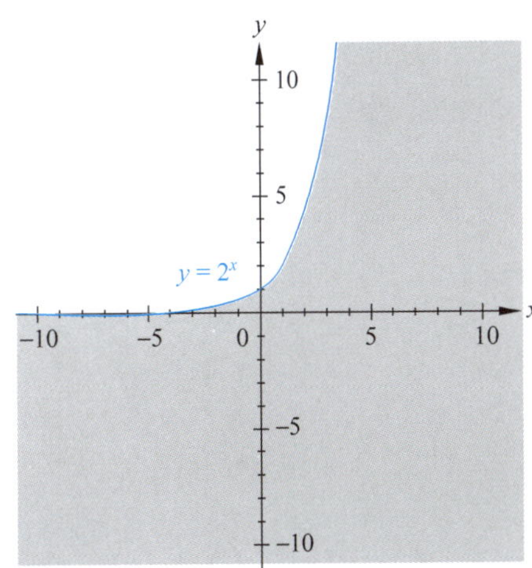

Figure 7-73

System of Functional Inequalities　函数不等式组

Definition: A collection of two or more functional inequalities with the same set of unknowns. The solution of a system of functional inequalities must satisfy every inequality in the system.

We represent the solution of a system of functional inequalities as a region in the graph. Though it can be represented by a piecewise-defined functional inequality, it is long and cumbersome.

Notation: $\begin{cases} \text{Functional Inequality 1} \\ \text{Functional Inequality 2.} \\ \cdots \end{cases}$

Properties: To find the solution, graph all the functional inequalities in the system. The shaded region that is common to **all** the functional inequalities in the system is the solution.

If none of the shaded regions is common to **all** the functional inequalities, we say the system has no solution.

Examples: 1. $\begin{cases} y > x \\ y < x \end{cases}$ has no solution. None of the points (x, y) satisfies both inequalities.

2. $\begin{cases} y \geqslant x \\ y \leqslant x \end{cases}$ has the solution $y = x$. The solution is a line instead of a 2-dimensional region.

3. To find the solution of $\begin{cases} y > 2x - 1 \\ y \geqslant 3x - 2 \end{cases}$:

Graphing the first inequality, we get the graph as shown in Figure 7-74.

Graphing the second inequality on the same coordinate plane, we get the graph as shown in Figure 7-75.

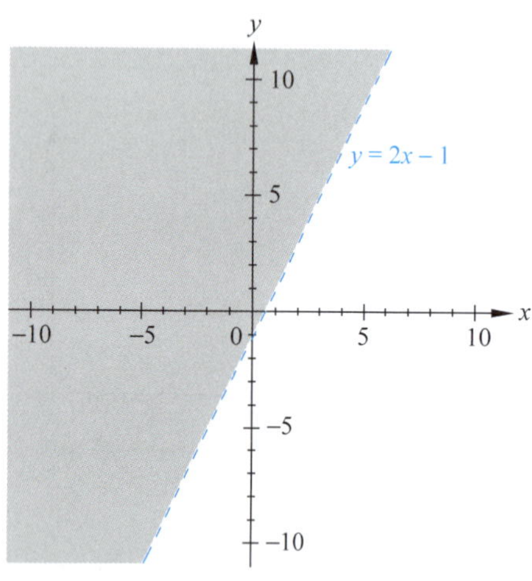

Figure 7-74

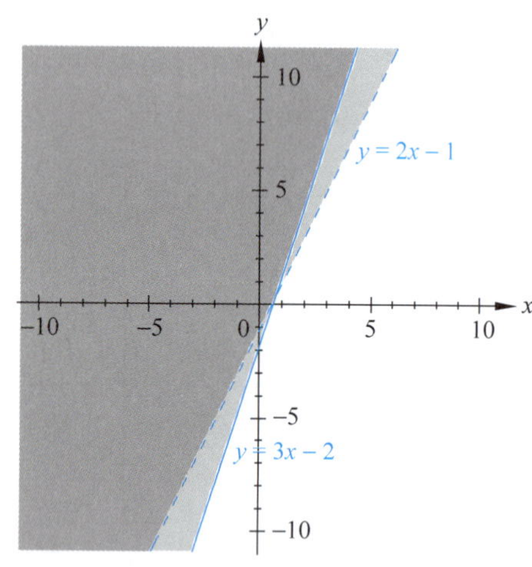

Figure 7-75

The region that has been shaded twice (darker region) is the solution.

The intersection of both lines is $(1, 1)$, as shown in Figure 7-76.

The piecewise-defined functional inequality is $f(x)\begin{cases} > 2x - 1, & x \leqslant 1 \\ \geqslant 3x - 2, & x > 1 \end{cases}$. Note that $(1, 1)$ is not a solution in this system—it does not satisfy the inequality $y > 2x - 1$.

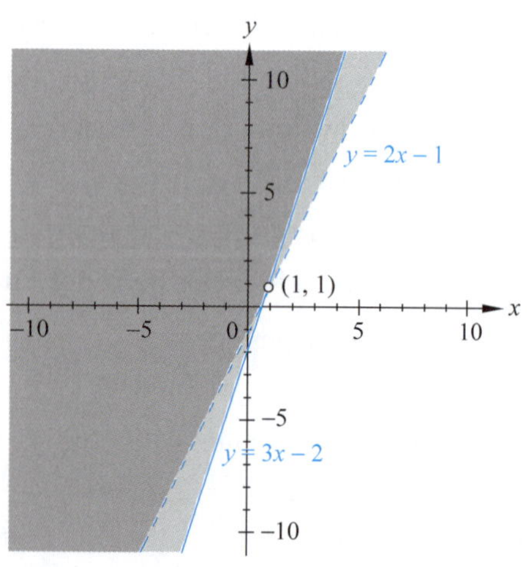

Figure 7-76

4. To find the solution of $\begin{cases} y > 2x \\ y \leqslant x^2 \\ y < 8 \end{cases}$:

Graphing the inequalities separately and taking the intersection of the shaded regions, we have the graph in Figure 7-77. The shaded region is the solution.

The two points of intersection are (0,0) and $(-2\sqrt{2},8)$, as shown in Figure 7-78. Note that these two points are not solutions of the system.

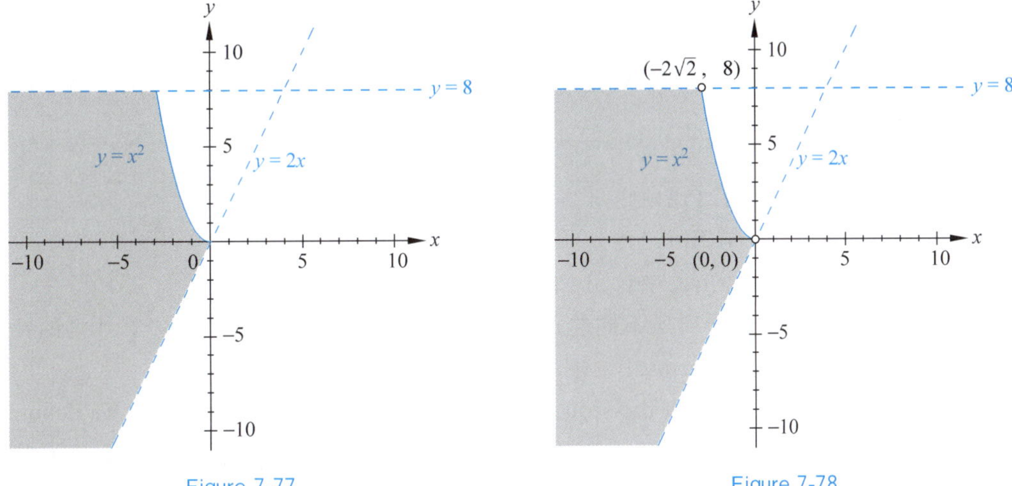

Figure 7-77　　Figure 7-78

The point (0,0) does not satisfy the inequality $y>2x$.

The point $(-2\sqrt{2},8)$ does not satisfy the inequality $y<8$.

The piecewise-defined functional inequality is $\begin{cases}2x<f(x)<8, & x\leqslant-2\sqrt{2}\\ 2x<f(x)\leqslant x^2, & -2\sqrt{2}<x<0\end{cases}$.

8. Sequences 数列

视频 18

8.1 Introduction 介绍

Sequence 数列

n. [ˈsikwəns]

Definition: An (ordered) array of numbers. Each number is known as a term. 数列是指一列有序的数字。每个数字称为项。

Notation: Terms are separated by commas: $a_1, a_2, a_3, \cdots$, for which a_1, a_2, and a_3 are the first, second, and third terms, respectively, and so on.

项以逗号隔开：$a_1, a_2, a_3, \cdots$，其中 a_1、a_2、a_3 分别为第一、第二、第三项，如此类推。

Properties: 1. Sequences can be both finite (if the last term exists) and infinite. 数列的项数可以为有限(若最后一项存在)或无限。

2. As opposed to sets, ordering and duplication matter in the sequence. For example,

   Sequence A: 1,2,3,4,5,6

   Sequence B: 1,1,2,2,3,4,5,6

   Sequence C: 6,5,4,3,2,1

   The three sequences are different from each other.

   But if

   Set A: $\{1,2,3,4,5,6\}$

   Set B: $\{1,1,2,2,3,4,5,6\}$

   Set C: $\{6,5,4,3,2,1\}$,

   then the three sets are the same.

   数列与集合有所不同的是,数列中的排序和重复的数字会影响数列的独特性。

3. If the ith term of a sequence is k, we denote it as $a_i = k$.

   For example, in the sequence 2,4,6,8,10, the 5th term is 10, so we denote this as $a_5 = 10$.

4. Sometimes the terms behave in a predictable pattern. The pattern of a sequence can be defined in two ways.

   有时候数列项有规律地出现。数列的这个规律可以用下列两种形式表示。

   (1) Explicitly: a_n = (an expression in terms of n).

   通项公式: a_n = 一个含有 n 的表达式。

   For example, in the infinite sequence $2,4,6,8,10,\cdots$, the nth term is $2n$. Thus the explicit rule for the sequence is $a_n = 2n$.

   This is similar to the function notation $f(n) = 2n$, for which $f(n)$ gives the nth term of the sequence. However, by convention, we will stick with $a_n = 2n$. 与函数表示法大同小异,但根据惯例,用 a_n 而不用$f(n)$表示第 n 个项。

   Advantage: Gives the nth term efficiently. That is, given n, can find a_n by arithmetic operations. 快速地给出第 n 个项——给出 n,能快速求出 a_n。

   (2) Implicitly/Recursively: $a_1 = \cdots$(the first term, or the first few terms); a_n = (an expression in terms of the terms before a_n).

   递推公式:给出第一个项或前几个项,a_n 的值用a_n 前面的项表示。

   For example, in the infinite sequence $2,4,6,8,10,\cdots$, the nth term is the sum of the $(n-1)$th term and 2. Thus the recursive rule for the sequence is $a_1 = 2$; $a_n = a_{n-1} + 2$ for all $n \geqslant 2$.

   Advantage: Sometimes patterns are more intuitive to be defined recursively than explicitly. 有时候递推公式比通项公式更直观。

5. Two common types of sequences are arithmetic sequences and geometric sequences. 两种常见的数列为等差数列和等比数列。

Examples: 1. In the sequence $10,20,30,40,50,\cdots$,

$a_1 = 10, a_2 = 20, a_3 = 30, \cdots$

Explicit: $a_n = 10n$.

Recursive: $a_1 = 10$; $a_n = a_{n-1} + 10$.

2. In the sequence $2,5,10,17,26,37,\cdots$,

$a_1=2, a_2=5, a_3=10, \cdots$

Explicit：$a_n=n^2+1$.

Recursive：$a_1=2$；$a_n=a_{n-1}+2n-1$ for all $n\geqslant 2$.

3. In the sequence 1,2,4,8,16,…,

   $a_1=1, a_2=2, a_3=4, \cdots$

   Explicit：$a_n=2^{n-1}$.

   Recursive：$a_1=1$；$a_n=2a_{n-1}$ for all $n\geqslant 2$.

4. Here is the famous Fibonacci Sequence：1,1,2,3,5,8,13,21,34,… 费波纳茨数列：

   $F_1=1, F_2=2, F_3=4, \cdots$

   Explicit：$F_n=\dfrac{(1+\sqrt{5})^n-(1-\sqrt{5})^n}{2^n\sqrt{5}}$. [Coming up with this is beyond the scope of Algebra 1.]

   Recursive：$F_1=1$；$F_2=1$；$F_n=F_{n-1}+F_{n-2}$ for all $n\geqslant 3$.

   Here, we use F_n instead of a_n for the nth term due to convention for the magical Fibonacci Sequence.　根据神奇的费波纳茨数列规律,用 F_n 表示第 n 个项。

Term　项

n. [tɜrm]

Definition：Each number in a sequence.　数列的每个数字。

Notation：Terms are numbered such that a sequence with n terms have the 1st term (denoted by a_1), 2nd term (denoted by a_2), 3rd term (denoted by a_3), … nth term (denoted by a_n).　在有 n 项的数列中,项均有编号。第一项为 a_1,第二项为 a_2,第三项为 a_3……第 n 项为 a_n。

8.2　Arithmetic Sequences　等差数列

Common Difference　公差

n. [ˈkɑmən ˈdɪfrəns]

Definition：The difference between two consecutive terms (latter minus former).

公差是指两个连续项之差(后项减前项)。

Properties：If the common difference in a sequence is constant, then the sequence is an arithmetic sequence.　若数列的公差为常值,则这个数列为等差数列。

Examples：1. In the sequence 1,5,8,4,7,…, the first four common differences, in order, are：

   $5-1=4$;

   $8-5=3$;

   $4-8=-4$;

   $7-4=3$.

2. In the sequence 4,7,10,13,16,…, the first four common differences, in order, are：

   $7-4=3$;

   $10-7=3$;

   $13-10=3$;

   $16-13=3$.

This sequence is an arithmetic sequence. 这个数列为等差数列。

3. In the sequence 1,3,9,27,81,…,the first four common differences,in order,are:

$3-1=2$;

$9-3=6$;

$27-9=18$;

$81-27=54$.

Arithmetic Sequence 等差数列

n. [ə'rɪθməˌtɪk 'sikwəns]

Definition: A sequence whose common difference is constant. 等差数列是指公差为常值的数列。

Properties: If d is the common difference, then the formula for finding the nth term is:

若 d 为公差,则求第 n 个项的公式为:

Explicit: $a_n = a_1 + d(n-1)$.

Recursive: a_1 is given; $a_n = a_{n-1} + d$ for all $n \geqslant 2$ (by definition).

Proof: To find the explicit rule, we can write out all the terms of the sequence in terms of the first term a_1 and the common difference d: $a_1, a_1+d, a_1+2d, a_1+3d, \cdots$. The nth term adds d "$(n-1)$ times". Thus $a_n = a_1 + d(n-1)$.

要证明等差数列的通项公式,可以把所有的项以首项 a_1 与公差 d 表示出来: $a_1, a_1+d, a_1+2d, a_1+3d, \cdots$第 n 个项加 d 加了$(n-1)$次,所以可得出 $a_n = a_1 + d(n-1)$。

Alternatively, we can treat this as a linear function $f(n)$, which gives the nth term. Since $f(1) = a_1$ and $f(2) = a_1 + d$, we can write the equation of the line that passes through $(1, a_1)$ and $(2, a_1+d)$. The slope is d (common difference). Substituting one point in the slope-intercept form, we get $a_1 = d(1) + b$, for which $b = a_1 - d$. Thus we have $f(n) = dn + (a_1 - d) = a_n = a_1 + d(n-1)$.

换一种方法,可以把等差数列当作线性函数 $f(n)$,从而求出第 n 个项。因为 $f(1) = a_1$ 和 $f(2) = a_1 + d$,根据点$(1, a_1)$和点$(2, a_1+d)$写出斜截式,得 $f(n) = dn + (a_1 - d) = a_n = a_1 + d(n-1)$。

Questions: 1. Write the pattern for each of the following arithmetic sequence explicitly and recursively.

(1) 4,8,12,16,20,…

(2) 1,7,13,19,25,…

(3) 100,93,86,79,72,…

2. If the 7th term of an arithmetic sequence is 55 and the 15th term is 111, what is the value of the 100th term?

Answers: 1. (1) $a_1 = 4$; $d = 4$.

Explicit: $a_n = 4 + 4(n-1)$.

Recursive: $a_1 = 4$; $a_n = a_{n-1} + 4$ for all $n \geqslant 2$.

(2) $a_1 = 1$; $d = 6$.

Explicit: $a_n = 1 + 6(n-1)$.

Recursive: $a_1 = 1$; $a_n = a_{n-1} + 6$ for all $n \geqslant 2$.

(3) $a_1 = 100$; $d = -7$.

Explicit: $a_n = 100 - 7(n-1)$.

Recursive: $a_1 = 100$; $a_n = a_{n-1} - 7$ for all $n \geqslant 2$.

2. We know that $a_7 = a_1 + d(7-1) = 55$ and $a_{15} = a_1 + d(15-1) = 121$. We have the system $\begin{cases} a_1 + 6d = 55 \\ a_1 + 14d = 111 \end{cases}$. Subtracting the first equation from the second, we get $8d = 56$, for which $d = 7$. It follows that $a_1 = 13$.

Therefore, the rule of this sequence is $a_n = 13 + 7(n-1)$. Thus $a_{100} = 13 + 7(100-1) = 706$.

8.3 Geometric Sequences 等比数列

Common Ratio 公比

n. [ˈkɑmən ˈreɪʃɪˌoʊ]

Definition: The quotient between two consecutive terms (latter divided by the former). 公比是指两个连续项之商(后项除以前项)。

Properties: If the common ratio in a sequence is constant, then the sequence is a geometric sequence. 若数列的公比为常值,则这个数列为等比数列。

Examples: 1. In the sequence 1,5,8,4,7,…, the first four common ratios, in order, are:

5/1 = 5;

8/5;

4/8 = 1/2;

7/4.

2. In the sequence 4,7,10,13,16,…, the first four common ratios, in order, are:

7/4;

10/7;

13/10;

16/13.

3. In the sequence 1,3,9,27,81,…, the first four common ratios, in order, are:

3/1 = 3;

9/3 = 3;

27/9 = 3;

81/27 = 3.

This sequence is a geometric sequence. 这个数列为等比数列。

Geometric Sequence 等比数列

n. [ˌdʒiəˈmetrɪk ˈsikwəns]

Definition: A sequence whose common ratio is constant. 公比为常值的数列。

Properties: If r is the common ratio, then the formula for finding the nth term is: 若 r 为公比,则求第 n 个项的公式为:

Explicit: $a_n = a_1 \cdot r^{n-1}$.

Recursive: a_1 is given; $a_n = a_{n-1} \cdot r$ for all $n \geqslant 2$ (by definition).

Proof: To find the explicit rule, we can write out all the terms of the sequence in terms of the first term a_1 and the common ratio r: $a_1, a_1 \cdot r, a_1 \cdot r^2, a_1 \cdot r^3, \cdots$. The nth term multiplies r "$(n-1)$ times". Thus $a_n = a_1 \cdot r^{n-1}$.

要证明等比数列的通项公式,可以把所有的项以首项 a_1 与公差 r 表示出来,即 $a_1, a_1 \cdot r$, $a_1 \cdot r^2, a_1 \cdot r^3, \cdots$ 第 n 个项乘 r 乘了$(n-1)$次。所以可得出 $a_n = a_1 \cdot r^{n-1}$。

Questions: 1. Write the pattern for each of the following geometric sequence explicitly and recursively.

(1) 2,4,8,16,32,⋯

(2) 12,36,108,324,972,⋯

(3) 1600,800,400,200,100,⋯

2. If the 3rd term of a geometric sequence is 56 and the 7th term is 896, and the common ratio is positive, what is the value of the 100th term?

Answers: 1. (1) $a_1 = 2$; $r = 2$.

Explicit: $a_n = 2 \times 2^{n-1} = 2^n$.

Recursive: $a_1 = 2$; $a_n = a_{n-1} \cdot 2$ for all $n \geqslant 2$.

(2) $a_1 = 12$; $r = 3$.

Explicit: $a_n = 12 \times 3^{n-1}$.

Recursive: $a_1 = 12$; $a_n = a_{n-1} \cdot 3$ for all $n \geqslant 2$.

(3) $a_1 = 1600$; $r = \frac{1}{2}$.

Explicit: $a_n = 1600 \times \left(\frac{1}{2}\right)^{n-1}$.

Recursive: $a_1 = 1600$; $a_n = a_{n-1} \cdot \frac{1}{2}$ for all $n \geqslant 2$.

2. We know that $a_3 = a_1 \cdot r^{3-1} = 56$ and $a_7 = a_1 \cdot r^{7-1} = 56$. We have the system $\begin{cases} a_1 \cdot r^2 = 56 \\ a_1 \cdot r^6 = 896 \end{cases}$. Dividing the second equation by the first, we get $r^4 = 16$, for which $r = 2$. It follows that $a_1 = 14$.

Therefore, the rule of this sequence is $a_n = 14 \times 2^{n-1}$.

Thus $a_{100} = 14 \times 2^{99}$.

9. Statistics Method—Calculations 统计方法——计算

9.1 Introduction 介绍

9.1.1 Basics 基础知识

Statistics 统计学

n. [stə'tɪstɪks]

Definition: A branch of mathematics that deals with the collection, analysis, interpretation, and presentation of masses of numerical data.　统计学是数学的一个分支,是专门针对大量数据的采集、分析、解释和介绍。

Data　数据

n.［ˈdeɪtə］

Definition：A number or value that relate to a particular subject.　数据是指跟对象有关的数值。

Examples：1. If we want to know the grade distribution of a class,then a data is the grade of one student.　若想测量一个班的成绩分布,则数据指的是一个学生的成绩。

2. If we want to measure the time distribution for a class of a one-mile run,then a data is the time for one student to run a mile.　若想测量一个班跑一英里(1 英里 = 1609.344 米)的时间分布,则数据指的是一个学生跑一英里的时间。

3. If we want to measure the height distribution of students in a class,then a data is the height of one student.　若想测量一个班的身高分布,则数据指的是一个学生的身高。

Data Set　数据集

n.［ˈdeɪtə set］

Definition：A collection of numbers or values that relate to a particular subject. Or,it is a collection of data.　数据集是指跟主题有关的数值组,也就是一组数据。

The number of data in a data set A is called the size of A.　数据集的数据个数称为集的大小。

Notations：List out all the data and separate each with a comma.　列出所有数据,以逗号隔开。

Properties：1. Duplication affects the uniqueness of a data set. For example,consider

A：1,2,3,3,4,5

B：1,2,3,4,5

A and B are different data sets.

重复的数据影响数据集的独特性。

2. Ordering does not affect the uniqueness of a data set. For example,consider

A：1,2,3,4,5

B：5,4,2,1,3

A and B are the same data set.

数据的排列不影响数据集的独特性。

Notes：1. Duplication affects the calculations of statistics. Thus duplication affects the uniqueness of a data set.　重复的数据影响统计数据的计算,所以重复的数据影响数据集的独特性。

2. Ordering does not affect the calculations of descriptive statistics. Thus ordering does not affect the uniqueness of a data set.　数据的排列不会影响统计数据的计算,所以数据的排列不影响数据集的独特性。

Examples：1. If we want to know the grade distribution of a class,then the data set is the collection of all grades of the students in the class.　若想测量一个班的成绩分布,数据集为所有学生成绩的汇集。

2. If we want to measure the time distribution for a class of a one-mile run,then the

data set is the collection of all mile-run times of the students in the class. 若想测量一个班跑一英里的时间分布,数据集为所有学生跑一英里时间的汇集。

3. If we want to measure the heights of students in a class, then the data set is the collection of all heights of the students in the class. 若想测量一个班的身高分布,数据集为所有学生身高的汇集。

9.1.2 Variables 变量

Variable 变量

n. ['veəriəbəl]

Definition: In statistics, a variable is a characteristic, a number, or a value that can be measured or counted. 在统计学中,变量为一个可被测量或统计的特征、数字或值。

We can divide every variable into one of the two categories: categorical variable or quantitative variable. 可以把变量分成两类:分类变量和定量变量。

Phrases: categorical~ 分类变量, quantitative~ 定量变量, independent~ 自变量, dependent~ 因变量, lurking~ 潜在变量, explanatory~ 解释变量, response~ 响应变量

Categorical Variable 分类变量

n. [ˌkætɪ'gɔrɪkəl 'veəriəbəl]

Definition: A variable that can take on one of the limited, and usually fixed, number of possible values. These values are qualitative, which is descriptive but has nothing to do with numbers. 在众多有限的值中取其一的变量。这些值是定性的——描述性的,与数量无关。

Examples: How does student feel about math? (Students must check one of these choices.)

(1) Very happy

(2) Happy

(3) Neutral

(4) Unhappy

(5) Very unhappy

Feeling about math is categorical. This question can be answered by one of the options from (1) through (5). These options are descriptive, but have nothing to do with numbers.

Quantitative Variable 定量变量

n. ['kwɑntɪˌteɪtɪv 'veəriəbəl]

Definition: A variable that can be measured in numeric or quantitative scale. Quantitative variables include ordinals, intervals, and ratios. Quantitative data are comparable using the numeric scale. 定量变量是指能用数字或数量测量的变量。定量变量包括数字、区间、比。定量数据能根据数值刻度进行对比。

Examples: 1. We are interested in finding the height distribution of adults in a community. 希望调查社区成年人的身高分布。

The height is the quantitative variable, which is a decimal. Heights are comparable

using the numeric scale.
身高是定量变量,用小数表示。身高能根据数值刻度进行对比。

2. On a 3-question quiz, if each question is worth one point, and there is no partial credit. We are interested in finding the grade distribution of the school.　在一个三问的测验里,每问值一分,不给部分分。我们想知道学校的成绩分布。
The possible score (one of 0,1,2,3) is the quantitative variable. The possible score is one of the four possibilities. These possibilities are descriptive and are relevant to numbers. Scores are comparable using the numeric scale.　可得的分数(0、1、2、3 中的一种)是一个定量变量。可得的分数为四个可能性之一。这些可能性有描述性,且跟量有关。分数能根据数值刻度进行对比。

9.2　One-Variable Statistics　单变量统计

视频 19

9.2.1　Descriptive Statistics　描述统计

Descriptive Statistics　描述统计

n. [dɪ'skrɪp.tɪv stə'tɪstɪks]

Definition: The process of organizing, analyzing, and describing statistical data through graphs or mathematical methods.　描述统计是指通过图表或数学方法,对数据资料进行整理、分析与描述的方法。

The most common tools of descriptive statistics are: mean, median, mode, range, minimum, quartile 1, quartile 3, maximum, variance, and standard deviation.　最常用的统计数据为:平均数、中位数、众数、极差、最小值、第一四分位数、第三四分位数、最大值、方差、标准差。

Mode　众数

n. [moʊd]

Definition: In a data set, the mode is the data that occurs the most number of times (at least twice).　众数是指在数据集中出现得最多的数据(至少两次)。

Properties: A data set can have any number of mode:
数据集可以有任意个众数:
0 modes—All data occurs the same number of times.
0 个众数——所有数据出现的次数相同。
1 mode—Only one data occurs the most number of times.
1 个众数——只有一个数据出现的次数最多。
2 or more modes—Otherwise.
2 个或更多个众数——其他情况。

Mean　平均数

n. [min]

Definition: The average of all numbers.

Notation: For a data set with n data, the mean $\bar{x}$ is calculated by: $\bar{x} = \frac{\sum_{i=1}^{n} x_i}{n} =$

$$\frac{\text{The sum of all numbers in the data set}}{\text{The number of data in the data set}}.$$

数据集的平均数定义为：$\frac{\text{所有数据的和}}{\text{数据的个数}}$。

Properties：1. From the equation in the notation section, we know that $n\bar{x}=\sum_{i=1}^{n}x_i$. The mean multiplied by the number of data equals the sum of the data.

通过移项可得出：数据的个数×平均数=所有数值的和。

2. Outliers may affect the mean drastically (Example 2). 离群值(极端值)能大幅影响平均数，见下面例2。

3. Sometimes the mean does not give a value that makes sense in the context, but it reflects the central tendency of the data.

有时候平均数并不表示一个实际的数量，但是它能反映出数据的集中趋势。

Examples：1. The test scores for a class of 10 students are as follows：

65,67,74,75,77,78,81,82,87,94

The sum of the scores is 780.

The mean score is 780/10=78.

2. The test scores for a class of 10 students are as follows：

7,70,85,86,87,87,87,90,90,91

The sum of the scores is 780.

The mean score is 780/10=78.

The student who got 85 is above average, but his score is the third lowest of the class!

The score of 7 pulls the average down by a ton.

Another method of calculating the mean involves dropping the lowest and highest scores first. After dropping the lowest and the highest, we get the test scores for the class are：70,85,86,87,87,87,90,90. The sum of these scores is 682. The mean is 682/8=85.25. The person who got 85 is now below average (and ranked #8 out of 10 students).

In short, 78 is the original mean, and 85.25 is the modified mean. In this scenario, the modified mean is better in describing the data because more scores are around 85.25.

78为原平均数；85.25为改进过的平均数。在这个场合中，改进过的平均数能更好地概括数据，因为很多分数都在85.25左右。

3. Jack and Jill each made a set of 10 free-throws 10 times. The data below shows the number of successes of each person for each of the 10 sets.

Jack：7,8,4,9,5,6,6,8,7,7

Jill：4,4,7,8,8,4,5,5,6,9

Jack's total number of successes is 67. The mean number of successes per set is 6.7.

Jill's total number of successes is 60. The mean number of successes per set is 6.

6.7 successes in one set does not make much sense (a free-throw is either a hit or a miss, and nothing in between). But the mean of 6.7 successes in one set shows that

Jack is better than Jill in free-throwing (because Jill only has 6 successes in one set, on average). Thus means are comparable and can be interpreted using the way above.

Minimum　最小值

n. ['mɪnəməm]

Definition: The minimum of a data set refers to the data with the lowest numerical value. There can be multiple copies of minimum in a data set.　数据集中最小的数据为最小值。一个数据集中能存在多个最小值。

Notation: The minimum of a data set A is denoted as $\min(A)$.

Examples: Consider the following data sets:

A: 1,2,3,6,7,8,10,11,18

B: 3,3,7,7,8,9,9

C: 1,4,5,6,7,8

D: 6,6,7,7,8,10,17,17

E: 5,6,8,7,2,0,1

$\min(A)=1$; $\min(B)=3$; $\min(C)=1$; $\min(D)=6$; $\min(E)=0$.

Maximum　最大值

n. ['mæksəməm]

Definition: The maximum of a data set refers to the data with the highest numerical value. There can be multiple copies of the maximum in a data set.　数据集中最大的数据为最大值。一个数据集中能存在多个最大值。

Notation: The maximum of a data set A is denoted as $\max(A)$.

Examples: Consider the following data sets:

A: 1,2,3,6,7,8,10,11,18

B: 3,3,7,7,8,9,9

C: 1,4,5,6,7,8

D: 6,6,7,7,8,10,17,17

E: 5,6,8,7,2,0,1

$\max(A)=18$; $\max(B)=9$; $\max(C)=8$; $\max(D)=17$; $\max(E)=8$.

Range　极差

n. [reɪndʒ]

Definition: The range of the data set is the difference between the maximum and minimum.　数据集中最大值与最小值的差。

Properties: Outliers affect the calculation of the range drastically. For example, in the data set 1, 50, 50, 53, 53, 50, 55, 100, most of the data are around 53. The range is $100-1=99$, which is large. If the data 1 and 100 were not there, the range would be $55-50=5$, which would be much smaller.　离群值能大幅影响极差。

Examples: Consider the following data sets:

A: 1,2,3,6,7,8,10,11,18

B：3,3,7,7,8,9,9

C：1,4,5,6,7,8

D：6,6,7,7,8,10,17,17

E：5,6,8,7,2,0,1

The range of A is $18-1=17$.

The range of B is $9-3=6$.

The range of C is $8-1=7$.

The range of D is $17-6=11$.

The range of E is $8-0=8$.

Median 中位数

n. [ˈmidiən]

Definition：The median of a **sorted** (arranged from least to greatest) data set is the number that is in the middle of the data set. 在从小到大排列的数据集中,中位数为数据集中间的数据值。

Notation："The median of a distribution is m" is sometimes written as "$M=m$". 中位数一般用 M 标记。

Properties：1. In a sorted data set：

在从小到大排列的数据集里：

- If the number of data is odd, the median is the middle number. To find the median, we simultaneously cross out **one** minimum and **one** maximum of the remaining data at a time, repeatedly until there is one number remaining. This number is known as the median. 若数据的个数为奇数,中位数是中间的数。要找出中位数,一次同时划去余下数据的最小值和最大值,直到余下一个数字。这个数字就为中位数。
- If the number of data is even, the median is the average of the two center numbers. To find the median, we simultaneously cross out **one** minimum and **one** maximum of the remaining data at a time, repeatedly until there are two numbers remaining. We take their average as the median. 若数据的个数为偶数,中位数是中间两个数的平均值。要找出中位数,一次同时划去余下数字的最小值和最大值,直到余下两个数字。这两个数字的平均数就为中位数。

2. To state Property 1 in an algebraic way: in a sorted data set with n data.

- If n is odd, the median is the $\left(\frac{n+1}{2}\right)$th data.
- If n is even, the median is the mean of the $\left(\frac{n}{2}\right)$th and $\left(\frac{n}{2}+1\right)$th data.

从代数的角度表达性质1。

3. Outliers do not affect the median. 离群值不影响中位数。

For example, in the data set 1, 68, 69, 70, 70, 72, we expect the "center" to be around 70.

The mean is $58.\overline{3}$ because the data 1 affects the calculation drastically.

The median is 69.5, which is close to our approximation 70.

Examples：Consider the following data sets：

A：1,2,3,6,7,8,10,11,18

B：3,3,7,7,8,9,9

C：1,4,5,6,7,8

D：6,6,7,7,8,10,17,17

E：5,6,8,7,2,0,1

To find the median of A, we get

$$\begin{aligned}\text{Median}(A) &= \text{median}(1,2,3,6,7,8,10,11,18)\\ &= \text{median}(2,3,6,7,8,10,11)\\ &= \text{median}(3,6,7,8,10)\\ &= \text{median}(6,7,8)\\ &= \text{median}(7)\\ &= 7.\end{aligned}$$

To find the median of B, we get

$$\begin{aligned}\text{Median}(B) &= \text{median}(3,3,7,7,8,9,9)\\ &= \text{median}(3,7,7,8,9)\\ &= \text{median}(7,7,8)\\ &= \text{median}(7)\\ &= 7.\end{aligned}$$

$\text{Median}(C) = \text{median}(1,4,5,6,7,8) = \cdots = \text{median}(5,6) = \text{mean}(5,6) = 5.5.$

$\text{Median}(D) = \text{median}(6,6,7,7,8,10,17,17) = \cdots = \text{median}(7,8) = \text{mean}(7,8) = 7.5.$

$\text{Median}(E) = \text{median}(5,6,8,7,2,0,1) = \text{median}(0,1,2,5,6,7,8) = \cdots = 5.$

Quartile　四分位数

n. [ˈkwɔːr.taɪl]

Definition：The quartiles are three points that divide the sorted data set (ascending order) into four equal groups.　四分位数是把从小到大排列的数据集平分为四个组的三个点。

Notation：Quartile 1 is denoted by $Q1$. Data that are lower than $Q1$ are in the bottom 25% of the distribution.　第一四分位数写作 $Q1$。比 $Q1$ 小的数据在数据集底部的 25%区域。

Quartile 2 is denoted by M, or simply called the median. Data that are lower than M are in the bottom 50% of the distribution.　第二四分位数写作 M，亦是中位数。比 M 小的数据在数据集底部的 50%区域。

Quartile 3 is denoted by $Q3$. Data that are higher than $Q3$ are in the top 25% of the distribution.　第三四分位数写作 $Q3$。比 $Q3$ 大的数据在数据集顶部的 25%区域。

Properties：Recall that in a sorted data set A (ascending order), suppose there are n data.　在从小到大排列的数据集 A 里，设有 n 个数据。

- If n is odd, the median is the $\left(\frac{n+1}{2}\right)$th data. 若 n 为奇数，中位数为第$\left(\frac{n+1}{2}\right)$个数据——中间的数据。
- If n is even, the median is the mean of the $\left(\frac{n}{2}\right)$th and $\left(\frac{n}{2}+1\right)$th data.　若 n 为偶

数，中位数为$\left(\frac{n}{2}\right)$和$\left(\frac{n}{2}+1\right)$个数据的平均数——中间两个数据的平均数。

1. To find $Q1$:
   要求出 $Q1$：$Q1$ 为中位数左边的数据的中位数。
   - If n is odd, $Q1$ is the median of the data set from the 1st through the $\left(\frac{n+1}{2}-1\right)$th data of A. In other words, $Q1$ is the median of the data that are on the left of the median.
   - If n is even, $Q1$ is the median of the data set from the 1st through the $\left(\frac{n}{2}\right)$th data of A.
2. To find $Q3$:
   要求出 $Q3$：$Q3$ 为中位数右边的数据的中位数。
   - If n is odd, $Q3$ is the median of the data set from the $\left(\frac{n+1}{2}+1\right)$th through the nth data of A.
   - If n is even, $Q3$ is the median of the data set from the $\left(\frac{n}{2}+1\right)$th through the nth data of A.
3. Since the process of finding quartiles is based on that of finding the median, outliers do not affect the calculations of quartiles. 因为求四分位数的过程是以求中位数的过程为依据的，离群值不影响四分位数的计算。

Examples: Consider the following data sets:

A: 1,2,3,6,7,8,10,11,18

B: 3,3,7,7,8,9,9

C: 1,4,5,6,7,8

D: 6,6,7,7,8,10,17,17

E: 5,6,8,7,2,0,1

For A: $Q1=\text{median}(1,2,3,6)=2.5$.
$Q2=\text{median}(A)=7$.
$Q3=\text{median}(8,10,11,18)=10.5$.

For B: $Q1=\text{median}(3,3,7)=3$.
$Q2=\text{median}(B)=7$.
$Q3=\text{median}(8,9,9)=9$.

For C: $Q1=\text{median}(1,4,5)=4$.
$Q2=\text{median}(C)=5.5$.
$Q3=\text{median}(6,7,8)=7$.

For D: $Q1=\text{median}(6,6,7,7)=6.5$.
$Q2=\text{median}(D)=7.5$.
$Q3=\text{median}(8,10,17,17)=13.5$.

For E: $Q1=\text{median}(0,1,2)=1$.
$Q2=\text{median}(0,1,2,5,6,7,8)=5$.

$Q3=\text{median}(6,7,8)=7.$

Interquartile Range（IQR）　四分位距

n.［ˌɪntəˌkwɔː. taɪl'reɪndʒ］

Definition：IQR = $Q3-Q1$.

Examples：A：1,2,3,6,7,8,10,11,18

B：3,3,7,7,8,9,9

C：1,4,5,6,7,8

D：6,6,7,7,8,10,17,17

E：5,6,8,7,2,0,1

According to the results in the example in quartile：

For A：IQR = 10.5 − 2.5 = 8.

For B：IQR = 9 − 3 = 6.

For C：IQR = 7 − 4 = 3.

For D：IQR = 13.5 − 6.5 = 7.

For E：IQR = 7 − 1 = 6.

Five-Number Summary　五数概括法

n.［fɑɪv 'nʌmbər 'sʌməri］

Definition：A description of a data set by showing five important statistics（in order）：

五数概括法是指按顺序展示五个重要的统计数据：

minimum,Quartile 1,median,Quartile 3,maximum

最小值,第一四分位数,中位数,第三四分位数,最大值

This summary of a data set shows the minimum and maximum,along with the bottom 25%,50%,and 75% cutoff values.　这个数据集的概括法展示了最小值、最大值和从数据集底部开始25%、50%、75%分隔点的数字。

Examples：Consider the following data sets：

A：1,2,3,6,7,8,10,11,18

B：3,3,7,7,8,9,9

C：1,4,5,6,7,8

D：6,6,7,7,8,10,17,17

E：5,6,8,7,2,0,1

According to the results in the example in quartile：

The five-number summary for A：1,2.5,7,10.5,18.

The five-number summary for B：3,3,7,9,9.

The five-number summary for C：1,4,5.5,7,8.

The five-number summary for D：6,6.5,7.5,13.5,17.

The five-number summary for E：0,1,5,7,8.

Outlier（of One-Variable Statistics）　离群值

n.［'aʊtˌlaɪ. ər］

Definition：A data is an outlier if it does not fit the general pattern of the data in the data set.

Outliers are particularly large or particularly small. 若一个数据不符合数据集里数据的总体规律,则称为离群值,也称极端值。离群值特别大或特别小。

Properties: The outliers can be either approximated subjectively (but sometimes they are hard to decide) or found exactly using the 1.5 IQR Rule. 离群值可以主观判定(但有时候又很难判定),或用1.5四分位距法则判定。

Questions: Approximate the outlier(s) for each of the following:

(1) 1,20,21,22,22,22,23,27,28,28,28,80

(2) 65,66,66,66,67,67,68,70,72,180,200

Answers: (1) 1,80.

(2) 180,200.

1.5 IQR Rule 1.5四分位距法则

Definition: A rule that states which data are outliers in a data set. 1.5四分位距法则是判断离群值的法则。

Properties: Data that are less than $Q1 - 1.5\text{IQR}$ or greater than $Q3 + 1.5\text{IQR}$ are outliers. 小于 $Q1 - 1.5\text{IQR}$ 或大于 $Q3 + 1.5\text{IQR}$ 的数据均为离群值。

Questions: Approximate the outlier(s) for each of the following:

(1) 1,20,21,22,22,22,23,27,28,28,28,80

(2) 65,66,66,66,67,67,68,70,72,180,200

Answers: (1) $Q1 = 21.5$, and $Q3 = 28$.

$\text{IQR} = 28 - 21.5 = 6.5$.

$Q1 - 1.5\text{IQR} = 11.75$, and $Q3 + 1.5\text{IQR} = 37.75$.

Data that are less than 11.75 or greater than 37.75 are outliers. Thus 1 and 80 are outliers.

(2) $Q1 = 66$, and $Q3 = 72$.

$\text{IQR} = 72 - 66 = 6$.

$Q1 - 1.5\text{IQR} = 57$, and $Q3 + 1.5\text{IQR} = 81$.

Data that are less than 57 or greater than 81 are outliers. Thus 180 and 200 are outliers.

Variance 方差

n. ['veəriəns]

Definition: Measurement of how close the data are deviated from the mean. In particular, it is the average of the squared differences from the mean. 方差是对数据偏离平均数的测量——与平均数之差平方的平均值。

Notation: For a data set with n data whose mean is $\bar{x}$, the variance s^2 is calculated by:

$$s^2 = \frac{1}{n-1}\sum_{i=1}^{n}(x_i - \bar{x})^2$$

$$= \frac{1}{\text{number of data} - 1} \cdot (\text{sum of the squares of the difference between each individual data and the mean})$$

This is known as the variance of the sample. 样本方差。

We sometimes use σ^2 for the variance, along with μ for the mean.

$$\sigma^2=\frac{1}{n}\sum_{i=1}^{n}(x_i-\mu)^2$$

This is known as the variance of the population. 总体方差。

Comparing these two calculations, the denominators are $n-1$ and n, respectively, due to the **degree of freedom**. This is beyond the scope of high school math. 两个方差计算方法的差别在于自由度,此知识点超出高中数学的范畴。

We will use the variance for the sample in this encyclopedia. 在本书中,会用到样本的方差。

Properties: For both population and sample variance:

对于总体方差和样本方差:

1. If all the data are identical in a data set, then the variance will be 0. 若数据集的所有数据相同,方差为0。

   The larger the range, the larger is the variance. 极差越大,方差越大。

2. Since outliers may affect the mean drastically, and the mean has a close connection with the variance, outliers may affect the variance drastically. For example, the data set of 4,4,4,4,4 has a mean of 4 and a variance of 0. Adding a data 100 into the set, the new data set is 4,4,4,4,4,100. The new mean is $120/6=20$ and the new variance is

   $$\frac{1}{6-1}[(4-20)^2+(4-20)^2+(4-20)^2+(4-20)^2+(4-20)^2+(100-20)^2]$$
   $$=\frac{1}{5}[(-16)^2+(-16)^2+(-16)^2+(-16)^2+(-16)^2+(80)^2]$$
   $$=1536.$$

   因为离群值会大幅度地影响平均数的计算,且平均数与方差紧密相连,所以离群值也会大幅度地影响方差的计算。

Notes: Property 1 is easy to show:

If all the data are identical, then the mean is one of the data.

$$s^2=\frac{1}{n-1}\sum_{i=1}^{n}(x_i-\bar{x})^2=\frac{1}{n-1}(0+0+\cdots+0)=0.$$

$$\sigma^2=\frac{1}{n}\sum_{i=1}^{n}(x_i-\mu)^2=\frac{1}{n}(0+0+\cdots+0)=0.$$

Questions: Calculate the variance for each of the following sample data set:

(1) 1,2,3,4,5,6,7

(2) 10,20,30,30,60

Answers: (1) $n=7$; $\bar{x}=4$.

$$s^2=\frac{1}{7-1}[(1-4)^2+(2-4)^2+(3-4)^2+(4-4)^2+(5-4)^2+(6-4)^2+(7-4)^2]$$
$$=\frac{1}{6}[(-3)^2+(-2)^2+(-1)^2+0^2+1^2+2^2+3^2]$$
$$=28/6$$
$$=14/3.$$

(2) $n=5$; $\bar{x}=30$.

$$\begin{aligned}s^2&=\frac{1}{5-1}[(10-30)^2+(20-30)^2+(30-30)^2+(30-30)^2+(60-30)^2]\\&=\frac{1}{4}[(-20)^2+(-10)^2+0^2+0^2+30^2]\\&=1400/4\\&=350.\end{aligned}$$

Standard Deviation 标准差

n. [ˈstændərd ˌdiːviˈeɪʃən]

Definition: Measurement of how close the data are deviated from the mean. In particular, it is the square root of the variance. 标准差是指对数据偏离平均数的测量——方差的平方根。

Notation: For samples, we use $\bar{x}$ for the mean, s^2 for the variance, and s for the standard deviation. The formula for the sample standard deviation is

对于样本,用$\bar{x}$ 表示平均数,s^2 表示方差,s 表示标准差。标准差的公式为

$$\begin{aligned}s&=\sqrt{\frac{1}{n-1}\sum_{i=1}^{n}(x_i-\bar{x})^2}\\&=\sqrt{s^2}\end{aligned}$$

= The square root of the sample variance.

样本标准差=样本方差的平方根。

For populations, we use μ for the mean, σ^2 for the variance, and σ for the standard deviation. The formula for the population standard deviation is

对于总体,用μ 表示平均数,σ^2 表示方差,σ 表示标准差。标准差的公式为

$$\begin{aligned}\sigma&=\sqrt{\frac{1}{n}\sum_{i=1}^{n}(x_i-\mu)^2}\\&=\sqrt{\sigma^2}\end{aligned}$$

= The square root of the population variance.

总体标准差=总体方差的平方根。

Properties: For both population and sample standard deviation:

对于总体标准差和样本标准差有如下属性:

1. The standard deviation is the square root of the variance. 标准差为方差的平方根。
2. If all the data are identical in the data set, then the standard deviation will be 0. 若数据集的所有数据相同,标准差为 0。

   The larger the range, the larger is the standard deviation. 极差越大,标准差越大。
3. Since outliers may affect the mean drastically, and the mean has connection with the standard deviation, outliers may affect the standard deviation drastically. For example, the data set of 4,4,4,4,4 has a mean of 4, a variance of 0, and a standard deviation of 0. Adding a data 100 into the set, the new data set is 4,4,4,4,4,100. The new mean is $120/6=20$; the new variance is

   $$\frac{1}{6-1}[(4-20)^2+(4-20)^2+(4-20)^2+(4-20)^2+(4-20)^2+(100-20)^2]$$

$$=\frac{1}{5}[(-16)^2+(-16)^2+(-16)^2+(-16)^2+(-16)^2+(80)^2]$$
$$=1536,$$

and the new standard deviation is $\sqrt{1536}\approx 39.192$.

因为离群值会大幅度地影响平均数的计算,且平均数与标准差紧密相连,所以离群值也会大幅度地影响标准差的计算。

Notes: Property 2 is easy to show:

If all the data are identical, then the mean is one of the data.

$$s^2=\frac{1}{n-1}\sum_{i=1}^{n}(x_i-\bar{x})^2=\frac{1}{n-1}(0+0+\cdots+0)=0.$$

$$s=\sqrt{s^2}=0.$$

$$\sigma^2=\frac{1}{n}\sum_{i=1}^{n}(x_i-\mu)^2=\frac{1}{n}(0+0+\cdots+0)=0.$$

$$\sigma=\sqrt{\sigma^2}=0.$$

Questions: Calculate the standard deviation for each of the following sample data set:

(1) 1,2,3,4,5,6,7

(2) 10,20,30,30,60

Answers: (1) According to the questions on variance, $s^2=14/3$. Therefore, $s=\sqrt{s^2}=\sqrt{14/3}\approx 2.160$.

(2) According to the questions on variance, $s^2=350$. Therefore, $s=\sqrt{s^2}=\sqrt{350}\approx 18.708$.

9.2.2 Frequencies　频数

Frequency　频数

n. ['frikwənsi]

Definition: The frequency of a data is the number of times it occurs in the data set.　在数据集中,一个数据的频数是指它出现的次数。

Examples: Josh wanted to know how many meals a person eats per day. He surveyed 20 random people. The sorted data set is as follows (each data represents the number of meals ate yesterday):

1,1,1,1,2,2,2,2,2,2,3,3,3,3,3,3,3,3,4,6

The frequency of 1 is 4; the frequency of 2 is 6; the frequency of 3 is 8; the frequency of 4 is 1; the frequency of 6 is 1.

Phrases: relative～相对频数

Relative Frequency　相对频数

n. ['relətɪv 'frikwənsi]

Definition: In a data set, the relative frequency of a data x is given by: $\frac{\text{frequency of } x}{\text{size of the data set}}$.

在数据集中,数据 x 的相对频数 $=\frac{x\text{ 的频数}}{\text{数据集的大小}}$。

Examples: Josh wanted to know how many meals a person eats per day. He surveyed 20 random people. The sorted data set is as follows (each data represents the number of meals ate yesterday):

1,1,1,1,2,2,2,2,2,2,3,3,3,3,3,3,3,3,4,6

Recall that the frequency of 1 is 4; the frequency of 2 is 6; the frequency of 3 is 8; the frequency of 4 is 1; the frequency of 6 is 1.

The relative frequency of 1 is 4/20 = 0.2; the relative frequency of 2 is 6/20 = 0.3; the relative frequency of 3 is 8/20 = 0.4; the relative frequency of 4 is 1/20 = 0.05; the relative frequency of 6 is 1/20 = 0.05.

Cumulative Frequency 累积频数

n. ['kjumjələtɪv 'frikwənsi]

Definition: The cumulative frequency of a data x is the number of data that are at or below x in the data set. 在数据集中,数据 x 的累计频数是小于或等于 x 的数据个数。

Examples: Josh wanted to know how many meals a person eats per day. He surveyed 20 random people. The sorted data set is as follows (each data represents the number of meals ate yesterday):

1,1,1,1,2,2,2,2,2,2,3,3,3,3,3,3,3,3,4,6

The cumulative frequency of 1 is 4.

The cumulative frequency of 2 is 10.

The cumulative frequency of 3 is 18.

The cumulative frequency of 4 is 19.

The cumulative frequency of 5 is 19.

The cumulative frequency of 6 is 20.

Cumulative Relative Frequency 累积相对频数

n. ['kjumjələtɪv 'relətɪv 'frikwənsi]

Definition: In a data set, the cumulative relative frequency of a data x is given by: $\frac{\text{cumulative frequency of } x}{\text{size of the dataset}}$.

在数据集中,数据 x 的累积相对频数 $=\frac{x\text{ 的累积频数}}{\text{数据集的大小}}$。

Examples: Josh wanted to know how many meals a person eats per day. He surveyed 20 random people. The sorted data set is as follows (each data represents the number of meals ate yesterday):

1,1,1,1,2,2,2,2,2,2,3,3,3,3,3,3,3,3,4,6

The cumulative relative frequency of 1 is 4/20 = 0.2.

The cumulative relative frequency of 2 is 10/20 = 0.5.

The cumulative relative frequency of 3 is 18/20 = 0.9.

The cumulative relative frequency of 4 is 19/20 = 0.95.

The cumulative relative frequency of 5 is 19/20 = 0.95.

The cumulative relative frequency of 6 is 20/20 = 1.

Frequency Distribution 频数分布

n. [ˈfrikwənsi ˌdɪstrəˈbjuʃən]

Definition：A summary showing the frequencies of different data or intervals of the data in a data set. 频数分布是指对于数据集中不同的数据和区间的频数概括。

Examples：Josh wanted to know how many meals a person eats per day. He surveyed 20 random people. The sorted data set is as follows (each data represents the number of meals ate yesterday)：

1,1,1,1,2,2,2,2,2,2,3,3,3,3,3,3,3,3,4,6

Using values, we have a table shown in Table 9-1.

Table 9-1

Number of Meals	1	2	3	4	5	6
Number of People	4	6	8	1	0	1

Using intervals (intervals can be done in many different ways), we have a table shown in Table 9-2.

Table 9-2

Number of Meals (inclusive)	0-1	2-3	4-5	6-7
Number of People	4	14	1	1

9.2.3 Summary 总结

1. For descriptive statistics：
   对于描述统计：
   - Measurements of the center：mean, median. 中心值的测量有：平均数、中位数。
   - Measurements of the spread：range, variance, standard deviation, and interquartile range. 分布的测量有：极差、方差、标准差、四分位距。
   - Measurements of frequency：mode. 频数的测量有：众数。
2. Outliers are extremely large or small data in a data set. They can be found using the 1.5 IQR Rule. 离群值是数据集中特别大或者特别小的数据，可以用1.5四分位距法则检测出来。
   - Measurements that are not resistant to outliers：mean, range, variance, standard deviation. 受离群值影响的测量有：平均数、极差、方差、标准差。
   - Measurements that are resistant to outliers：mode, median, quartiles. 不受离群值影响的测量有：众数、中位数、四分位数。
3. In a data set：
   在数据集中：
   - The frequency of a data is the number of times it occurs. 一个数据的频数是指它出现的次数。
   - The relative frequency of a data is the ratio of its frequency to the data set's size. 一个数据的相对频数是它的频数与数据集的大小之比。

- The cumulative frequency of a data is the number of times the data less than or equal to it occur. 一个数据的累计频数是小于或等于它的数据出现的次数。
- The cumulative relative frequency of a data is the ratio of its cumulative frequency to the data set's size. 一个数据的累积相对频数是它的累积频数与数据集的大小之比。

A frequency distribution is a summary showing the frequencies of different data or intervals of the data in a data set. 频数分布是数据集中对于不同的数据和区间的频数概括。

9.3 Visual Display for One-Variable Statistics 单变量统计图表

9.3.1 Visual Display for Data of a Categorical Variable 分类变量统计图表

Bar Graph 条形图

n. [ˈbɑr ˌɡræf]

Definition: A diagram that illustrates the frequencies of categories of a categorical variable using rectangles of equal widths but varying heights according to the frequencies. 条形图是指用矩形展示分类变量每个分类的频数的图表。这些矩形的宽相等,长则根据分类的频数而异。

Examples: John asked 100 people about the opinion of the following: "Roller coasters are fun". 34 people answered "strongly agreed"; 25 answered "agreed"; 10 answered "neutral"; 20 answered "disagreed", and 11 answered "strongly disagree".

The bar graph in Figure 9-1 shows the frequencies of these five categories.

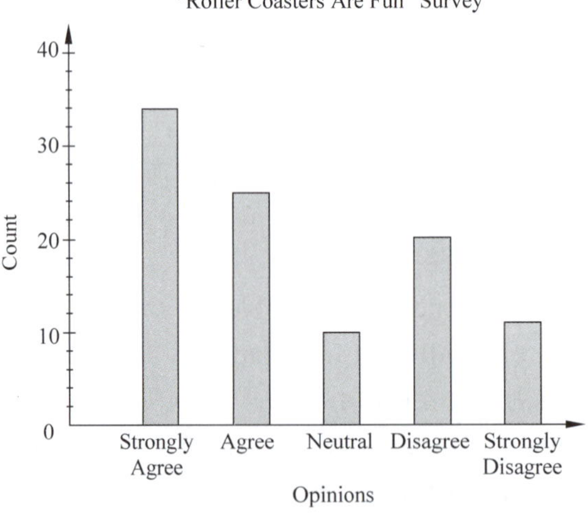

Figure 9-1

Pie Chart 饼图(圆形分格统计图表)

n. [ˈpaɪ ˌtʃɑrt]

Definition: A diagram that illustrates the relative frequencies of categories of a categorical variable, using sectors with varying degrees, according to the relative frequencies. 饼图是指用扇形展示分类变量每个分类的相对频数的图表。扇形的度数根据分类相对频数的不同而异。

Properties: Let $p\%$ be the relative frequency for a category. The number of degrees of this category's sector is $360° \cdot p\%$. 设 $p\%$ 为一个分类的相对频数。这个分类的扇形度数为 $360° \cdot p\%$。

Pie charts and bar graphs can be converted one from the other. 条形图和饼图可以互相转换。

Examples: John asked 100 people about the opinion of the following: "Roller coasters are fun". 34 people answered "strongly agreed"; 25 answered "agreed"; 10 answered "neutral"; 20 answered "disagreed", and 11 answered "strongly disagree".

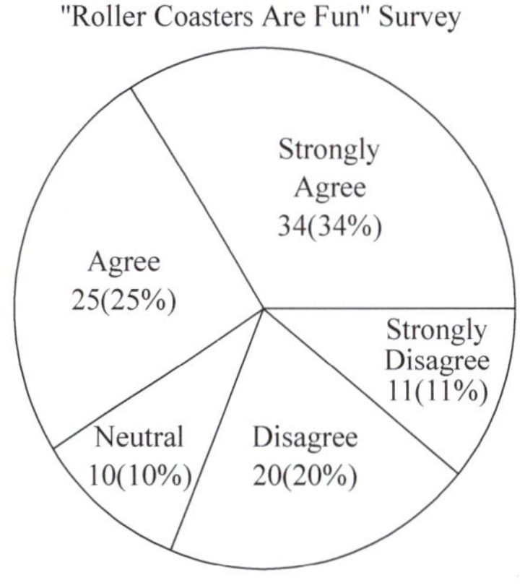

Figure 9-2

The bar graph in Figure 9-2 shows the frequencies and relative frequencies of these five categories.

34% of the people answered "strongly agreed". Thus the corresponding sector is $360° \cdot 34\% = 122.4°$.

25% of the people answered "agreed". Thus the corresponding sector is $360° \cdot 25\% = 90°$.

10% of the people answered "neutral". Thus the corresponding sector is $360° \cdot 10\% = 36°$.

20% of the people answered "disagreed". Thus the corresponding sector is $360° \cdot 20\% = 72°$.

11% of the people answered "strongly disagreed". Thus the corresponding sector is $360° \cdot 11\% = 39.6°$.

9.3.2 Visual Display for Data of a Quantitative Variable 定量变量统计图表

Table 9-3 summarizes the best graph for different characteristics of quantitative data set.
表 9-3 根据定量数据集的性质给出了最合理的图像展示。

Table 9-3 Histogram VS Dot Plot VS Stemplot 直方图、点图与茎叶图

Characteristics of Quantitative Data Set 定量数据集的性质	Small Spread & Small Data Set 分布小&数据集小	Large Spread & Small Data Set 分布大&数据集小	Small Spread & Large Data Set 分布小&数据集大	Large Spread & Large Data Set 分布大&数据集大
Best Fit of Graph 最合理的图像	Dot Plot or Stemplot 点图或茎叶图	Any 任何图像	Dot Plot 点图	Histogram 柱状图

Descriptions of Graphs of One-Variable Statistics 单变量统计图表描述

Definition: These descriptions apply to histograms, dot plots, stem-and-leaf plots(stemplots). 这些描述可应用到直方图、点图、茎叶图。

Vocabulary: We use the following four aspects to describe a graph: center, shape, spread, and outliers. 从以下 4 方面描述一个图表:中心值、形状、数据分布、离群值。

The center is the mean or the median of the data set (If outliers exists, use the median. Use either otherwise). 中心值是数据集的平均数或中位数(若存在离群值,则用中位数,否则用任何一个)。

The spread is the range of the data set (giving the minimum and maximum will be sufficient). 数据分布是指数据集的范围(给出最小值和最大值就足够了)。

The shape is one of no mode (no peak), unimodal (one peak), bimodal (two peaks), or multimodal (more than one peak), plus one of symmetric or skewed (left or right). We will focus on unimodal and bimodal. 形状有无峰、单峰、双峰或多峰,还有对称和非对称(左偏、右偏)。下面会集中介绍单峰与双峰的情况。

Figure 9-3 shows the case of unimodal, with the peak highlighted in blue. 图 9-3 展示了单峰的情况。峰以蓝色表示。

Figure 9-4 shows the case of bimodal, with the peaks highlighted in blue. 图 9-4 展示了双峰的情况。峰以蓝色表示。

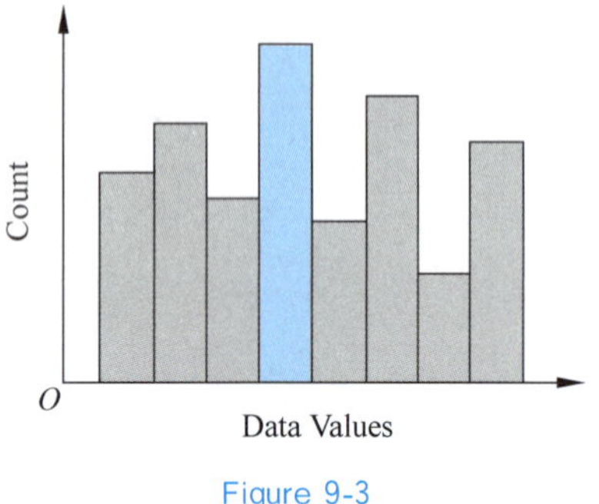

Figure 9-3

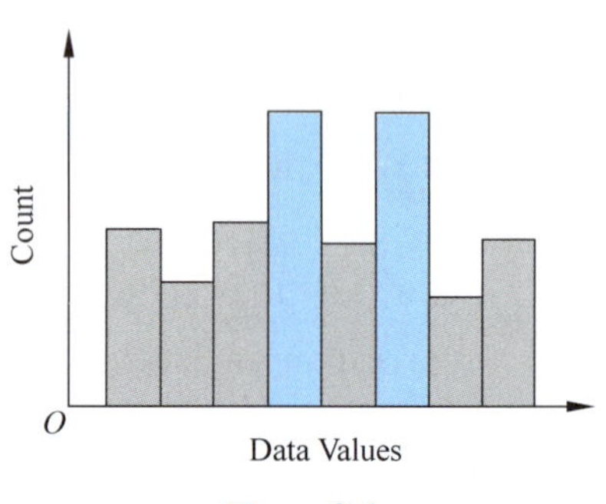

Figure 9-4

Being left-skewed means that data points generally fall on the higher side (above the mean) of the distribution, and there are so few data that are below the mean, as shown in Figure 9-5. 左偏的意思是在数据分布中,数据点大多数都高于平均数。很少低于平均数,如图 9-5 所示。

Being right-skewed means that data points generally fall on the lower side (below the mean) of the distribution, and there are so few data that are above the mean, as shown in Figure 9-6. 右偏的意思是在数据分布中,数据点大多数都低于平均数。很少高于平均数,如图 9-6 所示。

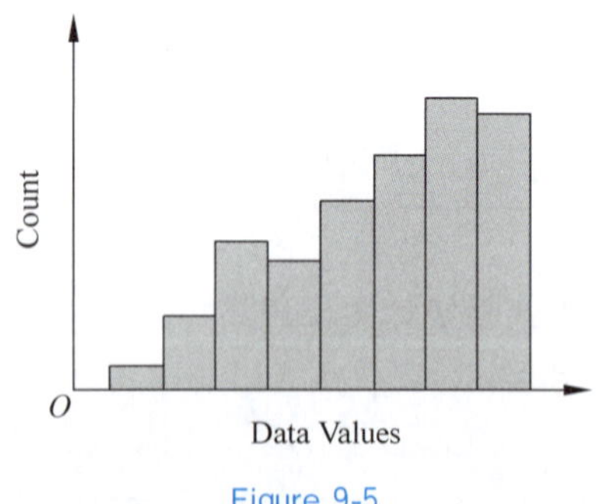

Figure 9-5

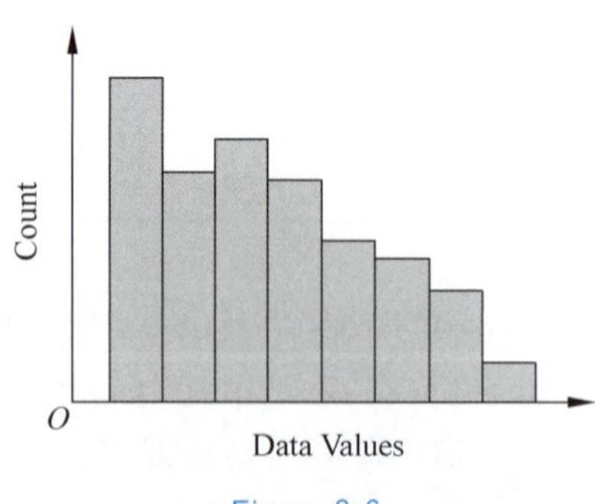

Figure 9-6

Being symmetric means that data points generally fall on both sides of the distribution evenly, as shown in Figure 9-7. 对称的意思是在数据分布中,数据点在平均数两边均匀分布,如图 9-7 所示。

See outliers (one-variable statistics) in Section 9.2.1. 见 9.2.1 节离群值(单变量统计)词条。

Properties: In a distribution, let $\bar{x}$ be the mean and M be the median. 在数据分布中,设 $\bar{x}$ 为平均数,M 为中位数。

- If the distribution is left-skewed, then $\bar{x}<M$, as shown in Figure 9-8.
  数据分布左偏的情况,如图 9-8 所示。

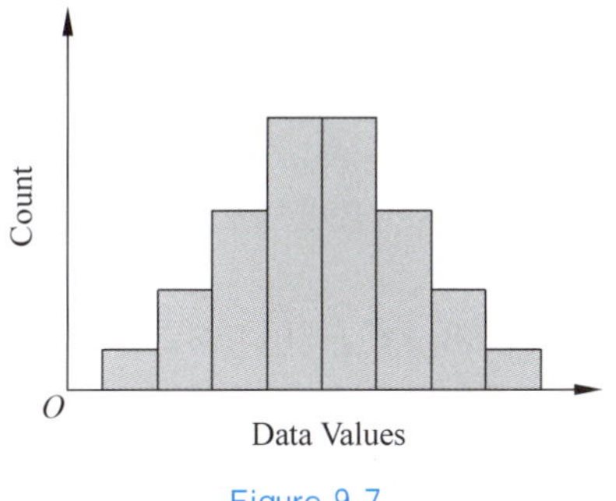

Figure 9-7

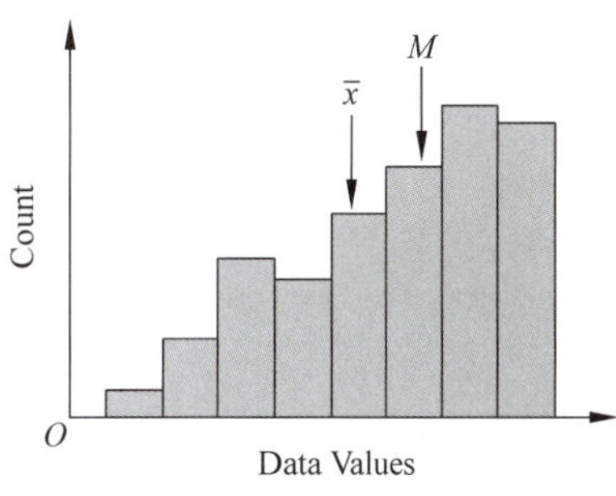

Figure 9-8

- If the distribution is right-skewed, then $\bar{x}>M$, as shown in Figure 9-9.
  数据分布右偏的情况,如图 9-9 所示。
- If the distribution is symmetric, then $\bar{x}\approx M$, as shown in Figure 9-10.
  数据分布对称的情况,如图 9-10 所示。

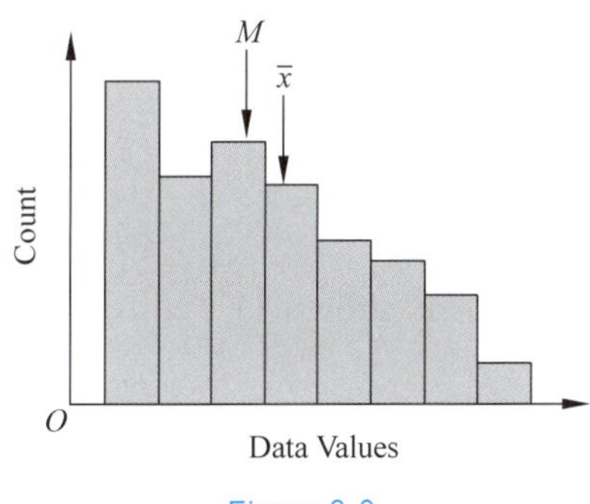

Figure 9-9

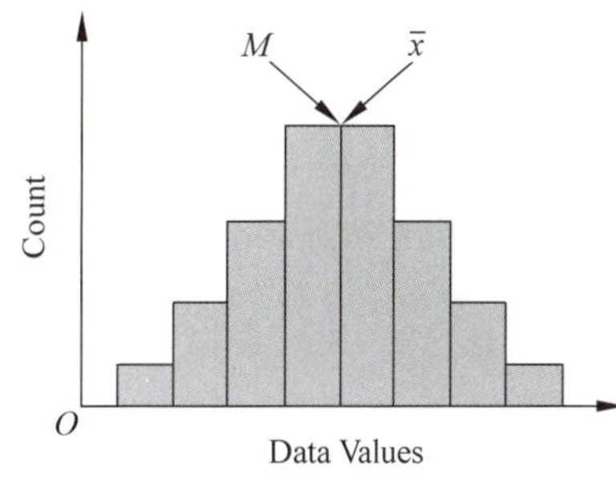

Figure 9-10

The median splits the area under the distribution in halves.
中位数等分数据分布下面的面积。

Histogram 直方图

n. [ˈhɪstəˌgræm]

Definition: A diagram that illustrates the frequencies of intervals of a quantitative variable using rectangles of equal widths but varying heights (depending on the frequencies of intervals).
直方图是指用矩形展示定量变量区间频数的图像。这些矩形的宽相等,长根据区间的频数的不同而异。

Figures 9-3 to 9-10 are histograms.

Properties: 1. Advantage: For data set with large size and large spread, displaying data using intervals is much easier and more efficient than plotting individual data points, as indicated in dot plot and stemplot.

优点：对于分布大的大数据集，用区间(即直方图)呈现数据比逐一呈现数据方便及有效。点图和茎叶图是逐一呈现数据的图像。

2. Disadvantage: For data set with small size or small number of possible values, displaying data using individual data points is more descriptive and specific, as shown in dot plot and stemplot.

缺点：对于分布小的小数据集，逐一呈现数据更能具体描述，即采用点图和茎叶图呈现数据的图像。

Examples: Jack measures the weights (rounded to the nearest pound) of 30 adults in his town and has the sorted data set as follows:

100,115,120,135,136,138,140,150,152,152,
156,156,157,157,158,160,162,165,170,175,
178,178,180,185,189,190,200,205,210,220

We can divide the intervals in several ways (there is no right or wrong answer, but we don't want too many or too few intervals).

Method I: Each interval has a length of 20, as shown in Figure 9-11.

Our convention is to include the left endpoint and exclude the right endpoint. For example, for the 100-120 interval, we include 100 but exclude 120.

Histograms come with descriptions.

Center: 159.

Spread: 100-220.

Shape: Unimodal and approximately symmetric.

Outliers: None.

Method II: Each interval has a length of 40, as shown in Figure 9-12.

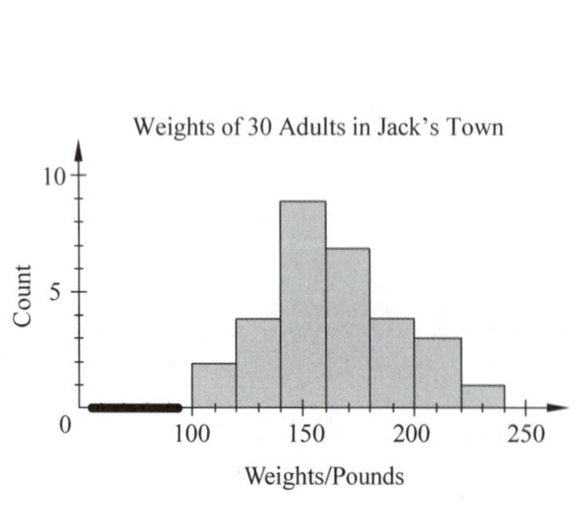

Figure 9-11

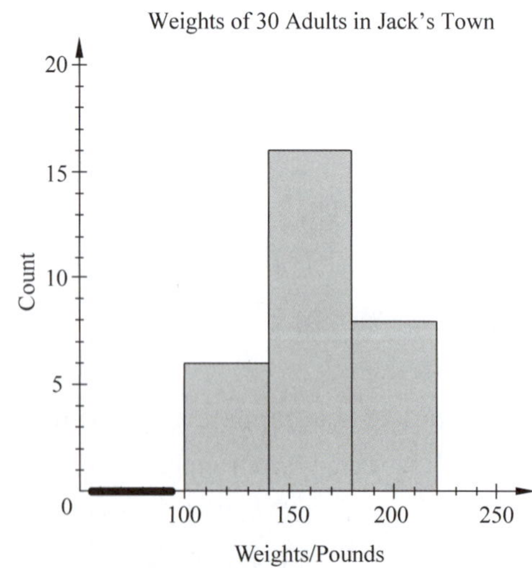

Figure 9-12

This is an example of histogram that has too few intervals.

Same descriptions as that for Figure 9-11.

And many more.

Dot Plot　点图

n. [dɑːt plɑːt]

Definition：A diagram that illustrates the frequencies of the data appeared in a quantitative variable using dots.　点图是指用点展示定量变量数据频数的图像。

Properties：1. Advantage：For a data set with a small size and a small spread，a dot plot can show each data's frequency. This reporting is very specific.

优点：对于分布小的小数据集，点图可展示每个数据的频数，很具体。

For a data set with a large size but a small spread，a dot plot still shows each data frequency. Alternatively we can manipulate the scale of the y-axis so that one unit on the y-axis may represent multiple occurrences of a data. It is similar to a bar graph，except the variable on the x-axis is quantitative instead of categorical，and we are using dots instead of bars for frequency.

对于分布小的大数据集，点图同样可展示每个数据的频数。可以变动 y 轴，使得 y 轴每个单位代表若干相同值的数据。点图类似于条形图，区别是条形图的 x 轴为分类变量，且用条形代表频数。点图的 x 轴为定量变量，且用点代表频数。

2. Disadvantage：For a large data set with a large spread，using histograms is far more effective than using dot plots because histogram divides each class using intervals.

缺点：对于分布大的大数据集，用直方图比用点图更有效，因为直方图用区间代表每个分类。

Examples：1. Eric wants to know how many books each person in his office read this month. Through survey，he gathered his data set as followed：

0,0,1,1,1,1,1,1,1,2,2,2,2,2,2,3,3,4,5,6

Since there are so few different values appeared (20 data points and 7 different values)，a dot plot is ideal for the visual display the distribution. Note that a description comes with a dot plot!

As shown in Figure 9-13，each dot represents a person.

Center：2.

Spread：0-6.

Shape：unimodal and right skewed.

Outliers：none.

2. John wanted to measure people's ability of thinking out of the box. He wrote a test of 10 multiple choice questions and posted it on the Internet (each question is worth 1 point，and there is no partial credit given). 150 people took this test，and the score distribution is as followed：

5 people got 0s；5 people got 1s；10 people got 2s；15 people got 3s；20 people got 4s；25 people got 5s；30 people got 6s；20 people got 7s；15 people got 8s；5 people got

9s；0 people got 10s.

As shown in Figure 9-14, each dot represents 5 people.

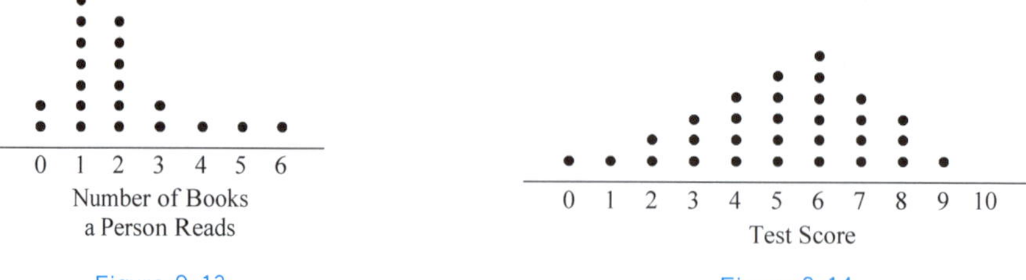

Figure 9-13　　Figure 9-14

Center：5.

Spread：0-9.

Shape：unimodal and left skewed.

Outliers：none.

It is similar to a bar graph, except that the variable on the x-axis is quantitative rather than categorical, and we are using dots instead of bars.

Stem-and-Leaf Plot/Stemplot 茎叶图

n. [stem ænd lif plɑt/stem.plɑt]

Definition：A diagram that illustrates all data appeared in a quantitative variable using a stem (usually records the digits except the last one, of a data) and leaf (usually records the last digit of a data). Stemplot comes with a key to provide interpretation of a combination of a stem and a leaf. 茎叶图是指用茎和叶展示定量变量的数据。茎通常记录数据除去最后一个数位的所有数位。叶通常记录最后一个数位。茎叶图通常都有图例，用于解释茎和叶的搭配。

Properties：1. Advantage：For data set with low spread and small size (usually less than 50), a stem-and-leaf plot/stemplot is ideal for modeling the distribution.

In addition, stemplot is ideal for comparing two 1-variable statistics using back-to-back stemplot.

To avoid having skyscraper-looking distribution (too many data of a particular stem), stemplots are flexible to split the stems using split stemplot.

优点：对于分布小的小数据集(数据数量小于 50)，茎叶图是描述数据分布的理想图像。此外，茎叶图能有效地对比两个单变量统计(用背靠背的茎叶图实现)。为了避免形状像摩天大楼的数据分布(一个茎里有太多的数据)，茎叶图可以灵活地分开茎(用分茎茎叶图)。

2. Disadvantage：For data set with high spread or large size, a stem-and-leaf plot is cumbersome to use since it records every data exactly.

缺点：对于分布大或数据多的数据集，用茎叶图记录每个数据是很困难的。

Examples：1. (Same example as that of histogram) 例题与柱状图的一样。

Jack measures the weights (rounded to the nearest pound) of 30 adults in his town and has the sorted data set as followed：

100,115,120,135,136,138,140,150,152,152,
156,156,157,157,158,160,162,165,170,175,
178,178,180,185,189,190,200,205,210,220

Stemplot comes with a description!

Figure 9-15 shows the stemplot of this data set.

Weights of 30 Adults (in Pounds)

Tens	Ones
10	0
11	5
12	0
13	5 6 8
14	0
15	0 2 2 6 6 7 7 8
16	0 2 5
17	0 5 8 8
18	0 5 9
19	0
20	0 5
21	0
22	0

Key: 13|5 represents 135

Figure 9-15

Center: 159.

Spread: 100-220.

Shape: unimodal and approximately symmetric.

Outliers: none.

2. Jill lives in the town next to Jack's. She measures the weights (rounded to the nearest pound) of 20 adults in her town and has the sorted data set as followed:

110,125,131,133,135,135,150,152,153,155,
155,155,165,168,175,180,182,182,195,210

Figure 9-16 shows the stemplot of this data set.

Weights of 20 Adults (in Pounds)

Tens	Ones
11	0
12	5
13	1 3 5 5
14	
15	0 2 3 5 5 5
16	5 8
17	5
18	0 2 2
19	5
20	
21	0

Key: 13|5 represents 135

Figure 9-16

Center: 155.

Spread: 110-210.

Shape: unimodal and slightly right skewed.

Outliers: none.

Note that even there are no weights in pounds in the 140s and 200s, it is still important to include these two stems.

We can make a back-to-back stemplot to compare the weights of the adults from both towns, as shown in Figure 9-17.

Weights of 50 Adults (in Pounds)

Jack's Town Ones	Tens	Jill's Town Ones
0	10	
5	11	0
0	12	5
8 6 5	13	1 3 5 5
0	14	
8 7 7 6 6 2 2 0	15	0 2 3 5 5 5
5 2 0	16	5 8
8 8 5 0	17	5
9 5 0	18	0 2 2
0	19	5
5 0	20	
0	21	0
0	22	

Key: 13|5 represents 135

Figure 9-17

Refer to the descriptions for both stemplot in Figures 9-15 and 9-16.

3. A grade distribution (out of 100) of a class of 30 students is as follows:

50,60,62,71,73,75,78,81,81,82,
82,83,85,85,86,87,87,87,88,88,
90,91,91,92,93,93,95,95,95,97

A regular stemplot will result a skyscraper-shaped distribution, as shown in Figure 9-18.

Test Scores

Tens	Ones
5	0
6	0 2
7	1 3 5 8
8	1 1 2 2 3 5 5 6 7 7 7 8 8
9	0 1 1 2 3 3 5 5 5 7

Key: 8|7 represents 87

Figure 9-18

Center: 86.5.

Spread: 50-97.

Shape: unimodal and left skewed.

Outliers：none.

So in this case，a split stemplot displays the result better，as shown in Figure 9-19.

Test Scores

Tens	Ones
5	0
5	
6	0 2
6	
7	1 3
7	5 8
8	1 1 2 2 3
8	5 5 6 7 7 7 8 8
9	0 1 1 2 3 3
9	5 5 5 7

Key: 8|7 represents 87

Figure 9-19

Note that the descriptions for the regular stemplot and the split stemplot are the same since they are displaying the same data set.

Box-and-Whisker Diagram/Boxplot　箱型图

n. [bɑks ænd ˈhwɪskər ˈdɑɪəˌɡræm/bɑks. plɑt]

Definition：A diagram that illustrates the five-number summary of a quantitative variable.　箱型图是展示定量变量的五数概括法的图表。

Notation：There are two types of boxplots：

有两种箱型图：

1. Regular boxplot，which illustrates the minimum，Q1，median，Q3，and maximum.　普通箱型图：展示最小值、第一四分位数、中位数、第三四分位数、最大值。
2. Modified boxplot，which illustrates the modified minimum，Q1，median，Q3，and the modified maximum，then plots the outliers separately.　改进箱型图：先找出改动过的最小值、第一四分位数、中位数、第三四分位数、改动过的最大值，然后逐一找出离群值。

   The modified minimum is the minimum after the outliers are excluded，and the modified maximum is the maximum after the outliers are excluded.　改动过的最小值(新最小值)是除去离群值后的最小值；改动过的最大值(新最大值)是除去离群值后的最大值。

Examples：A grade distribution (out of 100) of a class of 30 students is as followed：

10，30，62，62，63，65，67，67，68，69，
70，71，71，71，72，72，72，72，75，77，
77，77，78，79，79，79，81，81，98，99

min = 10；Q1 = 67；M = 72；Q3 = 78；max = 99.

IQR = Q3 − Q1 = 11.

Data that are less than Q1 − 1.5IQR = 50.5 are outliers.

Data that are greater than Q3 + 1.5IQR = 94.5 are also outliers.

We found out that 10,30,98,and 99 are outliers.

The boxplot is shown in Figure 9-20.

The modified boxplot is shown in Figure 9-21.

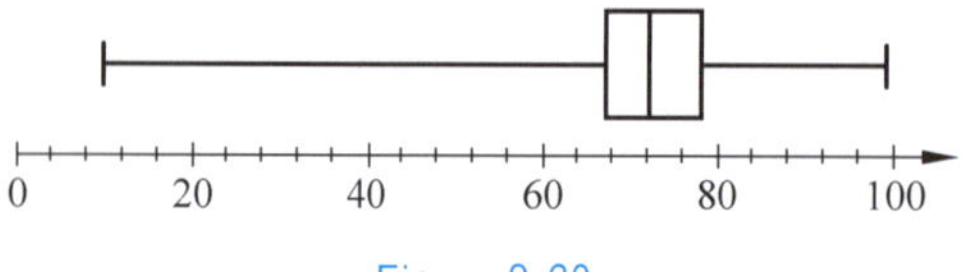

Figure 9-20

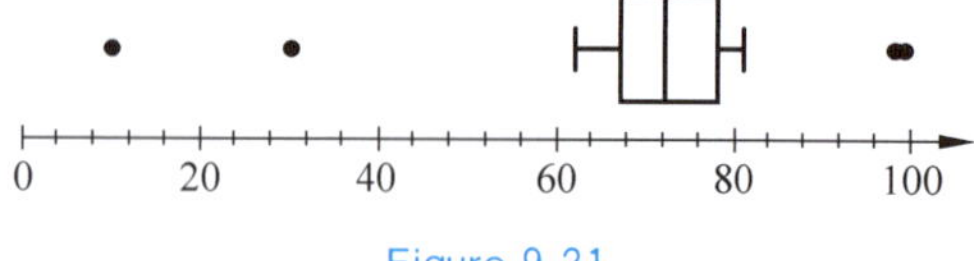

Figure 9-21

9.3.3 Summary 总结

1. Graphs that display categorical variables' data:
   展现分类变量数据的图表有:
   (1) Bar Graphs—display categories' frequencies.
   条形图——展现每个分类的频数。
   (2) Pie Charts—display categories' relative frequencies.
   饼图——展现每个分类的相对频数。
2. Graphs that display quantitative variables' data:
   展现定量变量数据的图表有:
   (1) Histogram—displays intervals of data. 直方图——展现不同区间的数据。
   (2) Dot Plot—displays frequencies of individual data.
   点图——展现每个数据的频数。
   (3) Stemplot—displays individual data. To compare data sets, use back-to-back stemplot. To avoid skyscraper-looking stemplots, split the stems.
   茎叶图——逐一展现数据。要对比数据集,用背靠背茎叶图。要避免形状像摩天大楼的茎叶图,把茎分开。
   (4) Boxplot—displays five-number summary of a data set. To distinguish the outliers, use modified boxplot (which modifies the minimum and maximum).
   箱型图——展现数据集的五数概括法。要区分开离群值,则用改进箱型图(改变原先的最小值和最大值)。

9.4 Two-Variable Statistics 双变量统计

Explanatory Variable and Response Variable 解释变量与响应变量

n. [ɪk'splænəˌtəri 'veəriəbəl] & n. [rɪ'spɑns 'veəriəbəl]

Definition: The explanatory variable is an analogue of the independent variable, and the response variable is an analogue of the dependent variable.
解释变量与自变量类似。响应变量与因变量类似。

Notes: Response variable is the central focus of the investigation, and the explanatory variable usually creates an effect on the response variable (The response variable changes as the explanatory variable changes).

We are interested in the relationship between explanatory and response variables (preferably causation).
响应变量是调研的重点。解释变量通常会影响响应变量(随着解释变量的变化,响应变量会变化)。
我们希望找出解释变量与响应变量的关系(最好为因果关系)。

Examples: 1. Jack wanted to know how the amount of study time affects a student's grade. He recorded the amount of study time for each student in his class along with the student's test score.
The explanatory variable is the amount of study time, and the response variable is the test score.
2. Josh wanted to know how much (in the scale from 1 to 10) that an on-campus college student feels homesick based on the distance between home and college.
The explanatory variable is the distance between home and college, and the response variable is the degree of homesickness.

Relationship between Explanatory and Response Variables　解释变量与响应变量之间的关系

Definition: One of causation, common response, and confounding.　解释变量与响应变量之间的关系可以为因果关系、共同作用关系、混杂关系中的一种。

Causation　因果关系

n. [kɔːˈzeɪ.ʃən]

Definition: The scenario that the response variable Y is **directly and solely** influenced by the explanatory variable X. In other words, X and Y have cause-and-effect relationship.
因果关系是指响应变量 Y 只会直接被解释变量 X 影响。换句话说,X 和 Y 是因果关系。
Figure 9-22 shows the situation of causation.

Figure 9-22

Examples: 1. Explanatory variable (X): the number of books bought.
Response variable (Y): the total cost of the books.
Scenario: Y increases as X increases.
Conclusion: X causes Y.
2. Explanatory variable (X): the height of the mother.
Response variable (Y): the height of the child.
Scenario: Y increases as X increases.
Conclusion: X causes Y.
3. Explanatory variable (X): the speed of traveling.
Response variable (Y): the distance traveled.
Scenario: Y increases as X increases.
Conclusion: X causes Y.

In each of the scenarios above, sometimes one argues that it is not a causation due to the presence of a lurking variable. See confounding.　在以上的每个情景中,有时候我们会争论潜伏变量是否存在。见混杂关系。

Common Response　共同作用关系

n. [ˈkɑmən rɪˈspɑns]

Definition: The scenario that the changes in both response variable Y and explanatory variable X is due to changes in another variable Z.

响应变量 Y 和解释变量 X 均被另一个变量 Z 影响的情况。

Figure 9-23 shows the situation of common response.

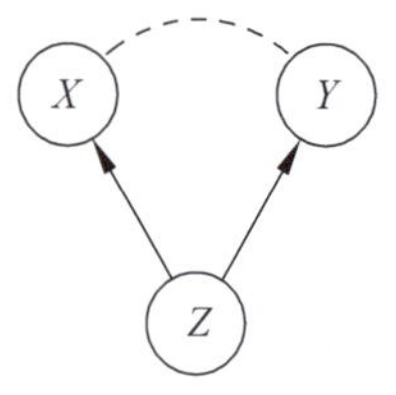

Figure 9-23

Examples: 1. Explanatory variable (X): the number of ice cream sold in a town.

Response variable (Y): the number of crimes in a town.

Z: the number of people in the town.

Scenario: Y increases as X increases.

It will be unsound to say that the number of ice cream sold in a town **causes** the number of crimes in a town.

Instead, this is a common response case: the changes in both X and Y is due to changes in another variable Z. Therefore,

Conclusion: Z causes X, and Z causes Y.

2. Explanatory variable (X): shoe size.

Response variable (Y): reading ability.

Z: age.

Scenario: Y increases as X increases.

It will be unsound to say X causes Y.

Conclusion: Z causes X, and Z causes Y.

Confounding　混杂关系

adj. [kənˈfaʊndɪŋ]

Definition: The scenario that the response variable Y is directly influenced by the explanatory X, but it is not certain whether there is a variable Z (**which is not explicitly measured**) that influences Y as well.

响应变量 Y 直接被解释变量 X 影响的情况，但是不确定是否存在同样影响 Y 的变量 Z(变量 Z 没有被明确测量)。

Variable Z is known as a lurking variable.　变量 Z 亦称潜伏变量。

Figure 9-24 shows the confounding situation.

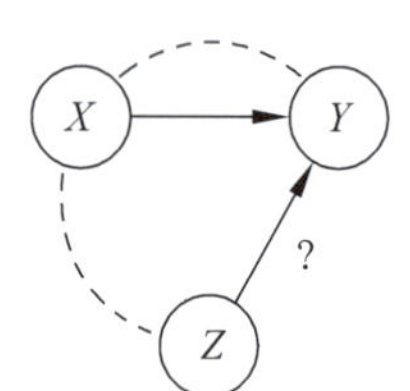

Figure 9-24

Examples: Examples 1-2 are from causation, and here it offers a view why some people interpret them as confounding (there is not really a right answer as long as the explanation is sound).

1. Explanatory variable (X): the number of books bought.

Response variable (Y): the total cost of the books.

Scenario: Y increases as X increases.

Conclusion: X causes Y.

Z: Other fees such as the increasing delivery fees.

We are not sure if Z causes Y as well. Thus the situation is confounding and Z is the lurking variable.

2. Explanatory variable (X): the height of the mother.

 Response variable (Y): the height of the child.

 Scenario: Y increases as X increases.

 Conclusion: X causes Y.

 Z: Daily habits of the child.

 We are not sure if Z causes Y as well. Thus the situation is confounding and Z is the lurking variable.

3. Explanatory variable (X): amount of video game time.

 Response variable (Y): grade on the exam.

 Scenario: Y decreases as X increases.

 Conclusion: X causes Y.

 Z: The ability of a student.

 We are not sure if Z causes Y as well. Thus the situation is confounding and Z is the lurking variable.

9.5 Visual Display for Two-Variable Statistics 双变量统计图表

9.5.1 Descriptions of Graphs of Two-Variable Statistics 双变量统计图表描述

Descriptions of Graphs of Two-Variable Statistics 双变量统计图表描述

Definition: These descriptions work for scatterplot. 这些描述是针对散点图的。

Vocabulary: We usually use the following four aspects to describe a scatterplot: form, association, strength (for linear models only), and outliers. 通常从以下几方面描述散点图:形状、关联性、强度、离群值。

The form is the appearance of the scatterplot (linear, or curve, or other irregular patterns). 散点图的形状有线性、曲线或其他形状。

Curve means it is not straight (can be a polynomial function with degree at least 2 or an exponential function, or other types of curves). 曲线暗示着是多项式函数(次数至少为2)或指数函数图像,或其他。

See association, strength and outliers. 见关联性、强度和离群值词条。

In addition, we also use correlation and influential points to describe scatterplots. 此外,也可以用相关系数和强影响点描述散点图。

Association 关联性

n. [əˌsoʊʃiˈeɪʃən]

Definition: One of negative association, zero association, and positive association.

关联性可以为负、零、正。

Let X be the explanatory variable and Y be the response variable.

设 X 为解释变量,Y 为响应变量。

- A negative association means that Y decreases as X increases (the general pattern of the graph). 负关联性指的是当 X 增加,Y 减少的图像总体规律。
- A zero association means that Y stays the same as X increases (the general pattern of the graph). 零关联性指的是当 X 增加,Y 不变的图像总体规律。
- A positive association means that Y increases as X increases (the general pattern of the graph). 正关联性指的是当 X 增加,Y 增加的图像总体规律。

Examples: 1. As shown in Table 9-4, this is a negative association. Y decreases as X increases.

Table 9-4

X	0	0	1	2	4	6	8	9	10
Y	100	98	80	86	75	71	67	70	55

2. As shown in Table 9-5, this is a zero association. Y stays the same as X increases.

Table 9-5

X	0	0	1	2	4	6	8	9	10
Y	100	100	100	98	96	100	102	104	100

3. As shown in Table 9-6, this is a positive association. Y increases as X increases.

Table 9-6

X	0	0	1	2	4	6	8	9	10
Y	1	3	10	20	18	35	60	57	100

Strength 强度

n. [streŋkθ]

Definition: The verbal description of how close the data points form a linear model. 强度是数据点的形状与直线接近程度的文字描述。

Strength has connection with correlation coefficient. 强度跟相关系数有关。

Examples: Figure 9-25 shows a strong linear relationship of Y against X.

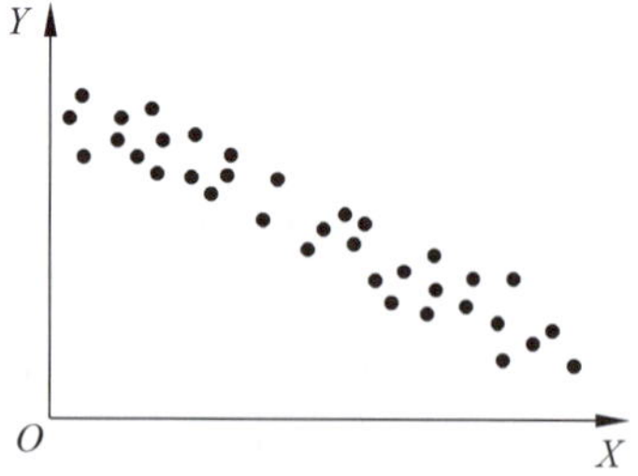

Figure 9-25

Figure 9-26 shows a moderate linear relationship of Y against X.

Figure 9-27 shows a weak linear relationship of Y against X.

Strong, moderate, and weak are descriptions of strength. We can add adverbs such as "moderately strong" and "moderately weak" when necessary.

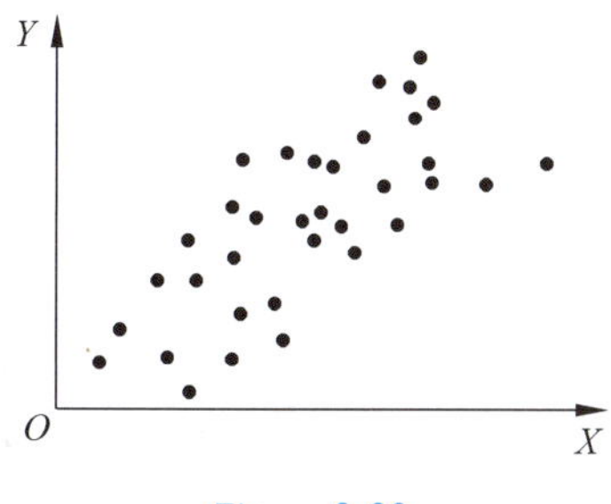

Figure 9-26

Figure 9-27

Correlation Coefficient 相关系数

n. [ˌkɔrəˈleɪʃən ˌkoʊɪˈfɪʃənt]

Definition: A value that describes the association and the strength. 相关系数是描述关联性和强度的值。

Notation: For pairs of values in explanatory variable X and response variable Y, the correlation coefficient r is given by

$$r=\frac{n\left(\sum xy\right)-\left(\sum x\right)\left(\sum y\right)}{\sqrt{\left[n\left(\sum x^2\right)-\left(\sum x\right)^2\right]\left[n\left(\sum y^2\right)-\left(\sum y\right)^2\right]}}.$$

n is the number of data (pair of values).

$\sum xy$ is the sum of the products of paired values.

$\sum x$ is the sum of the x-scores.

$\sum y$ is the sum of the y-scores.

$\sum x^2$ is the sum of the squares of the x-scores.

$\sum y^2$ is the sum of the squares of the y-scores.

Properties: 1. $-1\leqslant r\leqslant 1$.

2. r has the same sign as the association: $r<0$ if the association is negative; $r=0$ if the association is zero; $r>0$ if the association is positive.
   r 的正负值与关联性的正负值相同。
3. The closer $|r|$ is to 1, the stronger the strength is. If $|r|=1$, then the scatterplot of response variable VS explanatory variable is linear.
   $|r|$越接近1，强度越大。若$|r|=1$，则响应变量与解释变量的散点图为线性。
4. $r^2\cdot 100\%$ of the data points can be explained by the least-square regression line.
   $r^2\cdot 100\%$的数据点可以被最小二乘回归线解释。
5. Correlation does not imply causation. For example, let
   explanatory variable X = the number of ice cream sold in the town,
   response variable Y = the number of crimes in the town.
   If X and Y are positively and strongly correlated (Y increases as X increases, and the plot of Y VS X is almost linear.), it does not mean that X causes Y. Instead, this is a common response case. Let

Z = the number of people in the town.

X increases as Z increases, and Y increases as Z increases.

Z causes X and Z causes Y.

相关系数不代表因果关系。

Examples: We will not prove the properties since they are outside of the scope of high school math. Neither will we perform the calculation of the correlation. Nevertheless, we will show some sample scatterplots and approximate their correlation.

The correlation for the scatterplot in Figure 9-28 is approximately -0.9 (negative slope and strong linearity).

The correlation for the scatterplot in Figure 9-29 is approximately 0.8 (positive slope and moderately strong linearity).

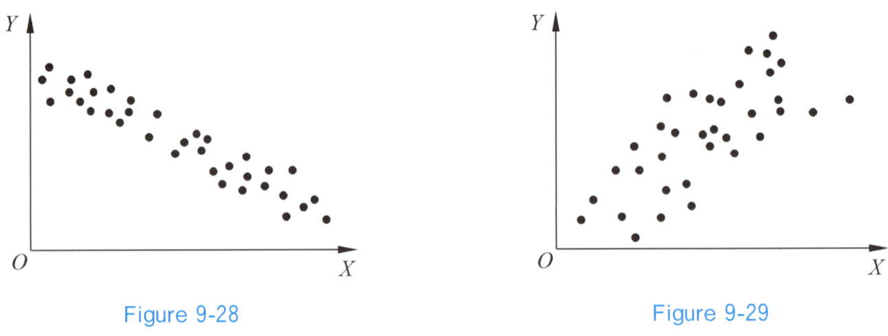

Figure 9-28　　Figure 9-29

The correlation for the scatterplot in Figure 9-30 is approximately 0.1 (positive slope and weak linearity).

The correlation for the scatterplot in Figure 9-31 is 0 (zero slope and no linearity at all).

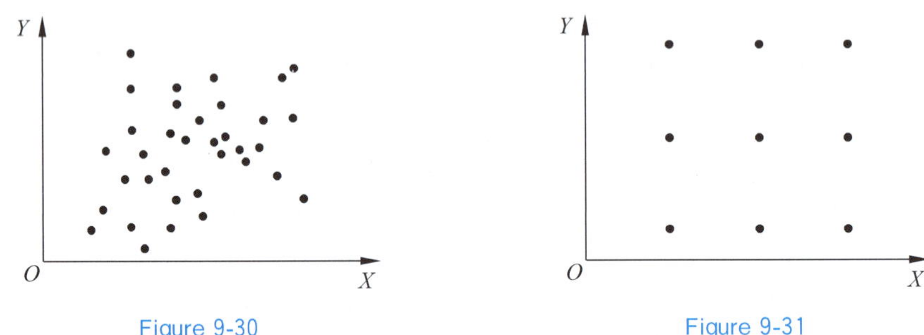

Figure 9-30　　Figure 9-31

Influential Point　强影响点

n. [ˌɪnfluˈenʃəl pɔɪnt]

Definition: A data point on the scatterplot whose removal will drastically change the slope of the least-square regression line.　强影响点是散点图上的一个点，若把它移除，最小二乘法回归线的斜率会有很大的变动。

Properties: 1. An influential point is not necessarily an outlier.　强影响点不一定为离群值。

2. We can use trial and error, or simply approximation to find an influential point.　用试错法或约算能找出强影响点。

Examples：1. As shown in Table 9-7，(5，70) is an influential point because it drastically changes the slope of the least-square regression line.

Table 9-7

X	0	1	2	3	4	5	6	7	8
Y	10	12	18	23	27	70	37	42	47

2. As shown in Table 9-8，(25，300) is an influential point because it drastically changes the slope of the least-square regression line.

Table 9-8

X	10	12	15	19	19	23	25	27	27
Y	50	59	76	100	98	125	300	153	145

3. As shown in Figure 9-32，the blue point is an influential point because it drastically changes the least-square regression line (LSRL).

Removing the influential point，the LSRL looks like this，as shown in Figure 9-33.

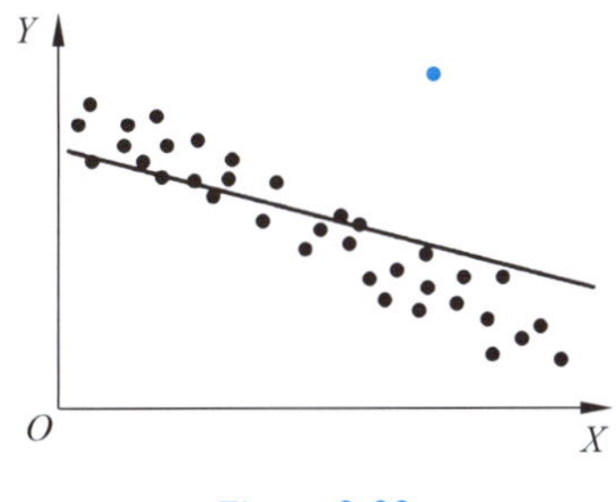

Figure 9-32

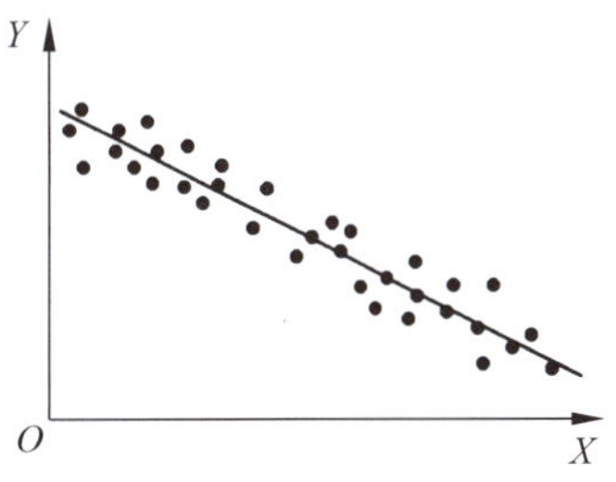

Figure 9-33

Outlier(of Two-Variable Statistics)　离群值

n. [ˈaʊtˌlaɪ.ər]

Definition：A data point on the scatterplot that lies far away from the majority of the data points.
离群值是散点图上离大部分数据点很远的点。

Properties：1. There is not a specified way to spot the outliers—the best way is by approximation.
没有固定的方法找离群值——唯一的一种方法是约算。

2. An outlier is not necessarily an influential point.　离群值不一定为强影响点。

Examples：1. As shown in Table 9-9，(100，9999) is an outlier because it lies far away from the majority of the data points. Moreover，(100，9999) is also an influential point. Its removal would result the rest of the data falls into a perfect linear pattern：$y=4x$.

Table 9-9

X	1	2	3	4	5	6	7	8	100
Y	4	8	12	16	20	24	28	32	9999

2. As shown in Table 9-10,(100,400) is an outlier because it lies far away from the majority of the data points. However,(100,400) is not an influential point. The data set as of now is a perfect linear model: $y=4x$. Removing the data point (100,400) does not change this fact. This illustrates Property 2.

Table 9-10

X	1	2	3	4	5	6	7	8	100
Y	4	8	12	16	20	24	28	32	400

3. As shown in Figure 9-34, the blue point is an outlier because it lies far away from the rest of the data points. However, it is not an influential point because its removal does not affect the least-square regression line.

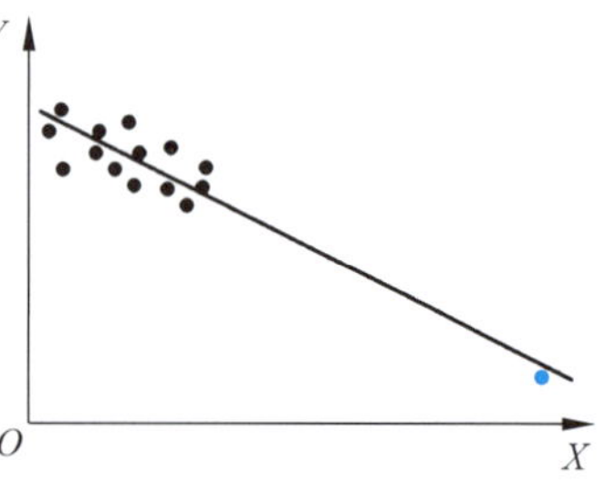

Figure 9-34

9.5.2 Scatterplots and Least-Square Regression Lines 散点图与最小二乘法回归线

Scatterplot 散点图

n. [skætər,plɑt]

Definition: A graph of plotted data points that shows the relationship between two quantitative variables. 通过数据点展现两个定量变量关系的图像。

Properties: The explanatory variable (independent variable) is plotted along the x-axis and the response variable (dependent variable) is plotted along the y-axis —same setup as that of a function's graph. 解释变量(自变量)沿着 x 轴变化,响应变量(因变量)沿着 y 轴变化——与函数图像的结构相同。

Examples: These examples come from explanatory variable & response variable. 这些例子与解释变量与响应变量的例子相同。

1. Jack wanted to know how the amount of study time affects a student's grade. He recorded the amount of study time for each student in his class along with the student's test score, as shown in Table 9-11.

   The explanatory variable is the amount of study time, and the response variable is the test score.

Table 9-11

Study Time/hrs	0	2	2	3	4	4	4	5	8
Grades Received	60	78	85	82	87	84	90	92	96

The scatterplot is shown in Figure 9-35.

Scatterplot comes with a description.

Form: linear.

Association: positive

Grades increase as study time increases.

Strength: moderately strong.

Outliers: The point for which the study time is 0 hours.

2. Josh wanted to know how much (in the scale from 1 to 10) that an on-campus college student feels homesick based on the distance between home and college.

The explanatory variable is the distance between home and college, and the response variable is the degree of homesickness.

The table is shown in Table 9-12.

Table 9-12

Distance from Home/km	100	200	300	400	500	600	600	650	1200
Degree of Homesickness	3	5	4	6	6	7	8	7	10

The scatterplot is shown in Figure 9-36.

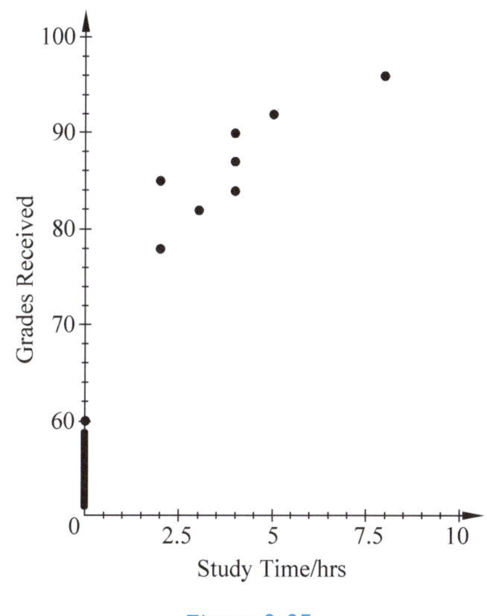

Figure 9-35

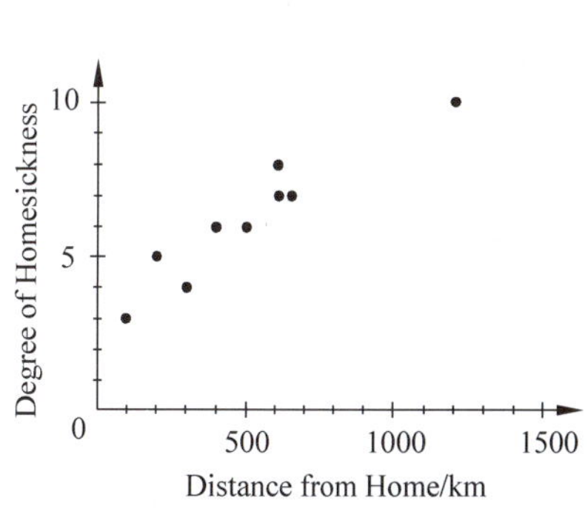

Figure 9-36

Form: linear.

Association: positive

Degree of homesickness increases as distance from home increases.

Strength: strong.

Outliers: The point for which the distance from home is 1200km.

Least-Square Regression Line (LSRL)　最小二乘法回归线

n. [list skweər rɪ'greʃən laɪn]

Definition: A line that fits the pattern of the data points in the scatterplot as closely as possible.

最小二乘法回归线是最符合散点图上点的规律的直线。

A less statistical name for LSRL is the best-fitting line.

Notation: $\hat{y} = a + bx$, for which a is the y-intercept and b is the slope. The variable $\hat{y}$ is the predicted y-value (points along the LSRL are predicted values).

最小二乘法回归线是可写作 $\hat{y} = a + bx$,其中 a 为 y 截距,b 为斜率,$\hat{y}$ 为 y 值的期望值(在最小二乘法回归线上)。

Properties: 1. For the explanatory variable X and the response variable Y:

对于解释变量 X 和响应变量 Y:

Let S_x be the standard deviation for the data set X, S_y be the standard deviation for the data set Y, r be the correlation between X and Y, $\overline{X}$ be the average of the data in X, $\overline{Y}$ be the average of the data in Y. We have

$$b = r\frac{S_y}{S_x}, \text{and } a = \overline{Y} - b\overline{X}.$$

设 S_x 和 S_y 分别为 X 和 Y 的标准差,r 为相关系数,$\overline{X}$ 和 $\overline{Y}$ 分别为 X 和 Y 数据的平均值,则有 $b = r\frac{S_y}{S_x}$, $a = \overline{Y} - b\overline{X}$。

2. The sum of all residuals of the scatterplot for the LSRL is 0 (and this is the reason why LSRL is the best-fitting line). 最小二乘法回归线上所有残差的和为 0。

3. $r^2 \cdot 100\%$ of the data can be explained by the LSRL $\hat{y} = a + bx$.

$r^2 \cdot 100\%$的数据可以被最小二乘法回归线$\hat{y} = a + bx$ 解释。

Notes: We will not prove Property 1 or 3 since it is outside of the scope of high school math.

Examples: We will skip the calculations in Property 1 since it is too cumbersome (it can be done in a graphing calculator efficiently). Instead, we will approximate the LSRL in our examples, as shown in Figures 9-37 and 9-38.

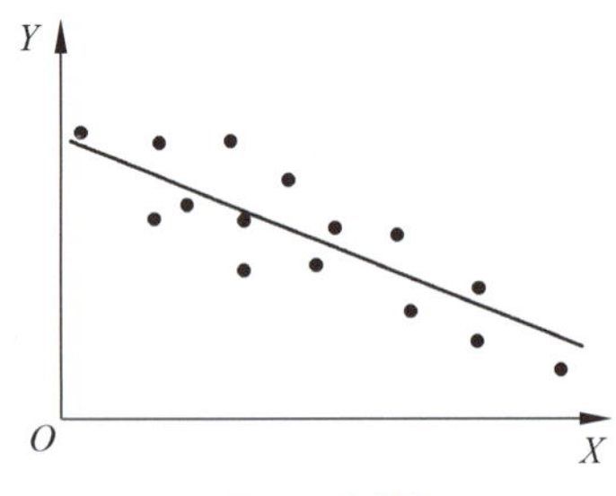

Figure 9-37

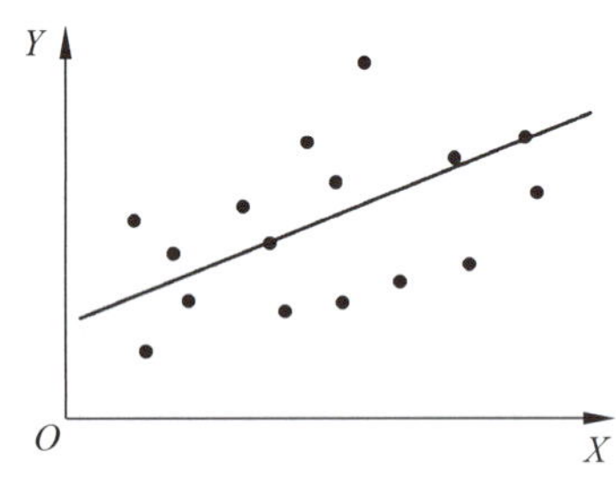

Figure 9-38

Residual 残差

n. [rɪ'zɪdʒuəl]

Definition: A point's residual is difference between the observed value and the predicted value of the response variable. 一个点的残差为响应变量中观察值与期望值的差。

Figure 9-39 illustrates this difference, d, from which is positive.

图 9-39 图解了这个差,d 是正数。

Points under the LSRL have negative residuals.

在 LSRL 下面的点的残差是负数。

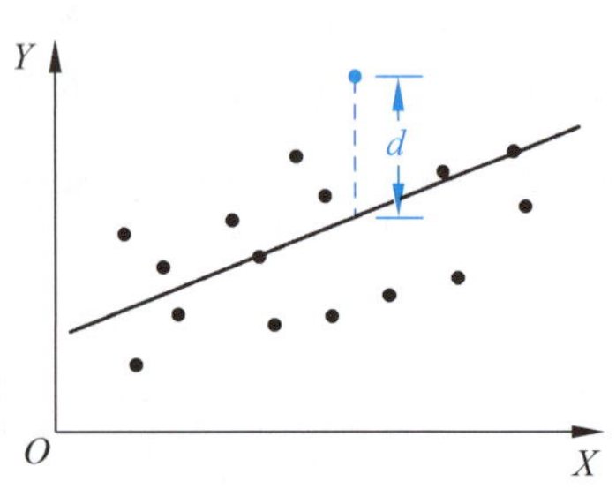

Figure 9-39

Notation: In a two-variable data set, suppose the LSRL is given by $\hat{y} = a + bx$, the residual of a point (x_0, y_0) is $y_0 - \hat{y}(x_0)$.

在双变量的数据集中,设最小二乘法回归线为 $\hat{y} = a +$

bx,点(x_0,y_0)的残差为 $y_0-\hat{y}(x_0)$。

Properties：The sum of residuals for the LSRL is 0, as the name LSRL implies.　LSRL 顾名思义,最小二乘法回归线的残差之和为 0。

Residual Plot　残差图

n. [rɪ'zɪdʒuəl plɑt]

Definition：A graph of plotted data points that shows the relationship between the explanatory variable and the residuals of the response variable.　残差图是指通过数据点,展现解释变量与残差(响应变量)关系的图像。

Properties：1. The LSRL corresponds to the x-axis. Data points that are above the x-axis imply positive residuals (observed > predicted), and the points below the x-axis imply negative residuals (observed < predicted).　x 轴对应最小二乘法回归线。在 x 轴之上的数据点代表正残差(观察值>期望值)。在 x 轴之下的数据点代表负残差(观察值<期望值)。

2. The residual plot shows good scatter if the data points are oscillating around the x-axis (the number of points that are above and below the x-axis are approximately equal, and no point has relatively large residuals).　若数据点在 x 轴徘徊,则残差图展现了理想的数据分布(在 x 轴之上和之下的数据点个数相近,且没有数据点有特别大的残差)。

3. The residual plot shares the same form as the corresponding scatterplot.　残差图与它对应的散点图形状相同。

Examples：1. Figure 9-40 shows a residual plot.

Form：linear.

Association：positive.

y increases as x increases.

Strength：moderately strong.

Outliers：none

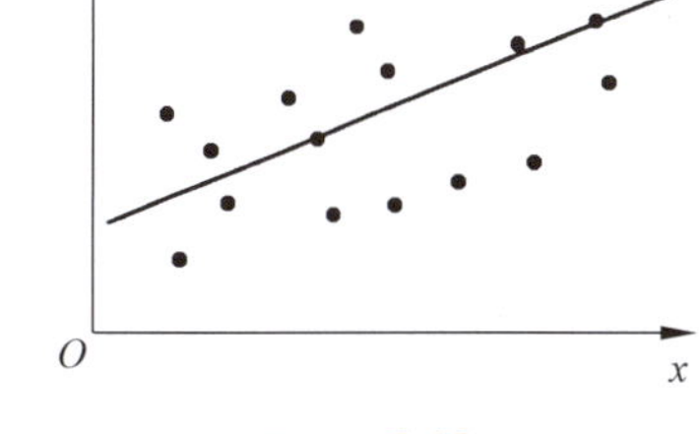

Figure 9-40

Figure 9-41 shows the residual plot from Figure 9-40. The residual plot shows good scatter (about half of the data points are above the LSRL and half are below). In addition, the data points are oscillating around the horizontal axis. This implies the linearity of the scatterplot. Neither exists any outliers or influential points.

2. Figure 9-42 shows a residual plot with good scatter, but based on the shape of the data points, it implies a curve on the scatterplot.

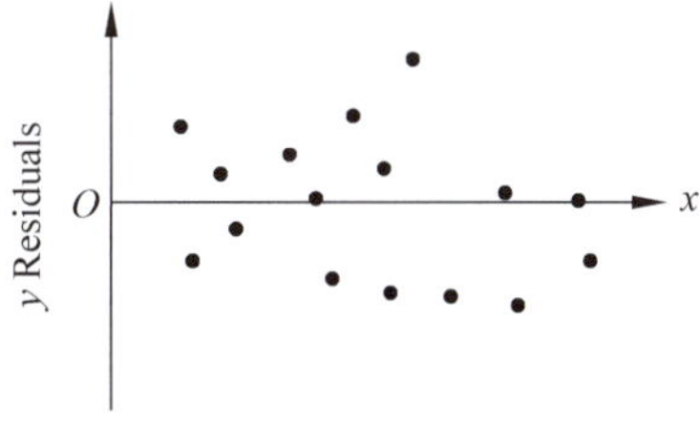

Figure 9-41

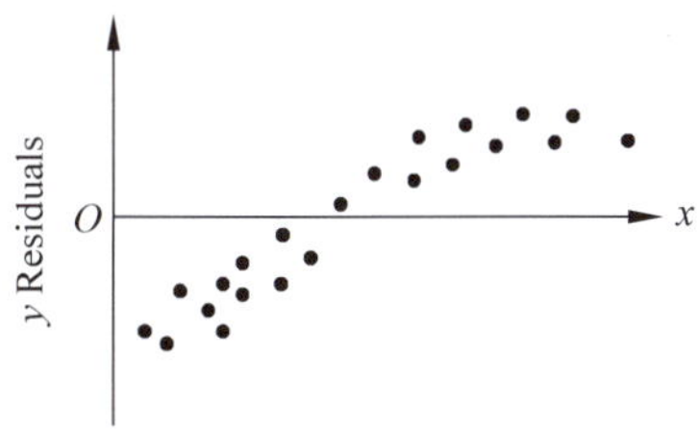

Figure 9-42

9.5.3 Extrapolation and Interpolation 外推和内推

Extrapolation 外推

n. [ɪkˌstræpə'leɪʃən]

Definition: The prediction of the value of the response variable evaluated at a value of the explanatory variable that is far beyond all the data points. 外推是指在所有数据点远处的解释变量值(沿着 x 轴)对响应变量值的预测。

Properties: Extrapolation is not very safe—sometimes it produces unsound results. 外推不是很安全——有时会给出不合理的结果。

Examples: As shown in Table 9-13, one extrapolation is that studying for 100 hours (suppose the exam is announced 5 days in advance) will result a grade about 500 points. This is unsound because a grade of 500 points will not be awarded. Secondly, the table suggests that once one studies for more than 4 hours, the grade will only change a little. Finally, studying for 100 hours will result over-study, which can backfire!

Table 9-13

Study Time/hrs	0	2	2	3	4	4	4	5	8
Grades Received	60	78	85	82	87	84	90	92	96

Interpolation 内推

n. [ˌɪn. tə. pə'leɪʃən]

Definition: The prediction of the value of the response variable evaluated at a value of the explanatory variable that is in the middle of the data points. 内推是指在数据点之间的解释变量值(沿着 x 轴)对响应变量值的预测。

Properties: Interpolation is much safer than extrapolation. The only dangerous scenario is the existence of influential points. 内推比外推安全,但要注意强影响点的存在。

Examples: As shown in Table 9-14, one interpolation is that studying for 6 hours will result a grade of 94. This is safe because there are no outliers or influential points in the scatterplot of this relation.

Table 9-14

Study Time/hrs	0	2	2	3	4	4	4	5	8
Grades Received	60	78	85	82	87	84	90	92	96

9.5.4 Summary 总结

1. To describe scatterplots of response variable Y VS explanatory variable X:
   描述响应变量 Y 与解释变量 X 的散点图:
   Form: linear/curve.
   形状:直线/曲线。
   Association: negative (Y decreases as X increases)/zero (Y remains as X increases)/positive (Y

increases as X increases).

关联性：负、零、正。

Strength：weak-strong,describing how closely the data points resemble to a line.

强度：弱-强,代表图像与直线的相似度。

Outliers：Data points that are far from the vast majority of the data points.

离群值：与绝大部分数据点相隔远的数据点。

Correlation：$-1\leqslant r\leqslant 1$. The sign of r describes association and $|r|$ describes strength.

相关系数：$-1\leqslant r\leqslant 1$。r 的正负符号代表关联性，$|r|$ 代表强度。

Influential Points：Data points whose removal affect calculations drastically.

强影响点：移除以后会大幅影响计算结果的点。

2. A scatterplot plots all the data points on a graph whose x-axis is the explanatory variable and whose y-axis is the response variable.　散点图以 x 轴为解释变量，y 轴为响应变量展示所有数据点。

   The LSRL is a line that best fits the pattern of data points on a scatterplot.　最小二乘法回归线是最符合数据点规律的直线。

   The residual of a point is difference between the observed value and the predicted value of the response variable.　一个点的残差为响应变量中观察值与期望值的差。

   A residual plot is a graph of plotted data points that shows the relationship between the explanatory variable and the residuals of the response variable. Its x-axis is the LSRL, and its y-axis is the residuals.　残差图是通过数据点，展现解释变量与残差(响应变量)关系的图像。它的 x 轴为最小二乘法回归线，y 轴为残差。

3. Review extrapolation and interpolation in Section 9.5.3.　复习 9.5.3 节外推和内推。

10. Statistics Method—Designing Studies 统计方法——研究设计

10.1 Introduction 介绍

10.1.1 Types of Studies 研究类型

Type of Studies 研究类型

Definition：See observational study and experiment.　研究类型有观察性研究和实验。

Observational Study 观察性研究

n. [ˌɒb.zəˈveɪ.ʃən.əl ˈstʌdi]

Definition：A type of study for which the individuals are observed and certain outcomes are measured. No treatment is deliberately imposed to attempt to affect the outcomes. 观察性研究是一种研究方式,即观察每个个体,测量某些结果。不会有意地实施实验手段影响结果。

Examples: 1. A researcher observes the studying habits from a class of students and measures their grades.
2. A researcher observes the amount of water plants have and compares their health conditions.
3. A educator observes the way math is taught in different schools and measures the difficulty of math classes of each school.

Experiment 实验

n. [ɪk'sperəmənt]

Definition: A controlled study for which the experimenter wants to know the cause-and-effect relationships between the explanatory and response variables. Treatments are deliberately imposed in an experiment. 实验是指被控制的研究,其中实验者想知道解释变量与响应变量的因果关系。这个研究有意地施以实验手段。

Properties: The two main characteristics of experiments are:
实验的两个主要特征是:
1. the method of dividing the experimental subjects. 对实验对象的分组方法。
2. the treatment each group of experimental subjects is assigned. 对每组实验对象分配的实验手段。

Golden Rule 1: To avoid experimental bias, all other explanatory variables besides the treatments must be kept the same for all experimental subjects.

黄金法则 1:要避免实验的偏差,除了实验手段外,所有其他解释变量在所有实验对象必须保持一致。

Golden Rule 2: Only one treatment can be administered in an experimental group. Otherwise the situation will be confounding.

黄金法则 2:对每个实验组只能用一个实验手段,否则会出现混杂关系的情况。

Examples: To illustrate Golden Rule 1:
Bob wanted to know if a fertilizer can help plants grow. He bought two identical plants (A and B). He fertilized Plant A and watered Plant B. However, he put A on the balcony and kept Plant B in his closet.
Two explanatory variables are being manipulated—the use of water/fertilizer and the location of the plants. If Plant A grew better than Plant B did, we had no idea whether the difference was due to the fertilizer or the location.
To illustrate Golden Rule 2:
A doctor wanted to know for seniors, (1) whether the habit of eating grapefruits regularly can lower the blood pressure and (2) whether taking a type of medicine can lower the blood pressure. He divided 300 seniors with high blood pressure into two equal groups. Group A ate grapefruits regularly and took the medicine. Group B did not receive any treatment. After a month, 120 members from Group A lowered their blood pressure, and 35 members from Group B lowered their blood pressure.

Explanatory Variable: the method used to lower blood pressure.

Response Variable: whether blood pressure is lowered.

Treatments: 1. the habit of eating grapefruits regularly.

2. the medicine.

We want to know whether the treatments have an effect.

It is evident that the treatments help lower blood pressure, but it is confounding because we do not know if one treatment or the other, or both help lower the blood pressure.

A better division of groups is to divide the 300 seniors with high blood pressures into 3 groups of 100. One group develops the habit of eating grapefruits regularly; one group takes the medicine; and one group does not receive any treatment. In this way there will be fewer lurking variables and the comparisons will be more persuasive.

Questions: For each of the following experiments, identify the explanatory variable, the response variable, and the treatment used.

1. A doctor wants to know if a certain type of new medicine can recover bad cold faster than a regular medicine. He divides 100 patients with bad cold into two equal groups. One group receives the new medicine and the other group resumes taking the regular medicine. After a week, the doctor measures the health conditions of the patients.
2. An automobile company wants to know which of the cars, A or B, is more popular. The developer asks 1000 people to drive model A, 1000 people to drive model B, and 1000 people to drive other models. He then records the satisfactory score for these 3000 people.
3. An elementary school teacher wants to know if a 2nd grader can learn multiplication by the multiplication table song faster than by the traditional memorization method. She decides to teach one of her classes (30 students) the multiplication song and another of her classes (30 students) the traditional memorization. At the end, she quizzes the students for multiplications.

Answers: 1. Explanatory Variable: the type of medicine a patient takes.

Response Variable: the health condition of a patient.

Treatment: the new medicine (We want to know whether it has an effect.).

2. Explanatory Variable: the car being driven.

Response Variable: the satisfactory score of a user.

Treatment: cars models A and B.

3. Explanatory Variable: the method of learning multiplication.

Response Variable: the quiz score.

Treatment: multiplication song (We want to know whether it has an effect.).

10.1.2 Vocabulary of Experiments 有关实验的术语

Treatment 实验手段

n. ['tritmənt]

Definition: An explanatory variable manipulated by the experimenter. We are interested to know whether a treatment has an influence on the experiment. See experiment in Section 10.1.1 for examples. 实验手段是指被实验者调整的解释变量。希望知道实验手段对实验的影响,见 10.1.1 节实验词条的例子。

Experimental Subject 实验对象

n. [ɪkˌsperə'mentəl 'sʌbdʒekt]

Definition: A member who is being studied in an experiment. 实验对象是指在实验中被研究的成员。

Group 分组

n. [grup]

Definition: A collection of experimental subjects who receive the same treatment (or no treatment). 分组是指用同样实验手段的(或者没有得到实验手段的)实验对象的汇集。

It is one of the experimental group or the control group. 分组包括实验组和控制组。

Properties: To ensure the fairness of the experiment, groups should have the same size, which refers to the number of experimental subjects. 为了实验的公正性,分组的大小(实验对象的个数)要相同。

Phrases: experimental～ 实验组, control～ 控制组

Experimental Group 实验组

n. [ɪkˌsperə'mentəl grup]

Definition: The group in an experiment that receives a treatment. 实验组是指在实验中得到实验手段的分组。

Properties: According to the Golden Rule 2 of experiment in Section 10.1.1, the experimental group can have exactly one treatment at a time. 根据 10.1.1 节实验词条的黄金法则 2,每个实验组只用一个实验手段。

Questions: For each of the following experiments, identify the experimental group.

1. A doctor wants to know if a certain type of new medicine and recover bad cold faster than a regular medicine. He divides 100 patients with bad cold into two equal groups. One group receives the new medicine and the other group resumes taking the regular medicine. After a week, the doctor measures the health conditions of the patients.
2. An automobile company wants to know which of the cars, A or B, is more popular. The developer asks 1000 people to drive model A, 1000 people to drive model B, and 1000 people to drive other models. He then records the satisfactory score for these 3000 people.
3. An elementary school teacher wants to know if a 2nd grader can learn multiplication by the multiplication table song faster than by the traditional memorization method. She decides to teach one of her classes (30 students) the multiplication song and another of her classes (30 students) the traditional memorization. At the end, she quizzes the students for multiplications.

Answers: 1. The group that receives the new medicine.

2. Groups that drive car models A and B.
3. The class for which the multiplication song is taught.

Control Group　控制组

n. [kən'troʊl grup]

Definition: The group in an experiment that does not receive a treatment. Control group serves as a benchmark for comparing the effects of the treatments.　控制组是指在实验中没有得到实验手段的分组。控制组是对比实验手段作用的基准。

Properties: Although subjects in the control group do not receive treatments, they may receive a placebo to ensure the fairness of the experiment.　虽然控制组的对象没有得到实验手段,但是他们可以得到安慰剂,用于确保实验的公平性。

Questions: For each of the following experiments, identify the control group.

1. A doctor wants to know if a certain type of new medicine and recover bad cold faster than a regular medicine. He divides 100 patients with bad cold into two equal groups. One group receives the new medicine and the other group resumes taking the regular medicine. After a week, the doctor measures the health conditions of the patients.
2. An automobile company wants to know which of the cars, A or B, is more popular. The developer asks 1000 people to drive model A, 1000 people to drive model B, and 1000 people to drive other models. He then records the satisfactory score for these 3000 people.
3. An elementary school teacher wants to know if a 2nd grader can learn multiplication by the multiplication table song faster than by the traditional memorization method. She decides to teach one of her classes (30 students) the multiplication song and another of her classes (30 students) the traditional memorization. At the end, she quizzes the students for multiplications.

Answers: 1. The group that receives the regular medicine.
2. The group that drives models other than A or B.
3. The class for which the traditional memorization is taught.

Placebo　安慰剂

n. [plə'siboʊ]

Definition: A fake/inactive treatment that deceives the experimental subjects in the control group as a treatment.　安慰剂是指伪/无作用的实验手段——但对控制组的实验对象它被谎称为实验手段。

Properties: 1. Placebos are widely used in blinded experiments, which increases the fairness of testing.
盲法实验经常用安慰剂来增强测试的公正性。
2. Subjects in the control group are unaware the facts that:
在控制组的实验对象不知道的情况有:
(1) They are in the control group. In fact, most of them will think they are in the experimental group.　他们在控制组,但很多人却认为自己在实验组。

(2) The treatment (placebo) they receive does not have any effect on the response variable, objectively. 客观来说,实验手段(安慰剂)对响应变量没有影响。

3. Placebos create the placebo effect. 安慰剂会产生安慰剂效应。

Examples: A doctor wanted to know whether drinking one particular type of juice before sleeping can help sleeping. He gathered 200 adults who had insomnia and divided them into two equal groups (A and B). Members in group A drank the juice before they slept, and members in group B drank iced lemonade (which the doctor claimed as a treatment) before they slept. After one month, the doctor compared the sleeping quality between these two groups.

Group B is the control group, and the lemonade here is the placebo. It is a "treatment" that does not have any real effect to cure insomnia, but the members of group B are unaware of this fact. Neither do they know that they are the control group. The experiment is designed this way to ensure fairness.

Placebo Effect 安慰剂效应

n. [plə'siboʊ ɪ'fekt]

Definition: The scenario that the members of the control group report that the placebo affects the response variable. 安慰剂效应是指控制组的组员汇报安慰剂影响响应变量的情况。

Properties: 1. The members in the control group are unaware the facts that they are in the control group and are using the placebo. On the other hand, most of them will think that they are in the experimental group. 控制组的组员不知道他们在控制组,也不知道他们在用安慰剂。相反,他们很多人会认为自己在实验组。

2. The placebo effect is due to psychological effects, which is out of the scope of high school mathematics. 安慰剂效应跟心理作用有关,这属于高中数学以外的知识。

Examples: A doctor wanted to know whether drinking one particular type of juice before sleeping can help sleeping. He gathered 200 adults who had insomnia and divided them into two equal groups (A and B). Members in group A drank the juice before they slept, and members in group B drank iced lemonade (which the doctor claimed as a treatment) before they slept. After one month, the doctor compared the sleeping quality between these two groups.

A placebo effect would be the case when most members from group B reporting that drinking iced lemonade helps sleeping.

10.2 Characteristics of Experimental Designs 实验设计的特征

Blinded Study 盲法研究

n. [blaɪndɪd 'stʌdi]

Definition: An experimental design for which the subjects do not know the groups they are in or the treatments (or placebo) they receive. 盲法研究是实验设计的一种,即实验对象不知道他们在哪一组,也不知道他们得到了什么实验手段(或者安慰剂)。

Properties: A blinded study reduces bias.　盲法研究减少了偏差。

Examples: A doctor wanted to know whether drinking one particular type of juice before sleeping can help sleeping. He gathered 200 adults who had insomnia and divided them into two equal groups (A and B). Members in group A drank the juice before they slept, and members in group B drank iced lemonade (which the doctor claimed as a treatment) before they slept. After one month, the doctor compared the sleeping quality between these two groups.

This is a blinded study if members of groups A and B do not know which drink (treatment or placebo) they are taking before they sleep.

Double-Blinded Study　双盲研究

n. [ˈdʌbəl blaɪndɪd ˈstʌdi]

Definition: An experimental design for which:

双盲研究也是实验设计的一种,有以下特点:

1. The subjects do not know the groups they are in or the treatments (or placebo) they receive.　实验对象不知道他们在哪组,不知道他们得到了什么实验手段(或者安慰剂)。
2. The experimenter does not know what particularly treatment each subject is receiving.　实验者不知道实验对象得到的实验手段。

Properties: Double-blinded study is the fairest method for the most circumstances, but there must be a third person keeping track of who is using which treatment.　双盲研究在大多数情况下都是最公正的,但是必须有第三个人记录谁在用哪种实验手段。

Examples: A doctor wanted to know whether drinking one particular type of juice before sleeping can help sleeping. He gathered 200 adults who had insomnia and divided them into two equal groups (A and B). Members in group A drank the juice before they slept, and members in group B drank iced lemonade (which the doctor claimed as a treatment) before they slept. After one month, the doctor compared the sleeping quality between these two groups.

This is a double-blinded study if both of the following are satisfied:

1. Members of groups A and B do not know which drink (treatment or placebo) they are taking before they sleep.
2. The doctor does not know who is drinking what—a third person keeps track of this information.

10.3　Types of Experimental Designs　实验设计的种类

Properties of Experimental Designs　实验设计性质

Properties: 1. The ultimate goal of an experiment is to present data that are as accurate and precise as possible.　实验的最终目的是尽可能准确和精密地呈现数据。

2. To reduce bias and variability as much as possible, a good experimental design involves:

要尽可能减少偏差和变异性,好的实验设计包括:

(1) Randomization—which subject is using which treatment is decided randomly instead of intentionally. Randomization can be done by the random number generator/random digit table.
随机选择——随机决定哪个实验对象有什么实验手段。可用随机数字生成器或随机数字表决定。

(2) Replication—performs the experiment several times because large data sets are always more reliable than small data sets.
重复——多次做实验,因为大数据集比小数据集可靠。

(3) Blocking—grouping experimental subjects according to their similarities (heights, grades, gender, …) that indirectly affect the response variable to ensure there are very few number of lurking variables.
区间分组——根据实验对象的相似度(身高、成绩、性别)进行分组。这些分组能间接影响响应变量,从而减少潜伏变量。

Independent Measure Design 独立测量设计

n. [ˌɪndɪˈpendənt ˈmeʒər dɪˈzaɪn]

Definition: An experimental design for which each subject tests exactly one condition of the independent variable (experiences exactly one treatment). 独立测量设计是实验设计的一种,即每个实验对象只用一个实验手段。

Properties: 1. Advantages:
优点:

(1) Each subject only needs to test one treatment in the experiment. So there are less complaints and fatigue (which helps the results).
每个实验对象只需要测试一个实验手段,从而减少实验对象的抱怨和由此产生的疲倦感,对结果有利。

(2) Easy to use randomization, which reduces the bias.
容易运用随机选择,从而减少偏差。

2. Disadvantages:
缺点:
Differences in individuals can be lurking variables, which creates bias. 每个个体的差别可以成为潜伏变量,从而导致偏差的产生。

Examples: A doctor wanted to know whether drinking one particular type of juice before sleeping can help sleeping. He gathered 200 adults who had insomnia and divided them into two equal groups (A and B). Members in group A drank the juice before they slept, and members in group B drank iced lemonade (which the doctor claimed as a treatment) before they slept. After one month, the doctor compared the sleeping quality between these two groups.
Each of the 200 subjects only needs to test one condition of the treatment (drinking the special juice or the lemonade before sleeping). The doctor can easily randomize the experiment.

Repeated Measure Design 重复测量设计

n. [rɪ'pitɪd 'meʒər dɪ'zaɪn]

Definition: An experimental design for which each subject tests all conditions of the independent variable (experiences all treatments), as opposed to independent measure design. 重复测量设计是实验设计的一种,即每个实验对象会用到所有的实验手段,与独立测量设计不同。

Even so, treatments must be imposed one at a time, otherwise it will be confounding. 尽管如此,实验手段必须逐个试用,否则会造成混杂关系。

Properties: 1. Advantages:

优点:

(1) Prevents the differences in individuals from creating bias. 防止因为个体的不同而产生偏差。

(2) Easy to use randomization, which reduces the bias. 容易运用随机选择,从而减少偏差的产生。

2. Disadvantages:

缺点:

Each subject needs to test all treatments in the experiment. So there are more complaints and fatigue (which can create bias). 每个实验对象在实验中需要测试所有实验手段,从而使实验对象产生抱怨并由此感到疲倦(会产生偏差)。

Examples: A doctor wanted to know whether drinking one particular type of juice before sleeping can help sleeping. He gathered 200 adults who had insomnia and divided them into two equal groups (A and B). Members in group A drank the juice before they slept, and members in group B drank iced lemonade (which the doctor claimed as a treatment) before they slept. **After one month, groups A and B switched treatments and the experiment continued for one more month.** Finally, the doctor compared the sleeping quality between these two groups.

Each of the 200 subjects needs to test both conditions of the treatment (drinking the special juice and the lemonade before sleeping). The doctor can easily randomize the experiment.

Matched Pair Design 配对设计

n. [ˌmætʃt peər dɪ'zaɪn]

Definition: An experimental design that groups experimental subjects with similar characteristics into pairs. For each pair, each subject receives a different treatment. For each pair, which subject uses what treatment can be decided by flipping a coin. 配对设计也是实验设计的一种,即根据相似的特征,把实验对象组成若干对。在每对里,每个实验对象得到不同的实验手段。在每对里,可抛硬币决定哪个实验对象用哪个实验手段。

Properties: 1. Advantages:

优点:

(1) Each subject only needs to test one treatment in the experiment. So there are less

complaints and fatigue (which helps the results). 每个实验对象在实验中只需要测试一种实验手段,从而减少抱怨和疲倦的产生,这对结果有利。

(2) Easy to use randomization, which reduces the bias. 容易运用随机选择,从而减少偏差。

(3) Combining this with the repeated measure design greatly reduces bias. 与重复测量设计结合,能更大地减少偏差。

2. Disadvantages:

缺点:

(1) It is very time consuming to group the subjects into pairs. 配对非常耗时。

(2) It is very difficult to make pairs according to similar characteristics (Unless they are identical twins!) 配对的过程很困难(除非实验对象是双胞胎)。

(3) If one subject drops out of the experiment, then the experimenter will lose all the data of that corresponding pair. 若一个实验对象退出实验,则实验者失去了那对实验对象的数据。

Examples: A doctor wanted to know whether drinking one particular type of juice before sleeping can help sleeping. He gathered 200 adults who had insomnia and divided them into 100 pairs. For each pair, the characteristics (gender, height, weight, level of insomnia) were generally similar. For each pair, one member drank the juice before he/she slept, and the other drank iced lemonade (which the doctor claimed as a treatment) before he/she slept (The doctor decides who receives what by flipping a coin). **After one month, the doctor compared the sleeping qualities between subjects in each pair.**

This example matches the definition of a matched pair design. A matched pair design with replication will be the same except replacing the bolded sentence with the following:

After one month, each subject switched treatment and continued for the second month. After these two months, the doctor compared the sleeping qualities between subjects in each pair.

10.4 Sampling 抽样

10.4.1 Introduction 介绍

Population 总体

n. [ˌpɑpjəˈleɪʃən]

Definition: A collection of subjects that the experimenter wants to make observations/inferences about. 总体是指实验者想观察/猜测的所有对象。

Questions: Identify the population from each of the following:

1. A professor wanted to know whether students in his class can improve their grades by going to the tutoring services. He selected 50 students who went to tutoring services regularly, and 50 students who did not receive tutoring and compare their grades.

2. Jack wanted to measure the length (in the number of words) of each paragraph in the

book **The Old Man and the Sea**. He measured the lengths of 30 paragraphs and tried to infer the pattern.

3. The school nurse wanted to measure the BMI values for the students in the 8th grade in the school. She measured the BMIs for forty 8th grade students and tried to observe the pattern for their heights.

Answers: 1. The population is all students in the class—the professor wants to observe their habits of going to tutoring services and their grades and infer whether tutoring services help their grades.

2. The population is all the paragraphs in the book **The Old Man and the Sea**—John wants to observe their lengths.

3. The population is all the 8th graders in the school. The school nurse wants to observe all their BMIs.

Sample　样本

n. ['sæmpəl]

Definition: A small part of the population that is intended to show the characteristics of the population.　样本是总体里的一小部分,用来展示总体的性质。

Phrases: sampling　抽样

Questions: Identify the sample from each of the following:

1. A professor wanted to know whether students in his class can improve their grades by going to the tutoring services. He selected 50 students who went to tutoring services regularly, and 50 students who did not receive tutoring and compare their grades.
2. Jack wanted to measure the length (in the number of words) of each paragraph in the book **The Old Man and the Sea**. He measured the lengths of 30 paragraphs and tried to infer the pattern.
3. The school nurse wanted to measure the BMI values for the students in the 8th grade in the school. She measured the BMIs for forty 8th grade students and tried to observe the pattern for their heights.

Answers: 1. The sample is the 100 students being selected to study—the professor wanted these 100 students to represent the overall trend of his class.

2. The sample is the 30 paragraphs being selected—John wanted these 30 paragraphs to represent the overall trend of the book.

3. The sample is the 40 students being selected to study—the nurse wanted these 40 students to represent the overall trend of the 8th grade.

10.4.2　Methods of Sampling with Probability　运用概率的抽样方法

Simple Random Sampling (SRS)　简单随机抽样

n. ['sɪmpəl 'rændəm 'sɑːmplɪŋ]

Definition: The basic sampling method where we select a smaller group (sample) from the larger group (population) in a way that every individual in the population has an equal chance to be included in the sample.　简单随机抽样是基础的抽样方法——从总体里选取一部分

作为样本,其中总体的每个个体被选中的概率是相同的。

Properties: 1. Advantage: It guarantees randomization. 优点:保证随机选择。

2. Disadvantage: There is some obvious sampling bias. Sometimes many individuals selected cannot accurately represent the population. 缺点:有很大的抽样偏差。有时被选中的很多个体都不能准确地代表总体。

Examples: A gym teacher wanted to know whether wearing a new type of sneakers can run faster. He gathered a sample of 200 students from the school (115 boys and 85 girls, using the simple random sampling method) and divided them into two groups (A and B). Students in group A wore the new type of sneakers, and students in group B wore the shoes they liked. Then the teacher compared the mile-run time for students in groups A and B.

SRS creates bias here—ideally we want 100 boys and 100 girls in the sample, and we want the individuals in our samples have mile-run times that are similar to that of an average student, but under SRS, every individual in the population has an equal chance of being selected and cannot guarantee that the teacher can select 100 boys and 100 girls. Neither can we guarantee that most of these 200 students have mile-run times similar to that of an average student.

Random Sampling 随机抽样

n. [ˈrændəm ˈsɑːmplɪŋ]

Definition: The basic sampling method where we select a smaller group (sample) from the larger group (population) in a way that **not** every individual in the population has an equal chance to be included in the sample. 随机抽样也是基础的抽样方法——从总体里选取一部分作为样本,其中总体的每个个体被选中的概率不是相同的。

Properties: It guarantees randomization and it can avoid the disadvantage of SRS. 确保随机选择,且能避免 SRS 的缺点。

Examples: A gym teacher wanted to know whether wearing a new type of sneakers can run faster. He gathered a sample of 200 students from the school using the random sample method and divided them into two groups (A and B). Students in group A wore the new type of sneakers, and students in group B wore the shoes they liked. Then the teacher compared the mile-run time for students in groups A and B.

Ideally we want 100 boys and 100 girls in the sample, and we want the individuals in our samples to have mile-run times that are similar to that of an average student. Random sampling guarantees the latter much better than the SRS does.

Systematic Sampling 等距抽样

n. [ˌsɪstəˈmætɪk ˈsɑːmplɪŋ]

Definition: Suppose we want a sample of size k. Below is the method of systematic sampling:

设选择一个大小为 k 的样本,等距抽样的方法如下:

List the individuals (suppose there are n of them, and $n>k$) from the population in an ordered manner. We will take every (n/k)th individual to be one subject in the sample.

列出总体的所有个体(设个体数为 n, $n>k$)并排列。选择每 n/k 个个体作为样本的一员。

Properties：1. Advantage：Often times，individuals selected can accurately represent the population.
优点：很多时候，被选中的个体能准确地代表总体。
2. Disadvantage：Less randomized than simple random sampling.
缺点：随机化比简单随机抽样小。

Examples：A gym teacher wanted to know whether wearing a new type of sneakers can run faster. He wanted to gather a sample of size 200 from the school population of 1200（500 are girls and 700 are boys）for his experiment. He ordered the students in the school so that ＃1-700 are boys，and ＃701-1200 are girls. He picked every student with a number that is a multiple of 6.

Ideally we want the sample to have 100 boys and 100 girls，**but this sampling method gives 116 boys and 84 girls，but it is better than SRS and has its own reason：** experimenters may argue that 116 boys and 84 girls is a good sample because the population has the ratio of boys to girls as 7∶5，and the sample distribution should be proportional to the population distribution.

Moreover，listing the students with similar characteristics（height，weight，gender，physical education grade）and using the systematic sampling gives a sample that is varied in characteristics，which better represents the population，but less randomized.

Stratified Random Sampling　分层随机抽样

n.［'strætəˌfaɪd 'rændəm 'sɑːmplɪŋ］

Definition：The sampling method that first divides the population into groups with similar characteristics（each group is called a stratum）. Then perform the simple random sampling for each stratum and combine the individuals selected from the strata to form a sample.　分层随机抽样的过程是：先根据相似的特征把总体分组(每个组为一个层)，然后对每层运用简单随机抽样，最后把这些选中的个体结合，成为最终样本。

Properties：1. Advantage：This provides a far better representation of the population in the previous methods. Smaller sample size under this type of sampling method can also represent the population well（smaller samples saves money and energy）.
优点：比前面的抽样方法更有代表性。用这种抽样方法，就算是小的样本也能很好地代表总体(小的样本能省钱省力)。
2. Disadvantage：It is too time-consuming as we need to do simple random sampling for each stratum.
缺点：对每层都要用简单随机抽样，很费时间。
3. If the strata are of different sizes，we have two solutions. Each has its advantage：
若层的大小不同，有两种方法。每种都有优点：
（1）Proportionate Stratified Random Sampling—the number of individuals selected from each stratum depends on the ratio of the size of that stratum to the size of the population.
均衡分层随机抽样——每层选择的个体数量取决于这层与总体的大小之比。
（2）Disproportionate Stratified Random Sampling—the number of individuals selected from each stratum is the same.

不均衡分层随机抽样——每层选择的个体数量相同。

Examples: A gym teacher wanted to know whether wearing a new type of sneakers can run faster. He wanted to gather a sample of size 200 from the school population of 1200 (500 are girls and 700 are boys) for his experiment. He divided the population into two strata: boys and girls.

The proportionate stratified random sampling method: Because the ratio of boys to girls is 7∶5 in the population, we want this to be the same in the sample. Therefore, we will pick $200(7/12)\approx 116$ boys and 84 girls from the corresponding strata using the simple random sampling method for each stratum.

The disproportionate stratified random sampling method: we will pick 100 boys and 100 girls from the corresponding strata using the simple random sampling method for each stratum.

Cluster Sampling 整群抽样

n. [ˈklʌstər ˈsɑːmplɪŋ]

Definition: The sampling method that first divides the population into separate groups (each group is called a cluster), then randomly picks some clusters to be included in the sample, and finally performs the simple random sampling for each selected cluster, and combines the individuals selected from these clusters to form a sample. 整群抽样的过程是：先把总体分组(每个组为一个群)，然后随机选出一些群作为潜在的样本，最后对每个被选中的群运用简单随机抽样，并把所有选中的个体结合，成为样本。

Properties: 1. The clusters must be of the same size. 群的大小必须相同。

2. This is the similar to stratified random sampling, except that we assume the experimental units have very similar characteristics (so we can divide them into several equal clusters without much consideration). We will determine the clusters that we want to include by probability, and perform SRS on each selected cluster. 整群抽样与分层随机抽样相似，假设实验对象在特征上很相似(从而不用做太多的考虑，就可以把它们分成若干不同的群)。可通过概率决定样本包括哪些群，然后对选上的群用简单随机抽样。

3. See advantages and disadvantages of stratified random sampling. 见分层随机抽样的优点和缺点。

Examples: Joe and Jack wanted to know the amount of electricity the households in Boston use. Joe performed the cluster sampling this way:

1. Divide Boston into districts (each district is called a cluster).
2. Randomly choose some districts to be included in the study.
3. SRS on households for each selected district.
4. Combine the data.

Joe believed that every district is more or less the same, thus he did step #2.

On the other hand, Jack does it using the stratified random sampling:

1. Divide Boston into districts (each district is called a stratum).
2. SRS on households for all districts.

3. Combine the data.

Jack believed that every district is different, thus he included all districts. Both Joe and Jack are reasonable.

10.4.3 Methods of Sampling without Probability　不运用概率的抽样方法

Convenience Sampling　便利抽样

n. [kən'vinjəns 'sɑːmplɪŋ]

Definition: A sampling method that does not depend on randomization or probability: it samples from subjects who are easy to reach.　便利抽样是不用随机选择的抽样方法，该方法从最容易联系到的个体中抽样。

Properties: 1. Advantages: Fast, easy, inexpensive.　优点：快速、简单、便宜。

2. Disadvantages: Whether the sample accurately represents the population is very doubtful. It creates experimental bias.　缺点：样本能否准确地代表总体要打个大的问号，会造成实验偏差。

Examples: A gym teacher wanted to know whether wearing a new type of sneakers can run faster **for a high school student**. He wanted to gather a sample of 50 best students in the classes he taught.

This is an example of convenience sampling—the gym teacher assumed that the students (more specifically, the designated students) in his school represent the high school students throughout the population (probably for the state, USA, or even the world). This is very questionable.

This type of samples is very easy to obtain.

10.4.4 Methods of Obtaining Data　获取数据的方法

Survey　调查

n. ['sɜrveɪ]

Definition: An investigation about the characteristics of a population. It is meant to collect data to form the sample and uses it to infer the population.　调查是指对总体特征的调研。它的目的是搜集数据组成样本，并且用来推测总体。

Examples: Rating an Iphone App, …

Census 普查

n. ['sensəs]

Definition: A survey that is given to every experimental subject. It is meant to collect data from the entire population.　普查是对每个实验对象的调查。它的目的是直接搜集整体的数据。

Examples: Population Census　人口普查。

10.4.5 Bias and Variability　偏差与变异

Figures 10-1 to 10-4 show the combination of high/low bias and high/low variability.

Figure 10-1 shows low bias and low variability (ideal model).

Figure 10-2 shows low bias and high variability.

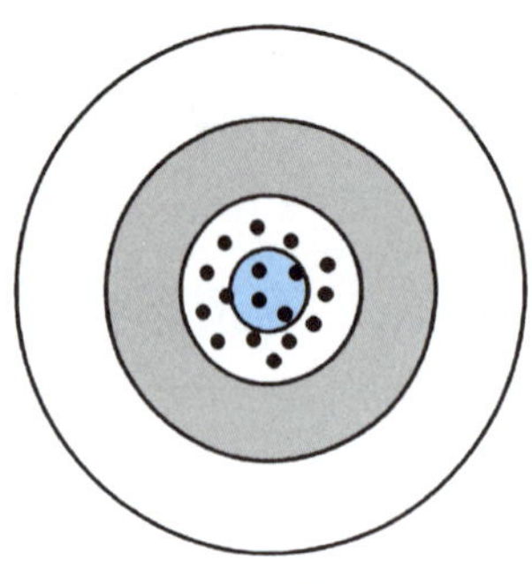
Figure 10-1

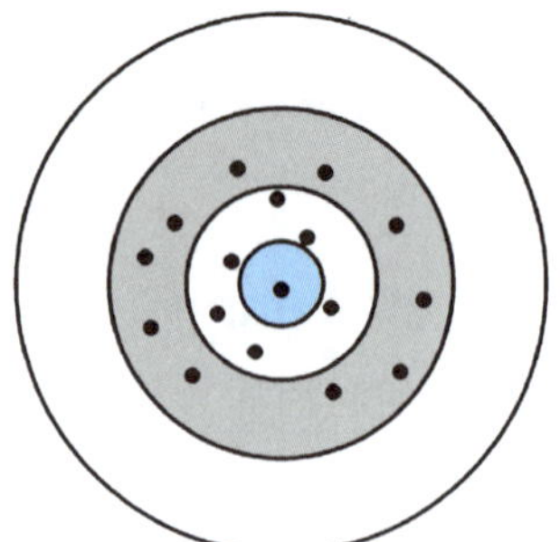
Figure 10-2

Figure 10-3 shows high bias and low variability.
Figure 10-4 shows high bias and high variability.

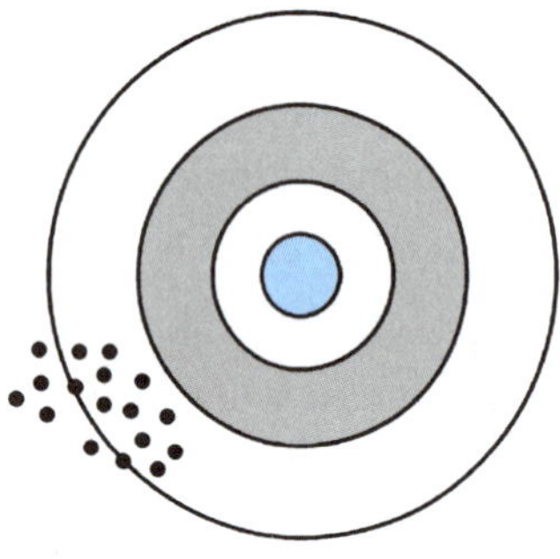
Figure 10-3

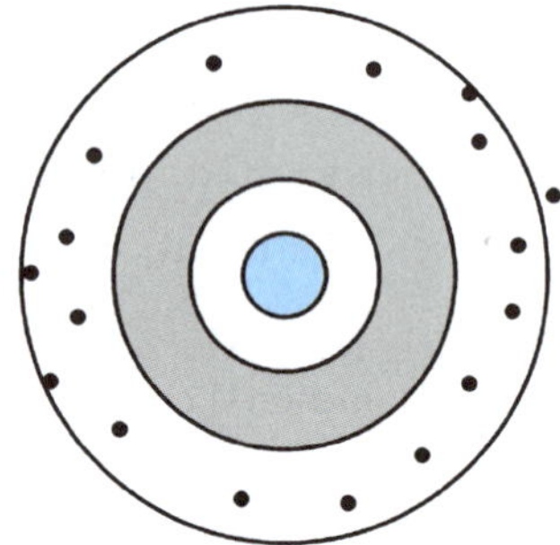
Figure 10-4

Bias 偏差

n. ['baɪəs]

Definition：The tendency of the sample systematically overestimating or underestimating the population，or the experiment is systematically in favor/against some experimental subjects. 偏差是指样本系统地高估/低估总体的趋向，或系统地有利于一些实验对象的趋向。

Accuracy measures how little the bias exists. 准确性测量可以测量偏差有多低。

Variability 变异

n. ['veəriəbəl]

Definition：The tendency of data in the sample having large differences from one another. 变异是指样本的数据互相之间差别大的趋势。

Precision measures how little variability exists. 精确性测量变异有多低。

Notes：Range，mean，variance，and standard deviation give information about the variability. 极差、平均数、方差、标准差能给出变异的信息。

10.4.6 Types of Experimental Bias 实验偏差的种类

Voluntary Response Bias 自愿回应偏差

n. ['vɑlən,teri rɪ'spɑns 'baɪəs]

Definition：The bias for which people participate in the experiment in a voluntary basis. This bias exists because usually these volunteers are people with strong opinions. 自愿回应偏差是指人们自愿性地参加实验所造成的偏差。这种偏差造成的原因是志愿者都有很强烈的意见。

Examples: Many applications on the smartphone ask whether the users want to rate the application or provide feedbacks. The users who do so usually have strong opinion about the application (either very good or very bad).

Nonresponse Bias 无回应偏差

n. [nɑnrɪ'spɑns 'bɑɪəs]

Definition: The bias for which the experimental subjects are unwilling or unable to be part of the experiment. This bias exists because the individuals sometimes refuse to answer sensitive questions, or they are unable to participate (not at home, not being able to pick up the phone at that time, …). 无回应偏差是指实验对象不愿意或者不能参与实验所造成的偏差。造成这种偏差的原因是志愿者拒绝回答敏感问题或不能参与调查(不在家,或不能接电话等)。

Examples: A research group wants to know whether each adult in the town has diabetes. The experimenter tries to call every household to find out. This has a nonresponse bias because not everyone is available to pick up the phone, or is willing to answer this question.

Undercoverage Bias 低覆盖面偏差

n. [ˌʌndər'kʌvərɪdʒ 'bɑɪəs]

Definition: The bias for which some groups of experimental subjects are left out in the study. Almost all convenience sampling has this type of bias. 低覆盖面偏差是指忽略一些组别的实验对象所造成的偏差。几乎所有的便利抽样都存在这种偏差。

Examples: A clothing company wants to know which belt is the most popular. The experimenter randomly surveys 200 males in the town. This is an undercoverage bias because the experimenter leaves out females in the study. And yes, there are belts for females.

11. Counting and Probability 计数与概率

11.1 Counting 计数

视频 20

11.1.1 Introduction 介绍

Enumeration Tree 枚举树

n. [ɪˌnumə'reɪ.ʃən tri]

Definition: The visual display that shows all combinations of doing two or more tasks. 枚举树是指

展示做两件或更多任务组合方法的图解。

A more elegant way that can achieve this is the Counting Principle. 计数原理是实现同样目的的捷径。

Figure 11-1 shows an enumeration tree for the combinations of doing two tasks.

Questions: 1. If there are 3 routes from city A to city B, and 2 routes from city B to city C, how many routes are there from city A to city C (assume that every route has to pass city B)?

2. If there are 3 different kinds of forks, 4 different kinds of spoons, and 5 different kinds of knives. How many ways are there to pick a dinnerware that consists of a fork, a spoon, and a knife?

Answers: 1. Figure 11-2 shows the enumeration tree for this question.

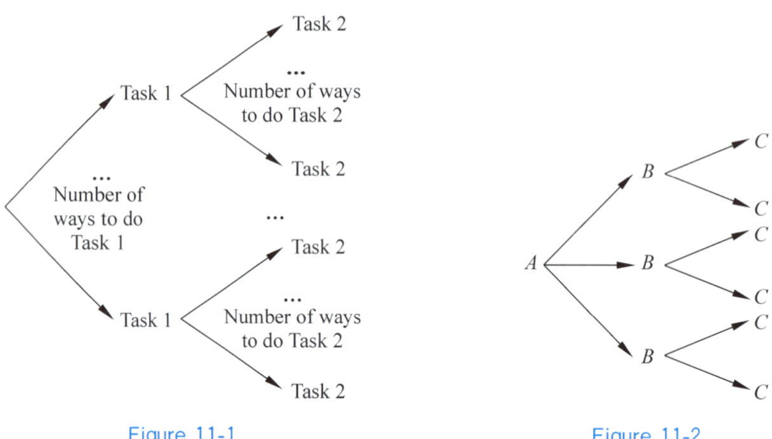

Figure 11-1　　Figure 11-2

The answer is 6. We count the number of rightmost branches.

2. The answer is 60. Come up with an enumeration tree yourself.

Counting Principle 计数原理

n. [kaʊntɪŋ 'prɪnsəpəl]

Definition: If there are m ways to do task A, and n ways to do task B, then there are mn combinations to do tasks A and B. 计数原理定义为：若有 m 种方法完成任务 A，n 种方法完成任务 B，则有 mn 种方法完成任务 A 和 B。

Properties: The generalized Counting Principle states that:

If there are m_1 ways to do task 1, m_2 ways to do task 2, m_3 ways to do task 3, ⋯, then there are $m_1 m_2 m_3 \cdots$ combinations to do all the tasks. 计数原理的一般性陈述。

Notes: This principle is based on the enumeration tree. 这个原理根据枚举树。

Questions: 1. If there are 3 routes from city A to city B, and 2 routes from city B to city C, how many routes are there from city A to city C (assume that every route has to pass city B)?

2. If there are 3 different kinds of forks, 4 different kinds of spoons, and 5 different kinds of knives. How many ways are there to pick a dinnerware that consists of a fork, a spoon, and a knife?

3. There are 6 symbols on the license plate. The first two symbols must be capital

letters. The rest of the symbols must be digits from 0 through 9. How many license plates are possible?

Answers：1. $3 \cdot 2 = 6$.

2. $3 \cdot 4 \cdot 5 = 60$.

3. $26 \cdot 26 \cdot 10 \cdot 10 \cdot 10 \cdot 10 = 6\ 760\ 000$.

Factorial　阶乘

n. [ˌfæk'tɔriəl]

Definition：For n is a positive integer, the factorial of n is the product of the first n positive integers.

对于正整数 n，n 的阶乘为前 n 个正整数的积。

The factorial of 0 is 1.

0 的阶乘为 1。

Notation：The factorial of n is written as $n!$.

Properties：1. $0! = 1$ (by definition).

2. For $n \geqslant 1, n! = 1 \times 2 \times 3 \times \cdots \times n = \prod_{i=1}^{n} i$.

3. For $n \geqslant 1, n! = n(n-1)!$.

4. Factorials in counting are based on the Counting Principle. It is a shorthand notation for some multiplication.　在计数中，阶乘的基础是计数原理。阶乘是某些乘法的简写。

Questions：1. Calculate the factorials of 0 through 10：$0!, 1!, 2!, \cdots, 10!$.

2. How many ways are there to order the letters A, B, C, and D?

3. Let n be an integer greater than 2, simplify each of the following：

(1) $\frac{n!}{(n-2)!}$

(2) $n! - (n-2)!$

Answers：1. $0! = 1$；$1! = 1$；$2! = 2$；$3! = 6$；$4! = 24$；$5! = 120$；$6! = 720$；$7! = 5040$；$8! = 40\ 320$；$9! = 362\ 880$；$10! = 3\ 628\ 800$.

2. By the Counting Principle, there are 4 ways to pick the first letter. After that, we have 3 ways to pick the second letter. After that, we have 2 ways to pick the third letter and 1 way to pick the fourth letter. Thus our answer is $4 \times 3 \times 2 \times 1 = 24$. Alternatively, using the factorial notation, our answer is $4! = 24$.

3. (1) $\frac{n!}{(n-2)!} = n(n-1)$.

(2) $n! - (n-2)! = n(n-1)(n-2)! - (n-2)! = (n^2 - n)(n-2)! - (n-2)! = (n^2 - n - 1)(n-2)!$.

Ordering　顺序

n. ['ɔːdərɪŋ]

Definition：The arrangement of objects.　顺序是指物体的排序。

Properties：In math applications, ordering sometimes matters and sometimes doesn't. If ordering matters, it is a permutation. If ordering does not matter, it is a combination. 在数学应用中，顺序有时候很重要。若考虑顺序，一组物体则是一个排列。若不考虑顺序，一组物体则是一个组合。

Examples: 1. We say "1,2,3,4,5" and "5,4,3,2,1" have different ordering.

2. We say "A,B,C,D,E" and "A,B,C,D,E" have the same ordering.

3. Here is an incident that the ordering matters: license plates. "ABC888" and "ACB888" are two different license plates for a car.

4. Here is an incident that the ordering does not matter: the items in shopping cart. It does not make a difference if one shops laptop first, then books, or shopping books first, then laptop.

11.1.2 Permutations and Combinations 排列与组合

Permutation 排列

n. [ˌpɜrmjuˈteɪʃən]

Definition: A selection of part or all of the members such that the ordering matters. 排列是指从给定个数的元素中取出部分或者所有的元素并进行排序，需要考虑元素的顺序。

Notation: The number of permutations of n objects taken r at a time is denoted by ${}_nP_r = n(n-1)\cdots(n-r+1)$[There are r factors.]$= \dfrac{n!}{(n-r)!}$.

If $n=r$, then ${}_nP_r = n!$ using the formula above.

If $n<r$, then ${}_nP_r = 0$ by the context. There are 0 possible arrangements.

If $r=0$, then ${}_nP_r = 1$ either by the formula or the context: there is only 1 way to take 0 objects from the n objects and arrange. This way is "not taking".

Proof: To prove why $n(n-1)\cdots$[There are r factors.]$=\dfrac{n!}{(n-r)!}$ for $n\geqslant r$:

$$
\begin{aligned}
n(n-1)\cdots[\text{There are } r \text{ factors.}] &= n(n-1)\cdots(n-r+1)\\
&= n(n-1)\cdots(n-r+1)\cdot\frac{(n-r)(n-r-1)\cdots 1}{(n-r)(n-r-1)\cdots 1}\\
&= \frac{n(n-1)\cdots(n-r+1)(n-r)(n-r-1)\cdots 1}{(n-r)(n-r-1)\cdots 1}\\
&= \frac{n!}{(n-r)!}.
\end{aligned}
$$

Questions: 1. How many ways are there to form four-digit numbers, for which the digits are distinct and nonzero?

2. How many ways are there to select 10 candidates to take up 6 different positions in the office?

Answers: 1. The ordering matters because 1234 is different from 4321.

By the Counting Principle: $9\cdot 8\cdot 7\cdot 6=3024$.

By permutation, ${}_9P_4 = 9\cdot 8\cdot 7\cdot 6=3024$.

Or, ${}_9P_4 = \dfrac{9!}{(9-4)!} = \dfrac{9!}{5!} = 9\cdot 8\cdot 7\cdot 6 = 3024$.

2. The ordering matters because "Person A is taking Position 1 and Person B is taking Position 2" is different from "Person A is taking Position 2 and Person B is taking Position 1".

By the Counting Principle: $10 \cdot 9 \cdot 8 \cdot 7 \cdot 6 \cdot 5 = 151\,200$.

By permutation, ${}_{10}P_6 = 10 \cdot 9 \cdot 8 \cdot 7 \cdot 6 \cdot 5 = 151\,200$.

Or, ${}_{10}P_6 = \dfrac{10!}{(10-6)!} = \dfrac{10!}{4!} = 10 \cdot 9 \cdot 8 \cdot 7 \cdot 6 \cdot 5 = 151\,200$.

Combination　组合

n. [ˌkɑmbəˈneɪʃən]

Definition: A selection of part or all of the members such that the ordering does not matter.　组合是指从给定个数的元素中仅仅取出部分或者所有的元素,不考虑元素的顺序。

Notation: The number of combinations of n objects taken r at a time is denoted by ${}_nC_r = \dfrac{n!}{r!\,(n-r)!} = \dfrac{{}_nP_r}{r!}$. From the permutation of n objects taken r at a time, we count each combination $r!$ times. Thus we have to divide out the repeats.

If $n = r$, then ${}_nC_r = 1$ using the formula above.

If $n < r$, then ${}_nC_r = 0$ by the context. There are 0 possible arrangements.

If $r = 0$, then ${}_nC_r = 1$ either by the formula or the context: there is only 1 way to choose 0 objects from the n objects. This way is "not choosing".

Questions: 1. (1) How many ways are there to pick 2 different flavors from 30 flavors of an ice cream shop to form an ice cream bowl?

(2) How many ways are there to pick 3 different flavors from 30 flavors of an ice cream shop to form an ice cream bowl?

2. How many ways are there to ask 4 out of 10 students to volunteer in community services?

Answers: 1. (1) By permutation, there are $30 \cdot 29 = 870$ different ways to pick 2 different flavors from 30, and the ordering matters.

However, in the context of this problem, ordering does not matter because chocolate-vanilla combination is the same as the vanilla-chocolate combination. We must divide out the number of times, which is 2, that we repeatedly counting each combination.

Thus our answer is $870/2 = 435$.

Using the combination formula, it is ${}_{30}C_2 = \dfrac{30!}{2!\,(30-2)!} = 435$.

(2) We have $30 \cdot 29 \cdot 28 = 24\,360$ ways to pick different triples such that the ordering matters. For each triple, we counted it $3! = 6$ times. Thus the answer is $24\,360/6 = 4060$.

Using the combination formula, it is ${}_{30}C_3 = \dfrac{30!}{3!\,(30-3)!} = 4060$.

2. The ordering does not matter because it does not make a difference who is called first.

Using combination formula, it is ${}_{10}C_4 = \dfrac{10!}{4!\,(10-4)!} = 210$.

11.2 Probability 概率

视频 21

11.2.1 Introduction 介绍

Outcome 结果

n. ['aʊt,kʌm]

Definition: A result of an experiment. 实验的一种结果。

Properties: Not every outcome in the experiment is equally likely. 实验中不是每一种结果的可能性都是相同的。

Examples: 1. Flipping a coin, we have 2 possible outcomes—heads or tails.

2. Rolling a die, we have 6 possible outcomes—1,2,3,4,5, or 6.

3. Randomly drawing a card from a standard 52-card deck, we have 52 possible outcomes, namely every card from the deck.

4. Randomly drawing a card and recording its value, we have 13 possible outcomes—A, 2,3,4,5,6,7,8,9,10,J,Q, or K.

5. Randomly drawing a marble from a bag with 40 red marbles and 60 green marbles and recording its color, we have 2 possible outcomes—red or green.

6. Flipping a coin twice, we have 4 possible outcomes: HH, HT, TH, TT. Ordering matters.

7. Rolling a die twice and recording the sum of outcomes, we have 11 possible outcomes: 2,3,4,5,6,7,8,9,10,11,12. **As shown in the property, not all outcomes here are equally likely. There are more than one way to get a sum of 7 (such as 2 + 5 or 3 + 4), but there is only one way to get a sum of 2 (1 + 1).**

Favorable Outcome 有利的结果

n. ['feɪvərəbəl 'aʊt,kʌm]

Definition: The outcome of interest. 有利的结果是指想要的结果。

Examples: 1. Flipping a coin, if we want it to be heads, then the favorable outcome is heads.

2. Rolling a die, if we want the outcome to be odd, then the favorable outcomes are 1,3,5.

Unfavorable Outcome 不利的结果

n. [ʌn'feɪvərəbəl 'aʊt,kʌm]

Definition: The opposite of favorable outcome. 不利的结果是指不想要的结果。

Examples: 1. Flipping a coin, if we want it to be heads, then the unfavorable outcome is tails.

2. Rolling a die, if we want the outcome to be odd, then the unfavorable outcomes are 2,4,6.

Sample Space 样本空间

['sæmpəl speɪs]

Definition: The set of **all possible** outcomes resulting from an experiment. 样本空间是指实验中包含所有结果的集。

Notation: The set for the sample space is named S.

Examples: 1. The sample space of flipping a coin is $S = \{\text{heads}, \text{tails}\}$.

2. The sample space of rolling a die is $S = \{1,2,3,4,5,6\}$.

3. The sample space for randomly drawing a card from a standard 52-card deck is $S = \{A♠, 2♠, 3♠, \cdots\}$, the set that consists of all 52 cards.
4. The sample space for randomly drawing a card and recording its value is $S = \{A, 2, 3, 4, 5, 6, 7, 8, 9, 10, J, Q, K\}$.
5. The sample space for randomly drawing a marble from a bag with 40 red marbles and 60 green marbles and recording its color is $S = \{\text{red}, \text{green}\}$.
6. The sample space for flipping a coin twice is $S = \{HH, HT, TH, TT\}$. Order matters.
7. The sample space for the sum of values of rolling two dice is $S = \{2, 3, 4, 5, 6, 7, 8, 9, 10, 11, 12\}$.

Event　事件

n. [ɪ'vent]

Definition: The set of one or more outcomes resulting from an experiment.　事件是指实验中包含一个或多个结果的集合。

Notation: {Outcome #1, Outcome #2, …}

Examples: 1. The event of rolling a 2 from a die is denoted by $\{2\}$.

2. The event of rolling an odd number from a die is denoted by $\{1, 3, 5\}$.
3. The event of randomly drawing a face card from the standard 52-card deck is denoted by $\{J, Q, K\}$ or $\{J, Q, K, A\}$, depending on whether Aces count as face cards.
4. The event of rolling a 7 from a die is denoted by $\{7\}$, which is an impossible event. The number 7 is known as an impossible outcome.
   不可能事件和不可能结果的例子。
5. The event of getting a head or a tail from flipping a coin is denoted by {head, tail}, which is a certain event. One of heads or tails is known as the certain outcome.
   必然事件和必然结果的例子。
6. The event of getting two heads when from flipping a coin twice is denoted by $\{HH\}$.

Complement　补集

n. ['kɑmpləˌment]

Definition: The complement of an event A is the set of all outcomes in the sample space that are **not** in A.　事件 A 的补集是在样本空间中,包含所有不在 A 里的结果的集。

Notation: The complement of an event A is denoted by A^C.

Properties: If A is a **possible** event,

1. $A \cap A^C = \varnothing$.
2. $A \cup A^C = S$.
3. $(A^C)^C = A$.

Examples: 1. The event of rolling a 2 from a die is denoted by $\{2\}$.
   Its complement is $\{1, 3, 4, 5, 6\}$.

2. The event of rolling an odd number from a die is denoted by $\{1, 3, 5\}$.
   Its complement is $\{2, 4, 6\}$.
3. The event of randomly drawing a face card from the standard 52-card deck is denoted

by {J,Q,K} or {J,Q,K,A},depending on whether Aces are counted as face cards.
Its complement is {A,2,3,4,5,6,7,8,9,10} or {2,3,4,5,6,7,8,9,10},depending on whether Aces are counted as face cards.

4. The event of rolling a 7 from a die is denoted by {7},which is an impossible event. The number 7 is known as an impossible outcome.
Its complement is {1,2,3,4,5,6},which is the sample space.

5. The event of getting a head or tail by filling a coin is denoted by {head,tail},which is a certain event. One of heads or tails is known as the certain outcome.
Its complement is $\varnothing$,which is the empty set.

Probability 概率

n. [ˌprɑbəˈbɪlɪti]

Definition: The likelihood of an event to happen. 概率是指事件发生的可能性。

Notation: The probability of event A is denoted by $P(A)$ or Prob(A).

Properties: 1. Let P be the probability of any event.
This inequality must hold: $0 \leqslant P \leqslant 1$.
概率用 P 表示,$0 \leqslant P \leqslant 1$。
If $P=0$,then the event is an impossible event. 如果 $P=0$,则为不可能事件。
If $P=1$,then the event is a certain event. 如果 $P=1$,则为必然事件。
If $0<P<1$,then the event is a possible event. 如果 $0<P<1$,则为可能事件。

2. Moreover,if all the elements in the sample space S are equally likely to happen,then

$$P(A)=\frac{\text{number of elements in } A}{\text{number of elements in } S}=\frac{\text{size of } A}{\text{size of } S}=\frac{\#\text{ of favorable outcomes}}{\#\text{ of total possible outcomes}}$$
$$=\frac{\#\text{ of favorable outcomes}}{\#\text{ of favorable outcomes}+\#\text{ of unfavorable outcomes}}.$$

There are advanced methods of calculating probability later.
此外,若在样本空间 S 中所有元素发生的概率相同,则

$$P(A)=\frac{A\text{ 的元素个数}}{S\text{ 的元素个数}}=\frac{A\text{ 的大小}}{S\text{ 的大小}}=\frac{\text{有利的结果的个数}}{\text{结果的个数}}=\frac{\text{有利的结果的个数}}{\text{有利的结果的个数}+\text{不利的结果的个数}}。$$

以后会介绍更高级的概率计算方法。

3. $P(S)=1$. 必然事件的概率为 1。事件在样本空间的概率为 1。

4. Let the sample space S be {outcome 1,outcome 2,$\cdots$,outcome n}, $\sum_{i=1}^{n} P(\text{outcome } i)=P(S)=1$.

5. For any event A in the sample space,$P(A)+P(A^{C})=1$.

Questions: 1. Rolling a fair die,what is the probability that we get a
(1) 3?
(2) prime number?
(3) 8?

2. Flipping a coin twice,what is the probability that we get
(1) two heads?
(2) different results?

3. Rolling a die twice,what is the probability that the sum of the outcomes is 7?

4. Picking a marble from a bag of red, green, and blue marbles only, if $P(\text{red}) = 1/5$, $P(\text{green}) = 3/10$, then what is $P(\text{blue})$?

Answers: 1. The sample space is $S = \{1,2,3,4,5,6\}$. **Every outcome from S is equally likely.**

(1) The event we want is $\{3\}$. Thus $P(3) = 1/6$.

(2) The event we want is $\{2,3,5\}$. Thus $P(\text{prime}) = 3/6 = 1/2$.

(3) The event we want is $\{8\}$, which is not in the sample space. Thus $P(8) = 0$, and it is an impossible event.

2. The sample space is $S = \{HH, HT, TH, TT\}$. **Every outcome from S is equally likely.**

(1) The event we want is $\{HH\}$. Thus $P(\text{two heads}) = 1/4$.

(2) The event we want is $\{HT, TH\}$. Thus $P(\text{different results}) = 2/4 = 1/2$.

3. The **sample space for the sum** is $S_{\text{sum}} = \{2,3,4,5,6,7,8,9,10,11,12\}$. **Since not every outcome from S is equally likely** (for example, there is only 1 way to get a 12, namely by rolling two 6's, but there are more than 1 way to get a 7, by rolling a 1 then a 6, or by rolling a 2 then a 5). The answer is not 1/11, and we must count carefully.

The **sample space for rolling a die twice** is $S_{\text{roll twice}} = \{11,12,13,14,15,16,21,22,23,24,25,26,31,32,33,34,35,36,41,42,43,44,45,46,51,52,53,54,55,56,61,62,63,64,65,66\}$, for which "35" means rolled a 3 then a 5. **There are 36 possible outcomes in this sample space, each of which is equally likely to occur.**

6 of these outcomes (namely 16, 25, 34, 43, 52, 61) have sum of 7.

Thus we have 6 favorable outcomes and 36 possible outcomes: $P(\text{sum} = 7) = 6/36 = 1/6$.

4. Since the sample space is $S = \{\text{red}, \text{green}, \text{blue}\}$, we have

$$P(\text{red}) + P(\text{green}) + P(\text{blue}) = 1$$
$$1/5 + 3/10 + P(\text{blue}) = 1$$
$$P(\text{blue}) = 1/2.$$

Theoretical Probability　理论概率

n. [ˌθiəˈretɪkəl ˌprɑbəˈbɪlɪti]

Definition: The probability that comes from calculation.　理论概率是指计算得到的概率。

Examples: See all the questions from probability.　见概率问题。

Experimental Probability　实验概率

n. [ɪkˌsperəˈmentəl ˌprɑbəˈbɪlɪti]

Definition: The probability that comes from actual experiment. The probability of an event A is given by $P(A) = \dfrac{\text{number of times it occurs in the experiment}}{\text{number of times the experiment is conducted}}$.

实验概率是指从实验中得到的概率。

11.2.2　Events　事件

Impossible Event　不可能事件

n. [ɪmˈpɑsəbəl ɪˈvent]

Definition: An event that never occurs.　不会发生的事件。

Properties: Let A be an impossible event,

1. $P(A) = 0$.

2. A is not a subset of the sample space.

Examples: 1. Rolling a 7 on a die.

2. Picking a white marble from a bag that only has red, green, and blue marbles.

Certain Event 必然事件

n. [ˈsɜrtən ɪˈvent]

Definition: An event that always occurs. 必会发生的事件。

Properties: Let A be a certain event,

1. $P(A)=1$.
2. A is the sample space.

Examples: 1. Rolling a number less than 7 on a die.

2. Picking a red, green, or blue marble from a bag that only has red, green, and blue marbles.

Possible Event 可能事件

n. [ˈpɑsəbəl ɪˈvent]

Definition: An event that sometimes occurs. 有时会发生的事件。

Properties: Let A be a possible event,

1. $0<P(A)<1$.
2. A is a proper subset of the sample space.
3. If $P(A)$ is close to 0, we say it is an unlikely event. 不太可能的情况。
   If $P(A)$ is close to 1, we say it is a likely event. 很有可能的情况。
4. Unlikely event is not impossible, and likely event is not certain. 不太可能并非不可能,很有可能并非必然。

Examples: 1. Winning a lottery ticket (unlikely event).

2. Rolling a 6 on a die (unlikely event).
3. Rolling an even number on a die.
4. Not rolling a 6 on a die (likely event).
5. Drawing a card from the standard deck and its suit is not hearts (likely event).

Mutually Exclusive Events/Disjoint Events 互斥事件

n. [ˈmjutʃuəli ɪkˈsklusɪv ɪˈvents/dɪsˈdʒɔɪnt ɪˈvents]

Definition: Events A and B are mutually exclusive if they cannot occur at the same time. 若事件 A 与 B 不会同时发生,则它们为互斥事件。

Events A and B are also called disjoint events.

Properties: Figure 11-3 shows the mutually exclusive events A and B in the sample space S.

图 11-3 展示了在样本空间 S 中 A 与 B 为互斥事件的情况。

$P(A\cap B)=0$.

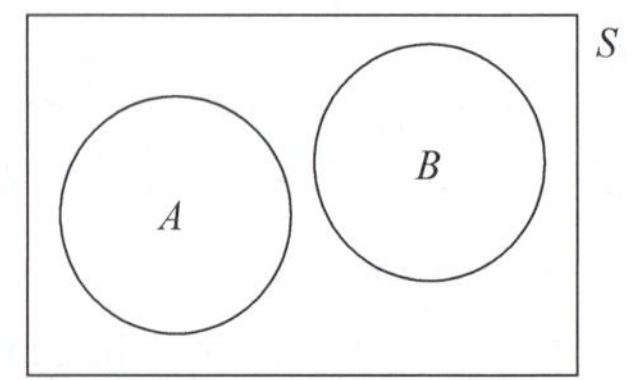

Figure 11-3

Examples: 1. Rolling a die:

A: Getting the number 3.

B: Getting the number 5.

You cannot get both 3 and 5 on one roll.

2. Flipping a coin:

A: Getting a head.

B: Getting a tail.

You cannot get both head and tail on one flip.

Complementary Events 对立事件

n. [ˌkɑːmˌpləˈmenˌtəːˌi ɪˈvents]

Definition: Events A and B are complementary events if they satisfy both conditions:

若符合以下条件,则事件 A 与 B 为对立事件:

1. A and B are mutually exclusive. A 与 B 互斥。
2. One of A or B will happen in every outcome. 每个结果要么是 A,要么是 B。

Properties: Figure 11-4 shows the complementary events A and B in the sample space S.

图 11-4 展示了在样本空间 S 中 A 与 B 为对立事件的情况。

This is a special case of mutually exclusive events.

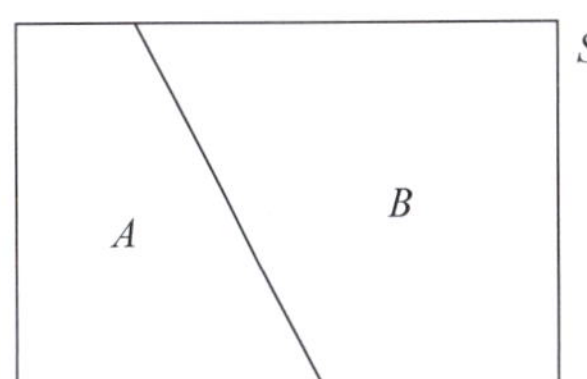

Figure 11-4

1. $P(A \cap B)=0$. Recall that A and B are mutually exclusive.
2. $P(A \cup B)=1$.
3. $P(A)+P(B)=1$. One of A or B must happen in every outcome.

Examples: 1. Rolling a die:

A: Getting an odd number.

B: Getting an even number.

You cannot get both odd and even on one roll.

Every roll is either an odd or even.

2. Flipping a coin:

A: Getting heads.

B: Getting tails.

You cannot get both heads and tails in one flip.

Every flip is either heads or tails.

11.2.3 Odds 赔率

Odds in Favor 有利赔率

n. [ɑdz ɪnˈfeɪvər]

Definition: The odds in favor is defined as

$$\frac{\#\text{ of favorable outcomes}}{\#\text{ of unfavorable outcomes}}=\frac{\#\text{ of favorable outcomes}}{\#\text{ of total outcomes}-\#\text{ of favorable outcomes}}$$

有利赔率是指有利的结果与不利的结果的数量之比。

Properties: 1. This is used to compare the numbers of favorable and unfavorable outcomes.

- If the odds in favor is greater than 1, then the "winning ratio" is more than 50%.
- If the odds in favor is 1, then the "winning ratio" is exactly 50%.

- If the odds in favor is less than 1, then the "winning ratio" is less than 50%.

2. If the number of unfavorable outcomes is 0, then the odds in favor is undefined.
   If the number of favorable outcomes is 0, then the odds in favor is 0.

Questions: The questions below are from those of probability in Section 11.2.1.

1. Rolling a fair die, what is the odds in favor that we get a
   (1) 3?
   (2) prime number?
   (3) 8?
2. Flipping a coin twice, what is the odds in favor that we get
   (1) two heads?
   (2) different results?

Answers: 1. (1) Since $P(3)=1/6$, its odds in favor is $1/5$.
(2) Since $P(2,3,\text{or }5)=1/2$, its odds in favor is $1/1$.
(3) Since $P(8)=0$, its odds in favor is $0/6=0$.
2. (1) Since $P(\text{two heads})=1/4$, its odds in favor is $1/3$.
(2) Since $P(\text{different results})=2/4=1/2$, its odds in favor is $1/1$.

Odds Against 不利赔率

n. [ɑdz əˈgenst]

Definition: The odds against is defined as

$$\frac{\#\text{ of unfavorable outcomes}}{\#\text{ of favorable outcomes}}=\frac{\#\text{ of unfavorable outcomes}}{\#\text{ of total outcomes}-\#\text{ of unfavorable outcomes}}$$

不利赔率是指不利的结果与有利的结果的数量之比。

Properties: 1. This is used to compare the numbers of unfavorable and favorable outcomes.

- If the odds against is greater than 1, then the "winning ratio" is less than 50%.
- If the odds against is 1, then the "winning ratio" is exactly 50%.
- If the odds against is less than 1, then the "winning ratio" is more than 50%.

2. If the number of favorable outcomes is 0, then the odds against is undefined.
   If the number of unfavorable outcomes is 0, then the odds against is 0.

Questions: The questions below are from those of probability in Section 11.2.1.

1. Rolling a fair die, what is the odds against that we get a
   (1) 3?
   (2) prime number?
   (3) 8?
2. Flipping a coin twice, what is the odds against that we get
   (1) two heads?
   (2) different results?

Answers: 1. (1) Since $P(3)=1/6$, its odds against is $5/1$.
(2) Since $P(2,3,\text{or }5)=1/2$, its odds against is $1/1$.
(3) Since $P(8)=0$, its odds against is undefined.
2. (1) Since $P(\text{two heads})=1/4$, its odds against is $3/1$.
(2) Since $P(\text{different results})=2/4=1/2$, its odds against is $1/1$.

11.2.4　Venn Diagrams　文氏图

Venn Diagram　文氏图

n. ['ven ˌdaɪəgræm]

Definition: A diagram that shows the relationship between two or more (finite) sets, or the probability model (the relationship between the sample space and a particular event), or the relationship between two quantities.　文氏图是展示两个或更多个集合之间关系的图表，它也是一个概率模型，用于展示样本空间和某个事件的关系或展示两个量的关系。

Notation: 1. To represent the relationship between two quantities, we use circles or ovals to denote the number of objects in the quantities. Figure 11-5 shows the illustration of the relationship between two quantities. The numbers inside the ovals represent the corresponding counts.

2. To represent the probability model, we use a rectangle to denote the sample space. Ovals inside the sample space represent the events. Figure 11-6 shows the illustration of the relationship between two events. The numbers inside the ovals represent the corresponding probabilities.

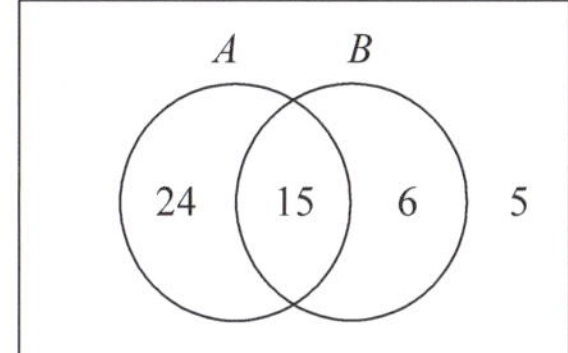

Figure 11-5

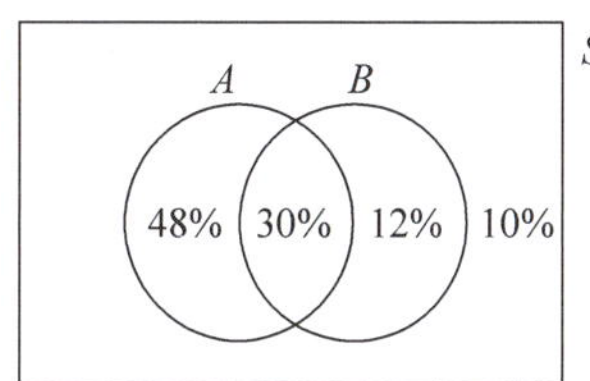

Figure 11-6

This is the same as Figure 11-5, except it is expressed in percents while Figure 11-5 expresses in counts.

3. To represent the relationship between two sets, we use circles or ovals to denote the sets. Figure 11-7 shows the illustration of the relationship between two sets. The numbers inside the ovals represent the sets' elements.

The sample space is all integers from 1 through 10.

Set A consists of all odd numbers.

Set B consists of all primes.

The intersection of A and B consists of all odd primes.

Properties: Each of the properties below illustrates the relationship between two sets, or two events (of a probability model), or two quantities, respectively. They are analogues of each other. (1) is for notations 1 and 3 (representing the counts), and (2) is for notation 2 (representing the probabilities), as shown in Figure 11-8.

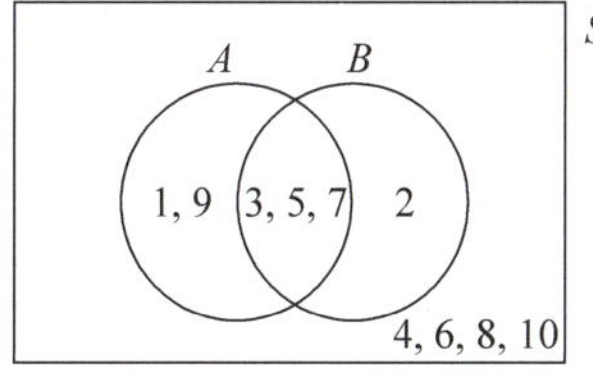

Figure 11-7

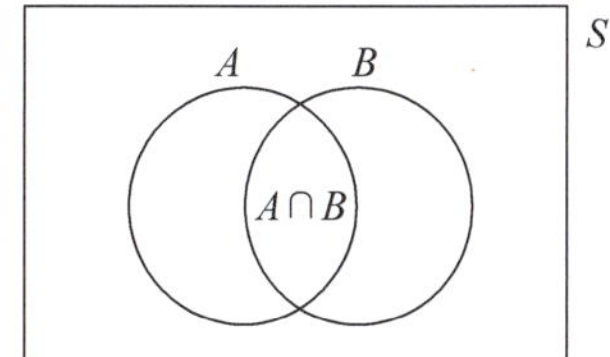

Figure 11-8

1. The general formulas:

(1) $|A\cup B|=|A|+|B|-|A\cap B|$.

(2) $P(A\cup B)=P(A)+P(B)-P(A\cap B)$.

2. If A and B are mutually exclusive events, then $A\cap B=\varnothing$, for which $|A\cap B|=0$ and $P(A\cap B)=0$. Therefore,

(1) $|A\cup B|=|A|+|B|$.

(2) $P(A\cup B)=P(A)+P(B)$.

3. If A and B are complements, then $A\cap B=\varnothing$, for which $|A\cap B|=0$ and $P(A\cap B)=0$, and $A\cup B=S$, for which $|A\cup B|=|S|$ and $P(A\cup B)=P(S)=1$. Therefore,

(1) $|A\cup B|=|A|+|B|=|S|$.

(2) $P(A\cup B)=P(A)+P(B)=P(S)=1$.

Questions: 1. If $A=\{2,3,5,7\}$ and $B=\{1,3,5,7,9\}$,

(1) Draw a Venn Diagram to represent their relationship.

(2) Calculate $A\cap B$ and $A\cup B$, then verify the formula $|A\cup B|=|A|+|B|-|A\cap B|$ is correct.

2. If $A=\{2,4,6,8,10\}$ and $B=\{1,3,5,7,9\}$,

(1) Draw a Venn Diagram to represent their relationship.

(2) Calculate $A\cap B$ and $A\cup B$, then verify the formula $|A\cup B|=|A|+|B|-|A\cap B|$ is correct.

3. If A = rolling a 6 on a die, B = rolling a 1 on a die,

(1) Draw a Venn Diagram to represent their relationship.

(2) Calculate $P(A)$, $P(B)$, $P(A\cap B)$, and $P(A\cup B)$, then verify the formula $P(A\cup B)=P(A)+P(B)-P(A\cap B)$ is correct.

4. If A = rolling an odd number on a die, B = rolling an even number on a die,

(1) Draw a Venn Diagram to represent their relationship.

(2) Calculate $P(A)$, $P(B)$, $P(A\cap B)$, and $P(A\cup B)$, then verify the formula $P(A\cup B)=P(A)+P(B)-P(A\cap B)$ is correct.

5. In the 100 households surveyed, 60 of which have cats and 70 of which have dogs. If 50 of which have both cats and dogs, how many of them have cats or dogs?

Answers: 1. (1) The Venn Diagram is shown in Figure 11-9.

(2) $A\cap B=\{3,5,7\}$, $A\cup B=\{1,2,3,5,7,9\}$.

$|A|=4$, $|B|=5$, $|A\cap B|=3$, and $|A\cup B|=6$.

$|A|+|B|-|A\cap B|=4+5-3=6=|A\cup B|$ is correct.

2. (1) The Venn Diagram is shown in Figure 11-10.

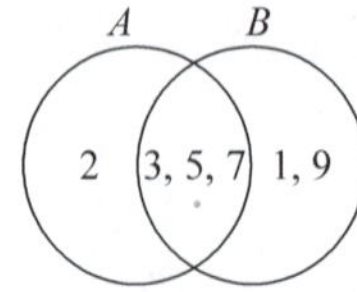

Figure 11-9

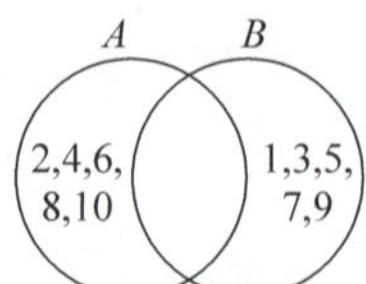

Figure 11-10

(2) $A \cap B = \varnothing$, $A \cup B = \{1,2,3,4,5,6,7,8,9,10\}$.

$|A| = 5$, $|B| = 5$, $|A \cap B| = 0$, and $|A \cup B| = 10$.

$|A| + |B| - |A \cap B| = 5 + 5 - 0 = 10 = |A \cup B|$ is correct.

Moreover, we can disregard $A \cap B$ because A and B are mutually exclusive.

3. (1) The Venn Diagram showing the elements is shown in Figure 11-11.

The Venn Diagram showing the counts is shown in Figure 11-12.

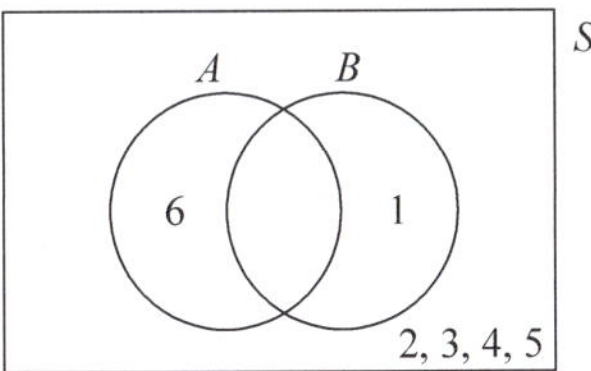

Figure 11-11

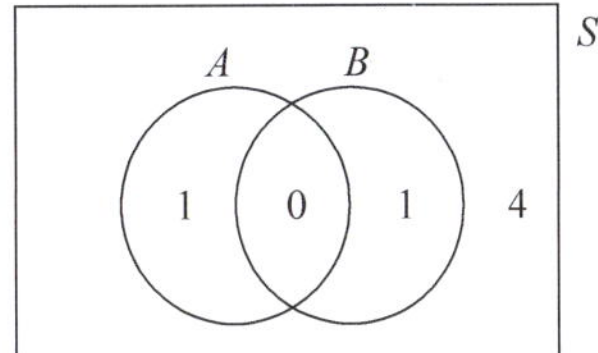

Figure 11-12

The Venn Diagram showing the probabilities is shown in Figure 11-13.

(2) $P(A) = P(B) = 1/6$, $P(A \cap B) = 0$, and $P(A \cup B) = 1/3$.

$P(A) + P(B) - P(A \cap B) = 1/6 + 1/6 - 0 = 1/3 = P(A \cup B)$ is correct.

Moreover, we can disregard $P(A \cap B)$ because A and B are mutually exclusive.

4. (1) The Venn Diagram showing the probabilities is shown in Figure 11-14.

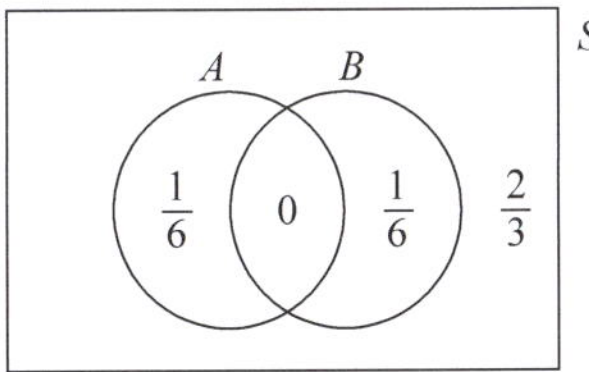

Figure 11-13

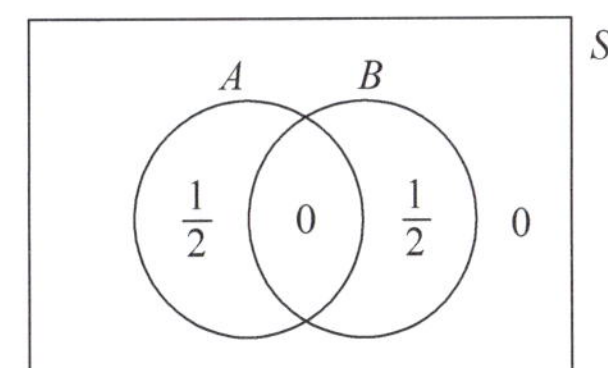

Figure 11-14

(2) $P(A) = P(B) = 1/2$, $P(A \cap B) = 0$, and $P(A \cup B) = 1$.

$P(A) + P(B) - P(A \cap B) = 1/2 + 1/2 - 0 = 1 = P(A \cup B)$ is correct.

Moreover, we can disregard $P(A \cap B)$ because A and B are mutually exclusive.

More precisely, they are complements for which $P(A \cup B) = 1$.

5. $|A|$: number of households that have cats $= 60$.

$|B|$: number of household that have dogs $= 70$.

$|A \cap B|$: number of households that have both cats and dogs $= 50$.

$|A \cup B|$: number of households that have cats or dogs $=$ unknown.

Using the formula $|A \cup B| = |A| + |B| - |A \cap B| = 60 + 70 - 50 = 80$.

The Venn Diagram in Figure 11-15 represents this relationship.

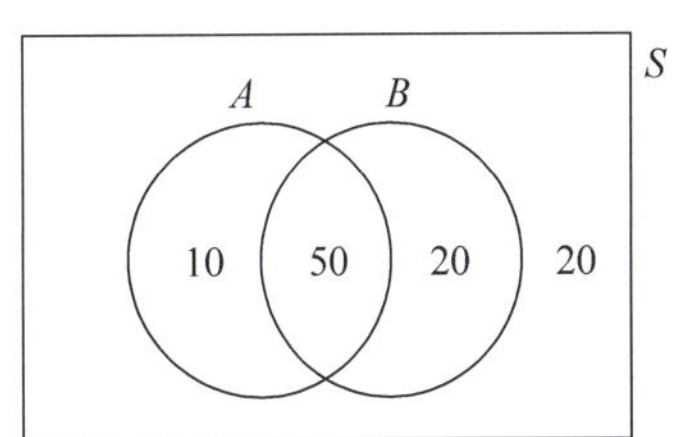

Figure 11-15

11.2.5 Independent Events and Dependent Events 独立事件与相关事件

Independent Events 独立事件

n. [ˌɪndɪˈpendənt ɪˈvents]

Definition: Events A and B are independent events if one's occurrence does not affect that of the other. 若事件 A 和 B 的发生互不影响,则它们为独立事件。

Properties: 1. $P(A\cap B)=P(A)\cdot P(B)$.

This formula is based on the Counting Principle.

2. In general, if $A_1, A_2, \cdots, A_n$ are independent events, then

$P(A_1\cap A_2\cap\cdots\cap A_n)=P(A_1)\cdot P(A_2)\cdot\cdots\cdot P(A_n)$.

Examples: 1. Rolling a fair die twice and find P(getting a 5, then a 3).

The outcome on the first roll does not influence the outcome of the second.

P(first roll is a 5) = 1/6.

P(second roll is a 3) = 1/6.

P(getting a 5, then a 3) = (1/6) · (1/6) = 1/36. This is the analogue of the Counting Principle. The numerator, 1, represents the number of ways to get a 5, then a 3. The denominator, 36, represents the number of possible outcomes (ordering matters).

2. From a bag of 4 red marbles and 6 green marbles, first randomly pick one marble and record its color, then put it back and randomly pick one marble again and record its color. We will find P(both green).

The outcome on the first pick does not influence the outcome of the second.

P(first pick is green) = 6/10.

P(second pick is green) = 6/10.

P(both green) = (6/10) · (6/10) = 36/100 = 9/25. This is the analogue of the Counting Principle. The unsimplified numerator, 36, represents the number of ways to pick two green marbles. The unsimplified denominator, 100, represents the number of ways to pick two marbles (ordering matters).

Questions: 1. Rolling a die twice (ordering matters), what is the probability that the outcomes have different parities (one odd and one even)?

2. Drawing four marbles from a bag of 4 red marbles and 6 green marbles, with replacement, what is the probability that the order of marbles we draw is green, green, red, green?

Answers: 1. Note that the outcomes are independent events.

P(different parities) = P(odd then even) + P(even then odd)
= P(1st roll odd)P(2nd roll even) +
P(1st roll even)P(2nd roll odd)
= (1/2) × (1/2) + (1/2) × (1/2)
= 1/2.

Alternatively, think of the problem this way. No matter what the first roll is, there are 3 favorable outcomes (out of the 6 possible outcomes) of the second roll. Therefore, our answer is 3/6 = 1/2.

2. (6/10) × (6/10) × (4/10) × (6/10) = 54/625.

Dependent Events　相关事件

n. [dɪ'pendənt ɪ'vents]

Definition: Events A and B are dependent events if one's occurrence affects that of the other.　若事件 A 和 B 的发生互相影响,则它们为相关事件。

Properties: 1. $P(A \cap B) = P(A) \cdot P(B|A)$.

$P(B|A)$ is read as "$P(B)$ given A happened".

If A and B are independent events, then $P(B|A) = P(B)$.

2. In general, if $A_1, A_2, \cdots, A_n$ are independent events,

$P(A_1 \cap A_2 \cap A_3 \cap \cdots \cap A_n) = P(A_1) \cdot P(A_2 | A_1) \cdot P(A_3 | A_1, A_2) \cdot \cdots \cdot P(A_n | A_1, A_2, A_3, \cdots, A_{n-1})$.

Examples: 1. Randomly picking two cards from a standard 52-deck without replacement, we want to get two 10's.

This is an example of dependent events. The outcome of the second card depends on the outcome of the first card.

P(first card is 10) = 4/52 = 1/13.

P(second card is 10) = 3/51 = 1/17. We assume that the first card picked is 10, and note that the deck now only has 51 cards since we are picking cards without replacement.

Therefore, our answer is (1/13) × (1/17) = 1/221.

2. From a bag of 4 red marbles and 6 green marbles, randomly pick two marbles without replacement. We will find P(both green).

The outcome on the first pick influences the outcome of the second pick.

P(first pick is green) = 6/10 = 3/5.

P(second pick is green) = 5/9. Assuming that the first pick is green, there are only 9 marbles left (5 of which are green).

P(both green) = (3/5) × (5/9) = 1/3.

Questions: 1. From a bag of 4 red marbles and 6 green marbles, randomly pick two marbles without replacement. What is the probability that they have different colors?

2. From a standard deck of 52 cards, randomly pick 13 without replacement. What is the probability that in order, their ranks are A, 2, 3, 4, 5, 6, 7, 8, 9, 10, J, Q, K?

Answers: 1. Note that the outcomes are dependent events since we are selecting without replacement.

$$\begin{aligned} P(\text{different color}) &= P(\text{1st red})P(\text{2nd green}|\text{1st red}) + \\ &\quad P(\text{1st green})P(\text{2nd red}|\text{1st green}) \\ &= (4/10) \times (6/9) + (6/10) \times (4/9) \\ &= 8/15. \end{aligned}$$

2. (4/52) × (4/51) × (4/50) × (4/49) × (4/48) × (4/47) × (4/46) × (4/45) × (4/44) × (4/43) × (4/42) × (4/41) × (4/40).

Part 2: Geometry

第2部分：

几　何

几何(geometry)一词由希腊语 geo(土地)和 metry(测量)两个词组合而成。几何学是人类在测量大自然的过程中形成的。公元前 300 年左右,希腊数学家欧几里得对几何原理进行汇总和拓展,完成论文集《几何原本》,书中他规定了一系列关于几何的定义、定理、公理和数学证明方法,构成了最早的几何学原理,对数学贡献巨大,他被后人尊称为"几何之父"。几何学又包括拓扑学、三角学、分形几何学、微分几何学和代数几何学等板块。

- 拓扑学(topology):是非常抽象的几何学分支,主要研究几何图形或空间在改变形状后还能保持一些基本性质的学科。拓扑学只考虑物体间的位置关系而不考虑它们的形状和大小,是要求学习者具有较高的逻辑推理能力和抽象思维能力的学科。
- 三角学(trigonometry):是以平面三角形和球面三角形的边和角的关系为基础进行研究,达到测量应用的学科。
- 分形几何学(fractal geometry):又称为大自然的几何学,是一门以不规则而又无限复杂、具备自相似结构的几何形态为研究对象的几何学,反映的是大自然复杂表面下的内在数学秩序。
- 微分几何学(differential geometry):主要是以分析方法来研究空间(微分流形)的几何性质。应用微分学来研究三维欧几里得空间里的曲线、曲面等图形性质的数学分支。差不多与微积分学同时起源于 17 世纪。
- 代数几何学(algebraic geometry):是用代数表达式描述几何物体的边和曲面的学科,是平面解析几何与三维空间解析几何的推广。

其中,三角形是几何的核心。几乎所有多边形的性质、圆的性质都派生于三角形。另外,证明是在几何中不可或缺的部分。利用公理严谨地证出引理和定理,从而可以举一反三证明出更为重要和有趣的命题。

通过这一部分的学习,读者能够熟悉几何的点、线、角等基本概念,并严谨地证明出它们的位置/大小/长度关系(第 12～14 章),然后学习三角形的各个部分及其性质以及直角三角形的三角学(第 15 章)。通过对三角形透彻的理解,可以融会贯通证明多边形与圆的重要定理,最后应用到立体图形上。

12. Introduction of Geometry 几何学介绍

12.1 Basics of Geometry 几何学基础

12.1.1 Types of Geometry 几何学种类

Geometry 几何学

n. [dʒiˈɑmətri]

Definition: The branch of mathematics that studies the properties, measurements, and relations of points, lines, planes, and solids.

几何学是数学的一个分支，专门介绍点、线、平面和立体的性质、测量、关系。

Phrases: plane～ 平面几何, solid～ 立体几何

Plane Geometry 平面几何

n. [pleɪn dʒiˈɑmətri]

Definition: The two-dimensional geometry. See also plane in Section 12.1.2 and coordinate plane in Chapter 6.

平面几何是指二维几何。见有关平面(12.1.2 节)和坐标平面(第 6 章)的介绍。

Solid Geometry 立体几何

n. [ˈsɑlɪd dʒiˈɑmətri]

Definition: The three-dimensional geometry. See also solids in Chapter 20.

立体几何是指三维几何。见第 20 章有关立体的介绍。

12.1.2 Basics 基础

Point 点

n. [pɔɪnt]

Definition: The representation of an exact location. In the coordinate plane, its location can be represented by an ordered pair.

点为精准位置的表示。在坐标平面上，点的位置用有序对表示。

Notation: A point is usually labelled with a capital letter.

点通常用大写字母标记。

Examples: Figure 12-1 shows the locations of different points.

The coordinates of point A are $(2,3)$.

The coordinates of point B are $(4,-3)$.

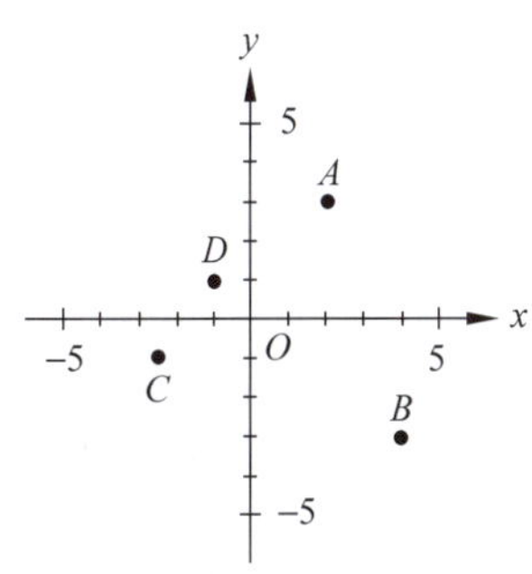

Figure 12-1

The coordinates of point C are $(-2.5,-1)$.

The coordinates of point D are $(-1,1)$.

Line 线

n. [laɪn]

Definition: A straight collection of points. Throughout this encyclopedia, we will assume that every line is a straight line, unless specified.

在本书中,除非特别指定,下面提到的"线"都为"直线"。

Notation: 1. A line is usually labelled with a lowercase letter, usually l.

线通常用小写字母标记,一般用 l。

2. If we know that a line passes through two points, A and B, we will denote this line by $\overleftrightarrow{AB}$, as shown in Figure 12-2.

若已知一条线经过两个点: A 和 B,则可以用 $\overleftrightarrow{AB}$ 标记这条线,如图 12-2 所示。

3. $\overleftrightarrow{AB}$ and $\overleftrightarrow{BA}$ refer to the same line.

4. Refer to Table A-5 in Appendix A for the notation of lines.

直线的标记方式参照附录 A 的表 A-5。

Properties: 1. A line has an indefinite length and an indefinite direction.

直线长度无法度量,方向无法度量。

2. Two different points determine a line.

两个不重合的点确定一条直线。

Examples: As shown in Figure 12-3, $\overleftrightarrow{CD}$, $\overleftrightarrow{DC}$, $\overleftrightarrow{DE}$, $\overleftrightarrow{ED}$, $\overleftrightarrow{CE}$, and $\overleftrightarrow{EC}$ are said to be the same line. Points C, D, and E are said to be collinear points because they lie on the same line.

如图 12-3 所示, $\overleftrightarrow{CD}$、$\overleftrightarrow{DC}$、$\overleftrightarrow{DE}$、$\overleftrightarrow{ED}$、$\overleftrightarrow{CE}$、$\overleftrightarrow{EC}$ 是同一条直线。点 C、D、E 是共线点,因为它们在同一条直线。

Figure 12-2　　Figure 12-3

Phrases: paralle lines 平行线, perpendicular lines 垂直线, ~of symmetry 对称轴

Plane 平面

n. [pleɪn]

Definition: A flat, two-dimensional surface that extends infinitely far.

The coordinate plane is a special and well-known example. Every point within this plane corresponds to a unique ordered pair.

平面是指一个平的无限长的表面。

坐标平面是一个众所周知的特殊平面。该平面的每个点都对应一个独特的有序对。

Properties: Three different noncollinear points determine a plane.

三个不重合的非共线点确定一个平面。

Examples：When three different noncollinear points，A，B，and C，are given，there is only one plane containing all of them. As shown in Figure 12-4，the plane contains A，B，and C.
给出三个不重合的非共线点，只有一个平面能包含它们全部。如图 12-4 所示的平面包含 A、B、C 三点。

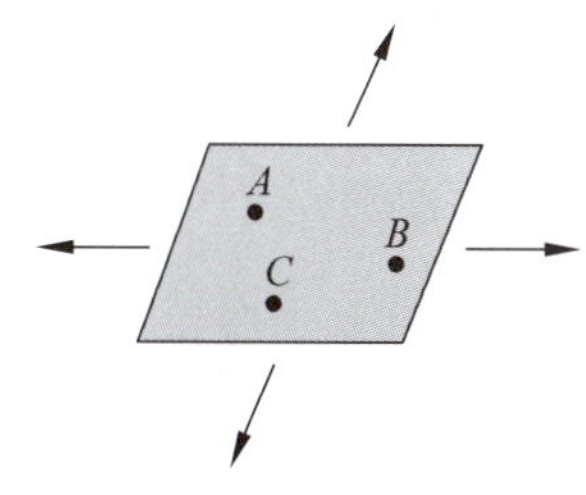

Figure 12-4

12.1.3　Dimensions　维度

Dimension　维度

n. [də'menʃən]

Definition：An extension in a given direction. We are living in the three-dimensional world. The three dimensions are called length，width，and height.
The dimensions are mutually perpendicular.
维度是指一个指定方向的延伸。我们生活在三维空间里。这三个维度的名称是长(度)、宽(度)、高(度)。
维度相互垂直。

Examples：1. A point is 0-dimensional because it has no length，width，or height. Figure 12-5 shows a dot.
点是零维图形，因为它没有长、宽或高。图 12-5 展示了一个点。

2. A line/ray/line segment is 1-dimensional because it has a length，but no width or height. Figure 12-6 shows a line，a ray，and a line segment，respectively.
直线/射线/线段是一维图形，因为它们有长，但没宽和高。图 12-6 分别展示了直线、射线和线段。

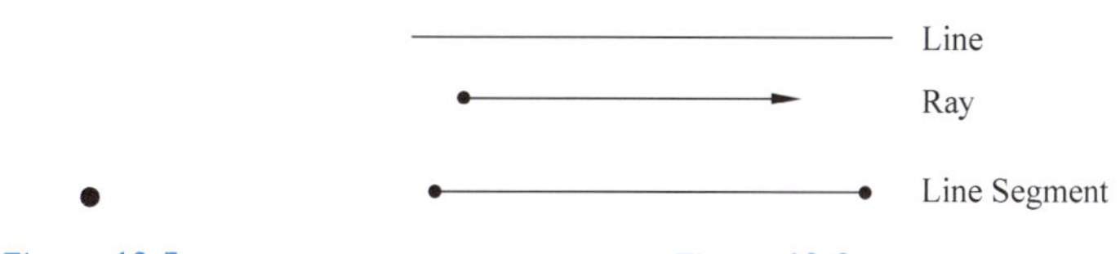

Figure 12-5　　Figure 12-6

3. A polygon is 2-dimensional because it has a length and a width，but no height. Figure 12-7 shows two polygons.
多边形是二维图形，因为它们有长和宽，但没有高。图 12-7 展示了两个多边形。

4. A solid is 3-dimensional because it has a length，a width，and a height. Figure 12-8 shows a solid.
立体是三维图形，因为有长、宽、高。图 12-8 展示了一个立体。

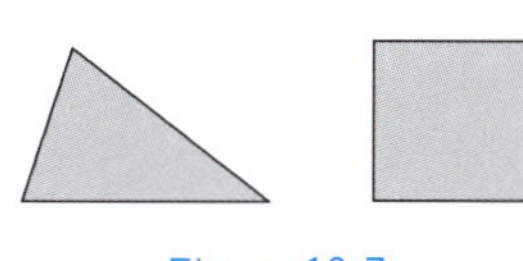
Figure 12-7

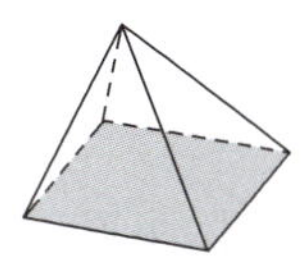
Figure 12-8

12.1.4 Geometry Shapes 几何图形

Shape 图形

n. [ʃeɪp]

Definition: The form of an object (how it is laid out in the space), bounded by line segments, curves, flat or curved surfaces. See 2D-shape/plane shape and 3D-shape/solid.

一个物体如何在空间中展现的形态,由线段、曲线、平面或曲面围成。见平面图形和立体图形。

2D-Shape/Plane Shape 平面图形

n. [tuˌdi ʃeɪp/pleɪn ʃeɪp]

Definition: A shape that has a length and a width, but no height or thickness. All plane shapes can be drawn within a plane.

平面图形是一个有长和宽的图形,但没有高(亦称厚度)。顾名思义,所有平面图形都能在同一个平面画出。

Properties: 1. The plane shapes can be categorized by the following:

Polygons—plane shapes that are bounded by three or more line segments. 多边形——用线段围成的平面图形。

Figure 12-9 shows four examples of polygons.

Circles—plane shapes that are 360° arcs. 圆形——360°弧。

Figure 12-10 shows a circle.

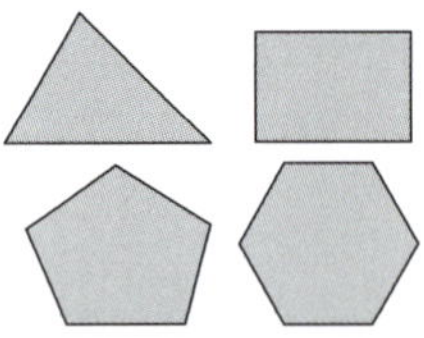

Figure 12-9

Figure 12-10

Sectors—plane shapes that are bounded by two radii of a circle and the included arc. 扇形——被圆的两条半径和夹弧围成的平面图形。

Figure 12-11 shows four examples of polygons.

Other irregular plane shapes, as shown in Figure 12-12.

Figure 12-11

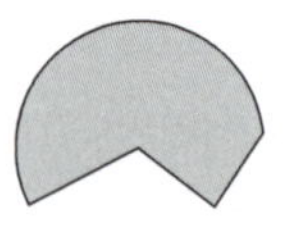

Figure 12-12

2. All plane shapes have perimeters and areas. 所有平面图形均有周长和面积。

3D-Shape/Solid 立体图形

n. [θriˌdi ʃeɪp/ˈsɑlɪd]

Definition: A shape that has a length, a width, and a height. All 3D-shapes/solids can be drawn within the *xyz*-plane.

立体图形是一个有长、宽、高的图形。所有立体图形都能在 *xyz* 平面画出。

Properties：1. The 3D-shapes/solids can be categorized by the following：

Polyhedrons—solids with flat faces, each of which is a polygon. Figure 12-13 shows three examples of polyhedrons.

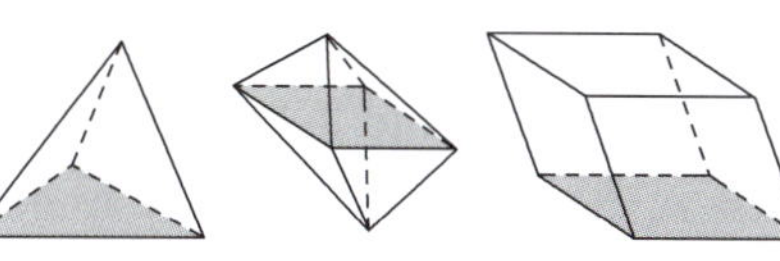

Figure 12-13

多面体——每个面都是平的多边形的立体图形，图 12-13 展示了三个多面体。

Common solids that involve circles：cylinders, circular cones, spheres. Figure 12-14 shows a cylinder, a cone, and a sphere respectively.

常用的含有圆的立体图形有圆柱、圆锥、球。图 12-14 分别展示了圆柱、圆锥、球形。

Other irregular 3D-shapes/solids, as shown in Figure 12-15.

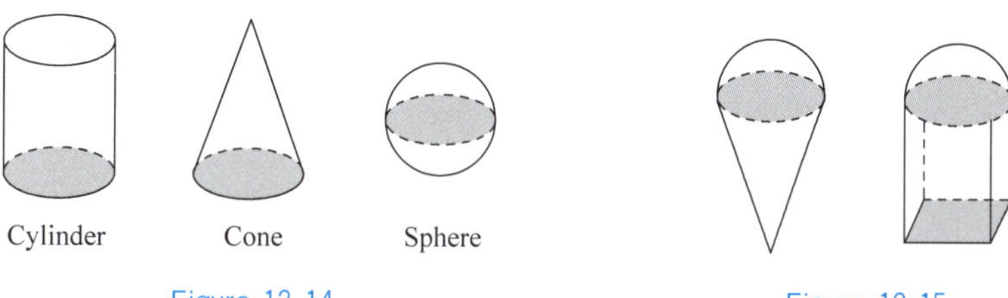

Figure 12-14　　Figure 12-15

2. All 3D-shapes/solids have surface areas and volumes. 所有立体图形均有表面积和体积。

12.1.5 Summary 总结

1. There are two types of geometry—plane geometry and solid geometry.

(1) Plane geometry studies the properties, measurements, and relations of points, lines, plane shapes within the same plane (up to 2-dimensional objects).

(2) Solid geometry studies properties, measurements, and relations of different planes and solids (up to 3-dimensional objects).

几何学有两种：平面几何与立体几何。

(1) 平面几何专门研究在一个平面上的点、线、面的性质、测量和关系(最多 2 个维度)。

(2) 立体几何专门研究不同的平面和立体的性质、测量、关系(最多 3 个维度)。

2. A dimension is an extension in a given direction. Our world is three-dimensional because it has a length, width, and height. Dimensions are mutually perpendicular.

(1) A point is zero-dimensional (no length, width, or height).

(2) A line/ray/line segment is one-dimensional (length only).

(3) A 2D shape or plane is two-dimensional (length and width only).

(4) A solid is three-dimensional (length, width, and height).

维度是一个指定方向的延伸。世界是三维的，因为有长(度)、宽(度)、高(度)。维度相互垂直。

(1) 点的维度是 0(没有长、宽、高)。

(2) 直线/射线/线段的维度是 1(只有长)。

(3) 平面图形和平面的维度是2(只有长和宽)。

(4) 立体的维度是3(有长、宽、高)。

3. A shape is a figure bounded by line segments, curves, flat or curved surface. It is one of 2D-shape (plane shape) and 3D-shape (solid).

(1) A 2D-shape that is only bounded by line segments is known as a polygon.

(2) A 3D-shape that is only bounded by polygons is known as a polyhedron.

图形是一个由线段、曲线、平面或曲面围成的形状。图形有两种:平面图形和立体图形。

(1) 只由线段围成的平面图形称为多边形。

(2) 只由多边形围成的立体图形称为多面体。

12.2 Endpoints, Lines, Rays, and Line Segments 端点、直线、射线与线段

12.2.1 Endpoints 端点

Endpoint 端点

n. [ˈendˌpɔɪnt]

Definition: A point that is located at the end of a ray or a line segment. 一个在射线或线段末端的点。

Examples: 1. As shown in Figure 12-16, points A and B are the endpoints of the line segment. 如图 12-16 所示,A 和 B 是线段的端点。

2. As shown in Figure 12-17, point C is the endpoint of the ray. 如图 12-17 所示,C 是射线的端点。

Figure 12-16 Figure 12-17

12.2.2 Lines, Rays, and Line Segments 直线、射线与线段

Line 直线

See line in Section 12.1.2.

Ray 射线

n. [reɪ]

Definition: A portion of a line that starts at one point and goes off to a particular direction indefinitely.

The starting point is called the endpoint.

直线的一部分——从一个点开始,对着一个方向无限延长。

射线的起点称为端点。

Notation: A ray whose endpoint is A and passes through B is denote by $\overrightarrow{AB}$, as shown in Figure 12-18. 端点为 A,经过 B 的射线写作 $\overrightarrow{AB}$,如图 12-18 所示。

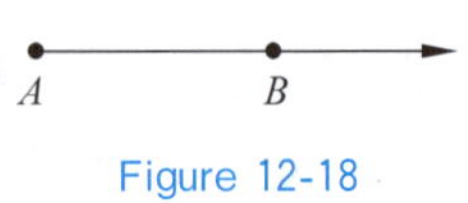

Figure 12-18

Refer to Table A-5 in Appendix A for the notation of rays.
射线的标记方式参照附录 A 的表 A-5。

Properties: 1. A ray has an indefinite length but a definite direction.
射线长度无法度量，但有确定的方向。

2. Two different points determine a ray, given that one of the point is the endpoint.
两个不重合的点(其中一个是端点)确定一条射线。

3. $\overrightarrow{AB}$ are $\overrightarrow{BA}$ are different rays. $\overrightarrow{AB}$和$\overrightarrow{BA}$是不同的射线。

Line Segment 线段

n. [ˈlaɪn ˌseɡmənt]

Definition: A portion of a line that starts at one point and ends at a different point.
直线的一部分——从一个点开始，到另一个点结束。
The starting point and ending point are called the endpoints.
线段的起点和终点称为端点。

Notation: 1. A line segment whose endpoints are A and B is denoted by $\overline{AB}$, as shown in Figure 12-19. 端点为 A 和 B 的线段写作 $\overline{AB}$，如图 12-19 所示。

2. To emphasize X is a point between $\overline{AB}$ (in other words, $\overline{AB}$ contains X), we denote $\overline{AB}$ as $\overline{AXB}$, as shown in Figure 12-20. 若要强调 X 是在 $\overline{AB}$ 间的一点，则可以把 $\overline{AB}$ 写成 $\overline{AXB}$，如图 12-20 所示。

Figure 12-19   Figure 12-20

3. Refer to Table A-5 in Appendix A for the notation of line segments.
线段的标记方式参照附录 A 的表 A-5。

Properties: 1. A line segment has a definite length but an indefinite direction.
线段有明确的长度，但是方向无法度量。

2. Two different points determine a line segment. 两个不重合的点确定一条线段。

3. The shortest path between two points, A and B, is $\overline{AB}$. 两点之间，线段最短。

4. The measurement of a line segment is length. 线段的测量称为长度。

5. $\overline{AB}$ and $\overline{BA}$ refer to the same line segment. $\overline{AB}$和$\overline{BA}$指的是同一条线段。

12.2.3 Measuring Line Segments 线段的测量

Length 长度

n. [leŋθ]

Definition: The length of a line segment is the distance between its endpoints. 线段的长度指的是两个端点间的距离。

Notation: The length of a line segment $\overline{AB}$ is denoted by AB. Note that the upper bar has disappeared from $\overline{AB}$ when we are talking about its length. 线段 $\overline{AB}$ 的长度写作 AB。注意，提及长度的时候，上画线没有了。

If $\overline{AB}$ and $\overline{CD}$ are equal in length, then we write $AB = CD$, or $\overline{AB} \cong \overline{CD}$. The latter reads as "$\overline{AB}$ and $\overline{CD}$ are congruent". 当我们表示$\overline{AB}$和$\overline{CD}$长度相等时,可以写作 $AB = CD$ 或 $\overline{AB} \cong \overline{CD}$。后者读作 $\overline{AB}$ 和 $\overline{CD}$ 全等。

Refer to Table A-5 in Appendix A for the notation of lengths.

长度的标记方式参照附录 A 的表 A-5。

Properties: 1. Some ways of measuring lengths:

(1) Metric system: km, m, cm, mm, …

(2) English system: mi, yd, ft, in, …

(3) The word "units" that represents the assumed unit. Usually we do not write out the word "units".

以上给出了一些长度单位制度。

2. Lengths are comparable and addible. 长度可比较,可相加。

3. **Segment Addition Postulate**: If C is a point on $\overline{AB}$, then $AB = AC + CB$. This is illustrated in Figure 12-21. 线段相加公理,如图 12-21 所示。

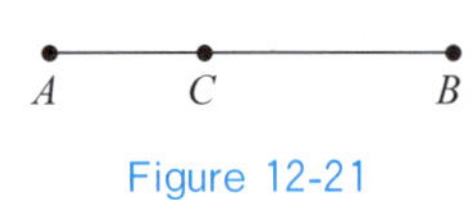

Figure 12-21

4. Trivial case: $AA = 0$. The length between point A and itself is 0.

5. The Distance Formula in Section 6.2 calculates the length of a line segment in the coordinate plane. 使用距离公式(6.2 节)可求出在坐标平面中线段的长度。

Questions: If A, B and C are collinear points in this order such that $AB = 7$ and $AC = 18$, what is the value of BC?

若 A、B、C 为共线点,满足 $AB = 7$ 和 $AC = 18$,则 BC 的值是多少?

Answers: According to Property 3, since B is a point on $\overline{AC}$, we have $AB + BC = AC$, so that $7 + BC = 18$, from which $BC = 11$.

答案的根据是线段相加定理(性质 3)。

Midpoint 中点

n. [ˈmɪdˌpɔɪnt]

Definition: M is the midpoint of a line segment $\overline{AB}$ if it satisfies both:

1. M is on $\overline{AB}$.

2. $AM = BM = AB/2$.

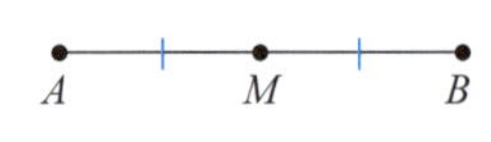

Figure 12-22

As shown in Figure 12-22, M is the midpoint of $\overline{AB}$.

Properties: The Midpoint Formula in Section 6.2 calculates the midpoint of a line segment in the coordinate plane. 使用中点公式(6.2 节)可求出在坐标平面中线段的中点。

Questions: 1. If M is a midpoint of $\overline{AB}$, and $AB = 8$, what is the value of AM?

2. If M is a midpoint of $\overline{AB}$, and $AM = 10$, what is the value of AB?

Answers: 1. $AM = AB/2 = 4$.

2. $AB = 2AM = 20$.

Line Segment Bisector 线段平分线

n. [ˈlaɪn ˌsegmənt baɪˈsektɚ]

Definition: A line segment bisector is a line/ray/line segment that cuts the line segment in halves. Putting it differently: Let M be the midpoint of $\overline{AB}$, any line/ray/line segment that

passes through M but not on $\overleftrightarrow{AB}$ is a line segment bisector of $\overline{AB}$.
Often time, of all possible line segment bisectors, we are interested in the perpendicular bisector.

若 M 是 $\overline{AB}$ 的中点，任何过 M 的直线/射线/线段(但不在 $\overleftrightarrow{AB}$ 上)都是 $\overline{AB}$ 的平分线。
在一条线段的众多平分线中，通常我们对垂直平分线感兴趣。

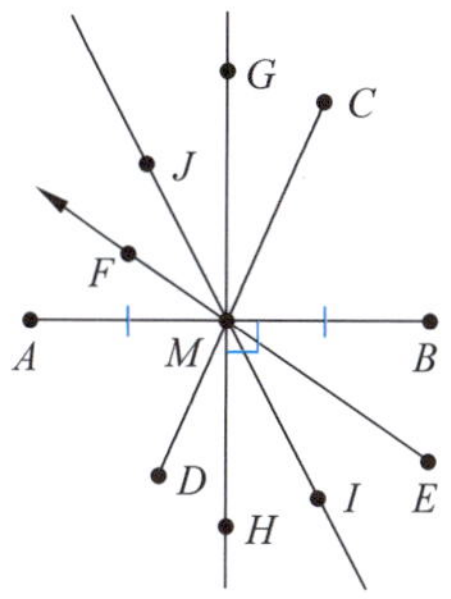

Examples: As shown in Figure 12-23, M is the midpoint of $\overline{AB}$.
$\overline{CD}$, $\overrightarrow{EF}$, $\overleftrightarrow{GH}$, and $\overleftrightarrow{IJ}$ are line segment bisectors of $\overline{AB}$. In particular, $\overleftrightarrow{GH}$ is a perpendicular bisector of $\overline{AB}$.
如图 12-23 所示，M 是 $\overline{AB}$ 的中点。
$\overline{CD}$、$\overrightarrow{EF}$、$\overleftrightarrow{GH}$、$\overleftrightarrow{IJ}$ 是 $\overline{AB}$ 的平分线。其中 $\overleftrightarrow{GH}$ 是 $\overline{AB}$ 的垂直平分线。

Figure 12-23

12.2.4 Measuring Rays 射线的测量

Direction 方向

n. [də'rekʃən]

Definition: The direction of a ray is the orientation the ray from its endpoint.
For now, we will only determine if two rays have the same direction.
射线的方向是它从端点的指向。现在只须判断两条射线的方向是否相同。

Examples: As shown in Figure 12-24, $\overrightarrow{AB}$ and $\overrightarrow{CD}$ have the same direction, and the other pairs of rays have different directions.
$\overrightarrow{AB}$ and $\overrightarrow{EF}$ have opposite directions.
$\overrightarrow{CD}$ and $\overrightarrow{EF}$ have opposite directions.
如图 11-24 所示，只有 $\overrightarrow{AB}$ 和 $\overrightarrow{CD}$ 的方向相同。
$\overrightarrow{AB}$ 和 $\overrightarrow{EF}$ 的方向相反。
$\overrightarrow{CD}$ 和 $\overrightarrow{EF}$ 的方向相反。

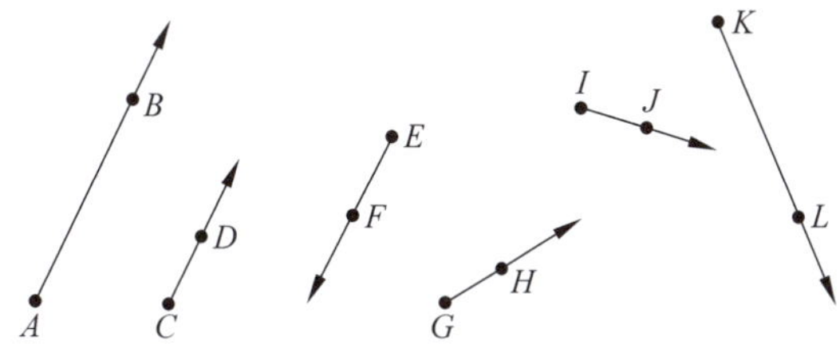

Figure 12-24

12.2.5 Summary 总结

Table 12-1 shows the properties of lines, line segments, and rays.

Table 12-1

Name 名称	Has Direction 是否有方向	Has Length 是否有长度
Line 直线	No	No
Line Segment 线段	No	Yes
Ray 射线	Yes	No

12.3 Angles 角

12.3.1 Introduction 介绍

Angle 角

n. [ˈæŋgəl]

Definition: The amount of turn between two rays (known as the arms/sides) that have a common endpoint (known as the vertex).

Figure 12-25 shows an angle with all parts labeled.

角是指两条具有共同端点的射线的转动空间大小。这两条射线叫作边,端点叫作顶点。图 12-25 展示了一个角并且标记边和顶点。

Notation: 1. As shown in Figure 12-26, points A and C are on different rays, and B is the rays' common endpoint. We will denote this angle by $\angle ABC$, or $\angle CBA$, **or simply $\angle B$.**

如图 12-26 所示,点 A 和 C 在不同的射线上。B 是射线的共同端点。这个角记作 $\angle ABC$ 或 $\angle CBA$ 或 $\angle B$。

2. As shown in Figure 12-27, points A, C, and D are on different rays, and B is the rays' common endpoint. We will notate the angle with double ticks as $\angle ABC$ or $\angle CBA$, but not $\angle B$ because $\angle B$ is ambiguous—we do not know whether it refers to $\angle ABC$ or $\angle CBD$ or $\angle ABD$.

如图 12-27 所示,点 A、C、D 在不同的射线上。B 是射线的共同端点。带有双画线的角记作 $\angle ABC$ 或 $\angle CBA$,但不能记作 $\angle B$,因为 $\angle B$ 太含糊了,可以指 $\angle ABC$ 或 $\angle CBD$ 或 $\angle ABD$。

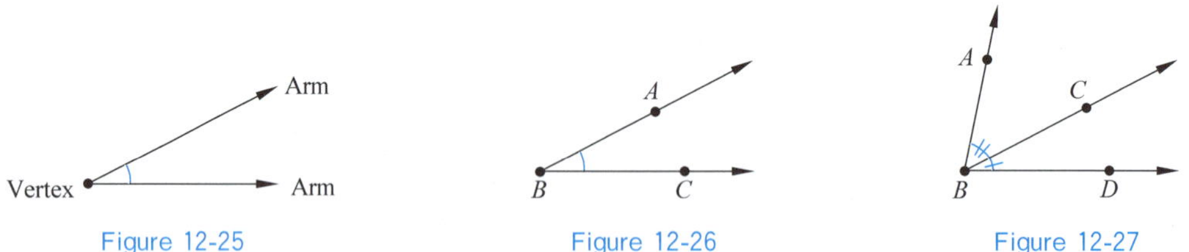

Figure 12-25　Figure 12-26　Figure 12-27

3. Refer to Table A-5 in Appendix A for the notation of angles.

角的标记方式参照附录 A 的表 A-5。

Properties: 1. Angles in geometry are measured in degrees or radians (which comes in Section 33.4).

角是用度数或弧度测量的。弧度将在 33.4 节介绍。

2. Each angle can be classified according to its measure, or the number of degrees. Such categories include: acute angle, right angle, obtuse angle, straight angle, reflex angle,

round angles.

根据度数，可以把角分为锐角、直角、钝角、平角、优角、周角。

3. Two angles sometimes have special relationship: congruent angles, adjacent angles, complementary angles, supplementary angles.

两个角有时有特殊关系，如等角、邻角、补角、余角。

4. As shown in Figure 12-28, the interior of an angle is the area between the rays that make up the angle, extending from the vertex to indefinitely.

如图 12-28 所示，角的内部是两条射线里面的区域。射线是从顶点无限延长的。

5. **Angle Addition Postulate**: If point B is in the interior of $\angle AOC$, then $m\angle AOB + m\angle BOC = m\angle AOC$, as shown in Figure 12-29.

角相加公理，如图 12-29 所示。

Examples: As shown in Figure 12-30, points B and C are at the interior of $\angle A$.

Points D and E are at the exterior of $\angle A$.

如图 12-30 所示，点 B 和 C 在 $\angle A$ 的内部。

点 D 和 E 在 $\angle A$ 的外部。

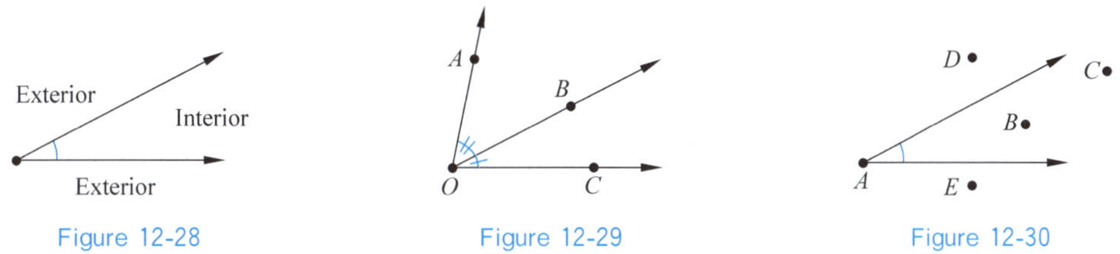

Figure 12-28　Figure 12-29　Figure 12-30

Phrases: From those mentioned in Properties 2 and 3.

12.3.2 Measurement 测量

Degree 度

n. [dɪ'gri]

Definition: A measurement unit of an angle. We will use a protractor to measure the number of degrees in an angle. 度是角的一种测量单位。一般用量角器测量角的度数。

A full rotation has 360 degrees. 一个全程旋转有 360°。

Notation: 1. To denote "d degrees", we write it as $d°$.

2. To express "$\angle ABC$ is d degrees", we write it as "$m\angle ABC = d°$". The letter m stands for the measure.

3. Refer to Table A-5 in Appendix A for the notation of degrees.

度的标记方式参照附录 A 的表 A-5。

Properties: Degrees is comparable and addible. 度数是可比较的和可相加的。

Examples: A circular clock is split into 60 equal parts (minutes), the minute hand rotates 6° per minute and 360° per hour. The hour hand rotates 360°/12 = 30° per hour. 一个圆形的钟被分成了 60 等份，每一等份为 1 分钟，则分针每分钟转动 6°，每小时转动 360°。时针每小时转动 30°。

Measure 测度/度数

n. [ˈmeʒər]

Definition: The measure of an angle is the number of degrees the angle has. 角的度数。

To express "$\angle ABC$ has d degrees", we write it as "$m\angle ABC = d°$". The letter m stands for the measure.

Refer to Table A-5 in Appendix A for the notation of measures.

测度的标记方式参照附录 A 的表 A-5。

Degree-Minute-Second (DMS) 度分秒表示法

n. [dɪˈgri ˈmɪnət ˈsekənd]

Definition: An alternate way of expressing the decimal measures of angles. DMS 是另一种表示角的度数的方法。

Notation: An angle that has d degrees m minutes s seconds is denoted as $d°\, m'\, s''$. We want $0 \leqslant m, s < 60$.

Refer to Table A-5 in Appendix A for the DMS notation.

度分秒标记法参照附录 A 的表 A-5。

Properties: 1. From DMS to decimal conversion: $d°\, m'\, s'' = \left(d + \frac{m}{60} + \frac{s}{3600}\right)°$. 下列公式给出了从度分秒表示法到小数表示法的转换 $d°\, m'\, s'' = \left(d + \frac{m}{60} + \frac{s}{3600}\right)°$。

2. From decimal to DMS conversion: Let w be the whole part and f be the fractional part of a number ($0 \leqslant f < 1$). To convert $(w + f)°$ to $d°\, m'\, s''$, $d = w$, $m = \lfloor 60f \rfloor$, and $s = 60(60f - \lfloor 60f \rfloor)$. 从小数表示法到度分秒表示法：度数和小数的整数部分相同。分数为小数表示法的小数部分与 60 的积向下取整。秒数为余下的部分(由分转换为秒)。

Proofs: 1. (Property 1) Recall that this is an analogue of the clock: think of the number of degrees as the hour. One minute is 1/60 of one hour, and one second is 1/3600 of one hour. 这与时钟类似：把度数试想为小时。1 分钟是 1/60 小时，1 秒钟是 1/3600 小时。

2. (Property 2) To satisfy $0 \leqslant m, s < 60$, we want d as large as possible. So $d = w$ is clear. After that, we want m as large as possible, thus $m = \lfloor 60f \rfloor$, and $s = 60(60f - \lfloor 60f \rfloor)$. 跟小学数学学到的时分秒到小数的转换法相同。

Questions: 1. Convert each of the following from DMS to decimal:

(1) $140°\, 30'\, 00''$

(2) $30°\, 20'\, 180''$

(3) $12°\, 24'\, 36''$

2. Convert each of the following from decimal to DMS:

(1) $30.5°$

(2) $54.321°$

(3) $12.345°$

Answers: 1. (1) $140°\, 30'\, 00'' = \left(140 + \frac{30}{60}\right)° = 140.5°$.

(2) $30°\,20'180'' = \left(30 + \frac{20}{60} + \frac{180}{3600}\right)° = 30.38\overline{3}°$.

(3) $12°\,24'36'' = \left(12 + \frac{24}{60} + \frac{36}{3600}\right)° = 12.41°$.

2. (1) We have $d = 30$, $m = \lfloor 60(0.5) \rfloor = 30$, and $s = 60(60(0.5) - \lfloor 60(0.5) \rfloor) = 0$. Thus our answer is $30°\,30'$.

(2) We have $d = 54$, $m = \lfloor 60(0.321) \rfloor = 19$, and $s = 60(60(0.321) - \lfloor 60(0.321) \rfloor) = 15.6$. Thus our answer is $54°\,19'15.6''$.

(3) We have $d = 12$, $m = \lfloor 60(0.345) \rfloor = 20$, and $s = 60(60(0.345) - \lfloor 60(0.345) \rfloor) = 42$. Thus our answer is $12°\,20'42''$.

12.3.3 Classifying an Angle 角的分类

Acute Angle 锐角

n. [ə'kjut 'æŋgəl]

Definition: An angle is an acute angle if its measure is greater than 0° and less than 90°.

To put it in mathematical notations, $\angle A$ is an acute angle if $0° < m\angle A < 90°$.

锐角是度数大于 0° 且小于 90° 的角。

Examples: Figure 12-31 shows some examples of acute angles.

Right Angle 直角

n. [raɪt 'æŋgəl]

Definition: An angle is a right angle if its measure is 90°.

To put it in mathematical notations, $\angle A$ is a right angle if $m\angle A = 90°$.

直角是度数为 90° 的角。

Notation: As shown in Figure 12-32, we mark right angles differently (using line segments, as opposed to using arcs). 如图 12-32 所示,我们会用线段(而不是弧线)标记直角。

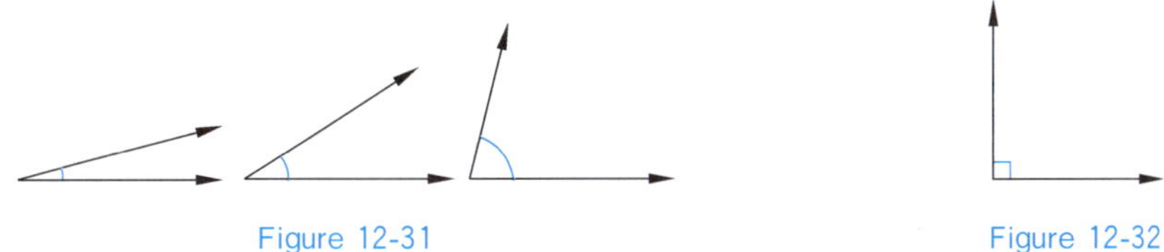

Figure 12-31 Figure 12-32

Properties: All right angles are congruent. 所有直角都是全等的。

Examples: The minute hand and hour hand form a right angle at 3 o'clock and at 9 o'clock. 在 3 点钟与 9 点钟的时候,分针与时针形成直角。

Obtuse Angle 钝角

n. [əb'tus 'æŋgəl]

Definition: An angle is an obtuse angle if its measure is greater than 90° and less than 180°.

To put it in mathematical notations, $\angle A$ is an obtuse angle if $90° < m\angle A < 180°$.

钝角是度数大于 90° 且小于 180° 的角。

Examples: Figure 12-33 shows some examples of obtuse angles.

Straight Angle 平角

n. [streɪt 'æŋgəl]

Definition: An angle is a straight angle if its measure is 180°.

To put it in mathematical notations, $\angle A$ is a straight angle if $m\angle A = 180°$.

平角是度数为 180° 的角。

Figure 12-34 shows an straight angle.

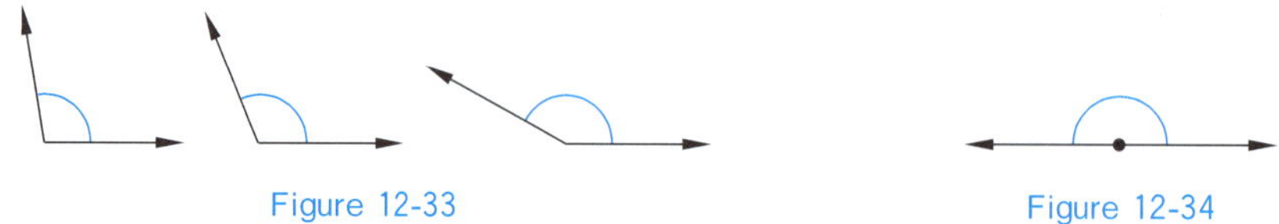

Figure 12-33　　Figure 12-34

Properties: 1. All straight angles are congruent. 所有平角都是全等的。

2. A straight angle is not a line! It is an angle whose sides are pointing at opposite directions. 平角并不是直线。它是两条边都指向对立方向的角。

Examples: The minute hand and hour hand form a straight angle at 6 o'clock. 在 6 点钟的时候，分针与时针形成平角。

Reflex Angle 优角

n. ['riˌfleks 'æŋgəl]

Definition: An angle is a reflex angle if its measure is greater than 180° and less than 360°.

To put it in mathematical notations, $\angle A$ is a reflex angle if $180° < m\angle A < 360°$.

优角是度数大于 180° 且小于 360° 的角。

Examples: Figure 12-35 shows some examples of reflex angles.

Round Angle 周角

n. [raʊnd 'æŋgəl]

Definition: An angle is a round angle if its measure is 360°.

To put it in mathematical notations, $\angle A$ is a round angle if $m\angle A = 360°$.

周角是度数为 360° 的角。

Figure 12-36 shows a round angle.

Figure 12-35　　Figure 12-36

Properties: 1. All round angles are congruent. 所有周角都是全等的。

2. A round angle is coincident to a 0°-angle. 周角和度数为 0° 的角完全重合。

Examples: The minute hand and hour hand form a round angle at 12 o'clock.

12.3.4 Relationship of Two or More Angles 两个或多个角的关系

Congruent Angles 等角

n. [kən'gruənt 'æŋgəls]

Definition: Two or more angles are congruent angles if their measures are the same. 若两个或多个角的度数相同，则称它们为等角。

Notation: To express "angles A and B are congruent", we write $\angle A \cong \angle B$. The symbol $\cong$ represents sameness without mentioning values.

$\cong$是全等符号(用来对比图形，但不用来对比数字)。

Figure 12-37 illustrates that $\angle A \cong \angle B$ and $\angle C \cong \angle D$. The single blue tick mark and double blue tick marks imply that the angles with the same number of tick marks are congruent.

图 12-37 展示了$\angle A \cong \angle B$ 和$\angle C \cong \angle D$。$\angle A$ 和$\angle B$，$\angle C$ 和$\angle D$ 是等角。

Properties: The statements "$\angle A \cong \angle B$" and "$m\angle A = m\angle B$" imply one another.

Adjacent Angles 邻角

n. [ə'dʒeɪsənt 'æŋgəls]

Definition: Two angles are adjacent angles if they have the common vertex and a common side but do not overlap. 若两个角的顶点相同且共用一条边但不重叠，则它们是邻角。

Examples: As shown in Figure 12-38, $\angle AOB$ and $\angle BOC$ are adjacent angles.

$\angle BOC$ and $\angle COD$ are adjacent angles.

$\angle AOB$ and $\angle BOD$ are adjacent angles.

$\angle AOC$ and $\angle COD$ are adjacent angles.

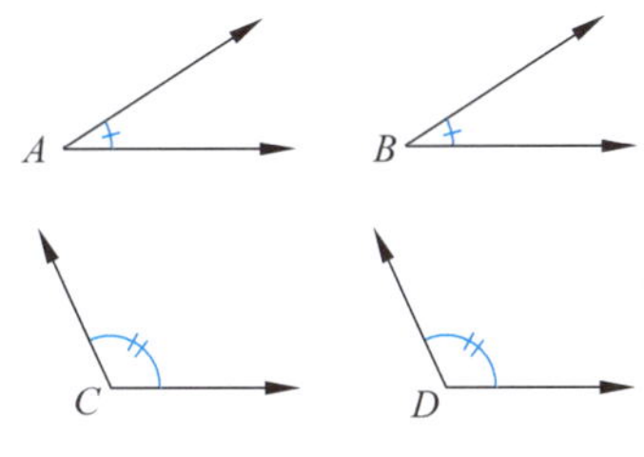

Figure 12-37

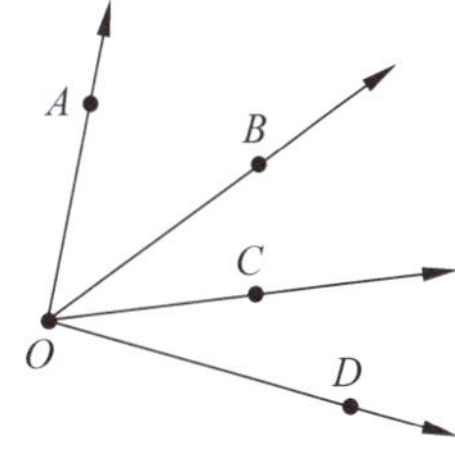

Figure 12-38

Complementary Angles 余角

n. [ˌkɑmplɪ'mentəri 'æŋgəls]

Definition: Two angles are complementary angles if their sum of measures is 90°. Each angle is the **complement** (abbr. of complementary angle) of the other. 若两个角的度数之和为90°，则它们互为余角，简称互余，即一个角是另外一个角的余角。

Properties: If $\angle A$ and $\angle B$ are complementary, and $\angle B$ and $\angle C$ are complementary, then $\angle A \cong \angle C$. 若$\angle A$ 和$\angle B$ 互余，$\angle B$ 和$\angle C$ 互余，则$\angle A \cong \angle C$。

Proofs: To prove the property, we are given that $m\angle A + m\angle B = 90°$ and $m\angle B + m\angle C = 90°$. By transitive, $m\angle A + m\angle B = m\angle B + m\angle C$, from which $m\angle A = m\angle C$, or $\angle A \cong$

$\angle C$. 可用互余的定义和传递性证明这个性质。

Examples: 1. As shown in Figure 12-39, $\angle AOB$ and $\angle BOD$ are complementary angles, and $\angle AOC$ and $\angle COD$ are complementary angles.

2. Complementary angles are not necessarily adjacent angles. 余角不一定是邻角。

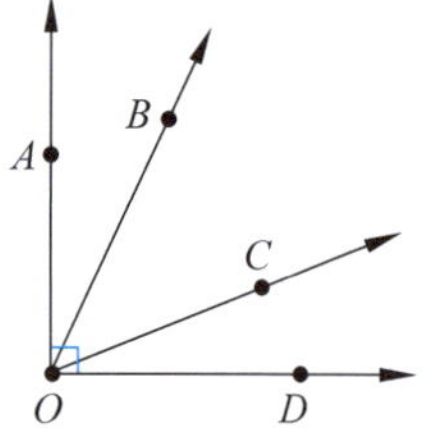

Figure 12-39

Questions: If $\angle A$ and $\angle B$ are complementary angles, and $m\angle A = 25°$, what is the value of $m\angle B$?

Answers: $m\angle B = 90° - m\angle A = 65°$.

Supplementary Angles 补角

n. [ˌsʌplə'mentəri 'æŋgəls]

Definition: Two angles are supplement angles if their sum of measures is 180°. Each angle is the **supplement** (abbr. of supplementary angle) of the other. 若两个角的度数之和为 180°，则它们互为补角，简称互补，即一个角是另外一个角的补角。

Properties: If $\angle A$ and $\angle B$ are supplementary, and $\angle B$ and $\angle C$ are supplementary, then $\angle A \cong \angle C$. 若$\angle A$ 和$\angle B$ 互补，$\angle B$ 和$\angle C$ 互补，则$\angle A \cong \angle C$。

Proofs: To prove the property, we are given that $m\angle A + m\angle B = 180°$ and $m\angle B + m\angle C = 180°$. By transitive, $m\angle A + m\angle B = m\angle B + m\angle C$, from which $m\angle A = m\angle C$, or $\angle A \cong \angle C$. 可用互补的定义和传递性证明这个性质。

Examples: 1. As shown in Figure 12-40, $\angle AOB$ and $\angle BOD$ are supplementary angles. $\angle AOC$ and $\angle COD$ are supplementary angles.

2. Supplementary angles are not necessarily adjacent angles. 补角不一定是邻角。

Questions: If $\angle A$ and $\angle B$ are supplementary angles, and $m\angle A = 120°$, what is $m\angle B$?

Answers: $m\angle B = 180° - m\angle A = 60°$.

Linear Pair 邻补角

n. ['lɪniər peər]

Definition: $\angle A$ and $\angle B$ is a linear pair if they are adjacent angles and supplementary angles. 若$\angle A$ 和$\angle B$ 是邻角并且互补，则称它们为邻补角。

Examples: As shown in Figure 12-41, $\angle AOB$ and $\angle BOD$ form a linear pair. $\angle AOC$ and $\angle COD$ also form a linear pair.

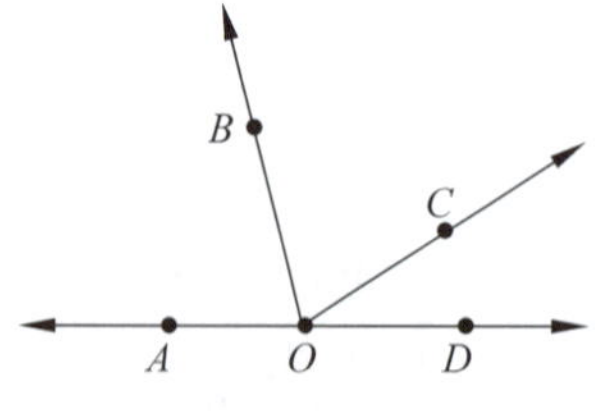

Figure 12-40

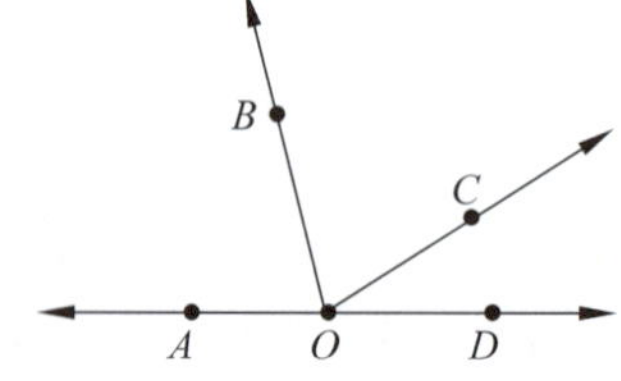

Figure 12-41

12.3.5 Summary 总结

1. (1) An angle is the amount of turn between two rays (known as the arms/sides) that have a

common endpoint (known as the vertex). 角是两条具有共同端点的射线的转动空间大小。这两条射线叫作边，端点叫作顶点。

(2) $\angle ABC$ can be named as $\angle B$ for short, when there is no ambiguity. 当不存在误解时，$\angle ABC$ 可被简写成 $\angle B$。

(3) Angles are measured and classified by degrees or radians. Angles are measured using protractors. 角可通过度数或弧度测量和分类。用量角器测量度数。

(4) The measures of angles can be represented by decimals or DMS Notation. See DMS Notation in Section 12.3.2 for their conversion. 角的度数可用小数或度分秒表示法表示。两者之间的转换见 12.3.2 节度分秒表示法的词条。

2. Table 12-2 shows the types of angles:

Table 12-2

Angle Category 角的种类	Measure (represented by d) 度数
Acute Angle 锐角	$0° < d < 90°$
Right Angle 直角	$d = 90°$
Obtuse Angle 钝角	$90° < d < 180°$
Straight Angle 平角	$d = 180°$
Reflex Angle 优角	$180° < d < 360°$
Round Angle 周角	$d = 360°$

3. Relationship of Two Angles (can satisfy zero or more of the following):
两角之间的关系：

- congruent 全等
- adjacent 相邻
- complementary 余角
- supplementary 补角
- linear pair 邻补角

12.4 Properties of Shapes 图形的性质

12.4.1 Optical Illusions 视觉幻象

Optical Illusion 视觉幻象

n. [ˈɑptɪkəl ɪˈluʒən]

Definition: A visual stimulus that is perceived by the eyes and then interpreted by the brain in a way that is different from the reality.

视觉幻象是指一个被眼睛感知后被大脑分析的视觉刺激物，分析结果通常与现实有差别。

Examples: As shown in Figure 12-42, the two blue line segments are equal in length, but many people think (at first glance) that the top line segment is shorter than the bottom one due to the black distracting arrows.

This is one of many examples.

如图 12-42 所示，两条蓝色线段的长度一样，但是许多人看了第一眼后会觉得上面的线段比下面的线段短，原因是被黑色箭头误导了。

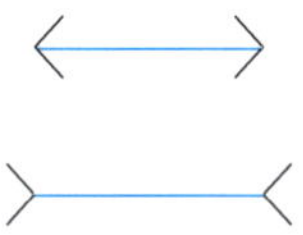

Figure 12-42

12.4.2 Symmetry Introduction 对称介绍

Symmetry 对称

n. [ˈsɪmɪtri]

Definition: The state for which one shape completely overlaps itself by flipping or turning it. The most common symmetries are line symmetry and rotational symmetry. 对称是指图形在翻转或旋转后与它本身完全重合的情景。最常见的对称是轴对称和旋转对称。

Center 中心

n. [ˈsentər]

Definition: The "middle point" of a shape in some sense (usually applied for shapes that have symmetry). The centers of circles and regular polygons are strictly defined.
中心是指图形(通常这些图形都对称)"中间的点"。圆和正多边形的中心有更加严谨的定义。

Examples: 1. As shown in Figure 12-43, the center of a circle is drawn.
图 12-43 画出了圆的中心。

2. As shown in Figure 12-44, the center of a regular pentagon is drawn.
图 12-44 画出了正五边形的中心。

3. As shown in Figure 12-45, the center of a polygon that has line symmetry is drawn. The lines of symmetry are drawn in blue dotted lines.
图 12-45 画出了一个带有线对称的多边形的中心。对称轴以蓝色虚线表示。

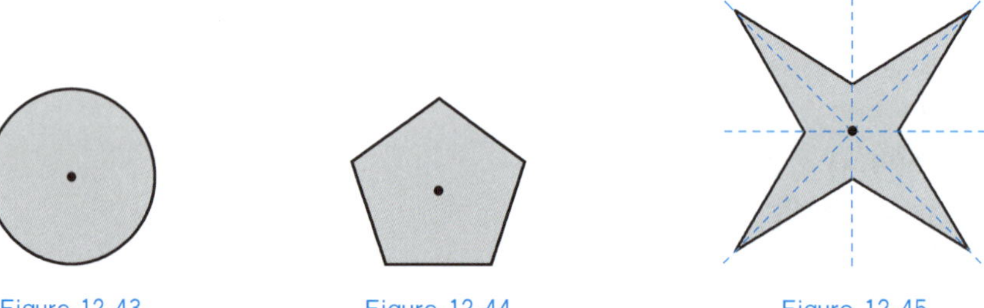

Figure 12-43 Figure 12-44 Figure 12-45

4. As shown in Figure 12-46, the center of an irregular shape that has line symmetry is drawn. 图 12-46 画出了一个带有线对称的不规则图形的中心。

5. Figure 12-47 shows a shape that does not have a center. 图 12-47 所示的图形没有中心。

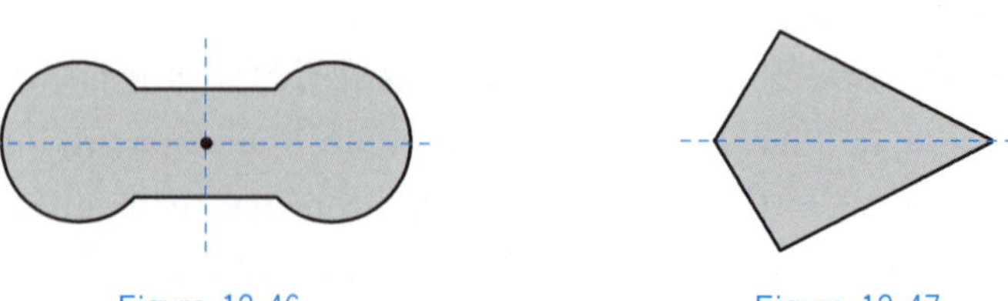

Figure 12-46 Figure 12-47

12.4.3 Types of Symmetry　对称种类

Line Symmetry　轴对称

n. [laɪn ˈsɪmɪtri]

Definition: If shape S is reflected/flipped over a line l and appears unchanged, then S has line symmetry, and l is known as a line of symmetry. See reflection in Section 18.2.2.

As shown in Figure 12-48 and Figure 12-49, the lines of symmetry of the two shapes are drawn in blue dotted lines.

若图形 S 沿着直线 l 反射后与 S 完全重合，则 S 就有轴对称，l 是对称轴，见 18.2.2 节的反射。

图 12-48 和图 12-49 中蓝色虚线是两个图形的对称轴。

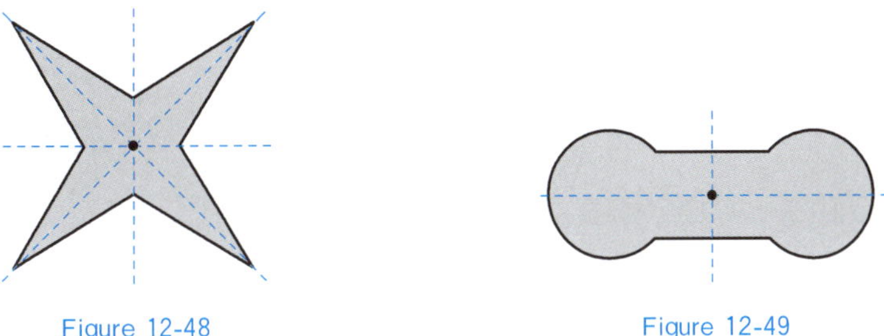

Figure 12-48　　Figure 12-49

Properties: The line(s) of symmetry must pass through the center of the shape.　对称轴必须经过图形的中心。

Examples: (1) An equilateral triangle has 3 lines of symmetry, as shown in Figure 12-50.　等边三角形有 3 条对称轴，如图 12-50 所示。

(2) A square has 4 lines of symmetry, as shown in Figure 12-51.　正方形有 4 条对称轴，如图 12-51 所示。

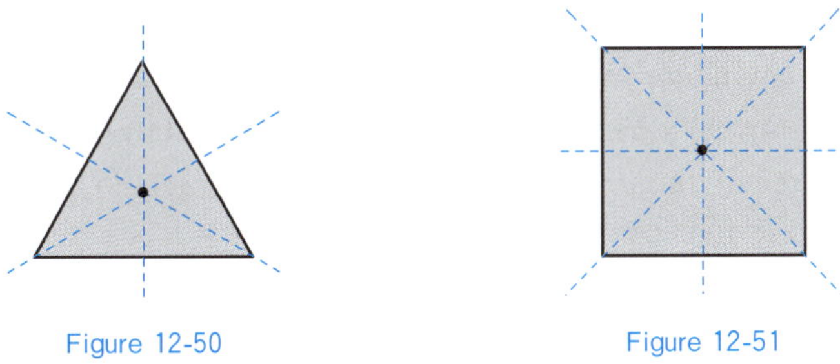

Figure 12-50　　Figure 12-51

(3) A regular pentagon has 5 lines of symmetry, as shown in Figure 12-52.　正五边形有 5 条对称轴，如图 12-52 所示。

(4) A regular n-gon has n lines of symmetry.

If n is odd, then a line connecting a vertex and the center (which also passes through midpoint of the vertex's opposite side) is a line of symmetry. There are n vertices, so there are n lines of symmetry.

If n is even, then there are $n/2$ lines of symmetry passing through the opposite vertices, and $n/2$ lines of symmetry joining the midpoints of the opposite sides. Refer to examples (1) through (3).

一个正 n 边形有 n 条对称轴。

若 n 为奇数,则任何一条连接顶点和中心的线(也经过这个顶点的对边)都是对称轴。因为有 n 个顶点,所以有 n 条对称轴。

若 n 为偶数,则有 $n/2$ 条对称轴通过两个对顶点,$n/2$ 条对称轴通过两条对边的中点。见例子(1)~(3)。

(5) A circle has infinitely many lines of symmetry, as shown in Figure 12-53. 圆有无数条对称轴,如图 12-53 所示。

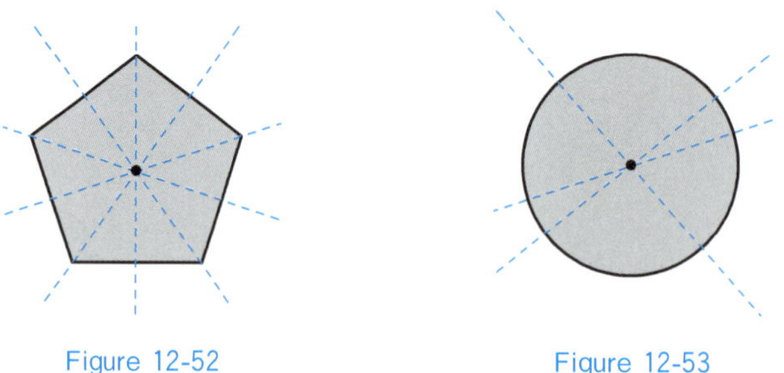

Figure 12-52　　Figure 12-53

Rotational Symmetry　旋转对称

n. [roʊ'teɪʃən 'sɪmɪtri]

Definition: If shape S is rotated/turned around its center by $d°$, where $d < 360$, and appears unchanged, then S has rotational symmetry. The minimum value of d is the angle of rotation. See rotation in Section 18.2.2. 若图形 S 沿着它的中心旋转 $d°$ ($d<360$) 后与 S 完全重合,则 S 有旋转对称。d 的最小值是图形的旋转角度,见 18.2.2 节的旋转。

Properties: 360° is an integer multiple of angle of rotation. 360°是旋转角度的整数倍数。

Examples: (1) An equilateral triangle's angle of rotation is 120°. 等边三角形的旋转角度是 120°。As shown in Figure 12-54, when $\triangle ABC$ rotates 120° counterclockwise around its center O, the image is $\triangle A'B'C'$. 如图 12-54 所示,当$\triangle ABC$ 沿着它的中心 O 逆时针旋转 120°,像是$\triangle A'B'C'$。

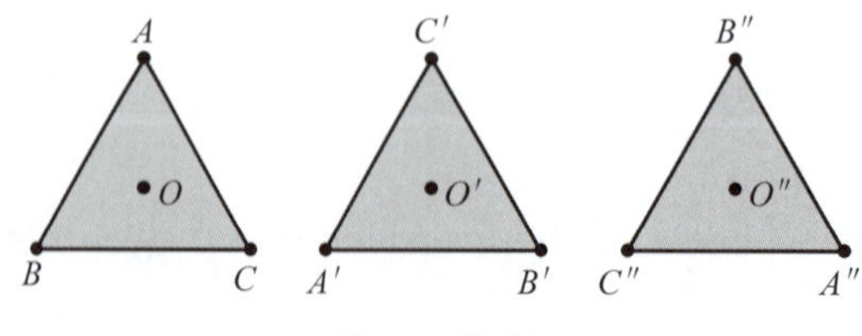

Figure 12-54

When $\triangle A'B'C'$ rotates 120° counterclockwise around its center O', the image is $\triangle A''B''C''$. 当$\triangle A'B'C'$沿着它的中心 O'逆时针旋转 120°,像是$\triangle A''B''C''$。

When $\triangle A''B''C''$ rotates 120° counterclockwise around its center O'', the image is $\triangle A'''B'''C'''$ (not drawn), which is the same as $\triangle ABC$. 当$\triangle A''B''C''$沿着它的中心O''逆时针旋转 120°，像是$\triangle A'''B'''C'''$（没有画出），这跟$\triangle ABC$的顶点方位是一模一样的。

(2) A square's angle of rotation is 90°. 正方形的旋转角度是 90°。

(3) A regular pentagon's angle of rotation is 72°. 正五边形的旋转角度是 72°。

(4) A regular n-gon's angle of rotation is 360°/n. 正 n 边形的旋转角度是 360°/n。

(5) A circle angle of rotation is infinitesimally small (approaches to 0°). 圆的旋转角度无穷小（接近 0°）。

12.4.4 Summary 总结

1. Symmetry is one of: (1) line symmetry, (2) rotational symmetry. 对称有两种：(1) 轴对称，(2) 旋转对称。

 (1) If shape S is reflected/flipped over a line l and appears unchanged, then S has line symmetry, and l is known as the line of symmetry. See reflection in Section 18.2.2. The line of symmetry must pass through the center of S. 若图形 S 沿着直线 l 反射后与 S 完全重合，则 S 就有轴对称，l 是对称轴，见 18.2.2 节的反射。对称轴必须经过 S 的中心。

 (2) If shape S is rotated/turned around its center by d°, where $d<360$, and appears unchanged, then S has rotational symmetry. The minimum value of d is the angle of rotation. See rotation in Section 18.2.2.

 360° is an integer multiple of angle of rotation. 若图形 S 沿着它的中心旋转 d°($d<360$)后与 S 完全重合，则 S 有旋转对称。d 的最小值是图形的旋转角度，见 18.2.2 节的旋转。

 360° 是旋转角度的整数倍数。

2. (1) A regular n-gon has n lines of symmetry. 正 n 边形有 n 条对称轴。

 (2) A regular n-gon's angle of rotation is 360°/n. 正 n 边形的旋转角度是 360°/n。

3. (1) A circle has infinitely many lines of symmetry. 圆有无数条对称轴。

 (2) A circle angle of rotation is infinitesimally small (approaches to 0°). 圆的旋转角度无穷小（接近 0°）。

12.5 Geometry Tool Kit 几何工具套装

12.5.1 Construction Tools 画图工具

Straightedge 直尺

n. [ˌstreɪtedʒ]

Definition: A tool that is used to transcribe straight lines, or measure the straightness of curves. 直尺是一个用来复制直线和测量曲线平直度的工具。

Properties: If a straightedge has even markings along the edge, then it is called a ruler, which can also be used to measure the lengths of the line segments. 若直尺的边缘有均匀的刻度，它就是刻度尺，能测量线段的长度。

Compass 圆规

n. ['kʌmpəs]

Definition: A tool that is used to construct circles and arcs. 圆规是用来画圆和弧的工具。

12.5.2 Other Tools 其他工具

Triangle Set 三角板

n. ['traɪˌæŋgəl set]

Definition: A set of educational tools that is used to study triangles. A triangle set consists of two triangles, one of which is 45° - 45° - 90°, and the other is 30° - 60° - 90°. 三角板是一套用来学习三角形的教学工具,包括两个三角尺,一个是 45° - 45° - 90° 三角尺,另一个是 30° - 60° - 90° 三角尺。

Protractor 量角器

n. [proʊ'træktər]

Definition: A tool that is used to measure angles. Protractors are transparent, and are usually semicircular (measures angles from 0° through 180°). 量角器是用来测量角度的工具,通常是透明的半圆形的,用来测量 0° ~180° 的角。

13. Mathematical Reasoning and Proofs 数学推理与证明

13.1 Introduction to Statements 命题的介绍

13.1.1 Types of Statements 命题的种类

(Logical) Statement 命题

n. [('lɑdʒɪkəl) 'steɪtmənt]

Definition: A sentence whose truth value is exactly one of true or false. See true statement and false statement. 命题是指可以判断真假的语句。见真命题和假命题的词条。

Phrases: true~ 真命题, false~ 假命题, conditional~ 条件命题, biconditional~ 双条件命题

True Statement 真命题

n. [tru 'steɪtmənt]

Definition: A sentence whose truth value is true. 判断为真的语句称为真命题。

Examples: 1. All dogs have 4 legs.

2. All right angles are 90°.

3. All triangles have three sides.

False Statement 假命题

n. [fɔls 'steɪtmənt]

Definition: A sentence whose truth value is false. 判断为假的语句称为假命题。

Examples: 1. All dogs have exactly 3 legs.

2. No person in the world is named "John".

3. All shirts are blue.

13.1.2 Quantifiers 量词

Quantifier 量词

n. ['kwɑntəˌfɑɪər]

Definition: One of universal quantifier or existential quantifier。

量词分为全称量词和存在量词。

Figure 13-1 shows the relationship between universal quantifiers and existential quantifiers—existential quantifiers are subcategories of universal quantifiers.

图 13-1 展示了全称量词与存在量词的关系。存在量词是全称量词的子类别。

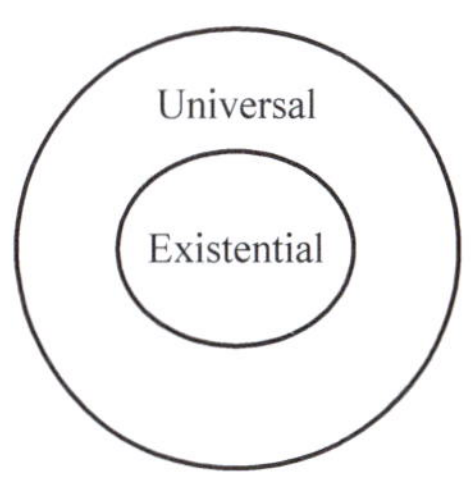

Figure 13-1

Universal Quantifier 全称量词

n. [junə'vɜrsəl 'kwɑntəˌfɑɪər]

Definition: A logic symbol that makes a statement about the set of values **such that all elements in the set** have a certain property as described. 全称量词是指在一个命题中描述的是一个集合里所有的元素都满足某个性质的逻辑符号。

Notation: The statement "$\forall a \in A$" is read as "for every element in set A" or "for all elements in set A".

写作: $\forall a \in A$。

读作: 所有在集合 A 里的元素。

Examples: 1. Suppose set $A=\{2,3,5,7,11,13\}$. We will form some true statements using the universal quantifier.

(1) For every element in A, it is a prime.

(2) All elements in A are positive.

(3) $\forall a \in A$ is an integer.

根据集合 $A=\{2,3,5,7,11,13\}$,可用全称量词表达一些真命题。

(1) 所有在集合 A 里的元素都是质数。

(2) 所有在集合 A 里的元素都是正数。

(3) 所有在集合 A 里的元素都是整数。

2. We will express the same sentence using the universal quantifier in different ways. Each of the sentences below has the same meaning. 以下命题的意思(所有狗都有 4 条腿)相同,只是全称量词用了不同的表达形式。

(1) All dogs have 4 legs.

(2) Every dog has 4 legs.

(3) $\forall d \in$ {all dogs} has 4 legs.

Existential Quantifier 存在量词

n. [ˌegzɪ'stenʃəl 'kwɑntəˌfɑɪər]

Definition: A logic symbol that makes a statement about the set of values **such that at least one element in the set** has a certain property as described. 存在量词是指在一个命题中描述的是一个集合里至少有一个元素满足某个性质的逻辑符号。

Notation: The statement "$\exists a \in A$" is read as "for some elements in set A" or "for at least one element in set A".

写作:$\exists a \in A$。

读作:集合 A 里存在某个/某些元素。

Examples: 1. Suppose set $A=\{1,2,3,4,5,6,7,8\}$. We will form some true statements using the existential quantifier.

(1) There exists a prime in A.

(2) At least one element in A is odd.

(3) $\exists a \in A$ is an integer.

On statement (3), we can change the existential quantifier to universal. What's true about universal must be true about existential.

根据集合 $A=\{1,2,3,4,5,6,7,8\}$,可用存在量词表达一些真命题。

(1) 某个在集合 A 里的元素是质数。

(2) 至少有一个在集合 A 里的元素是奇数。

(3) 集合 A 里存在一个元素是整数。

对于命题(3),可把存在量词改为全称量词。在含有全称量词的真命题中,量词改为存在量词,命题也必然为真。

2. We will express the same sentence using the existential quantifier in different ways. Each of the sentences below has the same meaning. 以下命题的意思(至少有一只狗是黑色的)相同,只是存在量词用了不同的表达形式。

(1) Some dogs are black.

(2) At least one dog is black.

(3) $\exists d \in$ {all dogs} is black.

13.1.3 Counterexamples 反例

Counterexample 反例

n. ['kaʊntərɪɡˌzɑːmpəl]

Definition: An example that proves a statement false. 反例是指可以证明一个假命题是假的例子。

Properties: 1. For a false statement that contains a universal quantifier, a counterexample is effective to prove the statement is false. 对一个带有全称量词的假命题,反例能有力地证明该命题是假的。

2. For a false statement that contains an existential quantifier, a counterexample does not exist, but the statement is false nevertheless. Showing that every subject

described in the statement is false is the only way to disprove such statement.　对一个带有存在量词的假命题，反例不存在，但是该命题仍然是假的。要证明这个命题是假的，只能用穷举法。

Examples：1. Statement：All dogs have exactly 3 legs.

The statement is false. To find a counterexample，simply find a dog with 4 legs.

2. Statement：No people in the world is named “John”.

The statement is false. “John Hancock” would be a counterexample.

3. Statement：All shirts are blue.

The statement is false. To find a counterexample，simply find a shirt that is not blue.

4. Statement：Some cats can fly (in real life).

The statement is false. In order to prove it is false，we cannot find a counterexample (we need to show that **every** cat cannot fly). See Property 2.

13.1.4　Conditional Statements　条件命题

Conditional Statement　条件命题

n. [kən'dɪʃənəl 'steɪtmənt]

Definition：A logical statement that consists of two parts：a hypothesis and a conclusion，written in the if-then form.

The hypothesis is the precondition.

The conclusion is a statement given that the hypothesis is true.

条件命题是指带有假设和结论的命题，用“若……则……”的形式表示。

假设就是前提条件；结论就是已知假设是正确时所得出的论断。

Notation：If (hypothesis)，then (conclusion).

Or，if p then q.

The hypothesis is usually written as p and the conclusion is usually written as q.

写作：若(假设)，则(结论)。

假设经常用 p 表示，结论通常用 q 表示。

Properties：1. To determine the truth value of a conditional statement，assume that the hypothesis is true and determine the truth value of the conclusion.　判断一个条件命题的真假值，首先假设前提条件是真的，再判断结论的真假。

2. Every statement can be written in the if-then form (conditional statement). Both the statement and its converted conditional statement have the same truth value.　所有命题都能被写成“若……则……”的条件命题形式。该命题和它被改写的条件命题的真假值一样。

Questions：Rewrite each of the following as a conditional statement and determine its truth value. If the statement is false，give a counterexample.　把下列命题改写成条件命题，并判断真假。若命题是假命题，给出一个反例。

1. All dogs have 4 legs.
2. All positive integers are greater than 1.
3. Some dogs are black.
4. Some cats can fly in real life.

Answers：1. If an animal is a dog，then it has 4 legs.

The statement is true.

2. If a number is a positive integer, then it is greater than 1.

   The statement is false. A counterexample would be 1.

3. If an animal is a dog, then it **may be** black.

   The statement is true, since we can find a black dog to prove it.

4. If an animal is a cat, then it may have the ability to fly.

   The statement is false, since we know that no cats can fly.

Biconditional Statement 双条件句

n. [ˌbaɪkən'dɪʃənəl 'steɪtmənt]

Definition: A logical statement asserting that the hypothesis depends on, and is dependent on, the conclusion. 双条件句是指"假设取决于和被取决于结论"的命题。

Notation: (Hypothesis) **if and only if** (conclusion).

Or, p if and only if q.

假设是真的**当且仅当**结论是正确的。

Properties: The biconditional statement "p if and only if q" is a combination of the two statements below.

(1) If p then q. (known as the "forward direction")

(2) If q then p. (known as the "backward direction")

To prove the biconditional statement is correct, need to show both (1) and (2) are correct.

双条件句是下面两个命题的结合。

(1) 若 p,则 q。(向前方向)

(2) 若 q,则 p。(向后方向)

证明一个双条件句是真的,需要证明(1)和(2)都是真的。

Questions: Rewrite each of the following biconditional statements as two conditional statements and determine its truth value. 把每个双条件句都改写成两个命题,并判断双条件句的真假。

1. An angle is a right angle if and only if its measure is 90°. 一个角是直角当且仅当它的度数是 90°。
2. An angle is an acute angle if and only if its measure is 30°. 一个角是锐角当且仅当它的度数是 30°。
3. An animal has four legs if and only if it is a dog. 一个动物是狗当且仅当它有四条腿。

Answers: 1. (1) If an angle is a right angle, then its measure is 90°. 若一个角是直角,则它的度数是 90°。

(2) If an angle's measure is 90°, then it is a right angle. 若一个角的度数是 90°,则它是一个直角。

Both statements (1) and (2) are true. So the biconditional statement is true.

因为命题(1)和(2)都是真的，所以双条件句是真的。

2. (1) If an angle is an acute angle, then its measure is 30°.　若一个角是锐角，则它的度数是 30°。

(2) If an angle's measure is 30°, then it is an acute angle.　若一个角的度数是 30°，则它是一个锐角。

The statement (1) is false, since a 60°-angle is a counterexample. Therefore, the biconditional statement is false.　命题(1)是假的，因为 60° 角是个反例。所以双条件句是假的。

3. (1) If an animal has four legs, then it is a dog.　若一个动物有四条腿，则它是狗。

(2) If an animal is a dog, then it has four legs.　若一个动物是狗，则它有四条腿。

The statement (1) is false, since a cat is a counterexample. Therefore, the biconditional statement is false.　命题(1)是假的，因为猫是个反例。所以双条件句是假的。

13.1.5 From One Statement to Another　从一个命题到另一个命题

Negation　否定

n. [nɪˈgeɪʃən]

视频 22

Definition: The negation of a statement P is a statement that says "not P".　命题 P 的否定所表达的命题为"不是 P"。

Notation: The negation of statement P is written as $\neg P$.

写作：$\neg P$。

读作：非 P。

Properties: 1. Exactly one of statements P and $\neg P$ is true, and the other is false.

P 和 $\neg P$ 两个命题一个是真，一个是假。

2. To negate a statement P, simply add the word "not" up front.

Or, to negate P more precisely, do the following:

(1) Change the quantifier contained in P from one to the other (from universal to existential, or from existential to universal).

(2) Write the opposite in the rest of the sentence.

否定一个命题，在前面加"不"字就可以。

更加精确地否定一个命题 P，可以按照以下方法操作：

(1) 改变量词(若是全称量词，则改成存在量词；若是存在量词，则改成全称量词)。

(2) 改动句子的余下部分，使其与原句相反。

3. For all statements P, $\neg(\neg P) = P$.　命题否定的否定和命题本身是相等的。

Examples: Although the truth values of negations are the same, there is more than one way to negate a statement P. Examples 1-2 illustrate Properties 1 and 2.　虽然命题的否定的真假值是相同的,但有多种方法否定命题 P。下面的例 1 和例 2 展示性质 1 和性质 2。

1. P: All dogs are red.　所有狗都是红色的。

   $\neg P$: **Not** all dogs are red.　不是所有狗都是红色的。

   $\neg P$: **Some** dogs are **not** red.　有的狗不是红色的。

   P is false, and $\neg P$ is true.　P 是假命题,$\neg P$ 是真命题。

2. P: Some marbles are green.　有的弹珠是绿色的。

   $\neg P$: **Not** some marbles are green.　弹珠没有绿色的。

   $\neg P$: **All** marbles are **not** green.　所有弹珠都不是绿色的。

   P is true, and $\neg P$ is false.　P 是真命题,$\neg P$ 是假命题。

Converse　逆命题

n. [ˈkɑnvɜrs]

Definition: The converse of a conditional statement P is a statement that is formed by switching the hypothesis and conclusion of P.　命题 P 的逆命题是指颠倒 P 的假设与结论的命题。

Notation: Conditional statement P: If p then q.

The converse of P: If q then p.

条件命题 P:若 p 则 q。

P 的逆命题:若 q 则 p。

Properties: Knowing the truth value of a conditional statement tells us nothing about the truth value of its converse!　条件命题的真假值与它的逆命题的真假值无关。

Questions: For each of the statements below, (1) Give the converse. (2) Give the truth values for both the statement and the converse.　对于下面的每一个命题,(1) 给出逆命题,(2) 给出它与它的逆命题的真假值。

1. Statement: If $m\angle A = 90°$, then $\angle A$ is a right angle.

   命题:若 $m\angle A = 90°$,则 $\angle A$ 是直角。

2. Statement: If $m\angle A = 30°$, then $\angle A$ is an acute angle.

   命题:若 $m\angle A = 30°$,则 $\angle A$ 是锐角。

3. Statement: If an integer is greater than 1, then it is a prime.

   命题:若一个整数大于 1,则它是质数。

4. Statement: If a positive integer is divisible by 2, then it is divisible by 3.

   命题:若一个正整数能被 2 整除,则它能被 3 整除。

Answers:

1. Converse: If $\angle A$ is a right angle, then $m\angle A = 90°$.

   Both the statement and the converse are true by definition of right angles.

2. Converse: If $\angle A$ is an acute angle, then $m\angle A = 30°$.

   The statement is true since the measure of any acute angle is between 0° and 90°, exclusive.

   The converse is false. One counterexample is $m\angle A = 60°$.

3. Converse: If an integer is a prime, then it is greater than 1.
   The statement is false. One counterexample is 4.
   The converse is true by the definition of primes.
4. Converse: If a positive integer is divisible by 3, then it is divisible by 2.
   The statement is false. One counterexample is 2.
   The converse is false. One counterexample is 3.

Contrapositive　对换句/逆否命题

n. [ˌkɒntrəˈpɒzɪtɪv]

Definition: The contrapositive of a conditional statement P is a statement that is formed by switching the hypothesis and conclusion of P, then negating both.　命题 P 的逆否命题是一个命题——把 P 的假设和结论分别换成结论的否定和假设的否定。

Notation: Conditional statement P: If p then q.
The contrapositive of P: If $\neg q$ then $\neg p$.
条件命题 P：若 p 则 q。
P 的逆否命题：若 $\neg q$ 则 $\neg p$.

Properties: The truth values of the conditional statement P and its contrapositive are the same. **Moreover, P and the contrapositive of P are equivalent statements.**　条件命题 P 与它的逆否命题真假值相同。它们是等价命题。

Proofs: One informal way is to think of "if p then q" as "p causes q". Therefore, if q does not happen, then p will not happen.　要证明为什么 P 和它的逆否命题是等价的，一个非正式的方法是把"若 p 则 q"想成"p 导致 q"。若 q 不发生，则 p 也不会发生。

Questions: For each of the statements below, (1) Give the contrapositive. (2) Give the truth values for both the statement and the contrapositive. These statements are the same as those in the question section in converse.　对于下面的每一个命题，(1) 给出逆否命题，(2) 给出它与它的逆否命题的真假值。下面的命题跟逆命题词条的问题是相同的。

1. Statement: If $m\angle A=90°$, then $\angle A$ is a right angle.
2. Statement: If $m\angle A=30°$, then $\angle A$ is an acute angle.
3. Statement: If an integer is greater than 1, then it is a prime.
4. Statement: If a positive integer is divisible by 2, then it is divisible by 3.

Answers:

1. Contrapositive: If $\angle A$ is **not** a right angle, then $m\angle A\neq 90°$.
   Both the statement and the contrapositive are true.
2. Contrapositive: If $\angle A$ is **not** an acute angle, then $m\angle A\neq 30°$.
   Both the statement and the contrapositive are true.
3. Contrapositive: If an integer is **not** a prime, then it is **not** greater than 1.
   Both the statement and the contrapositive are false.
   A counterexample for both is 4.
4. Contrapositive: If a positive integer is **not** divisible by 3, then it is **not** divisible by 2.
   Both the statement and the contrapositive are false.
   A counterexample for both is 2.

Inverse 否命题

n. [ˈɪnvɜrs]

Definition: The inverse of a conditional statement P is a statement that is formed by negating both the hypothesis and conclusion of P. 命题 P 的否命题是一个命题——把 P 的假设和结论分别换成假设的否定和结论的否定。

Notation: Conditional statement P: If p then q.

The inverse of P: If $\neg p$ then $\neg q$.

条件命题 P：若 p 则 q。

P 的否命题：若$\neg p$ 则 $\neg q$。

Properties: 1. The inverse of P is the contrapositive of the converse of P. Therefore, the inverse of P and the converse of P have the same truth value. P 的否命题就是 P 的逆命题的逆否命题。P 的否命题与 P 的逆命题的真假值相同。

2. Knowing the truth value of a conditional statement tells us nothing about the truth value of its inverse! 一个条件命题的真假值与它的否命题的真假值无关。

Proofs: To prove the properties, see the definitions and notations of converse and contrapositive. 要证明以上性质，见逆命题和逆否命题的定义和标记方式。

Questions: For each of the statements below, (1) Give the inverse, (2) Give the truth values for both the statement and the inverse. These statements are the same as those in the question section in converse. 对于下面的每一个命题，(1) 给出否命题，(2) 给出它与它的否命题的真假值。下面的命题跟逆命题词条的问题是相同的。

1. Statement: If $m\angle A=90°$, then $\angle A$ is a right angle.
2. Statement: If $m\angle A=30°$, then $\angle A$ is an acute angle.
3. Statement: If an integer is greater than 1, then it is a prime.
4. Statement: If a positive integer is divisible by 2, then it is divisible by 3.

Answers: 1. Inverse: If $m\angle A\neq 90°$, then $\angle A$ is **not** a right angle.

Both the statement and the inverse are true.

2. Inverse: If $m\angle A\neq 30°$, then $\angle A$ is **not** an acute angle.

The statement is true, but the inverse is false. A counterexample of the inverse is $m\angle A=40°$.

3. Inverse: If an integer is **not** greater than 1, then it is **not** a prime.

The statement is false. A counterexample would be 4.

The inverse is true.

4. Inverse: If a positive integer is **not** divisible by 2, then it is **not** divisible by 3.

Both the statement and the inverse are false. A counterexample of the statement would be 2, and a counterexample of the inverse is 3.

Table 13-1 shows the relationship between statement P and its negation, converse, contrapositive, and inverse.

表 13-1 展示了命题 P 与它的否定、逆命题、逆否命题、否命题之间的关系。

Table 13-1

Conditional Statement　条件命题	Notation　标记方式
Statement P　命题 P	If p then q.
Converse of P　命题 P 的逆命题	If q then p.
Contrapositive of P　命题 P 的逆否命题	If $\neg q$ then $\neg p$.
Inverse of P　命题 P 的否命题	If $\neg p$ then $\neg q$.

13.1.6　Summary　总结

1. A logical statement has a truth value, which classifies it to be one of "true statement" and "false statement".　逻辑命题有真假值，据此可分类成"真命题"或"假命题"。
2. There are two types of quantifiers: universal (for all) and existential (for some).　量词有两类：全称量词(所有)和存在量词(某个/某些)。
3. Counterexamples are powerful tools to disprove statements with universal quantifiers—one counterexample is sufficient to prove such statement to be false.　反例是证明带有全称量词的命题是假命题的有力工具。只需要一个反例就可以证明这种命题是假命题。
4. A conditional statement consists of two parts, hypothesis and conclusion. It is of the form "If (hypothesis), then (conclusion)" (if-then form). To determine a conditional statement's truth value, assume that the hypothesis is true, and determine the truth value of the conclusion.　条件命题包含两部分：假设和结论。它以"若(假设)，则(结论)"的方式出现("若—则"形式)。判断一个条件命题的真假，先假设前提条件是真的，再判断结论的真假。
5. A biconditional statement is of the form (hypothesis) **if and only if** (conclusion). A biconditional statement is true when the following two statements are true:
   (1) If hypothesis, then conclusion (forward direction).
   (2) If conclusion, then hypothesis (backward direction).
   双条件句以"(假设)当而且当(结论)"的形式出现。若一个双条件句是真命题，则以下两个命题都是真的。
   (1) 若(假设)，则(结论)。(向前方向)
   (2) 若(结论)，则(假设)。(向后方向)
6. Refer to Table 13-1 in Section 13.1.5.　见13.1.5节的表13-1。
   (1) A statement is equivalent to its contrapositive.　一个命题与它的逆否命题是等价的。
   (2) The converse of a statement P is equivalent to the inverse of P because they are contrapositives.　命题 P 的逆命题与 P 的否命题是等价的，因为它们互为逆否命题。

13.2　Postulates, Lemmas, Theorems, and Corollaries　公理、引理、定理与推论

13.2.1　Introduction　介绍

Postulate　公理

n. [ˈpɑstʃəˌleɪt]

Definition: A true statement without formal proofs.　公理是指没有正式证明的真命题。

Examples：1. **Line Segment Addition Postulate**：If C is a point on $\overline{AB}$，then $AB = AC + CB$，as shown in Figure 13-2.

线段相加公理，如图 13-2 所示。

2. **Angle Addition Postulate**：If point B is at the interior of $\angle AOC$，then $m\angle AOB + m\angle BOC = m\angle AOC$，as shown in Figure 13-3.

角相加公理，如图 13-3 所示。

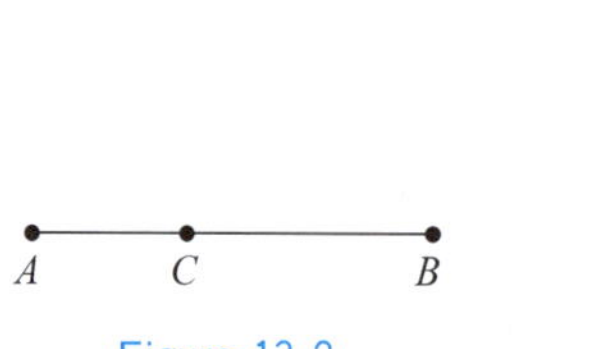

Figure 13-2

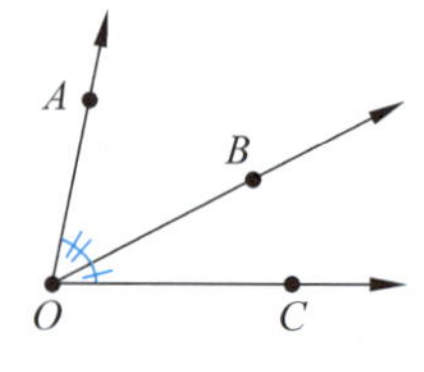

Figure 13-3

3. There is only one line to connect two different given points.

给出两个不重合的点，只有一条直线能连接它们。

Lemma 引理

n. [ˈlɛmə]

Definition：A proved proposition that helps to prove a larger result (theorem). 引理是已被证明的真命题，能用来证明定理。

Theorem 定理

n. [ˈθɪərəm]

Definition：A large result that is proved to be true. 定理是指已被证明的真命题，通常是比较重要的结论。

Phrases：Pythagorean～ 勾股定理

Corollary 推论

n. [ˈkɔːr.ə.ler.i]

Definition：A true result that requires little or no proof from a statement that is already proved. 推论是指能够"简单明了地"从已被证明的命题推出的论断。

Phrases：～of Pythagorean Theorem 勾股定理的推论

13.2.2 Summary 总结

Postulates，lemmas，theorems，and corollaries are true statements for which：

公理、引理、定理、推论是真命题：

1. Postulates do not need formal proofs，while the others do. 公理不需要证明，但是引理、定理、推论需要。
2. Lemma is a small result that helps to prove a large theorem. 引理是小的论述，用于证明大的定理。
3. Theorems are large results. 定理是大的结论。
4. Corollaries are derived from lemmas/theorems with a little or no effort. 推论可容易从引理和定理推断出。

13.3 Proofs 证明

13.3.1 Introduction 介绍

Proof 证明

n. [pruf]

Definition：A deductive argument for a true mathematical statement. The steps of a proof can be traced back to basic postulates or the other lemmas/theorems/corollaries. 证明是对真命题的演绎论证。证明的步骤可被追溯到基本的公理以及引理、定理、推论。

Properties：A complete proof consists of：

完整的证明包括：

Given—The conditions that are assumed to be true (preconditions of the proof).

已知条件——证明的前提条件，假设是真的。

Statements—The steps of the proof，one derived from the other，or from other postulates/lemmas/theorems/corollaries.

命题——证明的步骤，衍生自另一个命题或其他公理/引理/证明/推论。

Reasons—The explanation for each statement. The reason of a statement shows which statements/postulates/lemmas/theorems/corollaries that statement is derived from.

原因——对命题的解释，说明这个命题衍生自哪个命题/公理/引理/证明/推论。

Conclusion—Final step of the proof，which shows what we want to prove is true.

结论——证明的最后一步，展示了要证明的结论是正确的。

13.3.2 Proof Techniques 证明方法

13.3.2.1 Direct Proofs 直接证明

Direct Proof 直接证明

n. [də'rekt pruf]

Definition：A proof technique of proving the conclusions from the given using derived statements，without making any further assumptions. 直接证明是一种证明方法，即不用做任何假设，通过已知条件，使用派生命题证明结论。

Properties：1. A direct proof should prove the conclusion by deriving statements from given/other derived statements in the proof. Moreover，it never assumes the conclusion is true. 直接证明是通过已知条件和其他派生命题，证明最后的结论。直接证明从来不假设结论是真的。

2. There are three main appearances of a direct proof：two-column proof，paragraph proof，and a flowchart proof. They establish the same goal，and only differ in visual appearance. 三种常用的直接证明方法是两列式证明、自然段证明、流程图证明。它们的目标相同，只是表现方式不同。

Examples：We will show the two-column proof，paragraph proof，and a flowchart proof on the same questions in this section. 本节会对同样的问题给出两列式证明、自然段证明、流程图证明。

Two-Column Proof 两列式证明

n. [tu 'kɑləm pruf]

Definition: One method of direct proof that consists of two columns—statements (on the left) and reasons (on the right). Each reason backs up the corresponding statement. 两列式证明是直接证明的一种方法——通过两列(命题在左列,原因在右列)来证明命题。每个原因解释和它对应的命题。

Notation: Table 13-2 shows the chart for a two-column proof.

Table 13-2

Statements (S) 命题	Reasons (R) 原因
…	…

Questions: 1. Prove the statement: If $4(3x-2)+10=110$, then $x=9$.

2. Prove the statement: If $x\neq 0$, then $\frac{x^2}{x}=3x-x-x$.

Answers: 1. Table 13-3 answers Question 1.

Table 13-3

Statements (S) 命题	Reasons (R) 原因
1. $4(3x-2)+10=110$	1. Given
2. $4(3x-2)=100$	2. Subtract 10 from both sides
3. $3x-2=25$	3. Divide both sides by 4
4. $3x=27$	4. Add 2 to both sides
5. $x=9$	5. Divide both sides by 3

From step 3 can be $12x-8=100$ instead (by distributing the 4 in step 2). Either way, the proof is correct.

2. Table 13-4 answers Question 2.

Table 13-4

Statements (S) 命题	Reasons (R) 原因
1. $x\neq 0$	1. Given
2. $\text{LHS}=\frac{x^2}{x}=x$	2. Arithmetic, given that $x\neq 0$
3. $\text{RHS}=3x-x-x=x$	3. Combine like terms 合并同类项
4. $\frac{x^2}{x}=3x-x-x$	4. Transitive Property 传递性

Note that we never assume that $\frac{x^2}{x}=3x-x-x$ is true anywhere in our proof.

Paragraph Proof 自然段证明

n. ['pærəˌɡræf pruf]

Definition: One method of direct proof that consists of a paragraph, which presents a (derived) statement and the corresponding reason in a sentence. 自然段证明是直接证明的一种方法——通过自然段带出了命题和对应的原因。

Questions: 1. Prove the statement: If $4(3x-2)+10=110$, then $x=9$.

2. Prove the statement: If $x\neq 0$, then $\frac{x^2}{x}=3x-x-x$.

Answers: 1. From the given information $4(3x-2)+10=110$, subtracting 10 from both sides gives us $4(3x-2)=100$. Dividing both sides by 4, we have $3x-2=25$. Adding 2 to both sides leads to $3x=27$, and finally dividing both sides by 3 results $x=9$.

2. Since $x\neq 0$, LHS $=x$.

By combining like terms, RHS $=x$.

By transitive property, LHS = RHS, or $\frac{x^2}{x}=3x-x-x$.

Flowchart Proof 流程图证明

n. [ˈfloʊ.tʃɑːrt pruf]

Definition: One method of direct proof that consists of a flowchart, which shows the order of the derived statements from given to the conclusion. 流程图证明是直接证明的一种方法——通过流程图展示了已知条件的派生命题的顺序，直到得出最后的结论。

Questions: 1. Prove the statement: If $4(3x-2)+10=110$, then $x=9$.

2. Prove the statement: If $x\neq 0$, then $\frac{x^2}{x}=3x-x-x$.

Answers: 1. Figure 13-4 gives the proof for Question 1.

2. Figure 13-5 gives the proof for Question 2.

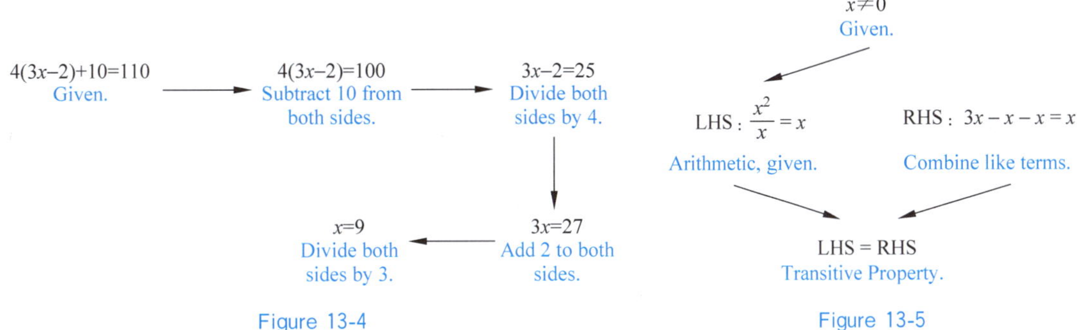

Figure 13-4 Figure 13-5

13.3.2.2 Mathematical Induction 数学归纳法

Mathematical Induction 数学归纳法

n. [mæθˈmætɪkəl ɪnˈdʌkʃən]

Definition: A proof technique that is used to prove a given statement about elements of a well-ordered set (each of its nonempty subsets has a least element), usually it is the set of all natural numbers. 数学归纳法是证明一个有关良序集元素的命题的方法（良序集即是一个集合：它的任意非空子集都有一个最小元素）。常用的良序集是自然数集。

Note: The set of all natural numbers is well-ordered because taking any of its nonempty subsets, such as $\{1,2,5\}$, $\{4,7,10,89\}$, $\cdots$, we can always find the least element in each subset. 自然数集是一个良序集,因为它的任意非空子集都有一个最小元素。

Properties: The mathematical induction on the set S consists of two steps:

1. Base Case: The statement is true when it is evaluated at the smallest element of S.
2. Inductive Case: Assume for every $n \in S$ the statement holds. Will use this assumption (sometimes along with the base case) to show that the statement holds for $n+1 \in S$.

Steps 1 and 2 generate that the statement holds for all elements in S.

Mathematical inductions work like dominos—when the first one falls, the rest of them fall one by one as well.

关于集合 S 的数学归纳法有两步:

1. 基本情况:当代入 S 的最小元素时命题成立。
2. 归纳情况:假设命题代入任意 $n \in S$ 时命题成立,则可以推导出代入 $n+1$ 时命题也成立(有时须借用基本情况)。

根据步骤 1 和 2 可以得出当命题代入任何 S 的元素时,命题均成立。

数学归纳法的原理就像玩多米诺骨牌——最前面一个倒了,后面的会一个接着一个倒。

Examples: For all positive integer n, $1+2+\cdots+n=\dfrac{n(n+1)}{2}$.

Proof by Induction:

1. Base Case: When $n=1$, LHS $=1$, and RHS $=\dfrac{1(1+1)}{2}=1$. The statement holds.
2. Inductive Case: Assume the proposition is true for any positive integer n. We will show that the proposition is true for $n+1$. Namely, we will show that $1+2+\cdots+n+(n+1)=\dfrac{(n+1)(n+2)}{2}$:

$$1+2+\cdots+n+(n+1)$$
$$=(1+2+\cdots+n)+(n+1)$$
$$=\frac{n(n+1)}{2}+(n+1) \qquad \text{Use the assumption.}$$
$$=\frac{(n+1)(n+2)}{2}. \qquad \text{Algebraic manipulation.}$$

13.3.2.3 Proofs by Contradiction 反证法

Proof by Contradiction 反证法

n. [pruf baɪ kɑntrə'dɪkʃən]

Definition: A proof technique for statement P by first assuming that P is false (so that $\neg P$ is true), and then showing that doing so arrives a contradiction.

The conclusion of the contradiction is that P is true.

When direct proofs run into a dead end, proofs by contradiction are very powerful tools.

反证法是一种证明命题的方法——先假设命题 P 是个假命题(则$\neg P$ 是个真命题),然后展示出这个假设的矛盾之处。

得出矛盾后的结论是:P 是个真命题。

当直接证明显得困难时,反证法是一个非常有力的工具。

Properties: Proof by contradiction relies on the fact that P and $\neg P$ always have opposite truth values. 反证法的原理是 P 和$\neg P$ 的真假值是对立的。

Examples: Proposition: $\sqrt{2}$ is irrational.

Proof: By contradiction, assume that $\sqrt{2}$ is rational. We will show that this assumption is false.

Therefore, $\sqrt{2}=\frac{p}{q}$, for which p and q are positive integers that are relatively prime.

Squaring both sides gives $2=\frac{p^2}{q^2}$, from which $p^2=2q^2$.

If p is odd, then the LHS is odd but the RHS is even, which arrives at a contradiction.

If p is even, then $p=2k$ for some positive integer. We have LHS $=p^2=(2k)^2=4k^2=2q^2=$ RHS.

From the above, $4k^2=2q^2$, for which $2k^2=q^2$. So q must be even (otherwise the LHS is even but the RHS is odd).

If p and q are both even, then GCF$(p,q)>1$, which contradicts the definition of relatively prime.

Therefore, assuming that $\sqrt{2}$ is rational arrives at a contradiction. We have shown that $\sqrt{2}$ is irrational.

命题：$\sqrt{2}$是无理数。

证明：假设$\sqrt{2}$是有理数。根据有理数的定义，$\sqrt{2}=\frac{p}{q}$，使得 p 和 q 为两个互质的正整数。

因此可以得到 $2=\frac{p^2}{q^2}$，即 $p^2=2q^2$。

p 和 q 的组合有以下情况，每一种情况都能被推出矛盾所在。

(1) p 是奇数——代入 $p^2=2q^2$ 以后，左边是奇数，但右边是偶数。

(2) p 是偶数——代入 $p^2=2q^2$ 以后，左边是偶数，q 必然也是偶数。故 GCF$(p,q)>1$。

因此，$\sqrt{2}$是有理数的假设不成立。故$\sqrt{2}$是无理数。证毕。

13.3.2.4 Proofs by Casework 分情况讨论法

Proof by Casework 分情况讨论法

n. [pruf baɪ ˈkeɪsˌwərk]

Definition: A proof technique that divides the statement into mutually exclusive cases, whose union covers all possibilities. Then prove that the statement holds for every case.

分情况讨论法是一种证明方法——先把命题分成若干种互斥的情况，这些情况的并集囊括所有的可能性，然后证明在每一种情况中命题都成立。

Examples: Proposition: For all positive integer n, $1+2+\cdots+n=\frac{n(n+1)}{2}$.

Proof: By casework. We will divide n into 2 cases: odd or even. These two cases are mutually exclusive, and they cover all possibilities of n.

Case 1: n is odd.

The average of each addend is $\frac{n+1}{2}$. There are n addends. Thus the sum is $\frac{n(n+1)}{2}$.

Case 2: n is even.

If we group 1 with n, 2 with $n-1$, …, then we will get $\frac{n}{2}$ pairs, each of which has a sum of $n+1$. Thus the sum on the LHS is $\frac{n(n+1)}{2}$.

命题：对于所有正整数 n，$1+2+\cdots+n=\frac{n(n+1)}{2}$。

证明：根据 n 的奇偶性对其进行分类。这两种情况是互斥的，而且囊括了 n 的所有可能值。对每一种情况都能证明命题成立。

13.3.2.5 Proofs by Contrapositive 逆否命题证明

Proof by Contrapositive 逆否命题证明

n. [pruf baɪ ˌkɒntrəˈpɒzɪtɪv]

Definition: A proof technique of showing the contrapositive of a statement P is true, then concluding that P is true. 逆否命题证明是一种证明命题的方法，即证明命题 P 的逆否命题是成立的，由此可得 P 是成立的。

Properties: Proof by contrapositive works because a statement always has the same truth value as its contrapositive. 因为一个命题和它的逆否命题的真假值相同，所以逆否命题证明是成立的。

Examples: Proposition: If a positive integer is not divisible by 2, then its ones digit is odd.

Proof: By contrapositive, if the ones digit of a positive integer n is even, then n is divisible by 2. If we show that the contrapositive is true, then we know that statement is true. If the ones digit of n is even, then $n=$ (A multiple of 10) $+2t$, for integer t such that $0\leqslant t\leqslant 4$. Since all multiples of 10 are divisible by 2, and $2t$ divisible by 2, n divisible by 2.

命题：若一个正整数不能被 2 整除，则它的个位数是奇数。

证明：该命题的逆否命题是若一个正整数 n 的个位数是偶数，则 n 能被 2 整除。若能证明逆否命题是真命题，则原命题也是真命题。若 n 的个位数是偶数，则 $n=$ (10 的倍数) $+2t$，t 为一个整数满足 $0\leqslant t\leqslant 4$。因为所有 10 的倍数都能被 2 整除，$2t$ 也能被 2 整除，所以 n 能被 2 整除。

13.3.3 Summary 总结

Summary 总结

1. Direct Proof—A proof technique of proving the conclusions from the given using derived statements, without making any further assumptions. Its appearances include:

(1) Two-Column Proof.

(2) Paragraph Proof.

(3) FlowChart Proof.

These appearances differ visually, but communicate the same idea.

直接证明：一种证明方法，即不用作任何假设，通过已知条件，使用派生命题证明结论。它有以下 3 种方式：

(1) 两列式证明。

(2) 自然段证明。

(3) 流程图证明。

这 3 种方式的表现形式有所不同,但所传达的信息是相同的。

2. Mathematical Induction—A proof technique that is used to prove a given statement about elements of a well-ordered set (each of its nonempty subsets has a least element), usually it is the set of all natural numbers.

Mathematical induction has 2 steps:

(1) Base Case: The statement is true when it is evaluated at the smallest element of S.

(2) Inductive Case: Assume for every $n \in S$ the statement holds. Will use this assumption (sometimes along with the base case) to show that the statement holds for $n+1 \in S$.

Steps (1) and (2) generate that the statement holds for all elements in S.

数学归纳法:证明一个有关良序集元素的命题的方法(良序集即是一个集合,它的任意非空子集都有一个最小的元素)。常用的良序集是自然数集。

数学归纳法分为两步:

(1) 基本情况:当代入 S 里的最小元素时,命题成立。

(2) 归纳情况:假设命题代入任意 $n \in S$ 时命题成立,则可以推导出代入 $n+1$ 时命题也成立(有时须借用基本情况)。

根据步骤(1)和(2)可以得出当命题代入任何 S 的元素时,命题均成立。

3. Proof by Contradiction—A proof technique for statement P by first assuming that P is false (so that $\neg P$ is true), and then showing that doing so arrives a contradiction, and therefore concludes that P is true.

反证法:证明命题的方法,即先假设命题 P 是个假命题(则$\neg P$ 是个真命题),然后展示出这个假设的矛盾之处,从而得出 P 是个真命题的结论。

4. Proof by Casework—A proof technique that divides the statement into mutually exclusive cases, whose union covers all possibilities. Then prove that the statement holds for every case.

分情况讨论法:证明命题的方法,即先把命题分成若干种互斥的情况。这些情况的并集囊括所有的可能性,然后证明在每一种情况中命题都成立。

5. Proof by Contrapositive—A proof technique of showing the contrapositive of a statement P is true, then concluding that P is true.

逆否命题证明:证明命题的方法,即证明命题 P 的逆否命题是成立的,由此可得 P 是成立的。

13.4 Congruence and Its Properties 全等及其性质

Congruent 全等

adj. [kən'gruənt]

Definition: Two line segments are congruent if their lengths are equal.

Two angles are congruent if their measures are equal.

若两条线段的长度相同,则它们全等。

若两个角的度数相同,则它们全等。

Notation: Figure 13-6 shows the notations in marking congruent line segments. Figure 13-7 shows the notations in marking congruent angles. We will use these notations throughout this encyclopedia. 图 13-6 展示了全等线段的标记。图 13-7 展示了全等角的标记。这些标记在本书中都会用到。

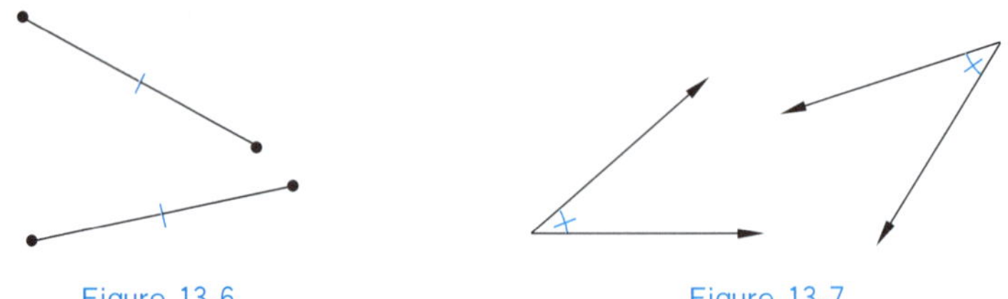

Figure 13-6　　Figure 13-7

Refer to Table A-5 in Appendix A for the notation of congruence.
全等的标记方式参照附录 A 的表 A-5。

Properties: An analogue of the properties of equation in algebra section.

1. $\angle A \cong \angle A$.
   $\overline{AB} \cong \overline{AB}$. Reflexive Property
2. If $\angle A \cong \angle B$, then $\angle B \cong \angle A$.
   If $\overline{AB} \cong \overline{CD}$, then $\overline{CD} \cong \overline{AB}$. Symmetric Property
3. If $\angle A \cong \angle B$ and $\angle B \cong \angle C$, then $\angle A \cong \angle C$.
   If $\overline{AB} \cong \overline{CD}$ and $\overline{CD} \cong \overline{EF}$ then $\overline{AB} \cong \overline{EF}$. Transitive Property

全等的性质跟本书代数部分的等式性质类似。

1. 反身性
2. 对称性
3. 传递性

14. Intersecting Lines and Parallel Lines 相交线与平行线

14.1 Relationship between Two Lines in a Plane 同一平面内两线的关系

Relationship between Two Lines in a Plane 在同一平面上两线之间的关系

Definition: The relationship between two lines in a plane is one of parallel, intersecting, or coincident. See Section 7.3.5.5 for reference. Here are their geometry definitions.
在同一平面里两线之间的关系可以是平行、相交、重合中的一种。可参考 7.3.5.5 节。现在给出它们的几何定义。

Parallel—Two lines that always have the same distance apart and never intersect, as shown in Figure 14-1.

平行——两线之间的距离总是一样，且不相交，如图 14-1 所示。

Intersecting—Two lines that have exactly one point in common, as shown in Figure 14-2.

相交——两线只有一个共同点，如图 14-2 所示。

Coincident—Two concurrent lines (infinitely many points in common), as shown in Figure 14-3.

重合——两条重叠的直线(有无数个共同点)，如图 14-3 所示。

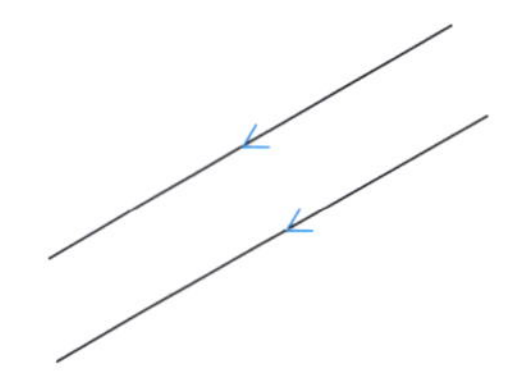
Figure 14-1

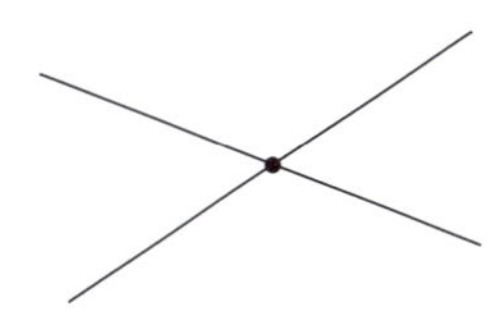
Figure 14-2

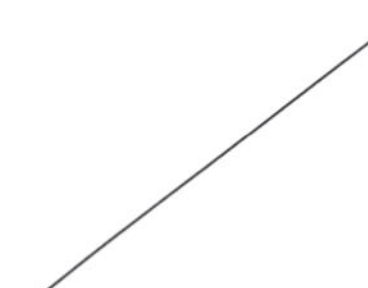
Figure 14-3

14.2 Properties of Intersecting Lines in a Plane 同一平面内相交线的性质

14.2.1 Introduction 介绍

Vertical Angles 对顶角

n. [ˈvɜrtɪkəl ˈæŋgəl]

Definition: The angles formed by intersecting lines that are opposite each other. As shown in Figure 14-4, ∠1 and ∠3 are vertical angles. ∠2 and ∠4 are vertical angles.

两个在相交线上对立的角。

如图 14-4 所示，∠1 与∠3 为对顶角。∠2 与∠4 为对顶角。

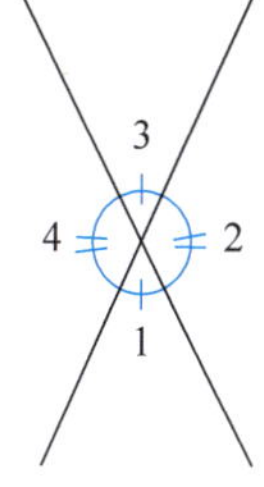

Figure 14-4

Properties: Vertical angles are congruent.

对顶角是全等的。

Proofs: Table 14-1 proves the property.

Table 14-1

Statements 命题	Reasons 原因
1. ∠2 and ∠1 form a linear pair ∠2 and ∠3 form a linear pair	1. Given
2. $m\angle 2 + m\angle 1 = 180°$ $m\angle 2 + m\angle 3 = 180°$	2. Definition of a linear pair 邻补角的定义
3. $m\angle 1 = 180° - m\angle 2$ $m\angle 3 = 180° - m\angle 2$	3. Subtract $m\angle 2$ from both sides in both equations
4. $m\angle 1 = m\angle 3$	4. Transitive Property 传递性
5. $\angle 1 \cong \angle 3$	5. Definition of congruent angles 全等角的定义

Questions: As shown in Figure 14-5 (not drawn in scale), if $m\angle 1=50°$, what are the values of $m\angle 2$, $m\angle 3$, and $m\angle 4$?

Answers: Since vertical angles are congruent, $m\angle 3=m\angle 1=50°$. Using the properties of vertical angles and linear pair, we have $m\angle 2=m\angle 4=180°-m\angle 1=130°$.

We have $m\angle 2=130°$, $m\angle 3=50°$, and $m\angle 4=130°$.

Perpendicular 垂直

adj. [ˌpərpənˈdɪkjələr]

Definition: Two lines are perpendicular if they intersect at right angles. Also refer to perpendicular in Section 7.3.5.5.

This is a special case of intersecting lines.

As shown in Figure 14-6, $m\angle 1=m\angle 2=m\angle 3=m\angle 4=90°$.

若两线相交形成直角,则两线垂直。可参考 7.3.5.5 节关于垂直的词条。

这是相交线的特殊情况。

Examples: The coordinate axes are perpendicular lines. 坐标轴是垂直线。

Note: See perpendicular in Section 7.3.5.5 for the equations of perpendicular lines in the coordinate plane. 在坐标平面上垂直线的等式,可参考 7.3.5.5 节关于垂直的词条。

Distance 距离

n. [ˈdɪstəns]

Definition: 1. The distance between points A and B is simply the AB, which is the length of $\overline{AB}$.

两点间的距离是指以两点为端点的线段的长度。

As shown in Figure 14-7, $AB=d$.

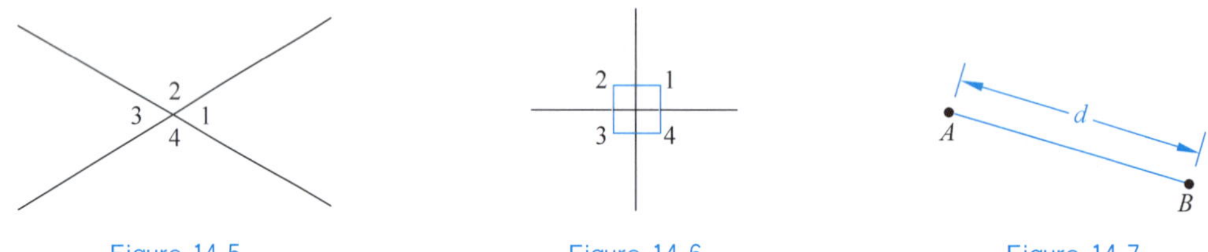

Figure 14-5 Figure 14-6 Figure 14-7

2. The distance between a point A and the line/ray/line segment/plane l is the length of $\overline{AB}$, for which B is on l, and $\overline{AB}\perp l$.

点 A 和直线/射线/线段/平面 l 的距离是线段 $\overline{AB}$ 的长度,其中 B 在 l 上且满足 $\overline{AB}\perp l$。

Figure 14-8 shows the distance between a point A and a line l.

图 14-8 展示了点 A 与直线 l 的距离。

Figure 14-9 shows the distance between a point A and a plane l.

图 14-9 展示了点 A 与平面 l 的距离。

3. The distance between parallel lines/rays/line segments/planes (l and m) is the length of the line segment $\overline{AB}$ for which A is on l and B is on m, such that $\overline{AB}\perp l$ and $\overline{AB}\perp m$, as shown from Figures 14-10 through 14-12. 平行的直线/射线/线段/

平面(l 和 m)的距离是线段 $\overline{AB}$ 的长度,其中 A 在 l 上,B 在 m 上,满足 $\overline{AB}\perp l$ 和 $\overline{AB}\perp m$,如图 14-10 至图 14-12 所示。

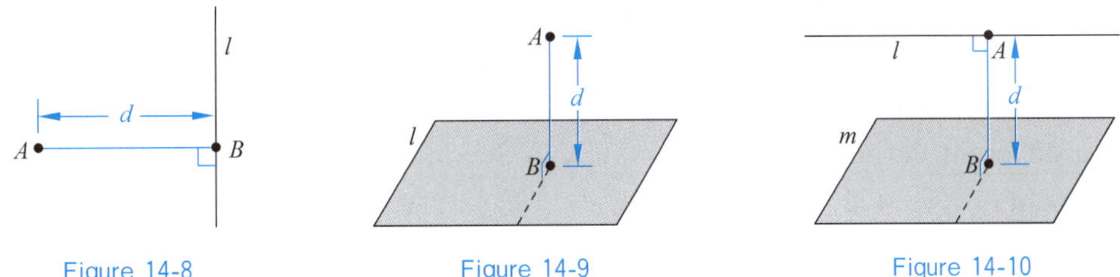

Figure 14-8　Figure 14-9　Figure 14-10

For points A, B, and C, if $AB = AC$, then we say A is equidistant from B and C, as shown in Figure 14-13.
在点 A、B、C 中,若 $AB = AC$,则说从 B 和 C 到 A 是等距的,如图 14-13 所示。

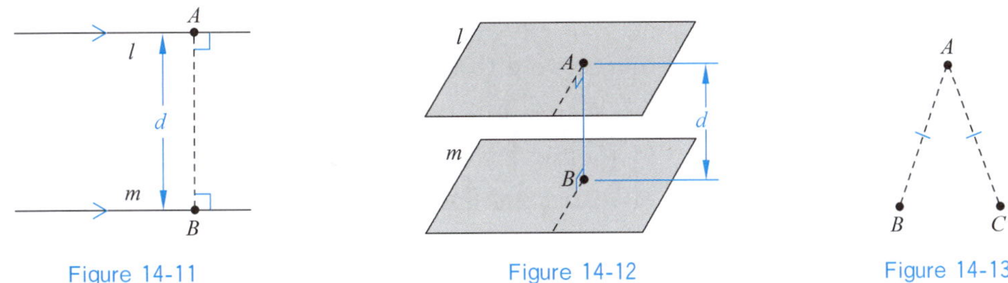

Figure 14-11　Figure 14-12　Figure 14-13

Properties: 1. Definition 2 relies on the postulate that assuming point A is not on line/ray/line segment l, if B is on l, then there is only one possibility for B to establish $\overline{AB}\perp l$. See **Perpendicular Postulate** from more properties for parallel and perpendicular lines (Section 14.3.3). 定义 2 是根据以下公理"设 A 不在直线/射线/线段 l 上。若 B 在 l 上,则 B 只有一种可能性,满足 $\overline{AB}\perp l$。"见 14.3.3 节更多平行线和垂直线的定理——**垂线公理**。

2. If point A is not on line/ray/line segment l and B is a point on l, then the value of AB minimizes when $\overline{AB}\perp l$, as shown in Figure 14-14.
The possible locations of B (denoted by B'), AB is minimized when $\overline{AB}\perp l$ (by the Pythagorean Theorem).
若 A 不在直线/射线/线段 l 上,B 在 l 上,则 AB 的最小值发生在 $\overline{AB}\perp l$ 的情况中,如图 14-14 所示。
在其他 B 的可能位置(以 B'表示),当 $\overline{AB}\perp l$ 时,AB 的值最小(根据勾股定理)。

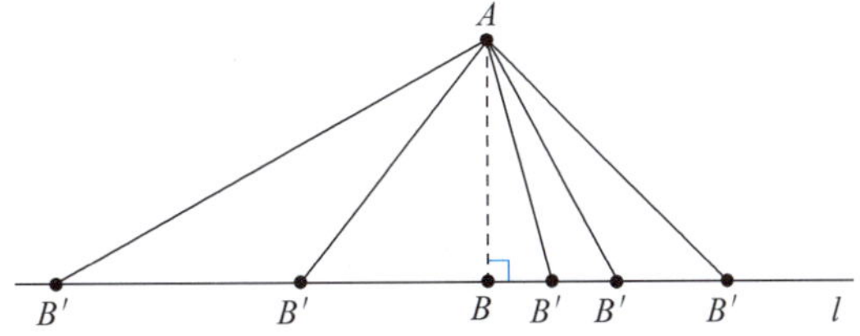

Figure 14-14

14.2.2 Summary 总结

1. Vertical angles are congruent. 对顶角全等。
2. Perpendicular lines intersect at right angles, and are special intersecting lines. 垂直线相交于直角，是特殊的相交线。
3. See different definitions about distance in Section 14.2.1 (the length of the perpendicular line segment). 见 14.2.1 节距离的不同定义(同垂直线段的长度)。

14.3 Properties of Parallel Lines in a Plane 同一平面内平行线的性质

14.3.1 Introduction 介绍

Transversal 截线

n. [træns'vɜrsəl]

Definition: A line that passes through two different lines on the same plane at two different points. 截线是指一条经过两条在同一平面内不重合的直线。截线与这两条直线的相交点均不同。

As shown in Figure 14-15, line t is a transversal of lines l and m.

As shown in Figure 14-16, each of lines l, m, and t is a transversal of the other two lines.

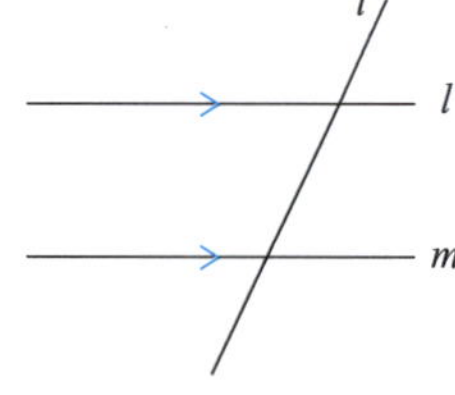

Figure 14-15

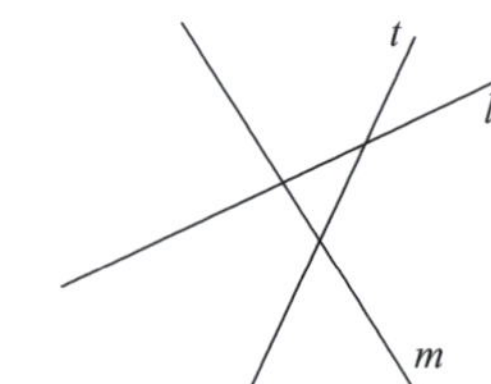

Figure 14-16

Properties: Transversals can check whether the two lines it passes through are parallel. 截线能检验两条线是否平行。

14.3.2 Relationship of Angles Formed by a Transversal Intersecting Two Lines 截线与两线相交后所形成的角关系

视频 23

As shown in Figure 14-17:

$\angle 1$ and $\angle 5$, $\angle 2$ and $\angle 6$, $\angle 3$ and $\angle 7$, and $\angle 4$ and $\angle 8$ are pairs of corresponding angles. 这四对角中，每一对互为同位角。

$\angle 3$ and $\angle 5$, and $\angle 4$ and $\angle 6$ are pairs of alternate interior angles. 这两对角中，每一对互为内错角。

$\angle 1$ and $\angle 7$, and $\angle 2$ and $\angle 8$ are pairs of alternate exterior angles. 这两对角中，每一对互为外错角。

$\angle 3$ and $\angle 6$, and $\angle 4$ and $\angle 5$ are pairs of consecutive interior angles. 这两对角中，每一对互为同旁内角。

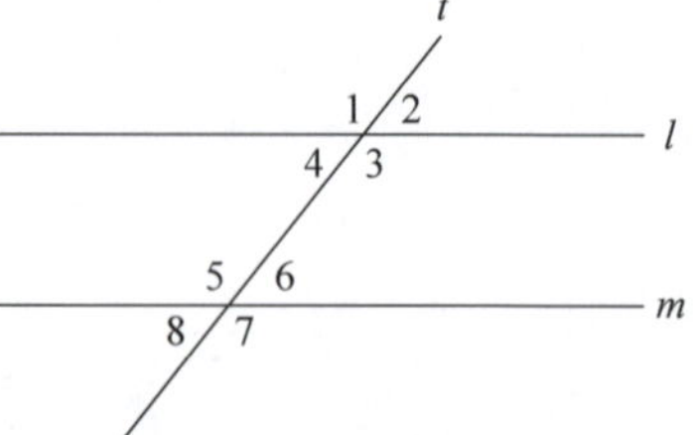

Figure 14-17

In this section, 14.3.2, we will be referring to Figure 14-17 for every vocabulary word.

Corresponding Angles　同位角

n. [kɔrə'spɑndɪŋ 'æŋgəls]

Definition: When lines l and m are crossed by the transversal t, the pair of angles in the matching corners is a pair of corresponding angles.　当截线 t 与直线 l 和 m 相交后，一对所对应的角互为同位角。

In Figure 14-17, ∠1 and ∠5, ∠2 and ∠6, ∠3 and ∠7, and ∠4 and ∠8 are pairs of corresponding angles.

Properties: 1. **Corresponding Angles Postulate**: If the transversal cuts two parallel lines, then the corresponding angles are congruent.　**同位角公理**：若截线相交两条平行线，则同位角全等。

2. **Converse of Corresponding Angles Postulate**: If the corresponding angles are congruent, then the transversal cuts two parallel lines.　**同位角公理的逆公理**：若同位角全等，则截线相交的是两条平行线。

Alternate Interior Angles　内错角

n. ['ɔltərˌneɪt ɪn'tɪəriər 'æŋgəls]

Definition: When lines l and m are crossed by the transversal t, the pair of angles on the opposite sides of t and between l and m is a pair of alternate interior angles.　当截线 t 与直线 l 和 m 相交后，一对在 t 对边的并在 l 和 m 之间的角互为内错角。

In Figure 14-17, ∠3 and ∠5, and ∠4 and ∠6 are pairs of alternate interior angles.

Properties: 1. **Alternate Interior Angles Theorem**: If the transversal cuts two parallel lines, then the alternate interior angles are congruent.　**内错角定理**：若截线相交两条平行线，则内错角全等。

2. **Converse of Alternate Interior Angles Theorem**: If the alternate interior angles are congruent, then the transversal cuts two parallel lines.　**内错角定理的逆定理**：若内错角全等，则截线相交的是两条平行线。

Proofs: All proofs refer to Figure 14-17.

1. Table 14-2 provides the proof for Property 1.

Proposition: If $l \parallel m$, then $\angle 3 \cong \angle 5$.

Table 14-2

Statements 命题	Reasons 原因
1. $l \parallel m$	1. Given
2. $\angle 3 \cong \angle 7$	2. **Corresponding Angles Postulate　同位角公理**
3. $\angle 7 \cong \angle 5$	3. Vertical angles are congruent　对顶角全等
4. $\angle 3 \cong \angle 5$	4. Transitive Property　传递性

2. Table 14-3 provides the proof for Property 2.

Proposition: If $\angle 3 \cong \angle 5$, then $l \parallel m$.

Table 14-3

Statements 命题	Reasons 原因
1. $\angle 3 \cong \angle 5$	1. Given
2. $\angle 5 \cong \angle 7$	2. Vertical angles are congruent　对顶角全等
3. $\angle 3 \cong \angle 7$	3. Transitive Property　传递性
4. $l \parallel m$	4. **Converse of Corresponding Angles Postulate　同位角公理的逆公理**

Alternate Exterior Angles　外错角

n. [ˈɔltərˌneɪt ekˈtɪəriər ˈæŋgəls]

Definition: When lines l and m are crossed by the transversal t, the pair of angles on the opposite sides of t but **not** between l and m is a pair of alternate exterior angles.　当截线 t 与直线 l 和 m 相交后，一对在 t 对边的并不在 l 和 m 之间的角互为外错角。

In Figure 14-17, $\angle 1$ and $\angle 7$, and $\angle 2$ and $\angle 8$ are pairs of alternate exterior angles.

Properties: 1. **Alternate Exterior Angles Theorem**: If the transversal cuts two parallel lines, then the alternate exterior angles are congruent.　**外错角定理**：若截线相交两条平行线，则外错角全等。

2. **Converse of Alternate Exterior Angles Theorem**: If the alternate exterior angles are congruent, then the transversal cuts two parallel lines.　**外错角定理的逆定理**：若外错角全等，则截线相交的是两条平行线。

Proofs: All proofs refer to Figure 14-17.

1. Table 14-4 provides the proof for Property 1.

Proposition: If $l \parallel m$, then $\angle 1 \cong \angle 7$.

Table 14-4

Statements 命题	Reasons 原因
1. $l \parallel m$	1. Given
2. $\angle 1 \cong \angle 5$	2. **Corresponding Angles Postulate　同位角公理**
3. $\angle 5 \cong \angle 7$	3. Vertical angles are congruent　对顶角全等
4. $\angle 1 \cong \angle 7$	4. Transitive Property　传递性

2. Table 14-5 provides the proof for Property 2.

Proposition: If $\angle 1 \cong \angle 7$, then $l \parallel m$.

Table 14-5

Statements 命题	Reasons 原因
1. $\angle 1 \cong \angle 7$	1. Given
2. $\angle 7 \cong \angle 5$	2. Vertical angles are congruent　对顶角全等
3. $\angle 1 \cong \angle 5$	3. Transitive Property　传递性
4. $l \parallel m$	4. **Converse of Corresponding Angles Postulate　同位角公理的逆公理**

Consecutive Interior Angles　同旁内角

n. [kənˈsekjətɪv ɪnˈtɪəriər ˈæŋgəls]

Definition: When lines l and m are crossed by the transversal t, the pair of angles on the same side of t and between l and m is a pair of consecutive interior angles. 当截线 t 与直线 l 和 m 相交后，一对在 t 同边的并在 l 和 m 之间的角互为同旁内角。

In Figure 14-17, $\angle 3$ and $\angle 6$, and $\angle 4$ and $\angle 5$ are pairs of consecutive interior angles.

Properties: 1. **Consecutive Interior Angles Theorem**: If the transversal cuts two parallel lines, then consecutive interior angles are supplementary. **同旁内角定理**：若截线相交两条平行线，则同旁内角互补。

2. **Converse of Consecutive Interior Angles Theorem**: If consecutive interior angles are supplementary, then the transversal cuts two parallel lines. **同旁内角定理的逆定理**：若同旁内角互补，则截线相交的是两条平行线。

Proofs: All proofs refer to Figure 14-17.

1. Table 14-6 provides the proof for Property 1.

Proposition: If $l \parallel m$, then $\angle 3$ and $\angle 6$ are supplementary.

Table 14-6

Statements 命题	Reasons 原因
1. $l \parallel m$	1. Given
2. $\angle 3$ and $\angle 4$ form a linear pair	2. Definition of a linear pair 邻补角的定义
3. $m\angle 3 + m\angle 4 = 180°$	3. Property of a linear pair 邻补角的性质
4. $\angle 4 \cong \angle 6$	4. **Alternate Interior Angles Theorem 内错角定理**
5. $m\angle 4 = m\angle 6$	5. Property of congruent angles 全等角的性质
6. $m\angle 3 + m\angle 6 = 180°$	6. Substitution 代入
7. $\angle 3$ and $\angle 6$ are supplementary	7. Definition of supplementary angles 补角的定义

2. Table 14-7 provides the proof for Property 2.

Proposition: If $\angle 3$ and $\angle 6$ are supplementary, then $l \parallel m$.

Table 14-7

Statements 命题	Reasons 原因
1. $\angle 3$ and $\angle 6$ are supplementary	1. Given
2. $m\angle 3 + m\angle 6 = 180°$	2. Definition of supplementary angles 补角的定义
3. $m\angle 6 = 180° - m\angle 3$	3. Subtract $m\angle 3$ from both sides
4. $\angle 3$ and $\angle 4$ form a linear pair	4. Definition of a linear pair 邻补角的定义
5. $m\angle 3 + m\angle 4 = 180°$	5. Property of a linear pair 邻补角的性质
6. $m\angle 4 = 180° - m\angle 3$	6. Subtract $m\angle 3$ from both sides
7. $m\angle 4 = m\angle 6$	7. Transitive Property 传递性
8. $\angle 4 \cong \angle 6$	8. Property of congruent angles 全等角的性质
9. $l \parallel m$	9. **Converse of Alternate Interior Angle Theorem 内错角定理的逆定理**

14.3.3 More Properties of Parallel and Perpendicular Lines 更多平行线与垂直线的性质

More Properties of Parallel and Perpendicular Lines 更多平行线与垂直线的性质

Properties: 1. **Parallel Postulate**: Given a line l and point P not on l, there is only one line that passes through P and is parallel to l. **平行公理**：给出直线 l 和一个不在 l 上的点 P，只有一条过 P 的直线平行于 l。

2. **Perpendicular Postulate**: Given a line l and point P not on l, there is only one line that passes through P and is perpendicular to l. **垂线公理**：给出直线 l 和一个不在 l 上的点 P，只有一条过 P 的直线垂直于 l。

   Figure 14-18 illustrates Properties 1 and 2.

3. **Transitivity of Parallel Lines**: For lines l, m, and n, if $l \parallel m$ and $m \parallel n$, then $l \parallel n$. **平行线的传递性**。

   As shown in Figure 14-19, we can conclude that all three lines are parallel.

4. **Two Perpendicular Lines Theorem**: In a plane, if two lines are perpendicular to the same line, then they are parallel to each other. **双垂线定理**：在同一平面上，若两条线都跟同一直线垂直，则这两条线平行。

   As shown in Figure 14-20, we can conclude that the solid lines are parallel. 如图 12-20 所示，可得出两条实线平行。

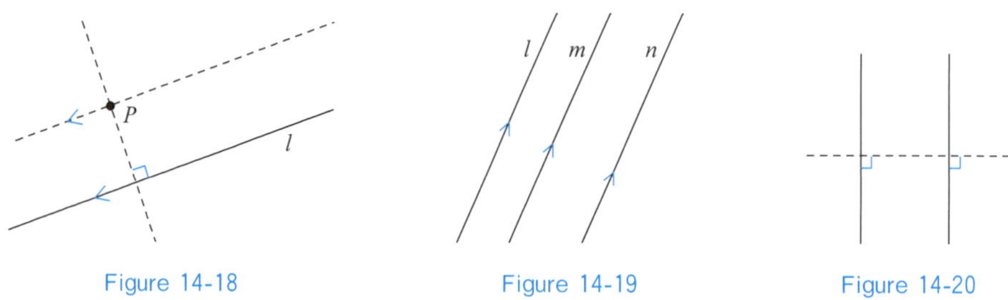

Figure 14-18　　Figure 14-19　　Figure 14-20

Proofs: To prove Property 4, use the converse of one of the following will do:
要证明性质 4，可用以下公理/定理的任意一个逆公理/逆定理：
Corresponding Angles Postulate　同位角公理
Alternate Interior Angles Theorem　内错角定理
Alternate Exterior Angles Theorem　外错角定理
Consecutive Interior Angles Theorem　同旁内角定理

14.3.4 Summary 总结

1. See **Section 14.3.2 Relationship between Two Angles Formed by a Transversal Intersecting Two Lines** for definitions. **定义见 14.3.2 节截线与两线相交后所形成的角关系**。

2. **Corresponding Angles Postulate and Its Converse**: The transversal cuts two parallel lines if and only if the corresponding angles are congruent. **同位角公理和逆公理**：截线相交两条平行线当且仅当同位角全等。

   The conclusion of the postulate is that the angles are congruent. 同位角公理的结论是角全等。
   The conclusion of the converse is that the lines are parallel. 同位角逆公理的结论是线平行。

4. **Theorems and Converses**: The transversal cuts two parallel lines if and only if the alternate interior angles are congruent; alternate exterior angles are congruent; consecutive interior angles are supplementary. 内错角、外错角、同旁内角定理和逆定理：截线相交两条平行线当且仅当内错

角全等；外错角全等；同旁内角互补。

The conclusion of the theorems is that the angles are congruent or supplementary. 上述 3 个定理的结论是角相等或互补。

The conclusion of the converses is that the lines are parallel. 上述 3 个定理的逆定理的结论是线平行。

14.4 Lines Not on the Same Plane 不在同一平面上的直线

Skew Lines 异面直线

n. [skju laɪns]

Definition: Two different lines are skew lines if they neither intersect nor parallel. 若两条不重合的直线既不相交也不平行，则它们互为异面直线。

Examples: In the cube as shown in Figure 14-21, some examples of skew lines are:

$\overleftrightarrow{AB}$ and $\overleftrightarrow{EF}$.

$\overleftrightarrow{DF}$ and $\overleftrightarrow{GH}$.

$\overleftrightarrow{BC}$ and $\overleftrightarrow{FG}$.

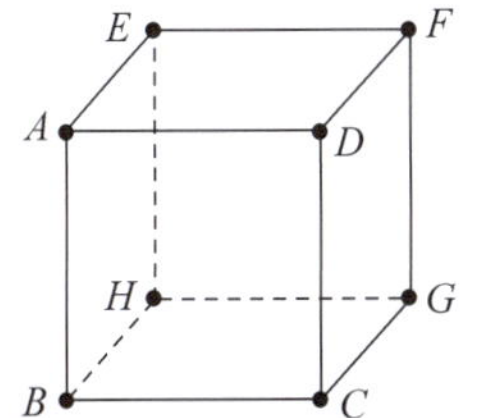

Figure 14-21

15. Triangles 三角形

15.1 Introduction 介绍

Triangle 三角形

n. [ˈtraɪˌæŋgəl]

Definition: A plane shape that is bounded by 3 line segments, each of which is called a side. 三角形是指被 3 条线段围成的平面图形，每条线段称为边。

Notation: Triangle ABC is denoted by $\triangle ABC$. The side-lengths that are opposite of $\angle A$, $\angle B$, and $\angle C$ are denoted by a, b, and c, respectively. Figure 15-1 shows $\triangle ABC$.

Refer to Table A-5 in Appendix A for the notation of triangles. 三角形的标记方式参照附录 A 的表 A-5。

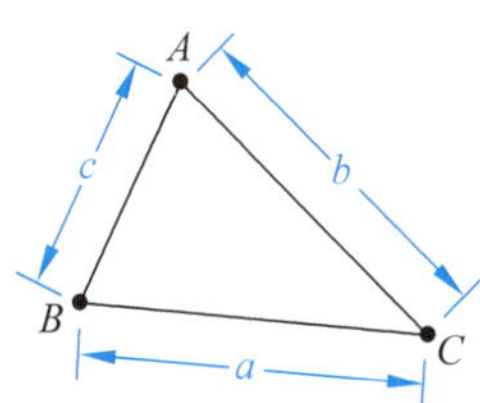

Figure 15-1

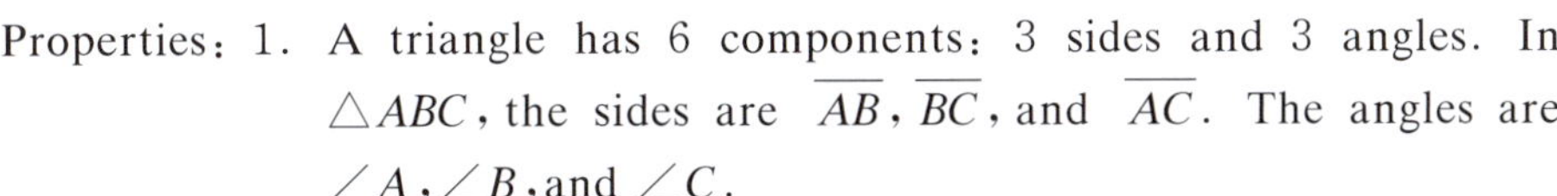

Properties: 1. A triangle has 6 components: 3 sides and 3 angles. In $\triangle ABC$, the sides are $\overline{AB}$, $\overline{BC}$, and $\overline{AC}$. The angles are $\angle A$, $\angle B$, and $\angle C$.

2. Classification of triangles according to angles: acute triangle, right triangle, obtuse triangle, equiangular triangles. Equiangular triangles are special acute triangles.

根据角分类,三角形分为锐角三角形、直角三角形、钝角三角形、等角三角形。等角三角形是特殊的锐角三角形。

3. Classification of triangles according to sides: scalene triangles, isosceles triangles, equilateral triangles. Equilateral triangles are special isosceles triangles.
根据边分类,三角形分为不等边三角形、等腰三角形、等边三角形。等边三角形是特殊的等腰三角形。

4. The sum of measures of the interior angles of every triangle is 180°. See interior angle in Section 15.3.1 for the proof. 任何三角形的内角度数之和为180°。证明见15.3.1节的内角的词条。
The sum of measures of the exterior angles of every triangle is 360°. See exterior angle in Section 15.3.2 for the proof. 任何三角形的外角度数之和为360°。证明见15.3.2节的外角的词条。

5. The sum of lengths of any two sides of a triangle is greater than that of the third side.
The difference of lengths of any two sides of a triangle is less than that of the third side.
三角形任意两边的长度之和大于第三边的长度。
三角形任意两边的长度之差小于第三边的长度。

6. The Pythagorean Theorem applies to all right triangles. 所有直角三角形均满足勾股定理。

7. To determine whether two triangles are congruent, use SSS, SAS, AAS, or ASA criteria. The HL Theorem applies on right triangles only. 判断两个三角形是否全等,用SSS(边边边)、SAS(边角边)、AAS(角角边)、ASA(角边角),或HL(斜边直角边)定理。HL定理是专门判断直角三角形的。

8. To determine whether two triangles are similar, use SSS, SAS, or AA criteria. 判断两个三角形是否相似,用SSS(边边边相似)、SAS(边角边相似)、AA(角角相似)。

9. The right triangle trigonometry solves a right triangle based on 3 pieces of given information (except the case when 3 angles are given), for which one piece is the right angle. Each piece of information refers to a side or an angle. 根据三个已知条件(其中一个是直角,且三个已知条件不能全是角,每一个条件指的是一条边或一个角),直角三角形的三角学能解直角三角形。

Phrases: acute～ 锐角三角形, right～ 直角三角形, obtuse～ 钝角三角形, equiangular～ 等角三角形, scalene～ 不等边三角形, isosceles～ 等腰三角形, isosceles right～ 等腰直角三角形, equilateral～ 等边三角形

15.2 Classification of Triangles 三角形的分类

15.2.1 Classification by Angles 按角分类

Acute Triangle 锐角三角形

n. [ə'kjut 'traɪˌæŋgəl]

Definition: A triangle with three acute angles. 锐角三角形是指三个角都是锐角的三角形。
Figure 15-2 shows an acute triangle.

Right Triangle　直角三角形

n. [raɪt 'traɪˌæŋgəl]

Definition：A triangle with one right angle. The sides that makes up the right angle are called the legs, and the remaining side (longest) is called the hypotenuse.　直角三角形是指有一个角是直角的三角形。包含直角的边叫直角边，另外一条边(最长)叫斜边。

Figure 15-3 shows a right triangle with its parts labelled.

Figure 15-2　　Figure 15-3

Properties：1. Every triangle can have at most one right angle.　一个三角形最多只能有一个直角。

2. **Right Triangle Property**：If T is a right triangle, then the two acute angles of T are complementary.　**直角三角形性质**：若 T 为直角三角形，则 T 的两个锐角互余。

3. **Converse of Right Triangle Property**：If two acute angles of triangle T are complementary, then T is a right triangle.　**直角三角形性质的逆命题**：若三角形 T 的两个锐角互余，则 T 是直角三角形。

Proofs：To prove the properties, note that the sum of measures of interior angles of any triangle is 180°. See interior angle in Section 15.3.1 for the proof.

要证明这些性质，注意三角形的内角之和为 180°，证明见 15.3.1 节内角的词条。

1. Having at least two right angles makes the sum of measures of interior angles is (180° + the measure of the third angle), which exceeds 180°.

   若至少有两个直角，则内角度数之和 = 180° + 第三个角的度数，也就是超过 180°。

2. We have (sum of measures of the two acute angles) + (measure of right angle) = 180°, from which (sum of measures of the two acute angles) + 90° = 180°. Therefore, the sum of measures of the two acute angles is 90°, which implies that the acute angles are complementary.

   两个锐角的度数之和 + 直角的度数 = 180°，因此"两个锐角的度数之和" + 90° = 180°，从而"两个锐角的度数之和" = 90°。故两个锐角互余。

3. If two acute angles of triangle T are complementary, then the sum of their measures is 90°. Thus, the measure of the remaining angle is 180° − 90° = 90°. Therefore, T is a right triangle by definition.

   若 T 的两锐角互余，则它们的度数之和为 90°。因此第三个角的度数为 180° − 90° = 90°。根据定义，T 是直角三角形。

Phrases：isosceles～　等腰直角三角形

Obtuse Triangle　钝角三角形

n. [əb'tus 'traɪˌæŋgəl]

Definition：A triangle with one obtuse angle.　有一个钝角的三角形。

Figure 15-4 shows an obtuse triangle. The obtuse angle is marked in blue.

Properties: A triangle can have at most one obtuse angle. 一个三角形最多只能有一个钝角。

Proofs: To prove the property, note that the sum of measures of interior angles of any triangle is 180°. See interior angle in Section 15.3.1 for the proof.

Having at least two obtuse angles makes the sum of measures of interior angles (greater than 180° + the measure of the third angle), which exceeds 180°.

要证明性质 3,注意三角形的内角之和为 180°,证明见 15.3.1 节内角的词条。

若至少有两个钝角,则内角度数之和≥180° + 第三个角的度数,也就是超过 180°。

Equiangular Triangle 等角三角形

n. [iːkwɪ'æŋgjʊlə 'trɑɪˌæŋgəl]

Definition: A triangle with three congruent angles, each of which measures 60° (The sum of measures of interior angles of any triangle is 180°). 三个角全等的三角形,每个角的度数为 60°(因为三角形的内角度数之和为 180°)。

Figure 15-5 shows an equiangular triangle.

Figure 15-4 Figure 15-5

Properties: Equiangular triangles are special acute triangles. 等角三角形是特殊的锐角三角形。

15.2.2 Classification by Sides 按边分类

Scalene Triangle 不等边三角形

n. ['skeɪliːn 'trɑɪˌæŋgəl]

Definition: A triangle that has three sides of different lengths. 不等边三角形是三边长度都不同的三角形。

Figure 15-6 shows a scalene triangle.

Isosceles Triangle 等腰三角形

n. [ɑɪ'sɑsəˌliz 'trɑɪˌæŋgəl]

Definition: A triangle that has two sides of equal lengths called the legs. The remaining side is called the base. 等腰三角形是指两边长度相等的三角形。这两边为腰,第三边为底。

Figure 15-7 shows an isosceles triangle with its parts labelled.

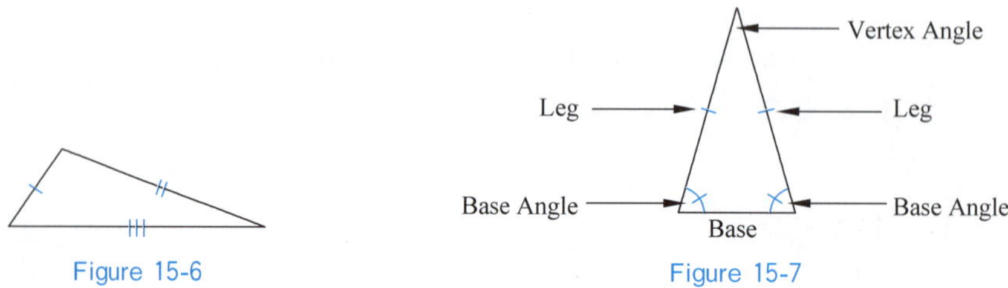

Figure 15-6 Figure 15-7

Properties：1. The angles opposite the legs (containing the base), called the base angles, are congruent.　对着腰的角(底是其中一边)叫作底角。底角全等。

2. The included angle of the legs, or the angle that is opposite the base, is called the vertex angle.　两腰的夹角(对着底的角)叫作顶角。

Proof：To prove the Property 1, see theorems and proofs of triangles in Section 15.5.3.　要证明性质 1,可参见 15.5.3 节的三角形的定理和证明。

Equilateral Triangle　等边三角形

n. [ˌikwəˈlætərəl ˈtraɪˌæŋgəl]

Definition：A triangle that has three sides of equal lengths. This is a special isosceles triangle.　等边三角形是指三边长度都相等的三角形。它是特殊的等腰三角形。

Figure 15-8 shows an equilateral triangle.

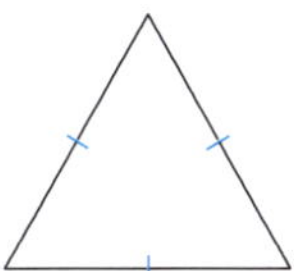

Figure 15-8

Properties：Equilateral triangles are equiangular. 等边三角形是等角的。

Proof：To prove the property, see theorems and proofs of triangles in Section 15.5.3.　要证明此性质,可参见 15.5.3 节的三角形的定理和证明。

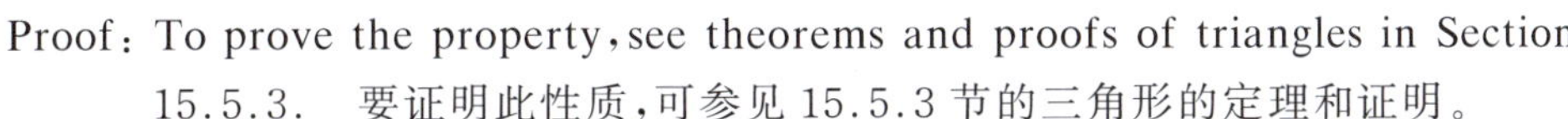

15.2.3　Special Triangles：Isosceles Right Triangles　特殊三角形：等腰直角三角形

Isosceles Right Triangle　等腰直角三角形

n. [aɪˈsɑsəˌliz raɪt ˈtraɪˌæŋgəl]

Definition：Refer to the definitions of isosceles triangle in Section 15.2.2 and right triangle in Section 15.2.1.　可参见 15.2.2 节的等腰三角形和 15.2.1 节的直角三角形的定义。

Figure 15-9 shows an isosceles right triangle.

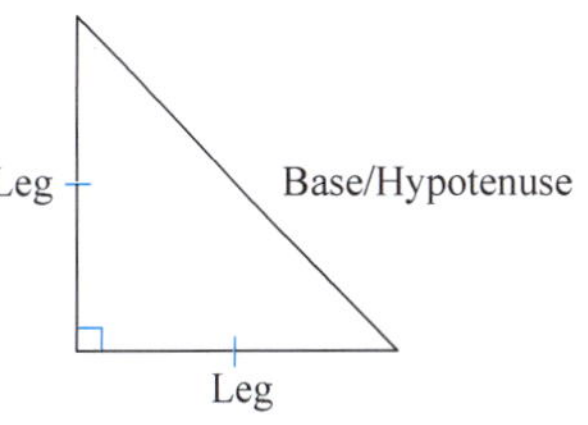

Figure 15-9

Properties：Refer to the properties of isosceles triangle in Section 15.2.2 and right triangle in Section 15.2.1.　性质见 15.2.2 节的等腰三角形和 15.2.1 节的直角三角形的词条。

15.2.4　Summary　总结

1. Classifying by angles, every triangle is one of：acute triangle, right triangle, or obtuse triangle. 按角分类,任意三角形必定是锐角三角形、直角三角形、钝角三角形中的一种。
   (1) Every triangle must have at least two acute angles.　任意三角形至少有两个锐角。
   (2) Equiangular triangles are special acute triangles (each angle is measured 60°).　等角三角形是特殊的锐角三角形(每个角的度数为 60°)。
2. Classifying by sides, every triangle is one of：scalene triangle or isosceles triangle.　按边分类,任意三角形必定是不等边三角形和等腰三角形中的一种。
   (1) Equilateral triangles are special isosceles triangles.　等边三角形是特殊的等腰三角形。
   (2) Equilateral triangles are equiangular, and vice versa.　等边三角形是等角的,反之亦然。
3. See theorems and proofs of triangles for more properties of isosceles triangles in Section 15.3.3. 有关等腰三角形的更多性质可参考 15.5.3 节的三角形的定理和证明。

15.3 Angles of a Triangle 三角形的角

15.3.1 Interior Angles 内角

Interior Angle 内角

n. [ɪn'tɪəriər 'æŋgəl]

Definition: The angle formed inside by any two adjacent sides. 内角是两条邻边所形成的角。

This definition also applies to polygons. 多边形的内角的定义与此定义相同。

The word "angle" of a polygon is assumed to be an interior angle, unless specified otherwise. 多边形的"角"可假设为内角,除非特别规定。

As shown in Figure 15-10, the three interior angles of a triangle are marked. 图 15-10 标出了三角形的三个内角。

Properties: The sum of measures of interior angles of every triangle is 180°. 三角形的内角度数之和为 180°。

Proof: To prove the property, for $\triangle ABC$ as shown in Figure 15-11, we draw line l containing A and parallel to $\overleftrightarrow{BC}$. There is only one way to draw l because of the **Parallel Postulate**.

要证明这个性质,如图 15-11 所示,在$\triangle ABC$ 中,过点 A 画出平行于 $\overleftrightarrow{BC}$ 的直线 l。因为**平行公理**,所以 l 只有一种画法。

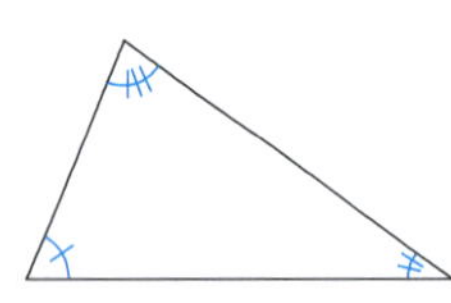

Figure 15-10

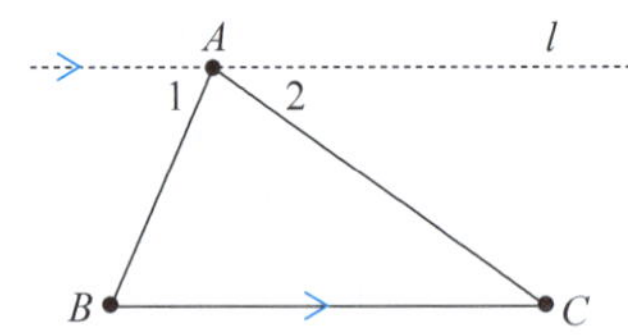

Figure 15-11

Table 15-1 gives the proof.

Table 15-1

Statements 命题	Reasons 原因
1. $l \parallel \overleftrightarrow{BC}$	1. Construction 画图
2. $\angle B \cong \angle 1$, $\angle C \cong \angle 2$	2. **Alternate Interior Angle Theorem 内错角定理**
3. $m\angle B = m\angle 1$, $m\angle C = m\angle 2$	3. Property of congruent angles 全等角的性质
4. $m\angle A + m\angle 1 + m\angle 2 = 180°$	4. Property of a straight angle and **Angle Addition Postulate** 平角的性质和**角相加公理**
5. $m\angle A + m\angle B + m\angle C = 180°$	5. Substitution 代入

As an extension, it is clear that the acute angles of a right triangle are complementary. 作为拓展,很明显直角三角形的两个锐角互余。

Questions: 1. In $\triangle ABC$, if $m\angle A = 30°$ and $m\angle B = 40°$, what is the value of $m\angle C$?

2. In right $\triangle ABC$, if the measure of one acute angle is 50°, what is the measure of the other acute angle? 在直角三角形 ABC 中，若一个锐角的度数为 50°，另一个锐角的度数是多少？

Answers: 1. $m\angle C = 180° - m\angle A - m\angle B = 110°$.

2. Since we know the measures of two of the angles, we can subtract their sum of measures from 180°. The sum of measures of two known angles is $50° + 90° = 140°$. Therefore, the measures of the other acute angle is $180° - 140° = 40°$. 已知两个角的度数，可以用 180° 减去这两个角的度数之和。这两个角的度数之和是 140°，故另一个锐角的度数就是 40°。

15.3.2 Exterior Angles 外角

Exterior Angle 外角

n. [ek'tɪəriər 'æŋgəl]

Definition: The angle formed when one side is extended beyond its adjacent sides. 外角是一条边和它邻边的延长线所形成的角。

This definition also applies to polygons. 多边形中内角的定义与此定义相同。

As shown in Figure 15-12, two sets of three exterior angles of a triangle are marked. 图 15-12 标出了两组三角形的三个外角。

A triangle has three exterior angles. Note that each black angle and its corresponding blue angle are congruent because of vertical angles. 三角形有三个外角。注意因为对顶角的关系，每个黑角和对应的蓝角是全等的。

Properties: 1. In every triangle or every other convex polygon, the sum of measures of an interior angle and its corresponding exterior angle is 180°. 在任意三角形或其他凸多边形中，内角和它对应的外角的度数之和为 180°。

As shown in Figure 15-13, the sum of measures of the two marked angles is 180°. 如图 15-13 所示，两个带标记的角度数之和为 180°。

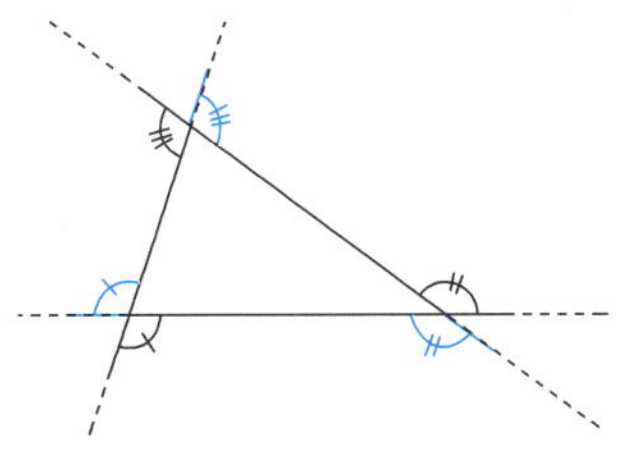

Figure 15-12

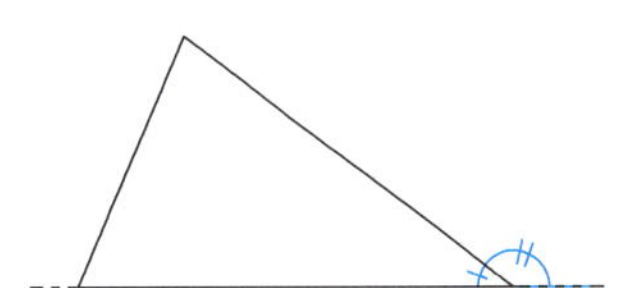

Figure 15-13

2. **Exterior Angle Theorem**: In every triangle, the sum of measures of two interior angles is equal to the measure of the exterior angle of the third angle. **外角定理**：在任意三角形中，两个内角的度数之和等于第三个角的外角度数。

As shown in Figure 15-14, the sum of measures of two black angles is equal to that of the blue angle. 如图 15-14 所示，两个黑角的度数之和等于蓝角的度数。

3. The sum of measures of exterior angles of every triangle is 360°. 任意三角形的外角

度数之和为 360°。

As shown in Figure 15-15, the sum of measures of the black angles is 360°, and the sum of measures of the blue angles is 360°. 如图 15-15 所示,黑角的度数之和为 360°,且蓝角的度数之和为 360°。

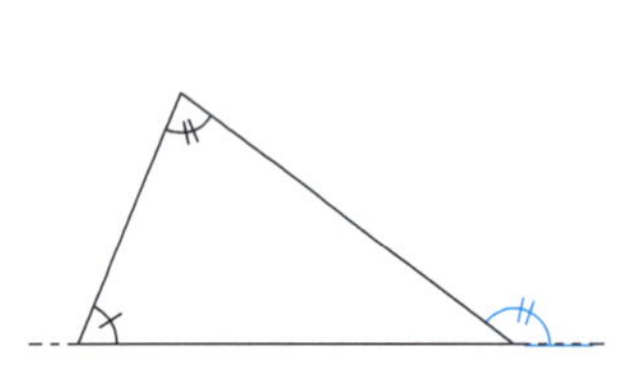

Figure 15-14

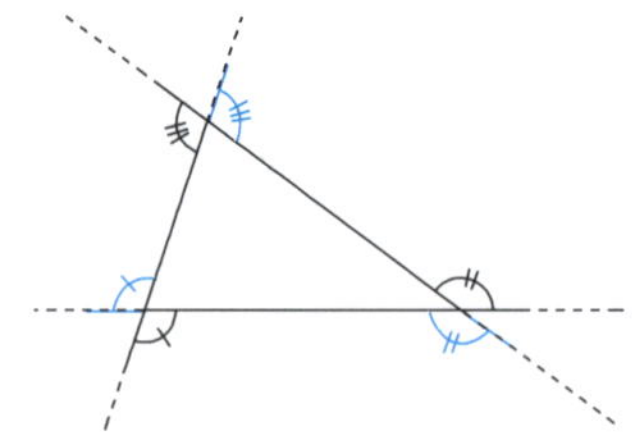

Figure 15-15

Proof: 1. Property 1 is based on the fact that the interior angle and its corresponding exterior angle form a linear pair.

性质 1 的根据是内角和它对应的外角形成邻补角。

2. To prove Property 2, as shown in Figure 15-16, in $\triangle ABC$, let $\angle C'$ be the exterior angle of C. We want to show that $m\angle A + m\angle B = m\angle C'$. Note that $m\angle A + m\angle B + m\angle C = 180°$ since the sum of measures of interior angles of every triangle is 180°. Also, $m\angle C + m\angle C' = 180°$ because of the linear pair. By transitive, $m\angle A + m\angle B + m\angle C = m\angle C + m\angle C'$, from which $m\angle A + m\angle B = m\angle C'$.

要证明性质 2,如图 15-16 所示,让$\angle C'$为 C 的外角。想证明 $m\angle A + m\angle B = m\angle C'$。因为三角形的内角之和等于 180°,所以 $m\angle A + m\angle B + m\angle C = 180°$。因为邻补角的关系,所以 $m\angle C + m\angle C' = 180°$。传递性给出了 $m\angle A + m\angle B + m\angle C = m\angle C + m\angle C'$,从而得知 $m\angle A + m\angle B = m\angle C'$。

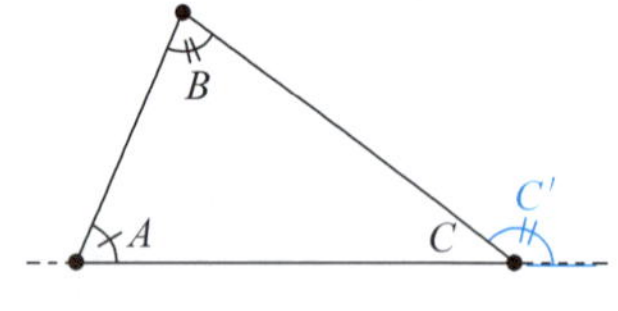

Figure 15-16

3. To prove Property 3, note that:

The sum of measures of the three straight angles is $180° \times 3 = 540°$.

The sum of measures of the three interior angles is 180°.

Therefore, the sum of measures of the three exterior angles is $540° - 180° = 360°$.

要证明性质 3,注意:

三个平角的度数之和是 $180° \times 3 = 540°$。

三个内角的度数之和是 180°。

三个外角的度数之和是 $540° - 180° = 360°$。

15.3.3 Summary 总结

In every triangle:

在任意三角形中:

1. The sum of measures of its interior angles is 180°. 内角的度数之和为 180°。

2. (1) The sum of measures of an interior angle and its corresponding exterior angle is 180°. This is

also true for any other convex polygons. 内角和它对应的外角的度数之和为 180°。对其他凸多边形也相同。

(2) Exterior Angle Theorem: The sum of measures of two interior angles is equal to the measure of the exterior angle of the third angle. 外角定理：两个内角的度数之和等于第三个角的外角度数。

(3) The sum of measures of its exterior angles is 360°. 外角度数之和为 360°。

15.4 Sides of a Triangle 三角形的边

Triangle Inequality Theorem 三角不等式定理

n. [ˈtraɪˌæŋɡəl ɪnɪˈkwɑlɪtɪ ˈθɪərəm]

Theorem: The sum of lengths of any two sides of a triangle is greater than the length of the third side. 定理：三角形任意两边的长度之和大于第三边的长度。

Corollary: The difference of lengths of any two sides of a triangle is less than the length of the third side. 推论：三角形任意两边的长度之差小于第三边的长度。

Proof: 1. For the theorem:

定理的证明：

If we can show that the sum of lengths of two shorter sides of every triangle is greater than the length of the longest side, then we are done. 若能证明较短的两边的长度之和大于第三边的长度，则证明完成。

Suppose in $\triangle ABC$, $AB < AC < BC$. We want to show that $AB + AC > BC$. 对于$\triangle ABC$，假设 $AB < AC < BC$。想证明 $AB + AC > BC$。

As shown in Figure 15-17, draw $\overline{AD}$, for which AD is the distance between A and $\overline{BC}$ (as we will introduce in Section 15.6.2.5, $\overline{AD}$ is an altitude of $\triangle ABC$). 如图 15-17 所示，画出 $\overline{AD}$，使得 AD 是 A 到 $\overline{BC}$ 的距离（将在 15.6.2.5 节介绍，$\overline{AD}$ 是$\triangle ABC$ 的高）。

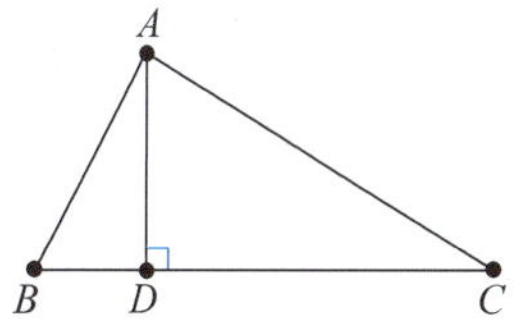

Figure 15-17

By the property of distance, note that BD is the distance from B to $\overline{AD}$, and $\overline{BD}$ is the shortest line segment from B to any point on $\overleftrightarrow{AD}$. Therefore, $AB > BD$. 根据距离的性质，BD 是 B 到 $\overline{AD}$ 的距离。$\overline{BD}$ 是从 B 到 $\overleftrightarrow{AD}$ 上任何一点的最短线段。所以 $AB > BD$。

Similarly, we can show that $AC > DC$. 同理，可得出 $AC > DC$。

Adding these two inequalities, we get $AB + AC > BD + DC$, for which $AB + AC > BC$, by Segment Addition Postulate. 两不等式相加可得出 $AB + AC > BD + DC$，用线段相加公理即得 $AB + AC > BC$。

2. For the corollary:

推论的证明：

If we can show that the difference of lengths of two longer sides of every triangle is less than the length of the shortest side, then we are done. 若能证明较长的两边的长度之差小于第三边的长度，则证明完成。

Suppose in $\triangle ABC$, $AB<AC<BC$. We want to show that $BC-AC<AB$.

We have shown that $AB+AC>BC$ is true. Subtracting AC from both sides gives us $AB>BC-AC$, from which $BC-AC<AB$.

Questions: Determine whether each of the following can be the side-lengths of a triangle. 判断以下每一组数是否可以成为一个三角形的边长。

(1) 4,5,6

(2) 1,2,3

Answers: (1) Yes.

(2) No, since $1+2>3$ is false.

15.5 Congruent Triangles 全等三角形

15.5.1 Introduction 介绍

Congruent Triangles 全等三角形

n. [kən'gruənt 'traɪˌæŋgəls]

Definition: Two triangles are congruent triangles if their corresponding sides and corresponding angles are congruent. 若两个三角形的对应边全等以及对应角全等,则它们为全等三角形。

Notation: If $\triangle ABC$ is congruent to $\triangle DEF$, then we write $\triangle ABC \cong \triangle DEF$.

读作: $\triangle ABC$ 和 $\triangle DEF$ 全等。

写作: $\triangle ABC \cong \triangle DEF$。

Properties: 1. If $\triangle ABC \cong \triangle DEF$, then all of the followings are true.

(1) $\overline{AB} \cong \overline{DE}$.

(2) $\overline{AC} \cong \overline{DF}$.

(3) $\overline{BC} \cong \overline{EF}$.

(4) $\angle BAC \cong \angle EDF$ (or simply $\angle A \cong \angle D$).

(5) $\angle ABC \cong \angle DEF$ (or simply $\angle B \cong \angle E$).

(6) $\angle ACB \cong \angle DFE$ (or simply $\angle C \cong \angle F$).

This is illustrated in Figure 15-18.

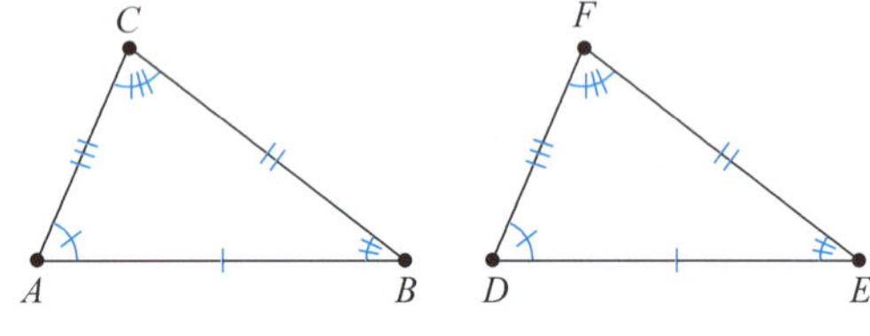

Figure 15-18

2. Property 1 illustrates CPCTC—Corresponding Parts of Congruent Triangles are Congruent. The converse of Property 1 is also true by the definition of congruent triangles. 性质1展现了全等三角形的对应部分全等的性质。根据全等三角形的定义,性质1的逆命题也成立。

3. We need at least three pieces of information (each of which is a side or angle) from each of the two triangles to determine whether they are congruent. 判断两个三角形是否全等,至少需要知道每个三角形的三个已知条件(边或角)。

4. Congruent triangles have the same perimeter and area. 全等三角形的周长相等,面积相等。

15.5.2 Determine Whether Two Triangles Are Congruent　判断两个三角形是否全等

SSS Congruence Postulate　边边边全等公理

Definition：If all corresponding sides of two triangles are congruent，then the two triangles are congruent. SSS is the abbreviation of Side-Side-Side.　若两个三角形所有对应的边全等，则这两个三角形全等。SSS 是边边边的缩写。

Examples：As shown in Figure 15-19，we can conclude that these two triangles are congruent.

SAS Congruence Postulate　边角边全等公理

Definition：If the two corresponding sides and the included angle of two triangles are congruent，then the two triangles are congruent. SAS is the abbreviation of Side-Angle-Side.　若两个三角形两条对应的边全等，并且夹角全等，则这两个三角形全等。SAS 是边角边的缩写。

Examples：As shown in Figure 15-20，we can conclude that these two triangles are congruent.

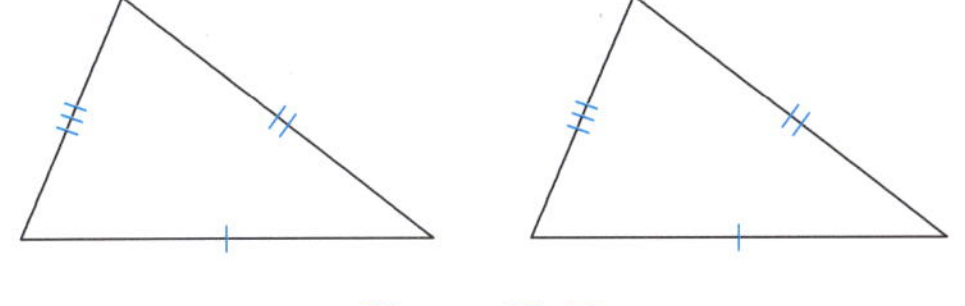

Figure 15-19

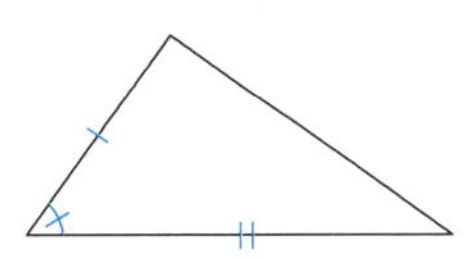

Figure 15-20

ASA Congruence Postulate　角边角全等公理

Definition：If the two corresponding angles and the included side of two triangles are congruent，then the two triangles are congruent. ASA is the abbreviation of Angle-Side-Angle.　若两个三角形两个对应的角全等，且夹边全等，则这两个三角形全等。ASA 是角边角的缩写。

Examples：As shown in Figure 15-21，we can conclude that these two triangles are congruent.

AAS Congruence Theorem　角角边全等定理

Definition：If the two corresponding angles and the non-included side of two triangles are congruent，then the two triangles are congruent. AAS is the abbreviation of Angle-Angle-Side.　若两个三角形两个对应的角全等，且非夹边全等，则这两个三角形全等。AAS 是角角边的缩写。

Proof：If we know the measures of two angles in a triangle，we will know the measure of the third angle because the sum of measures of interior angles of every triangle is 180°. This leads to the ASA Congruence Postulate.　若知道三角形中两角的度数，则能求出第三个角的度数（三角形的内角度数之和等于 180°）。这就变成了角边角的情况。

Examples：As shown in Figure 15-22，we can conclude that these two triangles are congruent.

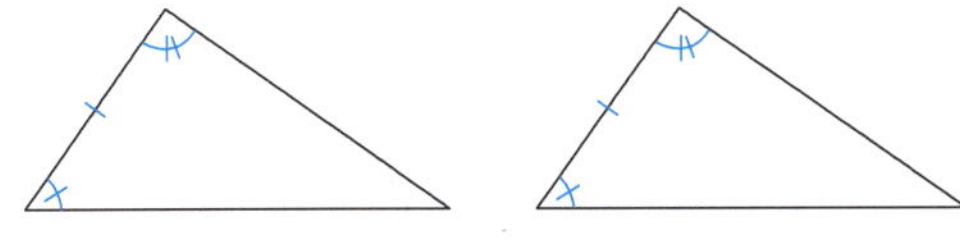

Figure 15-21

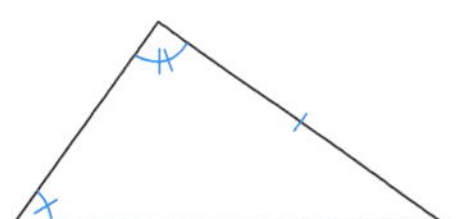

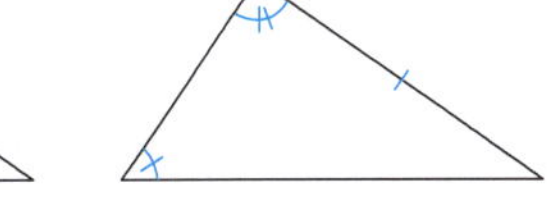

Figure 15-22

HL Congruence Theorem 斜边直角边全等定理

Definition：If two right triangles have congruent hypotenuses and one congruent leg, then the two triangles are congruent. HL is the abbreviation of Hypotenuse-Leg. 若两个直角三角形的斜边全等，其中一条直角边也全等，则这两个三角形全等。HL 是斜边直角边的缩写。

Proof：Proof of the HL Congruence Theorem：If two right triangles have congruent hypotenuses and one congruent leg, then we can use the Pythagorean Theorem and conclude that the other leg for both triangles are congruent. This leads to the SSS or SAS Congruence Postulate.

HL 定理的证明：若两个直角三角形的斜边全等，其中一条直角边也全等，则根据勾股定理可得知两个三角形的另一条直角边全等。现在就可用边边边或边角边得知它们全等了。

Examples：As shown below in Figure 15-23, we can conclude that these two triangles are congruent.

Combos That Cannot Conclude Two Triangles Are Congruent 不能断定两个三角形全等的组合条件

Definition：AAA (Angle-Angle-Angle)—Knowing that two triangles' all corresponding angles are congruent cannot conclude that the triangles are congruent. 角角角——即使已知两个三角形所有对应的角全等，也不能断定这两个三角形全等。

SSA (Side-Side-Angle)—Knowing that two triangles' corresponding sides and the corresponding non-included angle are congruent cannot conclude that the triangles are congruent. 边边角——已知两个三角形的两条对应的边全等，并且非夹角全等，也不能断定这两个三角形全等。

Proofs：1. To prove why AAA does not work, we give a counterexample：Each angle is measured 60° in all equilateral triangles, but there is more than one size of equilateral triangles, as shown in Figure 15-24. 证明为什么角角角不能断定两个三角形全等。反例如下：所有等边三角形角的度数都是 60°，但是有不止一种大小的等边三角形，如图 15-24 所示。

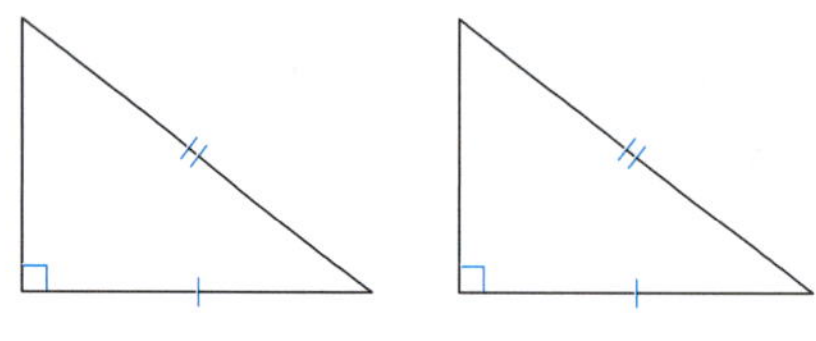

Figure 15-23

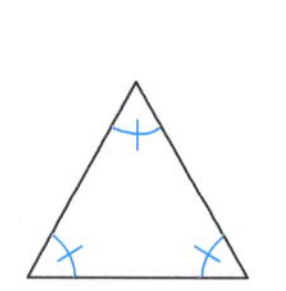

Figure 15-24

In fact, knowing any two triangles' all corresponding angles are congruent can only give information that they are similar to each other, not congruent. 其实知道两个三角形所有对应的角相等，只能断定这两个三角形相似，不能断定全等。

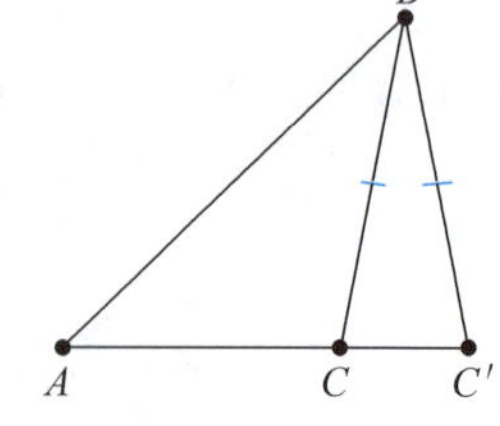

Figure 15-25

2. To prove why SSA does not work, consider the following counterexample. 要证明为什么边边角不能断定两个三角形全等，考虑以下的反例：

As shown in Figure 15-25, we know that $\angle A \cong \angle A$, $\overline{AB} \cong$

$\overline{AB}$, and $\overline{BC} \cong \overline{BC'}$, but we cannot conclude that $\triangle ABC \cong \triangle ABC'$.

15.5.3 Theorems, Proofs, and Applications of Triangles 三角形的定理、证明及应用

视频 24

Corresponding Parts of Congruent Triangles are Congruent (CPCTC) 全等三角形的对应部分全等

Properties: This is a very powerful tool to use in proving that two line segments or angles are congruent. See Property 1 of congruent triangles in Section 15.5.1. CPCTC 是一个证明两线段和两个角全等的有力工具。见 15.5.1 节全等三角形词条的性质 1。

Theorems and Proofs of Triangles 三角形的定理与证明

Theorems: 1. (1) **Base Angle Theorem**— In $\triangle ABC$, if $\overline{AB} \cong \overline{BC}$, then $\angle A \cong \angle C$ (The base angles of every isosceles triangle are congruent). **等边对等角定理**：在同一个三角形中，若两边全等，则这两边所对的角也全等(等腰三角形的底角全等)。

(2) **Corollary of Base Angle Theorem**—If a triangle is equilateral, then it is equiangular. **等边对等角定理的推论**：等边三角形是等角三角形。

2. (1) **Converse of Base Angle Theorem**—In $\triangle ABC$, if $\angle A \cong \angle C$, then $\overline{AB} \cong \overline{BC}$ (The legs of every isosceles triangle are congruent). **等边对等角定理的逆定理**，又称**等角对等边定理**：在同一个三角形中，若两角全等，则这两角所对应的边也全等(等腰三角形的腰全等)。

(2) **Corollary of Converse of Base Angle Theorem**—If a triangle is equiangular, then it is equilateral. 等角对等边定理的推论：**等角三角形是等边三角形。**

Proofs: 1. (1) To prove the **Base Angle Theorem** (Theorem 1(1)), let D be a point on $\overline{AC}$ such that $\overline{BD} \perp \overline{AC}$, as shown in Figure 15-26. Table 15-2 gives the proof.

Figure 15-26

Table 15-2

Statements 命题	Reasons 原因
1. $\overline{AB} \cong \overline{BC}$	1. Given
2. $\overline{BD} \perp \overline{AC}$	2. Construction 画图
3. $\overline{BD} \cong \overline{BD}$	3. Reflexive Property 反身性
4. $\angle BDA$ and $\angle BDC$ are right angles	4. Definition of perpendicular 垂直的定义
5. $\angle BDA \cong \angle BDC$	5. All right angles are congruent 所有直角全等
6. $\triangle BAD \cong \triangle BCD$	6. HL Congruence Theorem HL 定理
7. $\angle A \cong \angle C$	7. CPCTC 全等三角形的对应部分全等

(2) To prove the **Corollary of Base Angle Theorem** (Theorem 1(2)), we will use the **Base Angle Theorem** (Theorem 1(1)). Given that $\overline{AB} \cong \overline{AC} \cong \overline{BC}$ in $\triangle ABC$, we can conclude that $\angle A \cong \angle B \cong \angle C$, or $m\angle A = m\angle B = m\angle C$. We have

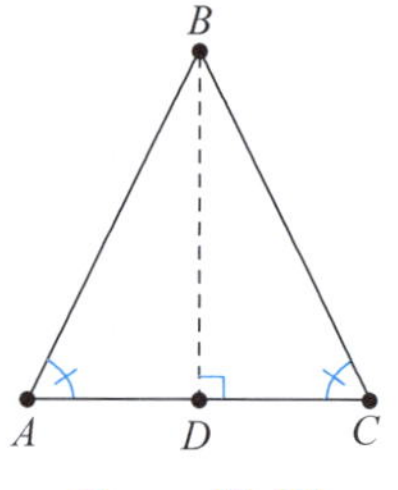

Figure 15-27

proved that the $\triangle ABC$ is equiangular. Moreover, since $m\angle A + m\angle B + m\angle C = 180°$, we have $m\angle A = m\angle B = m\angle C = 60°$.

2. (1) To prove the **Converse of Base Angle Theorem** (Theorem 2(1)), let D be a point on $\overline{AC}$ such that $\overline{BD} \perp \overline{AC}$, as shown in Figure 15-27.

Table 15-3 gives the proof.

Table 15-3

Statements 命题	Reasons 原因
1. $\angle A \cong \angle C$	1. Given
2. $\overline{BD} \perp \overline{AC}$	2. Construction 画图
3. $\overline{BD} \cong \overline{BD}$	3. Reflexive Property 反身性
4. $\angle BDA$ and $\angle BDC$ are right angles	4. Definition of perpendicular 垂直的定义
5. $\angle BDA \cong \angle BDC$	5. All right angles are congruent 所有直角全等
6. $\triangle BAD \cong \triangle BCD$	6. AAS Congruence Theorem 角角边
7. $\overline{AB} \cong \overline{BC}$	7. CPCTC: Recall that $\overline{AB}$ and $\overline{BA}$ are the same line segment 全等三角形的对应部分全等。注意 $\overline{AB}$ 和 $\overline{BA}$ 是同一条线段

(2) To prove the **Corollary of the Converse of Base Angle Theorem** (Theorem 2(2)), we will use the **Converse of Base Angle Theorem** (Theorem 2(1)). Given that $\angle A \cong \angle B \cong \angle C$ in $\triangle ABC$, we can conclude that $\overline{AB} \cong \overline{AC} \cong \overline{BC}$. We have proved that the $\triangle ABC$ is equilateral.

Applications of Congruence Theorems 全等三角形的应用

Properties: In the coordinate plane, the slopes of two perpendicular lines are opposite reciprocals. 在坐标平面上,两垂直线的斜率互为负倒数。

Proof: Given that lines $l \perp m$, we want to show that their slopes are opposite reciprocals. 已知 $l \perp m$,下面证明它们的斜率互为负倒数。

If l and m are horizontal and vertical lines, it is quite clear that their slopes are opposite reciprocals—the slope of the horizontal line is 0, and the slope of the vertical line is the opposite reciprocal of 0, which is undefined. Let's consider that case that neither of l and m is horizontal or vertical. 若 l 和 m 为横线和竖线,则它们的斜率很明显互为负倒数——横线的斜率是 0,竖线的斜率是 0 的负倒数,也就是无意义。现讨论非特殊情况。

Let C be the intersection of l and m. Construct horizontal line n passing C. Construct points A on l, B on n such that $\overline{AB} \perp n$. Construct point E on m such that $\overline{AC} \cong \overline{CE}$, and D on n such that $\overline{DE} \perp n$, as shown in Figure 15-28.

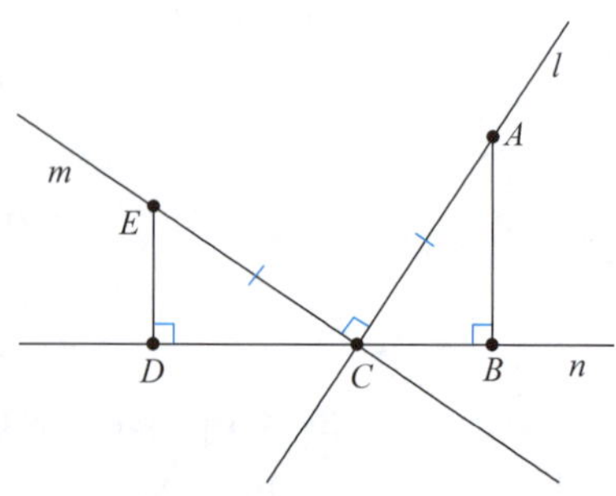

Figure 15-28

Table 15-4 gives the proof.

Table 15-4

Statements　命题	Reasons　原因
1. $l \perp m$	1. Given　已知条件
2. $\overline{AC} \cong \overline{CE}$	2. Construction　画图
3. $\angle ABC$ and $\angle CDE$ are right angles	3. $\overline{AB} \perp n$, $\overline{DE} \perp n$, and C is on n
4. $\angle ABC \cong \angle CDE$	4. All right angles are congruent　所有直角全等
5. $\angle BAC$ and $\angle ACB$ are complementary $\angle ACB$ and $\angle DCE$ are complementary	5. Definition of complementary angles　余角的定义
6. $\angle BAC \cong \angle DCE$	6. Property of complementary angles. See complementary angles　余角的性质。见余角
7. $\triangle ABC \cong \triangle CDE$	7. AAS Congruence Theorem　角角边
8. $\overline{AB} \cong \overline{CD}$, and $\overline{BC} \cong \overline{DE}$	8. CPCTC　全等三角形的对应部分全等
9. $AB = CD$, and $BC = DE$	9. Property of congruent line segments　全等线段的性质
10. Slope of $l = AB/BC$ Slope of $m = -DE/CD$	10. Definition of slopes　斜率的定义
11. Slope of $l = AB/BC$ Slope of $m = -BC/AB$	11. Substitution　代入
12. Slopes of l and m are opposite reciprocals	12. Definition of opposite reciprocals　负倒数的定义

15.5.4 Summary　总结

1. We need at least three pieces of information (each of which is a side or an angle) from each of the two triangles to determine whether they are congruent.　判断两个三角形是否全等,至少需要知道每个三角形的三个已知条件(边或角)。
   (1) Combos that can prove two triangles are congruent: SSS, SAS, ASA, AAS, HL.　能证明两个三角形全等的组合条件为边边边、边角边、角边角、角角边、HL。
   (2) Combos that cannot prove two triangles are congruent: AAA, SSA.　不能证明两个三角形全等的组合条件为角角角、边边角。
2. CPCTC: Corresponding Parts of Congruent Triangles are Congruent. This is a very powerful tool to prove two line segments or two angles are congruent from congruent triangles.　全等三角形的对应部分全等:一个从全等三角形中证明全等线段或全等角的有力工具。
3. Popular theorems/corollaries:
   常见的定理/推论:
   (1) **Base Angles Theorem.　等边对等角定理。**
   (2) If a triangle is equilateral, then it is equiangular.　等边三角形是等角三角形。
   (3) **Converse of the Base Angles Theorem.　等角对等边定理。**
   (4) If a triangle is equiangular, then it is equilateral.　等角三角形是等边三角形。

15.6 Special Line Segments of a Triangle 三角形的特殊线段

15.6.1 Introduction 介绍

Concurrent 共点

adj. [kən'kɜrənt]

Definition: Three or more lines/rays/line segments/curves are concurrent if they all share exactly one common point. This point is known as the point of concurrency. 当三条或更多直线/射线/线段/曲线共点,即它们都有一个共同点。这个点叫作交点。

If two lines/rays/line segments/curves share exactly one common point, we call that point "point of intersection", or simply intersection. 两条或更多直线/射线/线段/曲线有一个共同点,这个点叫作交点。

Point of intersection 和 point of concurrency 的意思都是交点。前者是形容两线的共同点;后者是形容三条或更多线的共同点。

As shown below in Figure 15-29, P is the point of intersection of two lines.

As shown below in Figure 15-30, P is the point of concurrency of four lines.

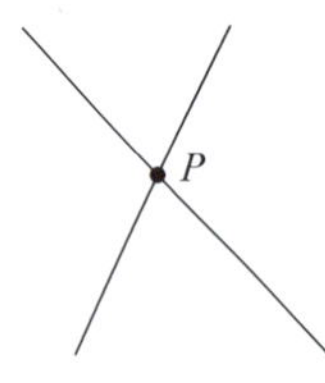

Figure 15-29

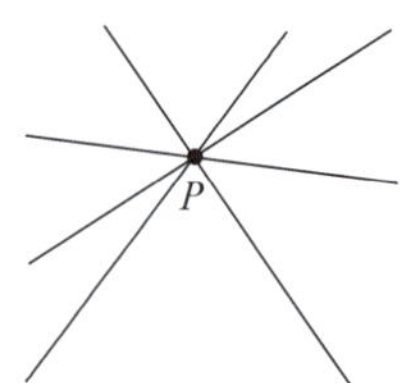

Figure 15-30

15.6.2 Special Line Segments 特殊线段

视频 25

15.6.2.1 Midsegments 中位线

Midsegment 中位线

n. [mɪd'segmənt]

Definition: A midsegment of a triangle T is a line segment whose endpoints are the midpoints of two sides of T, as shown in blue in Figure 15-31. 三角形 T 的中位线是一条端点为两边中点的线段。见图 15-31 的蓝色线段。

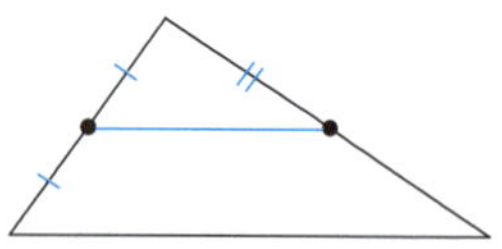
Figure 15-31

Properties: 1. **Midsegment Theorem**: In every triangle, a midsegment drawn from any two sides is parallel to the third side, with half of its length. **中位线定理**:在任意三角形中,连接两边的中位线与第三边平行,其长度是第三边的一半。

2. Three midsegments of a triangle T divide T into four smaller triangles with equal areas. 三角形 T 的所有中位线把 T 分成面积相等的 4 个小三角形。

Proofs: 1. Proof of Property 1: Without using similar triangles, the proof of the **Midsegment Theorem** will be incredibly long (but possible!). We will save it for later. See Theorem 3 of common theorems of similarity in Section 15.10.3 for the proof. 性

质 1 的证明：如果不用相似三角形，证明**中位线定理**是相当困难的。证明见 15.10.3 节“常见的相似定理”。

2. Proof of Property 2: Figure 15-32 labels the base b and height h for the large triangle T. Using similar triangles and **same base/altitude property**, we know that the base for each of the smaller triangles is half of that of T, and the height for each of the smaller triangles is also half of that of T. Therefore, each smaller triangle's area is 1/4 of that of T, as shown. 性质 2 的证明：图 15-32 标记了大三角形 T 的底 b 与高 h。用**相似三角形和等底/等高性质**，可得出每个小三角形的底长度是大三角形的底长度的一半。每个小三角形的高长度是大三角形的高长度的一半。因此，每个小三角形的面积是大三角形面积的 1/4。

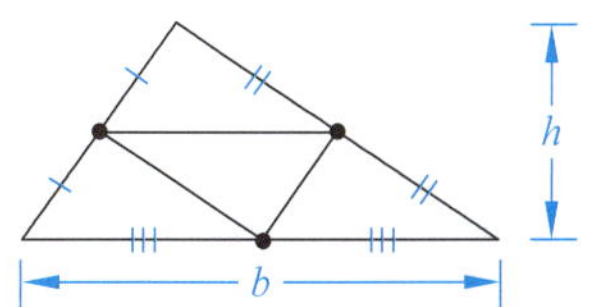

Figure 15-32

15.6.2.2 Perpendicular Bisectors 垂直平分线

Perpendicular Bisector 垂直平分线

n. [ˌpərpənˈdɪkjələr baɪˈsektɚ]

Definition: Suppose M is the midpoint of $\overline{AB}$. The perpendicular bisector l of $\overline{AB}$ is a line/ray/line segment that passes through M such that $l \perp \overline{AB}$. As shown in Figure 15-33, line l is the perpendicular bisector of $\overline{AB}$. 设 M 是 $\overline{AB}$ 的中点，$\overline{AB}$ 的垂直平分线 l 是一条经过 M 的直线/射线/线段，满足 $l \perp \overline{AB}$，如图 15-33 所示。

A perpendicular bisector is a special line segment bisector. 垂直平分线是特殊的线段平分线。

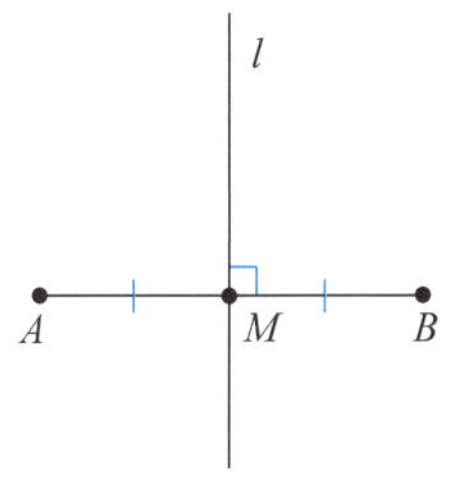

Figure 15-33

Properties: 1. **Perpendicular Bisector Theorem**: If P is a point on the perpendicular bisector of $\overline{AB}$, then $\overline{PA} \cong \overline{PB}$ (Any point on the perpendicular bisector of a line segment is equidistant to the endpoints of the line segment). **垂直平分线定理**：若 P 是 $\overline{AB}$ 的垂直平分线上的一点，则 $\overline{PA} \cong \overline{PB}$（一条线段垂直平分线上的任意一点离线段两端是等距的）。

2. **Converse of Perpendicular Bisector Theorem**: If $\overline{PA} \cong \overline{PB}$, then P is a point on the perpendicular bisector of $\overline{AB}$ (Any point equidistant to the endpoints of a line segment is on the perpendicular bisector of that line segment). **垂直平分线定理的逆定理**：若 $\overline{PA} \cong \overline{PB}$，则 P 是 $\overline{AB}$ 的垂直平分线上的一点（离一条线段两端等距的点都在线段的垂直平分线上）。

3. **Concurrency of Perpendicular Bisectors Theorem**: The perpendicular bisectors of the three sides of every triangle are concurrent. **垂直平分线共点定理**：任意三角形三条边的垂直平分线共点。

4. In a triangle, the point of concurrency of the perpendicular bisectors of the sides is known as the circumcenter, usually denoted by O, which is equidistant from each of the vertices of the triangle. A line segment whose endpoints are O and a vertex of the triangle is known as a circumradius. The circle with the circumcenter as the

center and the circumradius as a radius is known as the circumcircle. 三角形三边的垂直平分线的交点叫作外心，通常用 O 表示。它跟三角形的三个顶点都等距。端点 O 和三角形一个顶点的线段叫作外接圆半径。圆心为外心、半径为外接圆半径的圆叫作外接圆。

5. Knowing two perpendicular bisectors of a triangle immediately gives the third (which passes the intersection of the two perpendicular bisectors and the midpoint of the remaining side). 已知一个三角形两条边的垂直平分线就能立刻给出第三条边的垂直平分线(第三条垂直平分线经过两条垂直平分线的交点和第三边的中点)。

Proof: 1. Proof of Property 1: Let P be a point on the perpendicular bisector l of $\overline{AB}$, and l intersects $\overline{AB}$ at M. By the definition of perpendicular bisectors, M is the midpoint of $\overline{AB}$. 性质1的证明：P 为 $\overline{AB}$ 的垂直平分线 l 上的一点，M 为 l 和 $\overline{AB}$ 的交点。因为垂直平分线的定义，M 是 $\overline{AB}$ 的中点。We wish to show that $\overline{PA} \cong \overline{PB}$. If P is the same point as M, then we are done by the definition of midpoint. Now, consider that P is not the same point as M, as shown in Figure 15-34. 目标是证明 $\overline{PA} \cong \overline{PB}$。若 P 和 M 重合，则证毕(中点的定义)。若 P 和 M 不重合，如图 15-34 所示。

Figure 15-34

Table 15-5 gives the proof of Property 1.

Table 15-5

Statements 命题	Reasons 原因
1. P is a point on l and P is not on $\overline{AB}$. P 在 l 上但不在 $\overline{AB}$ 上	1. Given
2. $\overline{AM} = \overline{MB}$ $\angle AMP$ and $\angle BMP$ are right angles	2. Definition of perpendicular bisector 垂直平分线的定义
3. $\angle AMP \cong \angle BMP$	3. All right angles are congruent 所有直角全等
4. $\overline{PM} = \overline{PM}$	4. Reflexive Property 反身性
5. $\triangle AMP \cong \triangle BMP$	5. SAS Congruence Postulate 边角边
6. $\overline{PA} \cong \overline{PB}$	6. CPCTC 全等三角形的对应部分全等

2. Proof of Property 2: Suppose $\overline{PA} \cong \overline{PB}$. If P is on $\overline{AB}$, then P is the midpoint of $\overline{AB}$, which must be on the perpendicular bisector of $\overline{AB}$. Consider that P is not on $\overline{AB}$, as shown in Figure 15-35. 性质2的证明：假设 $\overline{PA} \cong \overline{PB}$。若 P 在 $\overline{AB}$ 上，P 是 $\overline{AB}$ 的中点，必然在 $\overline{AB}$ 的垂直平分线上。若 P 不在 $\overline{AB}$ 上，如图 15-35 所示。

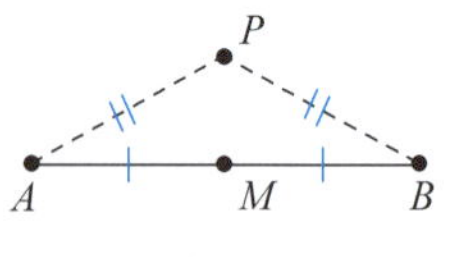

Figure 15-35

We wish to prove that P is on the perpendicular bisector of $\overline{AB}$. Let M be the midpoint of $\overline{AB}$. If we can prove that $\overleftrightarrow{PM}$ bisects $\overline{AB}$ and $\overleftrightarrow{PM} \perp \overline{AB}$, then we are done, by the definition of perpendicular bisectors. 目标是证明 P 在 $\overline{AB}$ 的垂直平分线上。设 M 为 $\overline{AB}$ 的中点。如果能证明 $\overleftrightarrow{PM}$ 对分 $\overline{AB}$ 并且 $\overleftrightarrow{PM} \perp \overline{AB}$，根据垂直平分线的定义，证明就完成了。

Table 15-6 gives the proof of Property 2.

Table 15-6

Statements　命题	Reasons　原因
1. $\overline{PA} \cong \overline{PB}$, and P is not on $\overline{AB}$	1. Given
2. $\overline{AM} \cong \overline{MB}$	2. Definition of midpoints　中点的定义
3. $\overleftrightarrow{PM}$ bisects $\overline{AB}$	3. The midpoint of $\overline{AB}$ is on $\overleftrightarrow{PM}$　$\overline{AB}$ 的中点在 $\overleftrightarrow{PM}$ 上
4. $\overline{PM} \cong \overline{PM}$	4. Reflexive Property　反身性
5. $\triangle AMP \cong \triangle BMP$	5. SSS Congruence Postulate　边边边
6. $\angle AMP \cong \angle BMP$	6. CPCTC　全等三角形的对应部分全等
7. $\angle AMP$ and $\angle BMP$ forms a linear pair. $\angle AMP$ 和 $\angle BMP$ 形成邻补角	7. Definition of linear pairs　邻补角
8. $\angle AMP$ and $\angle BMP$ are right angles. $\angle AMP$ 和 $\angle BMP$ 都是直角	8. Steps 6 and 7
9. $\overleftrightarrow{PM} \perp \overline{AB}$	9. Definition of perpendicular　垂直的定义
10. P is on the perpendicular bisector of $\overline{AB}$. P 在 $\overline{AB}$ 的垂直平分线上	10. Steps 3 and 9

3. Proof of Property 3: In $\triangle ABC$, suppose lines l, m, and n are perpendicular bisectors of $\overline{AB}$, $\overline{BC}$, and $\overline{AC}$ respectively, and O be the intersection of l and m, as shown in Figure 15-36.　性质 3 的证明：在$\triangle ABC$ 中，假设直线 l、m、n 分别是 $\overline{AB}$、$\overline{BC}$、$\overline{AC}$ 的垂直平分线，且 O 是 l 与 m 的交点，如图 15-36 所示。

   We are given that O is on both l and m, and wish to prove that O is on n. Property 1 states that since O is on l, $\overline{OA} \cong \overline{OB}$, and since O is on m, $\overline{OB} \cong \overline{OC}$. By transitive, $\overline{OA} \cong \overline{OC}$. By Property 2, O is on n.　知道 O 在 l 和 m 上，希望证明 O 在 n 上。因为 O 在 l 上，从性质 1 可得出 $\overline{OA} \cong \overline{OB}$。同时因为 O 在 m 上，$\overline{OB} \cong \overline{OC}$。根据传递性得出 $\overline{OA} \cong \overline{OC}$。由性质 2 可得 O 在 n 上。

   This also illustrates Property 5.

Circumcenter　外心

n. [ˈsɜrkəmˈsentər]

Definition: The point of concurrency of three perpendicular bisectors of a triangle. As shown in Figure 15-37, O is the circumcenter.　外心是指三角形的三条垂直平分线的交点。如图 15-37 所示，O 为外心。

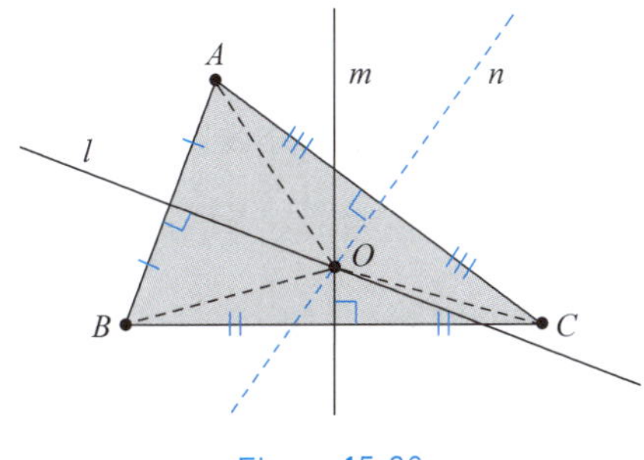

Figure 15-36

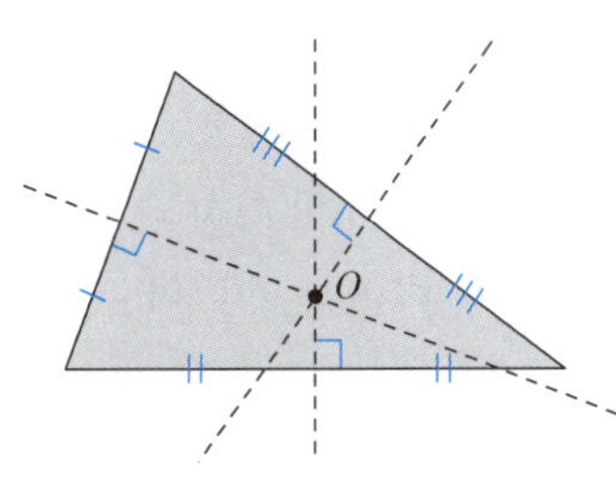

Figure 15-37

Notation: The circumcenter is denoted by the letter O. 外心通常用 O 表示。

Properties: 1. The circumcenter is equidistant from each of the vertices of the triangle. That is, for $\triangle ABC$, $\overline{OA} \cong \overline{OB} \cong \overline{OC}$. 外心离三角形的三个顶点等距。

2. As shown in Figure 15-38, the circumcenter of an acute triangle is inside the triangle. 如图 15-38 所示,锐角三角形的外心在三角形之内。

As shown in Figure 15-39, the circumcenter of a right triangle is the midpoint of the hypotenuse of the triangle. 如图 15-39 所示,直角三角形的外心在斜边的中点。

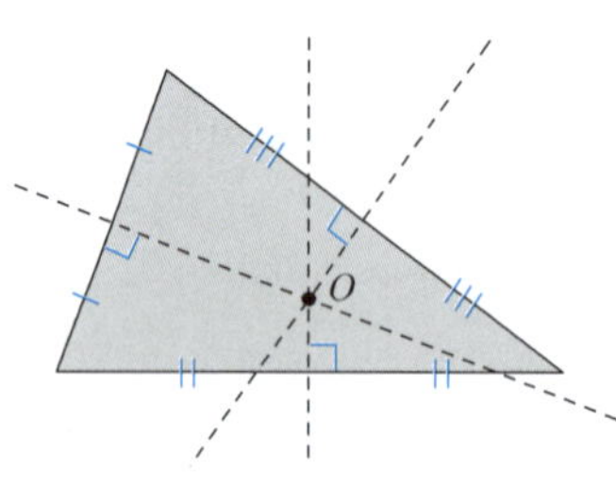

Figure 15-38

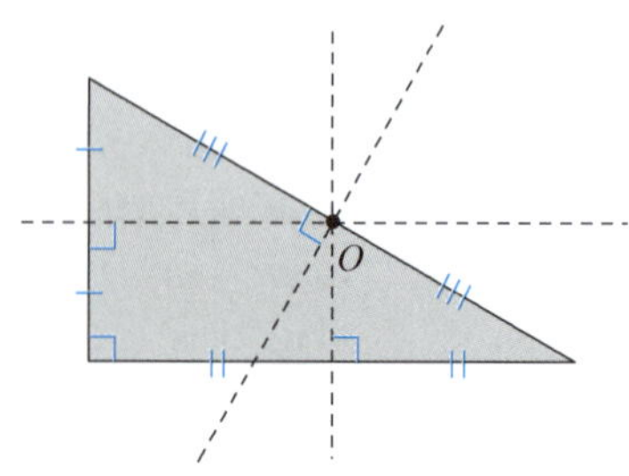

Figure 15-39

As shown in Figure 15-40, the circumcenter of an obtuse triangle is outside the triangle. 如图 15-40 所示,钝角三角形的外心在三角形之外。

Proof: 1. Proof of Property 1: We wish to show that $\overline{OA} \cong \overline{OB} \cong \overline{OC}$, in Figure 15-41.

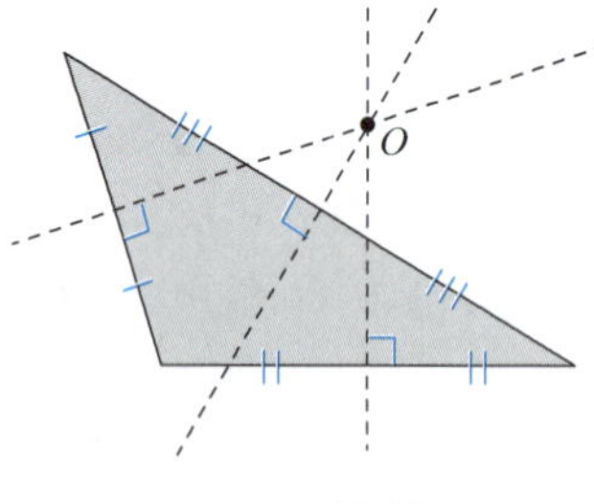

Figure 15-40

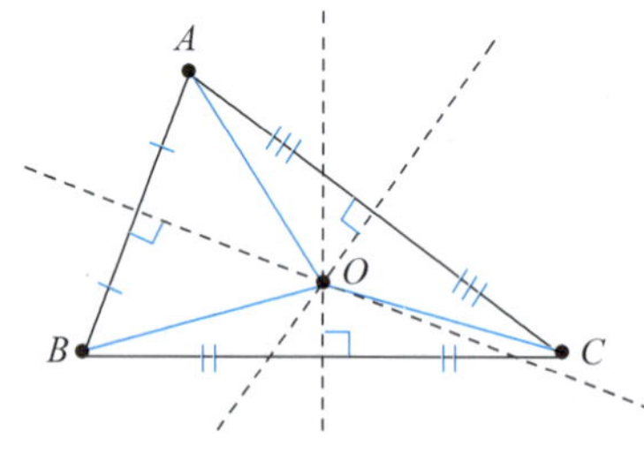

Figure 15-41

By the **Concurrency of Perpendicular Bisector Theorem**, perpendicular bisectors of the three sides of any triangle are concurrent. 见**垂直平分线共点定理**。

Since O is the point of concurrency of perpendicular bisectors, O is on each of the three perpendicular bisectors. Using the **Perpendicular Bisector Theorem**, we get $\overline{OA} \cong \overline{OB}$, $\overline{OB} \cong \overline{OC}$, and $\overline{OA} \cong \overline{OC}$, from which $\overline{OA} \cong \overline{OB} \cong \overline{OC}$ is true. 因为 O 是垂直平分线的交点,O 在三条垂直平分线上。用**垂直平分线定理**可得出 $\overline{OA} \cong \overline{OB}$,$\overline{OB} \cong \overline{OC}$,和 $\overline{OA} \cong \overline{OC}$。所以 $\overline{OA} \cong \overline{OB} \cong \overline{OC}$。

2. Proof of Property 2:

性质 2 的证明:

For now, start with the circle and observe. 开始观察圆。

This is a corollary from the **Inscribed Angle Theorem** in Section 19.1.5. **从圆周角定理**的推论可得出,见 19.1.5 节。

Circumcircle　外接圆

n. ['sɜrkəm'sɜrkəl]

Definition：The circumcircle of a triangle T is a circle that passes through all vertices of the T. Its center is the circumcenter, and its radius is the circumradius.　三角形 T 的外接圆是一个经过所有 T 的顶点的圆。它的圆心为外心，半径为外接圆半径。

As shown in Figure 15-42, the circumcircle of $\triangle ABC$ is drawn in blue, with center O. Note that $\overline{OA}$, $\overline{OB}$, and $\overline{OC}$ are radii of the circumcircle. They are called the circumradii of $\triangle ABC$.　如图 15-42 所示，$\triangle ABC$ 的外接圆已用蓝色画出，圆心为 O，$\overline{OA}$、$\overline{OB}$、$\overline{OC}$ 是外接圆的半径。

Circumradius　外接圆半径

n. ['sɜrkəm'reɪdiəs]

Definition：A radius of the circumcircle. One example would be a line segment whose endpoints are one vertex of a triangle and the triangle's circumcenter.　外接圆的半径，例如三角形的一个顶点到外心的线段。

As shown in Figure 15-43, some of the circumradii are shown in blue.

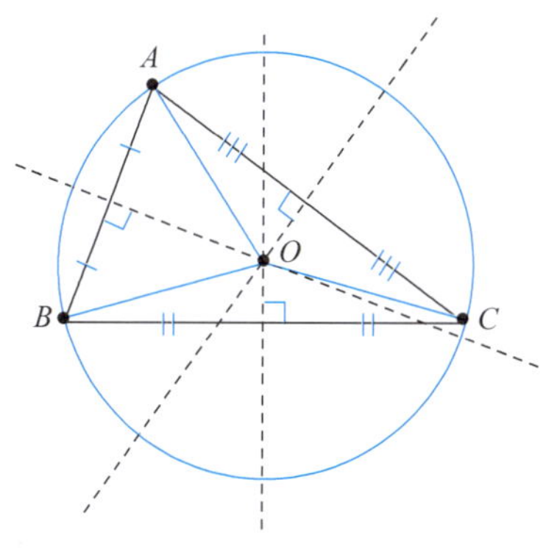

Figure 15-42

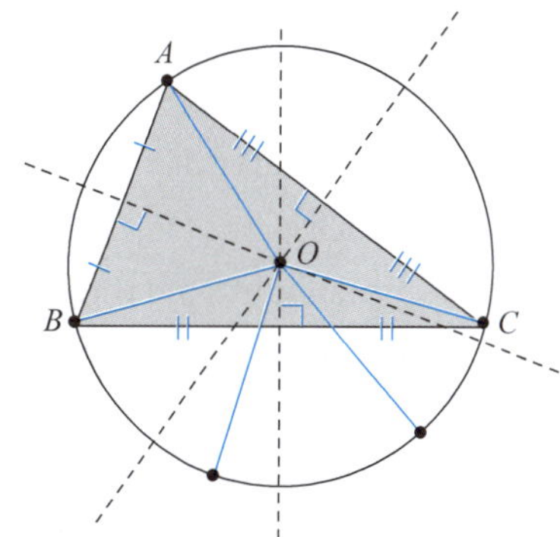

Figure 15-43

15.6.2.3　Angle Bisectors　角平分线

Angle Bisector　角平分线

n. ['æŋgəl baɪ'sektɚ]

Definition：The angle bisector of $\angle ABC$ is a line/ray/line segment passes through points B and P (B and P are different points), for which $\angle ABP \cong \angle CBP$, as shown in Figure 15-44. $\angle ABC$ 的角平分线是一条经过 B 和 P 的直线/射线/线段（B 和 P 不重合），满足 $\angle ABP \cong \angle CBP$，如图 15-44 所示。

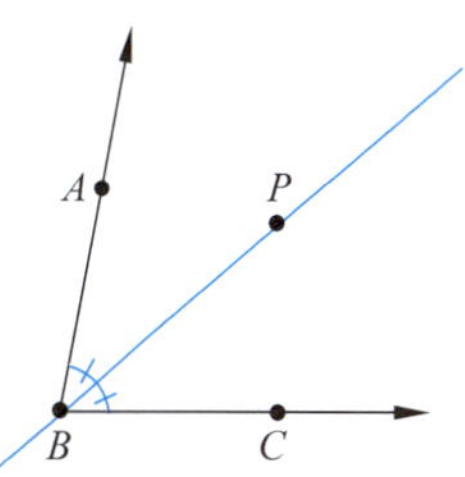

Figure 15-44

Properties：1. **Angle Bisector Property**: If P is a point on the angle bisector of $\angle ABC$, then P is equidistant from $\overline{AB}$ and $\overline{BC}$ (Any point on the angle bisector is equidistant from the sides of the angle).　**角平分线性质**：若 P 与 $\angle ABC$ 的角平分线上的一点，则 P 与 $\overline{AB}$ 和 $\overline{BC}$ 等距（任意在角平分线上的一点离角的两边等距）。

2. **Converse of Angle Bisector Property**: If P is a point equidistant from $\overline{AB}$ and $\overline{BC}$, then P is on the angle bisector of $\angle ABC$ (Any point equidistant from the sides of the angle is on the bisector of that angle). **角平分线性质的逆性质**:若 P 与 $\overline{AB}$ 和 $\overline{BC}$ 等距,则 P 是$\angle ABC$ 的角平分线上的一点(离角的两边等距的点都在角平分线上。)
3. **Concurrency of Angle Bisectors Theorem**: The angle bisectors of the three angles of every triangle are concurrent. **角平分线共点定理**:任意三角形三个角的角平分线共点。
4. In a triangle, the point of concurrency of the angle bisectors is known as the incenter, usually denoted by I, which is equidistant from each of the sides of the triangle. The line segment whose endpoints are I and a point on a side of the triangle is known as an inradius (must be perpendicular to that side). The circle with the incenter as the center and the inradius as a radius is known as the incircle. 三角形三角的角平分线的交点叫作内心,通常用 I 表示。它跟三角形的三边都等距。端点为 I 和三角形一条边上一点的线段叫作内切圆半径(必须与那条边垂直)。圆心为内心、半径为内切圆半径的圆叫作内切圆。
5. Knowing two angle bisectors in a triangle immediately gives the third (which passes the intersection of the two angle bisectors and the remaining vertex). 已知三角形的两条角平分线,就能给出第三条角平分线(第三条角平分线经过两条角平分线的交点和第三个角的顶点)。

Proof: 1. Proof of Property 1:

性质 1 的证明:

Let P be a point on the angle bisector of $\angle ABC$, and P is different from B (otherwise the case is trivial: P is equidistant from $\overline{AB}$ and $\overline{BC}$, for which the distance is 0), as shown in Figure 15-45. P 为$\angle ABC$ 平分线上的一点。P 不与 B 重合(否则就是一个特殊的情况:P 与 $\overline{AB}$ 和 $\overline{BC}$ 等距,距离是 0),如图 15-45 所示。

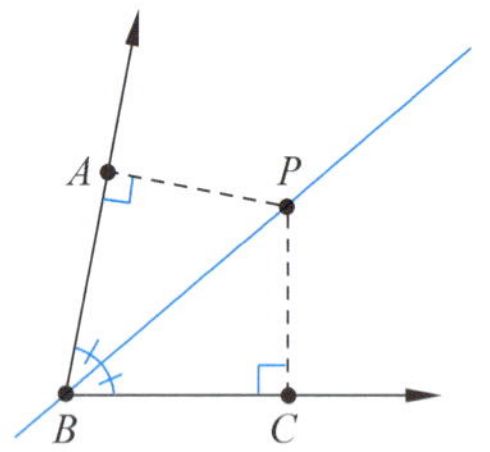

Figure 15-45

Suppose PA and PC are the distances from P to $\overline{AB}$ and $\overline{BC}$, respectively. We need to prove that $\overline{PA}\cong\overline{PC}$. 假设 PA 和 PC 分别是P 到 $\overline{AB}$ 的距离和P 到 $\overline{BC}$ 的距离。需要证明 $\overline{PA}\cong\overline{PC}$。

Table 15-7 gives the proof of Property 1.

Table 15-7

Statements 命题	Reasons 原因
1. P is a point on the angle bisector of $\angle ABC$. P 是$\angle ABC$ 平分线上的一点	1. Given
2. $\angle ABP\cong\angle CBP$	2. Definition of angle bisectors 角平分线的定义
3. $\angle PAB$ and $\angle PCB$ are right angles. $\angle PAB$ 和$\angle PCB$ 都是直角	3. Definition of distances 距离的定义
4. $\angle PAB\cong\angle PCB$	4. All right angles are congruent 所有直角全等
5. $\overline{PB}\cong\overline{PB}$	5. Reflexive Property 反身性
6. $\triangle PAB\cong\triangle PCB$	6. AAS Congruence Theorem 角角边
7. $\overline{PA}\cong\overline{PC}$	7. CPCTC 全等三角形的对应部分全等

2. Proof of Property 2:

性质 2 的证明：

Suppose PA and PC are the distances from P to $\overline{AB}$ and $\overline{BC}$, respectively. We know that $\overline{PA}\cong\overline{PC}$. 假设 PA 和 PC 分别是 P 到 $\overline{AB}$ 的距离和 P 到 $\overline{BC}$ 的距离。已知 $\overline{PA}\cong\overline{PC}$。

We will show that P is on the angle bisector of $\angle ABC$. If P is the same point as B, then the case is trivial, as the angle bisector of $\angle ABC$ must contains B. Consider that P is different from B, as shown in Figure 15-46. 证明 P 在$\angle ABC$ 的平分线上。若 P 和 B 重合，则是一个特殊情况——$\angle ABC$ 的平分线必会包含 P。假设 P 和 B 不重合，如图 15-46 所示。

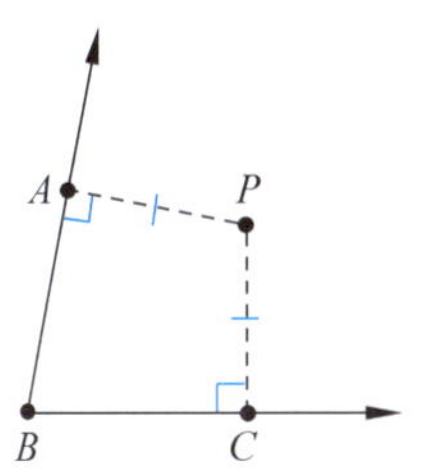

Figure 15-46

Table 15-8 gives the proof of Property 2.

Table 15-8

Statements 命题	Reasons 原因
1. $\overline{PA}\cong\overline{PC}$	1. Given
2. $\angle PAB$ and $\angle PCB$ are right angles $\angle PAB$ 和$\angle PCB$ 都是直角	2. Definition of distances 距离的定义
3. $\angle PAB\cong\angle PCB$	3. All right angles are congruent 所有直角全等
4. $\overline{PB}\cong\overline{PB}$	4. Reflexive Property 反身性
5. $\triangle PAB\cong\triangle PCB$	5. HL Congruence Theorem HL 定理
6. $\angle ABP\cong\angle CBP$	6. CPCTC 全等三角形的对应部分全等
7. P is a point on the angle bisector of $\angle ABC$	7. Definition of angle bisectors 角平分线的定义

3. Proof of Property 3: In $\triangle ABC$, suppose lines l, m, and n are the angle bisectors of $\angle A$, $\angle B$, and $\angle C$ respectively, and I be the intersection of l and m, as shown in Figure 15-47. 性质 3 的证明：在$\triangle ABC$ 中，设直线 l、m、n 分别是$\angle A$、$\angle B$、$\angle C$ 的垂直平分线，I 为 l 和 m 的交点，如图 15-47 所示。

We have I is on both l and m, and wish to prove that I is on n. Property 1 gives since I is on l, I is equidistant from $\overline{AB}$ and $\overline{AC}$, and since I is on m, I is equidistant from $\overline{AB}$ and $\overline{BC}$. Therefore, I is equidistant from $\overline{AC}$ and $\overline{BC}$ by transitive. By Property 2, I is on n. 可得出 I 在 l 和 m 上，想证明 I 在 n 上。因为 I 在 l 上，从性质 1 可得出 I 离 $\overline{AB}$ 和 $\overline{AC}$ 等距；因为 I 在 m 上，I 离 $\overline{AB}$ 和 $\overline{BC}$ 等距。由传递性，I 离 $\overline{AC}$ 和 $\overline{BC}$ 等距。根据性质 2，可得 I 在 n 上。

This also illustrates Property 5.

Incenter 内心

n. [ɪn'sentər]

Definition: The point of concurrency of three angle bisectors of a triangle. As shown in Figure 15-48, I is the incenter. 内心是指三角形三条角平分线的交点。如图 15-48 所示，I 为内心。

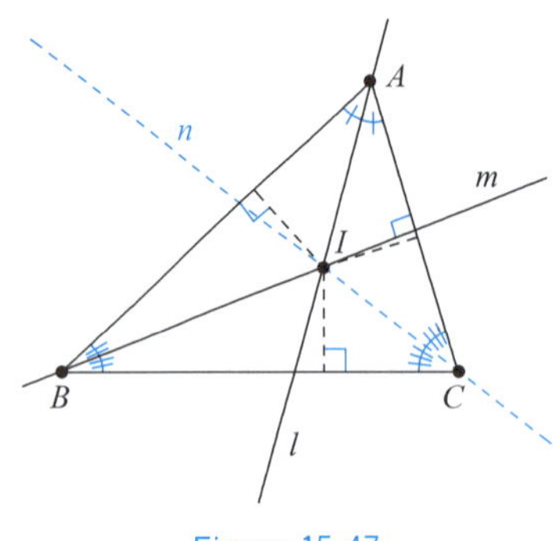

Figure 15-47

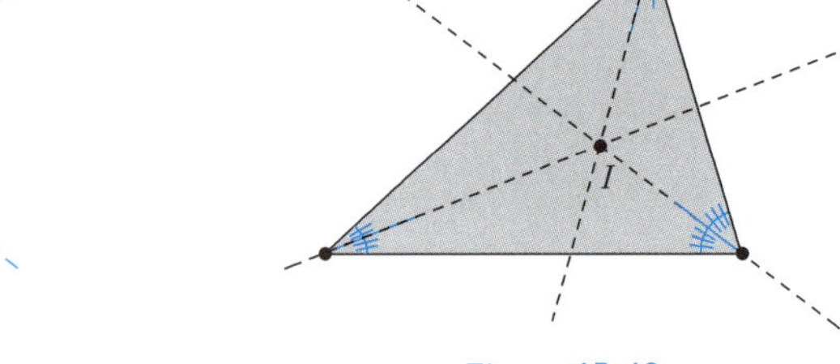

Figure 15-48

Notation：The incenter is denoted by the letter I.　内心一般用 I 表示。

Properties：1. The incenter is equidistant from each of the sides of the triangle.　三角形的内心离每一条边都等距。

2. The incenter of any triangle T is always inside T.　三角形 T 的内心永远在 T 之内。

Proof：We refer to Figure 15-49 to prove Property 1.

By the **Concurrency of Angle Bisectors Theorem**, angle bisectors of the three angles of any triangle are concurrent.　根据**角平分线共点定理**，三角形三个角的平分线共点。

Since I is the point of concurrency of angle bisectors, I is on each of the three angle bisectors. Using the **Angle Bisector Property**, we get I is equidistant from $\overline{AB}$, $\overline{BC}$, and $\overline{AC}$, from which the property is true.　因为 I 是角平分线的交点，I 在所有角平分线上。根据**角平分线性质**可得出 I 离 $\overline{AB}$、$\overline{BC}$、$\overline{AC}$ 等距。

Incircle　内切圆

n. [ɪn'sɜrkəl]

Definition：The incircle of a triangle T is a circle that is tangent to all sides of T. Its center is the incenter, and its radius is the inradius.　三角形 T 的内切圆是一个相切 T 三边的圆。

As shown in Figure 15-50, the incircle of $\triangle ABC$ is drawn in blue, with center I. Note that the blue line segments are radii of the incircle, or the inradii of $\triangle ABC$.　如图 15-50 所示，蓝色的是$\triangle ABC$ 的内切圆，圆心为 I。蓝色线段是内切圆的半径。

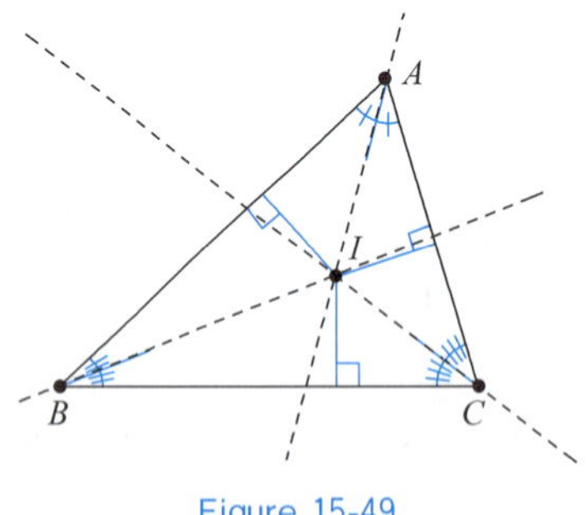

Figure 15-49

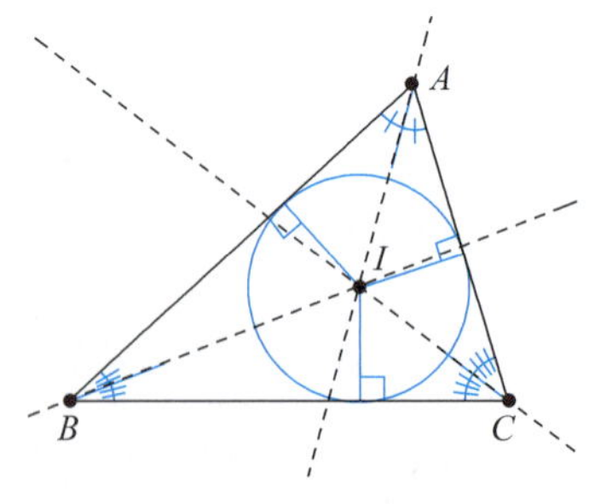

Figure 15-50

Inradius　内切圆半径

n. [ɪn'reɪdiəs]

Definition：A radius of the incircle. One example is a line segment whose endpoints are the incenter and one point on the side of the triangle, such that the inradius is perpendicular to that side.　内切圆的半径，例如端点为内心和三角形一边上的一点的线段(内切圆半径必须和那条边垂直)。

Some examples of inradii are drawn in blue, as shown in Figure 15-51.

Properties：Let s and r be the semiperimeter and the length of an inradius of $\triangle ABC$, respectively. The area of $\triangle ABC$ is sr.

设 s 和 r 分别为$\triangle ABC$ 半周长和内切圆半径的长度。$\triangle ABC$ 的面积为 sr。

Proof：As shown in Figure 15-52, the $[ABC]=[BIC]+[AIC]+[AIB]=\frac{1}{2}ar+\frac{1}{2}br+\frac{1}{2}cr=\frac{1}{2}(a+b+c)r=sr$.

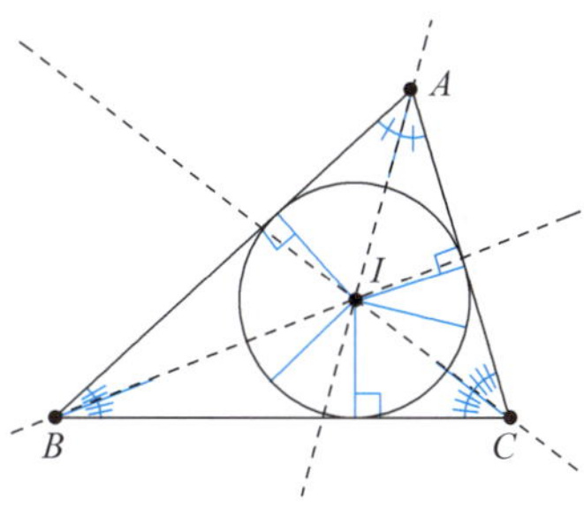

Figure 15-51

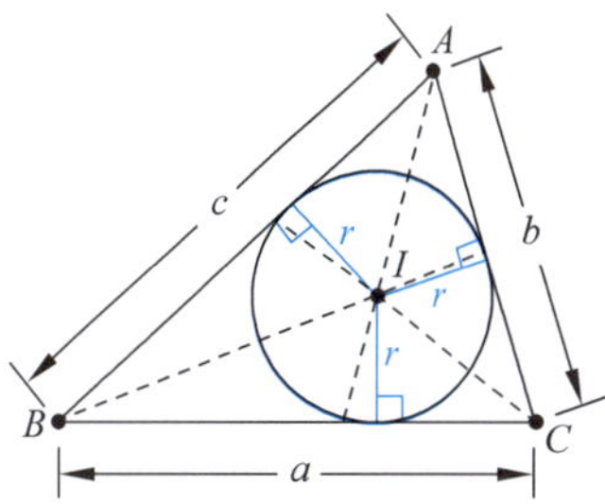

Figure 15-52

15.6.2.4　Medians　中线

Median　中线

n. ['midiən]

Definition：A median of a triangle is a line segment whose endpoints are a vertex and the midpoint of that vertex's opposite side.　三角形的中线是一条端点为一个顶点与它对边中点的线段。

As shown in Figure 15-53, $\overline{AD}$, $\overline{BE}$, and $\overline{CF}$ are the medians of $\triangle ABC$; G is the centroid of $\triangle ABC$.　如图 15-53 所示，$\overline{AD}$、$\overline{BE}$、$\overline{CF}$ 是$\triangle ABC$ 的中线；G 是$\triangle ABC$ 的重心。

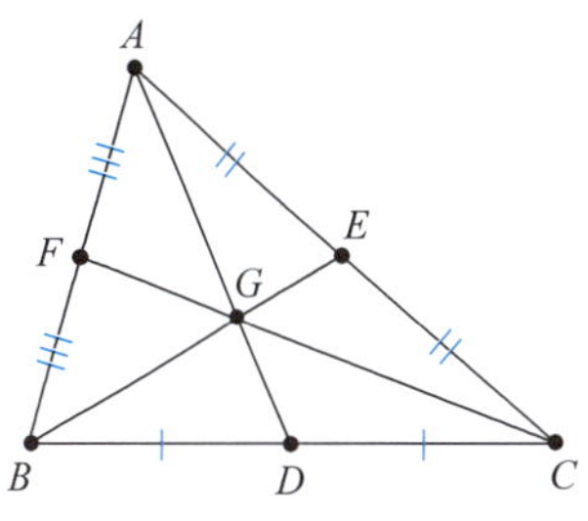

Figure 15-53

Properties：1. Three medians of a triangle T divide T into six smaller triangles with equal area.　三角形 T 的三条中线把 T 分割成 6 个面积相等的小三角形。

2. **Concurrency of Medians Theorem**：The three medians of any triangle are concurrent.　**中线共点定理**：三角形的三条中线共点。

3. The point of concurrency of the medians of a triangle T is known as the centroid of T.　三角形 T 中线的交点叫作 T 的重心。

4. Knowing two medians in a triangle immediately gives the third (which passes the intersection of the two medians and the remaining vertex).　已知三角形的两条中线，

即可给出第三条中线(第三条中线经过两条中线的交点和三角形的第三个顶点)。

5. The centroid divides each median's length into parts in the ratio 2 : 1. Within a median, the distance from a vertex to the centroid is twice of the distance from the centroid to the midpoint of the vertex's opposite side. 重心把一条中线的长度按照 2 : 1 的比例分开。在一条中线上，顶点到重心的距离是重心到顶点对边中心距离的两倍。
6. In a right triangle, the length of the median drawn to the hypotenuse is half of the length of the hypotenuse. 直角三角形中，到斜边的中线长度是斜边长度的一半。

Proof: 1. To prove Property 1, we need the **same base/altitude property**, see Property 3 of area in Section 21.1. 要证明定理 1，用**等底/等高性质**。见 21.1 节面积的性质 3。

As shown in Figure 15-54, let a, b, c, d, e, and f represent the areas of the smaller triangles.

By **same base/altitude property** (see Property 3 of area in Section 21.1) on each of the bases of the large triangle, we have $a = b$, $c = f$, and $d = e$, as shown in Figure 15-55.

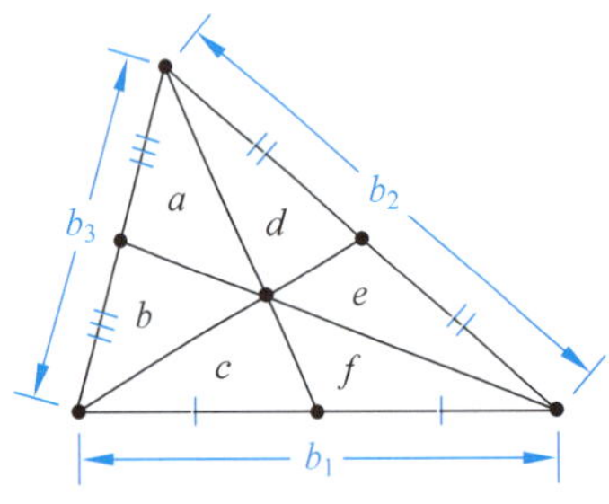

Figure 15-54

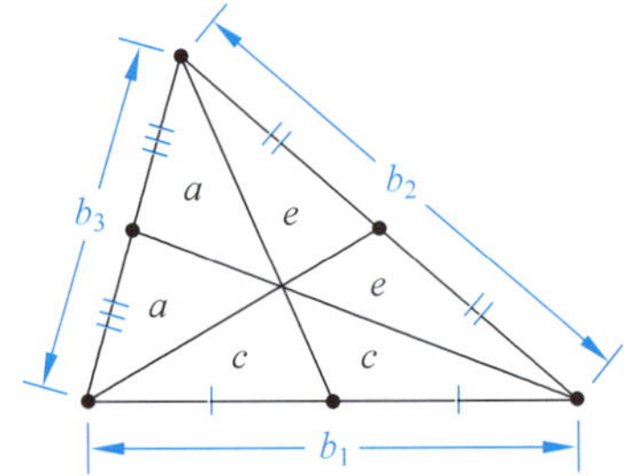

Figure 15-55

By **same base property** on b_3, we know that $a + 2c = a + 2e$, from which $c = e$. By **same base property** on b_2, we know that $2a + e = 2c + e$, from which $a = c$. Therefore, transitive gives $a = b = c = d = e = f$.

如图 15-54 所示，假设 a、b、c、d、e、f 是小三角形的面积，大三角形在每一个底用**等底/等高性质**(见 21.1 节面积的性质 3)可得出 $a = b$; $c = d$; $e = f$ (见图 15-55)。在 b_3 运用等底性质可得出 $a + 2c = a + 2e$，即 $c = e$。在 b_2 运用等底性质可得出 $2a + e = 2c + e$，即 $a = c$。根据传递性，$a = b = c = d = e = f$。

2. Property 2 is much more difficult to prove compared to the theorems about the concurrencies of the perpendicular bisectors and angle bisectors. We will assume it for now and will prove later. See Theorem 6 of common theorems of similarity in Section 15.10.3 for the proof. 性质 2 的证明比证明垂直平分线共点和角平分线共点困难得多。目前先假设是真的，并在以后证明。见 15.10.3 节“常见的相似定理(定理 6)”。
3. From Property 1 and the **same base/altitude property** (see Property 3 of area in Section 21.1), we have shown Property 5. 由于**等底/等高性质**(见 21.1 节面积的性质 3)的性质 1，已经证明了中线的性质 5。
4. Property 6 is a corollary from the **Inscribed Angle Theorem**. See Property 5 of

corollaries of circles in Section 19.1.5 for the proof. 性质 6 是**圆周角定理**的推论。证明见 19.1.5 节圆的推论性质 5。

Centroid 重心

n. ['sɛntrɔɪd]

Definition: The point of concurrency of three medians of a triangle. 重心是三角形三条中线的交点。

Notation: The centroid of a triangle is usually denoted by G. 三角形的重心一般用 G 表示。

As shown below in Figure 15-56, point G is the centroid. 如图 15-56 所示，G 是重心。

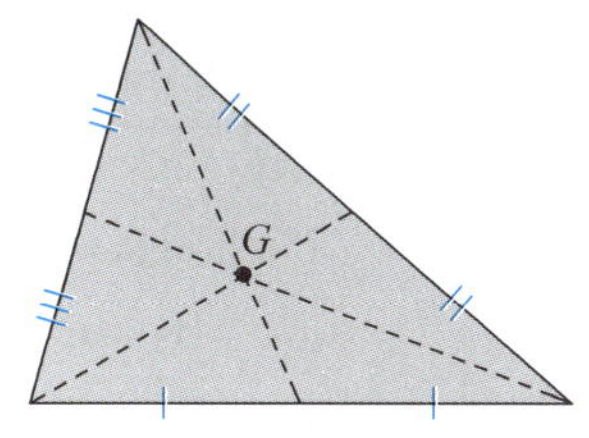

Figure 15-56

Properties: The centroid of a triangle T is always inside T. 三角形 T 的重心在 T 之内。

15.6.2.5 Altitudes 高线

Altitude 高线

n. ['æltɪˌtud]

Definition: An altitude of a triangle is a line segment whose endpoints are a vertex and a point on the opposite side (possibly extended), such that the altitude is perpendicular to that side. 三角形的高线是一条端点为一个顶点和它对边一点的线段(对边可能要延长)，高线必须与那条边垂直。

Figure 15-57 shows one altitude of an acute triangle. 图 15-57 是锐角三角形的一条高线。

Figure 15-58 shows one altitude of a right triangle. 图 15-58 是直角三角形的一条高线。

Figure 15-59 shows one altitude of an obtuse triangle. 图 15-59 是钝角三角形的一条高线。

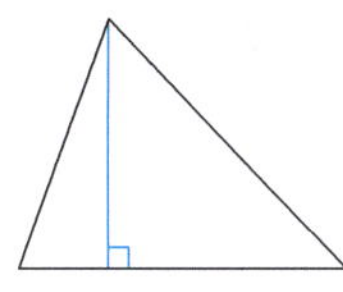
Figure 15-57

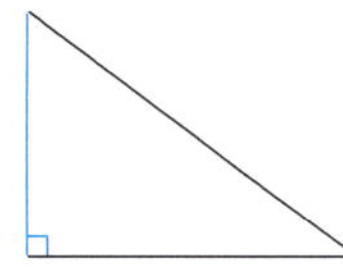
Figure 15-58

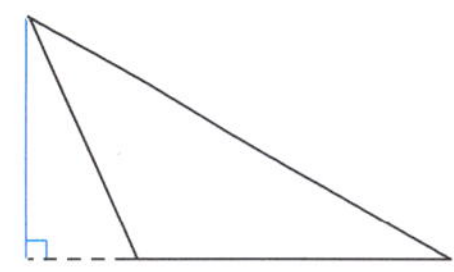
Figure 15-59

The length of an altitude is called the height. 高线的长度叫作高。

Properties: 1. **Concurrency of Altitudes Theorem**: The three altitudes (possibly extended) of any triangle are concurrent. 高线共点定理：三角形的三条高线(可能要延长)共点。

2. The point of concurrency of the altitudes in a triangle T is known as the orthocenter of T. 在三角形 T 中，高线的共点叫作 T 的垂心。

3. Knowing two altitudes of a triangle immediately gives the third (which passes the intersection of the two altitudes and the remaining vertex). 已知一个三角形的两条高线就能立刻给出第三条高线(第三条高线经过两条高线的交点和第三个顶点)。

Proof: To prove Property 1, let $\overline{AD}$, $\overline{BE}$, and $\overline{CF}$ be the altitudes of $\triangle ABC$. We want to prove that they all contain point H.

As shown in Figure 15-60, circumscribe $\triangle ABC$ with $\triangle PQR$ so that $\overline{AB}$, $\overline{BC}$, and $\overline{AC}$ are the midsegments of $\overline{PQ}$, $\overline{QR}$, and $\overline{PR}$, respectively, which implies that A, B, and C are the midpoints of $\overline{QR}$, $\overline{PR}$, and $\overline{PQ}$, respectively. By the **Midsegment Theorem**, we have $\overline{AB} \parallel \overline{PQ}$, $\overline{BC} \parallel \overline{QR}$, and $\overline{AC} \parallel \overline{PR}$. By the **Two Perpendicular Lines Theorem**, $\overline{AB}$ and $\overline{PQ}$ are perpendicular to $\overline{CF}$; $\overline{BC}$ and $\overline{QR}$ are perpendicular to $\overline{AD}$; $\overline{AC}$ and $\overline{PR}$ are perpendicular to $\overline{BE}$. 如图 15-60 所示,用$\triangle PQR$ 外接$\triangle ABC$,使得 $\overline{AB}$、$\overline{BC}$、$\overline{AC}$ 分别为 $\overline{PQ}$、$\overline{QR}$、$\overline{PR}$ 的中位线,也就是说 A、B、C 分别为 $\overline{QR}$、$\overline{PR}$、$\overline{PQ}$ 的中点。根据**中位线定理**,得出 $\overline{AB} \parallel \overline{PQ}$,$\overline{BC} \parallel \overline{QR}$,$\overline{AC} \parallel \overline{PR}$。根据**双垂线定理**,$\overline{AB}$和 $\overline{PQ}$ 垂直于 $\overline{CF}$;$\overline{BC}$和 $\overline{QR}$垂直于 $\overline{AD}$;$\overline{AC}$ 和 $\overline{PR}$垂直于 $\overline{BE}$。

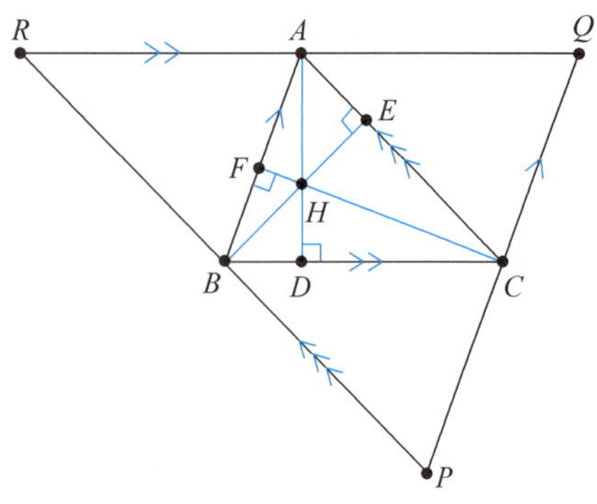

Figure 15-60

We have shown that $\overline{AD}$, $\overline{BE}$, and $\overline{CF}$ are the perpendicular bisectors of $\triangle PQR$. By the **Concurrency of Perpendicular Bisectors Theorem**, we know that they are concurrent at H, which means that the altitudes of $\triangle ABC$ are concurrent. 已经证明出了 $\overline{AD}$、$\overline{BE}$、$\overline{CF}$ 是$\triangle PQR$ 的垂直平分线。根据**垂直平分线共点定理**,知道它们相交于 H,也就是$\triangle ABC$ 的高线共点。

This also illustrates Property 3.

Orthocenter 垂心

n. [ˈɔːθəʊˌsɛntə]

Definition: The point of concurrency of the (possibly extended) altitudes of a triangle. 垂心是三角形中三条高线(可能要延长)的交点。

Properties: As shown in Figure 15-61, the orthocenter of an acute triangle is inside the triangle. 如图 15-61 所示,锐角三角形的垂心在三角形内。

As shown in Figure 15-62, the orthocenter of a right triangle is the vertex of the right angle. 如图 15-62 所示,直角三角形的垂心是直角的顶点。

As shown in Figure 15-63, the orthocenter of an obtuse triangle is outside the triangle. The extensions of the sides are drawn in solid lines and the extensions of the altitudes are drawn in dotted lines. 如图 15-63 所示,钝角三角形的垂心在三角形外面。边的延长用实线画出。高线的延长用虚线画出。

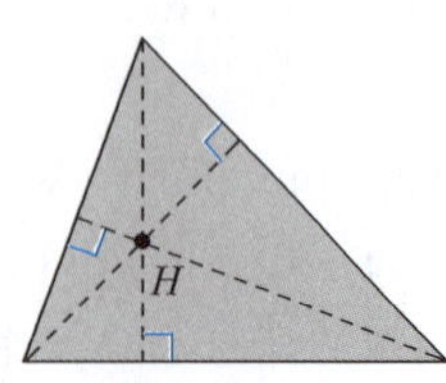

Figure 15-61

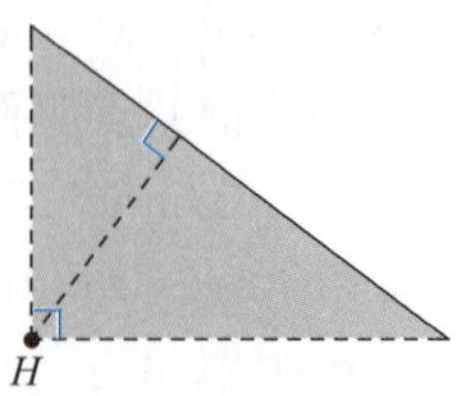

Figure 15-62

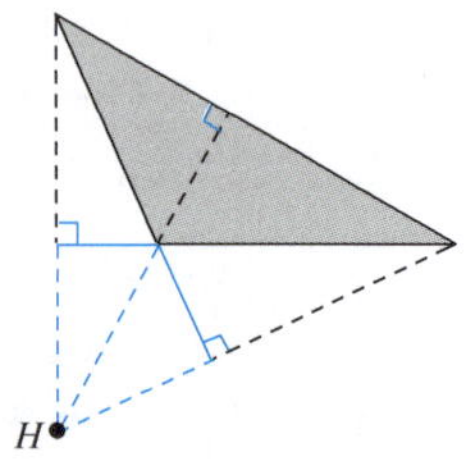

Figure 15-63

15.6.3 Summary 总结

1. “Intersecting” and “concurrent” are very similar in definitions.
   (1) “Intersecting” refers to two lines/rays/line segments have exactly one point in common. This point is known as the point of intersection, or simply intersection.
   (2) “Concurrent” refers to three or more lines/rays/line segments have exactly one point in common. This point is known as the point of concurrency.
2. Midsegments:
   中位线：
   (1) **Midsegment Theorem**: In every triangle, a midsegment drawn from any two sides is parallel to the third side, with half of its length. **中位线定理**：在任意三角形中，连接两边的中位线与第三边平行，长度是第三边的一半。
   (2) Three midsegments of a triangle T divide T into four smaller triangles with equal areas. 三角形 T 的三条中位线把 T 分成面积相等的 4 个小三角形。
3. Perpendicular Bisectors:
   垂直平分线：
   (1) **Perpendicular Bisector Theorem and Its Converse**: A point is on the perpendicular bisector of a line segment if and only if it is equidistant to the endpoints of the line segment. **垂直平分线定理及其逆定理**：一个点在一条线段的垂直平分线上当且仅当它与线段的两端等距。
   (2) **Concurrency of Perpendicular Bisectors Theorem**: The perpendicular bisectors of the three sides of every triangle are concurrent. **垂直平分线共点定理**：任意三角形的三条边的垂直平分线共点。
   (3) Of a triangle T:
   在三角形 T 中：
   Circumcircle: A circle that passes through three vertices of T.
   外接圆：通过三角形 T 的三个顶点的圆。
   Circumcenter: The center of the circumcircle; the point of concurrency of the perpendicular bisectors of T.
   The circumcenter of an acute triangle is inside the triangle.
   The circumcenter of a right triangle is at the midpoint of the hypotenuse.
   The circumcenter of an obtuse triangle is outside the triangle.
   外心：外接圆圆心，即 T 的三条垂直平分线的交点。
   锐角三角形的外心在三角形内。
   直角三角形的外心在斜边的中点。
   钝角三角形的外心在三角形外。
   Circumradius: The radius of the circumcircle; one example is a line segment whose endpoints are the circumcenter and one vertex of T.
   外接圆半径：一条端点是外心和 T 的一个顶点的线段。
4. Angle Bisectors:
   角平分线：

(1) **Angle Bisector Property and Its Converse**: A point is on the angle bisector if and only if it is equidistant from the sides of the angle. **角平分线性质及其逆性质**：一个点在一个角的平分线上当且仅当它与角的两边等距。

(2) **Concurrency of Angle Bisectors Theorem**: The angle bisectors of the three angles of every triangle are concurrent. **角平分线共点定理**：任意三角形的三个角的角平分线共点。

(3) Of a triangle T:
在三角形 T 中：
Incircle: A circle that is tangent to all three sides of T.
内切圆：与 T 的三条边相切的圆。
Incenter: The center of the incircle; the point of concurrency of angle bisectors of T, which is always inside T.
内心：内切圆圆心，即 T 的三条角平分线的交点，内心在 T 内。
Inradius: The radius of the incircle; one example is a line segment whose endpoints are the incenter and a point on one side of T, such that this line segment is perpendicular to that side.
内切圆半径：一条端点为内心和 T 的一边上的一点，使得内切圆半径与那条边垂直。

5. Medians:
中线：

(1) Three medians of a triangle T divide T into six smaller triangles with equal area. 三角形 T 的三条中线把 T 分割成六个面积相等的小三角形。

(2) **Concurrency of Medians Theorem**: The three medians of every triangle are concurrent. **中线共点定理**：三角形的三条中线共点。

(3) The centroid divides each median's length into parts in the ratio 2∶1. Within a median, the distance from a vertex to the centroid is twice of the distance from the centroid to the midpoint of the vertex's opposite side. 重心把一条中线的长度按照 2∶1 的比例分。在一条中线上，顶点到重心的距离是重心到顶点对边中心距离的两倍。

(4) The point of concurrency of the medians of a triangle T is known as the centroid of T, which is always inside T. 三角形 T 中线的交点叫作 T 的重心。重心在 T 内。

6. Altitudes:
高线：

(1) **Concurrency of Altitudes Theorem**: The three altitudes of every triangle are concurrent. **高线共点定理**：三角形的三条高线共点。

(2) The point of concurrency of the altitudes in a triangle T is known as the orthocenter of T.
The orthocenter of an acute triangle is inside the triangle.
The orthocenter of a right triangle is the vertex of the right angle.
The orthocenter of an obtuse triangle is outside the triangle.
在三角形 T 中，高线的共点叫作 T 的垂心。
锐角三角形的垂心在三角形内。
直角三角形的垂心是直角的顶点。
钝角三角形的垂心在三角形外。

15.7 More Properties of Triangles 三角形的其他性质

15.7.1 Stability of Triangles 三角形的稳定性

Stability of Triangles 三角形的稳定性

n. [stə'bɪlɪti əv 'traɪˌæŋgəl]

Definition: A triangle is fixed in shape (cannot be twisted, and its shape cannot be altered). The only way to alter the shape of a triangle is by breaking it. 三角形的形状稳定，一般不容易弯曲或变形。唯一使三角形变形的方法是破坏它。

Proof: As shown in Figure 15-64, to prove that a triangle T is stable, note that the uncommon endpoints (shown in black) of any two sides (shown in blue) are connected by the third side (shown in black), as shown below. 要证明三角形 T 是稳定的，首先留意到两条边(用蓝色表示)的非共端点(用黑色表示)被第三条边(用黑色表示)连接着，如图 15-64 所示。

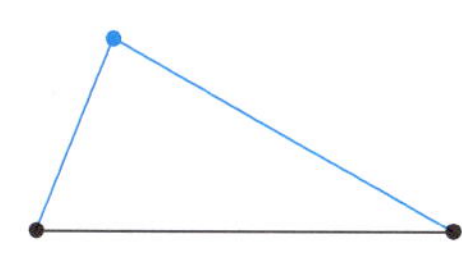

Figure 15-64

The third side cannot be shrunk /stretched /broken, and its length is fixed. Therefore, the included angle of those two sides is fixed. 第三条边不能被缩小/延伸/折断，故其长度是固定的。所以，两边的夹角也是固定的。

Applying similar arguments to the remaining sides and angles can conclude that all angles in T are fixed. Thus the T is stable. 同样的论点可应用到其他边和角上，可得出三角形 T 的所有角都是固定的。因此，T 有稳定性。

Examples: The triangular frame of a bicycle makes the bicycle itself stable.

15.7.2 Hinge Theorem and Its Converse 大角对大边定理与逆定理

Hinge Theorem 大角对大边定理

n. ['hɪndʒ 'θɪərəm]

Definition: If two sides of triangle T are congruent to two sides of triangle T', respectively, and the included angle of T is larger than that of T', then the third side of T is longer than that of T'. 若三角形 T 的两条边和 T' 的两条边全等，且 T 的夹角大于 T' 的夹角，则 T 的第三条边比 T' 的第三条边长。

Proof: As shown in Figure 15-65, we are given that $AB = DE$, $AC = DF$, $m\angle A > m\angle D$.

We want to show that $BC > EF$.

Construct point G in the interior of $\angle BAC$ so that $\triangle AGC \cong \triangle DEF$, as shown in Figure 15-66. 在$\angle BAC$ 的内部画出点 G，使得$\triangle AGC \cong \triangle DEF$，如图 15-66 所示。

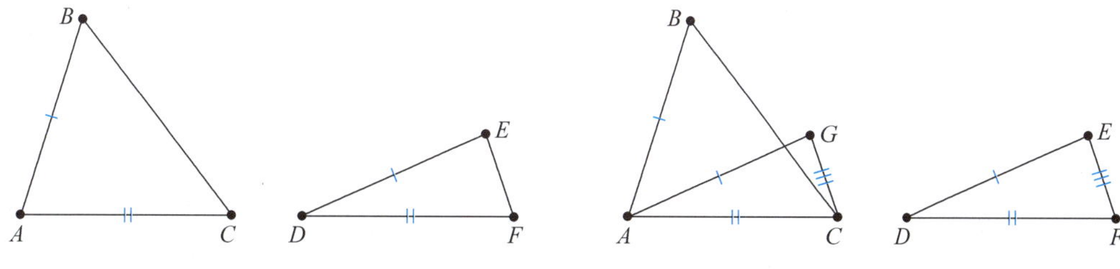

Figure 15-65 Figure 15-66

Draw the angle bisector of $\angle BAG$. Let M be the intersection of this angle bisector and $\overline{BC}$, as shown in Figure 15-67. 画出$\angle BAG$的角平分线,M为这条角平分线和$\overline{BC}$的交点,如图 15-67 所示。

From here, we can conclude that $\triangle AMB \cong \triangle AMG$ by SAS, from which we get $MB = MG$ by CPCTC. 根据边角边全等公理可以得到$\triangle AMB \cong \triangle AMG$。然后根据全等三角形的对应部分全等,得出$MB = MG$。

By **Triangle Inequality Theorem** on $\triangle CGM$, we get $CG < CM + MG$. Since $MB = MG$, substitution gives $CG < CM + MB$. 在$\triangle CGM$里运用**三角不等式定理**,可得出$CG < CM + MG$。因为$MB = MG$,代入可得出$CG < CM + MB$。

Segment Addition Postulate gives $BC = CM + MB$, from which it is clear that $BC > CG$. Since $CG = EF$, we have $BC > EF$. This finishes the proof. 根据**线段相加公理**,$BC = CM + MB$。显然,$BC > CG$。因为$CG = EF$,所以$BC > EF$。证毕。

Converse of the Hinge Theorem 大角对大边的逆定理——大边对大角定理

Definition: If two sides of triangle T are congruent to two sides of triangle T', respectively, and the third side of T is longer than that of T', then the included angle of T is larger than that of T'. 若三角形T的两边和T'的两边全等,且T的第三边比T'的第三边长,则T的夹角大于T'的夹角。

Proof: As shown in Figure 15-68, we are given that $AB = DE$, $AC = DF$, and $BC > EF$.

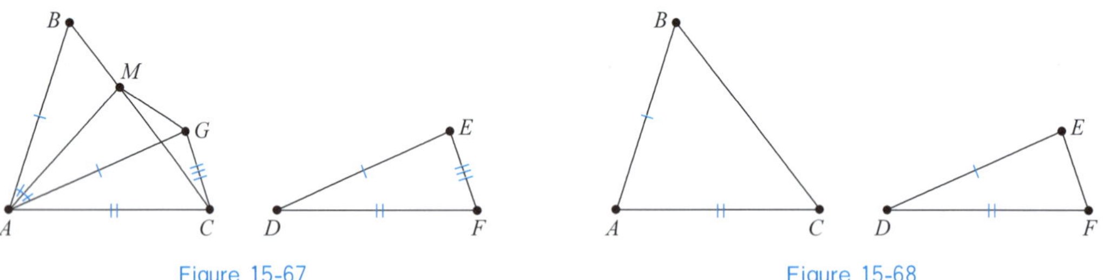

Figure 15-67 Figure 15-68

We want to show that $m\angle A > m\angle D$ by contradiction. 下面会用到反证法,证明$m\angle A > m\angle D$。

Suppose that $m\angle A > m\angle D$ is not true. We will have either $m\angle A = m\angle D$ or $m\angle A < m\angle D$. 若$m\angle A > m\angle D$是个假命题,则$m\angle A = m\angle D$或$m\angle A < m\angle D$是真命题。

If $m\angle A = m\angle D$, then we can conclude that $\triangle ABC \cong \triangle DEF$ by SAS, from which $BC = EF$ because of CPCTC. However, the given states that $BC > EF$, which is a contradiction. 若$m\angle A = m\angle D$,则根据边角边全等公理,$\triangle ABC \cong \triangle DEF$。根据全等三角形的对应部分全等,$BC = EF$。这与已知条件$BC > EF$矛盾。

If $m\angle A < m\angle D$, then by Hinge Theorem we can conclude that $BC < EF$. However, the given states that $BC > EF$, which is a contradiction. 若$m\angle A = m\angle D$,则根据大角对大边定理,$BC < EF$。这与已知条件$BC > EF$矛盾。

Therefore, it is true that $m\angle A > m\angle D$. 所以,$m\angle A > m\angle D$是真命题。

15.7.3 Corollaries 推论

Corollary of the Hinge Theorem 大角对大边定理的推论

Definition: Let a, b, and c be the side-lengths of triangle T, for which c is the length of the longest side.

If T is an acute triangle, then $a^2+b^2>c^2$.

If T is a right triangle, then $a^2+b^2=c^2$.

If T is an obtuse triangle, then $a^2+b^2<c^2$.

设 a、b、c 为三角形 T 的三条边的边长，且 c 是最长边边长。

若 T 是锐角三角形，则 $a^2+b^2>c^2$。

若 T 是直角三角形，则 $a^2+b^2=c^2$。

若 T 是钝角三角形，则 $a^2+b^2<c^2$。

Proof: Fix a and b. We will prove that if T is a right triangle, then $a^2+b^2=c^2$ in the **Pythagorean Theorem** in Section 15.9.1. The other two cases follow easily by the Hinge Theorem. 固定 a 和 b。若 T 为直角三角形，则 $a^2+b^2=c^2$（**勾股定理**，见 15.9.1 节）。另外两种情况在大角对大边定理中很容易得出。

Corollary of the Converse of the Hinge Theorem 大边对大角定理的推论

Definition: Let a, b, and c be the side-lengths of triangle T, for which c is the longest side.

If $a^2+b^2>c^2$, then T is an acute triangle.

If $a^2+b^2=c^2$, then T is a right triangle.

If $a^2+b^2<c^2$, then T is an obtuse triangle.

设 a、b、c 为三角形 T 的三条边边长，且 c 是最长边边长。

若 $a^2+b^2>c^2$，则 T 是锐角三角形。

若 $a^2+b^2=c^2$，则 T 是直角三角形。

若 $a^2+b^2<c^2$，则 T 是钝角三角形。

Proof: Fix a and b. We will prove that if $a^2+b^2=c^2$, then T is a right triangle in the **Converse of the Pythagorean Theorem** in Section 15.9.3. The other two cases are followed easily by the Converse of the Hinge Theorem. 固定 a 和 b。若 $a^2+b^2=c^2$，则 T 为直角三角形（**勾股定理的逆定理**，见 15.9.3 节）。另外两种情况在大边对大角定理中能够很容易得出。

15.7.4 Summary 总结

1. Stability of triangles: A triangle is fixed in shape (cannot be twisted, and its shape cannot be altered). The only way to alter the shape of a triangle is by breaking it. 三角形的形状稳定，不容易被弯曲或变形。唯一使三角形变形的方法是破坏它。
2. Hinge Theorem and Its Converse: If two sides of triangle T are congruent to two sides of triangle T', respectively, and the third side of T is longer than that of T' if and only if the included angle of T is larger than that of T'. 大角对大边定理与大边对大角定理：若三角形 T 的两条边和 T' 的两条边全等，则 T 的第三条边比 T' 的第三条边长当且仅当 T 的夹角大于 T' 的夹角。
3. Corollary of the Hinge Theorem and Its Converse: Let a, b, and c be the side-lengths of triangle

T, for which c is the longest side.

$a^2+b^2>c^2$ if and only if T is an acute triangle.

$a^2+b^2=c^2$ if and only if T is a right triangle.

$a^2+b^2<c^2$ if and only if T is an obtuse triangle.

大角对大边定理与大边对大角定理的推论：设 a、b、c 为三角形 T 的三条边边长，且 c 是最长边边长。

$a^2+b^2>c^2$ 当且仅当 T 是锐角三角形。

$a^2+b^2=c^2$ 当且仅当 T 是直角三角形。

$a^2+b^2<c^2$ 当且仅当 T 是钝角三角形。

15.8 More on Isosceles Triangles 更多等腰三角形知识点

See isosceles triangles in Section 15.2.2 for introduction. 参照 15.2.2 节等腰三角形的介绍。

Coincidence of Four Line Segments 四线合一

Definition：For isosceles $\triangle ABC$, with $\overline{AB}\cong\overline{AC}$, if point D is the midpoint of $\overline{BC}$, then $\overline{AD}$ is a perpendicular bisector, an angle bisector, a median, and an altitude of $\triangle ABC$. 在等腰$\triangle ABC$ 中，若 $\overline{AB}\cong\overline{AC}$，且点 D 是 $\overline{BC}$ 的中点，则 $\overline{AD}$ 同时是$\triangle ABC$ 的垂直平分线、角平分线、中线、高线。

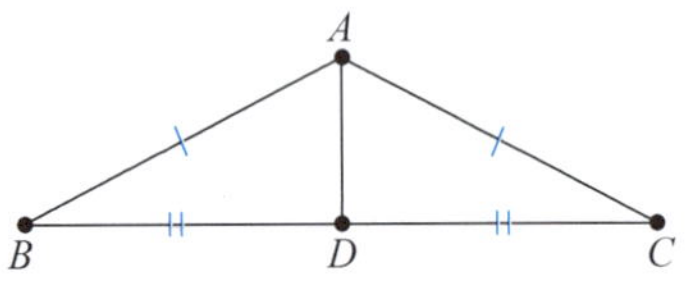

Figure 15-69

Proof：Refer to Figure 15-69.

Table 15-9 gives the proof for the coincidence of four line segments.

Table 15-9

Statements 命题	Reasons 原因
1. $\triangle ABC$ is isosceles, with $\overline{AB}\cong\overline{AC}$, and D is the midpoint of $\overline{BC}$ $\triangle ABC$ 是等腰三角形，$\overline{AB}\cong\overline{AC}$，点 D 是 $\overline{BC}$ 的中点	1. Given
2. $\overline{BD}\cong\overline{DC}$	2. Definition of midpoints 中点的定义
3. $\overline{AD}$ is a median of $\triangle ABC$ **$\overline{AD}$ 是$\triangle ABC$ 的中线**	**3. $\overline{BD}\cong\overline{DC}$ Property of medians 中线的性质**
4. $\overline{AD}\cong\overline{AD}$	4. Reflexive Property 反身性
5. $\triangle ABD\cong\triangle ACD$	5. SSS Congruence Postulate 边边边全等公理
6. $\angle ADB\cong\angle ADC$ $\angle BAD\cong\angle CAD$	6. CPCTC 全等三角形的对应部分全等
7. $\overline{AD}$ is an angle bisector of $\triangle ABC$ **$\overline{AD}$ 是$\triangle ABC$ 的角平分线**	**7. $\angle BAD\cong\angle CAD$. Property of angle bisectors 角平分线的性质**
8. $m\angle ADB+m\angle ADC=180°$	8. $\angle ADB$ and $\angle ADC$ form a linear pair 邻补角的性质

续表

Statements 命题	Reasons 原因
9. $m\angle ADB = m\angle ADC = 90°$	9. Steps 5 and 6
10. $\overline{AD} \perp \overline{BC}$	10. $\overline{AD}$ and $\overline{BC}$ intersect at right angles $\overline{AD}$和$\overline{BC}$相交于直角
11. $\overline{AD}$ is an altitude of $\triangle ABC$ $\overline{AD}$是$\triangle ABC$的高线	11. $\overline{AD} \perp \overline{BC}$ Property of altitudes 高线的性质
12. $\overline{AD}$ is a perpendicular bisector of $\triangle ABC$ $\overline{AD}$是$\triangle ABC$的垂直平分线	12. $\overline{AD} \perp \overline{BC}$ and $\overline{BD} \cong \overline{DC}$ Property of perpendicular bisectors 垂直平分线的性质

15.9 More on Right Triangles 更多直角三角形知识点

See right triangles in Section 15.2.1 for introduction. 参照 15.2.1 节直角三角形的介绍。

15.9.1 Pythagorean Theorem 勾股定理

视频 26

Pythagorean Theorem 勾股定理

n. [pəˌθæɡəˈriən ˈθɪərəm]

Definition: In every right triangle, the sum of the squares of the legs' lengths is equal to the square of the hypotenuse's length. Namely, if a, b, and c are the side-lengths of a right triangle, such that c is the length of the hypotenuse, then $a^2 + b^2 = c^2$. 在任何直角三角形中,两条直角边长度的平方和等于斜边长度的平方。换言之,若 a、b、c 分别为一个直角三角形的边长,其中 c 是斜边长度,那么必然满足 $a^2 + b^2 = c^2$。

Properties: 1. A Pythagorean Triple is an ordered triple (a, b, c), for which a, b, and c are integer side-lengths of a right triangle (usually $a < b$, or a is odd and b is even, depending on the convention), such that c is the length of the hypotenuse. 一个勾股数是一个整数三重序(a, b, c),使得 a, b, c 分别为一个直角三角形的边长(取决于作者惯例,通常指定 $a < b$ 或指定 a 为奇数,b 为偶数),其中 c 是斜边长度。

2. If (a, b, c) is a Pythagorean Triple, then (na, nb, nc), for every positive integer n, is also a Pythagorean Triple. 若(a, b, c)为勾股数,那么(na, nb, nc)也是勾股数(n 为正整数)。

Proofs: 1. Proof of the Pythagorean Theorem:

As shown in Figure 15-70, suppose a small square A inscribes a large square B. The white triangles that are formed are congruent (can be proved by AAS: Since $\angle 1$ and $\angle 2$ are complementary, $\angle 1$ and $\angle 3$ are complementary, $\angle 2 \cong \angle 3$ by transitive).

如图 15-70 所示,正方形 A 内接正方形 B,所形成的白色三角形全等(因为角角边全等公理:$\angle 1$ 和 $\angle 2$ 互补,$\angle 1$ 又和 $\angle 3$ 互补。从传递性可以得出$\angle 2 \cong \angle 3$)。

As shown in Figure 15-71, let c be the side-length of A, and $a + b$ be the side-length of B.

如图 15-71 所示,设 c 为 A 的边长,$a + b$ 为 B 的边长。

The area of B in terms of a and b: $(a + b)^2 = a^2 + 2ab + b^2$.

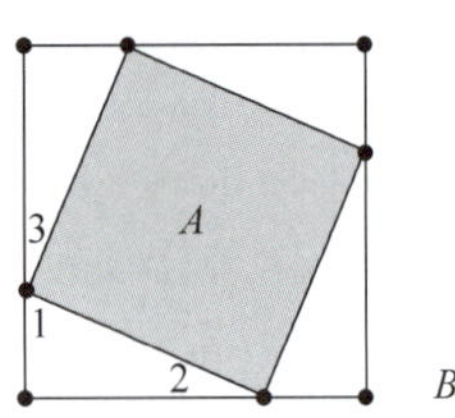

Figure 15-70

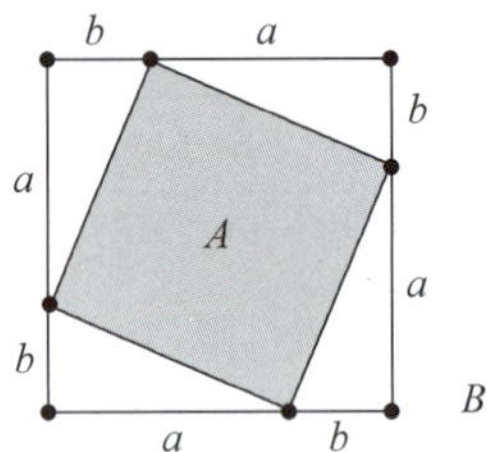

Figure 15-71

The area of B in terms of a, b, and c: $4\left(\frac{1}{2}ab\right)+c^2=2ab+c^2$.

Equating, we have $a^2+2ab+b^2=2ab+c^2$, from which $a^2+b^2=c^2$.

2. Proof of Property 2: We want to show that if $a^2+b^2=c^2$, then $(na)^2+(nb)^2=(nc)^2$. Note that LHS $=(na)^2+(nb)^2=n^2a^2+n^2b^2=n^2(a^2+b^2)=n^2c^2=(nc)^2=$ RHS.

Questions: 1. If a, b, and c are three sides of a right triangle, such that c is the hypotenuse,

(1) what is the value of c if $a=3$ and $b=4$?

(2) what is the value of b if $a=15$ and $c=17$?

若 a、b、c 为一个直角三角形的边长，其中 c 是斜边的长度，回答下面两个问题。

2. In right $\triangle ABC$, $m\angle B=90°$, $AB=7$, and $AC=25$. What is the value of BC?

Answers: 1. (1) $a^2+b^2=3^2+4^2=25=c^2$, from which $c=5$.

(2) We know that $a^2+b^2=c^2$. We have $15^2+b^2=17^2$, from which $225+b^2=289$, and $b^2=64$. Therefore, $b=8$.

2. We know that $\overline{AC}$ is the hypotenuse. $7^2+AB^2=25^2$, or $49+AB^2=625$, from which $AB^2=576$, and $AB=24$.

15.9.2 Special Right Triangles 特殊的直角三角形

Special Right Triangles 特殊的直角三角形

n. [ˈspeʃəl raɪt ˈtraɪˌæŋɡəls]

Definition: One of 45° − 45° − 90° triangle (isosceles right triangle) and 30° − 60° − 90° triangle.

见 45° − 45° − 90° 三角形(亦称为等腰直角三角形)和 30° − 60° − 90° 三角形。

As shown in Figure 15-72, the 45° − 45° − 90° is shown on the left, and the 30° − 60° − 90° is shown on the right.

45° − 45° − 90° Triangle/Isosceles Right Triangle 45° − 45° − 90° 三角形(又称等腰直角三角形)

Definition: A right triangle whose angle measures are 45°, 45°, and 90°. One example is shown in Figure 15-73.

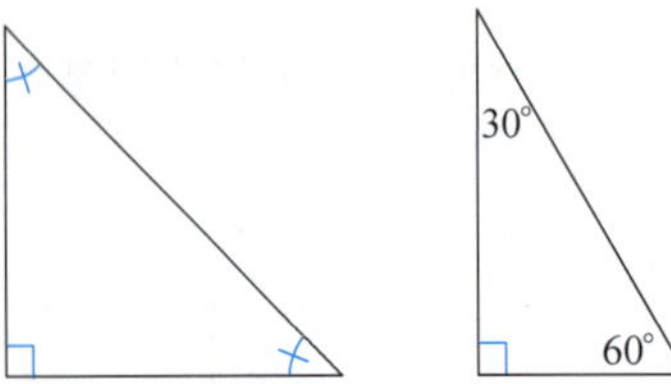

Figure 15-72

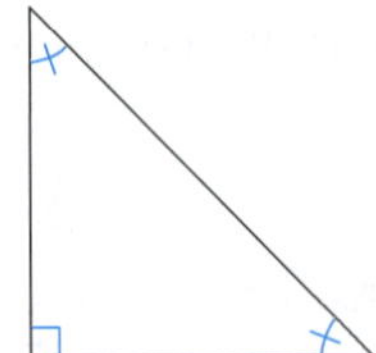
Figure 15-73

Properties: 1. By the Converse of Base Angle Theorem, a $45^\circ-45^\circ-90^\circ$ triangle is known as an isosceles right triangle. 用等角对等边定理，可得出 $45^\circ-45^\circ-90^\circ$ 都是等腰直角三角形。

2. The ratio of the side-lengths opposite the 45°, 45°, and 90° angles is $1:1:\sqrt{2}$, by the Pythagorean Theorem. 根据勾股定理，相对 45°、45°、90° 角的边长比例是 $1:1:\sqrt{2}$。

Question: If the length of a leg of an isosceles right triangle is 5, what is the length of the hypotenuse? 如果一个等腰直角三角形的直角边长度为5，那么斜边的长度是多少？

Answer: The length of the hypotenuse is $\sqrt{2}$ times the length of a leg. Thus our answer is $5\sqrt{2}$. 斜边长度是直角边长度的$\sqrt{2}$倍，所以答案是 $5\sqrt{2}$。

$30^\circ-60^\circ-90^\circ$ Triangle $30^\circ-60^\circ-90^\circ$ 三角形

Definition: A right triangle whose angle measures are 30°, 60°, and 90°. One example is shown in Figure 15-74.

Properties: 1. The ratio of the side-lengths opposite the 30°, 60°, and 90° angles is $1:\sqrt{3}:2$, by the Pythagorean Theorem. 根据勾股定理，相对 30°、60°、90° 角的边长比例是 $1:\sqrt{3}:2$。

2. From **Property 1**, note that the ratio of lengths of the shorter leg to the hypotenuse is $1:2$. 从**性质 1** 可得出较短的直角边与斜边的长度之比是 $1:2$。

Proof: Proof of **Property 1**: Note that an angle bisector of an equilateral triangle is also its perpendicular bisector (by the coincidence of four line segments). Clearly, the ratio of lengths of the shorter leg to the hypotenuse is $1:2$. By the Pythagorean Theorem, we'll obtain the ratio of lengths of the shorter leg to longer leg to hypotenuse to be $1:\sqrt{3}:2$. 要证明**性质 1**，注意一个等边三角形的一条角平分线同时也是一条垂直平分线（四线合一定理）。显然，较短的直角边与斜边的长度之比是 $1:2$。根据勾股定理可得出较短的直角边与较长的直角边与斜边的长度之比为 $1:\sqrt{3}:2$。

As shown in Figure 15-75, line l is both a perpendicular bisector and an angle bisector of the large equilateral triangle.

如图 15-75 所示，直线 l 同时是大等边三角形的垂直平分线和角平分线。

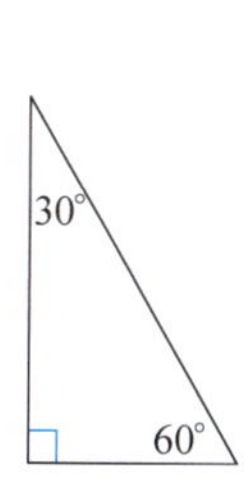

Figure 15-74

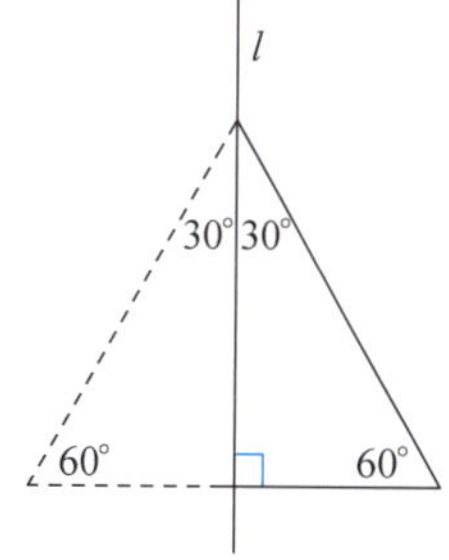

Figure 15-75

Question: If two sides of a $30^\circ-60^\circ-90^\circ$ triangle have lengths of 8 and 4, what is the length of the third side? 如果一个 $30^\circ-60^\circ-90^\circ$ 三角形的两边长度是 8 和 4，第三边的长度是多少？

Answer: Note that 4 is the length of the shorter leg, and 8 is the length of the hypotenuse

(**Property 2**). The length of the remaining leg is $4\sqrt{3}$. 根据**性质 2**,可得出 4 是较短的直角边长度,8 是斜边长度。剩下的直角边长度为 $4\sqrt{3}$。

15.9.3 Converse of the Pythagorean Theorem 勾股定理的逆定理

Converse of the Pythagorean Theorem 勾股定理的逆定理

Definition: In triangle T, if the sum of the squares of the lengths of the shorter sides is equal to the square of the length of the longest side, then T is a right triangle. Namely, if a, b, and c are the side-lengths of triangle T such that $a^2+b^2=c^2$, then T is a right triangle. 在三角形 T 中,如果直角边长度的平方和等于斜边长度的平方,那么 T 是直角三角形。换言之,若 a、b、c 分别为三角形 T 的边长,并满足 $a^2+b^2=c^2$,那么 T 是一个直角三角形。

Proof: Proof of the converse of the Pythagorean Theorem:

Suppose in $\triangle ABC$, we know that $a^2+b^2=c^2$. We want to show that $\triangle ABC$ is a right triangle.

Construct a right $\triangle A'B'C'$ so that $a'=a$, $b'=b$, and $c'=c$. From the Pythagorean Theorem, we know that $(a')^2+(b')^2=(c')^2$ from $a^2+b^2=c^2$.

Note that $\triangle ABC \cong \triangle A'B'C'$ by the SSS Congruence Postulate. Since $\triangle A'B'C'$ is a right triangle, $\triangle ABC$ is also a right triangle.

在$\triangle ABC$ 中,已知 $a^2+b^2=c^2$,证明$\triangle ABC$ 是直角三角形。

建立直角三角形 $A'B'C'$使得 $a'=a$, $b'=b$,和 $c'=c$。根据勾股定理,从 $a^2+b^2=c^2$ 可得知$(a')^2+(b')^2=(c')^2$。因为边边边全等公理,$\triangle ABC \cong \triangle A'B'C'$。因为$\triangle A'B'C'$是直角三角形,所以$\triangle ABC$ 也是直角三角形。

15.9.4 Pythagorean Trees 勾股树

Pythagorean Tree 勾股树

n. [pəˌθæɡəˈriən tri]

Definition: A tree that is formed by squares, for which special triangles (varying in sizes) are formed in the spaces enclosed by three squares (varying in sizes). This demonstrates the Pythagorean Theorem. 一棵以正方形画出来的树,三个正方形(大小不一)所包围的空间是一个特殊的直角三角形(大小不一)。勾股树展现了勾股定理。

Example: Figure 15-76 shows how two Pythagorean Trees are drawn procedurally.

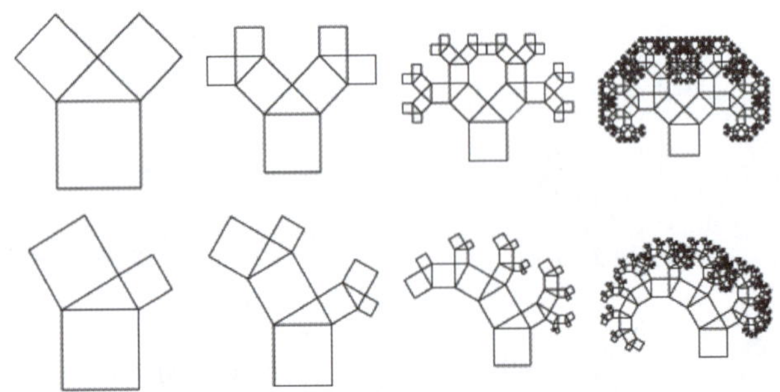

Figure 15-76

15.9.5　Summary　总结

1. All right triangles satisfy the Pythagorean Theorem: if a, b, and c are the side-lengths of a right triangle, such that c is the length of the hypotenuse, then $a^2+b^2=c^2$.　所有直角三角形都满足勾股定理：若 a、b、c 分别为一个直角三角形的边长，其中 c 是斜边长度，那么必然满足 $a^2+b^2=c^2$。
2. We can use the Converse of the Pythagorean Theorem to determine whether a triangle is a right triangle: In triangle T, if the sum of the squares of the lengths of the shorter sides is equal to the square of the length of the longest side, then T is a right triangle.　可以用勾股定理的逆定理判定一个三角形是否是直角三角形：在三角形 T 中，如果直角边长度的平方和等于斜边长度的平方，那么 T 是直角三角形。
3. Special right triangle—$45°-45°-90°$ triangle: It is also called the isosceles right triangle. The ratio of lengths of sides opposite the $45°$, $45°$, and $90°$ angles is $1:1:\sqrt{2}$.　特殊的直角三角形——$45°-45°-90°$ 三角形，也称为等腰直角三角形。相对 $45°$、$45°$、$90°$ 角的边长比例是 $1:1:\sqrt{2}$。
4. Special right triangle—$30°-60°-90°$ triangle: The ratio of lengths of sides opposite the $30°$, $60°$, and $90°$ angles is $1:\sqrt{3}:2$.　特殊的直角三角形——$30°-60°-90°$ 三角形，相对 $30°$、$60°$、$90°$ 角的边长比例是 $1:\sqrt{3}:2$。

15.10　Similarity　相似

15.10.1　Introduction　介绍

Review Ratio and Proportions in Section 2.3.　复习 2.3 节比和比例。

Similar Polygons　相似多边形

n. [ˈsɪmələr ˈpɑlɪˌgɑns]

Definition: Two polygons are similar polygons if the lengths of their corresponding line segments are proportional and their corresponding angles are congruent.　对于两个多边形，若对应的边的长度呈比例，对应的角全等，则它们是相似多边形。

Notation: If polygons P and P' are similar polygons, we write $P \sim P'$.

Congruent polygons are always similar.　全等多边形必然是相似多边形。

Refer to Table A-5 in Appendix A for the notation of similarity.

相似的标记方式参照附录 A 的表 A-5。

Examples: As shown in Figure 15-77, polygons $ABCD$ and $EFGH$ are similar polygons. We write $ABCD \sim EFGH$.

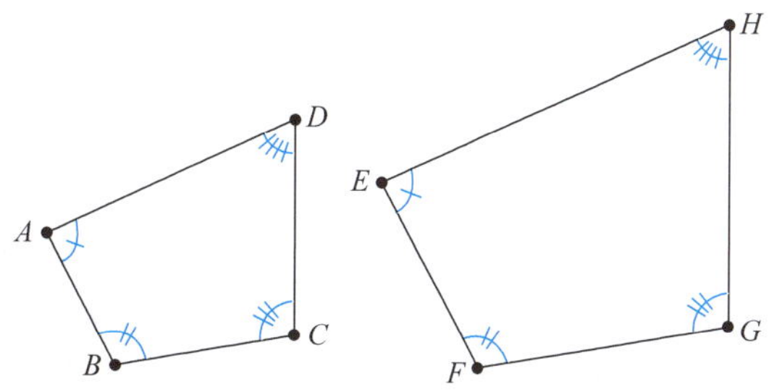

Figure 15-77

It satisfies:

1. $\frac{AB}{EF}=\frac{BC}{FG}=\frac{CD}{GH}=\frac{DA}{HE}$.

2. $\angle A\cong\angle E$, $\angle B\cong\angle F$, $\angle C\cong\angle G$, and $\angle D\cong\angle H$.

Ratio of Similitude 相似比

n. [ˈreɪʃɪˌoʊ əv səˈmɪl. ə. tuːd]

Definition: If polygons A and B are similar, then the ratio of similitude of A to B is the ratio of A's side-length to B's corresponding side-length. 若多边形 A 和 B 相似,则相似比为 A 的一条线段和 B 的对应线段的长度之比。

Examples: If polygons $ABCD$ and $EFGH$ are similar, then their ratio of similitude is $\frac{AB}{EF}=\frac{BC}{FG}=\frac{CD}{GH}=\frac{DA}{HE}$ (corresponding sides). 若多边形 $ABCD$ 和 $EFGH$ 相似,则它们的相似比为 $\frac{AB}{EF}=\frac{BC}{FG}=\frac{CD}{GH}=\frac{DA}{HE}$(对应的边之比)。

If polygons $ABCD$ and $EFGH$ are congruent, they are similar by definition, and the ratio of similitude is 1. 若多边形 $ABCD$ 和 $EFGH$ 全等,根据定义它们是相似多边形,相似比为 1。

Similar Triangles 相似三角形

n. [ˈsɪmələr ˈtraɪˌæŋgəls]

Definition: See similar polygons for the general definition. 相似三角形的一般定义见相似多边形的定义。

Notation: If $\triangle ABC$ and $\triangle DEF$ are similar, we write $\triangle ABC\sim\triangle DEF$.

Properties: 1. As shown in Figure 15-78, if $\triangle ABC\sim\triangle DEF$, then

(1) Lengths of corresponding line segments are proportional:
相似三角形对应的线段长度呈比例:
$\frac{AB}{DE}=\frac{BC}{EF}=\frac{AC}{DF}=\frac{\text{Height at } A}{\text{Height at } D}=\frac{\text{Height at } B}{\text{Height at } E}=\frac{\text{Height at } C}{\text{Height at } F}$.

(2) Corresponding angles are congruent:
相似三角形对应的角全等:
$\angle A\cong\angle D$, $\angle B\cong\angle E$, and $\angle C\cong\angle F$.

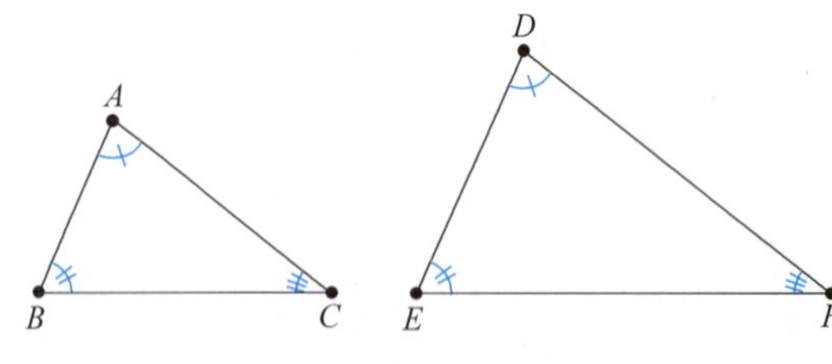

Figure 15-78

2. Congruent triangles are always similar. Their ratio of similitude is 1. 全等三角形必然是相似三角形,相似比为 1。

Questions: 1. If $\triangle ABC\sim\triangle DEF$, $m\angle A=40°$, $AB=12$, $AC=6$, and $DE=8$, what is the value of:

(1) $m\angle D$, and (2) DF?

2. If $\triangle ABC \sim \triangle DEF$, $m\angle A = 40°$, and $m\angle B = 60°$, what is the value of (1) $m\angle C$, (2) $m\angle D$, (3) $m\angle E$, and (4) $m\angle F$?

Answers: 1. (1) Since corresponding angles are congruent, $m\angle D = m\angle A = 40°$.

(2) From the ratio of similitude, $\frac{AB}{DE} = \frac{AC}{DF}$, from which $\frac{12}{8} = \frac{6}{DF}$, and $DF = 4$.

2. (1) $m\angle C = 180° - m\angle A - m\angle B = 80°$.

Since corresponding angles are congruent, we have:

(2) $m\angle D = m\angle A = 40°$.

(3) $m\angle E = m\angle B = 60°$.

(4) $m\angle F = m\angle C = 80°$.

15.10.2 Postulates and Theorems of Similarity 相似的公理和定理

AA Similarity Postulate 角角相似公理(两个角相等的两个三角形相似)

Definition: If two angles of one triangle are congruent to two angles of another, then the two triangles are similar. AA is the abbreviation of Angle-Angle. 若一个三角形的两个角与另一个三角形的两个角全等,则这两个三角形相似。AA 是角角的缩写。

Examples: As shown in Figure 15-79, we can conclude that $\triangle ABC \sim \triangle DEF$.

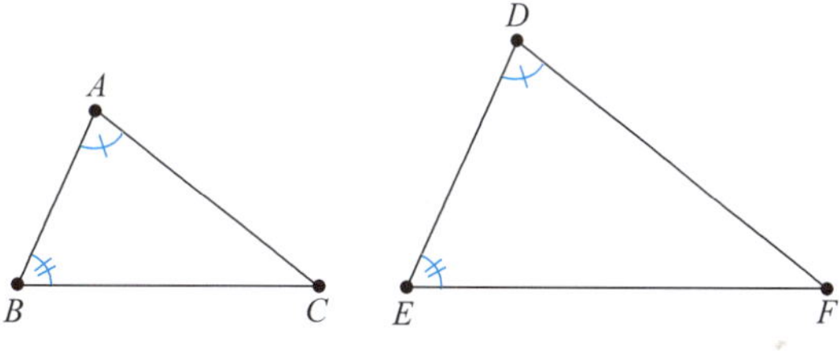

Figure 15-79

SAS Similarity Theorem 边角边相似定理(两条边对应成比例且夹角相等的两个三角形相似)

Definition: If two side-lengths of one triangle are proportional to two side-lengths of another, and the included angles are congruent, then the two triangles are similar. SAS is the abbreviation of Side-Angle-Side. 若一个三角形的两条边的边长与另一个三角形的两条边的边长成比例,且夹角相等,则这两个三角形相似。SAS 是边角边的缩写。

Proofs: As shown in Figure 15-80, we want to show that for $\triangle ABC$ and $\triangle DEF$, if $\frac{AB}{DE} = \frac{AC}{DF}$, and $\angle A \cong \angle D$, then $\triangle ABC \sim \triangle DEF$.

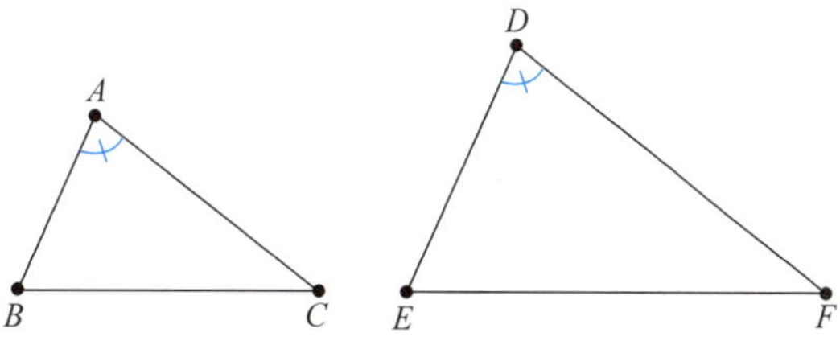

Figure 15-80

对 SAS 相似定理的证明:

If $AB = DE$, then $AC = DF$, and $\triangle ABC \cong \triangle DEF$ due to the SAS Congruence Postulate. Congruent triangles are always similar. 若 $AB = DE$,则 $AC = DF$,并且根据边角边全等公理,$\triangle ABC \cong \triangle DEF$。全等三角形必然相似。

Without the loss of generality, assume that $AB < DE$, so that $AC < DF$. 一般情况下,假设 $AB < DE$,可得出 $AC < DF$。

As shown in Figure 15-81, construct P on $\overline{DE}$ so that $DP = AB$, and Q on $\overline{DF}$ so that $\overline{PQ} \parallel \overline{EF}$. By the Corresponding Angles Postulate, $\angle DPQ \cong \angle DEF$. By the Reflexive Property, $\angle D \cong \angle D$. Therefore, $\triangle DPQ \sim \triangle DEF$ by AA Similarity Postulate. 如图 15-81 所示,在 $\overline{DE}$ 上取一点 P,使得 $DP = AB$;在 $\overline{DF}$ 上取一点 Q,使得 $\overline{PQ} \parallel \overline{EF}$。再根据同位角公理,$\angle DPQ \cong \angle DEF$。根据反身性,$\angle D \cong \angle D$。所以,根据角角相似公理,可得出$\triangle DPQ \sim \triangle DEF$。

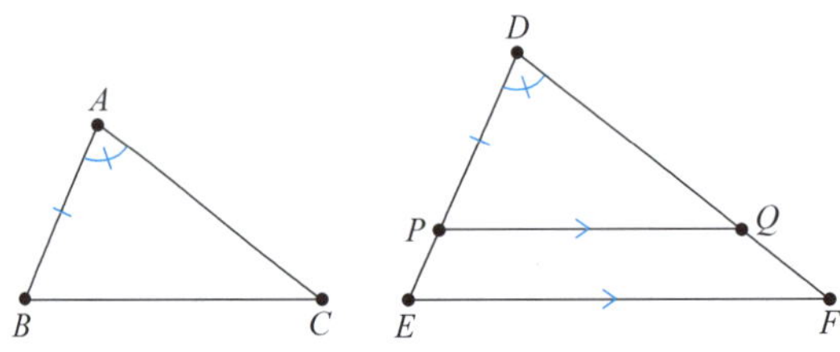

Figure 15-81

We located P so that $DP = AB$. By the ratio of similitude and substitution on $\triangle DPQ \sim \triangle DEF$, we have $\dfrac{DQ}{DF} = \dfrac{DP}{DE} = \dfrac{AB}{DE}$. Also, from the given information, we know that $\dfrac{AB}{DE} = \dfrac{AC}{DF}$. By transitive, $\dfrac{DQ}{DF} = \dfrac{AC}{DF}$, from which $DQ = AC$. 已取了一点 P 使得 $DP = AB$,$\triangle DPQ \sim \triangle DEF$,利用相似比和代入法,可得出$\dfrac{DQ}{DF} = \dfrac{DP}{DE} = \dfrac{AB}{DE}$。而且,由已知条件可知$\dfrac{AB}{DE} = \dfrac{AC}{DF}$。根据传递性,可得出$\dfrac{DQ}{DF} = \dfrac{AC}{DF}$,所以 $DQ = AC$。

We now have $\triangle ABC \cong \triangle DPQ$ by the SAS Congruence Postulate, since $AB = DP$, $\angle A \cong \angle D$, and $AC = DQ$. Since $\triangle DPQ \sim \triangle DEF$, we know that $\triangle ABC \sim \triangle DEF$. 因为 $AB = DP$,$\angle A \cong \angle D$,$AC = DQ$,根据边角边全等公理可得出$\triangle ABC \cong \triangle DPQ$。因为$\triangle DPQ \sim \triangle DEF$,所以$\triangle ABC \sim \triangle DEF$。

SSS Similarity Theorem 边边边相似定理(三条边对应成比例的两个三角形相似)

Definition: If the lengths of three sides of one triangle are proportional to the lengths of three sides of another, then the two triangles are similar. SSS is the abbreviation of Side-Side-Side. 若一个三角形的三条边的边长和另一个三角形的三条边的边长成比例,则这两个三角形相似。

Proofs: We want to show that for $\triangle ABC$ and $\triangle DEF$, if $\dfrac{AB}{DE} = \dfrac{AC}{DF} = \dfrac{BC}{EF}$, then $\triangle ABC \sim \triangle DEF$.

对边边边相似定理的证明:

If $AB=DE$, then $AC=DF$ and $BC=EF$, so $\triangle ABC \cong \triangle DEF$ due to SSS Congruence Postulate. Congruent triangles are always similar. 若 $AB=DE$, $AC=DF$, $BC=EF$, 则根据边边边全等公理, 可得 $\triangle ABC \cong \triangle DEF$, 全等三角形必然相似。

Without the loss of generality, assume that $AB<DE$, so that $AC<DF$ and $BC<EF$. 一般情况下, 假设 $AB<DE$, 则 $AC<DF$, $BC<EF$。

As shown in Figure 15-82, construct P on $\overline{DE}$ so that $DP=AB$, and Q on $\overline{DF}$ so that $DQ=AC$. By substitution on $\frac{AB}{DE}=\frac{AC}{DF}$, we have $\frac{DP}{DE}=\frac{DQ}{DF}$. By the Reflexive Property, $\angle D \cong \angle D$. Therefore, $\triangle DPQ \sim \triangle DEF$ by SAS Similarity Theorem. 如图 15-82 所示, 在 $\overline{DE}$ 上取一点 P, 使得 $DP=AB$; 在 $\overline{DF}$ 上取一点 Q, 使得 $DQ=AC$。由 $\frac{AB}{DE}=\frac{AC}{DF}$, 用代入法, 得出 $\frac{DP}{DE}=\frac{DQ}{DF}$。根据反身性, $\angle D \cong \angle D$。所以, 根据边角边相似定理, 可得 $\triangle DPQ \sim \triangle DEF$。

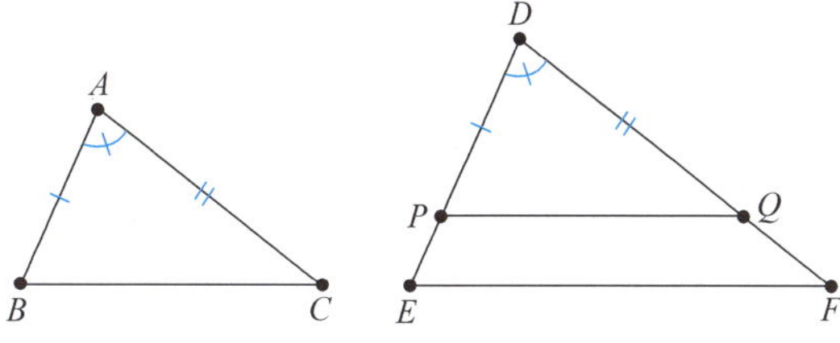

Figure 15-82

By the ratio of similitude, we have $\frac{DP}{DE}=\frac{DQ}{DF}=\frac{PQ}{EF}$. By substitution, we get $\frac{AB}{DE}=\frac{PQ}{EF}$ (1). 根据相似比的定义, $\frac{DP}{DE}=\frac{DQ}{DF}=\frac{PQ}{EF}$。用代入法可得出 $\frac{AB}{DE}=\frac{PQ}{EF}$ (1)。

It is given that $\frac{AB}{DE}=\frac{BC}{EF}$. By transitive with (1), $\frac{PQ}{EF}=\frac{BC}{EF}$, from which $PQ=BC$. 已知 $\frac{AB}{DE}=\frac{BC}{EF}$。由(1)根据传递性, $\frac{PQ}{EF}=\frac{BC}{EF}$, 可得出 $PQ=BC$。

We now have $\triangle ABC \cong \triangle DPQ$ by the SSS Congruence Postulate. Since $\triangle DPQ \sim \triangle DEF$, we know that $\triangle ABC \sim \triangle DEF$. 根据边边边全等公理, 得出 $\triangle ABC \cong \triangle DPQ$, 又因为 $\triangle DPQ \sim \triangle DEF$, 所以可知 $\triangle ABC \sim \triangle DEF$。

15.10.3 Common Theorems of Similarity 常见的相似定理

Common Theorems of Similarity 常见的相似定理

Theorems: 1. In $\triangle ABC$, if P is on $\overline{AB}$ and Q is on $\overline{AC}$ such that $\overline{PQ} \parallel \overline{BC}$, then (1) $\triangle APQ \sim \triangle ABC$ and (2) $\frac{AP}{PB}=\frac{AQ}{QC}$, as shown in Figure 15-83.

视频 27

2. If $\overline{AB} \parallel \overline{DE}$, and $\overline{AE}$ and $\overline{BD}$ intersect at C, such that C is between $\overline{AB}$ and $\overline{DE}$, then $\triangle ABC \sim \triangle EDC$, as shown in Figure 15-84.

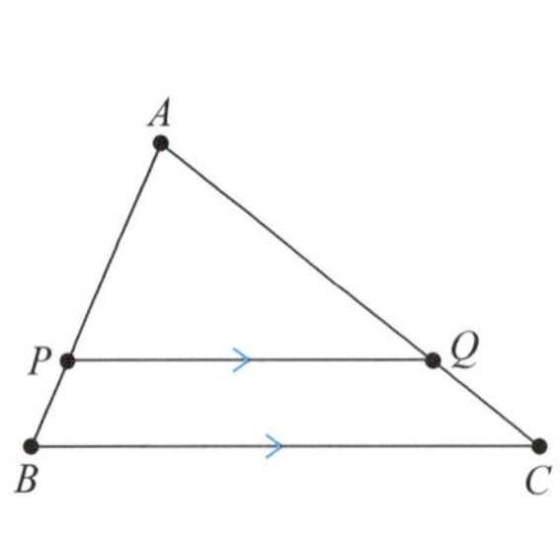

Figure 15-83

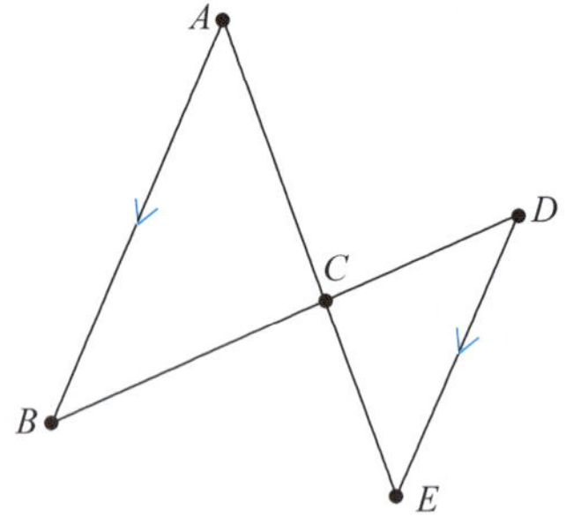

Figure 15-84

3. **Midsegment Theorem**: In every triangle, a midsegment drawn to any two sides is parallel to the third side, with half of its length. **中位线定理**：在任意三角形中，连接两边的中位线与第三边平行，且长度是第三边的一半。

   As shown in Figure 15-85, we can conclude that $\overline{PQ} \parallel \overline{AB}$, and $PQ = \frac{1}{2}AB$.

4. **Angle Bisector Theorem**: As shown in Figure 15-86, in $\triangle ABC$, if P is a point on $\overline{BC}$ and $\overline{AP}$ is an angle bisector of $\angle A$, then $\frac{AB}{BP} = \frac{AC}{CP}$. **角平分线定理**：在$\triangle ABC$中，假设$P$为$\overline{BC}$上一点，且$\overline{AP}$为$\angle A$的角平分线，则$\frac{AB}{BP} = \frac{AC}{CP}$。

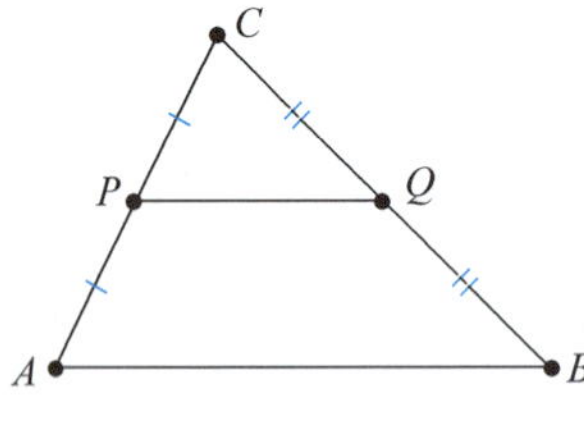

Figure 15-85

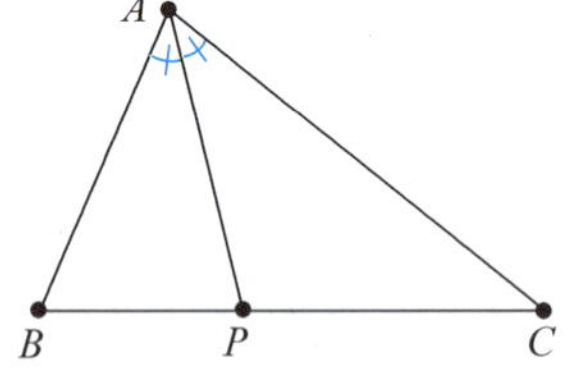

Figure 15-86

5. Figure 15-87 illustrates one common pattern of similar right triangles, from which we can conclude that $\triangle ABC \sim \triangle CDE$. 图 15-87 展示了相似直角三角形的一种常见规律：$\triangle ABC \sim \triangle CDE$。

6. **Concurrency of Medians Theorem**: The medians of every triangle are concurrent. **中线共点定理**：任意三角形的所有中线都是共点的。

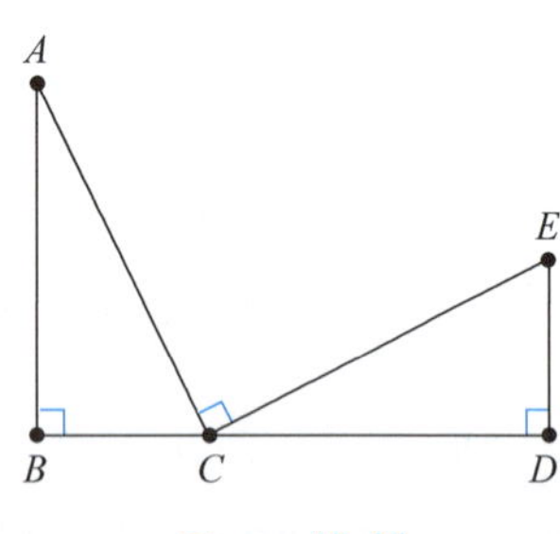

Figure 15-87

Proofs: Refer to the diagrams above.

1. To prove Theorem 1(1), since $\overline{PQ} \parallel \overline{BC}$, by the **Corresponding Angles Postulate**, $\angle APQ \cong \angle ABC$. By reflexive, $\angle A \cong \angle A$. Therefore, $\triangle APQ \sim \triangle ABC$ by AA Similarity Postulate. 要证明定理1(1)，因为$\overline{PQ} \parallel \overline{BC}$，所以根据**同位角公理**，$\angle APQ \cong \angle ABC$。又由于反身性，$\angle A \cong \angle A$。所以根据角角相似公理，可得$\triangle APQ \sim \triangle ABC$。

To prove Theorem 1(2), since $\triangle APQ \sim \triangle ABC$, the ratio of similitude implies that $\frac{AP}{AB}=\frac{AQ}{AC}$, from which $\frac{AB}{AP}=\frac{AC}{AQ}$, or $\frac{AP+PB}{AP}=\frac{AQ+QC}{AQ}$. From here, we have $1+\frac{PB}{AP}=1+\frac{QC}{AQ}$. Therefore, $\frac{PB}{AP}=\frac{QC}{AQ}$, from which $\frac{AP}{PB}=\frac{AQ}{AC}$. 要证明定理1(2),因为$\triangle APQ \sim \triangle ABC$,所以相似比为$\frac{AP}{AB}=\frac{AQ}{AC}$,也就是$\frac{AB}{AP}=\frac{AC}{AQ}$,即$\frac{AP+PB}{AP}=\frac{AQ+QC}{AC}$,能得出$1+\frac{PB}{AP}=1+\frac{QC}{AQ}$,所以有$\frac{PB}{AP}=\frac{QC}{AQ}$,满足$\frac{AP}{PB}=\frac{AQ}{AC}$。

2. To prove Theorem 2, since $\overline{AB} \parallel \overline{DE}$, by the **Alternate Interior Angles Theorem**, $\angle A \cong \angle E$ and $\angle B \cong \angle D$. Therefore, we have $\triangle ABC \sim \triangle EDC$ by AA Similarity Postulate. 要证明定理2,因为$\overline{AB} \parallel \overline{DE}$,所以根据**内错角定理**,可得$\angle A \cong \angle E$,$\angle B \cong \angle D$。根据角角相似公理,可得$\triangle ABC \sim \triangle EDC$。

3. To prove the **Midsegment Theorem**, let $\overline{PQ}$ be a midsegment of $\triangle ABC$ such that P is the midpoint of $\overline{AC}$ and Q is the midpoint of $\overline{BC}$. 要证明**中位线定理**,设$\overline{PQ}$为$\triangle ABC$的中位线,P为$\overline{AC}$的中点,Q为$\overline{BC}$的中点。

   We want to show that $\overline{PQ} \parallel \overline{AB}$ and $PQ=\frac{1}{2}AB$.

   To show $\overline{PQ} \parallel \overline{AB}$, by definition of midsegments $\frac{CP}{CA}=\frac{CQ}{CB}=\frac{1}{2}$. By reflexive, $\angle C \cong \angle C$. Therefore, we get $\triangle CPQ \sim \triangle CAB$ by SAS Similarity Theorem. By the properties of similar triangles, $\angle CPQ \cong \angle CAB$. By the **Converse of the Corresponding Angle Postulate**, $\overline{PQ} \parallel \overline{AB}$. 要证明$\overline{PQ} \parallel \overline{AB}$,根据中位线的定义,可得$\frac{CP}{CA}=\frac{CQ}{CB}=\frac{1}{2}$。由于反身性,$\angle C \cong \angle C$,所以根据边角边相似定理,可得$\triangle CPQ \sim \triangle CAB$。根据相似三角形的性质,可得$\angle CPQ \cong \angle CAB$。根据**同位角公理的逆公理**,得$\overline{PQ} \parallel \overline{AB}$。

   To show $PQ=\frac{1}{2}AB$, we know that $\triangle CPQ \sim \triangle CAB$ from $\overline{PQ} \parallel \overline{AB}$. So $\frac{CP}{CA}=\frac{CQ}{CB}=\frac{PQ}{AB}=\frac{1}{2}$, from which $2PQ=AB$, or $PQ=\frac{1}{2}AB$. 要证明$PQ=\frac{1}{2}AB$,从$\overline{PQ} \parallel \overline{AB}$可得知$\triangle CPQ \sim \triangle CAB$,所以$\frac{CP}{CA}=\frac{CQ}{CB}=\frac{PQ}{AB}=\frac{1}{2}$,可得出$2PQ=AB$,故$PQ=\frac{1}{2}AB$。

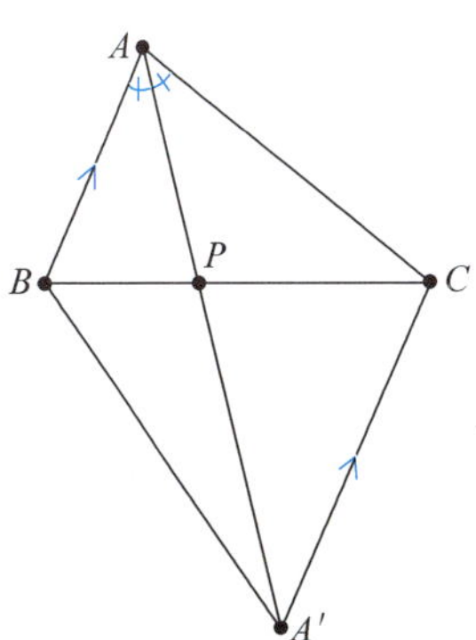

Figure 15-88

4. To prove Theorem 4, extend $\overline{AP}$ to A' so that $\overline{AB} \parallel \overline{A'C}$, as shown in Figure 15-88. By Theorem 2, $\triangle ABP \sim \triangle A'CP$. 要证明定理4,延长$\overline{AP}$到$A'$点,使得$\overline{AB} \parallel \overline{A'C}$。根据定理2,可得出$\triangle ABP \sim \triangle A'CP$。

   By the definition of angle bisectors, $\angle PAB \cong \angle PAC$. By the **Alternate Interior Angles Theorem**, $\angle PAB \cong \angle PA'C$. By transitive, $\angle PAC \cong \angle PA'C$. By the **Converse of the**

Base Angles Theorem, $AC = A'C$.
根据角平分线的定义,可得$\angle PAB \cong \angle PAC$,又根据**内错角定理**,得$\angle PAB \cong \angle PA'C$。由于传递性,可得$\angle PAC \cong \angle PA'C$。根据**等角对等边定理**,得 $AC = A'C$。

Since $\triangle ABP \sim \triangle A'CP$, by the ratio of similitude, $\frac{AB}{A'C} = \frac{BP}{CP}$. By substitution, we have $\frac{AB}{AC} = \frac{BP}{CP}$. Multiplying both sides by $\frac{AC}{BP}$ gives the result $\frac{AB}{BP} = \frac{AC}{CP}$. 因为$\triangle ABP \sim \triangle A'CP$,相似比为$\frac{AB}{A'C} = \frac{BP}{CP}$。用代入法可得出$\frac{AB}{AC} = \frac{BP}{CP}$。两边乘以$\frac{AC}{BP}$可得出$\frac{AB}{BP} = \frac{AC}{CP}$。

5. To prove Theorem 5, note that
$m\angle ACB + m\angle DCE = 90°$, from which $m\angle DCE = 90° - m\angle ACB$,
and $m\angle ACB + m\angle BAC = 90°$, from which $m\angle BAC = 90° - m\angle ACB$.
By transitive, we have $m\angle DCE = m\angle BAC$.
From here, we can conclude that $\triangle ABC \sim \triangle CDE$ by AA Similarity Postulate.

6. To prove the **Concurrency of Medians Theorem**, refer to Figure 15-89. 图 15-89 给出了**中线共点定理**的证明。
Suppose $\overline{AD}$ and $\overline{BE}$ are two medians of $\triangle ABC$, intersecting at G. Let points F and H be on $\overleftrightarrow{CG}$ such that F is the intersection of $\overleftrightarrow{CG}$ and $\overline{AB}$, and $CG = GH$. We want to show that $\overline{CGF}$ is a median of $\triangle ABC$ so that the medians are concurrent at G. 假设 $\overline{AD}$ 和 $\overline{BE}$ 是$\triangle ABC$ 的两条中线,相交于 G。设点 F 和 H 在 $\overleftrightarrow{CG}$ 上使得 F 是 $\overleftrightarrow{CG}$ 和 $\overline{AB}$ 的交点,并且 $CG = GH$。想证明 $\overline{CGF}$ 是$\triangle ABC$ 的中线,以便证明三条中线都交于 G。

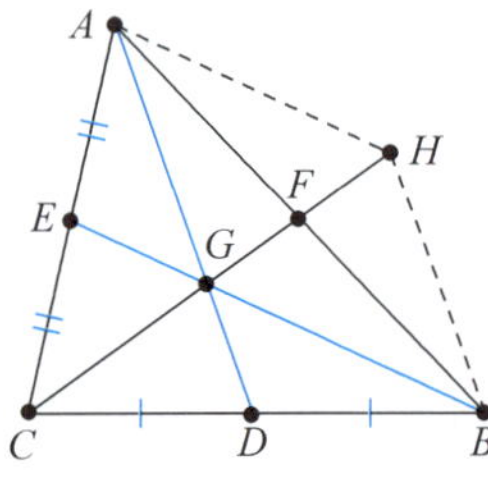

Figure 15-89

Clearly, $\overline{EG}$ is a midsegment of $\triangle ACH$. By the **Midsegment Theorem** we can conclude that $\overleftrightarrow{EG} \parallel \overline{AH}$, from which $\overline{GB} \parallel \overline{AH}$. 显然,$\overline{EG}$是$\triangle ACH$ 的中位线。根据**中位线定理**,可得出 $\overleftrightarrow{EG} \parallel \overline{AH}$,所以 $\overline{GB} \parallel \overline{AH}$。

Similarly, we can conclude that $\overleftrightarrow{DG} \parallel \overline{BH}$, from which $\overline{GA} \parallel \overline{BH}$. 同理,可得出 $\overleftrightarrow{DG} \parallel \overline{BH}$,所以 $\overline{GA} \parallel \overline{BH}$。
From here, we can conclude that $AGBH$ is a parallelogram. Since the diagonals of a parallelogram bisect each other (by Property 3 of parallelogram in Section 16.2.1), we have $AF = FB$, from which we can conclude that $\overline{CGF}$ is a median of $\triangle ABC$.
至此,可以得出 $AGBH$ 是平行四边形。因为对角线互相平分(见 16.2.1 节平行四边形的性质 3),可得出 $AF = FB$,也就是说 $\overline{CGF}$ 是$\triangle ABC$ 的中线。

15.10.4　Means　平均数

Arithmetic Mean　算术平均数

n. [ə'rɪθməˌtɪk min]

Definition: See "mean" in Section 9.2.1.　见 9.2.1 节"平均数"的词条。

Geometric Mean　几何平均数

n. [ˌdʒiə'metrɪk min]

Definition: The geometric mean of n numbers is the nth root of their product.

In geometry course, we will focus on the case when $n=2$.

n 个数的几何平均数为它们积的 n 次方根。

几何课会重点讲解 $n=2$ 的情况。

Notation: The geometric mean of $a_1, a_2, \cdots, a_n$ is $\sqrt[n]{a_1 a_2 \cdots a_n}$.

The geometric mean of two numbers, a and b, is simply $\sqrt{ab}$.

Properties: Below is an important result, AM-GM Inequality, the abbreviation for Inequality of Arithmetic and Geometric Means.

For a list of nonnegative numbers, the arithmetic mean is always greater than or equal to the geometric mean:

$$\frac{a_1 + a_2 + \cdots + a_n}{n} \geqslant \sqrt[n]{a_1 a_2 \cdots a_n}.$$

We will focus on $n=2$, from which $\frac{a+b}{2} \geqslant \sqrt{ab}$.

Proofs: We will prove the case when $n=2$.

To prove $\frac{a+b}{2} \geqslant \sqrt{ab}$, if we can show that $\frac{a+b}{2} - \sqrt{ab} \geqslant 0$, then we are done.

$$\begin{aligned}\frac{a+b}{2} - \sqrt{ab} &= \frac{a+b}{2} - \frac{2\sqrt{ab}}{2} \\ &= \frac{a - 2\sqrt{ab} + b}{2} \\ &= \frac{(\sqrt{a} - \sqrt{b})^2}{2} \\ &\geqslant 0\end{aligned}$$

From the last step, it is clear that the arithmetic and geometric means of two numbers are equal if and only if $a=b$.

Examples: In the example of similar triangles shown in Figure 15-90, we have $\triangle ABD \sim \triangle DBC \sim \triangle ADC$. We have this ratio of similitude: $\frac{AB}{BD} = \frac{DB}{BC}$. Since $BD = DB$, this is $\frac{AB}{BD} = \frac{BD}{BC}$, from which $BD = \sqrt{(AB)(BC)}$.

BD is the geometric mean of AB and BC.

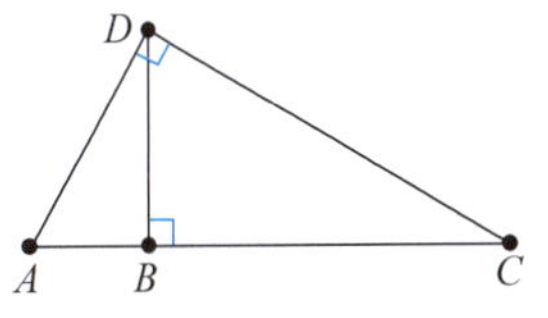

Figure 15-90

Now try to find the other sides that are the geometric mean of the other two sides.

Questions: What is the geometric mean of 3 and 48?

Answers: $\sqrt{3 \cdot 48} = \sqrt{144} = 12$.

15.10.5 Summary 总结

1. (1) Two polygons are similar polygons if the lengths of their corresponding line segments are proportional and their corresponding angles are congruent. Congruent polygons are similar. 对于两个多边形,若对应的边的长度成比例,对应的角全等,则它们是相似多边形。全等多边形是相似多边形。

   (2) If polygons A and B are similar, then the ratio of similitude of A to B is the ratio of A's side-length to B's corresponding side-length. The ratio of similitude of congruent polygons is 1. 若多边形 A 和 B 相似,则其相似比为 A 的一条线段与 B 的对应线段的长度之比。全等多边形的相似比为 1。

2. To prove two triangles are similar, use one of the following criteria:
   (1) AA Similarity Postulate. 角角相似公理。
   (2) SAS Similarity Theorem. 边角边相似定理。
   (3) SSS Similarity Theorem. 边边边相似定理。

3. See Common Theorems of Similarity in Section 15.10.3. 见 15.10.3 节的常见的相似定理。

4. Arithmetic mean of n numbers: $\frac{a_1 + a_2 + \cdots + a_n}{n}$.

   Arithmetic mean of two numbers: $\frac{a+b}{2}$.

   Geometric mean of n numbers: $\sqrt[n]{a_1 a_2 \cdots a_n}$.

   Geometric mean of two numbers: $\sqrt{ab}$.

15.11 Right Triangle Trigonometry 直角三角形三角学

15.11.1 Introduction 介绍

Trigonometry 三角学

n. [ˌtrɪgəˈnɑmitri]

Definition: A branch of mathematics that studies the relationship between sides and angles of triangles and the relevant functions of any angle. 三角学是数学的一个分支,专门研究三角形的边和角的关系以及与角相关的函数。

In this section, we will study the trigonometric functions of acute angles in a right triangle. Throughout Section 15.11, we will be referring to Figure 15-91 for the drawing of right $\triangle ABC$, with $m\angle C = 90°$. 本节学习直角三角形中锐角的三角函数。可参考图 15-91 直角三角形 ABC 的画法,图中 $m\angle C = 90°$。

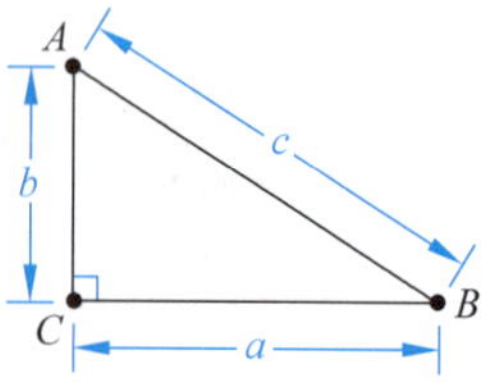

A
c
b
C
a
B

Figure 15-91

15.11.2　Parts of a Right Triangle　直角三角形部分

Opposite Side of an Acute Angle　锐角的对边

Definition：In right $\triangle ABC$, with $m\angle C = 90°$. The opposite side of $\angle A$ is $\overline{BC}$, whose length is a, and the opposite side of $\angle B$ is $\overline{AC}$, whose length is b.

在直角三角形 ABC 中，$m\angle C = 90°$。$\angle A$ 的对边是$\overline{BC}$，长度为 a。$\angle B$ 的对边是$\overline{AC}$，长度为 b。

Adjacent Side of an Acute Angle　锐角的邻边

Definition：In right $\triangle ABC$, with $m\angle C = 90°$. The adjacent side of $\angle A$ is $\overline{AC}$, whose length is b, and the adjacent side of $\angle B$ is $\overline{BC}$, whose length is a.

在直角三角形 ABC 中，$m\angle C = 90°$。$\angle A$ 的邻边是$\overline{AC}$，长度为 b。$\angle B$ 的邻边是$\overline{BC}$，长度为 a。

Note that the hypotenuse is not an adjacent side of an acute angle.　注意：斜边不为任何锐角的邻边。

Properties：In a right triangle, the opposite side of one acute angle is the adjacent side of the other.

在直角三角形中，一个锐角的对边等于另一个锐角的邻边。

15.11.3　Trigonometric Functions　三角函数

If we know the three pieces of information of a right triangle (besides 3 angles), then we know that there is a unique way to draw this right triangle.　已知一个直角三角形边或角的三个条件(除了已知三个角的条件外)，可以判断出这个直角三角形只有一种画法。

Trigonometric functions give us the ratio of lengths of two sides of a right triangle, based on the measure of an acute angle.　三角函数能根据一个直角三角形的锐角度数，给出两边的长度比。

Sine　正弦

n. [saɪn]

Definition：In a right triangle, the sine of an acute angle is defined as $\frac{\text{length of opposite}}{\text{length of hypotenuse}}$.　在直角三角形中，一个锐角的正弦的定义是它对边的长度和斜边的长度之比。

Notation：The sine of $\angle A$ is written as $\sin A$.

In right $\triangle ABC$ with $m\angle C = 90°$, $\sin A = a/c$, and $\sin B = b/c$.

Examples：In right $\triangle ABC$ with $m\angle C = 90°$, if $AC = 3$, $BC = 4$, and $AB = 5$, then $\sin A = a/c = 4/5$, and $\sin B = b/c = 3/5$.

Cosine　余弦

n. [ˈkoʊˌsaɪn]

Definition：In a right triangle, the cosine of an acute angle is defined as $\frac{\text{length of adjacent}}{\text{length of hypotenuse}}$.　在直角三角形中，一个锐角的余弦的定义是它邻边的长度和斜边的长度之比。

Notation：The cosine of $\angle A$ is written as $\cos A$.

In right $\triangle ABC$ with $m\angle C = 90°$, $\cos A = b/c$, and $\cos B = a/c$.

Properties: In right $\triangle ABC$ with $m\angle C=90°$, $\cos A=b/c=\sin B$, and $\cos B=a/c=\sin A$. 注：在直角三角形中，一个锐角的对边等于另一个锐角的邻边。

Examples: In right $\triangle ABC$ with $m\angle C=90°$, if $AC=3$, $BC=4$, and $AB=5$, then $\cos A=b/c=3/5$, and $\cos B=b/c=4/5$.

Tangent 正切

n. ['tændʒənt]

Definition: In a right triangle, the tangent of an acute angle is defined as $\frac{\text{length its opposite side}}{\text{length of its adjacent side}}$. 在直角三角形中，一个锐角的正切的定义是它对边的长度和邻边的长度之比。

Notation: The tangent of $\angle A$ is written as $\tan A$.

In right $\triangle ABC$ with $m\angle C=90°$, $\tan A=a/b$, and $\tan B=b/a$.

Properties: 1. For acute $\angle A$, $\tan A=\frac{\sin A}{\cos A}$, by the definition of sine and cosine functions: $\tan A=\frac{\text{length of opposite side of }\angle A}{\text{length of adjacent side of }\angle A}=\frac{\text{length of opposite side of }\angle A/\text{length of the hypotenuse}}{\text{length of adjacent side of }\angle A/\text{length of the hypotenuse}}=\frac{\sin A}{\cos A}$.

2. The tangent of one acute angle is the reciprocal of that of the other: $\tan A=\frac{\sin A}{\cos A}=\frac{\cos B}{\sin B}=\frac{1}{\frac{\sin B}{\cos B}}=\frac{1}{\tan B}$. 两个锐角的正切互为倒数。

Examples: In right $\triangle ABC$ with $m\angle C=90°$, if $AC=3$, $BC=4$, and $AB=5$, then $\tan A=a/b=4/3$, and $\tan B=b/a=3/4$.

15.11.4 Inverse Trigonometric Functions 反三角函数

Inverse trigonometric functions give us the measure of an acute angle in a right triangle, based on the ratio of lengths of two sides. 在直角三角形上，反三角函数能根据两边的长度比给出一个锐角的度数。

Inverse Trigonometric Functions 反三角函数

n. [ɪn'vɜrs ˌtrigənə'metrik 'fʌŋkʃəns]

Definition: The inverse functions for sine, cosine, and tangent are arcsine, arccosine, and arctangent, respectively. 反三角函数是指反正弦、反余弦、反正切函数。

Notation: The arcsine of x is written as $\arcsin x$.

The arccosine of x is written as $\arccos x$.

The arctangent of x is written as $\arctan x$.

Properties: In right $\triangle ABC$, with $m\angle C=90°$,

$\arcsin(\sin A)=m\angle A$ and $\arcsin(\sin B)=m\angle B$;

$\arccos(\cos A)=m\angle A$ and $\arccos(\cos B)=m\angle B$;

$\arctan(\tan A)=m\angle A$ and $\arctan(\tan B)=m\angle B$.

To explain one example in words, sin A represents $\frac{\text{length of } \angle A\text{'s opposite side}}{\text{length of the hypotenuse}}$; arcsin (sin A) represents the measure of the angle whose value of $\frac{\text{length of its opposite side}}{\text{length of the hypotenuse}}$ is sin A. Therefore, arcsin(sin A) = $m\angle A$ by the definition of sine.　sin A 代表的是 $\frac{\angle A\text{ 对边的长度}}{\text{斜边的长度}}$。arcsin(sin A)代表的是一个角的度数，这个角的对边长度与斜边长度之比是 sin A。根据正弦的定义，arcsin(sin A) = $m\angle A$。

Similar reasoning applies to the arccos and arctan functions.

反余弦、反正切函数同理。

Solving Right Triangles　解直角三角形

Definition: To solve a triangle, we need to find the information of the lengths of all sides and the measures of all angles. We will be using trigonometry.　解三角形需要解出所有边的长度和所有角的度数，会用到三角学。

We will learn how to solve a right triangle in this section.　本节学习如何解直角三角形。

Examples: 1. (Given the right angle and two sides.)　已知一个直角和两边，解直角三角形。

In right $\triangle ABC$ with $m\angle C = 90°$, $a = 3$, and $b = 4$:

We wish to find c, $m\angle A$, and $m\angle B$.

By the Pythagorean Theorem, $c = 5$.

$\sin A = a/c = 3/5$. Therefore, $m\angle A = \arcsin(\sin A) = \arcsin(3/5) \approx 36.870°$.

$\cos B = a/c = 3/5$. Therefore, $m\angle B = \arccos(\cos B) = \arccos(3/5) \approx 53.130°$.

We can get the same result by using different trigonometric functions.　用不同的三角函数可解出同样的结果，可谓是殊途同归。

2. (Given the right angle, an acute angle, and a side.)　已知一个直角、一个锐角、一边，解直角三角形。

In right $\triangle ABC$ with $m\angle C = 90°$, $m\angle A = 40°$, and $a = 8$:

We wish to find $m\angle B$, b, and c.

Clearly, $m\angle B = 50°$.

$\tan B = b/a$, from which $b = a \tan B = 8 \tan 50° \approx 9.534$.

$\sin A = a/c$. from which $c = a/\sin A = 8/\sin 40° \approx 12.446$.

15.11.5　Summary　总结

1. If we know the three pieces of information of a right triangle (besides 3 angles), then we know that there is a unique way to draw this right triangle.　已知一个直角三角形边或角的三个条件(除了三个角的条件外)，可以判断出这个三角形只有一种画法。

(1) Trigonometric functions give us the ratio of lengths of two sides of a right triangle, based on the measure of an acute angle.　三角函数能根据直角三角形的一个锐角的度数，给出两边的长度比。

(2) Inverse trigonometric functions give us the measure of an acute angle in a right triangle, based

on the ratio of lengths of two sides. 反三角函数能根据直角三角形的两边的长度比,给出锐角的度数。

2. In a right triangle, the opposite side of an acute angle is the same as the adjacent side of the other acute angle, and vice versa. 在一个直角三角形中,一个锐角的对边等于另一个锐角的邻边,反之亦然。

3. Trigonometric Functions: sin, cos, tan. See Section 15.11.3 for definitions. 三角函数包括正弦、余弦、正切函数。定义见 15.11.3 节。

   Note that:

   (1) $\tan x = \dfrac{\sin x}{\cos x}$.

   (2) $\tan x = \dfrac{1}{\tan(90° - x)}$.

   (3) $\sin x = \cos(90° - x)$.

   (4) $\cos x = \sin(90° - x)$.

4. Inverse Trigonometric Functions: arcsin, arccos, arctan. See Section 15.11.4 for definitions. 反三角函数包括反正弦、反余弦、反正切函数。定义见 15.11.4 节。

16. Quadrilaterals 四边形

Figure 16-1 shows the Venn diagram for types of quadrilaterals.
图 16-1 展示了四边形分类的文氏图。

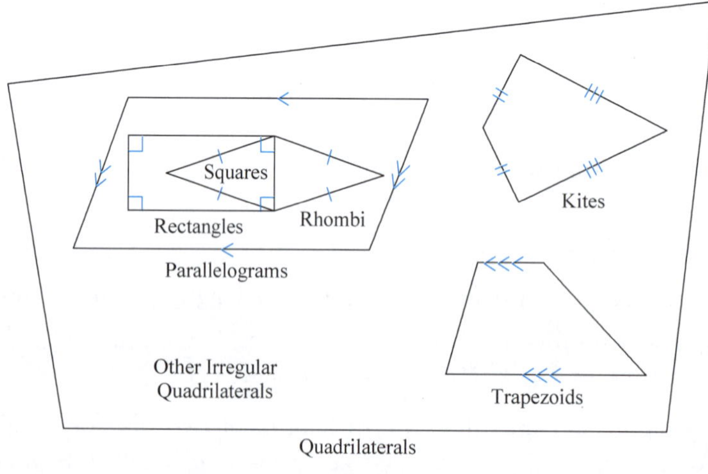

Figure 16-1

16.1 Introduction　介绍

Quadrilateral　四边形

n. [ˌkwɑdrəˈlætərəl]

Definition：A polygon with exactly four sides and exactly four angles/vertices.　四边形是指具有四条边和四个角/顶点的多边形。

Notation：We name a quadrilateral by listing the vertices in a consecutive order.　标记一个四边形，可按次序连续地列出它们的顶点。

For example，Figure 16-2 shows the quadrilateral $ABCD$，$BCDA$，$CDAB$，or $DABC$（depends on how one names it）.

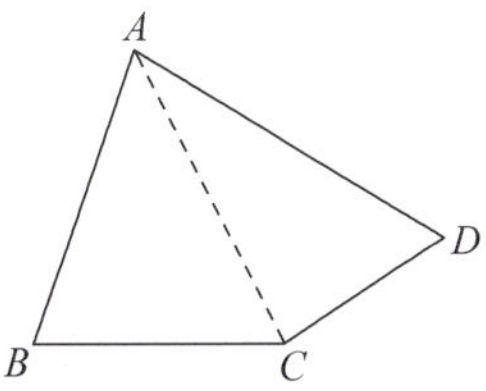

Figure 16-2

The quadrilateral $ABCD$ implies that（A and B），（B and C），（C and D），and（D and A）are adjacent vertices；（$\angle A$ and $\angle C$）and（$\angle B$ and $\angle D$）are opposite angles；$\overline{AB}$，$\overline{BC}$，$\overline{CD}$，and $\overline{DA}$ are sides；$\overline{AC}$ and $\overline{BD}$ are diagonals；（$\overline{AB}$ and $\overline{CD}$）and（$\overline{BC}$ and $\overline{DA}$）are opposite sides；（$\overline{AB}$ and $\overline{BC}$），（$\overline{BC}$ and $\overline{CD}$），and（$\overline{CD}$ and $\overline{DA}$）are adjacent sides. This is the same for $BCDA$，$CDAB$，or $DABC$.

四边形 $ABCD$ 暗示了 A 和 B、B 和 C、C 和 D、D 和 A 互为邻顶点；$\angle A$ 和$\angle C$、$\angle B$ 和$\angle D$ 互为对角；$\overline{AB}$、$\overline{BC}$、$\overline{CD}$、$\overline{DA}$ 为边；$\overline{AC}$ 和 $\overline{BD}$ 为对角线；$\overline{AB}$ 和 $\overline{CD}$、$\overline{BC}$ 和 $\overline{DA}$ 互为对边；$\overline{AB}$ 和 $\overline{BC}$、$\overline{BC}$ 和 $\overline{CD}$、$\overline{CD}$ 和 $\overline{DA}$ 互为邻边。这些条件在 $BCDA$、$CDAB$、$DABC$ 是相同的。

Properties：1. The main categories of quadrilaterals are：
四边形主要有：
（1）Parallelograms　平行四边形
（2）Trapezoids　梯形
（3）Kites　筝形
（4）Other　其他
Parallelograms and trapezoids must be convex.　平行四边形和梯形必然是凸四边形。

2. The sum of measures of interior angles of every quadrilateral is 360°.　四边形的内角度数之和为 360°。

Proof：To prove Property 2，constructing a diagonal as shown in Figure 16-3.　要证明性质 2，在任意四边形中画一条对角线，如图 16-3 所示。

We divide a quadrilateral into two non-overlapping triangles. Since the sum of measures of interior angles of a triangle is 180°，the sum of measures of interior angles of a quadrilateral is twice of that，which is 360°.　把一个四边形分成两个不交叠的三角形。因为三角形的内角度数之和为 180°，所以四边形的内角度数之和为它的 2 倍，也就是 360°。

Side　边

n. [saɪd]

Definition：A side of a polygon is a line segment whose endpoints are two adjacent vertices，as shown in Figure 16-4 in blue.　多边形的边是一条端点为两个邻顶点的线段，在图 16-4 中用蓝色表示。

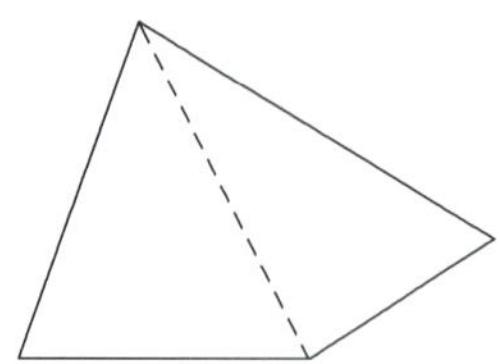

Figure 16-3

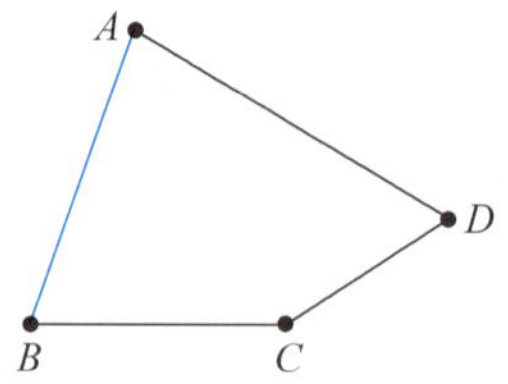

Figure 16-4

Examples: In quadrilateral $ABCD$, $\overline{AB}$, $\overline{BC}$, $\overline{CD}$, and $\overline{DA}$ are sides.

Diagonal 对角线

n. [daɪ'ægənəl]

Definition: A diagonal of a polygon is a line segment whose endpoints are two nonadjacent vertices (a line segment whose endpoints are two vertices, for which it is not a side), as shown in Figure 16-5 in blue. 多边形的对角线是一条端点为两个不相邻的顶点的线段(也就是一条端点为两个顶点的线段,且这条线段不是多边形的边),在图 16-5 中用蓝色表示。

Properties: The diagonals of concave polygons sometimes are drawn outside of the polygon. 凹多边形的对角线有时候在多边形之外。

As shown in Figure 16-6, diagonal $\overline{AC}$ of polygon $ABCDE$ (shaded) is drawn in blue.

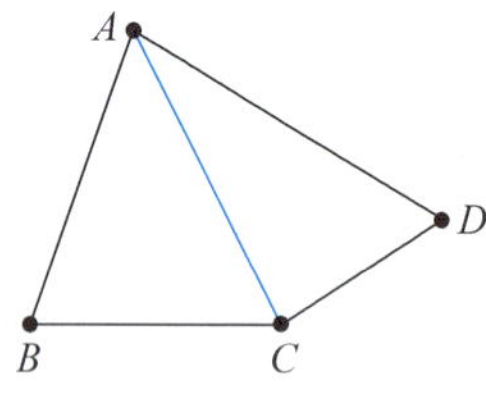

Figure 16-5

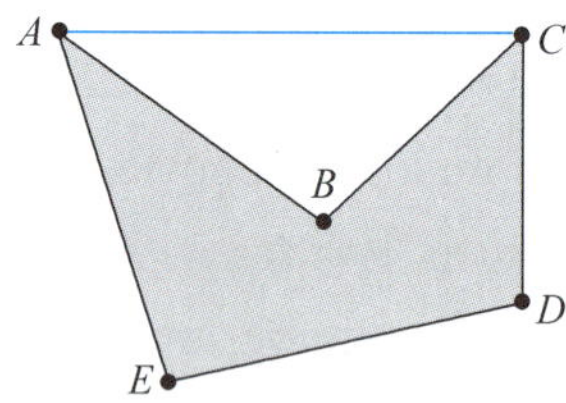

Figure 16-6

Examples: In quadrilateral $ABCD$, $\overline{AC}$ and $\overline{BD}$ are diagonals.

16.2 Parallelograms 平行四边形

视频 28

16.2.1 Introduction 介绍

Parallelogram 平行四边形

n. [ˌpærə'leləˌgræm]

Definition: A quadrilateral with two pairs of parallel sides. 有两组平行边的四边形称为平行四边形。

In parallelogram $ABCD$, $\overline{AB} \parallel \overline{CD}$ and $\overline{BC} \parallel \overline{AD}$, as shown in Figure 16-7.

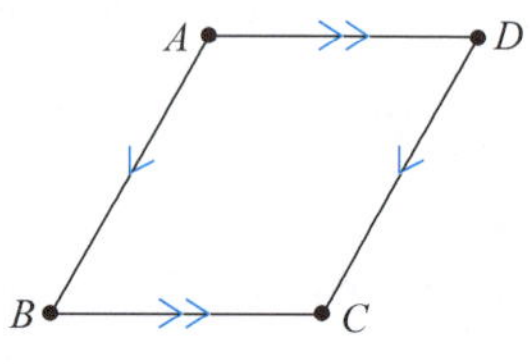

Figure 16-7

Properties: 1. Opposite angles of a parallelogram are congruent. 平行四边形的对角相等。

2. Opposite sides of a parallelogram are congruent. 平行四边形的对边相等。

3. Diagonals of a parallelogram bisect each other. 平行四边形的对角线互相平分。

Proofs: In the parallelogram $ABCD$:

在平行四边形 $ABCD$ 中：

1. To prove Property 1, we wish to show that $\angle A \cong \angle C$ and $\angle B \cong \angle D$. Since $\overline{AB} \parallel \overline{CD}$ and $\overline{BC} \parallel \overline{AD}$, by the **Consecutive Interior Angles Theorem**, $m\angle A + m\angle B = 180°$, from which $m\angle A = 180° - m\angle B$; and $m\angle B + m\angle C = 180°$, from which $m\angle C = 180° - m\angle B$. By transitive, $m\angle A = m\angle C$, or $\angle A \cong \angle C$. Similarly, we can prove that $\angle B \cong \angle D$. 要证明性质 1，须证出 $\angle A \cong \angle C$ 和 $\angle B \cong \angle D$。因为 $\overline{AB} \parallel \overline{CD}$，$\overline{BC} \parallel \overline{AD}$，用**同旁内角定理**可得出 $m\angle A + m\angle B = 180°$，也就是说 $m\angle A = 180° - m\angle B$；同样，$m\angle B + m\angle C = 180°$，也就是说 $m\angle C = 180° - m\angle B$。根据传递性，$m\angle A = m\angle C$，亦即 $\angle A \cong \angle C$。同理，可证出 $\angle B \cong \angle D$。
2. As shown in Figure 16-8, to prove Property 2, we wish to show that $\overline{AB} \cong \overline{CD}$ and $\overline{BC} \cong \overline{AD}$. Construct $\overline{BD}$. From Property 1, we know that $\angle A \cong \angle C$. By the **Alternate Interior Angles Theorem**, we know that $\angle ADB \cong \angle CBD$. By reflexive, we know that $\overline{BD} \cong \overline{BD}$. We know that $\triangle ADB \cong \triangle CBD$ by AAS, from which $\overline{AB} \cong \overline{CD}$ and $\overline{BC} \cong \overline{AD}$ by CPCTC. 如图 16-8 所示，要证明性质 2，须证出 $\overline{AB} \cong \overline{CD}$ and $\overline{BC} \cong \overline{AD}$。画出 $\overline{BD}$。从性质 1 知道 $\angle A \cong \angle C$。根据**内错角定理**，知道 $\angle ADB \cong \angle CBD$。用反身性可得出 $\overline{BD} \cong \overline{BD}$。用角角边全等公理可得出 $\triangle ADB \cong \triangle CBD$，因为全等三角形的对应部分全等，$\overline{AB} \cong \overline{CD}$ 和 $\overline{BC} \cong \overline{AD}$。
3. Suppose diagonals $\overline{AC}$ and $\overline{BD}$ intersect at E, as shown in Figure 16-9. To prove Property 3, we wish to show that $\overline{AE} \cong \overline{EC}$ and $\overline{BE} \cong \overline{ED}$. We have $\angle EAD \cong \angle ECB$ and $\angle EDA \cong \angle EBC$ by the **Alternate Interior Angles Theorem**, and $\overline{BC} \cong \overline{AD}$ by Property 2. From here, we get $\triangle EAD \cong \triangle ECB$ by ASA, from which $\overline{AE} \cong \overline{EC}$ and $\overline{BE} \cong \overline{ED}$ by CPCTC. 假设对角线 $\overline{AC}$ 和 $\overline{BD}$ 交于 E，如图 16-9 所示。要证明性质 3，须证出 $\overline{AE} \cong \overline{EC}$ 和 $\overline{BE} \cong \overline{ED}$。根据**内错角定理**，可得出 $\angle EAD \cong \angle ECB$ 和 $\angle EDA \cong \angle EBC$。根据性质 2，可得出 $\overline{BC} \cong \overline{AD}$。又由于角边角全等公理，可得 $\triangle EAD \cong \triangle ECB$。因为全等三角形的对应部分全等，所以 $\overline{AE} \cong \overline{EC}$ 和 $\overline{BE} \cong \overline{ED}$。

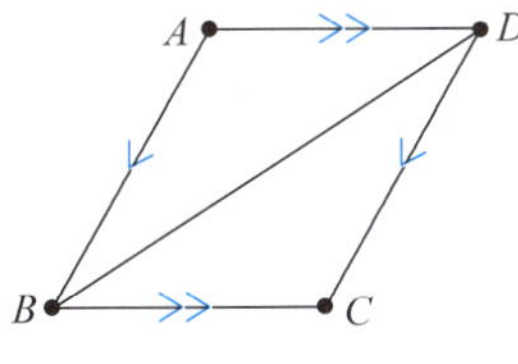

Figure 16-8

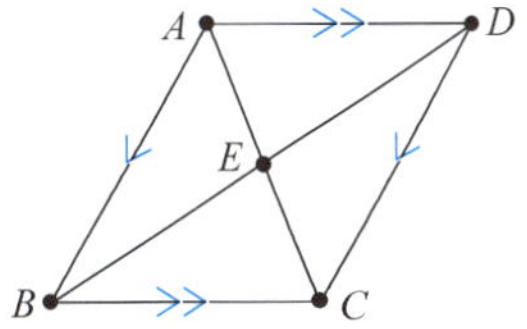

Figure 16-9

Questions: In parallelogram $ABCD$, if $AB = 5$, $BC = 8$, and $m\angle A = 70°$. What are values of (1) CD, (2) DA, (3) $m\angle B$, (4) $m\angle C$, (5) $m\angle D$?

Answers: (1) $CD = AB = 5$.

(2) $DA = BC = 8$.

(3) $m\angle B = 180 - m\angle A = 110°$.

(4) $m\angle C = m\angle A = 70°$.

(5) $m\angle D = m\angle B = 110°$.

Determining a Parallelogram 平行四边形的判定

Properties: In a quadrilateral:

在一个四边形中:

1. If the opposite sides are parallel, then it is a parallelogram. 若对边平行,则它是平行四边形。
2. If the opposite sides are congruent, then it is a parallelogram. 若对边全等,则它是平行四边形。
3. If two opposite sides are parallel and congruent, then it is a parallelogram. 若一组对边平行又全等,则它是平行四边形。
4. If the opposite angles are congruent, then it is a parallelogram. 若对角全等,则它是平行四边形。
5. If the diagonals bisect each other, then it is a parallelogram. 若对角线互相平分,则它是平行四边形。

Proofs: Suppose $ABCD$ is a quadrilateral.

设 $ABCD$ 是一个四边形。

1. Property 1 is by the definition of a parallelogram. 因为平行四边形的定义,可知上述性质 1 成立。
2. To prove Property 2, we wish to show that if $\overline{AB} \cong \overline{CD}$ and $\overline{BC} \cong \overline{AD}$, then $ABCD$ is a parallelogram. 要证明性质 2,即证明若 $\overline{AB} \cong \overline{CD}$ 且 $\overline{BC} \cong \overline{AD}$,则 $ABCD$ 是平行四边形。

   As shown in Figure 16-10, by reflexive, $\overline{AC} \cong \overline{AC}$. Therefore, by SSS, $\triangle ABC \cong \triangle CDA$, from which $\angle BAC \cong \angle DCA$ by CPCTC. By the **Converse of Alternate Interior Angles Theorem**, we get $\overline{AB} \parallel \overline{CD}$. We can show that $\overline{BC} \parallel \overline{AD}$ in a similar manner. By Property 1, we get that $ABCD$ is a parallelogram. 如图 16-10 所示,根据反身性,$\overline{AC} \cong \overline{AC}$。所以根据边边边全等公理,有$\triangle ABC \cong \triangle CDA$。又根据全等三角形的对应部分全等,可得$\angle BAC \cong \angle DCA$。根据**内错角定理的逆定理**可得出 $\overline{AB} \parallel \overline{CD}$。同样地,可证出 $\overline{BC} \parallel \overline{AD}$。根据性质 1,可得出 $ABCD$ 是平行四边形。
3. To prove Property 3, we wish to show that if $\overline{AB} \cong \overline{CD}$ and $\overline{AB} \parallel \overline{CD}$, then $ABCD$ is a parallelogram. 要证明性质 3,即证出若 $\overline{AB} \cong \overline{CD}$ 和 $\overline{AB} \parallel \overline{CD}$,则 $ABCD$ 是平行四边形。

   As shown in Figure 16-11, by reflexive, $\overline{AC} \cong \overline{AC}$. By the **Alternate Interior Angles Theorem**, $\angle BAC \cong \angle DCA$. Therefore, we have $\triangle BAC \cong \triangle DCA$ by SAS. By CPCTC, we get $\angle DAC \cong \angle BCA$. Using the **Converse of Alternate Interior Angles Theorem**, we show that $\overline{AD} \parallel \overline{BC}$. Therefore, $ABCD$ is a parallelogram by definition. 如图 16-11 所示,根据反身性,$\overline{AC} \cong \overline{AC}$。根据**内错角定理**,$\angle BAC \cong \angle DCA$。所以,根据边角边全等公理,可得$\triangle BAC \cong \triangle DCA$。根据全等三角形的对应部分全等,可得$\angle DAC \cong \angle BCA$。又根据**内错角定理的逆定理**,可得 $\overline{AD} \parallel \overline{BC}$。所以,根据定义,可知 $ABCD$ 是平行四边形。

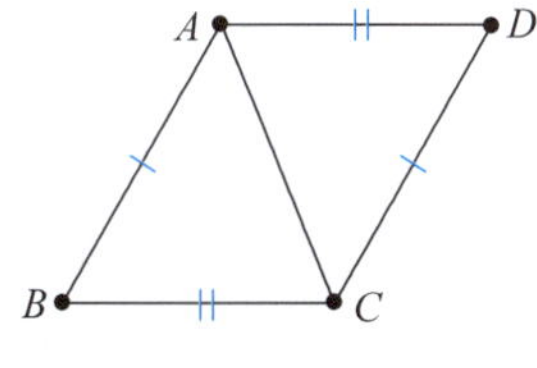

Figure 16-10

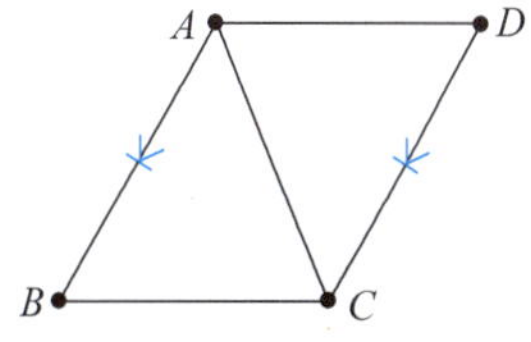

Figure 16-11

4. To prove Property 4, we wish to show that if $\angle A \cong \angle C$ and $\angle B \cong \angle D$, then $ABCD$ is a parallelogram.　要证明性质 4，即证出若 $\angle A \cong \angle C$ 和 $\angle B \cong \angle D$，则 $ABCD$ 是平行四边形。

   As shown in Figure 16-12, we know that $m\angle A + m\angle B + m\angle C + m\angle D = 360°$. By substitution, $2m\angle A + 2m\angle B = 360°$, from which $m\angle A + m\angle B = 180°$. By the **Converse of Consecutive Interior Angles Theorem**, we get that $\overline{AD} \parallel \overline{BC}$. Similarly, we can show that $\overline{AB} \parallel \overline{CD}$. By definition, $ABCD$ is a parallelogram.　如图 16-12 所示，知道 $m\angle A + m\angle B + m\angle C + m\angle D = 360°$。用代入法得出 $2m\angle A + 2m\angle B = 360°$，也就是说 $m\angle A + m\angle B = 180°$。根据**同旁内角定理的逆定理**，$\overline{AD} \parallel \overline{BC}$。同理，可得出 $\overline{AB} \parallel \overline{CD}$。根据定义，可得 $ABCD$ 是平行四边形。
5. To prove Property 5, suppose diagonals $\overline{AC}$ and $\overline{BD}$ intersect at E. We wish to show that if $\overline{AE} \cong \overline{EC}$ and $\overline{BE} \cong \overline{ED}$, then $ABCD$ is a parallelogram.　要证明性质 5，设对角线 $\overline{AC}$ 和 $\overline{BD}$ 相交于 E，即证出若 $\overline{AE} \cong \overline{EC}$ 和 $\overline{BE} \cong \overline{ED}$，则 $ABCD$ 是平行四边形。

   As shown in Figure 16-13, we know that $\angle AEB \cong \angle CED$ since vertical angles are congruent. Then we have $\triangle AEB \cong \triangle CED$ by SAS, from which $\angle EAB \cong \angle ECD$ by CPCTC. By the **Converse of Alternate Interior Angles Theorem**, we get that $\overline{AB} \parallel \overline{CD}$. Similarly, we can show that $\overline{AD} \parallel \overline{BC}$. By definition, $ABCD$ is a parallelogram.

   如图 16-13 所示，由对顶角全等可知 $\angle AEB \cong \angle CED$。根据边角边全等公理，可得到 $\triangle AEB \cong \triangle CED$。因为全等三角形的对应部分全等，所以 $\angle EAB \cong \angle ECD$。根据**内错角定理的逆定理**，可得出 $\overline{AB} \parallel \overline{CD}$。同理，可证出 $\overline{AD} \parallel \overline{BC}$。根据定义，可知 $ABCD$ 是平行四边形。

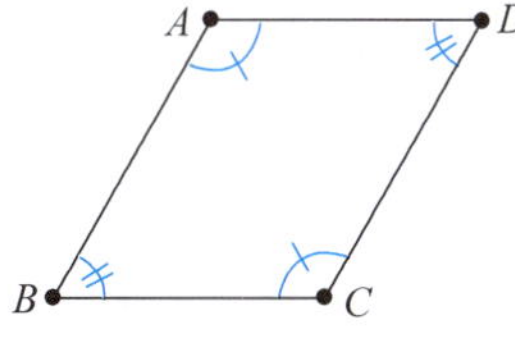

Figure 16-12

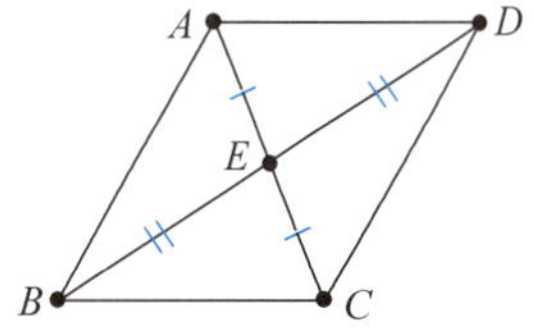

Figure 16-13

16.2.2　Subcategory—Rectangles　子类别——矩形

Rectangle　矩形

n. [ˈrektæŋɡəl]

Definition: An equiangular quadrilateral, as shown in Figure 16-14.　矩形是等角的四边形，如图 16-14 所示。

Properties：1. Rectangles are parallelograms. 矩形是平行四边形。

2. All angles of a rectangle are right angles. 矩形的所有角都是直角。

3. The diagonals of a rectangle are congruent. 矩形的对角线全等。

Figure 16-14

Proofs：1. To show Property 1, use Property 4 of determining a parallelogram in Section 16.2.1. 要证明矩形的性质 1，参照 16.2.1 节的平行四边形的判定性质 4.

Proofs 2 and 3 refer to rectangle $ABCD$. 证明 2 和 3 参照矩形 $ABCD$。

2. To prove Property 2, by the definition of equiangular, we have $m\angle A = m\angle B = m\angle C = m\angle D$. Since these four angles' measures sum up to 360°, each angle is measured 90°, which is a right angle. 要证明矩形的性质 2，根据矩形等角的定义，得到 $m\angle A = m\angle B = m\angle C = m\angle D$。因为这四个角的度数和为 360°，每个角的度数为 90°，也就是直角。

3. To prove Property 3, we need to show that $\overline{AC} \cong \overline{BD}$. By the Pythagorean Theorem, $AC^2 = AB^2 + BC^2$, and $BD^2 = BC^2 + CD^2$. Since opposite sides are congruent, we have $AB = CD$ and $AD = BC$. By substitution, $BD^2 = BC^2 + AB^2$. By transitive, $AC^2 = BD^2$. So, $AC = BD$, or $\overline{AC} \cong \overline{BD}$. 要证明矩形的性质 3，即需要证出 $\overline{AC} \cong \overline{BD}$。根据勾股定理，$AC^2 = AB^2 + BC^2$，且 $BD^2 = BC^2 + CD^2$。因为对边全等，可知 $AB = CD$ 和 $AD = BC$。用代入法可得出 $BD^2 = BC^2 + AB^2$。因为传递性，$AC^2 = BD^2$，也就是说，$AC = BD$，从而得出 $\overline{AC} \cong \overline{BD}$。

Questions：In rectangle $ABCD$, if $AB = 6$ and $BC = 8$, what are the values of (1) CD, (2) DA, (3) AC, (4) BD?

Answers：(1) $AB = CD = 6$ (true for any parallelogram).

(2) $DA = BC = 8$ (true for any parallelogram).

(3) $AC^2 = AB^2 + BC^2 = 100$, for which $AC = 10$.

(4) $BD = AC = 10$.

Determining a Rectangle 矩形的判定

Properties：1. If one angle of a parallelogram is a right angle, then the parallelogram is a rectangle. 若平行四边形的一个角是直角，则这个平行四边形为矩形。

2. If the diagonals of a parallelogram are congruent, then the parallelogram is a rectangle. 若平行四边形的对角线全等，则这个平行四边形为矩形。

3. If three angles of a quadrilateral are right angles, then the quadrilateral is a rectangle. 若四边形的三个角为直角，则这个四边形为矩形。

Proofs：1. In parallelogram $ABCD$, suppose $m\angle A = 90°$. To prove Property 1, we need to show that $ABCD$ is equiangular. 在平行四边形 $ABCD$ 中，设 $m\angle A = 90°$。要证明矩形的判定性质 1，即需要证出 $ABCD$ 等角。

Since opposite angles of a parallelogram are congruent, we have $m\angle C = 90°$. Since $m\angle A + m\angle B + m\angle C + m\angle D = 360°$, so $m\angle B + m\angle D = 180°$, from which $m\angle B = m\angle D = 90°$. We have shown that $m\angle A = m\angle B = m\angle C = m\angle D = 90°$, which is the definition of equiangular. 因为平行四边形的对角全等，可得到 $m\angle C = 90°$。因为

$m\angle A+m\angle B+m\angle C+m\angle D=360°$，所以 $m\angle B+m\angle D=180°$，也就是说 $m\angle B=m\angle D=90°$。现在已知 $m\angle A=m\angle B=m\angle C=m\angle D=90°$，也就是等角的定义。

2. In parallelogram $ABCD$, suppose $\overline{AC}\cong\overline{BD}$. To prove Property 2, we need to show that $ABCD$ is equiangular. 在平行四边形 $ABCD$ 中，设 $\overline{AC}\cong\overline{BD}$。要证明矩形的判定性质 2，即需要证明 $ABCD$ 等角。

   By reflexive, $\overline{CD}\cong\overline{CD}$. Since opposite sides of a parallelogram are congruent, $\overline{AD}\cong\overline{BC}$. Therefore, we have $\triangle ADC\cong\triangle BCD$ by SSS. Thus, by CPCTC we have $\angle ADC\cong\angle BCD$. By the **Consecutive Interior Angles Theorem**, we have $m\angle ADC+m\angle BCD=180°$. Therefore, $m\angle ADC=m\angle BCD=90°$. Since opposite angles of a parallelogram are congruent, we get $m\angle ADC=m\angle ABC=90°$, and $m\angle DAB=m\angle BCD=90°$. It is true that $m\angle ADC=m\angle ABC=m\angle DAB=m\angle BCD=90°$, for which $ABCD$ is equiangular. 因为反身性，$\overline{CD}\cong\overline{CD}$。因为平行四边形对边全等，$\overline{AD}\cong\overline{BC}$。所以，根据边边边全等公理可得到$\triangle ADC\cong\triangle BCD$。然后根据全等三角形的对应部分全等，$\angle ADC\cong\angle BCD$。根据**同旁内角定理**可得出 $m\angle ADC+m\angle BCD=180°$。所以 $m\angle ADC=m\angle BCD=90°$。因为平行四边形的对角全等，可得到 $m\angle ADC=m\angle ABC=90°$ 和 $m\angle DAB=m\angle BCD=90°$。已知 $m\angle ADC=m\angle ABC=m\angle DAB=m\angle BCD=90°$，也就是说 $ABCD$ 等角。

3. To prove Property 3, suppose in quadrilateral $ABCD$, $m\angle A=m\angle B=m\angle C=90°$. We wish to show that $ABCD$ is equiangular (which is a rectangle). 要证明矩形的判定性质 3，设在四边形 $ABCD$ 中，$m\angle A=m\angle B=m\angle C=90°$。即需要证出 $ABCD$ 等角(也就是一个矩形)。

   Since $m\angle A+m\angle B+m\angle C+m\angle D=360°$, it is clear that $m\angle D=90°$. We have shown that $ABCD$ is equiangular. 因为 $m\angle A+m\angle B+m\angle C+m\angle D=360°$，显然 $m\angle D=90°$。故 $ABCD$ 等角。

16.2.3 Subcategory—Rhombi 子类别——菱形

Rhombus 菱形

n. [ˈrɑmbəs]

Definition: An equilateral quadrilateral, as shown in Figure 16-15. 菱形是等边的四边形，如图 16-15 所示。

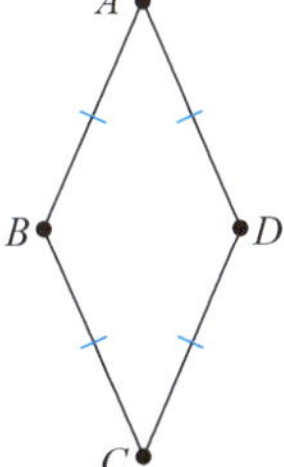

Figure 16-15

Properties: 1. A rhombus is a parallelogram. 菱形是平行四边形。

2. All sides of a rhombus are congruent. 菱形的所有边全等。

3. The diagonals of a rhombus are perpendicular, and each diagonal bisects two angles. 菱形的对角线垂直，且每条对角线平分两个角。

Proofs: 1. To show Property 1, use Property 2 of determining a parallelogram in Section 16.2.1. 要证明菱形的性质 1，参照 16.2.1 节的平行四边形的判定性质 2 即可。

2. Property 2 is by the definition of equilateral. 菱形的性质 2 根据菱形的定义可得出。

3. In Rhombus $ABCD$, to prove Property 3, we need to show:
   在菱形 $ABCD$ 中，要证明性质 3，则需要证明以下性质：

(1) $\overline{AC} \perp \overline{BD}$.

(2) $\angle CAB \cong \angle CAD$, $\angle ACB \cong \angle ACD$, $\angle BDA \cong \angle BDC$, and $\angle DBA \cong \angle DBC$.

To show (2), note that $\overline{AB} \cong \overline{BC} \cong \overline{CD} \cong \overline{DA}$. By reflexive, $\overline{AC} \cong \overline{AC}$. Therefore, $\triangle ABC \cong \triangle ADC$ by SSS. We have $\angle BAC \cong \angle DAC$ and $\angle BCA \cong \angle DCA$ by CPCTC. 要证出(2),留意到 $\overline{AB} \cong \overline{BC} \cong \overline{CD} \cong \overline{DA}$。根据反身性,$\overline{AC} \cong \overline{AC}$。所以,利用边边边全等公理,$\triangle ABC \cong \triangle ADC$。根据全等三角形的对应部分全等,可得到 $\angle BAC \cong \angle DAC$ 和 $\angle BCA \cong \angle DCA$。

Similarly, we can show that $\angle BDA \cong \angle BDC$, and $\angle DBA \cong \angle DBC$. 同理,可得出 $\angle BDA \cong \angle BDC$ 和 $\angle DBA \cong \angle DBC$。

To show (1), we will use the result we have for (2). As shown in Figure 16-16, suppose $\overline{AC}$ and $\overline{BD}$ intersect at E. By the definition of a rhombus, we have $\overline{AB} \cong \overline{AD}$. By (2) we have $\angle EAB \cong \angle EAD$. By reflexive we have $\overline{AE} \cong \overline{AE}$. Therefore, $\triangle ABE \cong \triangle ADE$ by SAS, for which $\angle BEA \cong \angle DEA$ by CPCTC. Since $\angle BEA$ and $\angle DEA$ are a linear pair, $m\angle BEA = m\angle DEA = 90°$. We have shown that the diagonals of a rhombus are perpendicular. 要证明(1),会用到(2)的结果。如图 16-16 所示。假设 $\overline{AC}$ 和 $\overline{BD}$ 交于 E。根据菱形的定义,$\overline{AB} \cong \overline{AD}$。根据(2)可得出 $\angle EAB \cong \angle EAD$。根据反身性可得出 $\overline{AE} \cong \overline{AE}$。因此,根据边角边全等公理,$\triangle ABE \cong \triangle ADE$。根据全等三角形的对应部分全等,$\angle BEA \cong \angle DEA$。因为 $\angle BEA$ 和 $\angle DEA$ 是邻补角,$m\angle BEA \cong m\angle DEA = 90°$,所以可证明菱形的对角线垂直。

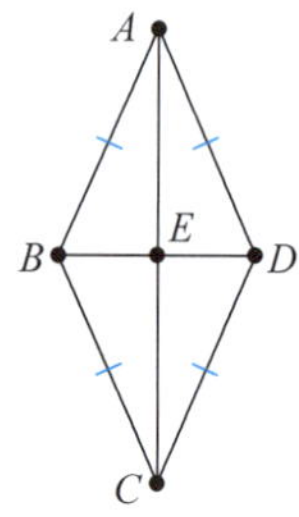

Figure 16-16

Questions: In rhombus $ABCD$, if $AB = 13$ and $AC = 24$, what are the values of: (1) BC, (2) CD, (3) DA, (4) BD?

Answers: $AB = BC = CD = DA = 13$. This answers parts (1) through (3).

To answer part (d), note that the diagonals of a parallelogram bisect each other, and the diagonals of a rhombus are perpendicular. Suppose $\overline{AC}$ and $\overline{BD}$ intersect at E, we have $AE = CE = \frac{1}{2}AC = 12$, $BE = DE = \frac{1}{2}BD$. By the Pythagorean Theorem, $AB^2 = BE^2 + AE^2$, for which $13^2 = BE^2 + 12^2$. We get $BE^2 = 25$ and $BE = 5 = \frac{1}{2}BD$. Therefore, $BD = 10$. 要解题(4),留意菱形的性质 3,并用勾股定理解出来。

Determining a Rhombus 菱形的判定

Properties: 1. If two adjacent sides of a parallelogram are congruent, then the parallelogram is a rhombus. 若平行四边形的两条邻边全等,则这个平行四边形是菱形。

2. If the diagonals of a parallelogram are perpendicular, then the parallelogram is a rhombus. 若平行四边形的对角线垂直,则这个平行四边形是菱形。

Proofs: In parallelogram $ABCD$:

在平行四边形 $ABCD$ 中:

1. To prove Property 1, we need to show that if $\overline{AB} \cong \overline{AD}$, then $ABCD$ is a rhombus. 要证明菱形的判定性质 1,即证明若 $\overline{AB} \cong \overline{AD}$,则 $ABCD$ 为菱形。

We are given that $\overline{AB}\cong\overline{AD}$. Since opposite sides of a parallelogram are congruent, we have $\overline{AB}\cong\overline{CD}$ and $\overline{BC}\cong\overline{AD}$. By transitive, $\overline{AB}\cong\overline{AD}\cong\overline{BC}\cong\overline{CD}$. Therefore, $ABCD$ is equilateral, which is the definition of a rhombus.　已知 $\overline{AB}\cong\overline{AD}$，根据平行四边形的对边全等，可得到 $\overline{AB}\cong\overline{CD}$ 和 $\overline{BC}\cong\overline{AD}$。根据传递性，$\overline{AB}\cong\overline{AD}\cong\overline{BC}\cong\overline{CD}$。所以，$ABCD$ 等边，即 $ABCD$ 为菱形。

2. To prove Property 2, we need to show that if $\overline{AC}\perp\overline{BD}$, then $ABCD$ is a rhombus. 要证明菱形的判定性质 2，即证明若 $\overline{AC}\perp\overline{BD}$，则 $ABCD$ 是菱形。

As shown in Figure 16-17, suppose $\overline{AC}$ and $\overline{BD}$ intersect at E. Since all right angles are congruent, we have $\angle BEA\cong\angle DEA$. Since diagonals bisect each other in a parallelogram, $\overline{BE}\cong\overline{ED}$. By reflexive, $\overline{AE}\cong\overline{AE}$. Therefore, we get that $\triangle ABE\cong\triangle ADE$ by SAS, from which $\overline{AB}\cong\overline{AD}$ by CPCTC. Using Property 1, we prove that $ABCD$ is a rhombus.　如图 16-17 所示，设 $\overline{AC}$ 与 $\overline{BD}$ 垂直于 E。因为所有直角全等，可得到 $\angle BEA\cong\angle DEA$。又因为平行四边形的对角线互相平分，$\overline{BE}\cong\overline{ED}$。根据反身性，$\overline{AE}\cong\overline{AE}$。根据边角边全等公理，可得到 $\triangle ABE\cong\triangle ADE$。由全等三角形的对应部分全等可得出 $\overline{AB}\cong\overline{AD}$。根据菱形的判定性质 1，可证明 $ABCD$ 是菱形。

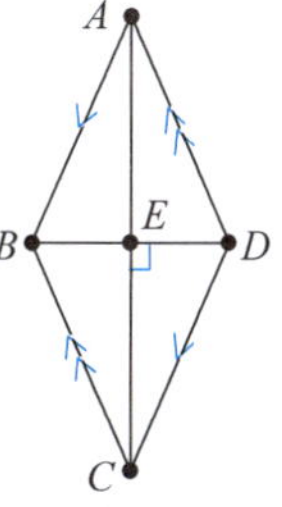

Figure 16-17

16.2.4　Subcategory of Subcategory—Squares　子类别的子类别——正方形

Square　正方形

n. [skweər]

Definition: A regular quadrilateral.　正方形是指正四边形。

Properties: By the definition of regular, a square is both equilateral and equiangular. Therefore, a square is also a rectangle and a rhombus. All of the properties from rectangle and rhombus apply to a square.　根据正多边形的定义，正方形等边并等角。所以，它既是矩形又是菱形。正方形满足矩形和菱形的所有性质。

16.2.5　Summary　总结

1. Refer to the Venn diagram at the beginning of the chapter.　可参照本章前面的文氏图。
2. A parallelogram is a quadrilateral whose opposite sides are parallel.　平行四边形是对边平行的四边形。
3. Rectangles, rhombi, and squares are special parallelograms.　矩形、菱形、正方形是特殊的平行四边形。
   (1) Rectangles are equiangular.　矩形等角。
   (2) Rhombi are equilateral.　菱形等边。
   (3) Squares are regular (both equiangular and equilateral). They are special rectangles and rhombi.　正方形是正多边形(等角以及等边)。它们是特殊的矩形和菱形。
4. See Section 16.2 for the properties of parallelogram, rectangle, rhombus, and square, along with the methods of determining a parallelogram, rectangle, rhombus, and square.　参见 16.2 节平行四边形、矩形、菱形、正方形的性质以及判定。

16.3 Non-Parallelograms 非平行四边形

16.3.1 Trapezoids 梯形

Trapezoid 梯形

n. [ˈtræpəˌzɔɪd]

Definition: A quadrilateral with exactly one pair of parallel sides. The parallel sides are called the bases, and the other two sides are called the legs, as shown in Figure 16-18. 梯形是指只有一组平行边的四边形。平行的边叫作底,其他两边叫作腰,如图 16-18 所示。

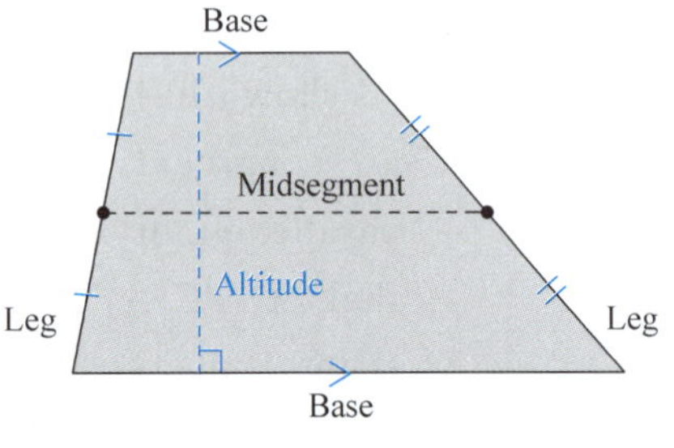

Figure 16-18

The line segment whose endpoints are the midpoints of both legs of a trapezoid is called the midsegment. 端点为两腰中点的线段叫中位线。

A line segment that connects the bases and perpendicular to the bases is called the altitude. 连接两底并且与其垂直的线段叫高线。

In an isosceles trapezoid, the legs are congruent, as shown in Figure 16-19. 等腰梯形的腰全等,如图 16-19 所示。

If a trapezoid has one right angle, then it is a right trapezoid, as shown in Figure 16-20 (By **Two Perpendicular Lines Theorem**, trapezoids that have one right angle must have two right angles.). 若梯形有一个直角,则该梯形称为直角梯形,如图 16-20 所示。(根据**双垂线定理**,有一个直角的梯形必有两个直角。)

Figure 16-19　　Figure 16-20

Properties: 1. An isosceles trapezoid has two pairs of congruent adjacent angles. 等腰梯形有两组全等的邻角。

2. The length of the midsegment of a trapezoid is the average of the lengths of the bases. 中位线的长度是两底长度的平均值。

Proofs: 1. To prove Property 1, we need to show that for the isosceles trapezoid $ABCD$, if $\overline{AB} \parallel \overline{CD}$, then $\angle ADC \cong \angle BCD$ and $\angle DAB \cong \angle CBA$. 要证明性质 1,即证明在等腰梯形 $ABCD$ 中,若 $\overline{AB} \parallel \overline{CD}$,则 $\angle ADC \cong \angle BCD$ 和 $\angle DAB \cong \angle CBA$。

As shown in Figure 16-21, since $ABCD$ is isosceles, the legs are congruent, or $\overline{AD} \cong \overline{BC}$. Without the loss of generality, let $\overline{AB}$ be the shorter base. Let $\overline{AE}$ and $\overline{BF}$ be the altitudes. Since all right angles are congruent, $\angle AED \cong \angle BFC$. Since $ABFE$ is a rectangle, $\overline{AE} \cong \overline{BF}$. Therefore, $\triangle AED \cong \triangle BFC$ by HL. We have $\angle ADE \cong \angle BCF$

by CPCTC. 如图 16-21 所示，因为 $ABCD$ 为等腰梯形，$\overline{AD} \cong \overline{BC}$。在一般情况下，设 $\overline{AB}$ 为较短的底，$\overline{AE}$ 和 $\overline{BF}$ 为高线。因为所有的直角全等，$\angle AED \cong \angle BFC$。因为 $ABFE$ 是矩形，所以 $\overline{AE} \cong \overline{BF}$。根据 HL 定理，$\triangle AED \cong \triangle BFC$。根据全等三角形的对应部分全等，可得到$\angle ADE \cong \angle BCF$。

Also, $\angle DAE \cong \angle CBF$ is by CPCTC too. Since $\overline{AE}$ and $\overline{BF}$ are the altitudes, we have $\angle EAB \cong \angle FBA$. Therefore, by substitution, $m\angle DAE + m\angle EAB = m\angle CBF + m\angle FBA$, or $m\angle DAB = m\angle CBA$ by **Angle Addition Postulate**. Therefore, we have $\angle DAB \cong \angle CBA$. 同样，根据全等三角形的对应部分全等，$\angle DAE \cong \angle CBF$。因为 $\overline{AE}$ 和 $\overline{BF}$ 均为高线，所以 $\angle EAB \cong \angle FBA$。通过代入法，$m\angle DAE + m\angle EAB = m\angle CBF + m\angle FBA$，也就是说通过**角相加公理**，$m\angle DAB = m\angle CBA$。因此，$\angle DAB \cong \angle CBA$。

2. To prove Property 2, as shown in Figure 16-22, the gray and white trapezoids are congruent. Together, they form a large parallelogram. Twice the length of the midsegment (shown in dotted line) is the sum of lengths of the bases of one of the trapezoids, Therefore, the length of the midsegment is the average of the lengths of both bases. 要证明性质 2，如图 16-22 所示，灰色和白色的梯形全等。它们合并起来形成一个大的平行四边形。梯形中线长度的 2 倍(虚线标出)等于一个梯形上底与下底的长度之和。所以，中位线的长度是上底和下底的平均长度。

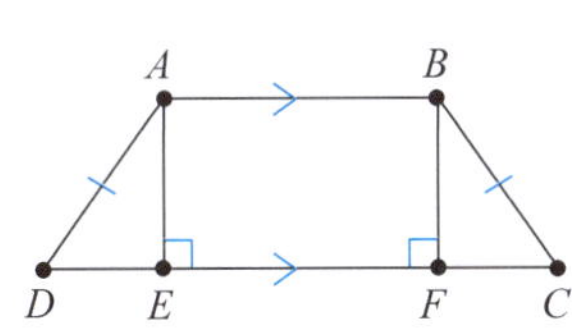

Figure 16-21

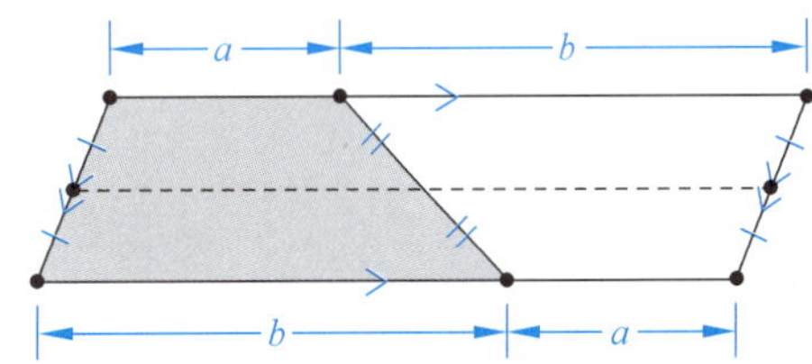

Figure 16-22

Phrases: isosceles～ 等腰梯形, right～ 直角梯形

Questions: If $ABCD$ is an isosceles trapezoid, with $\overline{AB} \parallel \overline{CD}$, $AB = 4$, the length of the altitude is 4, and $m\angle D = 60°$, then what are the values of (1) BC, (2) CD, (3) DA, (4) midsegment, (5) $m\angle C$, (6) $m\angle B$, and (7) $m\angle A$?

Answers: Let $\overline{AE}$ and $\overline{BF}$ be the altitudes. By the property of a $30° - 60° - 90°$ triangle, $DE = CF = \frac{4}{\sqrt{3}}$, and

(1) $BC = \frac{8}{\sqrt{3}}$.

(2) $CD = DE + EF + FC = 4 + \frac{8}{\sqrt{3}}$.

(3) $DA = BC = \frac{8}{\sqrt{3}}$.

(4) The length of the midsegment is $\frac{1}{2}(AB + CD) = 4 + \frac{4}{\sqrt{3}}$.

(5) $m\angle C = m\angle D = 60°$.

(6) $m\angle B = 180° - m\angle C = 120°$, by the **Consecutive Interior Angles Theorem**. **根据同旁内角定理**。

(7) $m\angle A = m\angle B = 120°$.

16.3.2 Kites 筝形

Kite 筝形

n. [kaɪt]

Definition: A quadrilateral that has exactly two pairs of congruent adjacent sides, as shown in Figure 16-23. 筝形是指一个只有两对全等邻边的四边形,如图 16-23 所示。

Properties: 1. A kite has exactly one pair of congruent angles. 筝形只有一对全等角。

2. One diagonal (possibly extended) of a kite bisects two angles and the other diagonal. 筝形的一条对角线(可能被延长)平分两个角,并且平分另一条对角线。

3. The diagonals (possibly extended) of a kite are perpendicular. 筝形的对角线互相垂直。

4. A kite can be concave. As shown in Figure 16-24, $\overline{AC}$ and $\overline{BD}$ (drawn in blue) are the diagonals of kite $ABCD$. 筝形可以是凹的。如图 16-24 所示,$\overline{AC}$ 和 $\overline{BD}$(用蓝线表示)是筝形 $ABCD$ 的对角线。

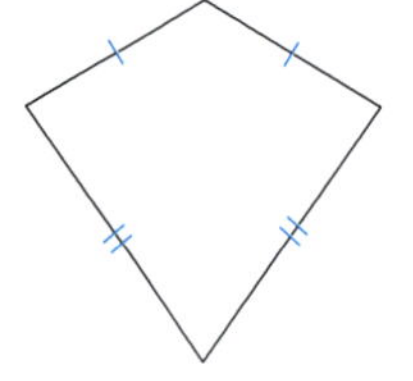

Figure 16-23

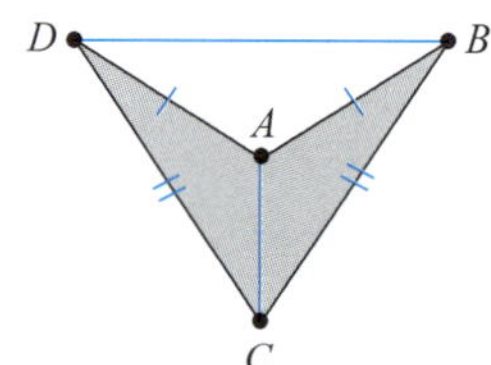

Figure 16-24

Proofs: 1. To prove Property 1, suppose in kite $ABCD$, $\overline{AB}\cong\overline{AD}$ and $\overline{CB}\cong\overline{CD}$. We want to show that $\angle B\cong\angle D$. 要证明性质 1,设在筝形 $ABCD$ 里,$\overline{AB}\cong\overline{AD}$ 和 $\overline{CB}\cong\overline{CD}$。需要证出$\angle B\cong\angle D$。

As shown in Figure 16-25, we have $\overline{AC}\cong\overline{AC}$ by reflexive. Therefore, we have $\triangle ABC\cong\triangle ADC$ by SSS. Thus we get $\angle B\cong\angle D$ by CPCTC. 如图 16-25 所示,根据反身性可得到 $\overline{AC}\cong\overline{AC}$。所以,根据边边边全等公理可得到$\triangle ABC\cong\triangle ADC$。因为全等三角形的对应部分全等,所以$\angle B\cong\angle D$。

The other pair of angles cannot be congruent. To prove why $m\angle BAD\neq m\angle BCD$, we can prove by contradiction. Suppose $m\angle BAD = m\angle BCD$. With $\angle B\cong\angle D$ as shown earlier, Property 4 of determining a parallelogram concludes that $ABCD$ is a parallelogram. However, a kite can never be a parallelogram, which arrives a contradiction. 另外一对角不能全等。用反证法证明为什么 $m\angle BAD\neq m\angle BCD$。设 $m\angle BAD = m\angle BCD$。先前得到了$\angle B\cong\angle D$。根据平行四边形的判定性质 4,$ABCD$ 是平行四边形。但是筝形不可能是平行四边形。故产生矛盾。

2. To prove Property 2, as shown in Figure 16-26, suppose the diagonals intersect at E.

We want to show that $\overline{AC}$ bisects $\angle BAD$, $\angle BCD$, and $\overline{BD}$. 要证明性质 2，如图 16-26 所示，假设对角线交于 E，须证明 $\overline{AC}$ 平分$\angle BAD$、$\angle BCD$、$\overline{BD}$。

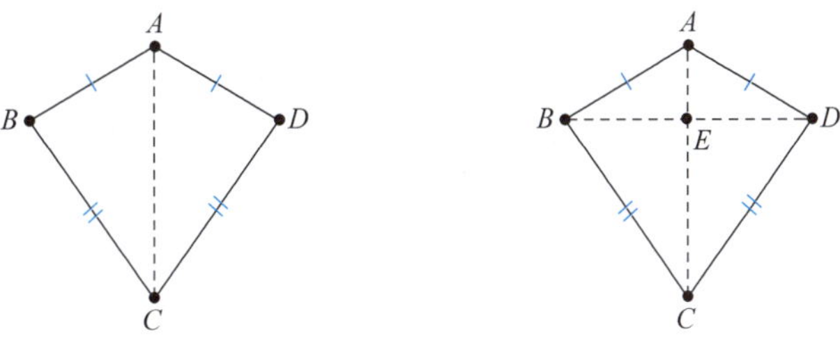

Figure 16-25 Figure 16-26

The proof of Property 1 shows that $\triangle ABC \cong \triangle ADC$ by SSS, so we can conclude that $\angle BAC \cong \angle DAC$ and $\angle BCA \cong \angle DCA$ by CPCTC. Here we have proved that $\overline{AC}$ bisects $\angle BAD$ and $\angle BCD$. 性质 1 的证明证出了由于边边边全等公理，$\triangle ABC \cong \triangle ADC$。所以可以因全等三角形的对应部分全等，得出$\angle BAC \cong \angle DAC$ 和$\angle BCA \cong \angle DCA$。由此证明了 $\overline{AC}$ 平分$\angle BAD$ 与$\angle BCD$。

We have $\overline{AE} \cong \overline{AE}$ by reflexive, so $\triangle ABE \cong \triangle ADE$ by SAS. It follows that by CPCTC, $\overline{BE} \cong \overline{DE}$. Here we have proved that $\overline{AC}$ bisects $\overline{BD}$. 根据反身性可得到 $\overline{AE} \cong \overline{AE}$，所以根据边角边全等公理，$\triangle ABE \cong \triangle ADE$。因为全等三角形的对应部分全等，$\overline{BE} \cong \overline{DE}$。由此证明了 $\overline{AC}$ 平分 $\overline{BD}$。

3. To prove Property 3, refer to the proof for Property 2. After concluding $\triangle ABE \cong \triangle ADE$ by SAS, it follows that $\angle AEB \cong \angle AED$ by CPCTC. Since these two angles form a linear pair, each of them must be a right angle. Therefore, we have proved that the diagonals of a kite are perpendicular. 要证明性质 3，参照性质 2 的证明。根据边角边全等公理得出$\triangle ABE \cong \triangle ADE$ 以后，因为全等三角形的对应部分全等，所以$\angle AEB \cong \angle AED$。因为两角是邻补角，每个角必须为直角，所以得出筝形的对角线相互垂直。

16.3.3 Summary 总结

1. A trapezoid is a quadrilateral with exactly one pair of parallel sides. The parallel sides are called the bases, and the other two sides are called the legs. 梯形是只有一组平行边的四边形。平行的边叫作底，其他两边叫作腰。
   (1) A trapezoid can have exactly 0 or 2 right angles. Trapezoids with two right angles are called right trapezoids. 梯形只有 0 个或 2 个直角。有 2 个直角的梯形叫直角梯形。
   (2) Trapezoids with congruent legs are called isosceles trapezoids. 腰全等的梯形叫等腰梯形。
   (3) The length of the midsegment of a trapezoid is the average of the sum of lengths of the bases. 中位线的长度是两底的平均长度。
2. A kite is a quadrilateral that has exactly two pairs of congruent adjacent sides. 筝形是只有两对全等邻边的四边形。

   A kite may be concave—the diagonals do not intersect. 筝形可以是凹四边形——对角线不相交。

3. (1) A kite has exactly one pair of congruent angles. 筝形只有一对全等角。

(2) One diagonal (possibly extended) of a kite bisects two angles and the other diagonal. 筝形的一条对角线（可能被延长）平分两个角，并且平分另一条对角线。

(3) The diagonals (possibly extended) of a kite are perpendicular. 筝形的对角线互相垂直。

17. Polygons 多边形

视频 29

17.1 Introduction 介绍

Polygon 多边形

n. [ˈpɒlɪˌɡɒn]

Definition: A plane shape that is bounded by three or more line segments. Each of these line segments is called a side. The intersection of two sides is called a vertex. Figure 17-1 shows a polygon—the blue line segments are sides and the black points are the vertices. 多边形是指被三条或更多条线段围成的平面图形。每条线段称为边。两条边的交点称为顶点。图 17-1 展示了一个多边形——蓝色线段为边，黑色点为顶点。

Notation: A polygon is named by writing down vertices consecutively. 要标记一个多边形，连续地依次标出各顶点即可。

For example, on the polygon in Figure 17-2, we will name it as one of $ABCDEF$, $BCDEFA$, $CDEFAB$, $DEFABC$, $EFABCD$, or $FABCDE$.

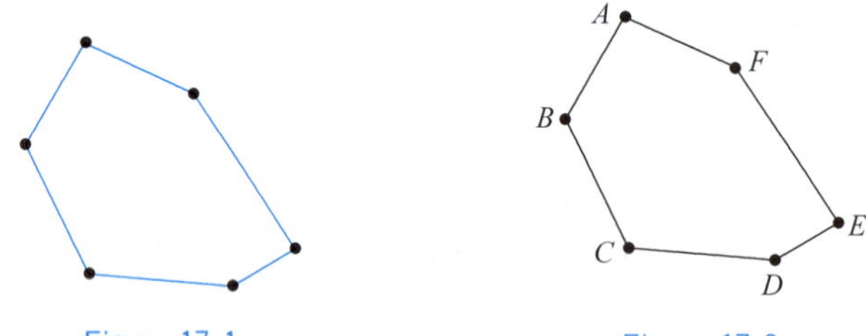

Figure 17-1　　Figure 17-2

Refer to Table A-5 in Appendix A for the notation of polygons.
多边形的标记方式参照附录 A 的表 A-5。

These names indicate that vertices (A and B), (B and C), (C and D), (D and E), (E and F), and (F and A) are adjacent vertices. 顶点的顺序决定了邻顶点。

Properties: Table 17-1 gives the names of the polygons according to their numbers of sides.

Table 17-1

Name 名称	Number of Sides 边数
Triangle 三角形	3
Quadrilateral 四边形	4
Pentagon 五边形	5
Hexagon 六边形	6
Heptagon 七边形	7
Octagon 八边形	8
Nonagon 九边形	9
Decagon 十边形	10
Dodecagon 十二边形	12
Icosagon 二十边形	20

A polygon with n sides is also called an n-gon. 有 n 条边的多边形亦称为 n 边形。

17.2 Describing Polygons 描述多边形

17.2.1 Convexity and Concavity 凹凸性

Convex Polygon 凸多边形

n. ['kɑnveks 'pɑlɪˌgɑn]

Definition: A convex polygon is a polygon that does not have a reflex angle. 凸多边形是指没有优角的多边形。

Properties: Choosing any two points in the interior of a convex polygon, the line segment whose endpoints are these two points are always completely inside the polygon. One example is shown in Figure 17-3. 在凸多边形内部任意选两点,端点为这两点的线段总是完全在多边形里面,如图 17-3 所示。

Concave Polygon 凹多边形

n. [kɑn'keɪv 'pɑlɪˌgɑn]

Definition: A concave polygon is a polygon that has at least one reflex angle. 凹多边形是指有优角的多边形。

Properties: In a concave polygon, there exists two points in the interior of the polygon such that the line segment (shown in blue) with these endpoints does not lie completely inside the polygon. One example is shown in Figure 17-4. 在凹多边形内部,存在两点使得端点为这两点的线段(用蓝色表示)没有完全在多边形里,如图 17-4 所示。

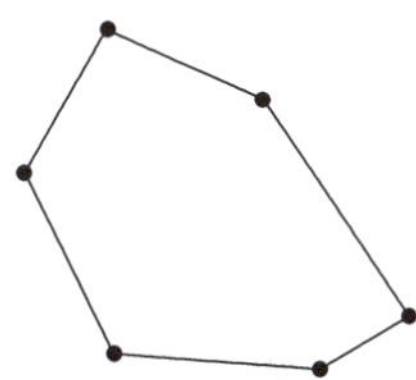

Figure 17-3

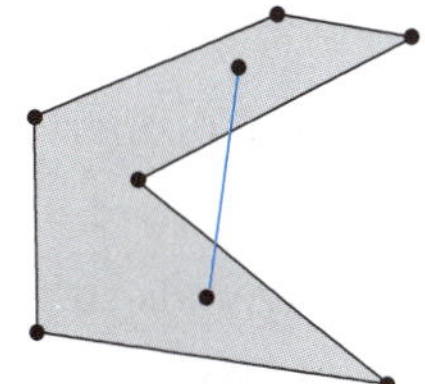

Figure 17-4

17.2.2 Regularity 规律性

Equiangular 等角的

adj. [iːkwɪˈæŋgjʊlə]

Definition：A polygon is equiangular if all of its angles are congruent. 所有角全等的多边形是等角的(等角多边形)。

Figure 17-5 shows two examples of equiangular polygons.

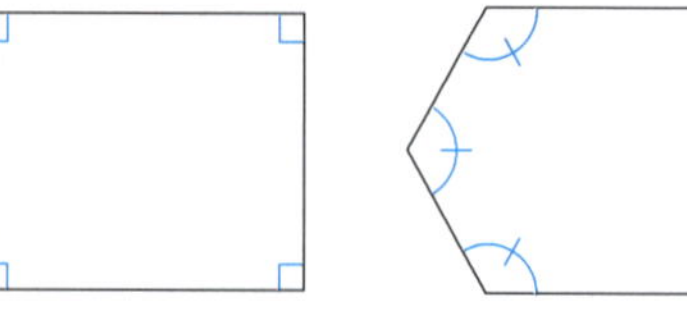

Figure 17-5

Equilateral 等边的

adj. [ˌikwəˈlætərəl]

Definition：A polygon is equilateral if all of its sides are congruent. 所有边全等的多边形是等边的(等边多边形)。

Figure 17-6 shows two examples of equilateral polygons.

Regular 正的

adj. [ˈregjələr]

Definition：A polygon is regular if it is equiangular and equilateral. 等角又等边的多边形是正的(正多边形)。

Figure 17-7 shows two examples of regular polygons.

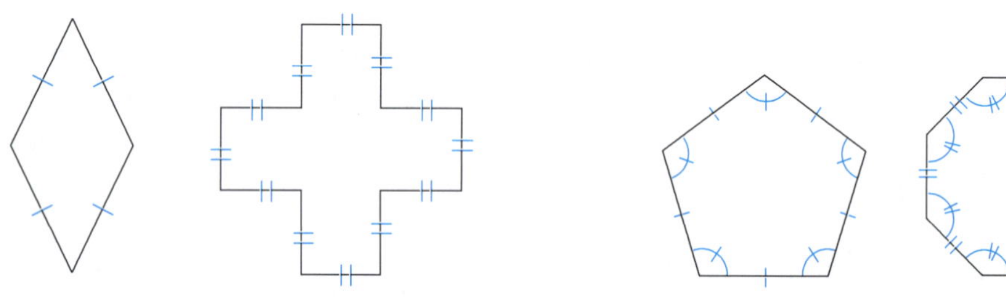

Figure 17-6

Figure 17-7

17.2.3 Diagonals 对角线

Diagonal 对角线

n. [daɪˈægənəl]

Definition：See diagonal in Section 16.1. 定义见 16.1 节四边形部分的对角线词条。

Properties：A convex n-gon has $n(n-3)/2$ diagonals. 凸 n 边形总共有 $n(n-3)/2$ 条对角线。

Proofs：To prove the property, note that for each vertex, there are $n-3$ other vertices (excluding itself and two adjacent vertices) that diagonals can be drawn. We have $n(n-3)$ diagonals so far. Since each diagonal is drawn twice (line segments "from A to B" and "from B to A" are consider the same), there are $n(n-3)/2$ diagonals for the convex n-gon. 要证明对角线的性质，留意到对于凸 n 边形的每个顶点，能连接 $n-3$ 个其他的顶点(除了它本身和它相邻的两个顶点)画出对角线，故可得到 $n(n-3)$条对角线。

因为每条对角线都画了两遍(从 A 到 B 的线段和从 B 到 A 的线段没有区别),所以凸 n 边形总共有 $n(n-3)/2$ 条对角线。

Questions: 1. How many diagonals does a convex heptagon have? 凸七边形有多少条对角线?

2. If a convex polygon has exactly 35 diagonals, how many sides does it have?

Answers: 1. Since a convex heptagon has 7 sides, it has $7(7-3)/2=14$ diagonals.

2. Let n be the number of sides. Since $n(n-3)/2=35$, or $n^2-3n-70=0$. We can factor the LHS so that $(n-10)(n+7)=0$, for which $n=10$ is the only reasonable answer.

17.2.4 Angles 角

Interior Angle 内角

n. [ɪn'tɪərɪər 'æŋgəl]

Definition: An interior angle of a polygon is an angle that is inside the polygon, as shown in Figure 17-8. 内角是指在多边形里面的角,如图 17-8 所示。

An n-gon has n interior angles. n 边形有 n 个内角。

Properties: 1. The sum of measures of interior angles of an n-gon is $(n-2)\times 180°$. n 边形的内角度数之和为 $(n-2)\times 180°$。

2. Each interior angle in a regular n-gon is measured $\frac{(n-2)\times 180°}{n}$. 正多边形的每个内角的度数是 $\frac{(n-2)\times 180°}{n}$。

Proofs: 1. To prove Property 1, from a vertex, we can draw diagonals from the remaining $n-3$ vertices (excluding the vertex itself and its two adjacent vertices) and partition the polygon into $n-2$ triangles. As shown in Figure 17-9, the diagonals are drawn in blue dotted lines. 从一个顶点到除了它本身和它相邻顶点的其他顶点,可以画出 $n-3$ 条对角线,并把多边形分成 $n-2$ 个三角形。如图 17-9 所示,对角线用蓝色虚线画出。

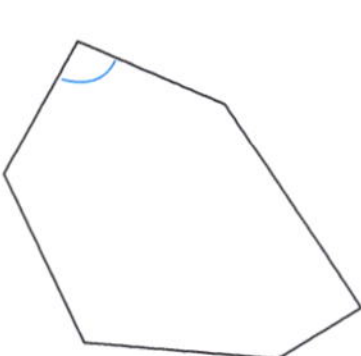

Figure 17-8

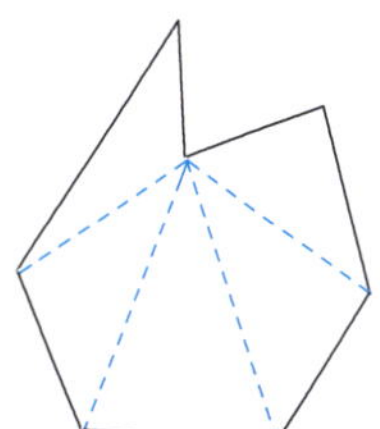

Figure 17-9

Since the sum of measures of interior angles of a triangle is 180°, we know that the sum of measures of interior angles of an n-gon is $(n-2)\times 180°$. 因为三角形内角的度数之和为 180°,由此可知 n 边形的内角度数之和为 $(n-2)\times 180°$。

The argument above applies to all convex n-gons. For concave n-gons, the argument sometimes works (as shown in Figure 17-9 already) and sometimes does not—to partition the n-gon into $n-2$ triangles, the diagonals we draw may not share the same vertex.

以上论点对所有凸 n 边形均有效。对凹 n 边形,以上论点有时有效(如图 17-9 所示),有时无效——要把 n 边形分割为 $n-2$ 个三角形,对角线也许没有共同顶点。

Nevertheless, the sum of measures of any n-gon is $(n-2)\times180°$.

不过,任何 n 边形的内角度数之和均为 $(n-2)\times180°$。

2. To prove Property 2, use Property 1. From Property 2, we can conclude that all regular polygons are convex. 用内角的性质 1 可证明内角的性质 2。从内角的性质 2 可得出所有正多边形都是凸多边形。

Questions: 1. What is the sum of measures of interior angles of a nonagon? 九边形的内角度数之和为多少?

2. What is the measure of an interior angle of a regular 18-gon? 正十八边形的一个内角度数为多少?

3. If an n-gon's sum of measures of interior angles is 2340°, what is the value of n? 若 n 边形的内角度数之和为 2340°,求 n 的值。

Answers: 1. Since a nonagon has 9 angles, the sum of measures of interior angles of a nonagon is $(9-2)\times180°=1260°$.

2. By Property 2, we have $\frac{(18-2)\times180°}{18}=160°$.

3. We have $(n-2)\times180°=2340°$, from which $n-2=13$, and $n=15$.

Exterior Angle 外角

n. [ek'tɪəriər 'æŋgəl]

Definition: An exterior angle of a convex polygon is the supplementary angle that forms a linear pair with the corresponding interior angle. As shown in Figure 17-10, two possible exterior angles of the same interior angle (drawn in black) are drawn in blue (the blue exterior angles are congruent due to vertical angles). 凸多边形的外角是它对应的内角的补角,跟这个内角形成邻补角。图 17-10 画出了一个内角(黑色)对应的两个外角(蓝色)。(因为对顶角的关系,蓝色的外角全等。)

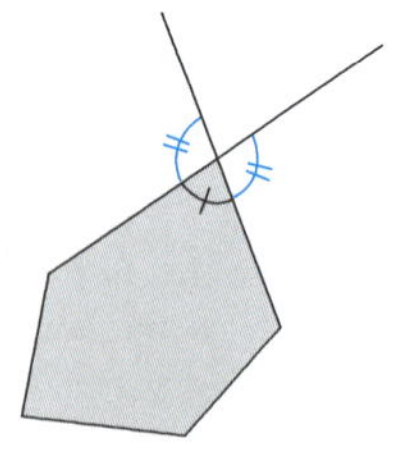

Figure 17-10

By definition, for convex polygons, the sum of measures of every interior angle and its corresponding exterior angle is 180°. 根据定义,对凸多边形而言,每个内角和它对应的外角度数之和为 180°。

An n-gon has n exterior angles. n 边形有 n 个外角。

Properties: The sum of measures of exterior angles of a convex n-gon is 360°. 凸 n 边形的外角度数之和为 360°。

Proofs: To prove the property, note that: 要证明外角的性质,注意:

The sum of measures of n straight angles is $n\cdot180°$.

n 个平角的度数之和为 $n\cdot180°$。

The sum of measures of interior angles of an n-gon is $(n-2)\cdot180°$.

n 边形的内角度数之和为 $(n-2)\times180°$。

Therefore, the sum of measures of exterior angles of an n-gon is $n\cdot180°-(n-2)\cdot$

$180° = 360°$.

所以，n 边形的外角度数之和为 $180° - (n-2) \cdot 180° = 360°$。

For example, as shown in Figure 17-11, the black angles are the interior angles of the gray polygon, and the blue angles are the exterior angles of the gray polygon. 举个例子，如图 17-11 所示，黑色的角是灰色多边形的内角；蓝色的角是灰色多边形的外角。

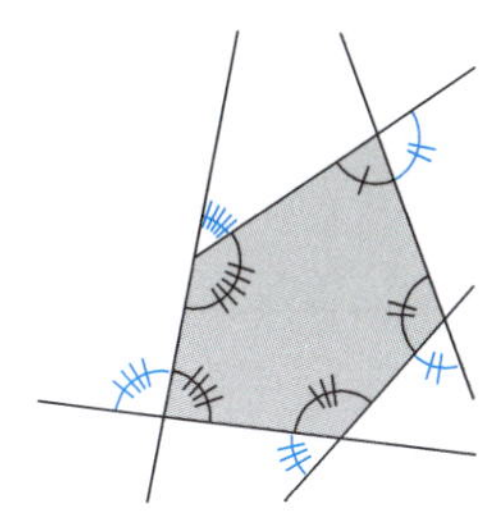

Figure 17-11

The sum of measures of all black and blue angles is $5 \cdot 180° = 900°$. 所有黑色角和蓝色角的度数之和为 $5 \cdot 180° = 900°$。

The sum of measures of black angles is $(5-2) \cdot 180° = 540°$. 黑色角的度数之和为$(5-2) \cdot 180° = 540°$。

The sum of measures of blue angles is $900° - 540° = 360°$. 蓝色角的度数之和为 $900° - 540° = 360°$。

Questions: If each interior angle in a regular polygon is measured 179°, how many sides does this polygon have? 若一个正多边形的每个内角度数为 179°，则这个多边形有多少条边？

Answers: Method Ⅰ: Solve the equation $\frac{(n-2) \cdot 180°}{n} = 179°$ (from Property 2 of interior angle) and get $n = 360$. 根据内角的性质 2 的等式。

Method Ⅱ: The question implies that each exterior angle is measured 1°. Therefore, $n = 360$.

Thinking about the exterior angles sometimes produces a much faster solution. 有时候从外角入手解决问题会更快。

17.2.5 Summary 总结

1. (1) A polygon is convex if it does not have any reflex angles. 若一个多边形没有优角，则它是凸多边形。
   (2) A polygon is concave if it has reflex angles. 若一个多边形有优角，则它是凹多边形。
2. (1) A polygon is equiangular if all of its angles are congruent. 所有角全等的多边形是等角多边形。
   (2) A polygon is equilateral if all of its sides are congruent. 所有边全等的多边形是等边多边形。
   (3) A polygon is regular if it is equiangular and equilateral. 等角又等边的多边形是正多边形。
3. A convex n-gon has $n(n-3)/2$ diagonals. 一个凸 n 边形有 $n(n-3)/2$ 条对角线。
4. (1) The sum of measures of interior angles of an n-gon is $(n-2) \cdot 180°$. n 边形的内角度数之和为$(n-2) \cdot 180°$。
   (2) Each interior angle in a regular n-gon is measured $\frac{(n-2) \cdot 180°}{n}$. 正多边形的每个内角的度数是$\frac{(n-2) \cdot 180°}{n}$。
5. The sum of measures of exterior angles of a convex n-gon is 360°. 凸 n 边形的外角度数之和为 360°。

18. Transformations 变换

18.1 Introduction 介绍

18.1.1 Basics 基础

Transformation 变换

n. [ˌtrænsfər'meɪʃən]

Definition: One of rigid transformation, or non-rigid transformation. 变换包括合同变换和非合同变换。

Translations, reflections, and rotations are called rigid transformations. 平移、反射、旋转属于合同变换。

Dilations are called non-rigid transformations. 位似变换属于非合同变换。

The original shape is called the preimage, usually denoted by $ABC\cdots$. 原图形叫作原像，通常用 $ABC\cdots$标记。

The shape after the transformation is called the image under transformation, denoted by $A'B'C'\cdots$, where A', B', C', $\cdots$ correspond to A, B, C, $\cdots$ on the preimage, respectively.

变换后的图形叫作像，通常用 $A'B'C'\cdots\cdots$标记。A'、B'、$C'\cdots\cdots$分别对应原像中的 A、B、$C\cdots\cdots$。

For each of the examples below, the preimage is colored in gray, and the image is colored in blue. 在下面各图中，灰色的图形是原像，蓝色的图形是像。

Figure 18-1 shows a translation. 平移

Figure 18-2 shows a reflection. 反射

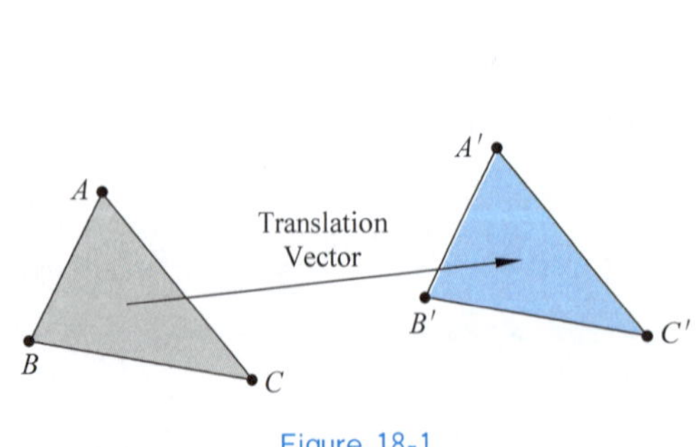

Figure 18-1

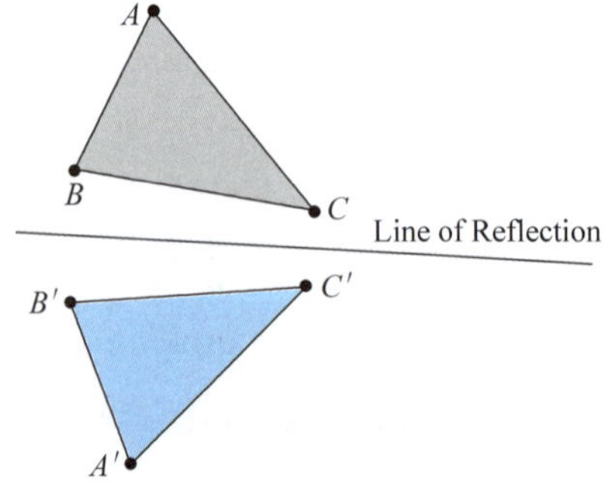

Figure 18-2

Figure 18-3 shows a rotation. 旋转

Figure 18-4 shows a dilation. 位似变换

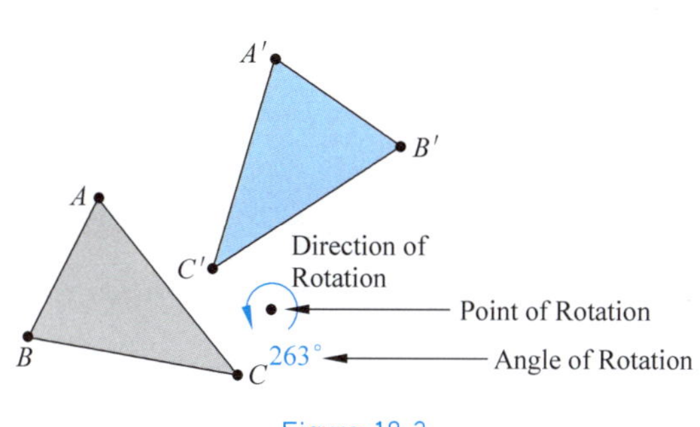

Figure 18-3

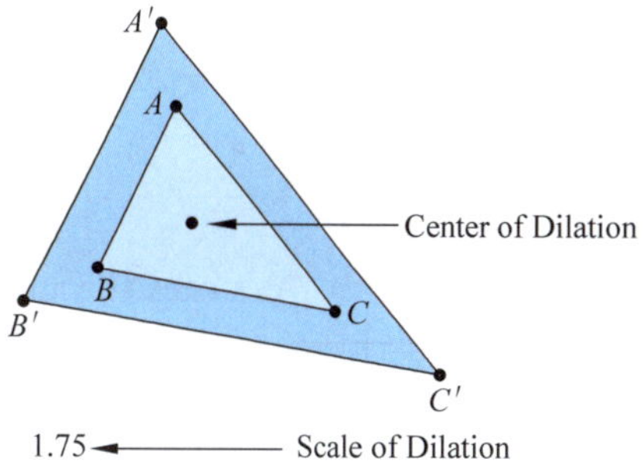

Figure 18-4

Vector　向量

n. ['vektər]

Definition: A directed line segment on the coordinate plane that has an initial point and a terminal point. One example is shown in Figure 18-5.　向量是指在坐标平面上有方向的线段。这条线段有始点和终点，如图 18-5 所示。

We say (2,3) is the initial point of this vector, and (9,7) is the terminal point of this vector.

(2,3)为初始点；(9,7)为终点。

Throughout this encyclopedia, we usually use position vectors—vectors whose initial point are at the origin, unless specified. One example is shown in Figure 18-6.　在本书中，除非特别指定，都会用位置向量——始点在坐标原点的向量，如图 18-6 所示。

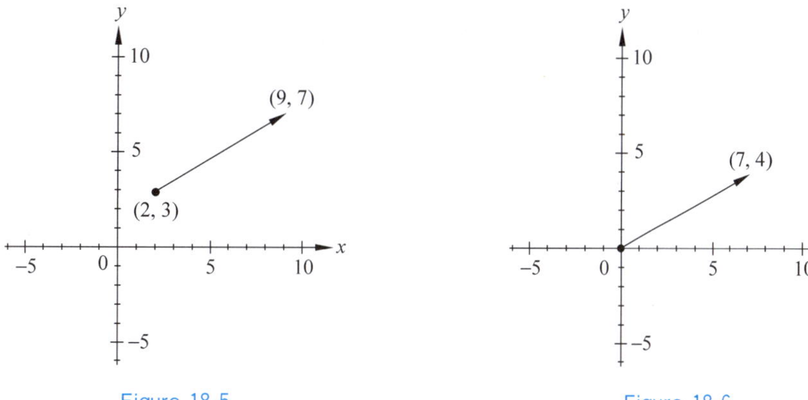

Figure 18-5　　Figure 18-6

Notation: A position vector $\boldsymbol{v}$ is denoted by its terminal point Q, written as followed:

位置向量 $\boldsymbol{v}$，用它的终点 Q 表示，写为

$$\boldsymbol{v}=(x\text{-coordinate of }Q, y\text{-coordinate of }Q)$$

$$\boldsymbol{v}=(Q\text{ 的 }x\text{ 坐标}, Q\text{ 的 }y\text{ 坐标})$$

Refer to Table A-5 in Appendix A for the notation of vectors.

向量的标记方式参照附录 A 的表 A-5。

Properties: 1. Vectors have both directions and magnitudes (length).　向量有确定的方向和量值

(长度)。

2. To find the magnitude of a vector $\boldsymbol{v}=(x_0,y_0)$, we have
要找出 $\boldsymbol{v}=(x_0,y_0)$ 的量值,根据以下公式:

$$|\boldsymbol{v}|=\sqrt{x_0^2+y_0^2}$$

$|\boldsymbol{v}|$ denotes the magnitude of $\boldsymbol{v}$. $|\boldsymbol{v}|$ 表示 $\boldsymbol{v}$ 的量值。
Refer to Table A-5 in Appendix A for the notation of the lengths of vectors.
向量长度的标记方式参照附录 A 的表 A-5。
This is an analogue of the Distance Formula. 与距离公式类似。

3. To add/subtract two vectors, add/subtract them component-wise: 要加/减两个向量,可加/减它们对应的部分。
If $\boldsymbol{v}_1=(x_1,y_1)$ and $\boldsymbol{v}_2=(x_2,y_2)$, then $\boldsymbol{v}_1+\boldsymbol{v}_2=(x_1+x_2,y_1+y_2)$ and $\boldsymbol{v}_1-\boldsymbol{v}_2=(x_1-x_2,y_1-y_2)$.

4. To multiply a vector with a constant (scalar multiple), multiply each component of the vector by that constant:
求一个向量和一个常数的积(标量倍数),则把向量里的每部分和常数相乘。
If $\boldsymbol{v}=(x_0,y_0)$ and k is a constant, then $k\boldsymbol{v}=(kx_0,\ ky_0)$.

5. If $\boldsymbol{v}$ is a vector and k is a constant, then $|k\boldsymbol{v}|=|k||\boldsymbol{v}|$.
Let $\boldsymbol{v}=(x_0,y_0)$. Note that $|k\boldsymbol{v}|=|(kx_0,ky_0)|=\sqrt{(kx_0)^2+(ky_0)^2}=\sqrt{k^2(x_0^2+y_0^2)}$ $=|k|\sqrt{x_0^2+y_0^2}=|k||\boldsymbol{v}|$.

Note: There are many other properties about vectors such as dot product and cross product, which we will not get into since they are out of the scope of high school geometry. 还有其他关于向量的性质,如向量点积和向量积。在高中几何课不会讲到。

Phrases: translation~ 平移向量

Questions: If $\boldsymbol{a}=(3,4)$ and $\boldsymbol{b}=(12,5)$, what are the values of: (1) $|\boldsymbol{a}|$, (2) $|\boldsymbol{b}|$, (3) $\boldsymbol{a}+\boldsymbol{b}$, (4) $\boldsymbol{a}-\boldsymbol{b}$, (5) $7\boldsymbol{a}-4\boldsymbol{b}$?

Answers: (1) $|\boldsymbol{a}|=\sqrt{3^2+4^2}=5$.

(2) $|\boldsymbol{b}|=\sqrt{12^2+5^2}=13$.

(3) $\boldsymbol{a}+\boldsymbol{b}=(3+12,4+5)=(15,9)$.

(4) $\boldsymbol{a}-\boldsymbol{b}=(3-12,4-5)=(-9,-1)$.

(5) $7\boldsymbol{a}-4\boldsymbol{b}=7(3,4)-4(12,5)=(21,28)-(48,20)=(-27,8)$.

Table 18-1 shows the properties of lines, line segments, rays, and vectors.

Table 18-1

1-D Shape 一维图形	Direction 方向	Length 长度
Line 直线	Indefinite 不确定	Indefinite 不确定
Line Segment 线段	Indefinite 不确定	Definite 确定
Ray 射线	Definite 确定	Indefinite 不确定
Vector 向量	Definite 确定	Definite 确定

18.1.2 Summary 总结

1. There are two types of transformations: rigid and nonrigid. 有两种变换：合同变换和非合同变换。
   Rigid transformations maintain the shape of the preimage, and nonrigid transformations do not. 合同变换能保持原像的图形不变，而非合同变换不能。
   (1) Rigid transformations include: translations, reflections, rotations. 合同变换包括平移、反射、旋转。
   (2) Nonrigid transformations include: dilations, and many other transformations that we won't cover in high school courses. 非合同变换包括位似变换，还有其他高中课程不会讲到的变换。
   Under every kind of transformation, the points $A', B', C', \cdots$ on the image correspond to the points $A, B, C, \cdots$ on the preimage, respectively. 像里的 A'、B'、C'……分别对应原像里的 A、B、C……
2. Some properties of vectors:
   变量的一些性质：
   (1) If $\boldsymbol{v} = (x_0, y_0)$, then $|\boldsymbol{v}| = \sqrt{x_0^2 + y_0^2}$. The expression $|\boldsymbol{v}|$ denotes the magnitude of $\boldsymbol{v}$.
   (2) If $\boldsymbol{v}_1 = (x_1, y_1)$ and $\boldsymbol{v}_2 = (x_2, y_2)$, then $\boldsymbol{v}_1 + \boldsymbol{v}_2 = (x_1 + x_2, y_1 + y_2)$ and $\boldsymbol{v}_1 - \boldsymbol{v}_2 = (x_1 - x_2, y_1 - y_2)$.
   (3) If $\boldsymbol{v} = (x_0, y_0)$ and k is a constant, then $k\boldsymbol{v} = (kx_0, ky_0)$.
   (4) If $\boldsymbol{v}$ is a vector and k is a constant, then $|k\boldsymbol{v}| = |k||\boldsymbol{v}|$.

18.2 Rigid Transformations 合同变换

视频 30

18.2.1 Introduction 介绍

Rigid Transformation 合同变换

n. [ˈrɪdʒɪd ˌtrænsfərˈmeɪʃən]

Definition: A transformation of the plane that preserves length so that the preimage and the image are congruent. 合同变换是指维持长度的变换，使得原像和像全等。
It is one of: translation, reflection, and rotation. 合同变换包括平移、反射、旋转。

18.2.2 Types of Rigid Transformations 合同变换的种类

Translation 平移

n. [trænsˈleɪ. ʃən]

Definition: A rigid transformation that moves a preimage a certain distance, without rotating or flipping it, as shown below in Figure 18-7. 平移是一种合同变换，是在不旋转或翻转的情况下移动原像，如图 18-7 所示。
A translation is determined by a translation vector, which has definite length and a direction. 平移向量决定一个平移。平移向量有明确的长度和方向。

Properties: In the coordinate plane, suppose a point on the preimage has coordinates (x, y). 在坐标平面上，假设在原像中有一点，坐标为(x, y)。
(1) Moving the preimage a units up, the translation vector is $(0, a)$. The corresponding point on the image would be $(x, y + a)$. 往上移动原像 a 格，平移向量是$(0, a)$。在

像中对应原像的点是$(x, y+a)$。

(2) Moving the preimage a units down, the translation vector is $(0, -a)$. The corresponding point on the image would be $(x, y-a)$. 往下移动原像 a 格,平移向量是$(0, -a)$。在像中对应原像的点是$(x, y-a)$。

(3) Moving the preimage b units left, the translation vector is $(-b, 0)$. The corresponding point on the image would be $(x-b, y)$. 往左移动原像 b 格,平移向量是$(-b, 0)$。在像中对应原像的点是$(x-b, y)$。

(4) Moving the preimage b units right, the translation vector is $(b, 0)$. The corresponding point on the image would be $(x+b, y)$. 往右移动原像 b 格,平移向量是$(b, 0)$。在像中对应原像的点是$(x+b, y)$。

This property is illustrated in Figure 18-8.

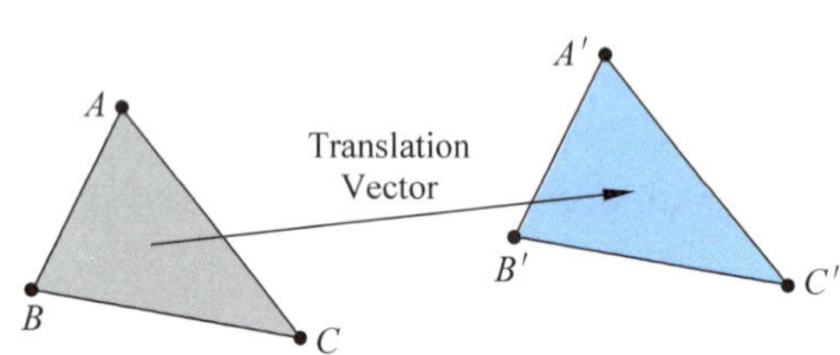

Figure 18-7

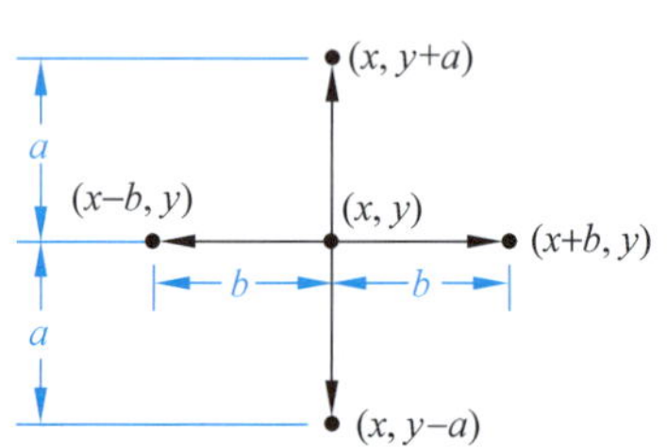

Figure 18-8

Questions: The coordinates of points A, B, C, and D are $(3,4)$, $(9,8)$, $(-3,-7)$, and $(-5,0)$, respectively. After translating the quadrilateral $ABCD$ 5 units up and 3 units left, what is (1) the translation vector, (2) the coordinates of the vertices of the image?

Answers: (1) $(-3,5)$: 5 units up and 3 units to the left.

(2) Suppose the image is $A'B'C'D'$.

$A' = (3-3, 4+5) = (0,9)$.

$B' = (9-3, 8+5) = (6,13)$.

$C' = (-3-3, -7+5) = (-6,-2)$.

$D' = (-5-3, 0+5) = (-8,5)$.

Reflection 反射

n. [rɪˈflekʃən]

Definition: A rigid transformation that flips the preimage about an axis, without translating or rotating it, as shown in Figure 18-9. 反射是一种合同变换,是指在不平移或旋转的情况下沿着一条轴翻转原像,如图 18-9 所示。

A reflection is determined by the axis of reflection or line of reflection. 根据反射轴可确定一个反射。

Properties: 1. In the coordinate plane, suppose a point on the preimage has coordinates (a, b). 在坐标平面里,假设在原像中有一点,坐标是(a, b)。

(1) Reflecting the preimage over the x-axis, the corresponding point on the image would be $(a, -b)$. 沿着 x 轴反射原像,在像中对应的点是$(a, -b)$。

(2) Reflecting the preimage over the y-axis, the corresponding point on the image would be $(-a, b)$. 沿着 y 轴反射原像，在像中对应的点是$(-a, b)$。

This property is shown in Figure 18-10.

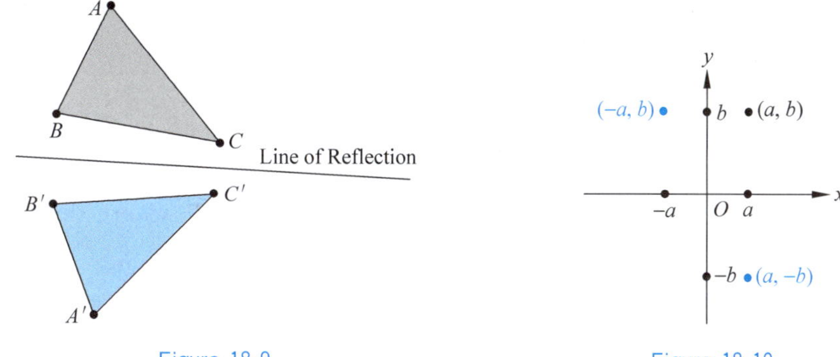

Figure 18-9

Figure 18-10

2. In a reflection, the distance from a point P in the preimage to the line of reflection is half the distance from P to its corresponding point, P', in the image, as shown in Figure 18-11. 在反射里，原像的一点 P 到反射轴的距离是 P 到 P' 的距离的一半。其中 P' 是在像中对应原像 P 的点，如图 18-11 所示。
3. A double reflection of a shape, whose lines of reflection are parallel, results a translation of the shape, as shown in Figure 18-12. Put it in another way, if the lines of reflection are parallel, then flipping a shape twice results a translation. 若两反射轴平行，一个图形的双反射的结果是一次平移，如图 18-12 所示。换言之，若两反射轴平行，则翻转两次相当于一次平移。

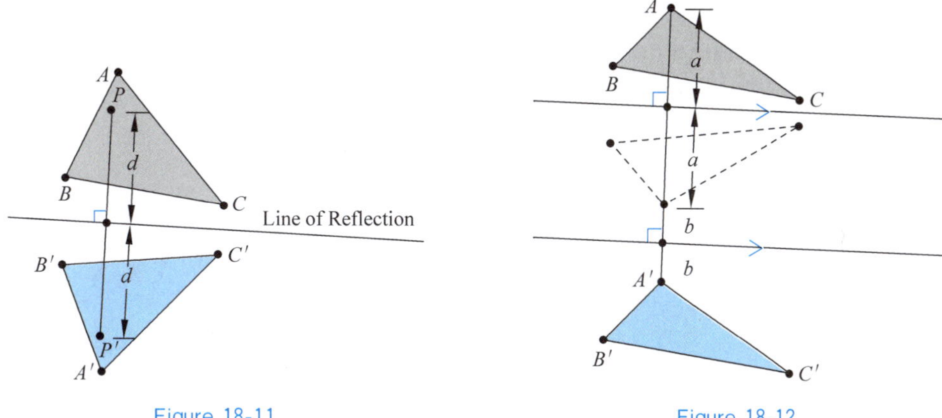

Figure 18-11

Figure 18-12

Moreover, the distance from a point P in the preimage to the corresponding point P' in the image is twice the distance between the lines of reflection, as shown above. 原像中点 P 到像中点 P' 的距离是两条反射轴的距离的 2 倍。

4. If the line of reflection passes through the center of the shape, and the image is coincident with the preimage, then by definition, the shape has a line symmetry, and

the line of reflection is a line of symmetry of the shape, as shown in Figure 18-13. 若反射轴经过图形的中心,且像和原像重合,则根据定义,这个图形有对称轴,且图形的反射轴就是对称轴,如图 18-13 所示。

Questions: 1. What is the image of the point (3,4) if it is:

(1) reflected over the x-axis?

(2) reflected over the y-axis?

2. As shown in Figure 18-14, two houses are located on points A and B. The transportation manager wants to build a bus station somewhere (call the location point S) on the blue line. To minimize $SA+SB$, where should S be located?

这是一种常见的求最短距离的问题,即将军饮马问题。

Figure 18-13　　Figure 18-14

Answers: 1. (1) (3, −4).

(2) (−3,4).

2. If we reflect A across the blue line and call the image A', The Perpendicular Bisector Theorem implies that $SA = SA'$. By substitution and the Triangle Inequality, $SA+SB = SA'+SB \leqslant A'B$. If A', S, and B are collinear, then $SA+SB$ is minimized since its length is $A'B$. Therefore, we will build S at the intersection of $\overline{A'B}$ and the blue line, as shown in Figure 18-15.

Rotation 旋转

n. [roʊ'teɪʃən]

Definition: A rigid transformation that turns the preimage about a point, without translating or flipping it, as shown in Figure 18-16. 旋转是一种合同变换,是指在不平移或翻转的情况下沿着一条线转动原像,如图 18-16 所示。

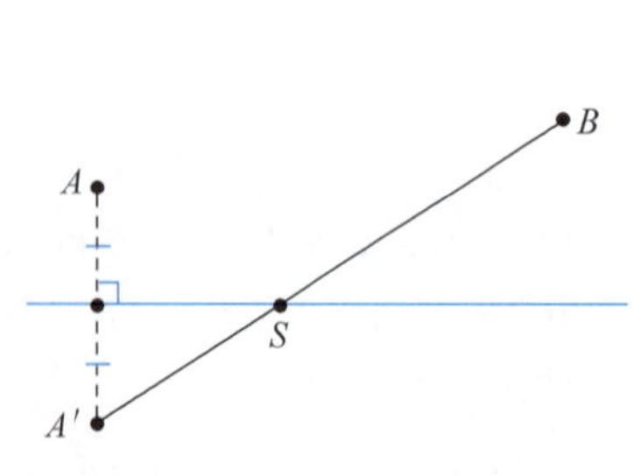

Figure 18-15

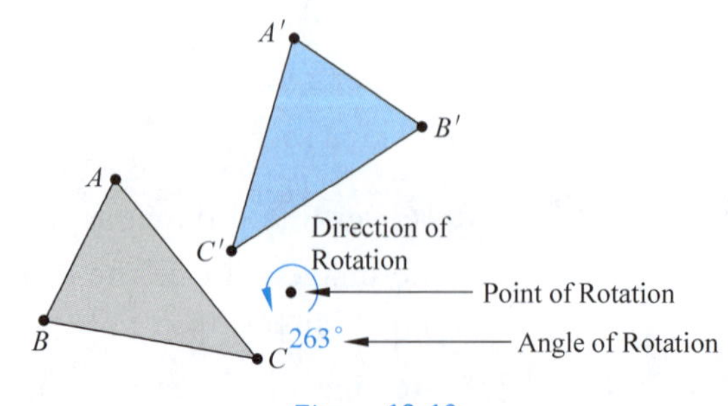

Figure 18-16

A rotation is determined by the point of rotation (which is also called the center of rotation), angle of rotation, and direction of rotation. 旋转点(亦称旋转中心)、旋转角度、旋转方向决定一个旋转。

Properties: 1. As shown in Figure 18-17, for every rotation, if O is the point of rotation, P is a point in the preimage, and P' is its corresponding point in the image, then $OP = OP'$ and $m\angle POP'$ = angle of rotation. The preimage (shown in gray) rotates 100° counterclockwise about O. The image is shown in blue. 如图 18-17 所示,在每个旋转中,若 O 为旋转点,P 为原像里的一点,P'为像中对应的点,则 $OP = OP'$,且 $m\angle POP'$是旋转角度。原像(灰色)沿着 O 逆时针旋转了 100° 形成像。像用蓝色表示。

2. If the point of rotation is the center of the shape, and the image is coincident with the preimage, then by definition, the shape has a rotational symmetry, as shown in Figure 18-18. 若旋转点是图形的中心,且像与原像重合。根据定义,这个图形有旋转对称,如图 18-18 所示。

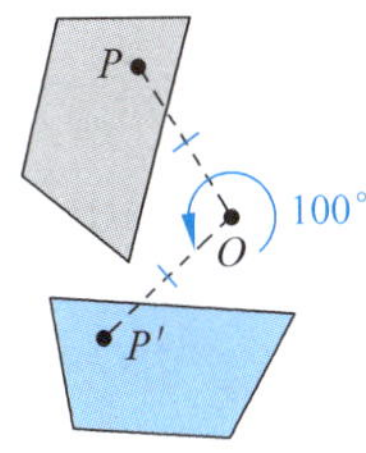

Figure 18-17

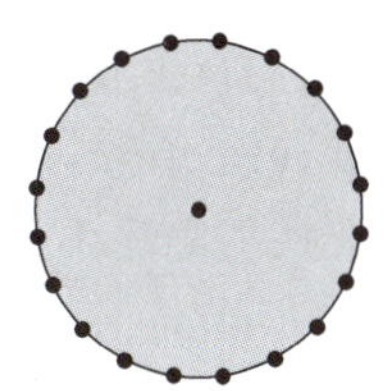
Figure 18-18

18.2.3 Order of Rigid Transformations 合同变换的顺序

Order of Rigid Transformations 合同变换的顺序

Properties: The order of rigid transformation combinations usually matters. One special exception is glide reflection. 合同变换的顺序跟结果有关系,例如滑移反射。

Examples: 1. The order of rigid transformations matters in this case. 合同变换的顺序很重要,下面举例说明。

From (4,9), if we translate 3 units up and 5 units to the right, we get (9,12). Reflecting the result across the y-axis, we end up with (−9,12). 从坐标(4,9)开始,若向上平移 3 格后向右平移 5 格,则得到(9,12)。再沿着 y 轴反射这个结果,得到了(−9,12)。

From (4,9), if we reflect across the y-axis first, we get (−4,9). Translating the result 3 units up and 5 units to the right, we end up with (1,12). 从坐标(4,9)开始,若沿 y 轴反射,则得到(−4,9);再向上平移 3 格后向右平移 5 格,得到(1,12)。

2. The order of rigid transformations does not matter in this case. 在下例中,合同变换的顺序不重要。

From (3,0), if we rotate 70° counterclockwise about the origin, then rotate the result 110° counterclockwise about the origin, we will end up with (−3,0). 从坐标(3,0)开始,若沿着坐标原点逆时针旋转 70°,然后沿坐标原点逆时针旋转 110°,则得到(−3,0)。

From (3,0), if we rotate 110° counterclockwise about the origin, then rotate the result 70° counterclockwise about the origin, we will end up with $(-3,0)$. 从坐标(3,0)开始,若沿着坐标原点逆时针旋转110°,然后沿着坐标原点逆时针旋转70°,则得到$(-3,0)$。

Glide Reflection 滑移反射

n. [glaɪd rɪ'flekʃən]

Definition: The rigid transformation combination that consists of a translation and a reflection, for which the translation vector is parallel to the line of reflection. 滑移反射是合同变换的组合,包含一次平移和一次反射,使得平移向量与反射轴平行。

As shown in Figure 18-19, the gray pre-image undergoes a translation and a reflection. When the translation vector is parallel to the line of reflection, performing the translation and reflection in either order results the same image.

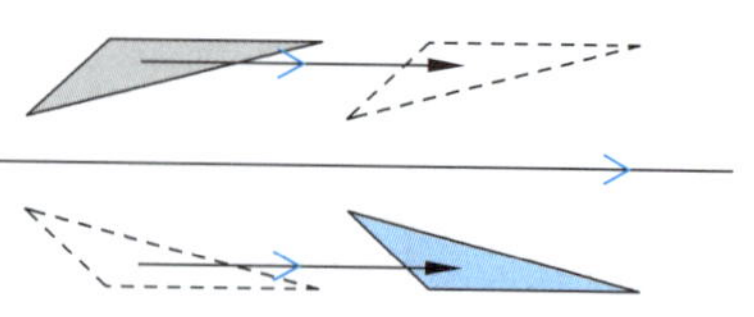

Figure 18-19

如图18-19所示,灰色的原像经历了一次平移与反射。

当平移向量与反射轴平行时,平移与反射的不同顺序所产生的像是相同的。

Properties: In a glide reflection, the order of rigid transformations does not matter. 在滑移反射中,合同变换的顺序不重要。

Questions: Perform the glide reflection on the point (5,7) below:

(1) Reflect across the line $y=5$ first, then translate 5 units to the right.

(2) Translate 5 units to the right first, then reflect across the line $y=5$.

Answers: (1) Reflecting across the line $y=5$, we get (5,3). Translating (5,3) 5 units to the right, we get (10,3), which is our image.

(2) Translating 5 units to the right, we get (10,7). Reflecting (10,7) across the line $y=5$, we get (10,3), which is our image.

18.2.4 Summary 总结

1. Rigid transformations:
合同变换:
   (1) A translation is determined by the translation vector. See translation for more properties in Section 18.2.2. 平移向量决定一次平移。更多性质见18.2.2节平移的词条。
   (2) A reflection is determined by the line of reflection. See reflection for more properties in Section 18.2.2. 反射轴决定一次反射。更多性质见18.2.2节反射的词条。
   (3) A rotation is determined by the point of rotation, angle of rotation, and the direction of rotation. See rotation for more properties in Section 18.2.2. 旋转点、旋转角度、旋转方向决定一次旋转。更多性质见18.2.2节旋转的词条。
2. The order of rigid transformation combinations usually matters. One exception is glide reflection. 合同变换的顺序跟结果有关系。一个例外是滑移反射。
3. A glide reflection is a rigid transformation combination that consists of a translation and a reflection, for which the translation vector is parallel to the line of reflection. 滑移反射是一个合同变换的组合,包含一个平移和一个反射,使得平移向量与反射轴平行。

18.3 Non-Rigid Transformations 非合同变换

18.3.1 Introduction 介绍

Non-Rigid Transformation 非合同变换

n. [nɑnˈrɪdʒɪd ˌtrænsfərˈmeɪʃən]

Definition: A transformation that changes the shape or size of the preimage. In high school geometry, we will only look at dilation. 非合同变换是指改变原像形状或大小的变换。在高中几何中,只会讲位似变换。

18.3.2 Dilations 位似变换

Dilation 位似变换

n. [daɪˈleɪʃən]

Definition: A non-rigid transformation that resizes the preimage proportionately but preserves shape, as shown in Figure 18-20. 位似变换是指成比例地改变原像大小(但保持原形状)的非合同变换,如图 18-20 所示。

A dilation is determined by the center of dilation and a scale of dilation. 位似中心和位似比决定一个位似变换。

Properties: Let (x, y) be a point on the preimage. If the center of dilation is (x_0, y_0) and the scale of dilation is k (for which $k>0$), then the corresponding point on the image is $(x_0, y_0)+k(x-x_0, y-y_0)$. 设$(x, y)$为原像上的一点。若位似中心在$(x_0, y_0)$,位似比为 k $(k>0)$,则在像中对应的点为$(x_0, y_0)+k(x-x_0, y-y_0)$。

Explanation: Suppose $P=(x, y)$ is a point on the preimage, P' is its corresponding point on the image, $O=(x_0, y_0)$ is the center of dilation, and k is the scale of dilation. To locate P', we start at O and travel OP units in the direction of $\overrightarrow{OP}$ for k times. Therefore, we have $P'=(x_0, y_0)+k(x-x_0, y-y_0)$. Since P' is located in the direction of $\overrightarrow{OP}$, points O, P, and P' are collinear.

解释:设 $P=(x, y)$是原像里的一点。P'是像中对应 P 的点,$O=(x_0, y_0)$是位似中心,k 是位似比。要求出 P',从 O 开始,沿着 $\overrightarrow{OP}$ 的方向走动 OP 格 k 次,得到了 $P'=(x_0, y_0)+k(x-x_0, y-y_0)$。因为 P'在 $\overrightarrow{OP}$ 的方向,所以 O、P、P'共线。

In other words, let P be a point in the preimage, P' be the corresponding point in the image, O be the center of dilation, and k be the scale of dilation. If $OP=d$, then $OP'=kd$. Or, $OP'=kOP$, as shown in Figure 18-21. 换个方式表达,设 P 是原像里的一点。P'是像中对应 P 的点,O 是位似中心,k 是位似比。若 $OP=d$,则 $OP'=kd$,或者说 $OP'=kOP$,如图 18-21 所示。

The preimage and image are similar, with k as the ratio of similitude. 原像和像相似,k 是相似比。

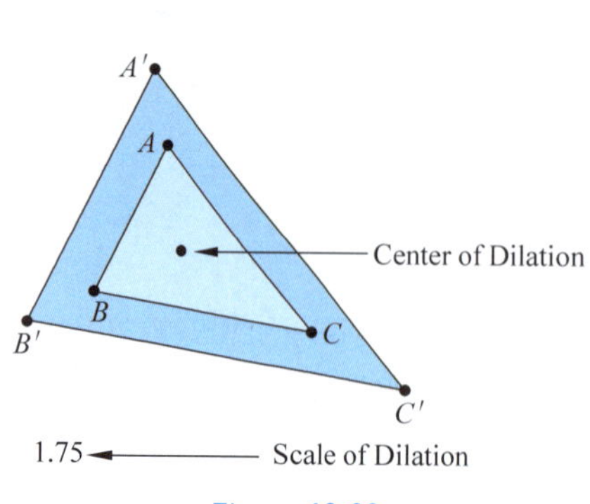

Figure 18-20

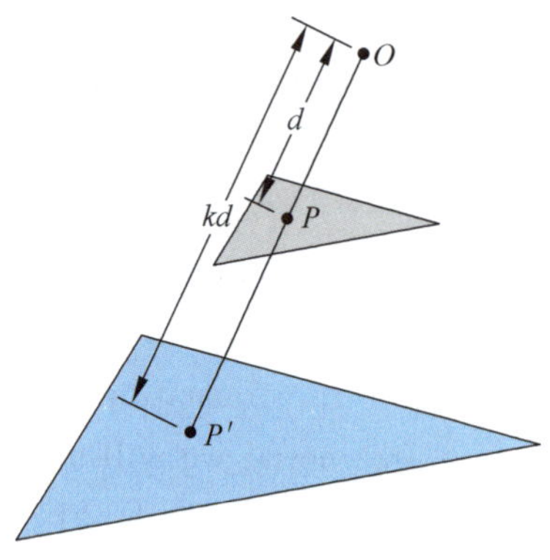

Figure 18-21

Questions: The coordinates of points A, B, C, and D are $(3,4)$, $(9,8)$, $(-3,-7)$, and $(-5,0)$, respectively. If the center of dilation is at $(2,2)$ and the scale of dilation is 3, what are the coordinates of the images of A, B, C, and D?

Answers: Using the formula from the property, we have

$A'=(2,2)+3(3-2,4-2)=(2,2)+3(1,2)=(5,8)$;

$B'=(2,2)+3(9-2,8-2)=(2,2)+3(7,6)=(23,20)$;

$C'=(2,2)+3(-3-2,-7-2)=(2,2)+3(-5,-9)=(-13,-25)$;

$D'=(2,2)+3(-5-2,0-2)=(2,2)+3(-7,-2)=(-19,-4)$.

19. Circles 圆

As shown in Figure 19-1:

在图 19-1 所示的圆中:

O is the center of the circle. O 为圆心。

$\overline{AB}$ is a chord of $\odot O$. $\overline{AB}$ 为弦。

$\overline{COD}$ is a diameter of $\odot O$, which is also a longest chord. $\overline{COD}$ 为直径,也是最长的弦。

$\overleftrightarrow{EF}$ is a secant of $\odot O$. $\overleftrightarrow{EF}$ 为割线。

$\overline{OP}$ is a radius of $\odot O$. $\overline{OP}$ 为半径。

l is a tangent of $\odot O$, with the point of tangency P. l 为切线,P 为切点。

$\overline{OP}\perp l$.

As shown in Figure 19-2:

在图 19-2 所示的圆中:

O is the center of the circle. O 为圆心。

$\angle AOB$ is a central angle. $\overset{\frown}{AB}$ is its intercepted arc. $\angle AOB$ 为圆心角,$\overset{\frown}{AB}$ 为其截弧。

$\angle CDE$ is an inscribed angle. $\overset{\frown}{CE}$ its intercepted arc.　$\angle CDE$ 为圆周角，$\overset{\frown}{CE}$ 为其截弧。

$\overset{\frown}{AB}$，$\overset{\frown}{BC}$，and $\overset{\frown}{CE}$ are some examples of minor arcs.　$\overset{\frown}{AB}$，$\overset{\frown}{BC}$ 和 $\overset{\frown}{CE}$ 均为劣弧。

$\overset{\frown}{AFB}$，$\overset{\frown}{BFC}$，and $\overset{\frown}{CFE}$ are some examples of major arcs.　$\overset{\frown}{AFB}$、$\overset{\frown}{BFC}$ 和 $\overset{\frown}{CFE}$ 均为优弧。

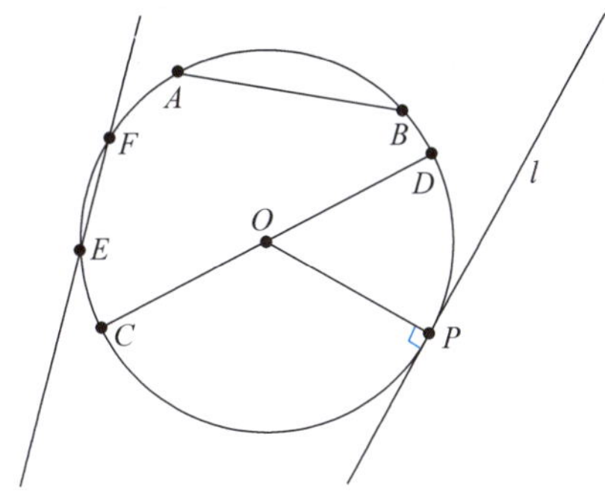

Figure 19-1

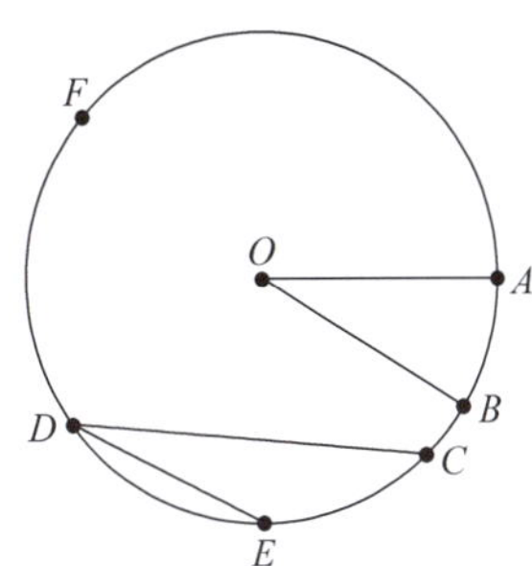

Figure 19-2

19.1　Introduction　介绍

视频 31

19.1.1　Definition of Circles　圆的定义

Circle　圆

n. ['sɜrkəl]

Definition：A collection of points that are equidistant from a given point. The distance is fixed.　圆是指离指定的点等距的点集。距离为固定值。

The given point is called the center，usually is denoted by O.　指定的点叫作圆心，通常用 O 表示。

A line segment whose endpoints are the center and one point on the circle is a radius. Its length is usually denoted by r. In the same circle，all radii are congruent by definition.　端点为圆心和圆上一点的线段叫作半径，其长度通常用 r 表示。圆内的所有半径相等。

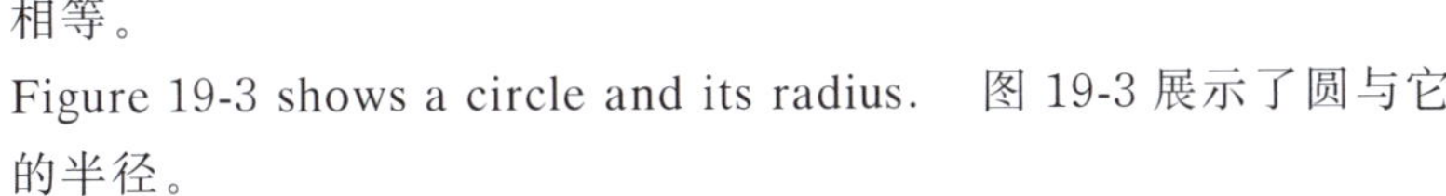

Figure 19-3 shows a circle and its radius.　图 19-3 展示了圆与它的半径。

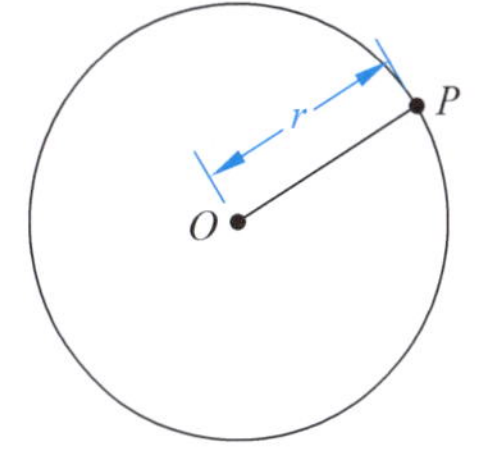

Figure 19-3

Notations：A circle whose center is O is denoted by "circle O" or "$\odot O$". 圆心为 O 的圆写作"圆 O"或"$\odot O$"。

Refer to Table A-5 in Appendix A for the notation of circles.

圆的标记方式参照附录 A 的表 A-5。

19.1.2　Two or More Circles on a Plane　同一平面上两个或多个圆

Concentric Circles　同心圆

n. [kən'sentrɪk 'sɜrkəls]

Definition：Two or more circles are concentric circles if they share the same center，as shown in

Figure 19-4. 若两个或多个圆的圆心相同,则称它们为同心圆,如图 19-4 所示。

Congruent Circles 等圆

n. [kən'gruənt 'sɜrkəls]

Definition: Two or more circles are congruent circles if their radii are congruent, as shown in Figure 19-5. 若两个或多个圆的半径相等,则称它们为等圆,如图 19-5 所示。

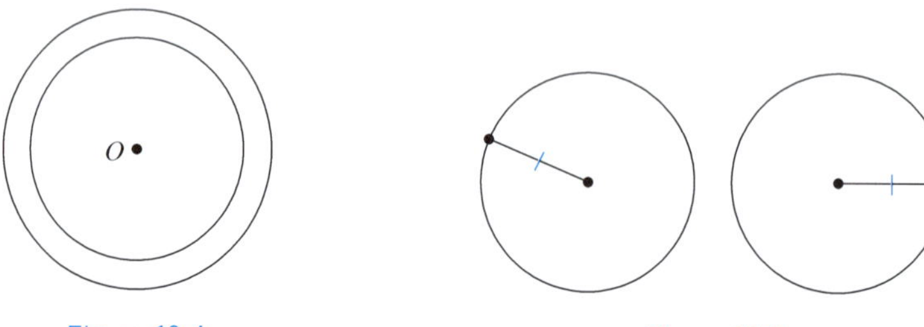

Figure 19-4 Figure 19-5

Concurrent Circles 同圆

n. [kən'kɜrənt 'sɜrkəls]

Definition: In a plane, two or more circles are concurrent circles if they are concentric and congruent, as shown in Figure 19-6. 若两个或多个圆是同心圆和等圆,则称它们为同圆,如图 19-6 所示。

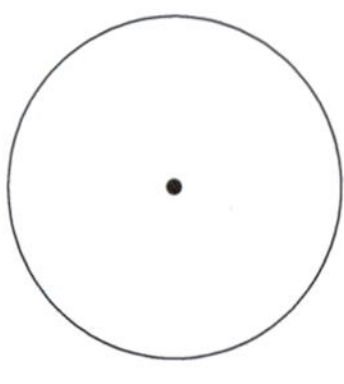

Figure 19-6

19.1.3 Chords and Diameters 弦和直径

Chord 弦

n. [kɔrd]

Definition: A chord of a circle is a line segment whose endpoints are both on the circle, as shown in Figure 19-7 in blue. 端点为圆周两点的线段称为弦,如图 19-7 蓝色所示。

Diameter 直径

n. [daɪ'æmɪtər]

Definition: A chord that passes through the center, as shown in Figure 19-8 in blue. 直径是过圆心的弦,如图 19-8 蓝色所示。

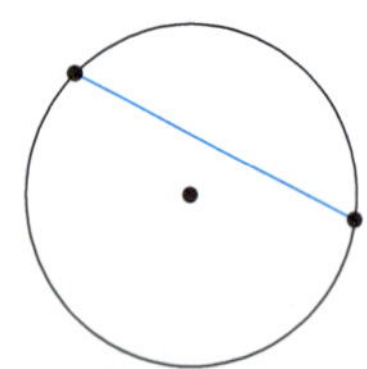

Figure 19-7

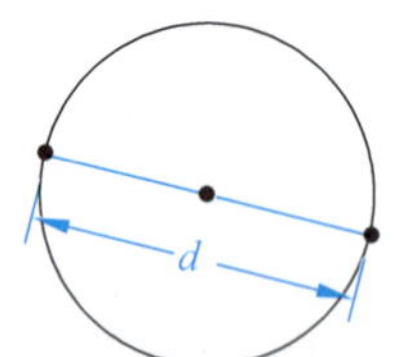

Figure 19-8

Notation: The length of a diameter is usually denoted by d. 直径的长度通常用 d 表示。

Properties: 1. In a circle, the longest chord is the diameter. 直径是圆里最长的弦。

2. In a circle, the length of a diameter is twice that of a radius. In short, $d=2r$. 在圆里,直径长度是半径长度的 2 倍,即 $d=2r$。

19.1.4 Arcs 弧

Arc 弧

n. [ɑrk]

Definition：A portion of the circumference of a circle，as shown in Figure 19-9. 弧是圆周的一部分，如图 19-9 所示。

Properties：An arc has a length and a measure. The circle itself is an arc，whose length is the circumference and whose measure is 360°. 弧有长度和度数。圆本身是一个弧，其长度是圆的周长，度数是 360°。

1. The length of the arc is given by
弧长的公式是

$$\frac{\text{measure of the arc}}{360^\circ}\cdot \text{Circumference}$$

$$\frac{\text{弧长的度数}}{360^\circ}\cdot \text{圆周长}$$

2. (1) Two arcs with the same measure can have different lengths. As shown in Figure 19-10，the blue arcs have the same measure，but different lengths. 相同度数的弧的长度可以不同，如图 19-10 所示，蓝色弧的度数相同，但长度不同。

(2) Two arcs with the same length can have different measures. As shown in Figure 19-11，the blue arcs have the same length，but different measures. 相同长度的弧的度数可以不同。如图 19-11 所示，蓝色弧的长度相同，但度数不同。

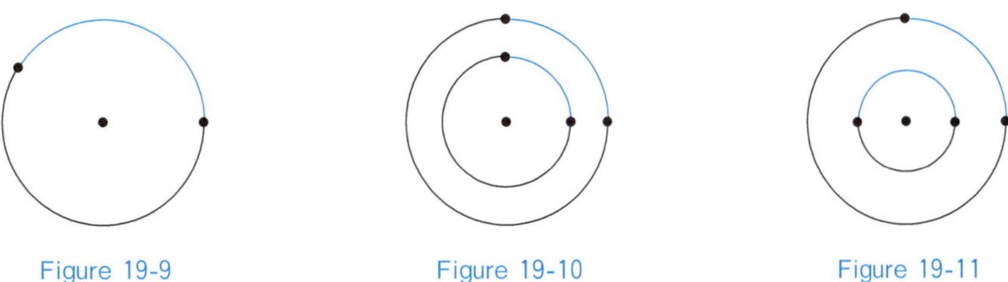

Figure 19-9　　Figure 19-10　　Figure 19-11

(3) In the same circle or congruent circles，the following statements are equivalent：
在同圆或等圆里，下面的命题等价：

① Two arcs have the same measure. 两弧的度数相同。

② Two arcs have the same length. 两弧的长度相同。

③ Two arcs are congruent. 两弧相等。

Examples：Figure 19-12 shows a 90°-arc，180°-arc，270°-arc，and 360°-arc respectively.

Phrases：minor～ 劣弧，major～ 优弧，congruent～ 等弧，intercepted～ 截弧

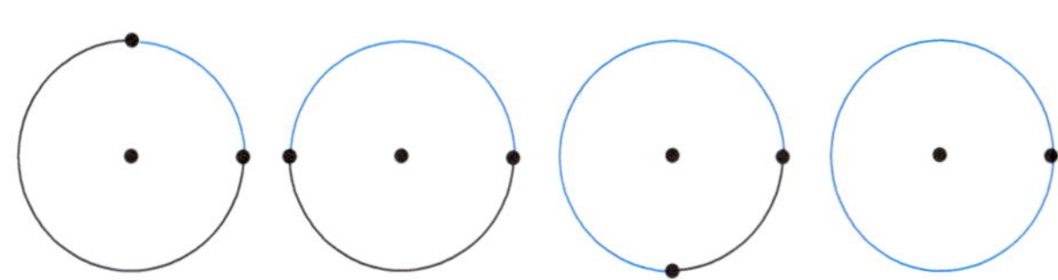

Figure 19-12

Minor Arc 劣弧

n. ['maɪnər ɑrk]

Definition: An arc whose measure is less than 180°. 劣弧是度数小于 180° 的弧。

Notation: To denote a minor arc with endpoints A and B, we write $\overset{\frown}{AB}$, as shown in Figure 19-13. 要标记一个端点为 A 和 B 的劣弧,可写作 $\overset{\frown}{AB}$,如图 19-13 所示。

Note that $\overset{\frown}{AB} \cong \overset{\frown}{BA}$.

Refer to Table A-5 in Appendix A for the notation of minor arcs.

劣弧的标记方式参照附录 A 的表 A-5。

Semicircle 半圆

n. ['semiˌsɜrkəl]

Definition: A 180° arc, as shown in Figure 19-14. 半圆是 180° 的弧,如图 19-14 所示。

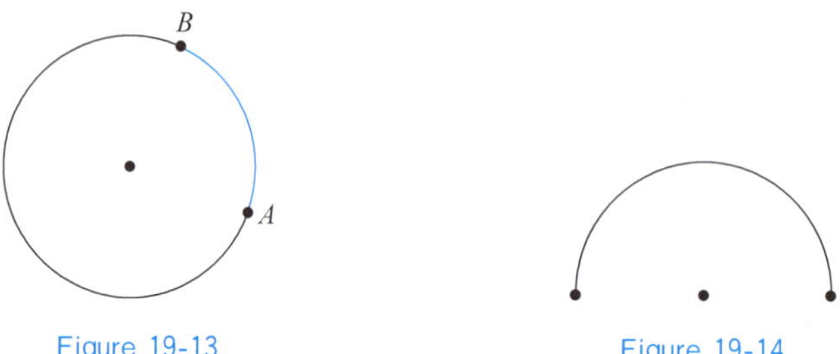

Figure 19-13

Figure 19-14

Notation: A semicircle adopts either the notation for a minor arc (two letters) or major arc (three letters), depending on conventions. 根据惯例,半圆可采用劣弧或优弧的表示方式。劣弧用两个字母表示,优弧用三个字母表示。

Major Arc 优弧

n. ['meɪdʒər ɑrk]

Definition: An arc whose measure is greater than 180°. 优弧是度数大于 180° 的弧。

Notation: To denote a major arc with endpoints A and B, let C be a point on the major arc and between A and B, we write $\overset{\frown}{ACB}$, as shown in Figure 19-15. 一个端点为 A 和 B,且经过 C 的优弧,可写作 $\overset{\frown}{ACB}$,如图 19-15 所示。

Note that $\overset{\frown}{ACB} \cong \overset{\frown}{BCA}$.

Refer to Table A-5 in Appendix A for the notation of major arcs.

优弧的标记方式参照附录 A 的表 A-5。

Congruent Arcs 等弧

n. [kən'gruənt ɑrks]

Definition: Two arcs are congruent if they have the same length and the same measure. 等弧是指两弧的长度相同,度数相同。

Examples: As shown in Figure 19-16, the blue arcs are congruent arcs. 如图 19-16 所示,蓝色的弧为等弧。

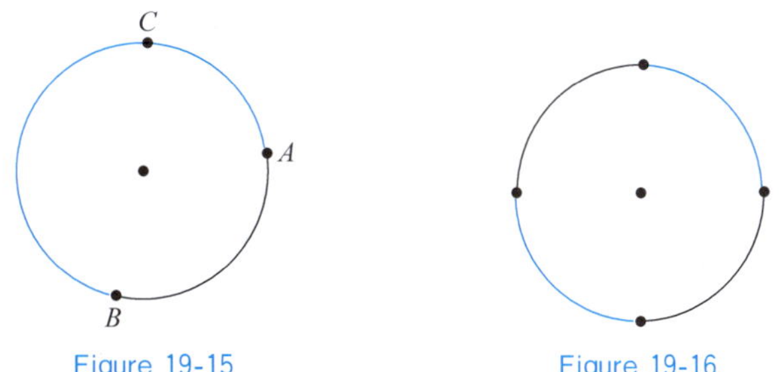

Figure 19-15　　Figure 19-16

19.1.5 Angles 角

Central Angle 圆心角

n. ['sentrəl 'æŋgəl]

Definition：An angle whose vertex is the center of the circle and whose sides are radii. 圆心角是顶点为圆心，两边为半径的角。

Figure 19-17 shows a central angle of a circle with its intercepted arc. 图 19-17 展示了圆的一个圆心角及其截弧。

Properties：The central angle has the same measure as its intercepted arc. 圆心角与它的截弧度数相同。

Inscribed Angle 圆周角

n. [ɪn'skraɪbd 'æŋgəl]

Definition：An angle whose vertex is on the circle and whose sides are chords. 圆周角是顶点在圆周上，两边为弦的角。

Figure 19-18 shows an inscribed angle of a circle with its intercepted arc. 图 19-18 展示了圆的一个圆周角及其截弧。

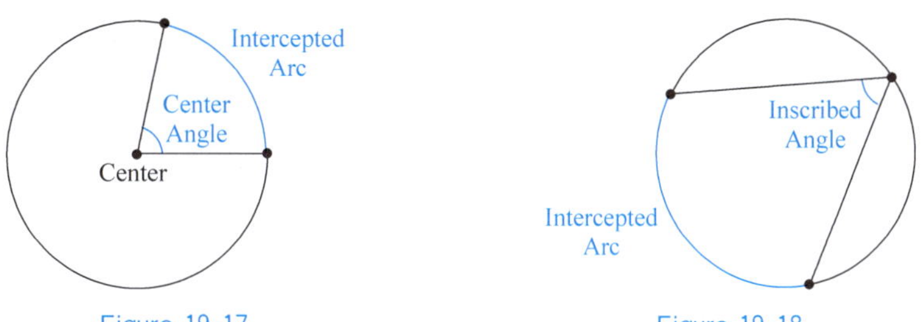

Figure 19-17　　Figure 19-18

Properties：**Inscribed Angle Theorem**：The measure of an inscribed angle is half the measure of the central angle that shares the same intercepted arc. **圆周角定理**：圆周角的度数是与它有同一个截弧的圆心角度数的一半。

Another version of the **Inscribed Angle Theorem**：The measure of an inscribed angle is half the measure of its intercepted arc. **圆周角定理**的另一种表述：圆周角的度数是它的截弧的度数的一半。

These versions are equivalent because of the property in central angles：The central angle has the same measure as its intercepted arc. 上面关于**圆周角定理**的两种表述等价，因为圆心角的性质，即圆心角和它的截弧度数相同。

Proofs: We will prove by casework regarding to the position of the center. For each case, let ψ be the measure of the inscribed angle, and θ be the measure of the central angle that shares the same intercepted arc as the inscribed angle does. For each case, we want to prove that $\theta=2\psi$. 可根据圆心的位置分情况讨论。对于每种情况,设 ψ 为圆周角的度数,θ 为与圆周角有共同截弧的圆心角的度数。对于每种情况,都需证明 $\theta=2\psi$。

The cases are listed in Figure 19-19.

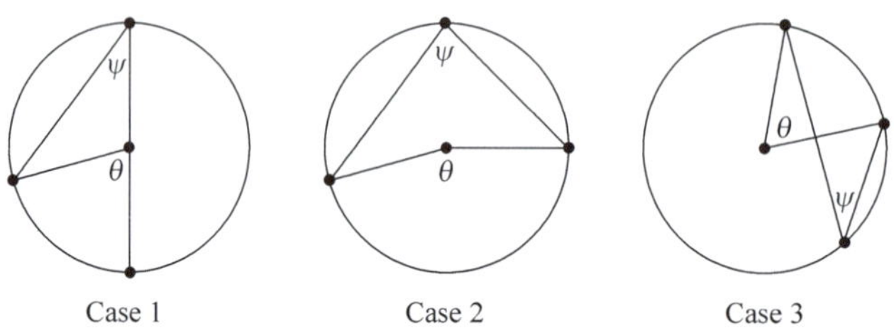

Figure 19-19

Case 1: The center lies on one side of the inscribed angle. 圆心在圆周角的一条边上。

As shown in Figure 19-20, since all radii are congruent, using the **Base Angle Theorem**, we can conclude that $m\angle OAB=m\angle OBA=\psi$. Using the **Exterior Angle Theorem**, we have $\theta=2\psi$. 如图 19-20 所示,因为半径全等,因此,根据**等边对等角定理**可得出 $m\angle OAB=m\angle OBA=\psi$。根据**外角定理**可得出 $\theta=2\psi$。

Case 2: The center is in the interior of the inscribed angle. 圆心在圆周角内部。

As shown in Figure 19-21, let $\psi=\psi_1+\psi_2$, and $\theta=\theta_1+\theta_2$.

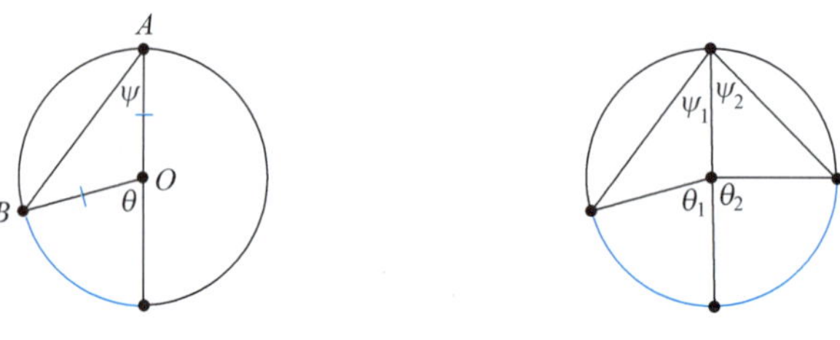

Figure 19-20 Figure 19-21

Use Case 1 twice and get $\theta_1=2\psi_1$ and $\theta_2=2\psi_2$. Therefore, $\theta=\theta_1+\theta_2=2\psi_1+2\psi_2=2(\psi_1+\psi_2)=2\psi$.

Case 3: The center is at the exterior of the inscribed angle. 圆心在圆周角外部。

As shown in Figure 19-22, using Case 1, we know that:

(1) $\theta'=2\psi'$.

(2) $\theta+\theta'=2(\psi+\psi')$.

Expanding the RHS of (2), we get $\theta+\theta'=2\psi+2\psi'$, from (1), we can simplify this to $\theta=2\psi$.

Semicircle 半圆形

n. [ˈsemiˌsɜrkəl]

Definition: A shape that is bounded by an 180°-arc and a line segment. This line segment is known as the diameter. 被 180° 的弧和一条线段围成的图形称为半圆形。这条线段称为直径。Figure 19-23 shows a semicircle.

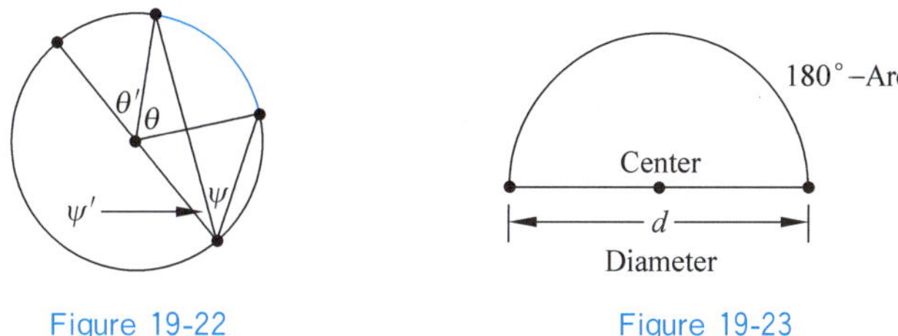

Figure 19-22 Figure 19-23

Notation: Like a circle, a semicircle is named by its center. 与圆相同,半圆是根据中心的符号命名的。

Corollaries of Circles 圆的推论

Corollaries: 1. Inscribed angles of congruent intercepted arcs are congruent, as shown in Figure 19-24. 有相等截弧的圆周角相等,如图 19-24 所示。

2. Any angle inscribed in a semicircle is a right angle, as shown in Figure 19-25. 内接半圆的圆周角是直角,如图 19-25 所示。

3. In a circle, the chord opposite any inscribed right angle is a diameter, as shown in Figure 19-26. 在圆的内接直角中,直角的对边是直径,如图 19-26 所示。

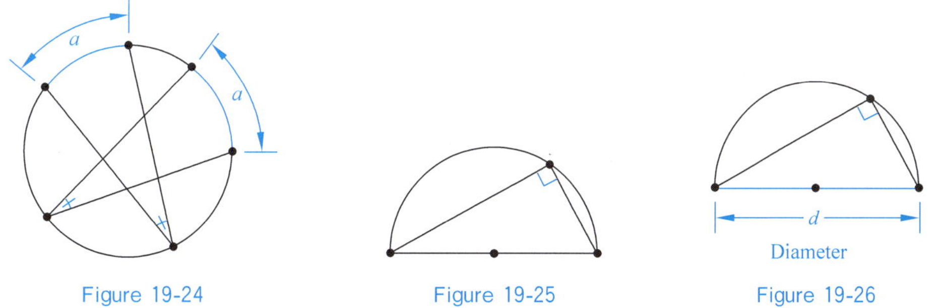

Figure 19-24 Figure 19-25 Figure 19-26

4. The hypotenuse of a right triangle T is a diameter of the circumcircle of T (Property 6 of median follows this). 直角三角形 T 的斜边是 T 的外接圆的直径(根据中线的性质 6)。

5. In a right triangle, the length of the median drawn to the hypotenuse is half of that of the hypotenuse. 在直角三角形中,到斜边的中线的长度是斜边长度的一半。

Proofs: Use the **Inscribed Angle Theorem** to prove each of Corollaries 1-4. Use Corollaries 2 and 3 to prove Corollary 5. 用**圆周角定理**证明圆的推论 1-4。用推论 2 和 3 证明推论 5。

Inscribed Polygon/Cyclic Polygon 圆内接多边形

n. [ɪn'skraɪbd 'pɑlɪˌgɑn]/['sɪklɪk 'pɑlɪˌgɑn]

Definition: An inscribed polygon/cyclic polygon is a polygon such that there exists a circle that contains all vertices of the polygon. 圆内接多边形是顶点均在同一个圆上的多边形。This circle is known as the circumcircle of a polygon. 这个圆叫作多边形的外接圆。

As shown below in Figure 19-27, the gray shape is an inscribed polygon/cyclic polygon. Its circumcircle is drawn. 如图 19-27 所示，灰色的图形是圆内接多边形，外接圆也已画出。

Properties: Opposite angles of a cyclic quadrilateral are supplementary. 圆内接四边形的对角互补。

Proofs: To prove the property, use the **Inscribed Angle Theorem**. 可用**圆周角定理**证明该性质。

Without the loss of generality, let $m\angle A > m\angle C$. As shown in Figure 19-28, note that $m\angle A = \frac{1}{2}m\widehat{BCD}$ and $m\angle C = \frac{1}{2}m\widehat{BD}$. Therefore, $m\angle A + m\angle C = \frac{1}{2}m\widehat{BCD} + \frac{1}{2}m\widehat{BD} = \frac{1}{2}(m\widehat{BCD} + m\widehat{BD}) = \frac{1}{2}(360°) = 180°$.

Similarly, $m\angle B + m\angle D = 180°$.

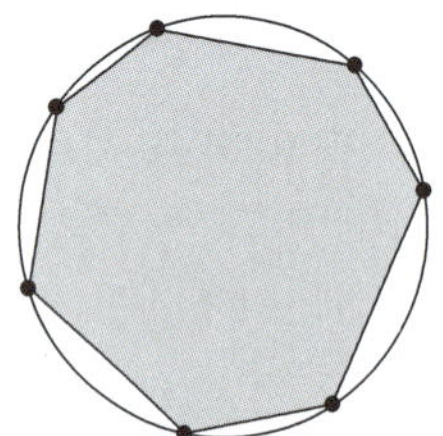
Figure 19-27

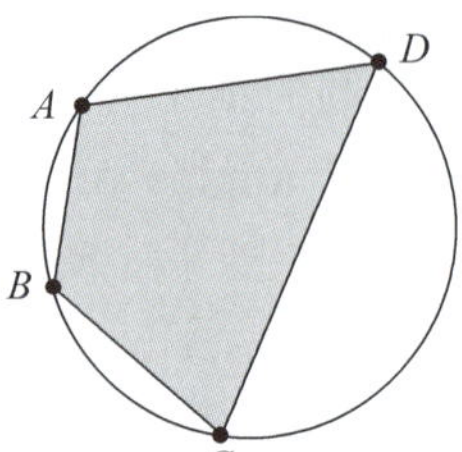

Figure 19-28

Questions: For cyclic quadrilateral $ABCD$, if $m\angle A = 150°$ and $m\widehat{ABC} = 220°$, what are the values of (1) $m\angle B$, (2) $m\angle C$, and (3) $m\angle D$?

Answers: We know that $m\widehat{AC} + m\widehat{ABC} = 360°$. So, $m\widehat{AC} = 140°$.

(1) $m\angle B = \frac{1}{2}m\widehat{AC} = \frac{1}{2}(140°) = 70°$.

(2) $m\angle C = 180° - m\angle A = 180° - 150° = 30°$.

(3) $m\angle D = 180° - m\angle B = 180° - 70° = 110°$.

19.1.6 Summary 总结

1. The location of the center and the length of a radius determine a circle. 圆心的位置和半径的长度决定一个圆。
2. A line segment whose endpoints are on the circle is called a chord. In particularly, diameters are the longest chords and are the only chords that pass through the center of the circle. The length of a diameter is twice that of a radius. 端点都在圆周上的线段叫作弦。直径是最长的弦，也是唯一通过圆心的弦。直径的长度是半径长度的 2 倍。
3. Relationship of two or more circles:
   两个或多个圆有如下关系：
   (1) concentric—share the same center. 同心
   (2) congruent—lengths of radii are the same. 全等
   (3) concurrent—concentric and congruent. 同圆
4. An arc is a part of the circumference, which has a length and a measure. The categories of arcs

according to measures are:
弧是圆周的一部分,有长度和度数。根据度数对弧分类如下:
(1) minor arc—measures less than 180°. 劣弧
(2) semicircle—measures 180°. 半圆
(3) major arc—measures more than 180°. 优弧
Two arcs are congruent if they have the same length and measure. 若长度相等且度数相等,则两弧全等。

5. A central angle is an angle whose vertex is the center of the circle and whose sides are two radii. An inscribed angle is an angle whose vertex is on the circle and whose sides are two chords. 圆心角是顶点为圆心,两边均为半径的角。圆周角是顶点在圆周上,两边均为弦的角。

Inscribed Angle Theorem: The measure of an inscribed angle is half the measure of the central angle that shares the same intercepted arc. **圆周角定理**:圆周角的度数是与它有同一个截弧的圆心角的度数的一半。

Many important results follow the **Inscribed Angle Theorem**. See corollaries of circles and inscribed polygons/cyclic polygons in Section 19.1.5. 很多重要的结论都是根据**圆周角定理**推导的。见19.1.5节圆的推论和圆内接多边形的词条。

19.2 Properties of Circles 圆的性质

19.2.1 Properties 性质

Symmetric Property 对称性质

n. [sɪ'mɛtrɪk]

Properties: 1. (1) A circle has line symmetry. 圆有轴对称性。

(2) A circle has infinitely many lines of symmetry. 圆有无数条对称轴。

Any line containing its center is a line of symmetry. Figure 19-29 shows some lines of symmetry in blue. 经过圆心的直线均为对称轴。图19-29用蓝线展示了若干条对称轴。

2. A circle has rotational symmetry around its center. The angle of rotation (see rotational symmetry in Section 12.4.3 for its definition) is infinitesimally small. 沿着圆心,圆有旋转对称性。旋转角度无限小。旋转角度的定义见12.4.3节的旋转对称。

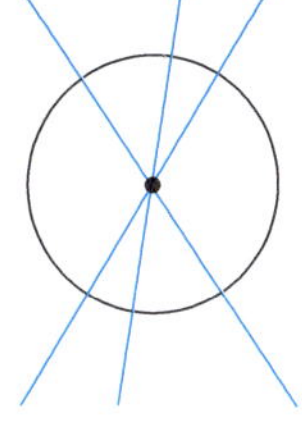
Figure 19-29

Theorem of Chord, Arc, Central Angles 弦、弧、圆心角定理

Theorem: In congruent circles, if two central angles are congruent, then
在等圆中,若两圆心角相等,则
(1) the arcs opposite the central angles are congruent. 圆心角相对的弧相等。
(2) the chords opposite the central angles are congruent. 圆心角相对的弦相等。
As shown in Figure 19-30, we can conclude that the blue line segments are congruent, and the blue arcs are congruent. 如图19-30所示,能得出蓝色线段相等,蓝色弧相等。

Proofs: Theorem (1) is by the property of the central angles. 定理(1)根据圆心角的性质可得出。
To show Theorem (2), as shown in Figure 19-31, since all radii in a circle are congruent,

we can show the two shaded triangles are congruent by SAS. Then by CPCTC we know that the chords (shown in blue) opposite the central angles are congruent.
要证明定理(2),如图 19-31 所示。因为圆的半径相等,根据边角边全等公理,可知两个灰色的三角形全等。根据全等三角形的对应部分全等,相对的弦(蓝色)相等。

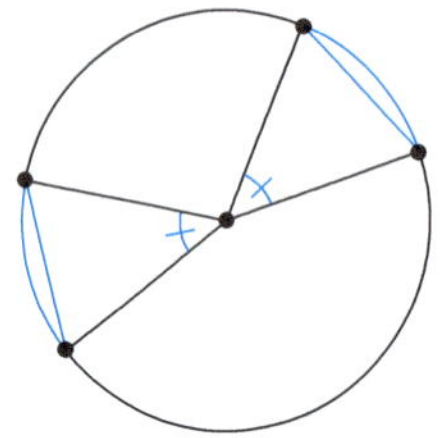
Figure 19-30

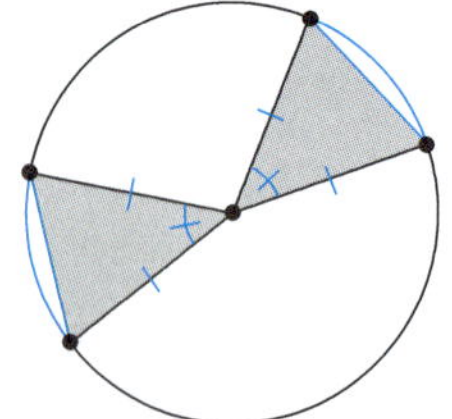
Figure 19-31

Corollaries of Chord, Arc, Central Angles 弦、弧、圆心角的推论

Corollaries: 1. In concurrent circles or congruent circles, if two arcs are congruent, then 在同圆或等圆中,若两弧相等,则

(1) the central angles opposite the arcs are congruent. 对应弧的圆心角相等。

(2) the chords opposite the arcs are congruent. 对应弧的弦相等。

2. In concurrent circles or congruent circles, if two chords are congruent, then 在同圆或等圆中,若两弦相等,则

(1) the central angles opposite the chords are congruent. 相对弦的圆心角相等。

(2) the major arcs opposite the chords are congruent. 相对弦的优弧相等。

(3) the minor arcs opposite the chords are congruent. 相对弦的劣弧相等。

Figure 19-32 shows the relationship between two congruent central angles, their intercepted arcs, and the opposite chords. 图 19-32 展示了两个相等圆心角、截弧与对应弦的关系。

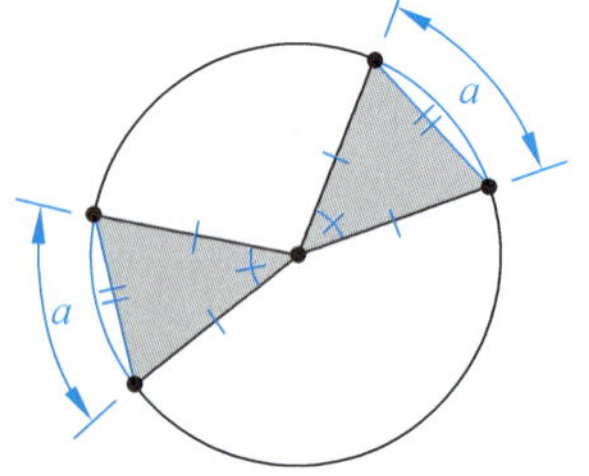

Figure 19-32

Proofs: Corollary 1(1) is based on definition, and Corollary 1(2) is based on theorem of chord, arc, central angles. We use SAS to show that the triangles are congruent. 推论 1(1)根据定义可得出;推论 1(2)根据弦、弧、圆心角的定理可得出,用边角边定理证出三角形全等。

Corollary 2 is based on theorem of chord, arc, central angles and definitions. We use SSS to show that the triangles are congruent. 推论 2 根据弦、弧、圆心角的定理及其定义可得出,用边边边全等公理证出三角形全等。

Perpendicular Chord Theorem 垂径定理

Definition: As shown in Figure 19-33, if a diameter is perpendicular to a chord, then it bisects the chord and its opposite arc. 如图 19-33 所示,若直径与弦垂直,则直径平分弦和弦的对弧。

Proofs: As shown in Figure 19-34, we wish to show that $\overline{EA} \cong \overline{EB}$, $\overset{\frown}{CA} \cong \overset{\frown}{CB}$, and $\overset{\frown}{DA} \cong \overset{\frown}{DB}$.

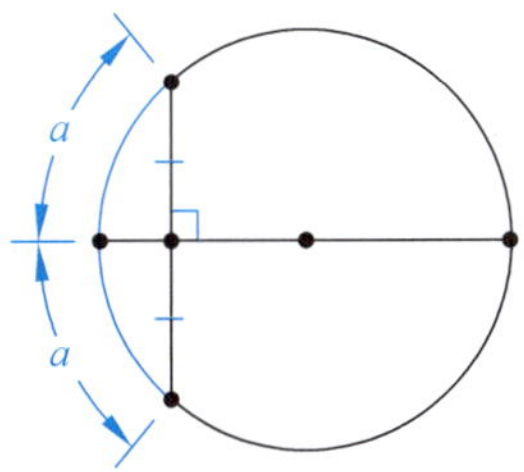

Figure 19-33

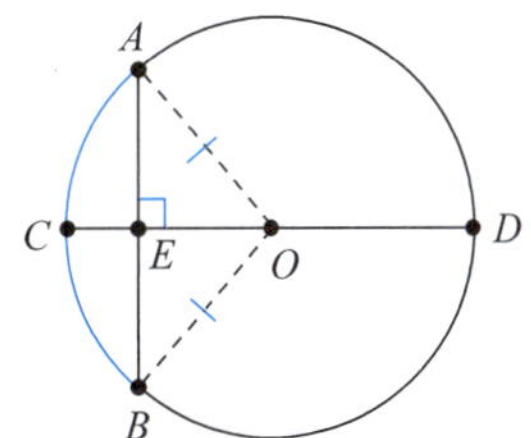

Figure 19-34

In $\odot O$, let $\overline{AB}$ be a chord and $\overline{CD}$ be a diameter such that $\overline{AB} \perp \overline{CD}$, and $\overline{AB}$ and $\overline{CD}$ intersect at E. By given, $\angle OEA \cong \angle OEB$. Since all radii are congruent, $\overline{OA} \cong \overline{OB}$. By reflexive, $\overline{OE} \cong \overline{OE}$. Therefore, $\triangle OEA \cong \triangle OEB$ by HL, from which we can conclude that $\overline{EA} \cong \overline{EB}$ by CPCTC. We have shown that the diameter bisects the chord.
在$\odot O$,设 $\overline{AB}$ 为弦,$\overline{CD}$ 为直径,满足 $\overline{AB} \perp \overline{CD}$,且 $\overline{AB}$ 与 $\overline{CD}$ 交于 E。根据已知条件,$\angle OEA \cong \angle OEB$。因为所有半径相等,所以 $\overline{OA} \cong \overline{OB}$。根据反身性,$\overline{OE} \cong \overline{OE}$。所以,根据 HL 定理给出$\triangle OEA \cong \triangle OEB$。因为全等三角形的对应部分全等,所以 $\overline{EA} \cong \overline{EB}$,可证出直径平分弦。
To show that the diameter bisects the arcs, note that $\angle AOE \cong \angle BOE$ by CPCTC, from which the opposite arcs are congruent: $\overset{\frown}{CA} \cong \overset{\frown}{CB}$. This implies $m\overset{\frown}{CA} = m\overset{\frown}{CB}$ so $180° - m\overset{\frown}{CA} = 180° - m\overset{\frown}{CB}$, from which $m\overset{\frown}{DA} = m\overset{\frown}{DB}$, so $\overset{\frown}{DA} \cong \overset{\frown}{DB}$. 要证出直径平分弧,因为全等三角形的对应部分全等,所以$\angle AOE \cong \angle BOE$,从而可得出 $\overset{\frown}{CA} \cong \overset{\frown}{CB}$。也就是说 $m\overset{\frown}{CA} = m\overset{\frown}{CB}$。所以 $180° - m\overset{\frown}{CA} = 180° - m\overset{\frown}{CB}$,从而 $m\overset{\frown}{DA} = m\overset{\frown}{DB}$,使得 $\overset{\frown}{DA} \cong \overset{\frown}{DB}$。

Corollary of the Perpendicular Chord Theorem　垂径定理的推论

Corollary: As shown in Figure 19-35, if a diameter bisects a chord, then it is perpendicular to the chord and bisects the opposite arc. 如图 19-35 所示,若直径平分一条弦,则它与弦垂直,并平分弦的对弧。

Proofs: As shown in Figure 19-36, we wish to show that $\overline{AB} \perp \overline{CD}$. 如图 19-36 所示,须证出 $\overline{AB} \perp \overline{CD}$。

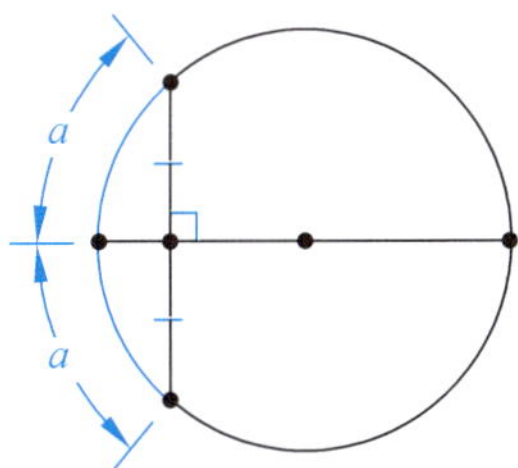

Figure 19-35

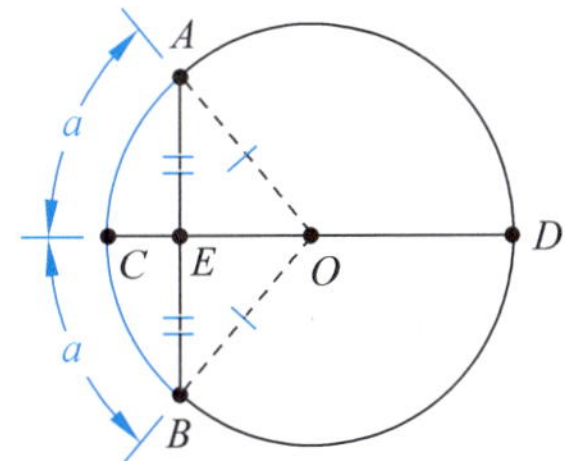

Figure 19-36

In $\odot O$, let $\overline{AB}$ be a chord and $\overline{CD}$ be a diameter such that $\overline{CD}$ bisects $\overline{AB}$, and $\overline{AB}$ and $\overline{CD}$ intersect at E. We are given that $\overline{EA} \cong \overline{EB}$. Since all radii are congruent, $\overline{OA} \cong \overline{OB}$. By reflexive, $\overline{OE} \cong \overline{OE}$. Therefore, $\triangle OEA \cong \triangle OEB$ by SSS. Then by CPCTC, we know $\angle OEA \cong \angle OEB$, from which $m\angle OEA = m\angle OEB$. Since $\angle OEA$ and $\angle OEB$ form a linear pair, we have $m\angle OEA + m\angle OEB = 180°$, from which $m\angle OEA = m\angle OEB = 90°$. Therefore, we have $\overline{AB} \perp \overline{CD}$.

在⊙O 里,设 $\overline{AB}$ 为弦,$\overline{CD}$ 为直径,使得 $\overline{CD}$ 平分 $\overline{AB}$,且 $\overline{AB}$ 与 $\overline{CD}$交于 E。已知 $\overline{EA} \cong \overline{EB}$。因为所有半径相等,所以 $\overline{OA} \cong \overline{OB}$。根据反身性,$\overline{OE} \cong \overline{OE}$。因此,根据边边边全等公理,$\triangle OEA \cong \triangle OEB$。根据全等三角形的对应部分全等,可知$\angle OEA \cong \angle OEB$,从而得 $m\angle OEA = m\angle OEB$。根据$\angle OEA$ 和$\angle OEB$ 是邻补角,可知 $m\angle OEA + m\angle OEB = 180°$。因此 $m\angle OEA = m\angle OEB = 90°$,可证明 $\overline{AB} \perp \overline{CD}$。

Use the second part of the proof in Perpendicular Chord Theorem to show that the diameter bisects the chord's opposite arcs. 用垂径定理可证明直径平分弦的对弧。

19.2.2 Summary 总结

1. (1) A circle has line symmetry. There are infinitely lines of symmetry. Any line passing through the center is a line of symmetry. 圆是轴对称图形,有无数条对称轴。任何一条经过圆心的直线均为对称轴。
   (2) A circle has rotational symmetry about its center. The angle of rotation is infinitesimally small. 圆是沿圆心旋转的对称图形。旋转角度无穷小。
2. Of following statements, one implies all the others. Refer to Figure 19-37. 下列命题中,由一个命题可知所有其他的命题,如图 19-37 所示。
   (1) The blue central angles are congruent. 蓝色的圆心角相等。
   (2) The blue arcs are congruent. 蓝色的弧相等。
   (3) The blue chords are congruent. 蓝色的弦相等。
3. Of following statements, one imply all the others. Refer to Figure 19-38. 下列命题中,由一个命题可知所有其他的命题,如图 19-38 所示。
   (1) The diameter is perpendicular to the chord. 直径和弦垂直。
   (2) The diameter bisects the chord. 直径平分弦。
   (3) The diameter bisects the blue arcs. 直径平分蓝色的弧。

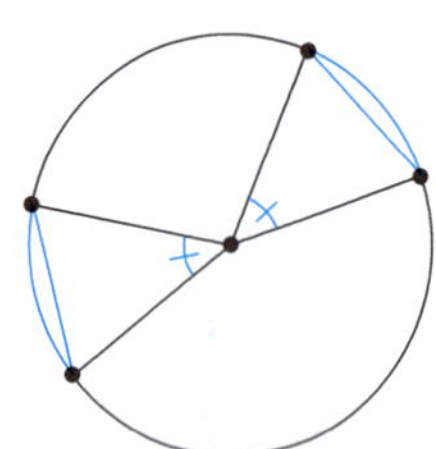

Figure 19-37

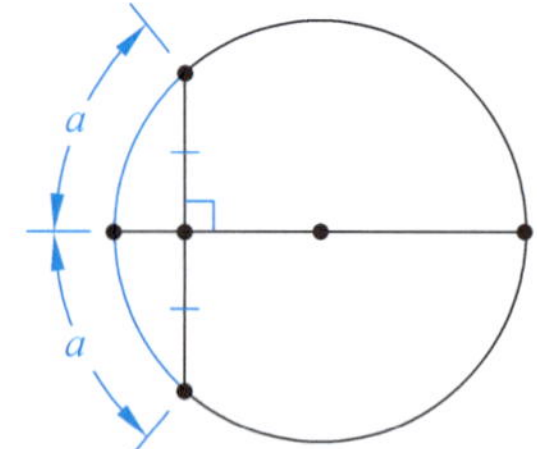

Figure 19-38

19.3 The Positional Relationship between a Point and a Circle 点与圆的位置关系

Positional Relationship between a Point and a Circle 点与圆的位置关系

Properties: 1. Suppose the length of a radius of ⊙O is r, and P is a point so that $OP = d$. 设⊙O 半径的长度为 r,P 是一点,满足 $OP = d$。

(1) P is outside of $\odot O$ if and only if $d>r$. P 在$\odot O$外当且仅当$d>r$。

(2) P is on $\odot O$ if and only if $d=r$. P 在$\odot O$的圆周上当且仅当$d=r$。

(3) P is inside $\odot O$ if and only if $d<r$. P 在$\odot O$内当且仅当$d<r$。

This is based on the definition of circles. 这些性质根据圆的定义得出。

These cases are shown in Figure 19-39.

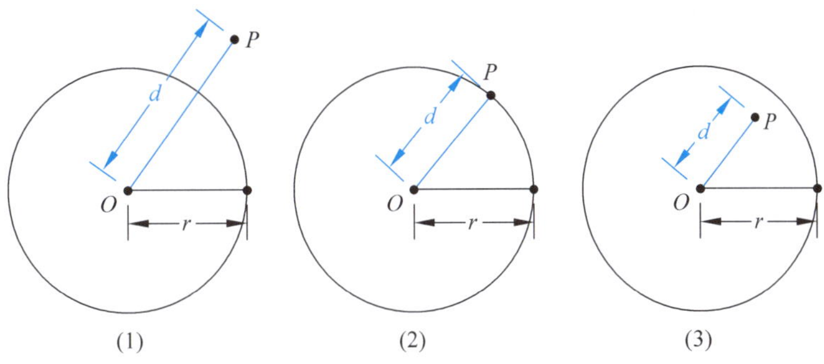

Figure 19-39

2. Three noncolinear points (A, B, and C) determine a circle, which is the circumcircle of $\triangle ABC$. 三个非共线点(A、B、C)能确定一个圆，也就是$\triangle ABC$的外接圆。

Questions: The center of $\odot O$ is at $(0,0)$, and the length of radius of $\odot O$ is 2. Determine the positional relationship for each of the following points and $\odot O$. 若$\odot O$的圆心在$(0,0)$，半径为2。判断下列各点和$\odot O$的位置关系。

(1) $A=(1,2/3)$

(2) $B=(\sqrt{3},1)$

(3) $C=(1.7,1.8)$

Answers: Using the Distance Formula:

(1) $OA=\sqrt{(1-0)^2+\left(\frac{2}{3}-0\right)^2}=\sqrt{\frac{13}{4}}<2$. Therefore, A is inside $\odot O$.

(2) $OB=\sqrt{(\sqrt{3}-0)^2+(1-0)^2}=\sqrt{4}=2$. Therefore, B is on $\odot O$.

(3) $OC=\sqrt{(1.7-0)^2+(1.8-0)^2}=\sqrt{6.13}>2$. Therefore, C is outside $\odot O$.

19.4 The Positional Relationship between a Line and a Circle 直线与圆的位置关系

视频 32

19.4.1 Introduction 介绍

Secant 割线

n. ['sikænt]

Definition: A line that intersects the circle at two points, as shown in Figure 19-40. 割线是指与圆有两个交点的直线，如图 19-40 所示。

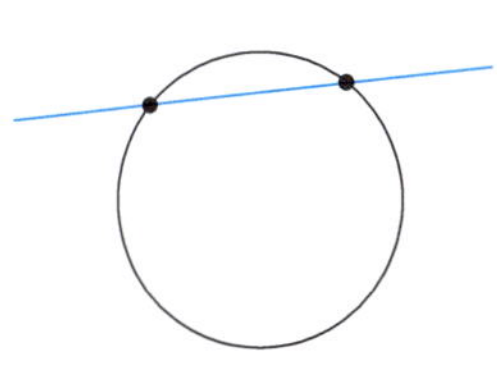

Figure 19-40

Tangent 切线、相切

n. /adj. ['tændʒənt]

Definition: (n.) A tangent of a circle is a line that intersects the circle at exactly one point. This point is known as the point of tangency. Figure 19-41 shows a tangent of a circle and the point of tangency. 切线是与圆只有一个交点的直线,交点称为切点。图 19-41 展示了圆的一条切线与切点。

(adj.) If a line/ray/line segment intersects a circle at only one point P, we say that the line/ray/line segment is tangent to the circle. P is known as the point of tangency. 若一条直线/射线/线段与圆只交于点 P,则这条直线/射线/线段与圆相切。P 是切点。

Properties: 1. **Tangent Theorem**: If a line l is a tangent of $\odot O$, then l is perpendicular to the radius of $\odot O$ that contains the point of tangency. Note that the point of tangency is an endpoint of that radius. **切线定理**:若直线 l 是$\odot O$ 的切线,则 l 垂直于$\odot O$ 中经过切点的半径。注意切点是半径的一个端点。

Figure 19-42 shows the relationship between the radius and the tangent. 图 19-42 展示了半径与切线的关系。

2. **Converse of Tangent Theorem**: If a line l intersects a radius' endpoint on $\odot O$ and is perpendicular to the radius, then l is a tangent of $\odot O$. **切线定理的逆定理**:若直线 l 与$\odot O$ 的一条半径的端点相交并与其垂直,则 l 是$\odot O$ 的切线。

3. If two line segments sharing one common endpoint are tangent to a circle, then the line segments are congruent, as shown in Figure 19-43. 若两条有一个共同端点的线段均和圆相切,则这两条线段等长,如图 19-43 所示。

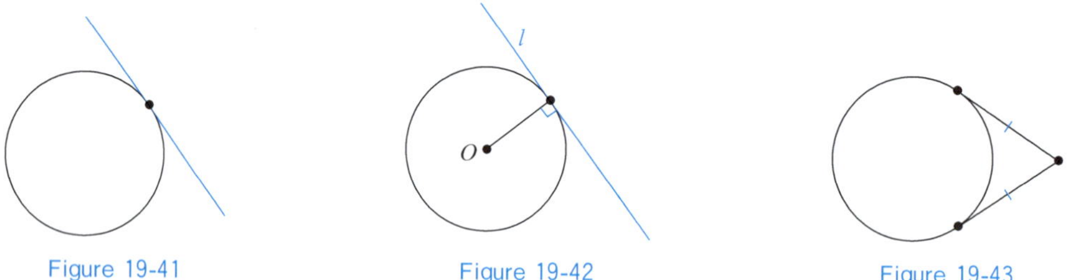

Figure 19-41 Figure 19-42 Figure 19-43

4. From any point A outside of $\odot O$, two line segments tangent to $\odot O$ can be drawn. If the points of tangency are B and C, then $\overline{OA}$ bisects $\angle BAC$, as shown in Figure 19-44. 从$\odot O$ 外的任意点 A 可画出两条与$\odot O$ 相切的线段。若切点为 B 和 C,则 $\overline{OA}$ 平分$\angle BAC$,如图 19-44 所示。

Proofs: 1. We will prove Property 1 by contradiction. 可用反证法证明性质 1。

As shown in Figure 19-45, suppose P is the point of tangency, and $\overline{OP}$ is not perpendicular to l. 如图 19-45 所示,设 P 为切点,且 $\overline{OP}$ 不与 l 垂直。

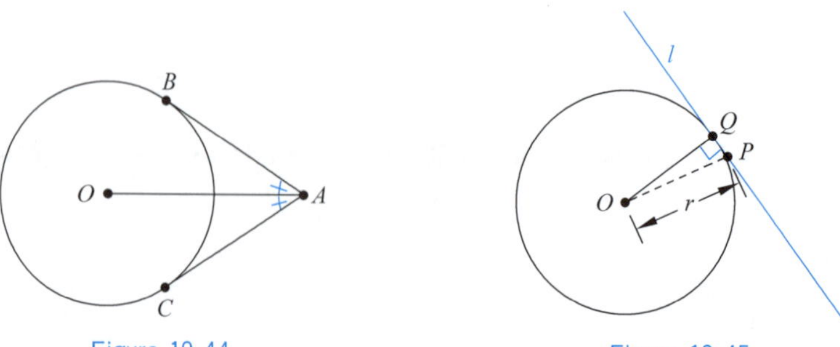

Figure 19-44 Figure 19-45

By the **Perpendicular Postulate**, there is a point Q on l such that $\overline{OQ}\perp l$. Clearly, P and Q are different points. By the positional relationship between a point and a circle, $OQ>r$ since l only intersects $\odot O$ at P, and Q is outside of $\odot O$. 根据垂线公理，在 l 上存在点 Q，满足 $\overline{OQ}\perp l$。显然，P 与 Q 不重合。因为点和圆的位置关系，$OQ>r$（原因是 l 只与$\odot O$ 交于 P，Q 在$\odot O$ 外）。

In $\triangle OPQ$, $\angle Q$ is a right angle. Therefore, $\overline{OP}$ is the hypotenuse. The length of the hypotenuse is r, but the length of one leg is $OQ>r$. Since a leg is always shorter than the hypotenuse in a right triangle, we arrive at a contradiction. 在$\triangle OPQ$ 里，$\angle Q$ 为直角。所以 $\overline{OP}$ 为斜边。斜边的长度是 r，但一条直角边的长度是 $OQ>r$。因为在直角三角形里，直角边必比斜边短，所以得出矛盾。

Therefore, the tangent is perpendicular to the radius that contains the point of tangency. 所以，切线与经切点的半径垂直。

2. We will prove Property 2 by contradiction. We know that l intersects with radius $\overline{OP}$ on $\odot O$ at P and $\overline{OP}\perp l$. We wish to prove l is a tangent of $\odot O$. 可用反证法证明性质 2。已知 l 与$\odot O$ 的半径 $\overline{OP}$ 交于 P，且 $\overline{OP}\perp l$。要证明 l 是$\odot O$ 的切线。

   As shown in Figure 19-46, suppose l intersects $\odot O$ at another point, Q, for which P and Q are different points. By the **Perpendicular Postulate**, we know that $\overline{OQ}$ is not perpendicular to l. In $\triangle OPQ$, $\angle P$ is a right angle. Therefore, $\overline{OQ}$ is the hypotenuse. Since $OQ>OP=r$, Q is outside of $\odot O$, contradicting the assumption that Q is on $\odot O$. Therefore, l intersects $\odot O$ at only one point. By definition, l is a tangent of $\odot O$. 如图 19-46 所示，设 l 与$\odot O$ 交于点 Q，使得 P 和 Q 不重合。根据**垂直公理**，$\overline{OQ}$ 与 l 不垂直。在$\triangle OPQ$ 里，$\angle P$ 是直角。所以，$\overline{OQ}$ 是直角边。因为 $OQ>OP=r$，Q 在$\odot O$ 外面，与先前的假设（Q 在圆周上）相矛盾。所以，l 只与$\odot O$ 交于一点。

3. We will use Property 1 to prove Property 3. As shown in Figure 19-47, we want to prove that $\overline{AB}\cong\overline{AC}$. 可用性质 1 证明性质 3。如图 19-47 所示，想证明 $\overline{AB}\cong\overline{AC}$。

   By reflexive, we have $\overline{OA}\cong\overline{OA}$. Since all radii are congruent, we have $\overline{OB}\cong\overline{OC}$. By Property 1, we have $\angle ABO\cong\angle ACO$. We can conclude that $\triangle ABO\cong\triangle ACO$ by HL. Using CPCTC, we get $\overline{AB}\cong\overline{AC}$.

   根据反身性，即 $\overline{OA}\cong\overline{OA}$。因为所有直径相等，所以 $\overline{OB}\cong\overline{OC}$。根据性质 1，可得到 $\angle ABO\cong\angle ACO$。根据 HL 定理可得出$\triangle ABO\cong\triangle ACO$。因为全等三角形的对应部分全等，所以 $\overline{AB}\cong\overline{AC}$。

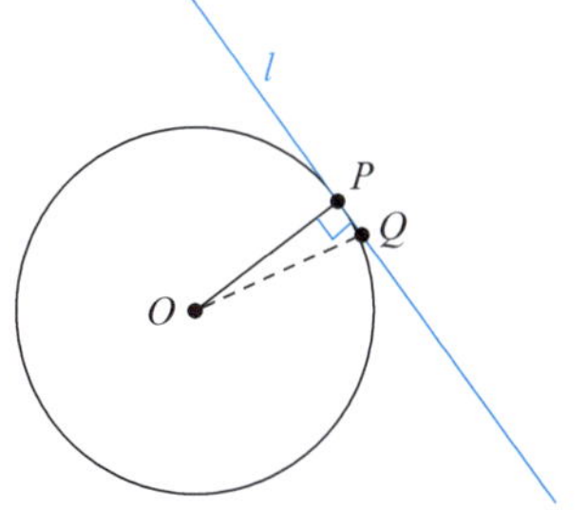

Figure 19-46

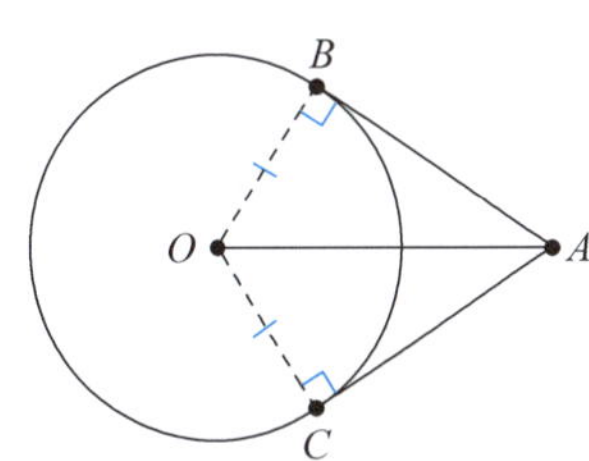

Figure 19-47

4. To prove Property 4, use the result from Property 3. Once we proved $\triangle ABO \cong \triangle ACO$, by CPCTC, we get $\angle BAO \cong \angle CAO$, from which it is true that $\overline{OA}$ bisects $\angle BAC$. 可用性质3证明性质4。证明了$\triangle ABO \cong \triangle ACO$以后,因为全等三角形的对应部分全等,所以$\angle BAO \cong \angle CAO$。从而可得$\overline{OA}$平分$\angle BAC$。

Questions: 1. As shown in Figure 19-48, what is the length of a radius of this circle?

2. As shown in Figure 19-49, what is the value of x?

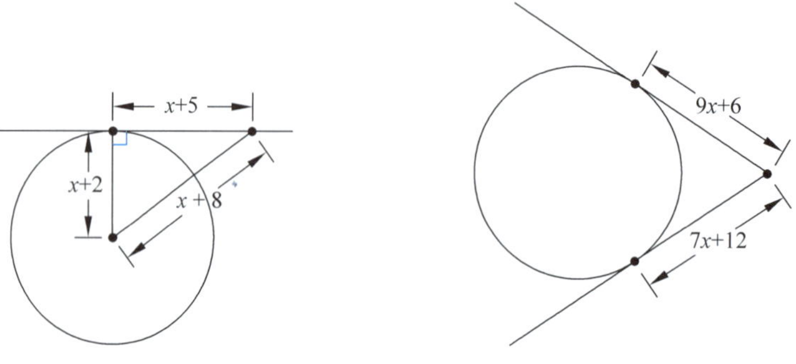

Figure 19-48　　Figure 19-49

Answers: (1) We have $(x+2)^2+(x+5)^2=(x+8)^2$, from which $2x^2+14x+29=x^2+16x+64$. Moving everything to the LHS, we get $x^2-2x-35=0$. Factoring, we have $(x-7)(x+5)=0$, and only $x=7$ is a reasonable solution. Therefore, the length of a radius is $x+2=9$.

(2) Setting the lengths of the line segments equal, we have $9x+6=7x+12$, from which $2x=6$, and $x=3$.

Positional Relationship between a Line and a Circle 直线与圆的位置关系

Properties: Suppose the length of a radius of $\odot O$ is r, and P is a point on line l such that the distance between O and l is $OP=d$. 设$\odot O$的半径长度为r,且P是直线l的一点,使得O和l的距离为$OP=d$。

(1) l does not intersect $\odot O$ if and only if $d>r$.
l不和$\odot O$相交当且仅当$d>r$。

(2) l is a tangent of $\odot O$ if and only if $d=r$.
l是$\odot O$的切线当且仅当$d=r$。

(3) l is a secant of $\odot O$ if and only if $d<r$.
l是$\odot O$的割线当且仅当$d<r$。

This is based on the positional relationship between a point and a circle, which is based on the definition of circles. 这些性质就是根据点和圆的位置关系得出的。点和圆的位置关系又是根据圆的定义得出的。

These cases are shown in Figure 19-50.

19.4.2 Summary 总结

1. A secant s of $\odot O$ intersects $\odot O$ at two points. The distance between O and s is less than the

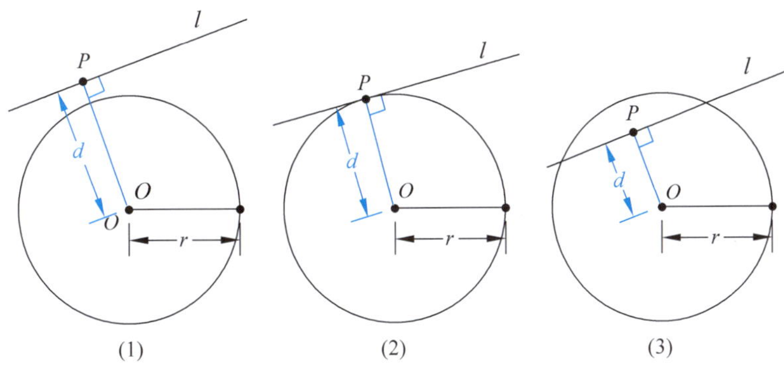

Figure 19-50

length of the ⊙O's radius.　割线 s 与⊙O 交于两点。O 和 s 的距离小于⊙O 的半径长度。

2. (1) A tangent t of ⊙O intersects ⊙O at one point P. The distance between O and t is equal to the length of the ⊙O's radius. In addition, t is the only line passing P and perpendicular to radius $\overline{OP}$.　切线 t 与⊙O 交于一点。O 和 t 的距离等于⊙O 的半径长度，且 t 是唯一一条过 P 且垂直于半径 $\overline{OP}$ 的直线。

(2) Suppose point P is on ⊙O and line l passes through P. The following statements are equivalent.　设 P 是⊙O 圆周上的一点。直线 l 经过 P。以下命题等价。

① l is a tangent of ⊙O.　l 是⊙O 的切线。

② $\overline{OP} \perp l$.

③ l intersects ⊙O at exactly one point, namely P.　l 只与⊙O 交于一点，P。

④ The distance between O and l is equal to r, which is the length of the radius of ⊙O.　O 到 l 的距离为 r，即⊙O 的半径长度。

(3) Properties 3 and 4 of tangents in Section 19.4.1.　见 19.4.1 节切线的性质 3 和 4。

19.5　Theorems of Circles　圆的定理

视频 33

Theorems of Circles　圆的定理

Theorems: 1. **Intersecting Chords Theorem**: If $\overline{AB}$ and $\overline{CD}$ are two chords of ⊙O, intersecting at point P, then $(AP)(PB)=(CP)(PD)$, as shown in Figure 19-51.

相交弦定理：若 $\overline{AB}$ 和 $\overline{CD}$ 是⊙O 的两条弦，交于点 P，则 $(AP)(PB)=(CP)(PD)$，见图 19-51。

2. **Chord Distance to Center Theorem**: If $\overline{AB}$ and $\overline{CD}$ are two congruent chords of ⊙O, then the distance from O to $\overline{AB}$ is the same as the distance from O to $\overline{CD}$, as shown in Figure 19-52.

弦心距定理：若 $\overline{AB}$ 和 $\overline{CD}$ 是⊙O 里两条等长的弦，则 O 到 $\overline{AB}$ 与 O 到 $\overline{CD}$ 的距离相等，见图 19-52。

3. **Parallel Lines Intercepted Arcs Theorem**: As shown in Figure 19-53, if $\overline{AB}$ and $\overline{CD}$ are two parallel chords of ⊙O, then $\overset{\frown}{AC} \cong \overset{\frown}{BD}$.

平行线截弧定理：如图 19-53 所示，若 $\overline{AB}$ 和 $\overline{CD}$ 是在⊙O 中两条平行的弦，则 $\overset{\frown}{AC} \cong \overset{\frown}{BD}$。

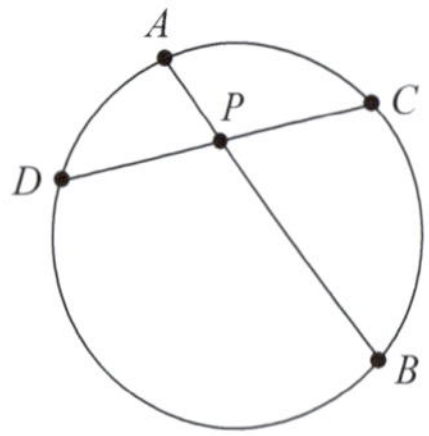

Figure 19-51

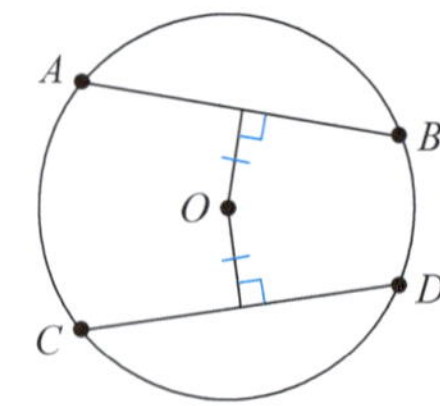

Figure 19-52

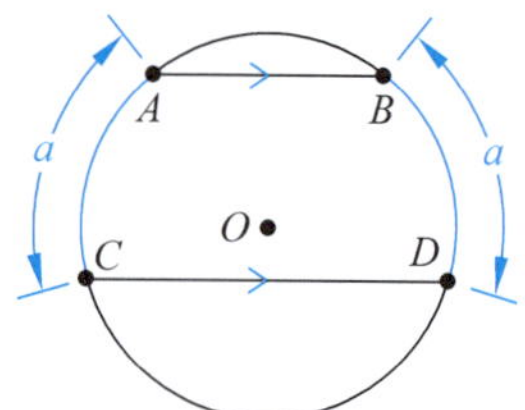

Figure 19-53

4. As shown in Figure 19-54, if $\overline{AB}$ and $\overline{CD}$ are two chords of $\odot O$ intersecting at point P, then $m\angle 1 = \frac{1}{2}(m\overset{\frown}{AC} + m\overset{\frown}{BD})$.

如图 19-54 所示,若 $\overline{AB}$ 和 $\overline{CD}$ 是$\odot O$ 的两条弦,交于点 P,则 $m\angle 1 = \frac{1}{2}(m\overset{\frown}{AC} + m\overset{\frown}{BD})$。

5. **Tangent-Chord Angle Theorem**: As shown in Figure 19-55, if line l is tangent to $\odot O$ at point P and $\overline{PA}$ is a chord of $\odot O$, then each of $m\angle 1$ and $m\angle 2$ is half the measure of its intercepted arc.

弦切角定理:如图 19-55 所示,若直线 l 与$\odot O$ 切于点 P,且 $\overline{PA}$ 是$\odot O$ 的弦,则 $m\angle 1$ 与 $m\angle 2$ 分别是它的截弧度数的一半。

6. As shown in Figure 19-56, if one secant intersects $\odot O$ at points A and B, and the other intersects $\odot O$ at points D and E, and the two secants intersect at point C (outside of $\odot O$), then $m\angle C = \frac{1}{2}(m\overset{\frown}{AD} - m\overset{\frown}{BE})$.

如图 15-56 所示,若一条割线与$\odot O$ 交于点 A 和 B,另外一条与$\odot O$ 交于点 D 和 E,且两条割线交于点 C(在$\odot O$ 外面),则 $m\angle C = \frac{1}{2}(m\overset{\frown}{AD} - m\overset{\frown}{BE})$。

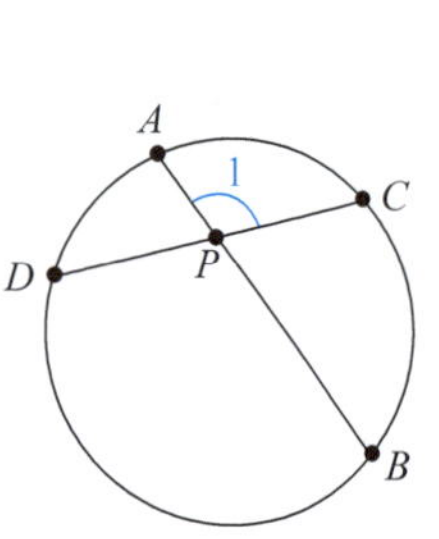

Figure 19-54

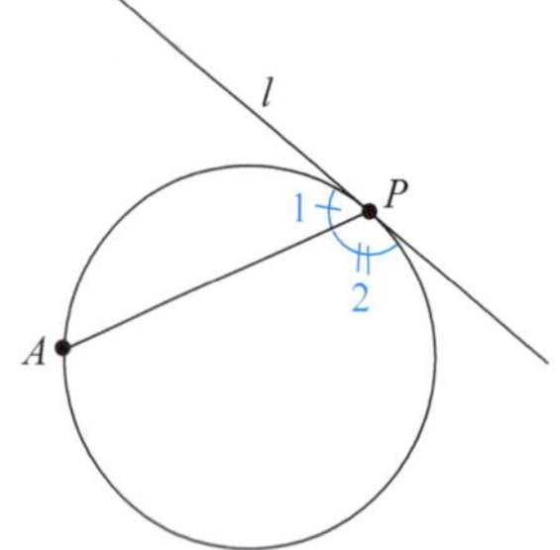

Figure 19-55

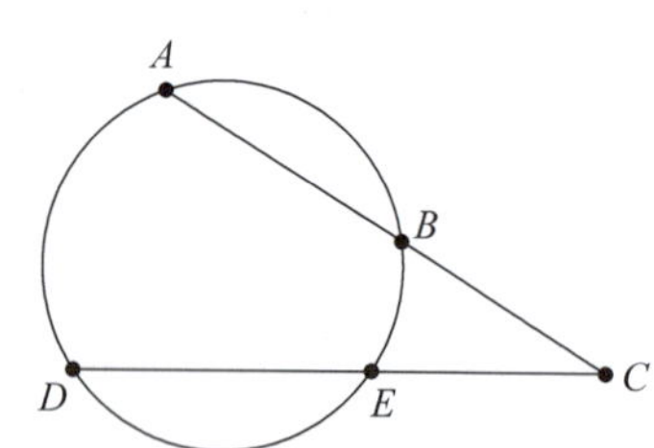

Figure 19-56

7. As shown in Figure 19-57, if one secant intersects $\odot O$ at points H and K, and one tangent intersects $\odot O$ at point J, and the tangent and secant intersect at point L (outside of $\odot O$), then $m\angle L = \frac{1}{2}(m\overset{\frown}{HJ} - m\overset{\frown}{JK})$.

如图 19-57 所示,若一条割线与$\odot O$ 交于点 H 和 K,一条切线与$\odot O$ 交于点 J,且切线与割线交于点 L(在$\odot O$ 外面),

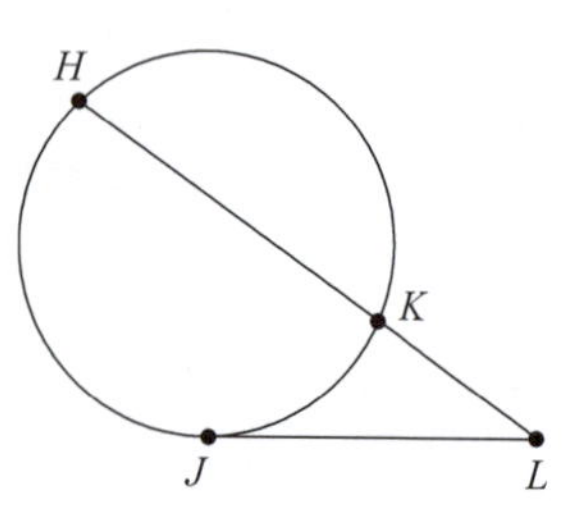

Figure 19-57

则 $m\angle L = \frac{1}{2}(m\overset{\frown}{HJ} - m\overset{\frown}{JK})$。

8. As shown in Figure 19-58, if one tangent intersects $\odot O$ at point B, and the other intersects $\odot O$ at point C, and the tangents intersect at point A, then $m\angle A = \frac{1}{2}(m\overset{\frown}{BDC} - m\overset{\frown}{BC})$.

   如图 19-58 所示，若一条切线交$\odot O$ 于 B，另一条交$\odot O$ 于 C，且两条切线交于点 A，则 $m\angle A = \frac{1}{2}(m\overset{\frown}{BDC} - m\overset{\frown}{BC})$。

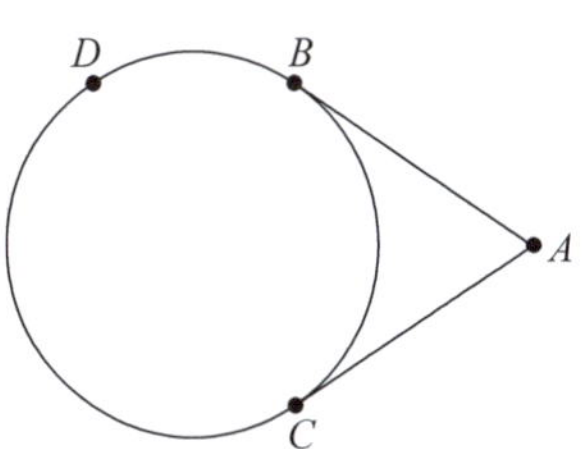

Figure 19-58

Proofs: 1. Proof of the **Intersecting Chords Theorem**:

相交弦定理的证明：

As shown in Figure 19-59, we have $\angle APD \cong \angle BPC$ by vertical angles. Using the **Inscribed Angle Theorem**, we get that $m\angle BAD = \frac{1}{2}m\overset{\frown}{BD} = m\angle BCD$, from which we can conclude that $m\angle BAD = m\angle BCD$ by transitive. From here, we have $\triangle APD \sim \triangle CPB$ by AA. Using the ratio of similitude, we get $\frac{AP}{CP} = \frac{PD}{PB}$, from which we get $(AP)(PB) = (CP)(PD)$.

如图 19-59 所示，根据对顶角相等，可得出$\angle APD \cong \angle BPC$。根据**圆周角定理**，可得出 $m\angle BAD = \frac{1}{2}m\overset{\frown}{BD} = m\angle BCD$。根据传递性，可得出 $m\angle BAD = m\angle BCD$。现在能根据角角相似公理，得出$\triangle APD \sim \triangle CPB$。根据相似比可得出$\frac{AP}{CP} = \frac{PD}{PB}$，从而$(AP)(PB) = (CP)(PD)$。

2. Proof of the **Chord Distance to Center Theorem**:

弦到圆心距离定理的证明：

As shown in Figure 19-60, we are given that in $\odot O$, $\overline{AB} \cong \overline{CD}$. Suppose points M and N are on $\overline{AB}$ and $\overline{CD}$ respectively, such that $\overline{OM} \perp \overline{AB}$ and $\overline{ON} \perp \overline{CD}$. We want to show that $\overline{OM} \cong \overline{ON}$.

如图 19-60 所示，已知在$\odot O$ 里，$\overline{AB} \cong \overline{CD}$。设点 M 和 N 分别在 $\overline{AB}$ 和 $\overline{CD}$ 上，使得 $\overline{OM} \perp \overline{AB}$，$\overline{ON} \perp \overline{CD}$，证明 $\overline{OM} \cong \overline{ON}$。

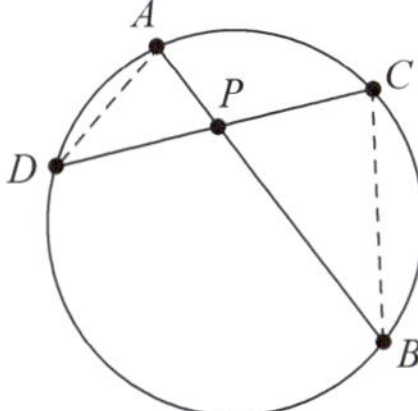

Figure 19-59

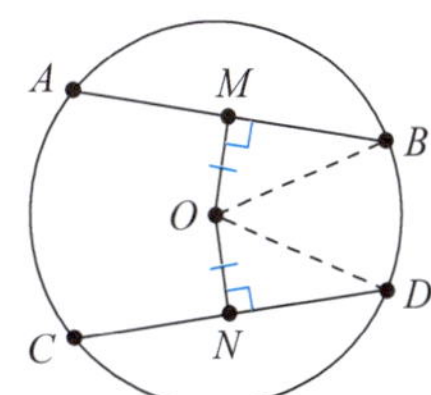

Figure 19-60

Clearly $\angle OMB \cong \angle OND$ since they are right angles. By the **Perpendicular Chord Theorem**, we get that $\frac{1}{2}AB = \frac{1}{2}CD = MB = ND$. Since all radii are congruent, we have $\overline{OB} \cong \overline{OD}$. Therefore, $\triangle OMB \cong \triangle OND$ due to HL, from which we can conclude $\overline{OM} \cong \overline{ON}$ by CPCTC.

因为都是直角,所以$\angle OMB \cong \angle OND$。根据**垂径定理**,得出$\frac{1}{2}AB = \frac{1}{2}CD = MB = ND$。根据半径相等,可得出$\overline{OB} \cong \overline{OD}$。根据 HL 定理,$\triangle OMB \cong \triangle OND$。因为全等三角形的对应部分全等,故$\overline{OM} \cong \overline{ON}$。

3. Proof of the **Parallel Lines Intercepted Arcs Theorem**:

平行线截弧定理的证明:

As shown in Figure 19-61, we wish to show that $\overset{\frown}{AC} \cong \overset{\frown}{BD}$. By the **Alternate Interior Angles Theorem**, we have $\angle BAD \cong \angle ADC$. Since we know that $m\angle BAD = \frac{1}{2}m\overset{\frown}{BD}$ and $m\angle ADC = \frac{1}{2}m\overset{\frown}{AC}$. Transitive gives $\frac{1}{2}m\overset{\frown}{AC} = \frac{1}{2}m\overset{\frown}{BD}$, for which $m\overset{\frown}{AC} = m\overset{\frown}{BD}$. Since the arcs are of the same circle, $\overset{\frown}{AC} \cong \overset{\frown}{BD}$.

如图 19-61 所示,证明$\overset{\frown}{AC} \cong \overset{\frown}{BD}$。用**内错角定理**可得出$\angle BAD \cong \angle ADC$。已知$m\angle BAD = \frac{1}{2}m\overset{\frown}{BD}$,$m\angle ADC = \frac{1}{2}m\overset{\frown}{AC}$,根据传递性,得出$\frac{1}{2}m\overset{\frown}{AC} = \frac{1}{2}m\overset{\frown}{BD}$,也就是说$m\overset{\frown}{AC} = m\overset{\frown}{BD}$。又因为是同圆的弧,可得出$\overset{\frown}{AC} \cong \overset{\frown}{BD}$。

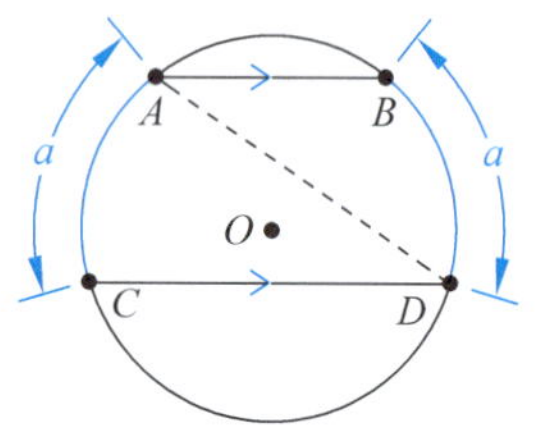

Figure 19-61

Note that by arc addition and subtraction, the following relationships are also true:

(1) $\overset{\frown}{AD} \cong \overset{\frown}{BC}$.

(2) $\overset{\frown}{ACD} \cong \overset{\frown}{BDC}$.

通过弧度加减法可得出以下关系式:

(1) $\overset{\frown}{AD} \cong \overset{\frown}{BC}$.

(2) $\overset{\frown}{ACD} \cong \overset{\frown}{BDC}$.

4. Proof of **Theorem 4**:

定理 4 的证明:

As shown in Figure 19-62, using the **Inscribed Angle Theorem**, we get $m\angle PBC = \frac{1}{2}m\overset{\frown}{AC}$ and $m\angle BCP = \frac{1}{2}m\overset{\frown}{BD}$. In $\triangle PBC$, by the **Exterior Angle Theorem**, we have $m\angle 1 = m\angle PBC + m\angle BCP = \frac{1}{2}m\overset{\frown}{AC} + \frac{1}{2}m\overset{\frown}{BD} = \frac{1}{2}(m\overset{\frown}{AC} + m\overset{\frown}{BD})$. Transitive gives $m\angle 1 = \frac{1}{2}(m\overset{\frown}{AC} + m\overset{\frown}{BD})$.

如图 19-62 所示,用**圆周角定理**可得出$m\angle PBC = \frac{1}{2}m\overset{\frown}{AC}$ 和$m\angle BCP = \frac{1}{2}m\overset{\frown}{BD}$。在

$\triangle PBC$ 中，根据**外角定理**可知 $m\angle 1 = m\angle PBC + m\angle BCP = \frac{1}{2}m\overset{\frown}{AC} + \frac{1}{2}m\overset{\frown}{BD} = \frac{1}{2}(m\overset{\frown}{AC} + m\overset{\frown}{BD})$。又由于传递性可得出 $m\angle 1 = \frac{1}{2}(m\overset{\frown}{AC} + m\overset{\frown}{BD})$。

5. Proof of the **Tangent-Chord Angle Theorem**:
弦切角定理的证明：
Note that $\angle 1$ and $\angle 2$ form a linear pair. If $m\angle 1 = m\angle 2 = 90°$, then it is clear that each of $m\angle 1$ and $m\angle 2$ is equal to half the measure of a semicircle $\left(\frac{1}{2}\times 180° = 90°\right)$. Now suppose that $m\angle 1 \neq m\angle 2$.
注意：$\angle 1$ 和 $\angle 2$ 互为邻补角。若 $m\angle 1 = m\angle 2 = 90°$，则很明显 $m\angle 1$ 和 $m\angle 2$ 各为半圆度数的一半$\left(\frac{1}{2}\times 180° = 90°\right)$。现在假设 $m\angle 1 \neq m\angle 2$。
Without the loss of generality, suppose that $m\angle 1 < 90° < m\angle 2$.
一般情况下，假设 $m\angle 1 < 90° < m\angle 2$。
As shown in Figure 19-63, note that $m\angle OPA = m\angle OAP = 90° - m\angle 1$. Therefore, $m\overset{\frown}{AP} = m\angle AOP = 180° - 2(90° - m\angle 1) = 2m\angle 1$, from which we can conclude $m\angle 1 = \frac{1}{2}m\overset{\frown}{AP}$.

Therefore, we have $m\angle 2 = 180° - m\angle 1 = 180° - \frac{1}{2}m\overset{\frown}{AP} = 180° - \frac{1}{2}(360° - m\overset{\frown}{ABP}) = \frac{1}{2}m\overset{\frown}{ABP}$.

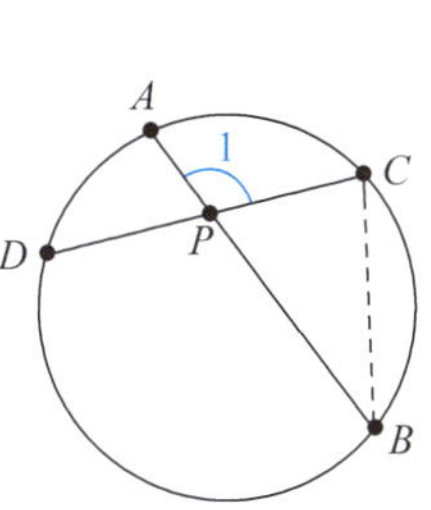

Figure 19-62

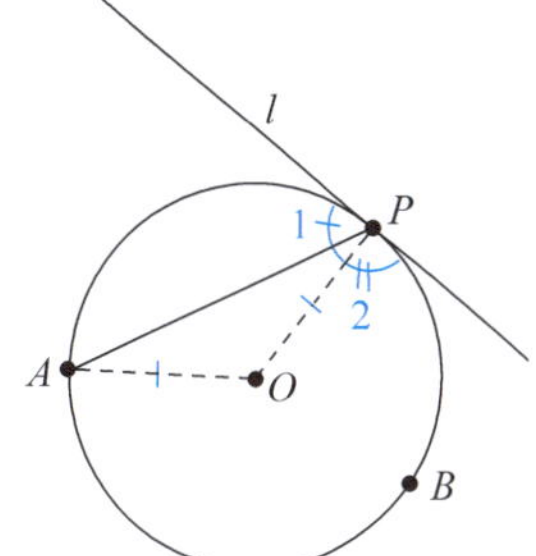

Figure 19-63

6. Proof of **Theorem 6**:
定理 6 的证明：

As shown in Figure 19-64, by the **Inscribed Angle Theorem** we have $m\angle AED = \frac{1}{2}m\overset{\frown}{AD}$ and $m\angle EAC = \frac{1}{2}m\overset{\frown}{BE}$. Therefore, $m\angle AEC = 180° - m\angle AED = 180° - \frac{1}{2}m\overset{\frown}{AD}$.

We have $m\angle C = 180° - m\angle AEC - m\angle EAC = 180° - (180° - \frac{1}{2}m\overset{\frown}{AD}) - \frac{1}{2}m\overset{\frown}{BE} =$

$\frac{1}{2}m\widehat{AD}-\frac{1}{2}m\widehat{BE}=\frac{1}{2}(m\widehat{AD}-m\widehat{BE})$.

根据**圆周角定理**证明。

7. Proof of **Theorem 7**:

定理 7 的证明:

As shown in Figure 19-65, by the **Inscribed Angle Theorem** we have $m\angle JKH=\frac{1}{2}m\widehat{HJ}$ so $m\angle JKL=180^\circ-m\angle JKH=180^\circ-\frac{1}{2}m\widehat{HJ}$. By the **Tangent-Chord Angle Theorem**, as in Theorem 5, we have $m\angle KJL=\frac{1}{2}m\widehat{JK}$. Therefore, we have $m\angle L=180^\circ-m\angle JKL-m\angle KJL=180^\circ-(180^\circ-\frac{1}{2}m\widehat{HJ})-\frac{1}{2}m\widehat{JK}=\frac{1}{2}m\widehat{HJ}-\frac{1}{2}m\widehat{JK}=\frac{1}{2}(m\widehat{HJ}-m\widehat{JK})$.

根据**圆周角定理**和**弦切角定理**证明。

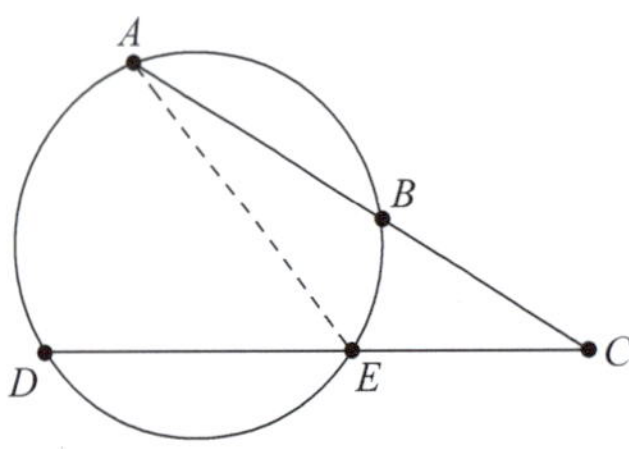

Figure 19-64

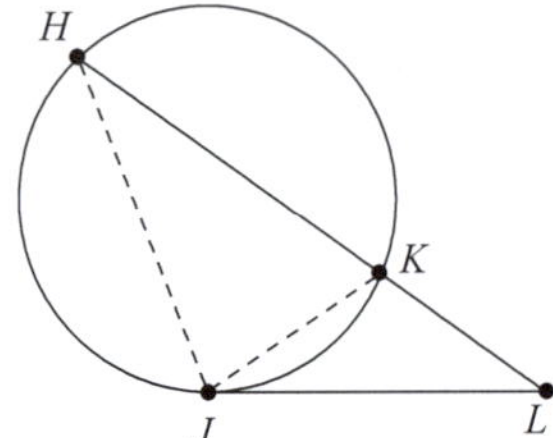

Figure 19-65

8. Proof of **Theorem 8**:

定理 8 的证明:

Note that $m\widehat{BC}+m\widehat{BDC}=360^\circ$.

As shown in Figure 19-66, by the **Tangent-Chord Angle Theorem**, as in Theorem 5, we have $m\angle B=m\angle C=\frac{1}{2}m\widehat{BC}$. Therefore, $m\angle A=180^\circ-m\angle B-m\angle C=180^\circ-\frac{1}{2}m\widehat{BC}-\frac{1}{2}m\widehat{BC}=180^\circ-m\widehat{BC}=\frac{1}{2}(m\widehat{BC}+m\widehat{BDC})-m\widehat{BC}=\frac{1}{2}(m\widehat{BDC}-m\widehat{BC})$.

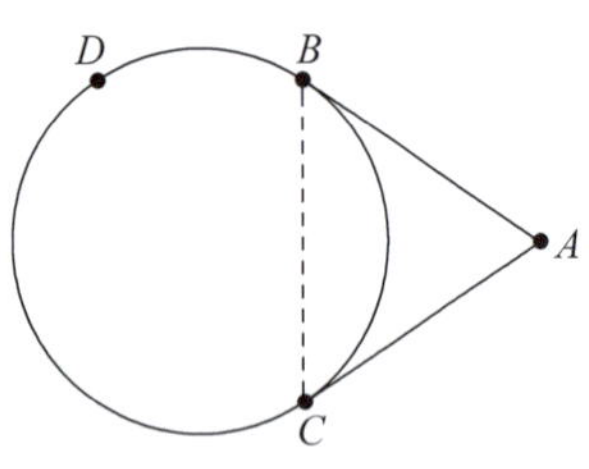

Figure 19-66

根据**弦切角定理**证明。

19.6 Common Tangents of Two Circles 两个圆的公切线

Common Tangent 公切线

n. [ˈkɑmən ˈtændʒənt]

Definition: A common tangent of two circles is a line that is tangent to both circles. It is one of common internal tangent and common external tangent. 两个圆的公切线是它们共同的切线,分为内公切线和外公切线。

Figure 19-67 shows one common tangent of two circles in blue.

Properties: Two different circles in the same plane can have 0,1,2,3,or 4 common tangents. 同一平面上两个不同的圆可有 0、1、2、3 或 4 条公切线。

Let n be the number of common tangents. 设 n 为公切线的条数。

Figures 19-68,19-69,19-70,19-71,and 19-72 show $n = 0,1,2,3,$and 4 respectively.

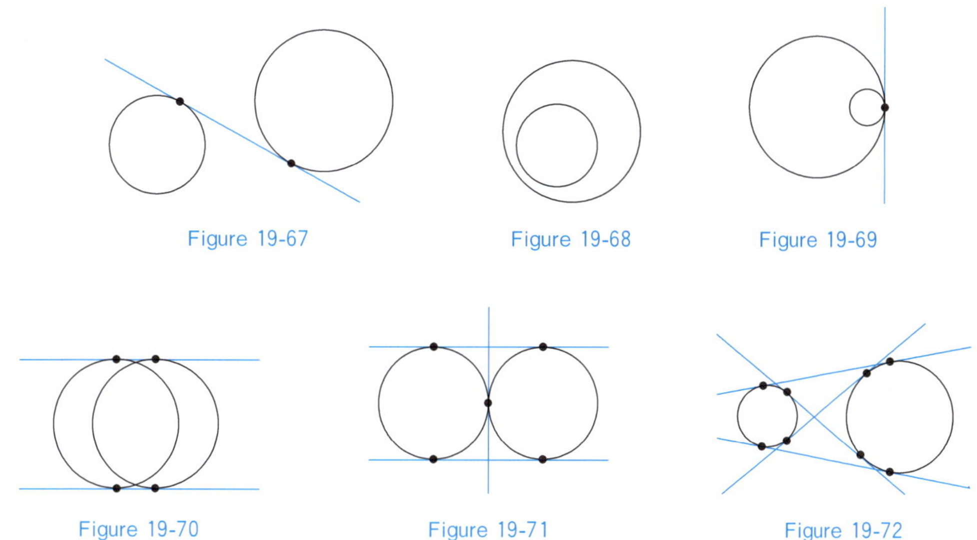

Figure 19-67 Figure 19-68 Figure 19-69

Figure 19-70 Figure 19-71 Figure 19-72

Common Internal Tangent 内公切线

n. [ˈkɑmən ɪnˈtɜrnəl ˈtændʒənt]

Definition: A common tangent that intersects the line segment whose endpoints are the centers of the two circles. 内公切线是指交于过两圆心的线段的公切线。

Figure 19-73 shows the common internal tangents of the two circles in blue.

Common External Tangent 外公切线

n. [ˈkɑmən ekˈtɜrnəl ˈtændʒənt]

Definition: A common tangent that does not intersect the line segment whose endpoints are the centers of the two circles.

外公切线是指不交于过两圆心的线段的公切线。

Figure 19-74 shows the common external tangents of the two circles in blue.

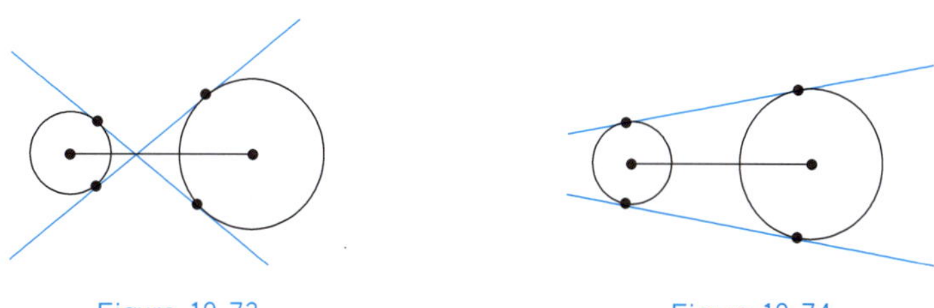

Figure 19-73 Figure 19-74

19.7 Regular Polygons and Circles 正多边形与圆

Parts of a Regular Polygon 正多边形的部分

Definition: Parts of a regular n-gon P.

正 n 边形 P 的部分定义如下。

Circumcircle: A circle that contains all of the P's vertices, as shown in Figure 19-75 in blue.

外接圆：圆周包含 P 所有顶点的圆，如图 19-75 蓝色所示。

Incircle: A circle that contains all of P's sides' midpoints, as shown in Figure 19-75 in blue.

内切圆：圆周包含 P 的所有边的中点的圆，如图 19-75 蓝色所示。

Center: The center of the P's circumcircle, also is the center of the P's incircle. In Figure 19-75, O is the center.

中心：P 的外接圆的圆心，也是 P 的内切圆的圆心。在图 19-75 中，O 为中心。

Radius: A line segment that whose endpoints are the center and a vertex of P, which is also a radius of the P's circumcircle. In Figure 19-75, $\overline{OA}$ and $\overline{OB}$ are radii.

外接圆半径：端点为中心和 P 的一个顶点的线段。在图 19-75 中，$\overline{OA}$ 和 $\overline{OB}$ 为外接圆半径。

Apothem: A line segment whose endpoints are the center and on one side of the P such that the line segment is perpendicular to that side. Apothem is also the radius of the incircle. In Figure 19-75, $\overline{OC}$ is an apothem.

边心距：端点为中心和 P 的一条边的中点的线段，也是内切圆的半径。在图 19-75 中，$\overline{OC}$ 为边心距。

Central Angle: The angle subtended at the center of P by one of its sides. In Figure 19-75, $\angle AOB$ is a central angle.

中心角：以 P 的一条边对向 P 的中心的角。在图 19-75 中，$\angle AOB$ 为中心角。

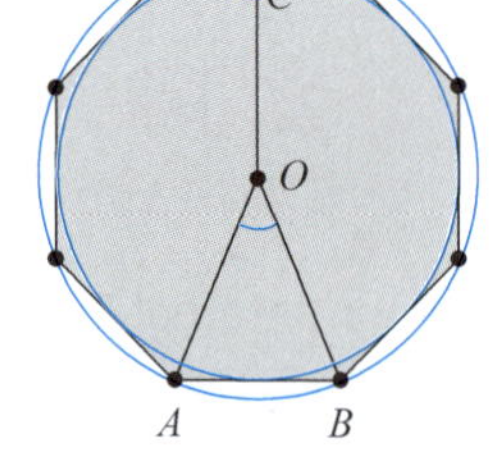

Figure 19-75

Properties: In a regular n-gon P:

在正 n 边形 P 中：

1. A central angle of an measured $360°/n$. 中心角的度数为 $360°/n$。
2. An apothem drawn to a side of P is intersects the midpoint of that side. 在 P 的一条边的边心距交于 P 这条边的中点。
3. All sides of P are tangent to the incircle. P 的所有边与内切圆相切。
4. Let s, a, and r be the lengths of the side, apothem, and radius of P, respectively. Their relationships are as followed:

   设 s、a、r 分别为 P 的边、边心距、外接圆半径，则它们的关系如下：

   (1) $\theta = \left(\dfrac{360°}{n}\right)/2$.

   (2) $\left(\dfrac{1}{2}s\right)^2 + a^2 = r^2$.

   (3) $\cos\theta = a/r$.

   (4) $\sin\theta = (s/2)/r$.

   These properties are illustrated in Figure 19-76. 这些性质如图 19-76 所示。

5. (1) P has line symmetry, and P has n lines of symmetry. See line symmetry in Section 12.4.3.
   P 具有轴对称性，有 n 条对称轴。见 12.4.3 节的轴对称。
   (2) P has rotational symmetry about its center. The angle of rotation is $360°/n$. See rotational symmetry in Section 12.4.3. P 具有中心旋转对称性。旋转角度是 $360°/n$。见 12.4.3 节的旋转对称。

Proofs: 1. To prove Property 1, as shown in Figure 19-77, all small triangles are congruent by SSS. Therefore, by CPCTC, all central angles are congruent. 要证明性质 1，如图 19-77 所示，根据边边边全等公理，所有小三角形全等。根据全等三角形的对应部分全等，可得所有中心角全等。

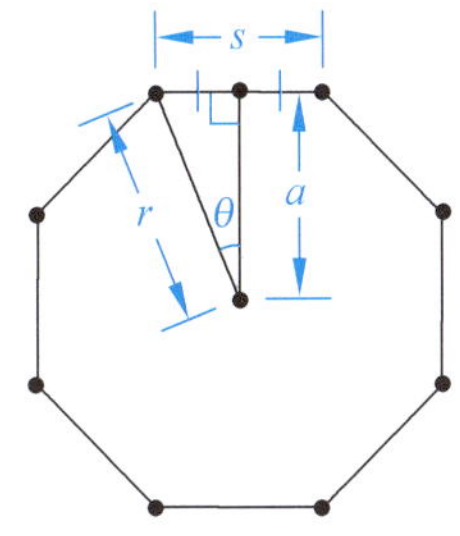

Figure 19-76

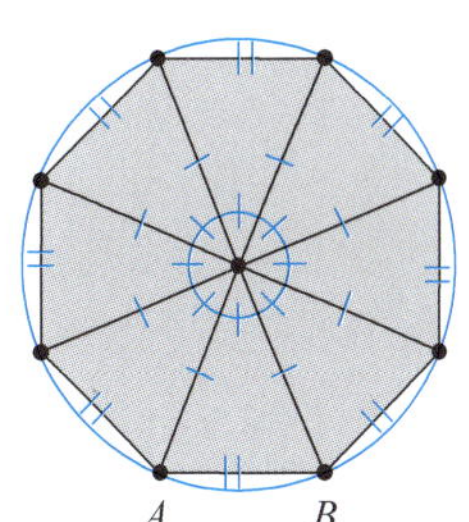

Figure 19-77

Note that a regular n-gon has n congruent central angles, and the sum of the measures of central angles is $360°$. Therefore, each central angle is measured $360°/n$.
注意一个正 n 边形有 n 个全等的中心角，它们的度数之和为 $360°$。因此每个中心角的度数是 $360°/n$。

2. To prove Property 2, suppose O is the center of P, $\overline{AB}$ is a side of P, and M is on $\overline{AB}$ such that $\overline{OM}\perp\overline{AB}$. We want to prove that M is the midpoint of $\overline{AB}$. 要证明性质 2，设 O 是 P 的中心，$\overline{AB}$ 是 P 的一条边，M 在 $\overline{AB}$ 上且 $\overline{OM}\perp\overline{AB}$。证明 M 是 $\overline{AB}$ 的中点。
As shown in Figure 19-78, reflexive gives $\overline{OM}\cong\overline{OM}$. Since all radii are congruent, we have $\overline{OA}\cong\overline{OB}$. Since $\angle OMA$ and $\angle OMB$ forms a linear pair and $\overline{OM}\perp\overline{AB}$ as given, we have $\angle OMA\cong\angle OMB$. Therefore, we can conclude that $\triangle OMA \cong \triangle OMB$ by HL, from which $\overline{AM}\cong\overline{BM}$ follows by CPCTC. So M is the midpoint of $\overline{AB}$.
如图 19-78 所示，由于反身性，则 $\overline{OM}\cong\overline{OM}$。根据所有半径相等，得出 $\overline{OA}\cong\overline{OB}$，又因为$\angle OMA$ 与$\angle OMB$ 互为邻补角，并且已知 $\overline{OM}\perp\overline{AB}$，可得$\angle OMA\cong\angle OMB$。根据 HL 定理，可知$\triangle OMA\cong\triangle OMB$。根据全等三角形的对应部分全等，得 $\overline{AM}\cong\overline{BM}$。因此，$M$ 是 $\overline{AB}$ 的中点。

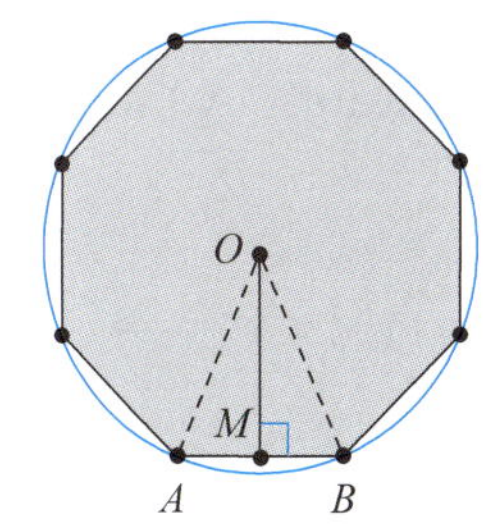

Figure 19-78

3. We will use Property 2 to prove Property 3. From Property 2, note that OM is the distance from O to $\overline{AB}$, and is the minimum length from O to any point on $\overline{AB}$. By the **Converse of Tangent Theorem** (Section 19.4.1), $\overline{AB}$ is tangent to the incircle. 可用性质 2 证明性质 3。在性质 2 中，注意 OM 是从 O 到 $\overline{AB}$ 的距离，此长度是 O 到 $\overline{AB}$

上任何点的长度中最小的。根据**切线定理的逆定理**(19.4.1 节),$\overline{AB}$ 与内切圆相切。

Questions: The side-length of a regular octagon P is 2.

设一个正八边形 P 的边长为 2。

(1) What is the measure of a central angle of P? 求 P 的中心角的度数。

(2) What is the length of a radius of P? 求 P 的半径。

(3) What is the apothem of P? 求 P 的边心距。

Answers: Using Property 3, we get $\theta = 360°/8/2 = 22.5°$, which is half the measure of the central angle.

(1) $360°/8 = 45°$.

(2) We have $\sin\theta = (s/2)/r$, from which $r = (s/2)/\sin\theta = (2/2)/\sin 22.5° \approx 2.613$.

(3) We have $\cos\theta = a/r$, from which $a = r\cos\theta = [(2/2)/\sin 22.5°]\cos 22.5° \approx 2.414$.

20. Solids 立体

20.1 Introduction 介绍

Face 面

n. [feɪs]

Definition: An individual surface of a polyhedron, as shown in Figure 20-1. 面是指多面体的一个表面,如图 20-1 所示。

Each face of a polyhedron is a polygon. 多面体的每个面均为多边形。

Edge 棱

n. [edʒ]

Definition: A line segment of a polyhedron that is the intersection of two faces, as shown in Figure 20-2. 棱是指多面体的一条线段,是两个面的交界,如图 20-2 所示。

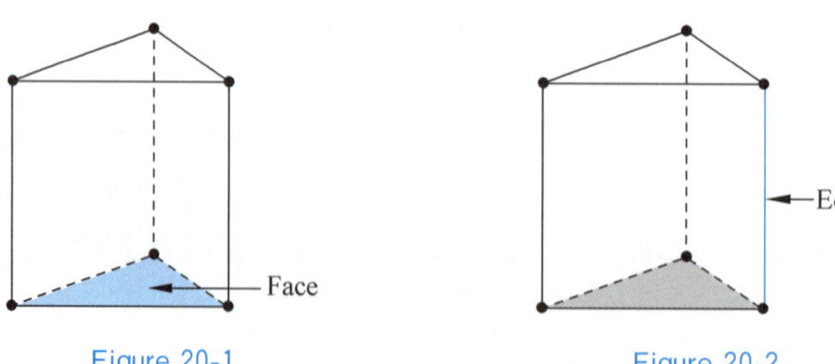

Figure 20-1　　Figure 20-2

Properties: The endpoints of an edge of a polyhedron P are the vertices of P. 多面体 P 的棱的端点就是 P 的顶点。

Vertex 顶点

n. [ˈvɜrteks]

Definition：A corner of a polyhedron (the intersection of three or more edges), as shown in Figure 20-3. 顶点是指多面体的三条或多条棱的交点，如图 20-3 所示。

Cross Section 横截面

n. [ˈkrɔsˌsekʃən]

Definition：The intersection of a solid and a plane. There are many different possibilities. 横截面是指立体和一个平面的交叠部分，有多种情况。

Examples：1. As shown in Figure 20-4, the cross section parallel to the bases of the prism is a blue triangle that is congruent to the bases. The plane that contains the cross section is shown in gray. 如图 20-4 所示，平行于棱柱的底的横截面是一个蓝色三角形，与两底全等，包含横截面的平面用灰色表示。

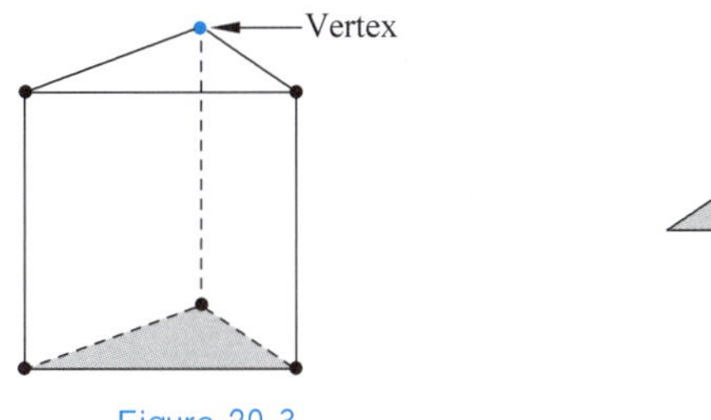

Figure 20-3

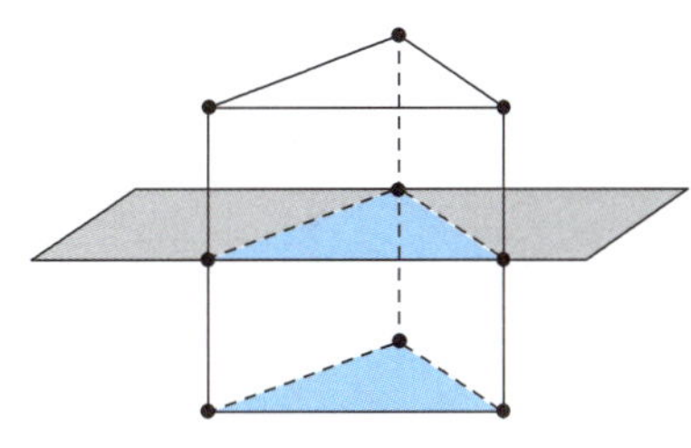
Figure 20-4

2. As shown in Figure 20-5, the cross section parallel to the face L of the prism is a blue rectangle that is similar to L. The plane that contains the cross section is shown in gray. 如图 20-5 所示，平行于棱柱的侧面 L 的横截面是一个蓝色矩形，与 L 相似。包含横截面的平面用灰色表示。

Net 展开图

n. [net]

Definition：A plane shape representing the unfolded form of a solid. 展开图代表一个展开的立体的平面图形。

Examples：A net of a die is shown in Figure 20-6. 图 20-6 是骰子的展开图。

A die is a cube. 骰子是立方体。

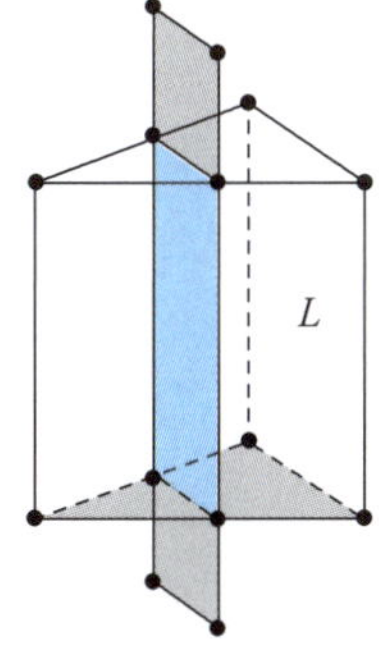

Figure 20-5

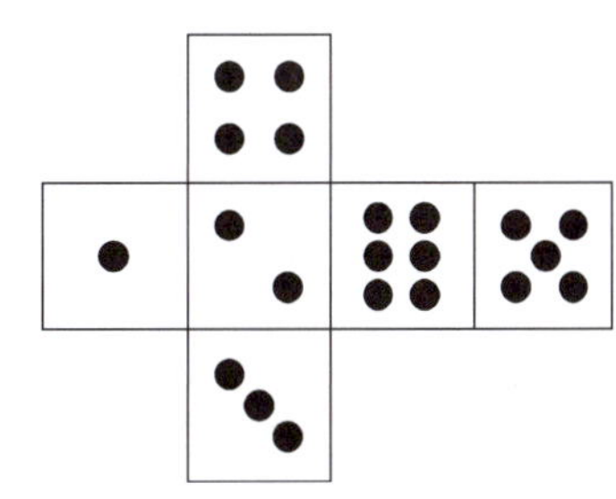
Figure 20-6

20.2 Polyhedrons 多面体

视频 34

20.2.1 Introduction 介绍

Polyhedron 多面体

n. [ˌpɑliˈhidrən]

Definition: A solid that is bounded by four or more faces, each of which is a polygon. 多面体是被四个或更多个面围成的立体，每个面均为多边形。

Special polyhedrons are prisms and pyramids. 棱柱和棱锥是特殊的多面体。

Examples: 1. Some examples of prisms are shown in Figure 20-7. 棱柱如图 20-7 所示。

2. Some examples of pyramids are shown in Figure 20-8. 棱锥如图 20-8 所示。

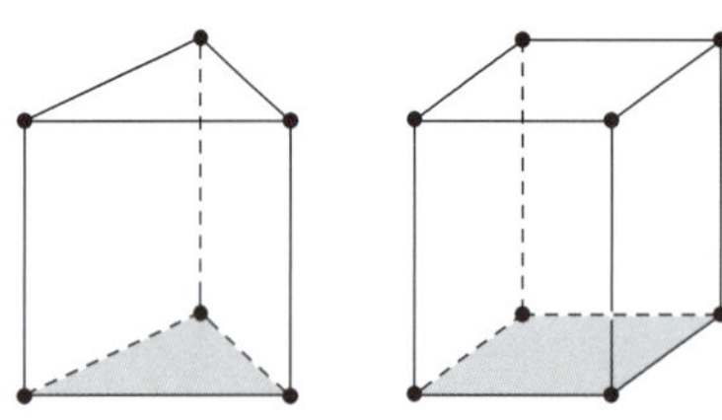
Figure 20-7

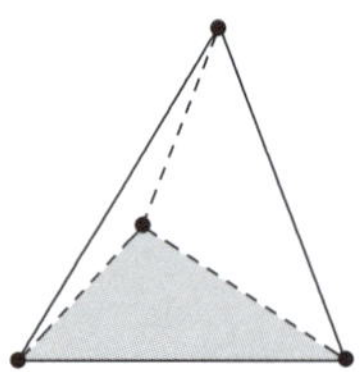
Figure 20-8

Phrases: List of polyhedrons by the number of faces:

根据面的个数多面体列表如下(根据后面的数字 n 可读作 n 面体)：

Tetrahedron (4), Pentahedron (5), Hexahedron (6), Heptahedron (7), Octahedron (8), Enneahedron (9), Decahedron (10), Hendecahedron (11), Dodecahedron (12), Tridecahedron (13), Tetradecahedron (14), Pentadecahedron (15), Hexadecahedron (16), Heptadecahedron (17), Octadecahedron (18), Enneadecahedron (19), Icosahedron (20)

Euler's Formula 欧拉公式

n. [ˈɔɪlərs ˈfɔrmjələ]

Definition: If a polyhedron has V vertices, E edges, and F faces, then it is true that $V+F=E+2$. 若一个多面体有 V 个顶点，E 条棱，F 个面，则 $V+F=E+2$。

Proofs: The proof of this formula is out of the scope of high school geometry. 欧拉公式的证明在高中几何知识范围之外。

20.2.2 Parts of Prisms and Pyramids 棱柱与棱锥的部分

Base 底

n. [beɪs]

The definition of a base is contained in both prism in Section 20.2.3 and pyramid in Section 20.2.4. 底的定义见 20.2.3 节棱柱与 20.2.4 节棱锥的定义。

Figure 20-9 shows the bases of a prism (left) and the base of a pyramid (right). 图 20-9 展示了棱柱的底(左)与棱锥的底(右)。

Lateral Face 侧面

n. ['lætərəl feɪs]

Definition: A face that is not a base.　不是底的面称为侧面。

Figure 20-10 shows a lateral face of a prism (left) and a lateral face of a pyramid (right).　图 20-10 展示了棱柱的一个侧面(左)与棱锥的一个侧面(右)。

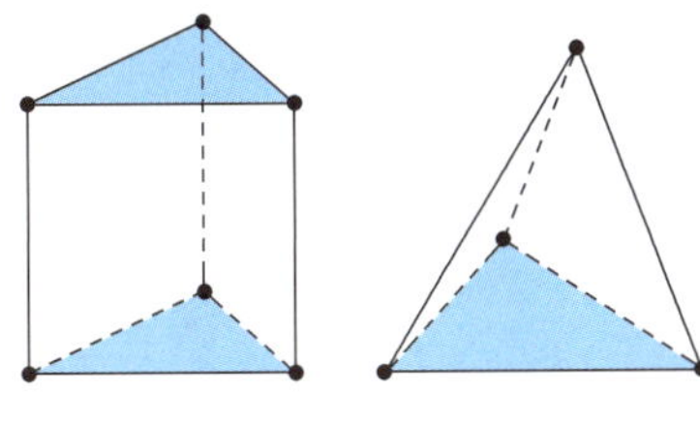

Figure 20-9

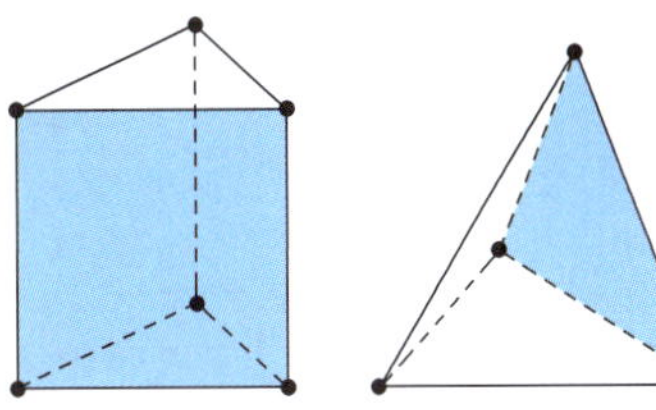

Figure 20-10

Lateral Edge 侧棱

n. ['lætərəl edʒ]

Definition: An edge that is not an edge of a base.　不是底上的棱称为侧棱。

Figure 20-11 shows a lateral edge of a prism (left) and a lateral edge of a pyramid (right).　图 20-11 展示了棱柱的一条侧棱(左)与棱锥的一条侧棱(右)。

20.2.3　Prisms　棱柱

Prism 棱柱

n. ['prɪzəm]

Definition: A polyhedron with two congruent parallel faces (called the bases, which can be any polygon), and the rest of the faces (called the lateral faces) are parallelograms.　棱柱是有两个面全等且平行的多面体(这两个面叫作底,可以是任意多边形)。其余的面(侧面)均为平行四边形。

For each prism, Figure 20-12 shows a base in gray and a lateral face in blue.　图 20-12 用灰色展示了棱柱的一个底,用蓝色展示了棱柱的一个侧面。

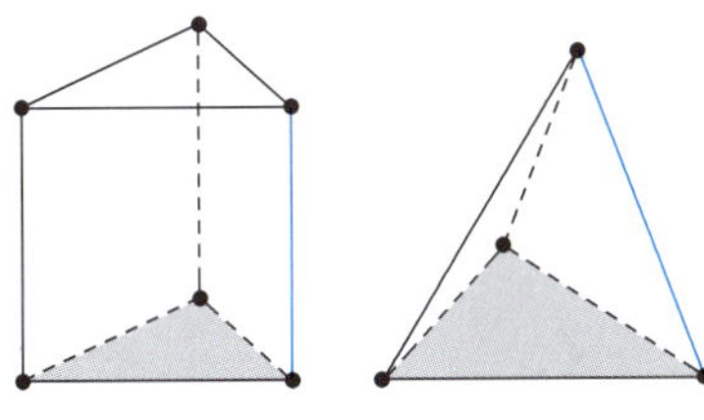

Figure 20-11

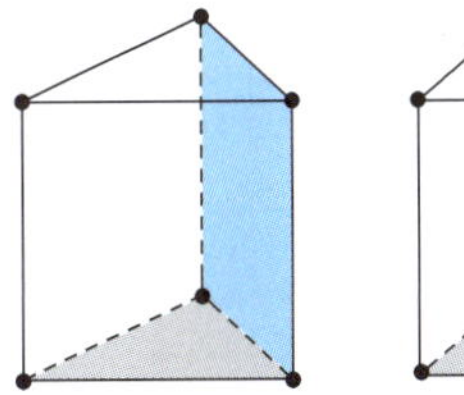

Figure 20-12

A prism is one of right prism and oblique prism.　棱柱是直棱柱和斜棱柱中的一种。

A regular prism is a prism with regular polygons as bases. According to our convention, a "regular prism" is implied to be a "regular right prism".　正棱柱是两底为正多边形的棱柱。根据惯例,"正棱柱"代表"正直棱柱"。

Properties: In a prism, any cross section parallel to those bases is congruent to the bases, as shown in Figure 20-13. 在棱柱里,任意与底平行的横截面与底全等,如图 20-13 所示。

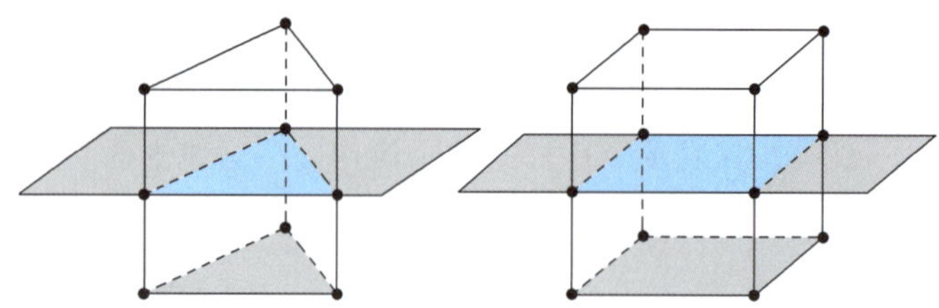

Figure 20-13

Phrases: triangular～: bases are triangles. 三棱柱。

quadrangular～: bases are quadrilaterals. 四棱柱。

rectangular～: bases are rectangles. 长方体。

pentagonal～: bases are pentagons. 五棱柱。

and so on.

Right Prism 直棱柱

n. [raɪt 'prɪzəm]

Definition: A prism whose lateral faces are all rectangles, as shown in Figure 20-14. 侧面均为矩形的棱柱为直棱柱,如图 20-14 所示。

Oblique Prism 斜棱柱

n. [oʊ'blik 'prɪzəm]

Definition: A prism with at least one lateral face that is not a rectangle, as shown in Figure 20-15. 斜棱柱是指至少有一个侧面不是矩形的棱柱,如图 20-15 所示。

Cube 立方体

n. [kjub]

Definition: A rectangular prism whose all faces are congruent squares, as shown in Figure 20-16. 立方体是指所有面都是全等正方形的直棱柱,如图 20-16 所示。

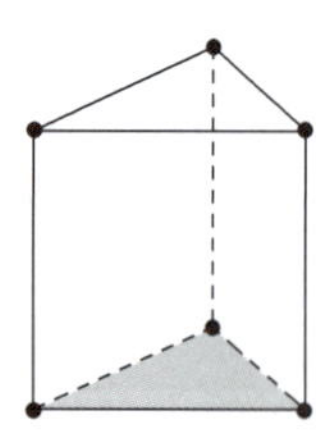

Figure 20-14

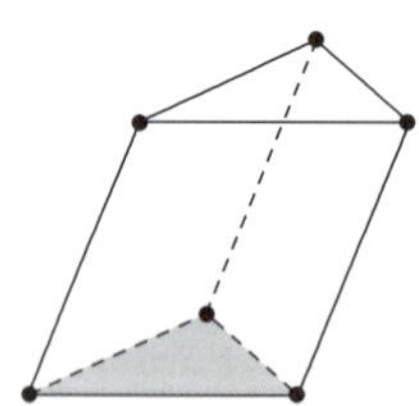

Figure 20-15

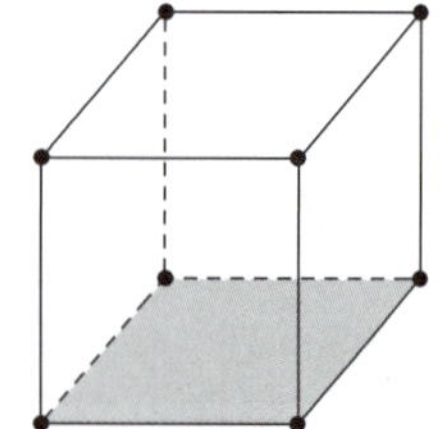

Figure 20-16

Altitude of a Prism 棱柱的高线

Definition: The altitude of a prism is a line segment connecting the planes of its bases (perpendicular to the bases). 棱柱的高线是指连接包含两底的平面的线段(垂直于两底)。

Notation: The length of the altitude of a prism is called the height, usually denoted by h. 棱柱高线的长度叫作高，通常用 h 表示。

Examples: Figure 20-17 shows altitudes for a right and an oblique prisms.

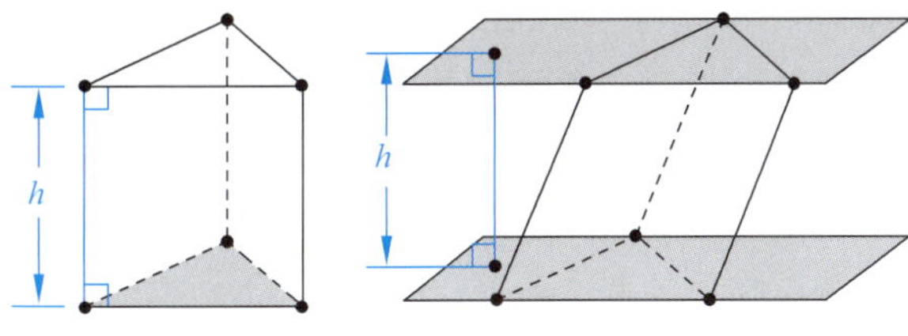

Figure 20-17

Note that a lateral edge of a right prism is an altitude. 正棱柱的侧棱本身就是一条高线。

20.2.4 Pyramids 棱锥

Pyramid 棱锥

n. ['pɪrəˌmɪd]

Definition: A polyhedron with a polygon (called the base) and triangles (called lateral faces) that meet at a common vertex (called the apex). Each lateral face shares an edge with the base, as shown in Figure 20-18. 棱锥是有一个多边形(底)以及若干三角形(侧面)的多面体。侧面的三角形相交于一个顶点(顶尖)。每个侧面与底有一条共同的棱，如图 20-18 所示。

A regular pyramid, as shown in Figure 20-19, is a pyramid whose base is a regular polygon and the line segment connecting the apex and the center of the base is perpendicular to the base. This line segment is called the altitude of a pyramid. A regular pyramid has a slant height. 如图 20-19 所示，正棱锥是一个底为正多边形，且连接顶尖和底中心的线段与底垂直的多面体。这线段叫作棱锥的高线。正棱锥有斜高。

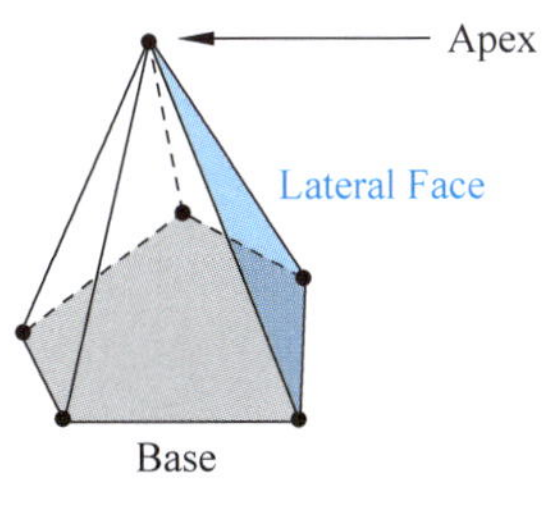

Figure 20-18

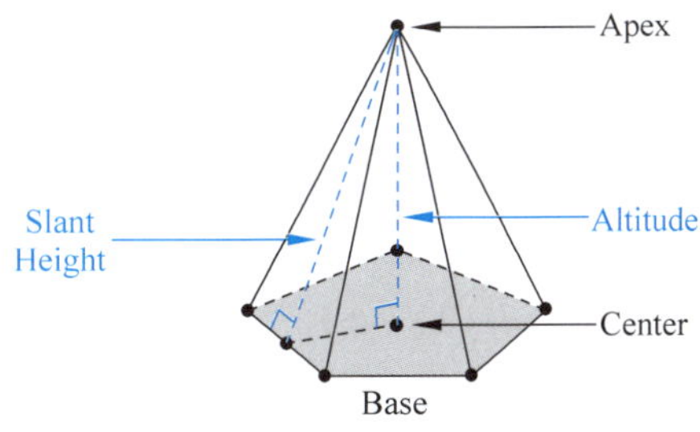

Figure 20-19

Properties: The lateral faces of a regular pyramid are congruent isosceles triangles. 正棱锥的侧面是全等的等腰三角形。

Proofs: As shown in Figure 20-20, we will use SAS to show that each triangle (shaded in gray) whose vertices are the apex, center of the base, and a vertex of the base, is congruent to the others. The dotted line segments are congruent because they are the radii of the base. Using CPCTC, we can show that all lateral edges (two of them drawn in blue) are congruent. So far we have shown that all lateral faces are isosceles triangles. 如图 20-20 所示,能用边角边全等公理证明每个灰色三角形(顶点为顶尖,另外两点为底中心、底的一个顶点)全等。虚线线段全等是因为它们都是底的外接圆半径。根据全等三角形的对应部分全等,可以证明所有侧棱(蓝色)全等,由此证明所有侧面都是等腰三角形。

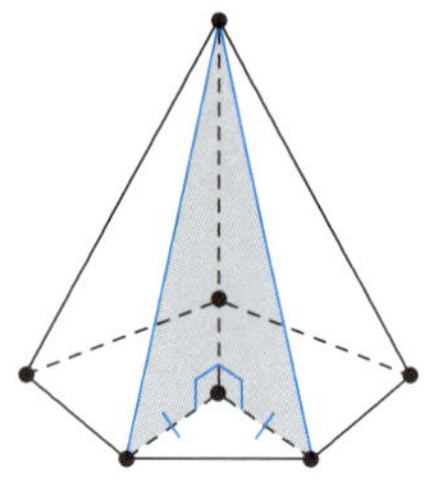

Figure 20-20

Finally, we will use SSS to show that all lateral triangle are congruent, recall that all sides of the base, which is a regular polygon, are congruent. 因为底是正多边形,底的边全等,用边边边全等公理可证出所有侧面三角形全等。

Altitude of a Pyramid 棱锥的高线

Definition: The line segment connecting its apex and one point of the plane containing the base, such that the altitude is perpendicular to the base. 棱锥的高线是指连接顶尖和包含底的平面上一点的线段,该线段垂直于底。

Notation: The length of the altitude of a pyramid is called the height, usually denoted by h. 棱锥高线的长度叫作高,通常用 h 表示。

Examples: Figure 20-21 shows the altitudes for two prisms.

Slant Height of a Regular Pyramid 正棱锥的斜高

n. [slænt haɪt …]

Definition: The distance between the apex and an edge of the base, as shown in blue in Figure 20-22. 正棱锥的斜高是指顶尖到底棱的距离,如图 20-22 所示用蓝色表示。

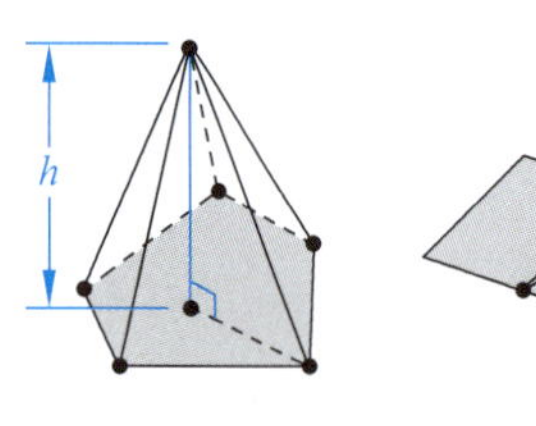

Figure 20-21

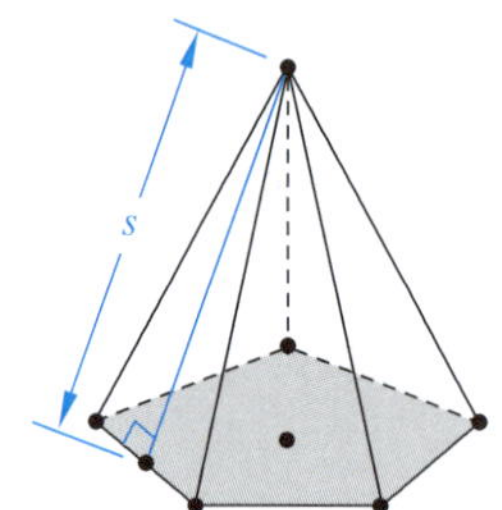

Figure 20-22

Notation: The slant height of a regular pyramid is usually denoted by s. 正棱锥的斜高通常用 s 表示。

Properties: 1. The line segment joining the apex and an edge of the base (whose length is the slant height) intersects the midpoint of that edge. 连接顶尖和底棱的线段(其长度为斜高)经过底棱中点。

2. In a regular pyramid, let h, a, and s be the height, base's apothem, and slant height, respectively. By the Pythagorean Theorem, $h^2+a^2=s^2$, as shown in Figure 20-23. 在

正棱锥里，h、a、s 分别为高、底的边心距、斜高。根据勾股定理，$h^2+a^2=s^2$，如图 20-23 所示。

Proofs: To prove Property 1, in the property of pyramid, we have shown that every lateral triangle is isosceles, as shown in Figure 20-24.

要证明性质 1，根据棱锥的性质，已经证出每个侧面的三角形均等腰，如图 20-24 所示。

The lateral triangle containing point P in Figure 20-24 is drawn in Figure 20-25. By the **Perpendicular Bisector Theorem** on the edge of the base, we know that point P is the midpoint of that edge since P is on its perpendicular bisector.

在图 20-24 中包含点 P 的三角形如图 20-25 所示。运用**垂直平分线定理**，因为点 P 在垂直平分线上，故 P 是那条棱的中点。

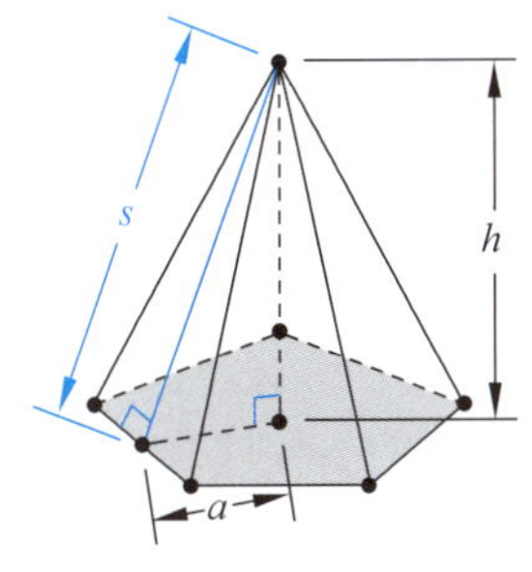

Figure 20-23

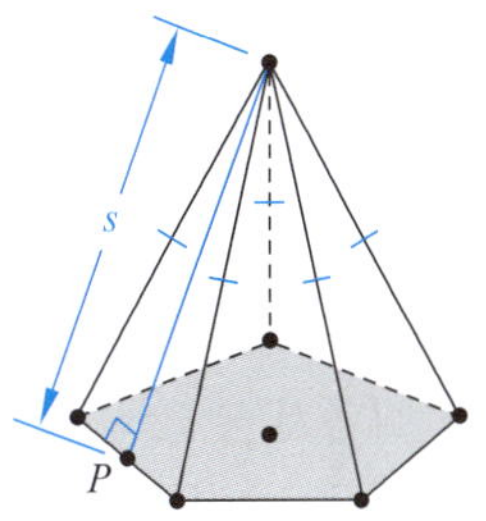

Figure 20-24

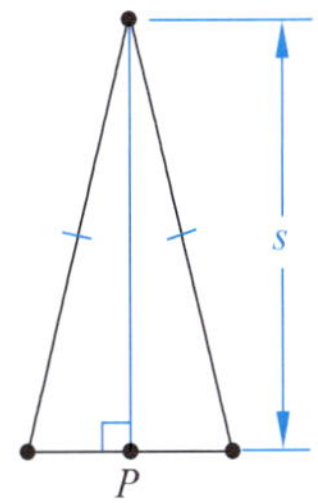

Figure 20-25

20.2.5 Summary 总结

1. Polyhedrons are the only solids that have faces, edges, and vertices (see their definitions in Section 20.1). The relationship between the numbers of faces (F), edges (E), and vertices (V) is explained by the Euler's Formula: $V+F=E+2$. 多面体是唯一一种有面、棱、顶点的立体(见 20.1 节的定义)。
2. Prisms and pyramids are special polyhedrons. 棱柱和棱锥是特殊的多面体。
3. A prism is one of right prism and oblique prism. Regular prisms belong to right prisms. 棱柱包括直棱柱和斜棱柱。正棱柱属于直棱柱。
4. A regular pyramid is a pyramid with a regular polygon as the base, and the line segment connecting the apex and the base's center is perpendicular to the base (This line segment is called the altitude.). In addition, in any regular pyramid R: 正棱锥是底为正多边形，且连接顶尖和底中心的线段与底垂直的棱锥(这条线段叫作高线)。另外，在任意正棱锥 R 里，有：
   (1) The lateral faces of R are congruent isosceles right triangles. R 的侧面是全等的等腰三角形。
   (2) R has a slant height (the distance between the apex and an edge of the base). R 有斜高(顶尖和一条底棱的距离)。
   (3) $h^2+a^2=s^2$, where h, a, and s are the height, base's apothem, and slant height, respectively. $h^2+a^2=s^2$。 h、a、s 分别是高、底的边心距、斜高。

20.3 Non-Polyhedrons 非多面体

视频 35

20.3.1 Cylinders 圆柱

Cylinder 柱体

n. ['sɪləndər]

Definition: A solid with two congruent and parallel closed curves as bases, joining by a smooth curved surface.

柱体是以两个全等且平行的密闭曲线为底,且由平滑的曲面连接两个底的立体。

Figure 20-26 shows two cylinders.

A circular cylinder's bases are congruent circles. A circular cylinder can be formed by rotating a rectangle 360° about one of its sides.

圆柱的底是等圆。矩形沿着一条边旋转 360° 可形成一个圆柱。

We will refer cylinder as circular cylinder, unless specified.

除非特别指定,通常所说的柱体为圆柱。

Figure 20-27 shows the parts of a cylinder.

图 20-27 展示了圆柱的各部分。

A cylinder is one of right cylinder and oblique cylinder.

圆柱是直圆柱和斜圆柱的一种。

Right Cylinder 直圆柱

n. [raɪt 'sɪləndər]

Definition: A cylinder for which the line joining the centers of the bases is perpendicular to the bases, as shown in Figure 20-28.

直圆柱是指连接两底圆心的直线与两底垂直的圆柱,如图 20-28 所示。

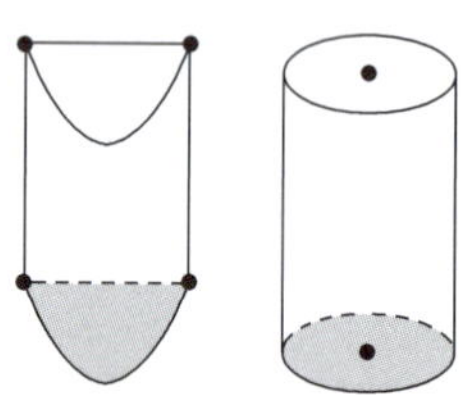
Figure 20-26

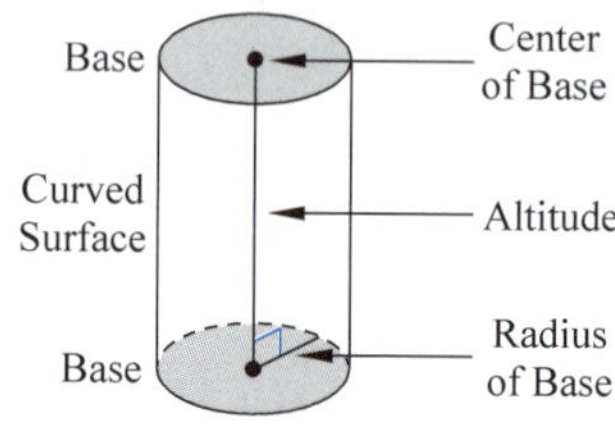

Figure 20-27

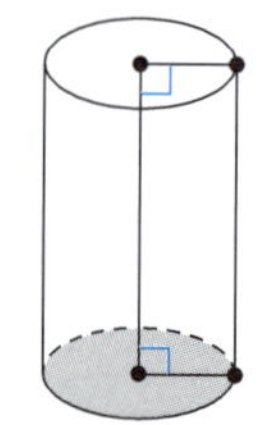
Figure 20-28

Oblique Cylinder 斜圆柱

n. [oʊ'blik 'sɪləndər]

Definition: A cylinder for which the line joining the centers of the bases is not perpendicular to the bases, as shown in Figure 20-29.

斜圆柱是指连接两底圆心的直线不与两底垂直的圆柱,如图 20-29 所示。

Altitude of a Cylinder 圆柱的高线

Definition: The line segment connecting the planes containing its bases(perpendicular to the bases).

圆柱的高线是指连接包含两底平面的线段(垂直于两底)。

Notation：The length of the altitude of a cylinder is called the height，usually denoted by h.　圆柱高线的长度叫作高，通常用 h 表示。

Examples：Figure 20-30 shows the altitudes of two cylinders in blue.　图 20-30 中蓝色线展示了两个圆柱的高线。

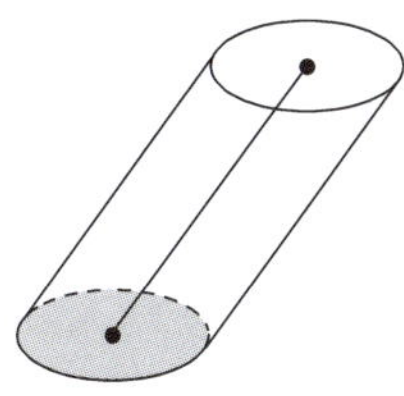

Figure 20-29

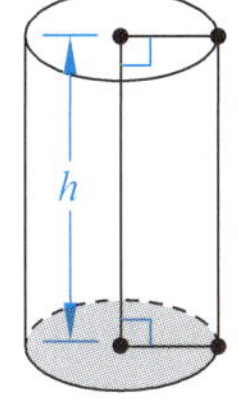

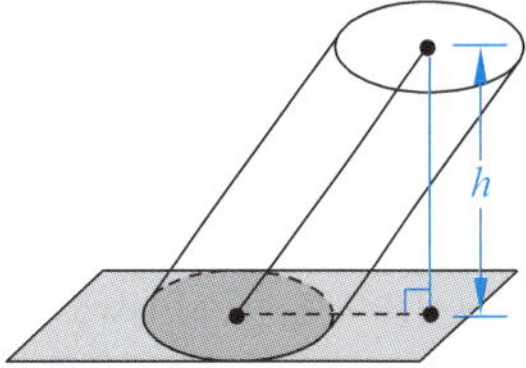

Figure 20-30

20.3.2　Cones　圆锥

Cone　锥体

n. [koʊn]

Definition：A solid with a closed curve as the base and a curved surface that connects the base to the vertex. This vertex is known as the apex.　锥体是以一个密闭曲线为底，且由一个平滑曲面连接底和顶点的立体。这个顶点又叫作顶尖。

Figure 20-31 shows two cones.

A circular cone's base is a circle.　圆锥的底是圆。

We will refer a cone as a circular cone，unless specified.　除非特别指定，通常所说的锥体为圆锥。

Figure 20-32 shows the parts of a cone.　图 20-32 展示了圆锥的各部分。

A cone is one of right cone and oblique cone.　圆锥包括直圆锥和斜圆锥。

Right Cone　直圆锥

n. [raɪt koʊn]

Definition：A cone for which the line connecting its apex and the center of the base is perpendicular to the base，as shown in Figure 20-33. Right cone has a slant height.　直圆锥是指连接顶尖和底圆心的线段与底垂直的圆锥，如图 20-33 所示。直圆锥有斜高。

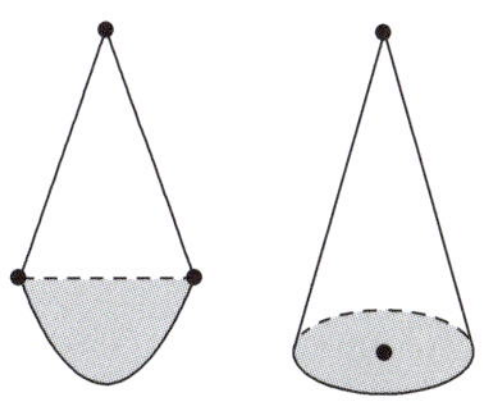

Figure 20-31

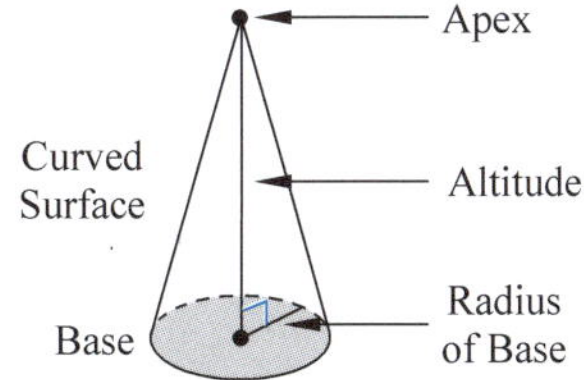

Figure 20-32

Figure 20-33

Oblique Cone　斜圆锥

n. [oʊ'blik koʊn]

Definition：A cone for which the line connecting its apex and the center of the base is not perpendicular to the base, as shown in Figure 20-34. 斜圆锥是指连接顶尖和底圆心的线段不与底垂直的圆锥,如图 20-34 所示。

Altitude of a Cone 圆锥的高线

Definition: The line segment connecting its apex and one point of the plane containing the base, such that the altitude is perpendicular to the base. 圆锥的高线是指连接顶尖和包含底的平面的线段(垂直于底)。

Notation: The length of the altitude of a cone is called the height, usually denoted by h. 圆锥的高线叫作高,通常用 h 表示。

Examples: Figure 20-35 shows the altitudes of two cones in blue. 图 20-35 用蓝色线展示了两个圆锥的高线。

Slant Height of a Right Cone 直圆锥的斜高

n. [slænt haɪt …]

Definition: The distance between the apex and a point on the circumference of the base, as shown in Figure 20-36. 直圆锥的斜高是指顶尖到底的圆周的距离,如图 20-36 所示。

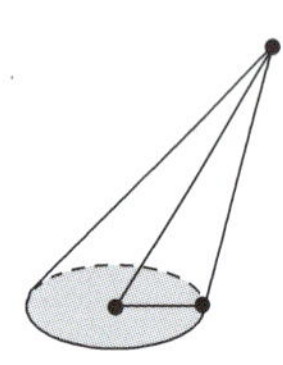

Figure 20-34

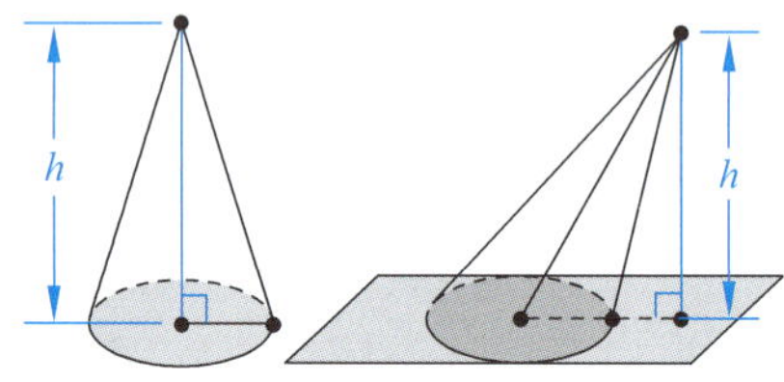

Figure 20-35

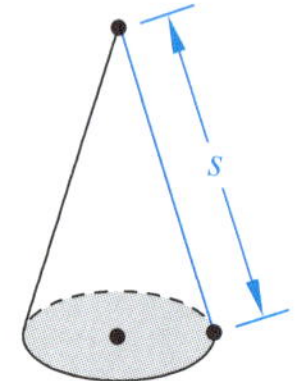

Figure 20-36

Notation: The slant height of a right cone is usually denoted by s.

Properties: In a right cone, let h, r, and s be the height, radius length, and slant height, respectively. By the Pythagorean Theorem, $h^2 + r^2 = s^2$, as shown in Figure 20-37. 在直圆锥中,h、r、s 分别为高、半径、斜高。根据勾股定理,则得 $h^2 + r^2 = s^2$,如图 20-37 所示。

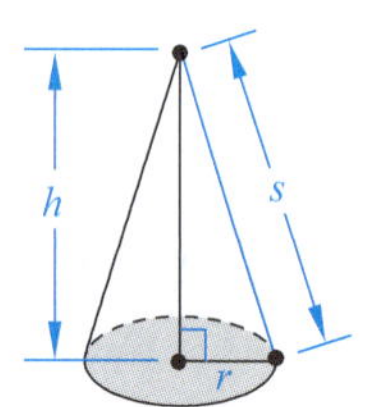

Figure 20-37

20.3.3 Spheres 球

Sphere 球

n. [sfɪər]

Definition: The set of points in the 3-dimensional space that is equidistant to one given point. The distance is given. The given point is known as the center of the sphere. 在三维空间中,与一个指定的点等距(距离指定)的点集称为球。指定的点叫作球心。

We say one point is "on the sphere" if it is on the surface of the sphere. 一个点"在球上"表示它在球的表面上。

The line segment connecting the center and one point on the sphere is the radius of the sphere. 连接球心和球上一点的线段叫作球的半径。

The parts of sphere are shown in Figure 20-38, many of which are analogues to those of a circle. 球的各部分标记如图 20-38 所示,很多都与圆类似。

A cross section of a sphere is a circle, as shown in the gray. 球的横截面为圆,如灰色所示。

O is the center of the sphere. O 为球心。

$\overline{OP}$ is a radius of the sphere. $\overline{OP}$ 为半径。

l is a tangent of the sphere, from which P is the point of tangency. l 为切线，P 为切点。

$\overline{AOB}$ is a diameter of the sphere. $\overline{AOB}$ 为直径。

s is a secant of the sphere. s 为割线。

$\overline{CD}$ is a chord of the sphere. $\overline{CD}$ 为弦。

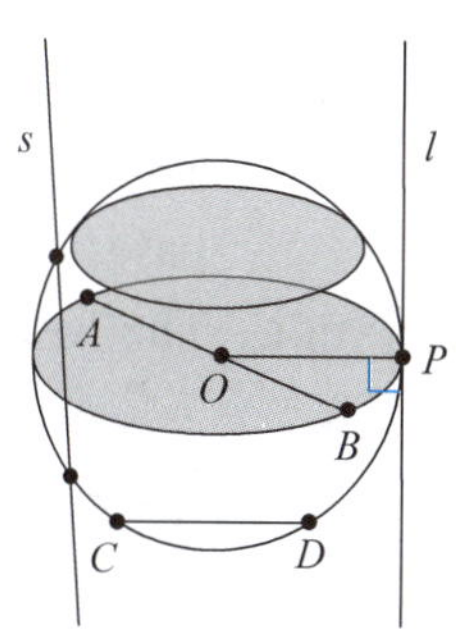

Figure 20-38

Notation：A sphere is named by its center, for which usually we denote that by O. 球是以球心命名的，一般称作 O。

Properties：Every cross section of a sphere is a circle. 球的横截面是圆。

20.3.4 Pyramidal and Conical Frusta 棱锥与圆锥的平截头体

Frustum 平截头体

n. [ˈfrʌs. təm]

Definition：A frustum of a pyramid or cone is a portion of the solid that lies between one or two parallel planes cutting it. 棱锥或圆锥的平截头体是这个立体的一部分，在一个或两个切割平面的中间。

The bases of a frustum are parallel, and we will assume that the top base is smaller than the bottom base. 假设上底比下底小，平截头体的底互相平行。

Figure 20-39 shows a pyramidal frustum. 图 20-39 展示了棱锥的平截头体。

Figure 20-40 shows a right conical frustum. 图 20-40 展示了直圆锥的平截头体。

Figure 20-41 shows a conical frustum. 图 20-41 展示了圆锥的平截头体。

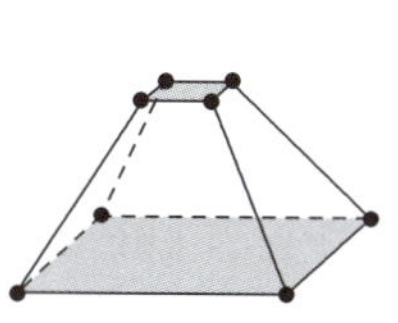
Figure 20-39

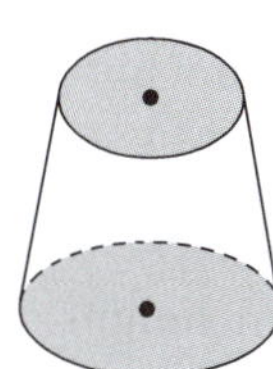
Figure 20-40

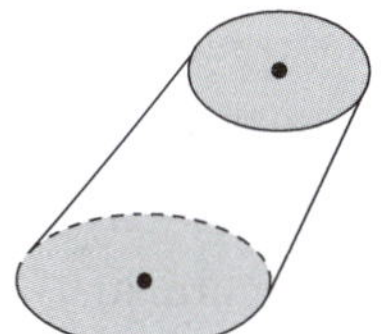
Figure 20-41

The parts of a pyramidal frustum is shown in Figure 20-42.
图 20-42 展示了棱锥的平截头体的各部分。

The parts of a right conical frustum is shown in Figure 20-43.
图 20-43 展示了直圆锥的平截头体的各部分。

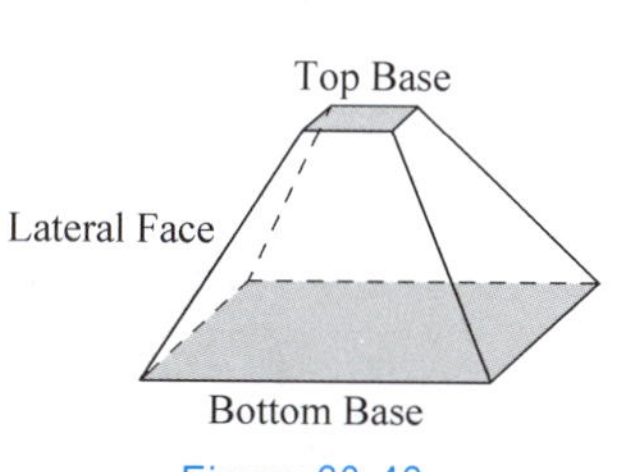

Figure 20-42

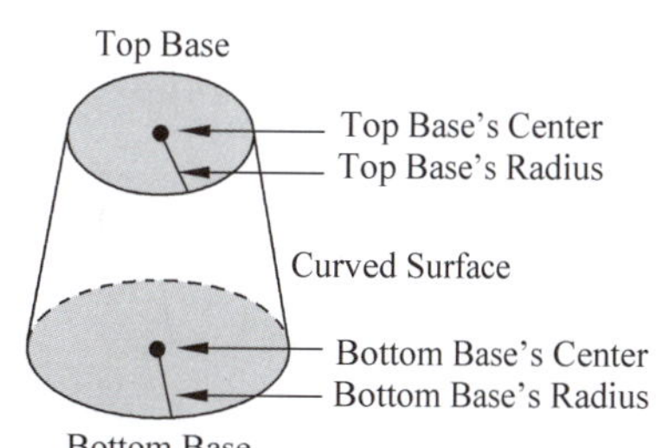

Figure 20-43

Altitude of a Frustum 平截头体的高线

Definition: The line segment connecting the planes containing its bases (perpendicular to the bases). 平截头体的高线是指连接包含两底的平面的线段(垂直于两底)。

Notation: The length of the altitude of a frustum is called the height, usually denoted by h. 平截头体高线的长度叫作高,通常以 h 表示。

Examples: Figure 20-44 shows an altitude of a pyramidal frustum.
图 20-44 展示了棱锥的平截头体的高线。
Figure 20-45 shows an altitude of a right conical frustum.
图 20-45 展示了直圆锥的平截头体的高线。
Figure 20-46 shows an altitude of a conical frustum.
图 20-46 展示了圆锥的平截头体的高线。

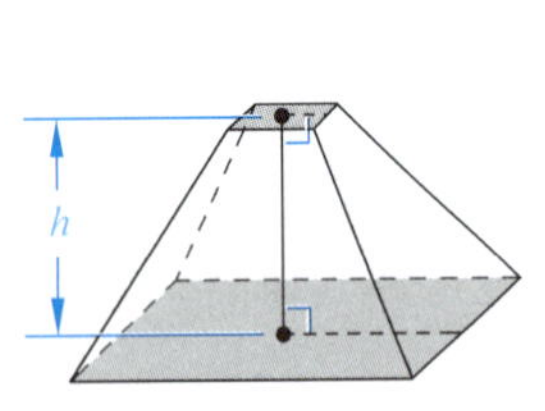

Figure 20-44

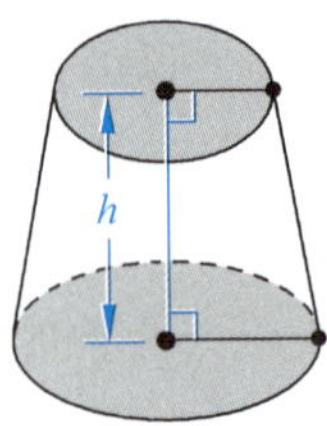

Figure 20-45

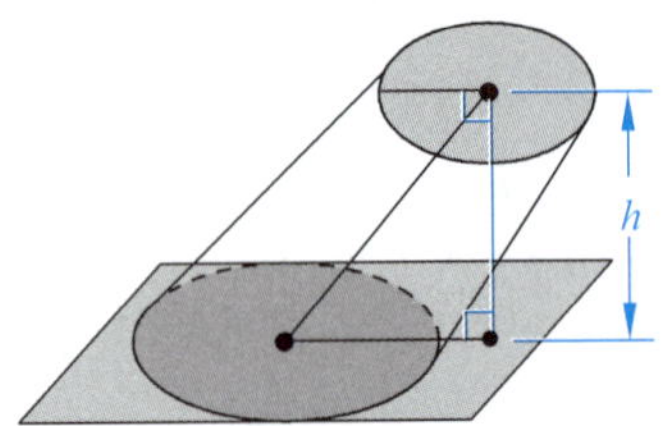

Figure 20-46

Slant Height of a Right Conical Frustum 正圆锥平截头体的斜高

n. [slænt haɪt …]

Definition: The slant height of a right conical frustum is PP', for which P and P' are points on the circumferences of the top and bottom bases, respectively, and $\overline{PP'}$ is along the curved surface of the frustum. 正圆锥平截头体的斜高是 PP'。其中 P 和 P' 分别在上底和下底的圆周上,使得 $\overline{PP'}$ 是沿着平截头体的曲面的线段。

Figure 20-47 shows the slant height of a right conical frustum.

Notation: The slant height of a right conical frustum is usually denoted by s. 正圆锥平截头体的斜高通常用 s 表示。

Slant Height of a Regular Pyramidal Frustum 正棱锥平截头体的斜高

n. [slænt haɪt …]

Definition: The distance between two edges in the bases such that these two edges are the edges of the same lateral face. Or, it is the height of a lateral face. 两底中两条棱的距离(这两条棱是同一个侧面的)。换言之,它是侧面的高。

Figure 20-48 shows the slant height of a regular pyramid.

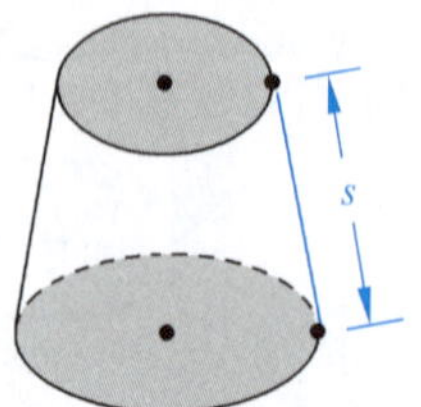

Figure 20-47

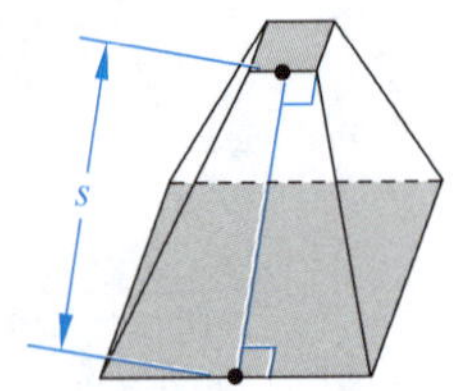

Figure 20-48

Notation: The slant height of a frustum of a regular pyramid is usually denoted by s. 正棱锥平截头体的斜高通常用 s 表示。

20.3.5 Summary 总结

1. A cylinder is one of right cylinder and oblique cylinder. 圆柱包括直圆柱和斜圆柱。
2. A cone is one of right cone and oblique cone. A right cone has a slant height, and satisfies $h^2 + r^2 = s^2$, where h, r, and s are the height, base's radius' length, and slant height respectively. 圆锥包括直圆锥和斜圆锥。直圆锥有斜高,并满足 $h^2 + r^2 = s^2$,其中 h、r、s 分别为高、底的半径长度、斜高。
3. Spheres are analogues of circles. 球与圆类似。
4. Among all frusta, only right conical frusta and regular pyramidal frusta have slant heights. 在所有平截头体中,只有直圆锥的平截头体和正棱锥的平截头体有斜高。

20.4 The Relationship between Two Solids 两个立体的关系

Congruent Solids 全等立体

n. [kənˈgruənt ˈsɑlɪds]

Definition: Two solids are congruent solids if their shape and size are the same. 若两个立体的形状和大小相同,则它们是全等立体。

Figure 20-49 shows congruent prisms and Figure 20-50 shows congruent cones. 图 20-49 展示了全等棱柱,图 20-50 展示了全等圆锥。

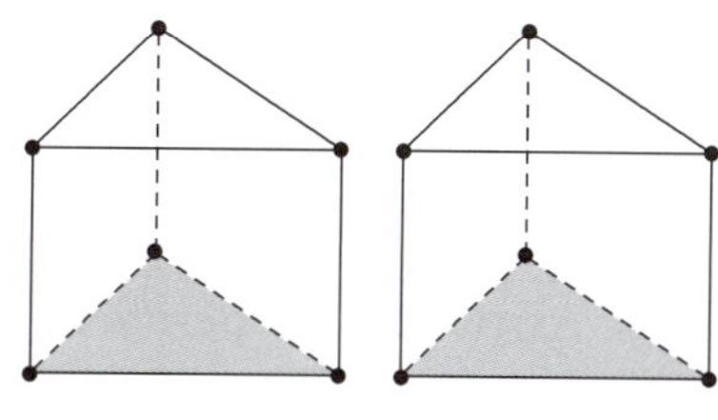

Figure 20-49

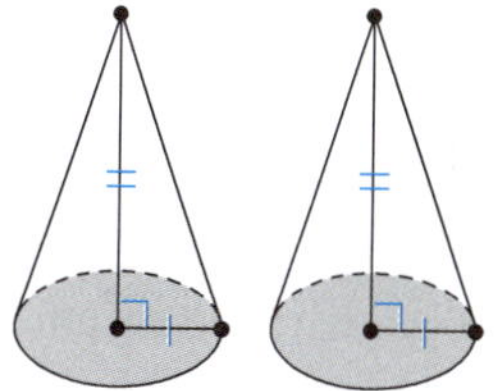

Figure 20-50

Similar Solids 相似立体

n. [ˈsɪmələr ˈsɑlɪds]

Definition: Two solids are similar solids if they have the same shape and proportional dimension. 若两个立体的形状相同,维度成比例,则它们是相似立体。

The factor by which the sizes of solids differ is known as the scale factor, which is an analogue of ratio of similitude of similar polygons. 立体大小差别的因子叫作比例因子,与相似多边形的相似比类似。

Properties: 1. Congruent solids are similar solids. Their scale factor is 1. 全等立体是相似立体,比例因子为 1。

2. All spheres are similar solids. 所有球都是相似立体。

Examples: Figure 20-51 shows similar prisms and Figure 20-52 shows similar cones. 图 20-51 展示了相似棱柱,图 20-52 展示了相似圆锥。

In Figure 20-52, we have, $\frac{h}{r}=\frac{H}{R}$.

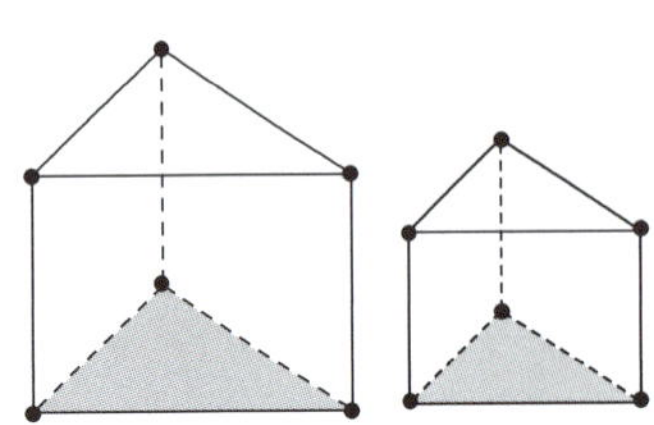

Figure 20-51

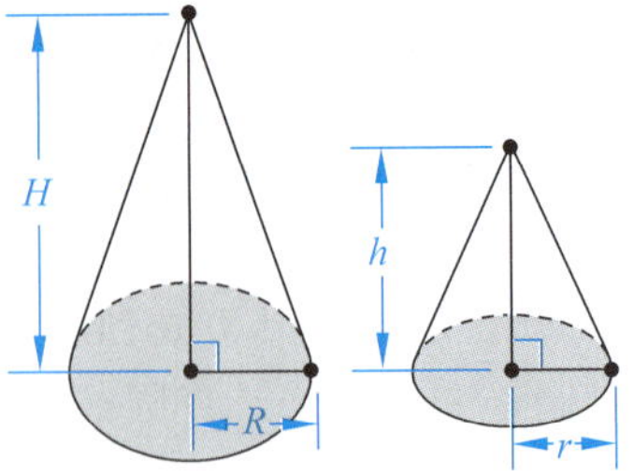

Figure 20-52

21. Calculations in Geometry 几何计算

视频 36

21.1 Perimeters and Areas of Plane Shapes 平面图形的周长与面积

Perimeter 周长

n. [pə'rɪmɪtər]

Definition: The distance around a plane shape. In particular, the perimeter of a polygon is the sum of lengths of its sides. 周长是指平面图形边界的距离。对于多边形,周长是它所有边的长度之和。

Refer to Table A-1 in Appendix A for the perimeter formulas of plane shapes. 平面图形的周长公式参照附录 A 的表 A-1。

Notation: The perimeter of a plane shape is denoted by P. 平面图形的周长用 P 表示。

The circumference, or the perimeter of a circle, is denoted by C. 圆周(圆的周长)用 C 表示。

Properties: 1. The perimeters of congruent plane shapes are equal. 全等平面图形的周长相等。

2. Perimeter is the sum of lengths of line segments/curves, which is one-dimensional. Therefore, 因为周长是线段/曲线的长度总和,是一维的,所以,

(1) The ratio of perimeters of two similar polygons is equal to their ratio of similitude. 两个相似多边形的周长之比等于它们的相似比。

(2) The ratio of circumferences of two circles is equal to their ratio of radii's lengths. 两个圆的圆周之比等于它们的半径长度之比。

Area　面积

n. [ˈeəriə]

Definition: The amount of two-dimensional space that a plane shape contains, or, the number of unit squares that takes to cover the entire shape.　面积是指平面图形所占的二维空间(能覆盖整个图形的单位正方形个数)。

Refer to Table A-2 in Appendix A for the area formulas of plane shapes.　平面图形的面积公式参照附录 A 的表 A-2。

Notation: The area of a plane shape is denoted by A.　平面图形的面积用 A 表示。

The area of a polygon $ABC\cdots$ is denoted by $[ABC\cdots]$.　多边形 $ABC\cdots$的面积用$[ABC\cdots]$表示。

Refer to Table A-5 in Appendix A for the notation of areas.　面积的标记方式参照附录 A 的表 A-5。

Properties: 1. The areas of congruent plane shapes are equal. 全等平面图形的面积相等。

2. Area is the sum of two-dimensional space. Therefore,　面积是二维空间之和。所以可知：

(1) The ratio of areas of two similar polygons is equal to the square of their ratio of similitude.　两个相似多边形的面积之比等于它们相似比的平方。

(2) The ratio of areas of two circles is equal to the square of their ratio of radius lengths.　两个圆的面积之比等于它们半径之比的平方。

3. **Same Base/Same Altitude Property:**

等底/等高性质：

(1) If the lengths of bases of triangles A and B are the same, and the height of triangle A is k times that of triangle B, then the area of A is k times that of B. 若三角形 A 和 B 的底的长度相同，且 A 的高是 B 的高的 k 倍，则 A 的面积是 B 的面积的 k 倍。

(2) If the heights of triangles A and B are the same, and the length of the base of triangle A is k times that of triangle B, then the area of A is k times that of B. 若三角形 A 和 B 的高相同，且 A 的底的长度是 B 的底的长度的 k 倍，则 A 的面积是 B 的面积的 k 倍。

The property also holds for parallelograms and trapezoids. For trapezoids, we may want to rename this property as **Same Sum-of-Bases/Same Altitude Property.**　该性质也可应用到平行四边形和梯形中。在梯形中，该性质可称为**等底长度和/等高性质**。

Proofs: To prove Property 3(1), suppose the lengths of the bases are b, the heights of triangle B and A are h and kh, respectively. We have:

要证明性质 3(1)，设底的长度均为 b。三角形 B 和三角形 A 的高分别为 h 和 kh。得到：

$$\text{Area of Triangle } A = \frac{1}{2}b(kh) = k\left(\frac{1}{2}bh\right).$$

$$\text{Area of Triangle } B = \frac{1}{2}bh.$$

It satisfies that Area of Triangle $A = k$(Area of Triangle B).

To prove Property 3(2), suppose the heights are h, the lengths of the bases of triangle B and A are b and kb, respectively. We have:
要证明性质 3(2),设高均为 h。三角形 B 和三角形 A 底的长度分别为 b 和 kb。得到:

Area of Triangle $A = \frac{1}{2}(kb)h = k\left(\frac{1}{2}bh\right)$.

Area of Triangle $B = \frac{1}{2}bh$.

It satisfies that Area of Triangle $A = k$(Area of Triangle B).

Similarly, using the area formulas can show this property for parallelograms and trapezoids. 同理,用面积公式可证明平行四边形和梯形的这个性质。

21.2 Surface Areas and Volumes of Solids 立体图形的表面积与体积

Surface Area 表面积

n. [ˈsɜː.fɪs ˌer.i.ə]

Definition: The total area of a solid's surface. In particular, the surface area of a polyhedron is the sum of areas of its faces. 立体图形的表面积是指立体表面的面积。对于多面体来说,表面积是它所有面的面积之和。

Refer to Table A-3 in Appendix A for the surface area formulas of solids. 立体的表面积公式参照附录 A 的表 A-3。

Notation: The surface area of a solid is denoted by SA. 立体的表面积用 SA 表示。

Properties: 1. The surface areas of congruent solids are equal. 全等立体的表面积相等。

2. Surface area is the sum of two-dimensional space. Therefore,
表面积是二维空间之和。所以,

(1) The ratio of surface areas of two similar solids is equal to the square of their scale factor. 两个相似立体的表面积之比等于它们的比例因子之比的平方。

(2) The ratio of surface areas of two spheres is equal to the square of the ratio of their radius lengths. 两个球的表面积之比等于它们的半径之比的平方。

Volume 体积

n. [ˈvɑljəm]

Definition: The amount of three-dimensional space inside a solid, or, the number of unit cubes that takes to cover the entire solid. 体积是指立体图形所占的三维空间(能覆盖整个立体的单位立方体个数)。

Refer to Table A-4 in Appendix A for the volume formulas of solids. 立体的体积公式参照附录 A 的表 A-4。

Notation: The volume of a solid is denoted by V. 立体的体积用 V 表示。

Properties: 1. The volumes of congruent solids are equal.

全等立体的体积相等。

2. Volume is the sum of three-dimensional space. Therefore,
体积是三维空间之和。所以,

(1) The ratio of volumes of two similar solids is equal to the cube of their scale factor. 两个相似立体的体积之比等于它们的比例因子的立方。

(2) The ratio of volumes of two spheres is equal to the cube of the ratio of their radius lengths. 两个球的体积之比等于它们的半径之比的立方。

Part 3: Pre-Calculus

第3部分：

微积分初步

本部分内容为微积分的基础知识。微积分（calculus）产生于17世纪，是高等数学中研究函数的微分（differentiation）和积分（integration），以及有关概念和应用的数学分支。微积分描述的是连续的变化，无论是现实世界的运动，还是纯抽象的代数曲线，这些变化都是无穷小的瞬时变化。微积分的发明堪称人类智慧的结晶，被称为“数学创造的最有效的科学研究工具”。

牛顿和莱布尼茨两位科学巨匠之间的“到底谁先发明了微积分”之争直到他们离世都没有得出分晓。现在我们称微积分公式为“牛顿-莱布尼茨公式”，不知道这对“冤家”看到后人把他们的名字写在一起会做何感想。

通过这一部分的学习，学生对函数的概念会更加熟悉，本部分内容包括但不局限于函数运算、因式分解、多项式除法、多项式函数、有理函数、倒数函数、反函数、对数函数，最后会介绍微积分的极限与导数。导数与上述的连续变化密切相关。此外，本部分也介绍了代数的一些重要课题，与数学竞赛、大学数学分析密不可分，包括但不局限于三维空间、矩阵、复数、圆锥曲线、级数、概率和高等三角函数。

22. Systems of Linear Inequalities with Two Variables 二元一次不等式组

Review Functional Inequalities in Section 7.6. 复习 7.6 节函数不等式。

Linear Programming 线性规划

n. [ˈlɪniər ˈproʊ.græm.ɪŋ]

Definition: The process of writing multiple linear inequalities related to some situation and finding the optimal value (minimum or maximum) of a linear objective function. 线性规划是指对一个环境列出若干线性不等式,并求出目标线性函数最值的步骤。

Each inequality is called a constraint. 每条不等式称为约束条件。

The region bounded by the graphs of the inequalities is called the feasible region. 被不等式图像围成的区域称为可行域。

The objective function is called the optimization equation. It must be linear. 目标函数必须为线性。

The maximum or minimum value of the optimization equation always occurs at one of the vertices of the feasible region. 目标函数的最值必然发生在可行域中的一个顶点上。

Examples: As a tutor in English and mathematics, Bob offers two types of services. He allots 40 minutes for an English lesson and 60 minutes for a math lesson. He cannot tutor more than 6 math lessons per day. Every day he has 10 hours available for lessons. If an English lesson costs \$ 60 and a math lesson costs \$ 75, what is a combination of numbers of English and math lessons that will maximize Bob's income per day?

Let e be the number of English lessons and m be the number of math lessons.

We have $\begin{cases} e \geqslant 0 \\ m \geqslant 0 \\ m \leqslant 6 \\ 40e + 60m \leqslant 600 \end{cases}$.

Our optimization equation is $I(e, m) = \$60e + \$75m$, from which we want to find its maximum value.

We graph the system of equations as shown in Figure 22-1.

We find the vertices (e, m) to be $(0,0)$, $(0,6)$, $(6,6)$, and $(15,0)$.

We plug each vertex into our optimization equation, as shown in Table 22-1:

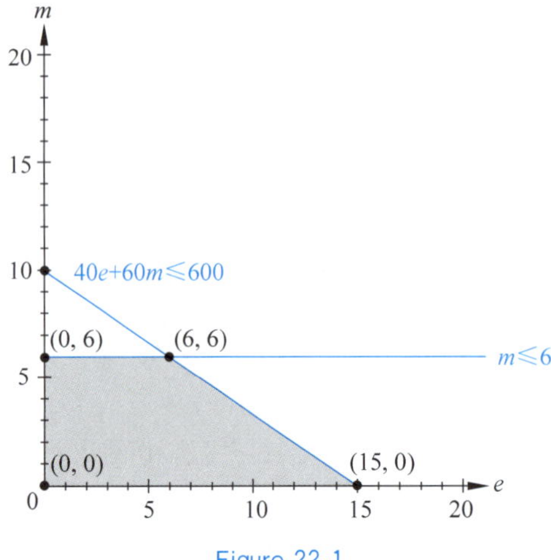

Figure 22-1

Table 22-1

Input (e,m) 输入	Output $I(e,m)=\$60e+\$75m$ 输出
(0,0)	\$0
(0,6)	\$450
(6,6)	\$810
(15,0)	\$900

We know that (15, 0) produces the maximum output. In other words, tutoring 15 English lessons per day produces the maximum income.

23. The Three-Dimensional Space 三维空间

23.1 Introduction 介绍

Three-Dimensional Space 三维空间

n. [θri də'menʃənəl speɪs]

Definition: The space determined by three mutually perpendicular axes: x-axis, y-axis, and z-axis.

三维空间是被三条互相垂直的轴(x 轴、y 轴、z 轴)确定的空间。

Compared to the coordinate plane, the three-dimensional space has one extra dimension (z-axis), which represents the height of a plane. 与坐标平面比起来,三维空间多了一个维度(z 轴),代表一个平面的高。

Many terminologies from the coordinate plane are analogues for those of the three-dimensional space. 坐标平面与三维空间的很多术语均类似。

The three-dimensional space is shown in Figure 23-1.

Properties：1. Each point on the three-dimensional plane can be specified by an ordered triple of numbers. 三维空间上的每个点都能用有序三元组表示。

2. (1) The xy-plane is a plane that contains the x-axis and y-axis and is perpendicular to the z-axis, as shown in Figure 23-2.

xy 平面包含 x 轴和 y 轴，与 z 轴垂直，如图 23-2 所示。

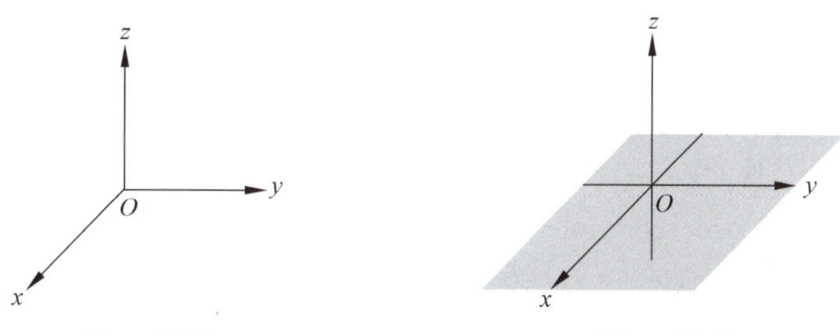

Figure 23-1 Figure 23-2

(2) The xz-plane is a plane that contains the x-axis and z-axis and is perpendicular to the y-axis, as shown in Figure 23-3.

xz 平面包含 x 轴和 z 轴，与 y 轴垂直，如图 23-3 所示。

(3) The yz-plane is a plane that contains the y-axis and z-axis and is perpendicular to the x-axis, as shown in Figure 23-4.

yz 平面包含 y 轴和 z 轴，与 x 轴垂直，如图 23-4 所示。

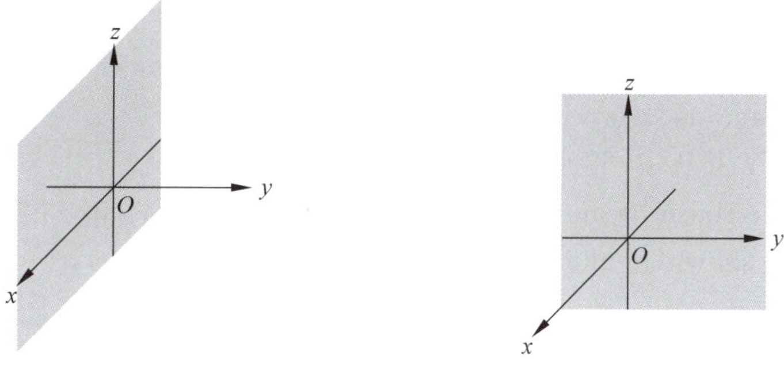

Figure 23-3 Figure 23-4

Coordinate Axis 坐标轴

n. [koʊ'ɔrdənˌeɪt 'æksəs]

Definition：Refer to the x-axis, the y-axis, or the z-axis. 坐标轴指 x 轴、y 轴或 z 轴。

Coordinate 坐标

n. [koʊ'ɔrdənˌeɪt]

Definition：Refer to the x-coordinate, the y-coordinate, or the z-coordinate. 坐标指 x 坐标、y 坐标或 z 坐标。

The coordinates of a point are represented by an ordered triple. 每个点的坐标都可以用有序三元组表示。

Ordered Triple 有序三元组

n. [ˈɔːrdəd ˈtrɪpəl]

Definition: The representation of the location (as a point) on the three-dimensional space. 有序三元组是点在三维空间上的位置的表达方法。

Notation: (x, y, z), in which x represents the x-coordinate of the point, y represents the y-coordinate of the point, and z represents the z-coordinate of the point. 在(x,y,z)中，x代表点的x坐标，y代表点的y坐标，z代表点的z坐标。

If point A is located at (a,b,c), then we also say A has coordinates of (a,b,c). 若A的位置在(a,b,c)，即A的坐标为(a,b,c)。

Properties:

1. Each ordered triple (a,b,c) on the coordinate plane is located in the intersection of the planes $x=a$, $y=b$, and $z=c$. 三维空间上的有序三元组(a,b,c)的位置为平面上$x=a$、$y=b$、$z=c$的交点。
2. Points along the x-axis have y-coordinate and z-coordinate equal to 0. 沿着x轴的点的y坐标与z坐标均为0。
3. Points along the y-axis have x-coordinate and z-coordinate equal to 0. 沿着y轴的点的x坐标与z坐标均为0。
4. Points along the z-axis have x-coordinate and y-coordinate equal to 0. 沿着z轴的点的x坐标与y坐标均为0。
5. The origin's has coordinates of $(0,0,0)$. 坐标原点的坐标为$(0,0,0)$。
6. Figure 23-5 shows how to plot the blue point (a,b,c). 图23-5展示了如何画出蓝色的点(a,b,c)。
   We first locate the point $(a,b,0)$ from the xy-plane, as shown in the gray rectangle.
   首先须从xy平面找到点$(a,b,0)$，如灰色矩形所示。
   We finish by moving the point $(a,b,0)$ c units up, as shown in the rectangle with vertices $(0,0,0)$, $(a,b,0)$, (a,b,c), $(0,0,c)$.
   最后把$(a,b,0)$上移c格，如顶点为$(0,0,0)$、$(a,b,0)$、(a,b,c)、$(0,0,c)$的矩形所示。

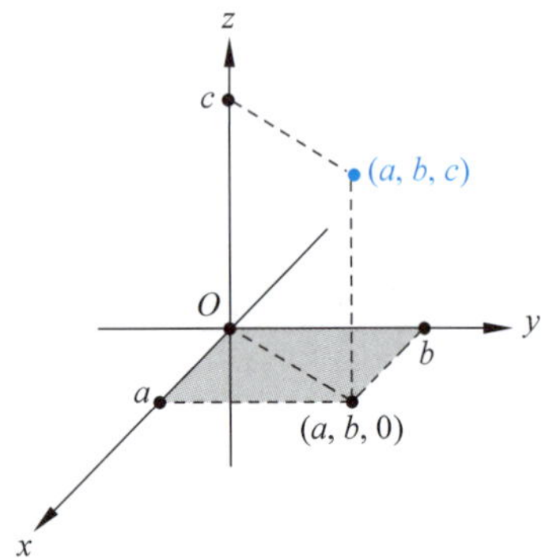

Figure 23-5

Point 点

n. [pɔɪnt]

Definition: The visual representation on the three-dimensional space of an ordered triple. 点为有序三元组在三维空间上的视觉表示法。

Origin 坐标原点

n. [ˈɔrədʒɪn]

Definition: The intersection of the coordinate axes. The origin has $(0,0,0)$ as the coordinates. 坐标原点是指坐标轴的交点，其坐标为$(0,0,0)$。

Intercept 截距

n. [ˌɪntərˈsept]

Definition: The number on a coordinate axis in which the graph intersects that axis. 截距是指图像与坐标轴相交的坐标数字。

For a graph in the three-dimensional space:

对于在三维空间的图像有:

The x-intercept is the x-coordinate of the point that the graph intersects the x-axis. At this point, the y and z-coordinates are 0.

x 截距是图像与 x 轴交点的 x 坐标。在这点上,y 和 z 坐标均为 0。

The y-intercept is the y-coordinate of the point that the graph intersects the y-axis. At this point, the x and z-coordinates are 0.

y 截距是图像与 y 轴交点的 y 坐标。在这点上,x 和 z 坐标均为 0。

The z-intercept is the z-coordinate of the point that the graph intersects the z-axis. At this point, the x and y-coordinates are 0.

z 截距是图像与 z 轴交点的 z 坐标。在这点上,x 和 y 坐标均为 0。

Phrases: x-intercept, y-intercept, z-intercept

Questions: What are the intercepts of the plane $5x+3y-2z=30$?

Answers: To find the x-intercept, set y and z to 0:

$$5x+3(0)-2(0)=30$$
$$5x=30$$
$$x=6.$$ The x-intercept is 6.

To find the y-intercept, set x and z to 0:

$$5(0)+3y-2(0)=30$$
$$3y=30$$
$$y=10.$$ The y-intercept is 10.

To find the z-intercept, set x and y to 0:

$$5(0)+3(0)-2z=30$$
$$-2z=30$$
$$z=-15.$$ The z-intercept is -15.

23.2 Formulas 公式

Midpoint Formula 中点公式

n. [ˈmɪdˌpɔɪnt ˈfɔrmjələ]

Definition: Given two points A and B, the Midpoint Formula gives the midpoint of $\overline{AB}$. 中点公式给出端点为两个点的线段的中点。

Notation: Suppose point A has the coordinates (x_1, y_1, z_1) and point B has the coordinates (x_2, y_2, z_2), the Midpoint Formula is $M=\left(\frac{x_1+x_2}{2}, \frac{y_1+y_2}{2}, \frac{z_1+z_2}{2}\right)$, for which $AM=BM$.

Properties: 1. Note that the x, y and z coordinates of M are independent of each other. 计算中点

时,坐标之间相互独立。

2. Switching points (x_1, y_1, z_1) and (x_2, y_2, z_2) leads to the same result. This is due to the Commutative Property of Addition. 根据加法交换律,把两点互换后代入公式,结果相同。

Questions: What is the coordinates of the midpoint of $\overline{AB}$ for each of the following if:

(1) $A=(5,9,4)$ and $B=(3,1,8)$

(2) $A=(-7,-2,-10)$ and $B=(-11,-16,-4)$

Answers: (1) $\left(\frac{5+3}{2},\frac{9+1}{2},\frac{4+8}{2}\right)=(4,5,6)$.

(2) $\left(\frac{-7+(-11)}{2},\frac{-2+(-16)}{2},\frac{-10+(-4)}{2}\right)=(-9,-9,-7)$.

Distance Formula 距离公式

n. [ˈdɪstəns ˈfɔrmjələ]

Definition: Given two points A and B, the Distance Formula gives the length of $\overline{AB}$. 距离公式给出端点为两个点的线段的长度。

Notation: Suppose point A has coordinates (x_1, y_1, z_1) and point B has coordinates (x_2, y_2, z_2), the Distance Formula is $\sqrt{(x_2-x_1)^2+(y_2-y_1)^2+(z_2-z_1)^2}$. Sometimes it is also written in $\sqrt{(\Delta x)^2+(\Delta y)^2+(\Delta z)^2}$, for which Δx denotes the change in x, Δy denotes the change in y, and Δz denotes the change in z. All of Δx, Δy, and Δz are nonnegative.

Properties: 1. This formula is based on applying Pythagorean Theorem several times. 这个公式源于多次运用勾股定理。

For two points $A=(x_1, y_1, z_1)$ and $B=(x_2, y_2, z_2)$ on the xyz-plane, we can construct a rectangular prism for which $\overline{AB}$ is the long diagonal, as shown in Figure 23-6.

给出 xyz 平面的两点 A 与 B,要求它们的距离,先画一个长方体,使得 $\overline{AB}$ 为对角线,如图 23-6 所示。

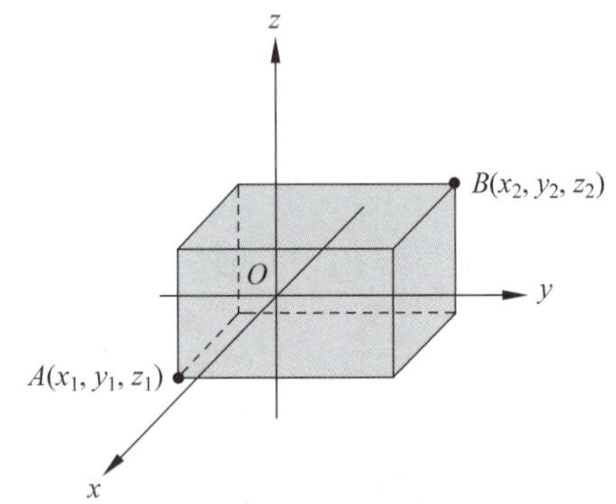

Figure 23-6

If exactly one of the coordinates are the same in A and B, then the rectangular prism degenerates into a rectangle. This reduces to the Pythagorean Theorem on a two-dimensional plane.

若 A 与 B 的其中一个坐标相同,则长方体退化为矩形。距离可用勾股定理求出。

If exactly two of the coordinates are the same in A and B, then the rectangular prism degenerates into a line segment parallel to the third axis. The distance is simply the difference between the third coordinate.

若 A 与 B 的其中两个坐标相同,则长方体退化为平行于第三个坐标轴的线段。距离为第三个坐标的差。

If all three of the coordinates are the same in A and B, then the rectangular prism degenerates into a point, from which the distance is 0.

若 A 与 B 的所有坐标相同,则长方体退化为一个点,距离为 0。

As shown in Figure 23-7, we know that $AB^2 = AD^2 + BD^2$, which is $AB^2 = AD^2 + (\Delta z)^2$. By Pythagorean Theorem, $AD^2 = (\Delta x)^2 + (\Delta y)^2$. Therefore, $AB^2 = AD^2 + (\Delta z)^2 = (\Delta x)^2 + (\Delta y)^2 + (\Delta z)^2$, from which $AB = \sqrt{(\Delta x)^2 + (\Delta y)^2 + (\Delta z)^2}$.

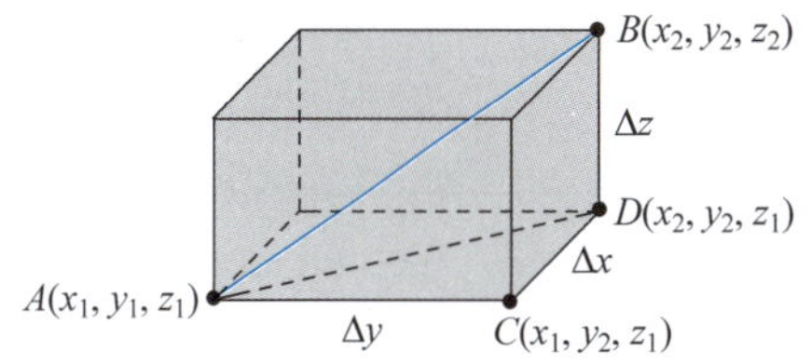

Figure 23-7

2. Switching points (x_1, y_1, z_1) and (x_2, y_2, z_2) leads to the same result. This is due to the fact that the squares of a number and its opposite are equal: $a^2 = (-a)^2$. 把点互换后代入公式,结果相同,因为一个数的平方与它相反数的平方相等。

Note that the following formulas are equivalent.

以下公式等价:

$AB = \sqrt{(x_2 - x_1)^2 + (y_2 - y_1)^2 + (z_2 - z_1)^2}$.

$AB = \sqrt{|x_2 - x_1|^2 + |y_2 - y_1|^2 + |z_2 - z_1|^2}$.

$AB = \sqrt{(\Delta x)^2 + (\Delta y)^2 + (\Delta z)^2}$.

Questions: What is the distance for each of the following pairs of points?

(1) (1,3,8) and (2,6,3)

(2) (−5,−3,4) and (5,−8,−2)

Answers: (1) $\sqrt{(2-1)^2 + (6-3)^2 + (3-8)^2} = \sqrt{1^2 + 3^2 + (-5)^2} = \sqrt{35}$.

(2) $\sqrt{(5-(-5))^2 + (-8-(-3))^2 + (-2-4)^2} = \sqrt{10^2 + (-5)^2 + (-6)^2} = \sqrt{161}$.

23.3 Planes 平面

Equation of a Plane 平面的方程式

Definition: The equation $ax + by + cz = d$, for which a, b, c, and d are constants.

Properties: 1. Special cases of a, b, and c:

(1) If a and b are both 0, then we have $cz = d$, or $z = d/c$.

In other words, z is a constant. This is a plane perpendicular to the z-axis.

当 a 与 b 均为 0,得出来的等式是 $cz = d$ 或 $z = d/c$,即 z 等于一个常数。这个平面与 z 轴垂直。

(2) If a and c are both 0, then we have $by = d$, or $y = d/b$.

In other words, y is a constant. This is a plane perpendicular to the y-axis.

当 a 与 c 均为 0,得出来的等式是 $by = d$ 或 $y = d/b$,即 y 等于一个常数。这个平面与 y 轴垂直。

(3) If b and c are both 0, then we have $ax = d$, or $x = d/a$.

In other words, x is a constant. This is a plane perpendicular to the x-axis.

当 b 与 c 均为 0,得出来的等式是 $ax = d$ 或 $x = d/a$,即 x 等于一个常数。这个平

面与 x 轴垂直。

2. Equation of planes that contain two coordinate axes:
   包含两条坐标轴的平面的方程式:
   (1) Equation of the xy-plane: $z=0$. $(a,b,d=0,c\neq 0)$
   (2) Equation of the xz-plane: $y=0$. $(a,c,d=0,b\neq 0)$
   (3) Equation of the yz-plane: $x=0$. $(b,c,d=0,a\neq 0)$
3. We can graph planes by intercepts:
   可用截距画图。

   To graph $ax+by+cz=d$, note the x-intercept is $\frac{d}{a}$, the y-intercept is $\frac{d}{b}$, and the z-intercept is $\frac{d}{c}$.

   Locate the points $\left(\frac{d}{a},0,0\right)$, $\left(0,\frac{d}{b},0\right)$, and $\left(0,0,\frac{d}{c}\right)$ and draw a plane passing through these three points. Note that three non-collinear points determine a plane.

   在方程式 $ax+by+cz=d$ 中, x 截距为$\frac{d}{a}$, y 截距为$\frac{d}{b}$, z 截距为$\frac{d}{c}$。

   找出点$\left(\frac{d}{a},0,0\right)$、$\left(0,\frac{d}{b},0\right)$、$\left(0,0,\frac{d}{c}\right)$,画一个过这三点的平面。注意三个非共线点确定一个平面。
4. Coming up with an equation for a plane requires cross product of vectors, which will be covered in college calculus. 求出平面的等式需要用到向量积,这会在大学微积分课程提及。

Examples: To graph the plane $12x+15y+20z=60$, we locate the points $(5,0,0)$, $(0,4,0)$, and $(0,0,3)$, then draw a plane connecting these points, as shown in Figure 23-8.

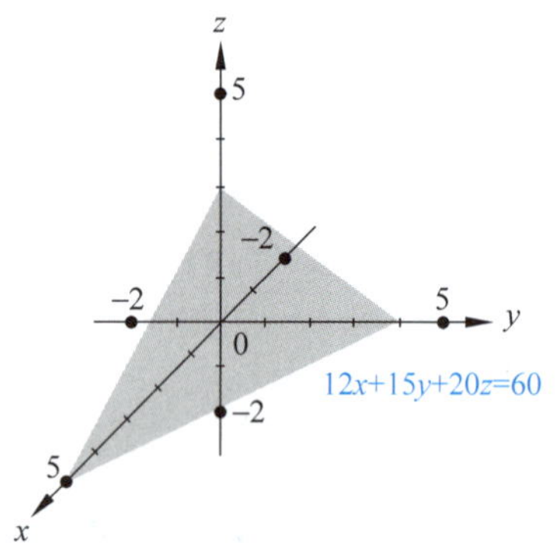

Figure 23-8

23.4 Systems of Linear Equations with Three Unknowns 三元一次方程组

视频 37

Warning: Solving system of linear equations with three or more unknowns by graphing or substitution is extremely cumbersome. We will skip it and stick to elimination in the future.

说明:用画图法或代入消元法解三元或更多元一次方程组太复杂了,一般会忽略这两种方法,而用加减消元法,且以后都会用到。

Solving Systems of Linear Equations with Three Unknowns by Elimination 加减消元法

Definition: This is based on Solving Systems of Linear Equations with Two Unknowns by Elimination. 解三元一次方程组的加减消元法基于解二元一次方程的加减消元法。

Properties: Suppose you have a system of three linear equations: Equations A, B, and C. 假设现在有三元一次方程组:方程 A、B、C。

1. Group two equations. Use elimination to eliminate one variable. Call the resulting equation Equation D.
先把两个方程式结为一组，用加减消元法消去一个未知数，得出的方程式称为 D。
2. Of Equations A, B, and C, group another pair of equations (different from Step 1). Use elimination to cancel out the variable that is eliminated in Step 1. Call the resulting equation Equation E.
在方程 A、B、C 中，把两个方程式结为另一组（与第 1 步的方程组不同）。用加减消元法消去在第 1 步里消去的那个未知数。得出的方程式称为 E。
3. Group Equations D and E. Solve this system of linear equations with two unknowns using elimination.
把方程 D 和 E 结为一组，用加减消元法解二元一次方程组。
4. Substitute the result from Step 3 back to any of Equations A, B, and C to find the third unknown.
把第 3 步的结果代入 A、B、C 的任何一条方程，解第三个未知数。

Examples: 1. Suppose we want to solve the following system of equations:

$$\begin{cases} A: x+2y-3z=-4 \\ B: x-3y+z=-2 \\ C: x+3y+5z=22 \end{cases}.$$

We note that x is the easiest to eliminate.

Step 1: Group equations A and B. Use elimination to cancel out x.

$$\begin{cases} A: x+2y-3z=-4 \\ B: x-3y+z=-2 \end{cases}.$$

$A-B$ gives $5y-4z=-2$, which is Equation D.

Step 2: Group equations A and C. Use elimination to cancel out x.

$$\begin{cases} A: x+2y-3z=-4 \\ C: x+3y+5z=22 \end{cases}.$$

$C-A$ gives $y+8z=26$, which is Equation E.

Step 3: Group equations D and E. Use elimination.

$$\begin{cases} D: 5y-4z=-2 \\ E: y+8z=26 \end{cases}.$$

Double Equation D:

$$\begin{cases} 10y-8z=-4 \\ y+8z=26 \end{cases}.$$

Adding equations gives $11y=22$, from which $y=2$.

Substituting $y=2$ in any of the equations involving y and z only, we have $z=3$.

Substituting $y=2$ and $z=3$ in any of Equations A, B, or C gives $x=1$.

Therefore, our answer is $(x,y,z)=(1,2,3)$.

2. Suppose we want to solve the following system of equations:

$$\begin{cases} A: 7x + y - 3z = 16 \\ B: 2x + 4y - 5z = 19 \\ C: 3x - 2y + 2z = -2 \end{cases}.$$

We note that y is the easiest to eliminate.

Step 1: Group equations A and B. Use elimination to cancel out y.

$$\begin{cases} A: 7x + y - 3z = 16 \\ B: 2x + 4y - 5z = 19 \end{cases}.$$

Multiply Equation A by 4:

$$\begin{cases} 28x + 4y - 12z = 64 \\ 2x + 4y - 5z = 19 \end{cases}.$$

Subtracting equations gives $26x - 7z = 45$, which is Equation D.

Step 2: Group equations A and C. Use elimination to cancel out y.

$$\begin{cases} A: 7x + y - 3z = 16 \\ C: 3x - 2y + 2z = -2 \end{cases}.$$

Multiply Equation A by 2:

$$\begin{cases} 14x + 2y - 6z = 32 \\ 3x - 2y + 2z = -2 \end{cases}.$$

Adding equations gives $17x - 4z = 30$, which is Equation E.

Step 3: Group equations D and E. Use elimination.

$$\begin{cases} D: 26x - 7z = 45 \\ E: 17x - 4z = 30 \end{cases}.$$

Multiply Equation D by 4 and Equation E by 7.

$$\begin{cases} 104x - 28z = 180 \\ 119x - 28z = 210 \end{cases}.$$

Subtracting the equations, we get $-15x = -30$, from which $x = 2$.

Substituting $x = 2$ in any of the equations involving x and z only, we have $z = 1$.

Substituting $x = 2$ and $z = 1$ in any of Equations A, B, or C gives $y = 5$.

Therefore, our answer is $(x, y, z) = (2, 5, 1)$.

Geometric Interpretation of Systems of Linear Equations with Three Unknowns 三元一次方程组的几何分析

Review: Recall that in a system of linear equations with two unknowns (x and y), each equation is an equation of **a line in the coordinate plane**. Two lines in the coordinate plane have three scenarios:

1. Parallel—the system has no solution.
2. Intersecting—the system has exactly one solution.
3. Coincidence—the system has infinitely many solutions.

回顾二元一次方程组(未知数为 x 和 y),每个方程所代表的是线的方程式。在坐标平面上,两条线有以下 3 种情况:

1. 平行——方程组无解。
2. 相交——方程组仅有一解。
3. 重合——方程组有无限多个解。

Definition：This is an analogue of the review.

In a system of linear equations with three unknowns (x, y, and z), each equation is an equation of **a plane in the three-dimensional space**. Three planes in the three-dimensional space have the following scenarios:

1. No point of concurrency—the system has no solution, as shown in Figures 23-9, 23-10, and 23-11.
2. One point of concurrency—the system has exactly one solution, as shown in Figure 23-12.
3. A line of concurrency—the system has infinitely many solutions, as shown in Figure 23-13.
4. Coincidence—the system has infinitely many solutions, as shown in Figure 23-14.

与解二元一次方程组类似。

在三元一次方程组(未知数为 x、y、z)中，每个方程所代表的是平面方程式。在三维空间上，三个平面有以下情况：

1. 没有共点——方程组无解，如图 23-9、图 23-10、图 23-11 所示。
2. 一个共点——方程组有一解，如图 23-12 所示。
3. 一条共线——方程组有无限多个解，如图 23-13 所示。
4. 重合——方程组有无限多个解，如图 23-14 所示。

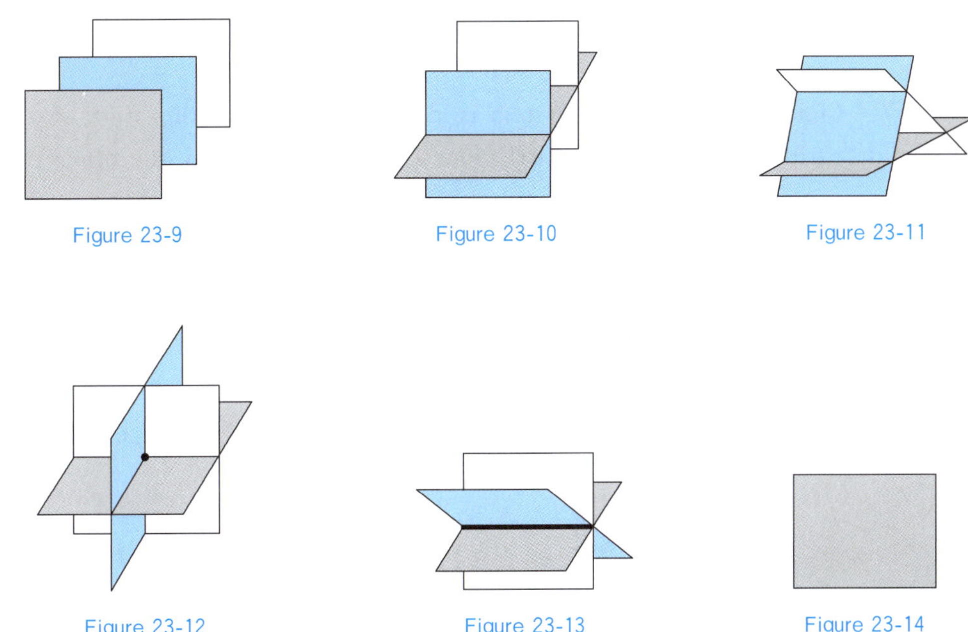

Figure 23-9　Figure 23-10　Figure 23-11

Figure 23-12　Figure 23-13　Figure 23-14

Examples：1. The system $\begin{cases} x+2y+3z=100 \\ 4x+5y+6z=200 \\ 9x+9y+9z=500 \end{cases}$ has no solution.

Subtracting the first equation from the second gets $3x+3y+3z=100$. Multiplying this result by 3 gives $9x+9y+9z=300$.

This contradicts the third equation in the system, $9x+9y+9z=500$. Therefore, the system has no solution.

2. Refer to the examples from **Solving Systems of Linear Equations with Three Unknowns by Elimination** for systems with one solution. The solution is a point in the *xyz*-plane.

3. The system $\begin{cases} x+y+z=1 \\ x+3y+4z=5 \\ x+7y+10z=13 \end{cases}$ has infinitely many solutions.

   Subtracting the first equation from the second, the second from the third, we get $\begin{cases} 2y+3z=4 \\ 4y+6z=8 \end{cases}$. These two equations are scalar multiples of each other, which has infinitely many solutions.

   Any ordered triple (x, y, z) that satisfies $2y+3z=4$ will be a solution in the *xyz*-plane. The solutions form the line $2y+3z=4$ in the *xyz*-plane.

   We will parametrize the solutions.

   Let $y=r$, it follows $z=\frac{4-2r}{3}$ and $x=-\frac{r+1}{3}$. Our solution is $\left(-\frac{r+1}{3}, r, \frac{4-2r}{3}\right)$.

4. The system $\begin{cases} x+y+z=100 \\ 2x+2y+2z=200 \\ 5x+5y+5z=500 \end{cases}$ has infinitely many solutions.

   Each equation in the system is a scalar multiple of the other two. So, any ordered triple (x, y, z) that satisfies one equation will satisfy the other two. The solutions form the plane $x+y+z=100$ in the *xyz*-plane.

   We will parametrize the solutions. Let's work with the first equation: $x+y+z=100$.

   Let $x=r$ and $y=s$, we have $z=100-r-s$. Our solution is $(r, s, 100-r-s)$.

24. Matrices 矩阵

24.1 Introduction 介绍

Matrix 矩阵

n. [ˈmeɪtrɪks]

Definition: A rectangular array of numbers arranged in rows and columns. 矩阵是按照行列排序的

矩形数阵。

Notation：$\begin{bmatrix} a_{11} & a_{12} & \cdots & a_{1n} \\ a_{21} & a_{22} & \cdots & a_{2n} \\ \vdots & \vdots & \ddots & \vdots \\ a_{m1} & a_{m2} & \cdots & a_{mn} \end{bmatrix}$

Above shows a matrix with m rows and n columns. Referring by dimension, this is an $m \times n$ matrix.

a_{ij} is the element that is on the ith row and the jth column.

上面展示了 m 行 n 列的矩阵。

a_{ij} 指的是在第 i 行第 j 列的元素。

Properties：1. For an $m \times n$ matrix $\boldsymbol{M}$ (m and n are positive integers)：

- If $m = n$, then $\boldsymbol{M}$ is a square matrix.　方块矩阵(方阵)
- If $m = 1$, then $\boldsymbol{M}$ is a row matrix.　行矩阵
- If $n = 1$, then $\boldsymbol{M}$ is a column matrix.　列矩阵

2. For an $n \times n$ matrix $\boldsymbol{N}$ (n is a positive integer)：

The main diagonal consists of all elements in the diagonal running from the top-left to the bottom-right.　矩阵的主对角线包括从矩阵左上到右下的对角线的元素。

Algebraically, the main diagonal is the collection of entries a_{ij}, from which $i = j$.

$$\boldsymbol{N} = \begin{bmatrix} a_{11} & a_{12} & \cdots & a_{1n} \\ a_{21} & a_{22} & \cdots & a_{2n} \\ \vdots & \vdots & \ddots & \vdots \\ a_{n1} & a_{n2} & \cdots & a_{nn} \end{bmatrix}$$

- If all entries below the main diagonal are 0's, then $\boldsymbol{N}$ is an upper triangular matrix.　上三角矩阵

  Algebraically, upper triangular matrices satisfy that all $a_{ij} = 0$, from which $i > j$.

$$\begin{bmatrix} a_{11} & a_{12} & \cdots & a_{1n} \\ 0 & a_{22} & \cdots & a_{2n} \\ 0 & 0 & \ddots & \vdots \\ 0 & 0 & 0 & a_{nn} \end{bmatrix}$$

- If all entries above the main diagonal are 0's, then $\boldsymbol{N}$ is a lower triangular matrix.
  下三角矩阵

  Algebraically, lower triangular matrices satisfy that all $a_{ij} = 0$, from which $i < j$.

$$\begin{bmatrix} a_{11} & 0 & 0 & 0 \\ a_{21} & a_{22} & 0 & 0 \\ \vdots & \vdots & \ddots & 0 \\ a_{n1} & a_{n2} & \cdots & a_{nn} \end{bmatrix}$$

- If $\boldsymbol{N}$ is both an upper triangular matrix and a lower triangular matrix, then it is a diagonal matrix.　对角矩阵

  Algebraically, diagonal matrices satisfy that all $a_{ij} = 0$, from which $i \neq j$.

$$\begin{bmatrix} a_{11} & 0 & 0 & 0 \\ 0 & a_{22} & 0 & 0 \\ 0 & 0 & \ddots & 0 \\ 0 & 0 & 0 & a_{nn} \end{bmatrix}$$

- If $\boldsymbol{N}$ is a diagonal matrix and the main diagonals are all 1's, then N is an identity matrix. 单位矩阵

  Algebraically, identity matrices satisfy that all $a_{ij} = 0$, from which $i \neq j$, and all $a_{ij} = 1$, form which $i = j$.

$$\begin{bmatrix} 1 & 0 & 0 & 0 \\ 0 & 1 & 0 & 0 \\ 0 & 0 & \ddots & 0 \\ 0 & 0 & 0 & 1 \end{bmatrix}$$

3. If all entries of a matrix are 0's, then this matrix is called the zero matrix. 零矩阵

Examples: 1.

$$\begin{bmatrix} 1 & 2 & 3 & 4 \\ 5 & 6 & 7 & 8 \\ 9 & 10 & 11 & 12 \end{bmatrix}$$

The dimension of the matrix above is 3×4. It has 3 rows and 4 columns.

2.

$$\begin{bmatrix} 1 & 2 & 3 \\ 4 & 5 & 6 \\ 7 & 8 & 9 \end{bmatrix}$$

The dimension of the matrix above is 3×3. This is a square matrix.

The main diagonal consists of entries 1, 5, and 9.

Questions: What are all the entries of a 4×5 matrix that satisfies $a_{ij} = i + j$?

Answers:

$$\begin{bmatrix} 2 & 3 & 4 & 5 & 6 \\ 3 & 4 & 5 & 6 & 7 \\ 4 & 5 & 6 & 7 & 8 \\ 5 & 6 & 7 & 8 & 9 \end{bmatrix}$$

24.2 Matrix Operations 矩阵运算

Matrix Addition and Subtraction 矩阵加法与减法

Definition: To add/subtract two matrices, add/subtract them entry-wise.

两个矩阵加减,是将矩阵中对应的元素进行加减。

Properties: 1. In order to add/subtract two matrices, their dimensions must be the same. 要对两个矩阵进行加法或减法,两矩阵的维度必须相同。

2. Written algebraically, suppose $\boldsymbol{A}$, $\boldsymbol{B}$, and $\boldsymbol{C}$ are matrices such that $\boldsymbol{A} + \boldsymbol{B} = \boldsymbol{C}$. It must be true that $a_{ij} + b_{ij} = c_{ij}$ for every entry.

Questions：If possible，add/subtract each of the following.

1. $\begin{bmatrix}1 & -2 & 3\\4 & -5 & 6\\7 & 8 & -9\end{bmatrix}+\begin{bmatrix}3 & -3 & 3\\-3 & 3 & -3\\3 & -3 & 3\end{bmatrix}$

2. $\begin{bmatrix}3 & 4\\5 & 6\\7 & 8\end{bmatrix}-\begin{bmatrix}1 & -1 & 3\\-3 & -6 & -3\\3 & -4 & 9\end{bmatrix}$

3. $\begin{bmatrix}1 & 2\\-3 & -4\\5 & 6\\-7 & -8\end{bmatrix}-\begin{bmatrix}-1 & -2\\3 & 4\\6 & 5\\-8 & -7\end{bmatrix}$

Answers：1. $\begin{bmatrix}1 & -2 & 3\\4 & -5 & 6\\7 & 8 & -9\end{bmatrix}+\begin{bmatrix}3 & -3 & 3\\-3 & 3 & -3\\3 & -3 & 3\end{bmatrix}=\begin{bmatrix}1+3 & -2+(-3) & 3+3\\4+(-3) & -5+3 & 6+(-3)\\7+3 & 8+(-3) & -9+3\end{bmatrix}=\begin{bmatrix}4 & -5 & 6\\1 & -2 & 3\\10 & 5 & -6\end{bmatrix}$

2. We cannot subtract the matrices since they have different dimensions.

$\begin{bmatrix}3 & 4\\5 & 6\\7 & 8\end{bmatrix}$ is a 3×2 matrix.

$\begin{bmatrix}1 & -1 & 3\\-3 & -6 & -3\\3 & -4 & 9\end{bmatrix}$ is a 3×3 matrix.

3. $\begin{bmatrix}1 & 2\\-3 & -4\\5 & 6\\-7 & -8\end{bmatrix}-\begin{bmatrix}-1 & -2\\3 & 4\\6 & 5\\-8 & -7\end{bmatrix}=\begin{bmatrix}2 & 4\\-6 & -8\\-1 & 1\\1 & -1\end{bmatrix}$

Product of a Matrix with a Constant 矩阵与常数的积

Definition：To find the product of a matrix and a constant，multiply every entry of the matrix by that constant. 要求出矩阵与常数的积，把矩阵的每个元素都乘上该常数。

Questions：Evaluate the following：

$3\cdot\begin{bmatrix}2 & 4 & -5\\-1 & 7 & -3\\6 & -7 & 3\end{bmatrix}$

Answers：$3\cdot\begin{bmatrix}2 & 4 & -5\\-1 & 7 & -3\\6 & -7 & 3\end{bmatrix}=\begin{bmatrix}3\cdot 2 & 3\cdot 4 & 3\cdot(-5)\\3\cdot(-1) & 3\cdot 7 & 3\cdot(-3)\\3\cdot 6 & 3\cdot(-7) & 3\cdot 3\end{bmatrix}=\begin{bmatrix}6 & 12 & -15\\-3 & 21 & -9\\18 & -21 & 9\end{bmatrix}$

Dot Product 点积

n. [dɑt 'prɑdəkt]

Definition：An algebraic operation that takes in two equal-length sequence，then computes the products of corresponding entries，and finally sum up those products. 点积是一种代数

运算,即先输入两个等长的数列,再计算同位元素的积,最后把这些积加起来。

The result of a dot product is a number. 点积的结果是一个数字。

Notation: The dot product of sequences $\boldsymbol{A}=(a_1,a_2,\cdots,a_n)$ and $\boldsymbol{B}=(b_1,b_2,\cdots,b_n)$ is written as

$$\begin{aligned}\boldsymbol{A}\cdot\boldsymbol{B}&=(a_1,a_2,a_3,\cdots,a_n)\cdot(b_1,b_2,b_3,\cdots,b_n)\\&=a_1b_1+a_2b_2+a_3b_3+\cdots+a_nb_n\\&=\sum_{i=1}^{n}a_ib_i.\end{aligned}$$

Examples: 1. The dot product of (1,2,3,4) and (5,6,7,8) is

$$\begin{aligned}(1,2,3,4)\cdot(5,6,7,8)&=1(5)+2(6)+3(7)+4(8)\\&=5+12+21+32\\&=70.\end{aligned}$$

2. The costs of a math textbook, a science textbook, and an English textbook are \$35, \$45, and \$40, respectively.

Ben buys 4 math textbooks, 3 science textbooks, and 5 English textbooks. We can find the total price using the dot product.

The sequences below are organized as (math, science, English).

Unit price sequence (in dollars): (35,45,40).

Quantity sequence: (4,3,5)

The total price is

$$\begin{aligned}(35,45,40)\cdot(4,3,5)&=35(4)+45(3)+40(5)\\&=140+135+200\\&=475\text{ dollars}.\end{aligned}$$

Matrix Multiplication 矩阵乘法

Definition: To multiply two matrices $\boldsymbol{A}$ and $\boldsymbol{B}$, the entry on the ith row jth column of the product is the dot product of the ith row of $\boldsymbol{A}$ and the jth column of $\boldsymbol{B}$. 两个矩阵 $\boldsymbol{A}$ 和 $\boldsymbol{B}$ 相乘,积的 i 行 j 列的元素是 $\boldsymbol{A}$ 的第 i 行与 $\boldsymbol{B}$ 的第 j 列的点积。

Notation: To multiply two matrices

$$\boldsymbol{A}=\begin{bmatrix}a_{11}&a_{12}&\cdots&a_{1n}\\a_{21}&a_{22}&\cdots&a_{2n}\\\vdots&\vdots&\ddots&\vdots\\a_{m1}&a_{m2}&\cdots&a_{mn}\end{bmatrix}\text{ and }\boldsymbol{B}=\begin{bmatrix}b_{11}&b_{12}&\cdots&b_{1p}\\b_{21}&b_{22}&\cdots&b_{2p}\\\vdots&\vdots&\ddots&\vdots\\b_{n1}&b_{n2}&\cdots&b_{np}\end{bmatrix}$$

$$\boldsymbol{AB}=\boldsymbol{C}.$$

$$c_{ij}=\sum_{k=1}^{n}a_{ik}b_{kj}\ .$$

Properties: 1. To multiply two matrices $\boldsymbol{A}$ and $\boldsymbol{B}$, the number of columns in $\boldsymbol{A}$ must equal to the number of rows in $\boldsymbol{B}$, since dot product is the operation of two equal-length sequences.

矩阵 $\boldsymbol{A}$ 和 $\boldsymbol{B}$ 相乘,$\boldsymbol{A}$ 的行数必须等于 $\boldsymbol{B}$ 的列数,因为点积是两个长度相等的数列的运算。

In short, we can multiply $\boldsymbol{A}$ and $\boldsymbol{B}$ only if the dimension of $\boldsymbol{A}$ is $m\times n$ and the dimension of $\boldsymbol{B}$ is $n \times p$.

2. It follows that from Property 1, $\boldsymbol{AB} \neq \boldsymbol{BA}$ for some matrices $\boldsymbol{A}$ and $\boldsymbol{B}$.

Questions: Evaluate $\boldsymbol{AB}$ for each of the following when possible:

1. $\boldsymbol{A} = \begin{bmatrix} 1 & 2 & 3 \\ 4 & 5 & 6 \end{bmatrix}$

$\boldsymbol{B} = \begin{bmatrix} 2 & 6 \\ 1 & -1 \\ -2 & -3 \end{bmatrix}$

2. $\boldsymbol{A} = \begin{bmatrix} 1 & 2 & 3 \\ 4 & 5 & 6 \end{bmatrix}$

$\boldsymbol{B} = \begin{bmatrix} -3 & -2 & -1 \\ -6 & -5 & -4 \end{bmatrix}$

3. $\boldsymbol{A} = \begin{bmatrix} 1 & 2 \end{bmatrix}$

$\boldsymbol{B} = \begin{bmatrix} 3 \\ 4 \end{bmatrix}$

4. $\boldsymbol{A} = \begin{bmatrix} 3 \\ 4 \end{bmatrix}$

$\boldsymbol{B} = \begin{bmatrix} 1 & 2 \end{bmatrix}$

Answers: 1. $\boldsymbol{AB} = \begin{bmatrix} 1 & 2 & 3 \\ 4 & 5 & 6 \end{bmatrix} \begin{bmatrix} 2 & 6 \\ 1 & -1 \\ -2 & -3 \end{bmatrix} = \begin{bmatrix} 1(2)+2(1)+3(-2) & 1(6)+2(-1)+3(-3) \\ 4(2)+5(1)+6(-2) & 4(6)+5(-1)+6(-3) \end{bmatrix} = \begin{bmatrix} -2 & -5 \\ 1 & 1 \end{bmatrix}$.

2. The dimension of $\boldsymbol{A}$ is 2×3, while the dimension of $\boldsymbol{B}$ is 2×3. It is incompatible to find $\boldsymbol{AB}$.

3. $\boldsymbol{AB} = \begin{bmatrix} 1 & 2 \end{bmatrix} \begin{bmatrix} 3 \\ 4 \end{bmatrix} = \begin{bmatrix} 1(3)+2(4) \end{bmatrix} = \begin{bmatrix} 11 \end{bmatrix}$.

4. $\boldsymbol{AB} = \begin{bmatrix} 3 \\ 4 \end{bmatrix} \begin{bmatrix} 1 & 2 \end{bmatrix} = \begin{bmatrix} 3(1) & 3(2) \\ 4(1) & 4(2) \end{bmatrix} = \begin{bmatrix} 3 & 6 \\ 4 & 8 \end{bmatrix}$.

24.3 Determinant of a (Square) Matrix （方块）矩阵的行列式

24.3.1 Introduction 介绍

Determinant 行列式

n. [dɪ'tɜː. mɪ. nənt]

Definition: A special number obtained from a square matrix that reveals many of the matrix's properties. 行列式是指从方阵得到的数字，可以展示这个方阵的很多性质。

Note that only square matrices have determinants. 只有方阵有行列式。

Notation: The determinant of matrix $\boldsymbol{A}$ is denoted by $\det(\boldsymbol{A})$ or $|\boldsymbol{A}|$.

Properties: 1. For a 1×1 matrix $[a]$, the determinant is simply a.

2. For a 2×2 matrix $\begin{bmatrix} a & b \\ c & d \end{bmatrix}$, the determinant is $a|d| - b|c| = ad - bc$.

3. For a 3×3 matrix $\begin{bmatrix} a & b & c \\ d & e & f \\ g & h & i \end{bmatrix}$, the determinant is $a\begin{vmatrix} e & f \\ h & i \end{vmatrix} - b\begin{vmatrix} d & f \\ g & i \end{vmatrix} + c\begin{vmatrix} d & e \\ g & h \end{vmatrix}$.

$\begin{vmatrix} e & f \\ h & i \end{vmatrix}$ is the minor of a. $\begin{vmatrix} e & f \\ h & i \end{vmatrix}$是 a 的子式。

$\begin{vmatrix} d & f \\ g & i \end{vmatrix}$ is the minor of b. $\begin{vmatrix} d & f \\ g & i \end{vmatrix}$是 b 的子式。

$\begin{vmatrix} d & e \\ g & h \end{vmatrix}$ is the minor of c. $\begin{vmatrix} d & e \\ g & h \end{vmatrix}$是 c 的子式。

4. In general, for an $n\times n$ matrix $\begin{bmatrix} a_{11} & a_{12} & \cdots & a_{1n} \\ a_{21} & a_{22} & \cdots & a_{2n} \\ \vdots & \vdots & \ddots & \vdots \\ a_{n1} & a_{n2} & \cdots & a_{nn} \end{bmatrix}$, in order to calculate the determinant, we can use the formula $a_{11}M_{11} - a_{12}M_{12} + a_{13}M_{13} - a_{14}M_{14} + \cdots + \operatorname{sgn}(a_{1n})a_{1n}M_{1n}$, where “sgn” is the sign function, defined by

$$\operatorname{sgn}(a_{ij}) = \begin{cases} 1, & i+j \text{ is even} \\ -1, & i+j \text{ is odd} \end{cases}.$$

Above we have $i=1$, for which

$$\operatorname{sgn}(a_{1j}) = \begin{cases} 1, & 1+j \text{ is even} \\ -1, & 1+j \text{ is odd} \end{cases}.$$

M_{11} denotes the minor of a_{11}, and so on.

In fact, we can select any ith row and use the formula $\sum\limits_{\substack{1\leqslant i\leqslant n \\ j=1}}^{j=n} (\operatorname{sgn}(a_{ij})a_{ij}M_{ij}) = \operatorname{sgn}(a_{i1})a_{i1}M_{i1} + \operatorname{sgn}(a_{i2})a_{i2}M_{i2} + \cdots + \operatorname{sgn}(a_{in})a_{in}M_{in}$

or select any ith column and use the formula $\sum\limits_{\substack{1\leqslant i\leqslant n \\ j=1}}^{j=n} (\operatorname{sgn}(a_{ji})a_{ji}M_{ji}) = \operatorname{sgn}(a_{1i})a_{1i}M_{1i} + \operatorname{sgn}(a_{2i})a_{2i}M_{2i} + \cdots + \operatorname{sgn}(a_{ni})a_{ni}M_{ni}$.

Both produce the same result.

Notes: For more proofs and patterns of the determinant, refer to a college linear algebra book. Proofs are out of scope of high school mathematics.

Questions: What is the determinant for each of the following?

1. $\boldsymbol{A} = [5]$

2. $\boldsymbol{B} = \begin{bmatrix} -1 & 2 \\ -3 & 4 \end{bmatrix}$

3. $\boldsymbol{C} = \begin{bmatrix} 1 & -2 & 3 \\ -1 & 2 & 3 \\ 1 & 2 & -3 \end{bmatrix}$

Answers: 1. $|\boldsymbol{A}| = 5$.

2. $|\boldsymbol{B}| = -1\cdot 4 - 2\cdot(-3) = -4+6 = 2$.

3. $|\boldsymbol{C}| = 1\begin{vmatrix} 2 & 3 \\ 2 & -3 \end{vmatrix} - (-2)\begin{vmatrix} -1 & 3 \\ 1 & -3 \end{vmatrix} + 3\begin{vmatrix} -1 & 2 \\ 1 & 2 \end{vmatrix} = 1(-12) + 2(0) + 3(-4) = -24.$

24.3.2 Applications 应用

24.3.2.1 Solving Systems of Linear Equations Using Cramer's Rule 用克拉默法则解线性方程组

Cramer's Rule 克拉默法则

n. [kreɪmərs rul]

Definition: An approach that uses determinants to solve systems of linear equations. 克拉默法则是一种用行列式解线性方程组的方法。

Properties: 1. For the following system of linear equations:

$$\begin{cases} ax + by = e \\ cx + dy = f \end{cases},$$

the solution is $x = \dfrac{\begin{vmatrix} e & b \\ f & d \end{vmatrix}}{\begin{vmatrix} a & b \\ c & d \end{vmatrix}}$ and $y = \dfrac{\begin{vmatrix} a & e \\ c & f \end{vmatrix}}{\begin{vmatrix} a & b \\ c & d \end{vmatrix}}$, from which $\begin{vmatrix} a & b \\ c & d \end{vmatrix} \neq 0$.

2. For the following system of linear equations

$$\begin{cases} ax + by + cz = j \\ dx + ey + fz = k, \\ gx + hy + iz = l \end{cases}$$

The solution is $x = \dfrac{\begin{vmatrix} j & b & c \\ k & e & f \\ l & h & i \end{vmatrix}}{\begin{vmatrix} a & b & c \\ d & e & f \\ g & h & i \end{vmatrix}}$, $y = \dfrac{\begin{vmatrix} a & j & c \\ d & k & f \\ g & l & i \end{vmatrix}}{\begin{vmatrix} a & b & c \\ d & e & f \\ g & h & i \end{vmatrix}}$, $z = \dfrac{\begin{vmatrix} a & b & j \\ d & e & k \\ g & h & l \end{vmatrix}}{\begin{vmatrix} a & b & c \\ d & e & f \\ g & h & i \end{vmatrix}}$, and $\begin{vmatrix} a & b & c \\ d & e & f \\ g & h & i \end{vmatrix} \neq 0$.

Notes: 1. To illustrate the Cramer's Rule, we can solve the system of linear equations $\begin{cases} ax + by = e \\ cx + dy = f \end{cases}$ using elimination:

Multiplying the first equation by d and the second by b, we have

$$\begin{cases} adx + bdy = de \\ bcx + bdy = bf \end{cases}.$$

Subtracting equations gives

$$(ad - bc)x = de - bf.$$

We now get

$$x = \frac{de - bf}{ad - bc} = \frac{\begin{vmatrix} e & b \\ f & d \end{vmatrix}}{\begin{vmatrix} a & b \\ c & d \end{vmatrix}}.$$

Substituting, we have

$$y=\frac{af-ce}{ad-bc}=\frac{\begin{vmatrix} a & e \\ c & f \end{vmatrix}}{\begin{vmatrix} a & b \\ c & d \end{vmatrix}}.$$

2. Same process applies to show that the Cramer's Rule works for systems of linear equations with three unknowns.

   The proof of Cramer's Rule is out of the scope of high school mathematics, but will appear in college linear algebra books.

Questions: Use Cramer's Rule to solve the system of equations below.

$$\begin{cases} -3x+2y=19 \\ 5x+7y=-11 \end{cases}$$

Answers: $x=\frac{\begin{vmatrix} 19 & 2 \\ -11 & 7 \end{vmatrix}}{\begin{vmatrix} -3 & 2 \\ 5 & 7 \end{vmatrix}}=\frac{155}{-31}=-5$ and $y=\frac{\begin{vmatrix} -3 & 19 \\ 5 & -11 \end{vmatrix}}{\begin{vmatrix} -3 & 2 \\ 5 & 7 \end{vmatrix}}=\frac{-62}{-31}=2$.

24.3.2.2 Solving Systems of Linear Equations Using Inverse Matrices 用逆矩阵解线性方程组

Identity Matrix 单位矩阵

n. [aɪ'dentɪti 'meɪtrɪks]

Definition: A square matrix whose entries on the main diagonal ($a_{11}, a_{22}, a_{33}, \cdots$) are 1's and the other entries are 0's. 单位矩阵是指主对角的元素为 1,其余元素为 0 的方阵。

$$\begin{bmatrix} 1 & 0 & 0 & 0 \\ 0 & 1 & 0 & 0 \\ 0 & 0 & \ddots & 0 \\ 0 & 0 & 0 & 1 \end{bmatrix}$$

Notation: An $n \times n$ identity matrix is denoted by $\boldsymbol{I}_n$.

Properties: 1. Multiplying any matrix $\boldsymbol{M}$ with the identity matrix (if compatible) gives $\boldsymbol{M}$.

2. Multiplying the identity matrix with any matrix $\boldsymbol{M}$ (if compatible) gives $\boldsymbol{M}$.

Inverse Matrix 逆矩阵

n. ['ɪnvɜrs 'meɪtrɪks]

Definition: Matrices $\boldsymbol{A}$ and $\boldsymbol{A}^{-1}$ are inverses if $\boldsymbol{A}\boldsymbol{A}^{-1}=\boldsymbol{A}^{-1}\boldsymbol{A}=\boldsymbol{I}$, from which $\boldsymbol{I}$ is an identity matrix.

$\boldsymbol{A}$ and $\boldsymbol{A}^{-1}$ must be square matrix.

Notation: The inverse matrix of $\boldsymbol{A}$ is denoted by $\boldsymbol{A}^{-1}$.

Properties: Due to the scope of high school mathematics, we will only be covering the inverse matrices for 2×2 matrices.

The inverse matrix for $\boldsymbol{A}=\begin{bmatrix} a & b \\ c & d \end{bmatrix}$ is $\boldsymbol{A}^{-1}=\frac{1}{ad-bc}\begin{bmatrix} d & -b \\ -c & a \end{bmatrix}$, where $\det(\boldsymbol{A})=ad-bc\neq0$.

Proofs: To show why $\boldsymbol{A}^{-1}=\frac{1}{ad-bc}\begin{bmatrix} d & -b \\ -c & a \end{bmatrix}$, suppose we want to solve for $\boldsymbol{A}\boldsymbol{A}^{-1}=$

$$\begin{bmatrix} a & b \\ c & d \end{bmatrix} A^{-1} = \begin{bmatrix} 1 & 0 \\ 0 & 1 \end{bmatrix}.$$

Suppose $A^{-1} = \begin{bmatrix} w & x \\ y & z \end{bmatrix}$, from which we want to solve for w, x, y, and z.

We know that $AA^{-1} = \begin{bmatrix} a & b \\ c & d \end{bmatrix} A^{-1} = \begin{bmatrix} a & b \\ c & d \end{bmatrix}\begin{bmatrix} w & x \\ y & z \end{bmatrix} = \begin{bmatrix} aw+by & ax+bz \\ cw+dy & cx+dz \end{bmatrix} = \begin{bmatrix} 1 & 0 \\ 0 & 1 \end{bmatrix}$.

We have the two systems of equations: $\begin{cases} aw+by=1 \\ cw+dy=0 \end{cases}$, which solves for w and y, and $\begin{cases} ax+bz=0 \\ cx+dz=1 \end{cases}$, which solves for x and z.

Expressing w, x, y, and z in terms of a and b proves the result.

Questions: 1. Determine if each pair of matrices are inverses below:

(1) $A = \begin{bmatrix} 4 & 4 \\ -1 & 2 \end{bmatrix}$ and $B = \begin{bmatrix} \frac{1}{4} & \frac{1}{4} \\ -1 & \frac{1}{2} \end{bmatrix}$

(2) $C = \begin{bmatrix} 1 & 2 & 3 \\ 3 & 2 & 1 \\ 3 & 1 & 1 \end{bmatrix}$ and $D = \begin{bmatrix} -1/8 & -1/8 & 1/2 \\ 0 & 1 & -1 \\ 3/8 & -5/8 & 1/2 \end{bmatrix}$

2. What is the inverse matrix for $M = \begin{bmatrix} 2 & 3 \\ 5 & 7 \end{bmatrix}$?

Answers: 1. (1) $AB = \begin{bmatrix} 4 & 4 \\ -1 & 2 \end{bmatrix}\begin{bmatrix} \frac{1}{4} & \frac{1}{4} \\ -1 & \frac{1}{2} \end{bmatrix} = \begin{bmatrix} -3 & 3 \\ -\frac{9}{4} & \frac{3}{4} \end{bmatrix}$, which is not the identity matrix.

Therefore, A and B are not inverses.

(2) $CD = \begin{bmatrix} 1 & 2 & 3 \\ 3 & 2 & 1 \\ 3 & 1 & 1 \end{bmatrix}\begin{bmatrix} -1/8 & -1/8 & 1/2 \\ 0 & 1 & -1 \\ 3/8 & -5/8 & 1/2 \end{bmatrix} = \begin{bmatrix} 1 & 0 & 0 \\ 0 & 1 & 0 \\ 0 & 0 & 1 \end{bmatrix}$, which is the identity matrix. Therefore, C and D are inverses.

2. $M^{-1} = \frac{1}{2\times7-3\times5}\begin{bmatrix} 7 & -3 \\ -5 & 2 \end{bmatrix} = -\begin{bmatrix} 7 & -3 \\ -5 & 2 \end{bmatrix} = \begin{bmatrix} -7 & 3 \\ 5 & -2 \end{bmatrix}$.

Solving Systems of Linear Equations Using Matrices 用矩阵解线性方程组

Definition: We will use inverse matrices to solve systems of linear equations. 在这里会采用逆矩阵解线性方程组。

Since we only learned finding inverse matrices of 2×2 matrices, we will only be solving systems of linear equations with two variables. 因为只学过如何求 2×2 矩阵的逆矩阵，所以只解二元一次方程组。

Properties: Suppose we want to solve for the system $\begin{cases} ax+by=e \\ cx+dy=f \end{cases}$:

Let $\boldsymbol{A}=\begin{bmatrix}a & b\\ c & d\end{bmatrix}$, $\boldsymbol{X}=\begin{bmatrix}x\\ y\end{bmatrix}$, and $\boldsymbol{B}=\begin{bmatrix}e\\ f\end{bmatrix}$, we have $\boldsymbol{AX}=\begin{bmatrix}a & b\\ c & d\end{bmatrix}\begin{bmatrix}x\\ y\end{bmatrix}=\begin{bmatrix}ax+by\\ cx+dy\end{bmatrix}=\begin{bmatrix}e\\ f\end{bmatrix}=\boldsymbol{B}$.

To solve the equation $\boldsymbol{AX}=\boldsymbol{B}$, we multiply both sides by $\boldsymbol{A}^{-1}$:

$$\begin{aligned}\boldsymbol{AX}&=\boldsymbol{B}\\ \boldsymbol{A}^{-1}(\boldsymbol{AX})&=\boldsymbol{A}^{-1}(\boldsymbol{B})\\ (\boldsymbol{A}^{-1}\boldsymbol{A})\boldsymbol{X}&=\boldsymbol{A}^{-1}\boldsymbol{B}\\ \boldsymbol{X}&=\boldsymbol{A}^{-1}\boldsymbol{B}.\end{aligned}$$

It is important to note that matrix multiplication is not commutative. When we multiply both sides by $\boldsymbol{A}^{-1}$, we must have $\boldsymbol{A}^{-1}$ on the left for both sides. Getting $\boldsymbol{X}=\boldsymbol{BA}^{-1}$ would not be correct. 矩阵的乘法没有交换律。当两边乘以 $\boldsymbol{A}^{-1}$ 时,均需要把 $\boldsymbol{A}^{-1}$ 放置在左侧。$\boldsymbol{X}=\boldsymbol{BA}^{-1}$ 不是正确答案。

Questions: Use matrices to solve the following system:

$$\begin{cases}4x+3y=11\\ -3x+5y=-30\end{cases}$$

Answers: We know that $\boldsymbol{A}=\begin{bmatrix}4 & 3\\ -3 & 5\end{bmatrix}$, $\boldsymbol{X}=\begin{bmatrix}x\\ y\end{bmatrix}$, and $\boldsymbol{B}=\begin{bmatrix}11\\ -30\end{bmatrix}$.

We have $\boldsymbol{AX}=\boldsymbol{B}$, from which $\begin{bmatrix}4 & 3\\ -3 & 5\end{bmatrix}\begin{bmatrix}x\\ y\end{bmatrix}=\begin{bmatrix}11\\ -30\end{bmatrix}$.

The inverse of $\begin{bmatrix}4 & 3\\ -3 & 5\end{bmatrix}$ is $\frac{1}{29}\begin{bmatrix}5 & -3\\ 3 & 4\end{bmatrix}$.

Multiplying both sides of the equation by the inverse on the left, we get $\frac{1}{29}\begin{bmatrix}5 & -3\\ 3 & 4\end{bmatrix}\begin{bmatrix}4 & 3\\ -3 & 5\end{bmatrix}\begin{bmatrix}x\\ y\end{bmatrix}=\frac{1}{29}\begin{bmatrix}5 & -3\\ 3 & 4\end{bmatrix}\begin{bmatrix}11\\ -30\end{bmatrix}$, from which $\begin{bmatrix}x\\ y\end{bmatrix}=\frac{1}{29}\begin{bmatrix}145\\ -87\end{bmatrix}=\begin{bmatrix}5\\ -3\end{bmatrix}$.

24.3.3 Summary 总结

1. (1) For a 1×1 matrix $[a]$, the determinant is simply a.

   (2) For a 2×2 matrix $\begin{bmatrix}a & b\\ c & d\end{bmatrix}$, the determinant is $a|d|-b|c|=ad-bc$.

   (3) For a 3×3 matrix $\begin{bmatrix}a & b & c\\ d & e & f\\ g & h & i\end{bmatrix}$, the determinant is $a\begin{vmatrix}e & f\\ h & i\end{vmatrix}-b\begin{vmatrix}d & f\\ g & i\end{vmatrix}+c\begin{vmatrix}d & e\\ g & h\end{vmatrix}$.

2. Review Solving Systems of Linear Equations Using Cramer's Rule in Section 24.3.2.1. 复习 24.3.2.1 节的"用克拉默法则解线性方程组"。

3. For any square matrix $\boldsymbol{M}$, it is true that $\boldsymbol{MM}^{-1}=\boldsymbol{M}^{-1}\boldsymbol{M}=\boldsymbol{I}$, from which $\boldsymbol{M}^{-1}$ is the inverse of $\boldsymbol{M}$, and $\boldsymbol{I}$ is the identity matrix. 逆矩阵的积为单位矩阵。

The inverse matrix of $\boldsymbol{A}=\begin{bmatrix} a & b \\ c & d \end{bmatrix}$ is $\boldsymbol{A}^{-1}=\frac{1}{ad-bc}\begin{bmatrix} d & -b \\ -c & a \end{bmatrix}$.

4. To solve the linear system $\boldsymbol{AX}=\boldsymbol{B}$, we have $\boldsymbol{X}=\boldsymbol{A}^{-1}\boldsymbol{B}$.

25. Complex Numbers　复数

25.1 Introduction　介绍

视频 38

We know that the equation $x^2=-1$ has no real solutions. This chapter will uncover the solutions for this type of equations.

Imaginary Unit　虚数单位

n. [ɪ'mædʒəˌneri 'juːnɪt]

Notation: The imaginary unit is denoted by i.

The definition for i is $\sqrt{-1}$, for which $\mathrm{i}^2=-1$.

Properties: 1. Imaginary unit cannot be located on the real number line.　虚数单位不在实数轴上。

2. The powers of i appear in cycles:

i 的次幂按如下规律循环：

$\mathrm{i}^1=\mathrm{i}$	$\mathrm{i}^5=\mathrm{i}$	$\mathrm{i}^9=\mathrm{i}$	$\mathrm{i}^{13}=\mathrm{i}$
$\mathrm{i}^2=-1$	$\mathrm{i}^6=-1$	$\mathrm{i}^{10}=-1$	$\mathrm{i}^{14}=-1$
$\mathrm{i}^3=-\mathrm{i}$	$\mathrm{i}^7=-\mathrm{i}$	$\mathrm{i}^{11}=-\mathrm{i}$	$\mathrm{i}^{15}=-\mathrm{i}$
$\mathrm{i}^4=1$	$\mathrm{i}^8=1$	$\mathrm{i}^{12}=1$	$\mathrm{i}^{16}=1$

And so on.

In general, for i^n:

If n is 0 mod 4 (leaves the remainder of 0 when divided by 4), then $\mathrm{i}^n=1$.

If n is 1 mod 4, then $\mathrm{i}^n=\mathrm{i}$.

If n is 2 mod 4, then $\mathrm{i}^n=-1$.

If n is 3 mod 4, then $\mathrm{i}^n=-\mathrm{i}$.

The sum of four consecutive powers of i is 0.

i 的四个连续次幂之和为 0。

3. Review rules of radicals:

$\sqrt{a}\cdot\sqrt{b}=\sqrt{ab}$, when $a\geqslant 0$ and $b\geqslant 0$.

However, when $a<0$ or $b<0$, we must individually express each of $\sqrt{a}$ and $\sqrt{b}$ in terms of i, then multiply.

Warning: $\sqrt{-4}\cdot\sqrt{-9}\neq\sqrt{(-4)(-9)}$!!!

LHS $=\sqrt{-4}\cdot\sqrt{-9}=\sqrt{(-1)(4)}\cdot\sqrt{(-1)(9)}=2\sqrt{-1}\cdot 3\sqrt{-1}=2i\cdot 3i=6i^2=6(-1)=-6$.

RHS $=\sqrt{(-4)(-9)}=\sqrt{36}=6$.

Questions: 1. If i represents the imaginary unit, what is the value of

$$i^1+i^2+i^3+\cdots+i^{2015}+i^{2016}?$$

2. Simplify each of the following:

(1) $\sqrt{-8}\cdot\sqrt{-18}$.

(2) $\sqrt{-8\cdot(-18)}$.

Answers: 1. Note that the values of powers of i behaves periodically.

$i^1+i^2+i^3+i^4=i+(-1)+(-i)+1=0$.

$i^5+i^6+i^7+i^8=i+(-1)+(-i)+1=0$.

…

The sum of four consecutive powers of i is 0.

Since 2016 is a multiple of 4, the sum is 0.

2. (1) $\sqrt{-8}\cdot\sqrt{-18}=\sqrt{(-1)(8)}\cdot\sqrt{(-1)(18)}=\sqrt{8}\sqrt{-1}\cdot\sqrt{18}\sqrt{-1}=\sqrt{8}i\cdot\sqrt{18}i=\sqrt{144}i^2=12(-1)=-12$.

(2) $\sqrt{-8\cdot(-18)}=\sqrt{144}=12$.

Complex Number 复数

n. [kəm'pleks 'nʌmbər]

Definition: A number of the form $a+bi$, from which a and b are real constants, and i is the imaginary unit.

复数是指以 $a+bi$ 的形式表示的数。a 和 b 为实数常值, i 为虚数单位。

Notation: $z=a+bi$.

We denote $\text{Re}(z)$ as the real part of z.

We denote $\text{Im}(z)$ as the real coefficient of the imaginary part of z.

$\text{Re}(z)=a$.

$\text{Im}(z)=b$, **not bi**.

Properties: 1. All numbers are complex numbers.

All complex numbers are located in the complex plane.

Every complex number has a norm.

所有数都是复数。

所有复数都在复平面上。

所有复数都有范数。

2. For complex number $z=a+bi$:

If $b\neq 0$, then we say z is an imaginary number. 虚数情况。

If $a=0$ and $b\neq 0$, then we say z is a pure imaginary number. 纯虚数情况。

If $b=0$, then we say z is a real number. 实数情况。

Figure 25-1 shows the relationship among complex numbers, real numbers, imaginary numbers, and pure imaginary numbers.

图 25-1 展示了复数、实数、虚数、纯虚数的关系。

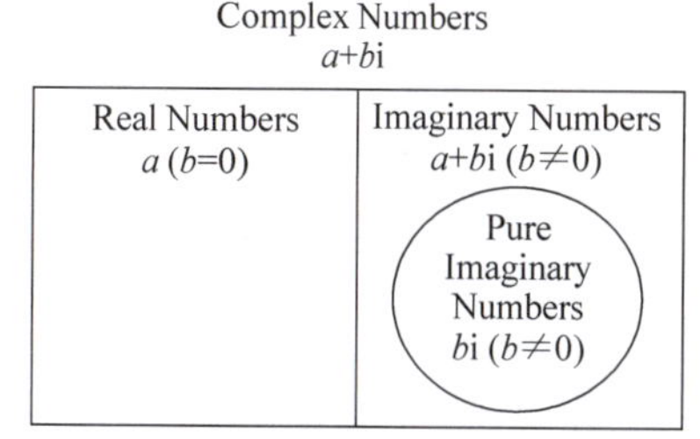

Figure 25-1

3. In operations between complex numbers, treat the i as a variable and combine like terms. 在进行复数运算时，把 i 当作变量，可以合并同类项。

Questions: Simplify each of the following in the form of $a + bi$.

(1) $(5+3i)+(3-6i)$

(2) $(4-5i)-(-7+8i)$

(3) $(-7+3i)(10-2i)$

(4) $\dfrac{4+10i}{-2}$

Answers: (1) $(5+3i)+(3-6i)=5+3i+3-6i=(5+3)+(3i-6i)=8-3i$.

(2) $(4-5i)-(-7+8i)=4-5i+7-8i=(4+7)+(-5i-8i)=11-13i$.

(3) $(-7+3i)(10-2i)=-7(10)+(-7)(-2i)+(3i)(10)+(3i)(-2i)=-70+14i+30i+6=-64+44i$.

(4) $\dfrac{4+10i}{-2}=\dfrac{4}{-2}+\dfrac{10i}{-2}=-2-5i$.

25.2 Complex Plane 复平面

Complex Plane 复平面

n. [kəm'pleks pleɪn]

Definition: A geometric representation of complex numbers, as shown in Figure 25-2. 复数的几何表示方法，如图 25-2 所示。

The horizontal axis is called the real axis, or a-axis.

水平轴为实数轴，又称为 a 轴。

The vertical axis is called the imaginary axis, or b-axis. It represents a complex number's imaginary part's real coefficient.

垂直轴为虚数轴，又称为 b 轴。它代表复数的虚数部分的实数系数。

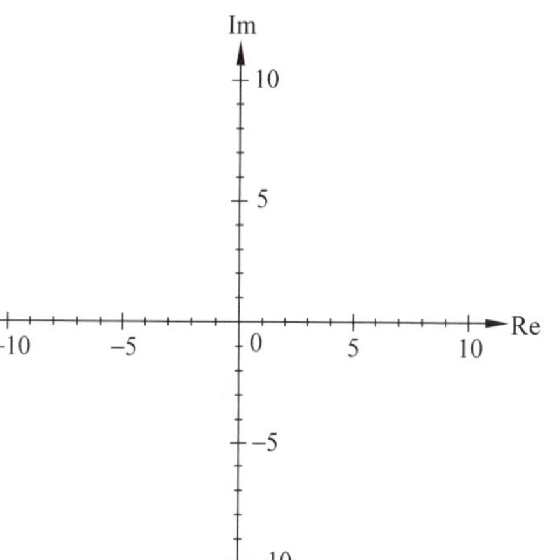

Figure 25-2

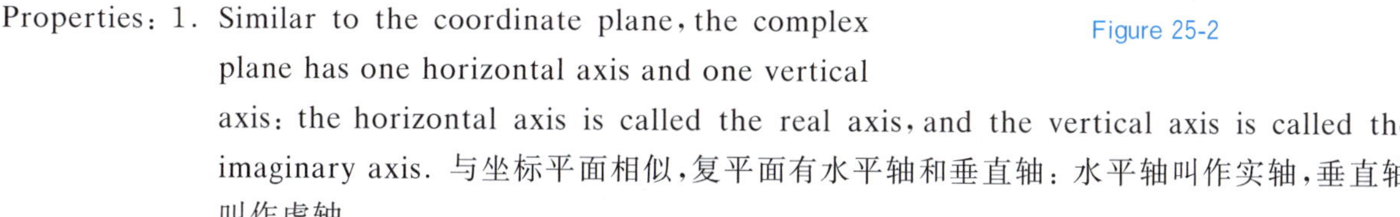

Properties: 1. Similar to the coordinate plane, the complex plane has one horizontal axis and one vertical axis: the horizontal axis is called the real axis, and the vertical axis is called the imaginary axis. 与坐标平面相似，复平面有水平轴和垂直轴：水平轴叫作实轴，垂直轴叫作虚轴。

2. Every point on the complex plane can be represented by a complex number. 复平面的每个点都能用复数表示。

   For a point P, if a and b are the horizontal and vertical distances from P to the origin, respectively, then P can be written as $a+bi$, as shown in Figure 25-3.

   对于点 P,若 P 离原点的横距离和纵距离分别为 a 和 b,则 P 可以写作 $a+bi$,如图 25-3 所示。

Examples: Figure 25-4 shows some complex numbers in the complex plane.

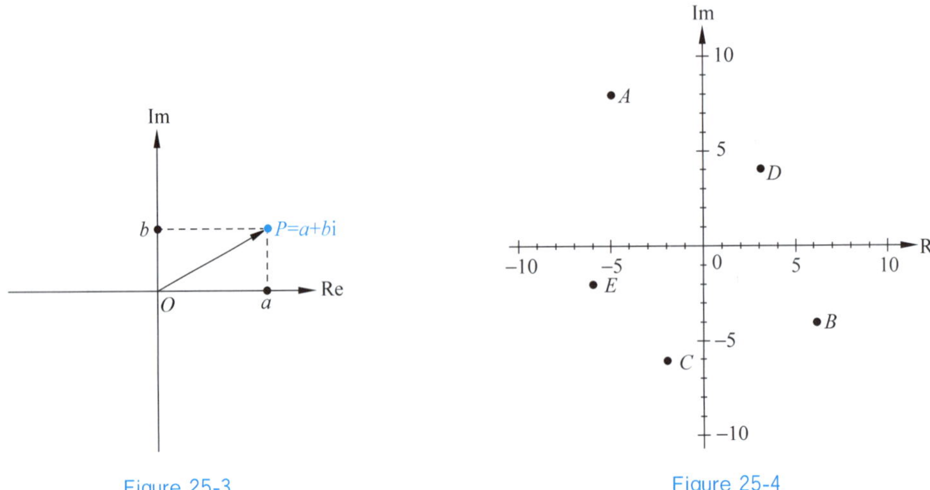

Figure 25-3

Figure 25-4

For complex numbers A, B, C, D, and E, we have:

$A=-5+8i$;

$B=6-4i$;

$C=-2-6i$;

$D=3+4i$;

$E=-6-2i$.

Norm 范数

n. [nɔrm]

Definition: The norm of a complex number z is the distance from the origin to the point representing z in the complex plane. 复数 z 的范数是复平面上与 z 对应的点到原点的距离。

The norm of a complex number is also known as its absolute value. 复数的范数也称为它的绝对值。

Notation: The norm of a complex number z is written as $|z|$.

Properties: The formula for the norm of $z=a+bi$ is $|z|=\sqrt{a^2+b^2}$.

This formula is based on the Pythagorean Theorem and/or the Distance Formula. 根据勾股定理和距离公式可得出范数公式。

Questions: What is the norm for each of the following?

(1) $3+4i$

(2) $4-i$

(3) 0

Answers：(1) $|3+4\mathrm{i}|=\sqrt{3^2+4^2}=\sqrt{9+16}=\sqrt{25}=5$.

(2) $|4-\mathrm{i}|=\sqrt{4^2+(-1)^2}=\sqrt{16+1}=\sqrt{17}$.

(3) $|0|=0$.

Conjugate Complex Numbers　共轭复数

n. [ˈkɑndʒəˌgeɪt kəmˈpleks ˈnʌmbərs]

Definition：Two complex numbers are conjugate complex numbers (or simply conjugates) when their real parts are equal and the imaginary parts are opposites.　两个复数若实部相等，虚部相反，则互为共轭复数。

Notation：$z=a+b\mathrm{i}$ and $\overline{z}=a-b\mathrm{i}$ are conjugates of each other.

The conjugate of z is denoted by $\overline{z}$.

Properties：1. For $z=a+b\mathrm{i}$：

(1) If $a=0$, then $\overline{z}=-b\mathrm{i}$.

(2) If $b=0$, then $\overline{z}=a=z$.

Complex conjugate and irrational conjugate are analogues.　复数的共轭与无理数的共轭类似。

2. On the complex plane, the conjugate of a complex number z is the image when z is reflected across the real axis, as shown in Figure 25-5.　在复平面上，复数 z 的共轭是 z 沿着实轴反射的像，如图 25-5 所示。

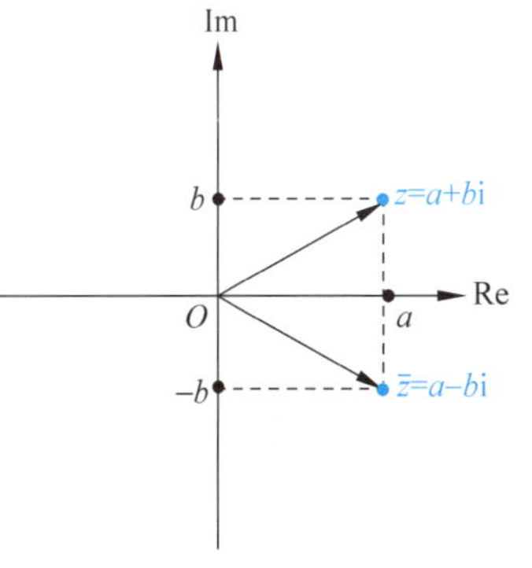

Figure 25-5

3. For $z=a+b\mathrm{i}$ and conjugate $\overline{z}=a-b\mathrm{i}$, their product is a real number：

$z\overline{z}=(a+b\mathrm{i})(a-b\mathrm{i})=a^2-(b\mathrm{i})^2=a^2-b^2\mathrm{i}^2=a^2+b^2$.

We can use complex conjugates to divide complex numbers and express the result in the form of $a+b\mathrm{i}$.

复数与其共轭的积为实数。

可以用复数共轭做复数的除法，并且把得数写成 $a+b\mathrm{i}$ 的形式。

4. From Property 3, note that the product of complex conjugates is the same as the square of the norm of these complex numbers：$(a+b\mathrm{i})(a-b\mathrm{i})=a^2+b^2=\sqrt{a^2+b^2}^{\,2}=|a+b\mathrm{i}|^2=|a-b\mathrm{i}|^2$.

由性质 3 可知复数与其共轭的积与这些复数的范数平方相同。

Questions：1. What is the conjugate for each of the following?

(1) $3+8\mathrm{i}$

(2) $4-\mathrm{i}$

(3) $-9\mathrm{i}$

(4) 7

2. Write each of the following in the form of $a+bi$.

(1) $\frac{4+3i}{-2i}$

(2) $\frac{7-2i}{3+5i}$

Answers: 1. (1) $3-8i$.

(2) $4+i$.

(3) $9i$.

(4) 7.

2. (1) We multiply both the numerator and the denominator by the conjugate of the denominator so that the denominator becomes a real number.

$\frac{4+3i}{-2i}=\frac{(4+3i)(2i)}{(-2i)(2i)}=\frac{8i+6i^2}{-4i^2}=\frac{8i-6}{4}$.

From here, we can write the fraction in the form of $a+bi$.

$\frac{8i-6}{4}=\frac{8i}{4}-\frac{6}{4}=2i-\frac{3}{2}=-\frac{3}{2}+2i$.

(2) $\frac{7-2i}{3+5i}=\frac{(7-2i)(3-5i)}{(3+5i)(3-5i)}=\frac{21-35i-6i+10i^2}{9-25i^2}=\frac{21-35i-6i-10}{9+25}=\frac{11-41i}{34}=\frac{11}{34}-\frac{41}{34}i$.

25.3 Solving Quadratic Equations with Imaginary Roots 解有虚数根的一元二次方程

Complete the Square Review 配方法回顾

Definition: See complete the square method for solving quadratic equations in Section 3.5.2. 定义参照3.5.2节解一元二次方程的配方法。

Questions: Solve each of the following:

(1) $x^2+16=0$

(2) $x^2+6x+18=0$

(3) $x^2-10x+26=0$

Answers: (1) $x^2+16=0$

$x^2=-16$

$x=\pm\sqrt{-16}$

$x=\pm 4\sqrt{-1}$

$x=\pm 4i$.

(2) $x^2+6x+18=0$

$x^2+6x+9=-9$

$(x+3)^2=-9$

$x+3=\pm\sqrt{-9}$

$$x+3=\pm 3\sqrt{-1}$$
$$x+3=\pm 3\mathrm{i}$$
$$x=-3\pm 3\mathrm{i}.$$

(3) $x^2-10x+26=0$

$$x^2-10x+25=-1$$
$$(x-5)^2=-1$$
$$x-5=\pm\sqrt{-1}$$
$$x-5=\pm\mathrm{i}$$
$$x=5\pm\mathrm{i}.$$

Quadratic Formula Review　一元二次方程求根公式回顾

Definition: See Quadratic Formula for solving quadratic equations in Section 3.5.2.　定义参照3.5.2节解一元二次方程的求根公式词条。

Properties: To solve $ax^2+bx+c=0$ ($a\neq 0$):

We have derived that $x=\dfrac{-b\pm\sqrt{b^2-4ac}}{2a}$.

The discriminant is b^2-4ac:

判别式的情况：

- If $b^2-4ac>0$, then the equation has two different real roots.
  两个不同的实数根的情况。
- If $b^2-4ac=0$, then the equation has one real root.
  一个实数根的情况。
- If $b^2-4ac<0$, then the equation has two different imaginary roots.
  两个不同的虚数根的情况。

We will focus on the third case here.

Questions: The questions are the same as those from Complete the Square Review.

(1) $x^2+16=0$

(2) $x^2+6x+18=0$

(3) $x^2-10x+26=0$

Answers: (1) We have $a=1$, $b=0$, and $c=16$.

$$x=\frac{-0\pm\sqrt{0^2-4(1)(16)}}{2(1)}=\frac{\pm\sqrt{-64}}{2}=\frac{\pm 8\mathrm{i}}{2}=\pm 4\mathrm{i}.$$

(2) We have $a=1$, $b=6$, and $c=18$.

$$x=\frac{-6\pm\sqrt{6^2-4(1)(18)}}{2(1)}=\frac{-6\pm\sqrt{-36}}{2}=\frac{-6\pm 6\mathrm{i}}{2}=3\pm 3\mathrm{i}.$$

(3) We have $a=1$, $b=-10$, and $c=26$.

$$x=\frac{10\pm\sqrt{(-10)^2-4(1)(26)}}{2(1)}=\frac{10\pm\sqrt{-4}}{2}=\frac{10\pm 2\mathrm{i}}{2}=5\pm\mathrm{i}.$$

26. Polynomial Functions 多项式函数

26.1 Function Operations 函数运算

Basic Operations 基本运算

Table 26-1 shows the definitions and examples of basic functional operations.

Table 26-1

Operation	Definition	Example when $f(x)=2x+1$ and $g(x)=6x-5$
Addition	$(f+g)(x)=f(x)+g(x)$	$(2x+1)+(6x-5)=8x-4$
Subtraction	$(f-g)(x)=f(x)-g(x)$	$(2x+1)-(6x-5)=-4x+6$
Multiplication	$(f\cdot g)(x)=f(x)\cdot g(x)$	$(2x+1)(6x-5)=12x^2-4x-5$
Division	$\left(\frac{f}{g}\right)(x)=\frac{f(x)}{g(x)}, g(x)\neq 0$	$\frac{2x+1}{6x-5}, x\neq\frac{5}{6}$

Function Composition 复合函数

n. [ˈfʌŋkʃn ˌkɑːmpəˈzɪʃn]

Definition: The application of one function to the result of another function to produce a third function. 复合函数是指在一个函数的结果上应用另一个函数,从而产生第三个函数。

Notation: $(f\circ g)(x)=f(g(x))$.

First, we apply function g on input x, then we apply function f to the result $g(x)$.

Properties: On many cases, $f(g(x))\neq g(f(x))$.

Questions: 1. For each of the following, find $f(g(x))$ and $g(f(x))$.

(1) $f(x)=3x, g(x)=2x$

(2) $f(x)=5x-1, g(x)=3x$

(3) $f(x)=4x-3, g(x)=2^x$

2. What is the value of $f(g(7))$ for each of the following?

(1) $f(x)=2x, g(x)=9$

(2) $f(x)=3x-7, g(x)=5x+3$

(3) $f(x)=8x, g(x)=2^x-1$

Answers: 1. (1) $f(g(x))=f(2x)=3(2x)=6x$.

$g(f(x))=g(3x)=2(3x)=6x$.

(2) $f(g(x))=f(3x)=5(3x)-1=15x-1$.

$g(f(x))=g(5x-1)=3(5x-1)=15x-3$.

Note that $f(g(x))\neq g(f(x))$ here.

(3) $f(g(x))=f(2^x)=4(2^x)-3$.

$g(f(x))=g(4x-3)=2^{4x-3}$.

Note that $f(g(x))\neq g(f(x))$ here.

2. (1) $f(g(x))=f(9)=2(9)=18$. This is a constant function.

Therefore, $f(g(7))=18$.

Alternatively, we can find $f(g(7))$ directly:

$f(g(7))=f(9)=2(9)=18$.

(2) $f(g(x))=f(5x+3)=3(5x+3)-7=15x+2$.

Therefore, $f(g(7))=15(7)+2=107$.

Alternatively, we can find $f(g(7))$ directly:

$f(g(7))=f(5(7)+3)=f(38)=3(38)-7=107$.

(3) $f(g(x))=f(2^x-1)=8(2^x-1)$.

Therefore, $f(g(7))=8(2^7-1)=8(127)=1016$.

Alternatively, we can find $f(g(7))$ directly:

$f(g(7))=f(2^7-1)=f(127)=8(127)=1016$.

26.2 Polynomial Divisions 多项式除法

视频 39

26.2.1 Introduction 介绍

Rational Expression 有理式

n. [ˈræʃənəl ɪkˈspreʃən]

Definition: A rational expression is either an integral expression or a fraction. 有理式是整式与分式的统称。

Notation: A/B, where A and B are integral expressions:

If B does not contain a variable, then A/B is an integral expression.

若 B 不含变量，则 A/B 是整式。

If B contains a variable, then A/B is a fraction.

若 B 含有变量，则 A/B 是分式。

Long Division 长除法

n. [lɔŋ dɪˈvɪʒən]

Definition: An algorithm of dividing polynomials, i.e. representation of a rational expression as the sum of a polynomial and a rational expression, for which the denominator in the resulting rational expression has a larger degree. 长除法是指除以多项式的算法，也就是把一个分式改写成多项式与分式的和，使得分式分母的次数大于分子的次数。

Notation: For polynomials $A(x)$ and $B(x)$, when $A(x)$ is divided by $B(x)$ using the long division, we have

多项式 $A(x)$ 除以多项式 $B(x)$ 的写法为

$$\frac{A(x)}{B(x)}=Q(x)+\frac{R(x)}{B(x)}$$

$Q(x)$ is the quotient, which is a polynomial.

$Q(x)$ 为商式，是多项式。

$R(x)$ is the remainder, a polynomial whose degree is less than the degree of $B(x)$.

$R(x)$ 为余式，是次数小于 $B(x)$ 的次数的多项式。

Properties: The algorithm is an analogue of the long division algorithm of numbers. 这个算法与数字的长除法类似。

In the long division algorithm with numbers, the remainder must be less than the divisor. 在数字的长除法中,余数必须比除数小。

In this algorithm, the remainder's degree must be less than that of the divisor. 在这个算法中,余式的次数必须比除式的次数小。

Examples: Figure 26-1 shows the long division of $\dfrac{3x^3+2x^2-5x+7}{x^2-x+1}$.

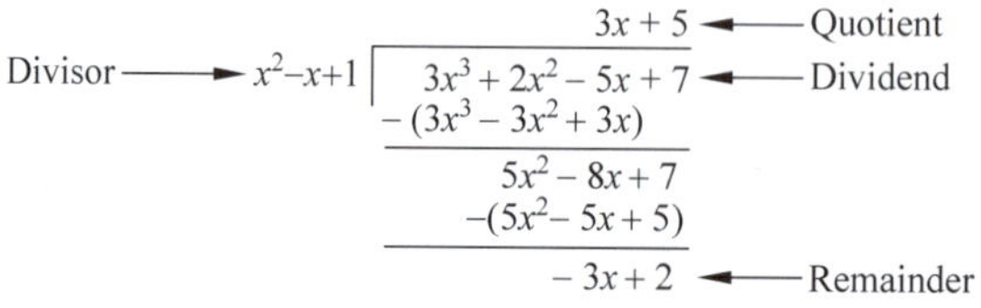

$$\begin{array}{r|l} & 3x+5 \quad \leftarrow \text{Quotient} \\ \text{Divisor} \rightarrow x^2-x+1 & 3x^3+2x^2-5x+7 \quad \leftarrow \text{Dividend} \\ & -(3x^3-3x^2+3x) \\ \hline & 5x^2-8x+7 \\ & -(5x^2-5x+5) \\ \hline & -3x+2 \quad \leftarrow \text{Remainder} \end{array}$$

Figure 26-1

In the quotient, we line up the terms with the dividend's terms by like terms. For example, we line up the quotient's $3x$ term with the dividend's $-5x$ term.

在商式里,根据与被除式的同类项进行对齐。举例如下:商式的项 $3x$ 与被除式的项 $-5x$ 对齐。

We write $\dfrac{3x^3+2x^2-5x+7}{x^2-x+1}=3x+5+\dfrac{-3x+2}{x^2-x+1}$.

Questions: When $x=1000$, round $\dfrac{x^4+3x^2}{x^2+6}$ to the nearest integer.

Answers: Using long division, $\dfrac{x^4+3x^2}{x^2+6}=x^2-3+\dfrac{18}{x^2+6}$.

When $x=1000$, $\dfrac{18}{x^2+6}$ is clearly less than 0.5.

Rounding to the nearest integer, this is $x^2-3=1000^2-3=999997$.

26.2.2 Theorems Related to Long Division 有关长除法的定理

Factor Theorem 因式定理

n. [ˈfæktər ˈθɪərəm]

Definition: A polynomial $P(x)$ has a factor $(x-k)$ if and only if $P(k)=0$.

We also call k a root for $P(x)$.

因式定理:$(x-k)$为 $P(x)$的因式,当且仅当 $P(k)=0$,称 k 为$P(x)$的根。

Proof: (1) Statement: If a polynomial $P(x)$ has a factor $(x-k)$, then $P(k)=0$.

Proof: By definition, we have $P(x)=(x-k)$(other polynomial), from which $P(k)=(k-k)$(other polynomial) $=0$(other polynomial) $=0$.

(2) Statement: If a polynomial $P(x)$ satisfies that $P(k)=0$, then $P(x)$ has a factor $(x-k)$.

Proof: From long division, we can rewrite as $P(x)=(x-k)Q(x)+R(x)$, where $Q(x)$ is the polynomial quotient and $R(x)$ is the polynomial remainder.

根据长除的定义，多项式 $P(x)$ 可改写为 $P(x)=(x-k)Q(x)+R(x)$，其中 $Q(x)$ 为商式，$R(x)$ 为余式。

We are given that $P(k)=(k-k)Q(k)+R(k)=R(k)=0$. Since the polynomial remainder is 0, we have $P(x)=(x-k)Q(x)$, from which we know that $P(x)$ has a factor $(x-k)$.

代入 $x=k$ 可得出余式为 0。根据定义，$P(x)$ 有 $(x-k)$ 这个因式。

Questions: What is the polynomial $P(x)$ with the least number of degree such that $P(x)$ has roots of 1, 2, 3, 4, and 5, and $P(0)=-720$?

Answers: Since $P(x)$ has roots of 1, 2, 3, 4, and 5, we know that $(x-1)$, $(x-2)$, $(x-3)$, $(x-4)$, and $(x-5)$ are factors of $P(x)$.

The least number of degree is 5.

We have $P(x)=a(x-1)(x-2)(x-3)(x-4)(x-5)$, where a is the leading coefficient.

Substituting, we get $P(0)=a(0-1)(0-2)(0-3)(0-4)(0-5)=-120a=-720$, from which $a=6$.

Therefore, our answer is $P(x)=6(x-1)(x-2)(x-3)(x-4)(x-5)$.

Remainder Theorem 余式定理

n. [rɪ'meɪndər 'θɪərəm]

Definition: The remainder of the division of a polynomial $P(x)$ by a linear polynomial $x-k$ is equal to $P(k)$.

多项式 $P(x)$ 除以 $x-k$ 的余式为 $P(k)$。

Proofs: From long division, we can rewrite as $P(x)=(x-k)Q(x)+R(x)$, where $Q(x)$ is the polynomial quotient and $R(x)$ is the polynomial remainder. 根据长除的定义，多项式 $P(x)$ 可改写为 $P(x)=(x-k)Q(x)+R(x)$，其中 $Q(x)$ 为商式，$R(x)$ 为余式。

By substitution, $P(k)=(k-k)Q(k)+R(k)=R(k)$.

Note that the divisor is $x-k$ and the remainder is $R(x)$. The degree of $R(x)$ must be less than that of $x-k$. So, the degree of $R(x)$ is 0, which means that $R(x)$ is a constant.

注意除式为 $x-k$，余式为 $R(x)$。因为 $R(x)$ 的次数必须比 $x-k$ 的次数小，所以 $R(x)$ 的次数为 0，即 $R(x)$ 为一个常值。

Notes: The Factor Theorem is a corollary of the Remainder Theorem. 因式定理是余式定理的推论。

Questions: What is the remainder of $\dfrac{x^3+2x^2+3x+4}{x-5}$?

Answers: Suppose $P(x)=x^3+2x^2+3x+4$. We know that the remainder of $\dfrac{P(x)}{x-5}$ is $P(5)=194$.

26.2.3 Rational Root Theorem 有理根定理

Rational Root Theorem 有理根定理

n. ['ræʃənəl rʊt 'θɪərəm]

Definition: Given a polynomial $P(x)=a_nx^n+a_{n-1}x^{n-1}+\cdots+a_1x+a_0$ with integral coefficients, and $a_n \neq 0$. The Rational Root Theorem states that if $P(x)$ has a rational root $r = \pm\frac{p}{q}$, where p and q are relatively prime positive integers, then p is a divisor of a_0 and q is a divisor of a_n.

多项式 $P(x)=a_nx^n+a_{n-1}x^{n-1}+\cdots+a_1x+a_0$,其中系数为整数,且 $a_n\neq 0$。根据有理根定理,若 $P(x)$有有理根 $r=\pm\frac{p}{q}$,其中 p 和 q 为互质的正整数,则 p 为 a_0 的因数,且 q 为 a_n 的因数。

Proofs: Given that $\frac{p}{q}$ is a rational root of $f(x)=a_nx^n+a_{n-1}x^{n-1}+\cdots+a_1x+a_0$, where the coefficients are integers. We wish to show that $p \mid a_0$ and $q \mid a_n$. Since $\frac{p}{q}$ is a root, it follows that

$$0=a_n\left(\frac{p}{q}\right)^n+a_{n-1}\left(\frac{p}{q}\right)^{n-1}+\cdots+a_1\left(\frac{p}{q}\right)+a_0.$$

Multiplying both sides by q^n gives

$$0=a_np^n+a_{n-1}p^{n-1}q+\cdots+a_1pq^{n-1}+a_0q^n.$$

Taking this modulo p, we have

$$0\equiv 0+0+\cdots+0+a_0q^n \pmod{p}.$$

We know that $a_0q^n\equiv 0 \pmod{p}$. Since p and q are relatively prime, we have $p \mid a_0$. Similarly, taking $0=a_np^n+a_{n-1}p^{n-1}q+\cdots+a_1pq^{n-1}+a_0q^n$ mod q results that $a_np^n\equiv 0 \pmod{q}$. Since p and q are relatively prime, we have $q \mid a_n$.

Questions: List all possible rational roots for the polynomial

$$P(x)=4x^4+7x^3+6x^2-8x-9.$$

Answers: All possible rational roots are of the form $\pm\frac{p}{q}$, where p is a divisor of 9 and q is a divisor of 4.

The possible rational roots are: $\pm 1, \pm 3, \pm 9, \pm\frac{1}{2}, \pm\frac{3}{2}, \pm\frac{9}{2}, \pm\frac{1}{4}, \pm\frac{3}{4}, \pm\frac{9}{4}$.

26.2.4 Summary 总结

1. A rational expression $\frac{A(x)}{B(x)}$ can be written as $Q(x)+\frac{R(x)}{B(x)}$ using long division. The degree of $R(x)$ must be less than that of $B(x)$. 根据长除法定义,可把分式$\frac{A(x)}{B(x)}$改写为 $Q(x)+\frac{R(x)}{B(x)}$,其中 $R(x)$的次数必须小于 $B(x)$的次数。
2. Factor Theorem: A polynomial $P(x)$ has a factor $(x-k)$ if and only if $P(k)=0$.
   因式定理
3. Remainder Theorem: The remainder of the division of a polynomial $P(x)$ by a linear polynomial $x-k$ is equal to $P(k)$.
   余式定理

4. Review Rational Root Theorem in Section 26.2.3.
复习 26.2.3 节的有理根定理。

26.3　Factoring Completely　完全分解

视频 40

Synthetic Division　综合除法

n. [sɪn'θetɪk dɪ'vɪʒən]

Definition: A quick method of factoring a polynomial, based on Rational Root Theorem and Factor Theorem.
综合除法是根据有理根定理和因式定理的快速因式分解法。

Procedure: To divide $P(x)=a_nx^n+a_{n-1}x^{n-1}+\cdots+a_1x+a_0$ by $x-k$, we use synthetic division.

Figure 26-2 shows the synthetic division for $\dfrac{ax^n+bx^{n-1}+cx^{n-2}+\cdots}{x-k}$.

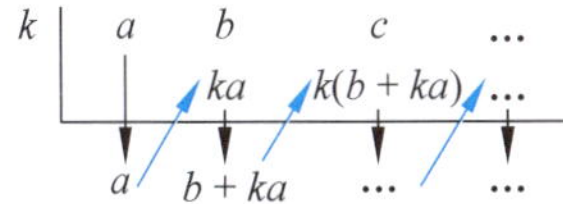

Figure 26-2

Vertical pattern (black arrows): Add terms.

Horizontal pattern (blue arrows): Multiply by k.

Remember from the Remainder Theorem that the remainder of $P(x)$ divided by $x-k$ is equal to $P(k)$. 根据余式定理，$P(x)$ 除以 $x-k$ 的余式与 $P(k)$ 相等。

We can only use synthetic division when the divisor polynomial is linear. The reason is that synthetic division is based on the Remainder Theorem. 综合除法只对线性多项式除式有效，因为综合除法是根据余式定理得来的。

Notes: Why does synthetic division work?

We can compare the methods of long division and synthetic division.

Figure 26-3 shows the long division for $\dfrac{4x^3+2x^2-3x+8}{x-2}$.

Figure 26-4 shows the synthetic division for $\dfrac{4x^3+2x^2-3x+8}{x-2}$.

$$
\begin{array}{r|l}
 & 4x^2+10x+17 \\ \hline
x-2 & 4x^3+2x^2-\ \ 3x+\ 8 \\
 & -(4x^3-8x^2) \\ \hline
 & 10x^3-\ \ 3x \\
 & -(10x^2-20x) \\ \hline
 & 17x+\ 8 \\
 & -(17x-34) \\ \hline
 & 42
\end{array}
$$

Figure 26-3

$$
\begin{array}{r|rrrr}
2 & 4 & 2 & -3 & 8 \\
 & & 8 & 20 & 34 \\ \hline
 & 4 & 10 & 17 & 42
\end{array}
$$

Figure 26-4

Looking at the bottom row in the synthetic division, the last number is the remainder and the first three numbers are the coefficients of the quotient. Looking at the second row in the synthetic division, the numbers correspond to $-8x^2$, $-20x$, and -34 in the steps of the long division. We will analyze how this happens:

观察综合除法的最后一行：最后的数字是余式，且前三个数是商式的系数。观察综合除法的第二行，与数字 8、20、34 对应的分别是长除法中间步骤的 $-8x^2$、$-20x$、-34 三个项。

分析如下：

As we are dividing $4x^3+2x^2-3x+8$ by $x-2$, we want the degree of the quotient to be $3-1=2$. **The leading coefficient of the quotient is clearly 4. That's how we have the 4 on the last row in the synthetic division.** Note: the previous bolded sentences would not be true if the divisor were not in the form of $x-k$.

当 $4x^3+2x^2-3x+8$ 除以 $x-2$ 时，商的次数为 $3-1=2$。**商的最高次项系数显然为 4。这就是综合除法中最后一行 4 的由来。**注：若除式的形式不是 $x-k$，加粗的句子就不成立了。

We multiply $4x^2$ by $x-2$, the x^2-term's coefficient of the product is -8. As we subtract the product from $4x^3+2x^2-3x+8$, the x^2-term's coefficient is $2-(-8)=10$. And clearly, when that difference is divided by $x-2$, the x-term's coefficient in the quotient is 10. This is what the next column in the synthetic division indicates: $2+8=10$. The 10 represents the coefficient of the second term in the quotient, namely the x-term.

把 $4x^2$ 乘上 $x-2$，积的 x^2 项系数为 -8。当 $4x^3+2x^2-3x+8$ 减去这个积时，x^2 的系数为 $2-(-8)=10$。显然，当这个差除以 $x-2$ 时，商的 x 项系数为 10。这也是综合除法下一列所表示的：$2+8=10$。10 代表的是商第二个项的系数，也就是 x 项的系数。

And so on.

以此类推。

The generalized method in proving the synthetic division uses the exact same idea.

证明综合除法的泛化方法运用了相同的思想。

Solving Polynomial Equations 解多项式方程

Procedure: To solve the equation of the form $P(x)=0$, from which $P(x)$ is a polynomial:

1. Find all possible rational roots based on the Rational Root Theorem. 根据有理根定理，找出所有有理根的可能值。
2. Test each rational root using the Synthetic Division. 用综合除法测试有理根。
3. Factor out a linear factor based on the result of the Synthetic Division, and repeat the process, until the resulting polynomial is easy enough to factor (quadratic or able to be factored by grouping). 根据综合除法的结果，分解出一个线性因式。对新得出的多项式重复此步骤，直到新得出的多项式容易分解(二次或者可以分组分解)。
4. Factor out everything and find the roots. 完全分解，并求出根。

Questions: What are all roots of the polynomial $2x^4-9x^3+37x-30=0$?

Answers: We use the Rational Root Theorem to find all possible rational roots: $\pm1, \pm2, \pm3, \pm5, \pm6, \pm10, \pm15, \pm30, \pm\frac{1}{2}, \pm\frac{3}{2}, \pm\frac{5}{2}, \pm\frac{15}{2}$.

We then test the rational roots one by one. Of course, a strong number sense allows one to recognize that 1 is a root. We perform synthetic division as shown in Figure 26-5:

The quotient is $2x^3-7x^2-7x+30$.

We can rewrite the equation as $(x-1)(2x^3-7x^2-7x+30)=0$.

We use the same procedure on the quotient, as shown in Figure 26-6.

The quotient is $2x^2-x-10$.

We can rewrite the equation as $(x-1)(x-3)(2x^2-x-10)=0$.

At this point, the resulting polynomial is easy enough to factor. We can either keep going with the synthetic division or use what we learned in quadratic equations to factor.

Figure 26-7 gives the complete synthetic division for this problem.

1	2	−9	0	37	−30
		2	−7	−7	30
	2	−7	−7	30	0

Figure 26-5

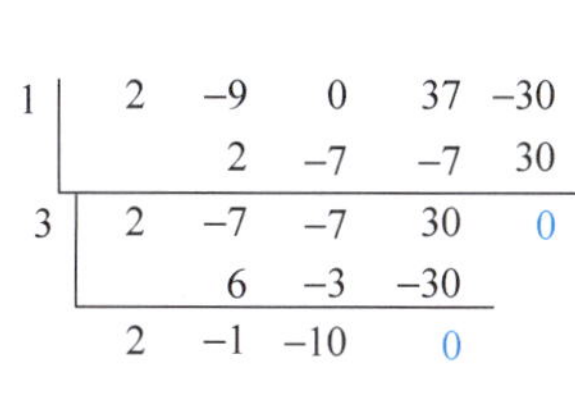

Figure 26-6

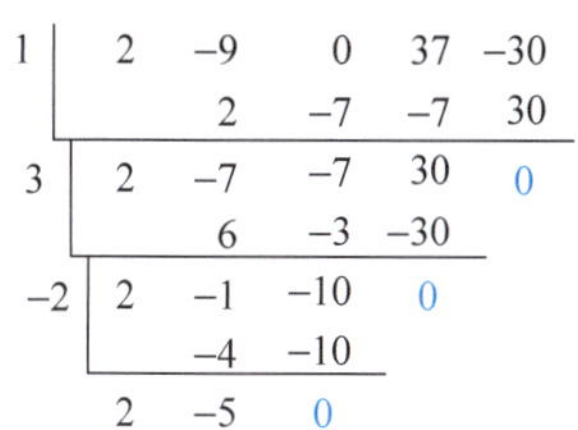

Figure 26-7

The quotient is $2x-5$.

We can rewrite the equation as $(x-1)(x-3)(x+2)(2x-5)=0$.

Therefore, our roots are $x=1,3,-2,5/2$.

26.4 Theorems of Algebra 代数定理

Fundamental Theorem of Algebra 代数基本定理

n. [ˌfʌndəˈmentəl ˈθɪərəm ʌv ˈældʒəbrə]

Definition: If $f(x)$ is a polynomial of degree n, where $n>0$, then f has at least one zero in the complex number system.

若 $f(x)$是一个次数为 n 的多项式,其中 $n>0$,则 f 至少有一个复数零值。

Proofs: The proof of this theorem requires topology and mathematics analysis, which is out of the scope of high school mathematics.

这个定理的证明需要用到拓扑学和数学分析。这些知识点都在高中数学范围之外。

Linear Factorization Theorem 线性分解定理

n. [ˈlɪniər ˌfæktəraɪzˈzeɪʃən ˈθɪərəm]

Definition: If $f(x)$ is a polynomial of degree n, where $n>0$, then f has exactly n linear factors:

$$f(x)=a_n(x-c_1)(x-c_2)\cdots(x-c_n),$$

where $c_1, c_2, \cdots, c_n$ are complex numbers.

若 $f(x)$是一个次数为 n 的多项式,其中 $n>0$,则 f 有 n 个线性因式,$c_1, c_2, \cdots, c_n$ 均为复数。

Proofs: Using the Fundamental Theorem of Algebra, we know that f has at least one zero in the complex number system. Call this zero c_1. We rewrite f as

$$f(x)=(x-c_1)f_1(x).$$

If the degree of $f_1(x)$ is greater than zero, applying the Fundamental Theorem of Algebra again gives

$$f(x)=(x-c_1)(x-c_2)f_2(x),$$

and so on.

Clearly, the degree of f is n; the degree of f_1 is $n-1$; the degree of f_2 is $n-2$. The degree of f_i is $n-i$. Applying the Fundamental Theorem of Algebra n times results that

$$f(x)=a_n(x-c_1)(x-c_2)\cdots(x-c_n),$$

where a_n is the leading coefficient of $f(x)$.

对一个 n 次多项式$f(x)$运用 n 次代数基本定理,每次分解一个线性因式,新的多项式的次数比原多项式的次数少 1。

运用 n 次代数基本定理可得出$f(x)=a_n(x-c_1)(x-c_2)\cdots(x-c_n)$,其中 a_n 为$f(x)$的最高次项系数。

26.5 Characteristics of a Polynomial Function and Dominance of Functions 多项式函数的特征与主导函数

Seven Characteristics of a Polynomial Function 多项式函数的 7 个特征

Characteristics: For a polynomial function $f(x)$ with degree n and leading coefficient a:

对 n 次且最高次项系数为 a 的多项式函数$f(x)$,有以下 7 个特征:

1. The graph of f is continuous. f 的图像是连续的。
2. All turns, if any, of the graph of f are smooth, unlike the sharp turns of the graphs of absolute value functions. 所有的弯都是平滑的,不像绝对值函数图像的尖转弯。
3. The graph of f has at most $n-1$ extrema. f 的图像至多有 $n-1$ 个极值点。
4. The graph of f has at most n x-intercepts. f 的图像至多有 n 个 x 截距。
5. For n is odd:

   If $a>0$, then the graph of f falls to the left and rises to the right, as shown in Figure 26-8. 若 $a>0$,则 f 的图像往左向下运动、往右向上运动,如图 26-8 所示。

   If $a<0$, then the graph of f rises to the left and falls to the right, as shown in Figure 26-9. 若 $a<0$,则 f 的图像往左向上运动、往右向下运动,如图 26-9 所示。

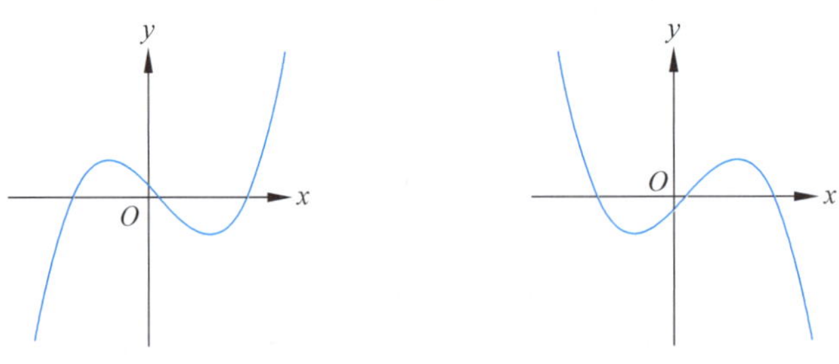

Figure 26-8 Figure 26-9

6. For n is even:

   If $a>0$, then the graph of f rises both to the left and right, as shown in Figure 26-10. 若 $a>0$,则 f 的图像两边向上运动,如图 26-10 所示。

   If $a<0$, then the graph of f falls both to the left and right, as shown in Figure 26-11. 若 $a<0$,则 f 的图像两边向下运动,如图 26-11 所示。
7. The graph of f has exactly one y-intercept. f 的图像仅有一个 y 截距。

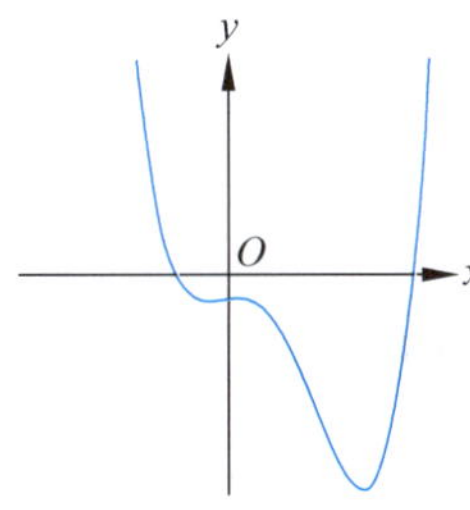

Figure 26-10

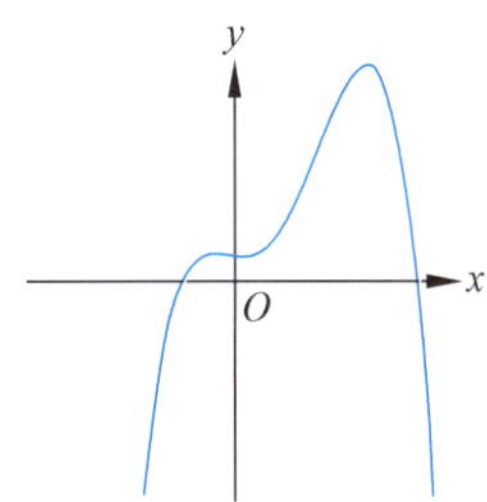

Figure 26-11

Dominance of Functions　主导函数

Definition: For functions f and g, we say f dominates g if $\frac{f(x)}{g(x)} \to \pm\infty$ as $x \to +\infty$ or as $x \to -\infty$.

Dominance decides the end behavior.　主导函数决定了末端情况。

Properties: 1. The following types of functions are ordered from dominance as $x \to \infty$:

(1) Factorial Functions: $x!$ (x is an integer.)　阶乘函数

(2) Exponential Functions: b^x ($|b|>1$.)　指数函数

(3) Power Functions: x^n ($n>0$)　幂函数

(4) Logarithmic Functions: $\log_u x$ ($u>1$)　对数函数

2. For power functions: When $a>b>0$, x^a dominates x^b as $x \to \pm\infty$.

3. Using Property 2, for polynomial function $f(x) = a_nx^n + a_nx^n + \cdots + a_0$, the leading term, a_nx^n, is the dominating term, which affects the behavior of the output as $x \to +\infty$ or $x \to -\infty$ regardless of the other terms. 多项式函数的主导项 a_nx^n 决定当$x \to +\infty$或 $x \to -\infty$时 $f(x)$的趋势。其他项与 $f(x)$的趋势无关。

We have Characteristics 5 and 6 from Seven Characteristics of A Polynomial Function:

- For odd n:

  If $a_n<0$, then $a_nx^n \to \infty$ as $x \to -\infty$. Since a_nx^n is the dominating term, we know that $f(x) \to \infty$ as $x \to -\infty$. We refer this as rising to the left. Similarly, $f(x) \to -\infty$ as $x \to \infty$. We refer this as falling to the right.

  If $a_n>0$, then $f(x) \to -\infty$ as $x \to -\infty$. We refer this as falling to the left. In addition, $f(x) \to \infty$ as $x \to \infty$. We refer this as rising to the right.

- For even n:

  If $a_n<0$, then $f(x) \to -\infty$ as $x \to \pm\infty$. We refer this as falling to the left and falling to the right.

  If $a_n>0$, then $f(x) \to \infty$ as $x \to \pm\infty$. We refer this as rising to the left and rising to the right.

Questions: What is the end behavior for each of the following?

1. $f(x) = 2x^4 - 3x^2 + 5$

2. $g(x) = -7x^5 + 8x^4 + 9$

Answers: 1. $f(x) \to \infty$ as $x \to \pm\infty$. The graph of f rises to both left and right.

2. $f(x)\to\infty$ as $x\to-\infty$. The graph of f rises to the left.
   $f(x)\to-\infty$ as $x\to\infty$. The graph of f falls to the right.

27. Domain Restrictions and Inverse Functions 定义域限制与反函数

27.1 Domain Restrictions 定义域限制

Domain Restriction 定义域限制

n. [doʊ'meɪn rɪ'strɪkʃən]

Definition: The domain restriction of a function $f(x)$ is the restriction that the domain is not the set of all real numbers. i.e. If input x takes a certain value a, $f(a)$ will be undefined.
函数 f 的定义域限制是指"定义域不能为实数集"。当输入 x 为 a 值时,$f(a)$为无意义。

Properties: So far, the two types of the domain restrictions we encountered are:

1. The denominator cannot be 0. 分母不能为 0。
2. The radicand of an (even)th root cannot be negative. 偶次方根的被开方数不能为负数。

Questions: Identify the domain restrictions for each of the following.

(1) $f(x)=\dfrac{2017}{x^2-5x-14}$

(2) $g(x)=\sqrt{5x-3}$

(3) $h(x)=\dfrac{x^3-5x}{\sqrt{x^2-4}}$

(4) $m(x)=\dfrac{\sqrt{5x-3}}{\sqrt{x^2-4}}$

Answers: (1) The denominator $x^2-5x-14$ cannot be 0.
$x^2-5x-14=(x-7)(x+2)$ is 0 when $x=7$ or -2.
Therefore, the domain of f is the set of all real numbers except $x=7$ or -2.

(2) The radical is the square root (even root). The radicand $5x-3$ cannot be less than 0.
$5x-3$ is less than 0 when $x<3/5$.
Therefore, the domain of g is the set of all real numbers for which $x\geqslant 3/5$.

(3) The radical is the square root (even root). The radicand x^2-4 cannot be 0 (because it is in the denominator) or negative.

So, the radicand x^2-4 must be positive.

To solve $x^2-4=(x+2)(x-2)>0$, note that the zeros are -2 and 2.

We want both $x+2$ and $x-2$ to have the same sign so that their product is positive.

If $x<-2$, then both of them are negative. Their product is positive.

If $-2<x<2$, then $x+2$ is positive, but $x-2$ is negative. Their product is negative.

If $x>2$, then both of them are positive.

Thus we have $x<-2$ or $x>2$.

Therefore, the domain of h is the set of all real numbers such that $x<-2$ or $x>2$.

(4) Note that $h(x)$ and $m(x)$ share the same denominator, which has the same story.

From the denominator, we know that $x<-2$ or $x>2$.

From the numerator, we know that $x>3/5$, which is the same story for $g(x)$.

Both numerator and denominator have to follow the restriction rule, so we take the intersection of ($x<-2$ or $x>2$) and $x>3/5$.

Therefore, the domain of h is the set of all real numbers such that $x<-2$ or $x>2$.

27.2 Inverse Functions　反函数

27.2.1 Introduction　介绍

Inverse Function　反函数

n. [ɪnˌvɜrs ˌfʌŋkʃən]

Definition: If functions f and g are inverses of each other, then it is true that $f(g(x)) = g(f(x))=x$. In other words, f and g undo each other, and their composition function maps x back to x itself.

若 f 和 g 互为反函数，则有 $f(g(x))=g(f(x))=x$。即 f 和 g 相互抵消，并且把 x 仍映射为 x。

Notation: If g is the inverse function for f, then we write g as f^{-1}.

Properties: If g is the inverse function for f:

若 g 是 f 的反函数：

1. As shown in Figure 27-1:

   One's domain is the other's range, and vice versa.

   In other words, for $f(g(x))$, the range of g is the domain of f.

   For $g(f(x))$, the range of f is the domain of g.

   如图 27-1 所示：一个的定义域为另一个的值域，反之亦然。

   即对于 $f(g(x))$，g 的值域为 f 的定义域。

   对于 $g(f(x))$，f 的值域为 g 的定义域。

2. As shown in Figure 27-2, f and g undo each other, regardless of which is performed first.

   $f(g(x)) = x$.

   $g(f(x)) = x$.

   如图 27-2 所示，不管函数的运算顺序，f 和 g 相互抵消。

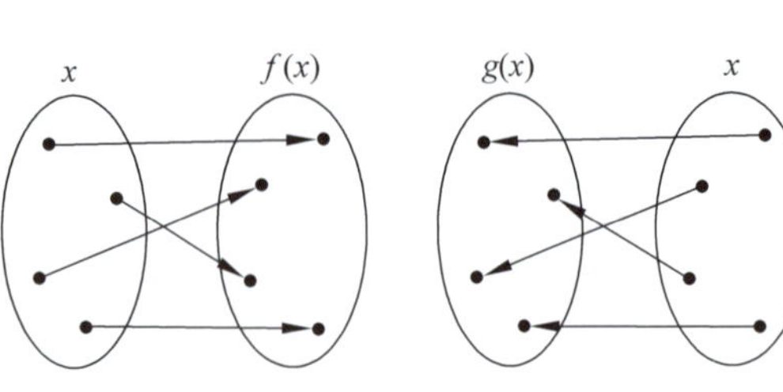

Figure 27-1

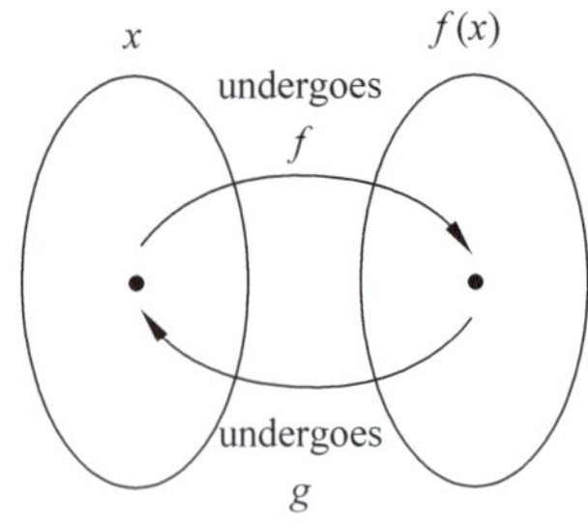

Figure 27-2

3. Given $f(x)$, in order to find its inverse $g(x)$, all we need to do is the following: On the equation $f(x)=\cdots x\cdots$, we replace $f(x)$ with x and x with $g(x)$. We need to solve for $g(x)$, which is also called $f^{-1}(x)$.
   若给定一个关于 x 的函数$f(x)$,要求出其反函数 $g(x)$,则只需做以下变换:
   原式中的 $f(x)$替换为 x,x 替换为$g(x)$。将 $g(x)$整理即可,也称 $g(x)$为 $f^{-1}(x)$。
4. See the graphs of inverse functions for more properties.
   更多性质见反函数的图像。

Questions: 1. For each of the following, identify whether they are inverse functions.

(1) $f(x) = 5x \qquad g(x) = \dfrac{1}{5}x$

(2) $j(x) = 4x-6 \quad k(x) = \dfrac{1}{4}x+6$

2. If possible, find the inverse function for each of the following:

(1) $\{(1, 100), (2, 200), (3, 300), (4, 400)\}$

For every ordered pair (x, y) in this set, x is the input and y is the output.

(2) $f(x) = 5x+3$

(3) $g(x) = -x-4$

(4) $h(x) = \sqrt[3]{x-6}$

(5) $j(x) = \sqrt{x}$

(6) $k(x) = x^2$

3. Dan loves to go to gym! He pays \$80 for the one-time registration fee and \$7 for each visit.

(1) Write the equation of the total cost in dollars, $f(x)$, as a function of number of visits, x.

(2) What does $f^{-1}(x)$ mean?

(3) Write the equation of $f^{-1}(x)$.

Answers: 1. (1) $f(g(x)) = 5\left(\dfrac{1}{5}x\right) = x$.

$$g(f(x)) = \frac{1}{5}(5x) = x.$$

f and g are inverse functions.

(2) $j(k(x))=4\left(\frac{1}{4}x+6\right)-6=x+24-6=x+18\neq x$.

j and k are not inverse functions.

2. (1) By definition, the inverse function is $\{(100, 1), (200, 2), (300, 3), (400, 4)\}$.

(2) To find the inverse function for $f(x) = 5x+3$:

By Property 3, we will replace $f(x)$ with x and x with $f^{-1}(x)$, then solve for $f^{-1}(x)$:

$$x = 5f^{-1}(x)+3$$

$$f^{-1}(x) = \frac{x-3}{5}.$$

We can verify inverses by checking $f(f^{-1}(x)) = x$ and $f^{-1}(f(x)) = x$.

(3) To find the inverse function for $g(x) = -x-4$:

By Property 3, we will replace $g(x)$ with x and x with $g^{-1}(x)$, then solve for $g^{-1}(x)$:

$$x = -g^{-1}(x)-4$$

$$-g^{-1}(x) = x+4$$

$$g^{-1}(x) = -x-4.$$

We can verify inverses by checking $g(g^{-1}(x)) = x$ and $g^{-1}(g(x)) = x$.

Note that in this case, $g(x)=g^{-1}(x)$! Why?

(4) $x=\sqrt[3]{h^{-1}(x)-6}$

$h^{-1}(x)=x^3+6$.

(5) $x=\sqrt{j^{-1}(x)}$

$j^{-1}(x)=x^2$.

But note that for $j(x)=\sqrt{x}$, the domain is the set of all nonnegative numbers and the range is the set of all nonnegative numbers.

By Property 1, we have $j^{-1}(x)=x^2$, **but the domain must be restricted to the set of all nonnegative numbers** and the range is the set of all nonnegative numbers.

(6) $x=(k^{-1}(x))^2$

$k^{-1}(x)=\pm\sqrt{x}$.

k^{-1} is not a function because the same input cannot go with two different outputs! For example, $k^{-1}(4)=\pm\sqrt{4}=\pm 2$.

What kind of functions do not have inverses?

The answer is in graphs of inverse functions.

3. (1) $f(x)=7x+80$.

(2) f^{-1} is the inverse function of f, which undoes the action of f.

$f(x)$: When the number of visits x is given, finds the total cost, in dollars.

$f^{-1}(x)$: When the total cost, x dollars, is given, finds the number of visits.

(3) $x=7f^{-1}(x)+80$

$$f^{-1}(x)=\frac{x-80}{7}.$$

Graphs of Inverse Functions 反函数的图像

Properties: The graphs of inverse functions are symmetric about the line $y=x$.

反函数的图像的对称轴为 $y=x$。

Proofs: As shown in Figure 27-3, suppose $P=(x_0, y_0)$ is a point on the graph of f. By definition of inverse functions, the corresponding point for the graph of f^{-1} is $P'=(y_0, x_0)$.

If P is on $y=x$, then $x_0=y_0$ and $P=P'$, and trivially, P and P' are symmetric about $y=x$.

如图 27-3 所示,假设 $P=(x_0, y_0)$是函数 f 图像的一点,那么根据反函数的定义,$P'=(y_0, x_0)$是函数 f^{-1} 图像的对应点。

若 P 在 $y=x$ 上,则 P 与 P'共点,也就是对称的特殊情况。

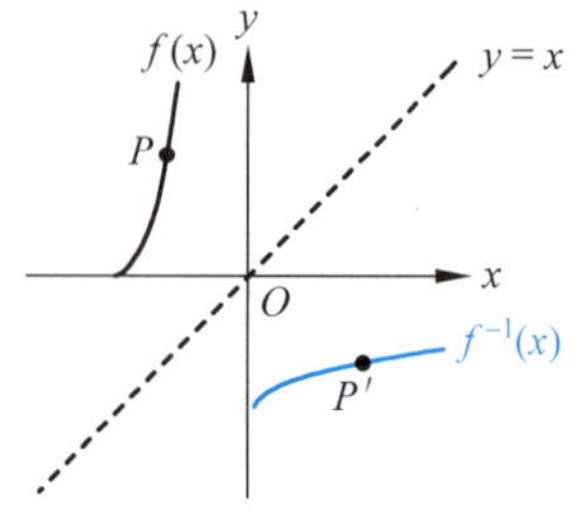

Figure 27-3

Otherwise, in order to prove that the line $y=x$ is the line of reflection of the graphs, we want to show that $y=x$ is the perpendicular bisector of $\overline{PP'}$.

To show that $y=x$ is perpendicular to $\overline{PP'}$, note that their slopes are opposite reciprocals:

The slope of $y=x$ is 1.

The slope of $\overline{PP'}$ is $\dfrac{x_0-y_0}{y_0-x_0}=-1$.

To show that $y=x$ bisects $\overline{PP'}$, we want to show that $y=x$ passes the midpoint of $\overline{PP'}$.

The midpoint of $\overline{PP'}$ is $\left(\dfrac{x_0+y_0}{2}, \dfrac{x_0+y_0}{2}\right)$, which is on $y=x$.

We have proved that for every point P on the graph of f (corresponds to P' on the graph of f^{-1}), the line $y=x$ is the perpendicular bisector of $\overline{PP'}$. Therefore, graphs of f and f^{-1} are symmetric about the line $y=x$.

否则,需要证明 $y=x$ 是$\overline{PP'}$的垂直平分线。

即必须证出 $y=x$ 与$\overline{PP'}$垂直,并且 $y=x$ 平分$\overline{PP'}$。

上述证明证出了$\overline{PP'}$与 $y=x$ 的斜率互为相反倒数(垂直),且 $y=x$ 必经$\overline{PP'}$中点(平分)。

结论是,对于 f 的图像的每个点 P(在 f^{-1} 的图像的对应点为 P'),$y=x$ 都是$\overline{PP'}$的垂直平分线。可以得出反函数图像对称于 $y=x$。

27.2.2 One-to-One Functions 一对一函数

One-To-One Function 一对一函数

n. ['wʌntə'wʌn 'fʌŋkʃən]

Definition: A function for which every element of the range corresponds to exactly one element of the domain. 一对一函数是指值域里每个值都只对应定义域里一个值的函数。

Review function's definition in Section 7.1.2. 回顾 7.1.2 节函数的定义。

Properties: 1. Only one-to-one functions have inverse functions. 只有一对一函数有反函数。

2. One-to-one functions must pass both the horizontal line test and the vertical line test. 一对一函数必须通过垂线测试和水平线测试。

The vertical line test determines whether the graph is a function.　垂线测试检测图像是否为函数。

The horizontal line test determines whether the function is one-to-one.　水平线测试检测函数是否为一对一。

Proofs: To prove Property 1, for inverse functions f and f^{-1}, recognize that the domain of f is the range of f^{-1}, and the range of f is the domain of f^{-1}.

If one output of f goes with two or more inputs of f, f is still a function, but in f^{-1}, it becomes one input goes with two or more outputs, which means f^{-1} is not a function, or, the inverse function for f does not exist.

要证明性质1，注意在一对反函数 f 和 f^{-1} 中，一个函数的定义域为另一个函数的值域，反之亦然。若在 f 中一个输出对应多个输入，则 f 仍是函数，但在 f^{-1} 中一个输入对应多个输出，即 f^{-1} 不是函数，或者 f 不存在反函数。

Questions: For each of the following relations, each ordered pair (a, b) represents the input is a and the output is b. Determine whether it is a one-to-one function.

(1) $\{(1,10),(2,20),(3,30),(4,40),(5,50)\}$

(2) $\{(1,10),(2,10),(3,44),(4,55),(5,77)\}$

(3) $\{(1,16),(2,81),(3,23),(3,37),(5,71)\}$

Answers: (1) Yes.

(2) No. The output 10 goes with both inputs 1 and 2.

(3) No. The relation is not even a function. Input 3 goes with both outputs 23 and 37.

Horizontal Line Test　水平线测试

n. [ˌhɔrə'zɑntəl lɑɪn test]

Definition: A visual way to determine whether a function is one-to-one.　水平线测试是通过视觉判断函数是否为一对一函数的方法。

Properties: 1. If two or more points of the same graph lies in the same horizontal line, then the function is not one-to-one.　若图像中两个或更多的点在同一条水平线上，则这个函数不是一对一函数。

2. Functions that do not pass the horizontal line test are not one-to-one functions and do not have an inverse functions.　通不过水平线测试的函数都不是一对一函数，都没有反函数。

Notes: If two or more points of the same graph lies in the same horizontal line, then the graph violates the definition of a one-to-one function: each element in the range can only map to one element in the domain.

若两个或更多的点在图像的同一条水平线上，则图像违反了一对一函数的定义：值域中的每个值只与定义域中的一个值对应。

Examples: As shown in Figure 27-4, f is a one-to-one function. The graph of its inverse function is drawn in blue.

As shown in Figure 27-5, g is not a one-to-one function. The graph of its “inverse” is drawn in blue. Its inverse is not a function because it fails the vertical line test, as g fails the horizontal line test.

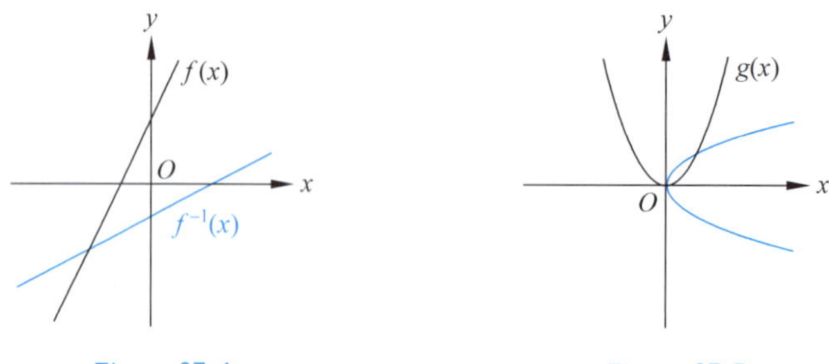

Figure 27-4　　Figure 27-5

27.2.3　Summary　总结

1. If functions f and g are inverses of each other, then it is true that $f(g(x)) = g(f(x)) = x$. In other words, f and g undo each other, and their composition function maps x back to x itself. 若 f 和 g 互为反函数,则有 $f(g(x)) = g(f(x)) = x$。即 f 和 g 相互抵消,并且把 x 仍映射为 x。

   For inverse functions f and g, one's domain is the other's range, and vice versa.

   In other words, for $f(g(x))$, the range of g is the domain of f.

   For $g(f(x))$, the range of f is the domain of g.

   对于互反函数 f 和 g,一个的定义域为另一个的值域,反之亦然。

   即对于 $f(g(x))$,g 的值域为 f 的定义域。

   对于 $g(f(x))$,f 的值域为 g 的定义域。

   Given $f(x)$, in order to find its inverse $g(x)$, all we need to do is the following:

   On the equation $f(x) = \cdots x \cdots$, we replace $f(x)$ with x and x with $g(x)$. We need to solve for $g(x)$, which is also called $f^{-1}(x)$.

   若给定一个关于 x 的函数 $f(x)$,要求出其反函数 $g(x)$,则只需作以下变换:

   原式中的 $f(x)$ 替换为 x,x 替换为 $g(x)$。将 $g(x)$ 整理即可,也称 $g(x)$ 为 $f^{-1}(x)$。

2. For inverse functions f and f^{-1}, their graphs are symmetric about the line $y = x$.　函数与它的反函数的图像关于直线 $y = x$ 对称。

3. Only one-to-one functions have inverse functions. For function f, if two or more inputs go with the same output, then f^{-1} will not be a function as its two or more outputs go with the same input.　只有一对一函数有反函数。若函数 f 至少有两个输入对应同一个输出,则 f^{-1} 将不会是一个函数,因为它至少有两个输出对应同一个输入。

   A function must pass the vertical line test.　函数必须通过垂线测试。

   A one-to-one function must pass the horizontal line test in addition.　一对一函数必须另外通过水平线测试。

   The horizontal line test checks whether a function is one-to-one.　水平线测试检验一个函数是否一对一。

28. Exponential and Logarithmic Functions 指数函数与对数函数

Review exponential functions in Section 7.4.2.　回顾 7.4.2 节指数函数。

28.1 Irrational Number e　无理数 e

e

n. [i]

Definition: The value of $\left(1+\frac{1}{n}\right)^n$ as $n \to +\infty$.　当 n 趋向无穷大时，$\left(1+\frac{1}{n}\right)^n$ 的极限值为 e。

The value of e is approximately 2.71828.

e 是一个趋近 2.71828 的值。

Properties: Suppose P is the invested principal, n is the number of compoundings per year, and r is the annual interest rate, the balance after t years is $A = P\left(1+\frac{r}{n}\right)^{nt}$.

Suppose n increases without bound, we call this continuous compounding: We can rewrite the formula in terms of e.

Let $m = n/r$. We can rewrite:

若 P 为投资本金，n 为每年的复利次数，r 为年利率，则 t 年后余额为 $A = P\left(1+\frac{r}{n}\right)^{nt}$。

设 n 没有限制地增加，这个现象叫作连续复利：下面用 e 改写有关 A 的公式。设 $m = n/r$，可得

$$\begin{aligned} A &= P\left(1+\frac{r}{n}\right)^{nt} \\ &= P\left(1+\frac{r}{mr}\right)^{mrt} \\ &= P\left(1+\frac{1}{m}\right)^{mrt} \\ &= P\left[\left(1+\frac{1}{m}\right)^{m}\right]^{rt} \\ &= P\mathrm{e}^{rt}. \end{aligned}$$

28.2 Logarithms　对数

视频 41

Logarithm　对数

n. [ˈlɔgəˌrɪðəm]

Definition: An operation that finds the exponent when given a base and a power.　对数是指已知底

数和幂,算出指数的运算。

Notation: $\log_b a = c$ if and only if $b^c = a$, for which $a>0$, $b>0$, and $b \neq 1$.

b is known as the base. b 为底数。

When we omit the b in our notation, the logarithm takes the assumed base, 10. 当省略不写 b 时,对数的底数为假定的 10。

When the base is e, we write ($\ln a$) instead of ($\log_e a$). The number e is known as the natural base. 当底数为 e 时,$\log_e a$ 可改写成 $\ln a$。e 为自然底数。

Properties: 1. Special cases of logarithms:

(1) $\log_b 1 = 0$, since $b^0 = 1$.

(2) $\log_b b = 1$, since $b^1 = b$.

2. Logarithms are the inverse operation of powers. 对数是次幂的逆运算。

(1) $\log_b(b^a) = a$.

(2) $b^{\log_b a} = a$.

The following properties are based on the properties of exponents. 以下性质跟指数的性质有关。

3. (1) Logarithm Product Rule: $\log_b u + \log_b v = \log_b(uv)$.

(2) Logarithm Quotient Rule: $\log_b u - \log_b v = \log_b \dfrac{u}{v}$.

4. Logarithm Power Rule:

$\log_b(u^v) = v\log_b u$.

5. Logarithm Base-Change Rule: 换底定理

$$\log_b u = \frac{\log_c u}{\log_c b}.$$

6. Logarithm Base-Switch Rule:

$$\log_b u = \frac{1}{\log_u b}.$$

Notes: 1. Proof of Property 2(1):

By the definition of logarithms, $\log_b b^a$ gets the exponent such that when b is raised to that exponent, the result is b^a. Clearly, this exponent is a.

Proof of Property 2(2):

Let $\log_b a = x$. By definition, we have $b^x = a$.

$\text{LHS} = b^{\log_b a} = b^x = a = \text{RHS}$.

2. Proof of Property 3(1):

Suppose $\log_b u = x$ and $\log_b v = y$.

By definition, we have $b^x = u$ and $b^y = v$.

$\text{LHS} = \log_b u + \log_b v = x + y$.

$\text{RHS} = \log_b(uv) = \log_b(b^x b^y) = \log_b(b^{x+y}) = x + y$.

Therefore, LHS = RHS.

Proof of Property 3(2):

Suppose $\log_b u = x$ and $\log_b v = y$.

By definition, we have $b^x = u$ and $b^y = v$.

$\text{LHS} = \log_b u - \log_b v = x - y$.

$\text{RHS} = \log_b \dfrac{u}{v} = \log_b \left(\dfrac{b^x}{b^y}\right) = \log_b (b^{x-y}) = x - y$.

Therefore, LHS = RHS.

3. Proof of Property 4:

Suppose $\log_b u = x$.

By definition, we have $b^x = u$.

$\text{LHS} = \log_b (u^v) = \log_b [(b^x)^v] = \log_b [b^{xv}] = xv$.

$\text{RHS} = v\log_b u = vx = xv$.

Therefore, LHS = RHS.

4. Proof of Property 5:

Let $\log_c u = x$ and $\log_c b = y$, which means that $c^x = u$ and $c^y = b$.

$\text{RHS} = \dfrac{x}{y}$.

$\text{LHS} = \log_b u = \log_{c^y} (c^x) = x \log_{c^y} c = x\left(\dfrac{1}{y}\right) = \dfrac{x}{y}$.

Therefore, LHS = RHS.

5. Proof of Property 6:

From Property 5, set $c = u$.

$\text{LHS} = \log_b u = \dfrac{\log_c u}{\log_c b} = \dfrac{\log_u u}{\log_u b} = \dfrac{1}{\log_u b} = \text{RHS}$.

Questions: 1. Evaluate each of the following.

(1) $\log_2 16$

(2) $\log_3 243$

(3) $\log_{17} 1$

(4) $\log_{256} 2$

(5) $\log_{20} \dfrac{1}{400}$

2. Evaluate each of the following.

(1) $\log_{10} 4 + \log_{10} 25$

(2) $\log_2 384 - \log_2 3$

(3) $\log_3 81^{25}$

3. If x and y are positive numbers such that $x \neq 1$ and $y \neq 1$, evaluate the following.

$$\log_x y \cdot \log_y x$$

4. What is the value of x if $(-3)^x = 81$?

Answers: 1. We evaluate each of following using the definition of logarithms.

(1) $\log_2 16 = 4$.

(2) $\log_3 243 = 5$.

(3) $\log_{17} 1 = 0$, recall that $a^0 = 1$ for all $a \neq 0$.

(4) $\log_{256} 2 = \dfrac{1}{8}$, recall that $\sqrt[8]{256} = 2$.

(5) $\log_{20}\dfrac{1}{400}=-2$, recall that $20^{-2}=\dfrac{1}{20^2}=\dfrac{1}{400}$.

2. We evaluate each of following using the properties of logarithms.

(1) $\log_{10}4+\log_{10}25=\log_{10}(4\cdot 25)=\log_{10}100=2$.

(2) $\log_2 384-\log_2 3=\log_2\dfrac{384}{3}=\log_2 128=7$.

(3) $\log_3 81^{25}=25\log_3 81=25(4)=100$.

3. $\log_x y\cdot\log_y x=\dfrac{\log_b y}{\log_b x}\cdot\dfrac{\log_b x}{\log_b y}=1$.

4. Since $(-3)^4=81$, $x=4$.

Note that we cannot say $x=\log_{-3}81=4$ because the base of a logarithm must be a positive number other than 1.

28.3 Logarithmic Functions 对数函数

Logarithmic Function 对数函数

n. [ˈlɔgəˌrɪðəmɪk ˈfʌŋkʃən]

Definition: See the definition of logarithm in Section 28.2. 见 28.2 节对数的定义。

Properties: 1. Since logarithms are inverse operations of powers, logarithmic functions are inverse functions of exponential functions:

因为对数是次幂的逆运算,所以对数函数是指数函数的反函数:

$f(x)=b^x$ and $g(x)=\log_b x$ are inverse functions, for which b is positive and $b\neq 1$.

2. For the graph of $y=\log_b x$:

- The domain is all positive numbers. 定义域为所有正数。
- The range is all real numbers. 值域为所有实数。
- The vertical asymptote is at $x=0$. 垂直渐近线为 $x=0$。
- The x-intercept is 1. x 截距为 1。

Figure 28-1 shows the graph of $y=\log_b x$.

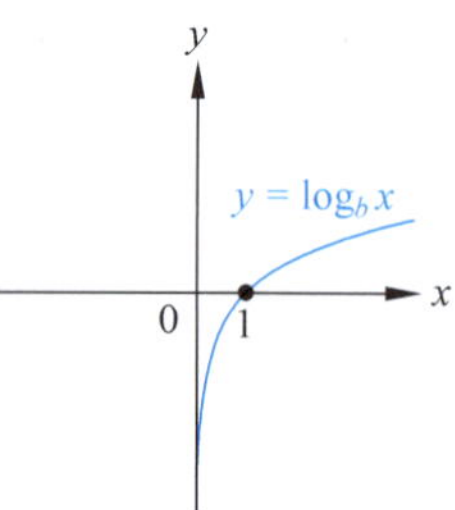

Figure 28-1

Notes: To show that f and g are inverse functions in Property 1:

证明性质 1 中的 f 和 g 互为反函数。

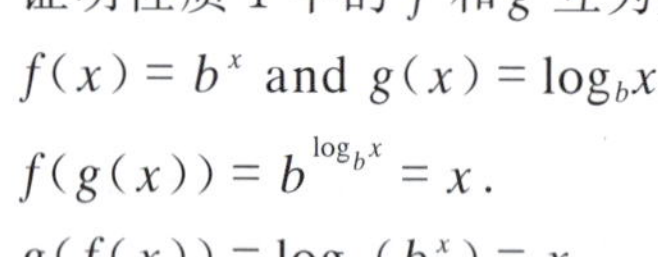

$f(x)=b^x$ and $g(x)=\log_b x$

$f(g(x))=b^{\log_b x}=x$.

$g(f(x))=\log_b(b^x)=x$.

Solving Exponential and Logarithmic Equations 解指数与对数方

Properties: Recall that powers and logarithms are inverse operations.

Suppose a and b are positive constants such that $b\neq 1$.

(1) To solve for x in $b^x=a$, take the logarithm with base b on both sides:

$$\begin{aligned} b^x &= a \\ \log_b(b^x) &= \log_b a \\ x &= \log_b a. \end{aligned}$$

(2) To solve for x in $\log_b x = a$, exponentiate both sides with base b:

$$\begin{aligned}\log_b x &= a\\ b^{\log_b x} &= b^a\\ x &= b^a.\end{aligned}$$

Questions: 1. Solve each exponential equation below.

(1) $7^{x+6} = 343$

(2) $6(4^{2x-5}) + 7 = 391$

(3) $(-5)^x = -3125$

2. Solve each logarithmic equation below.

(1) $5\log_4(3x-2) = 15$

(2) $6\log_3(6x+3) + 5 = 23$

Answers: 1. (1)

$$\begin{aligned}7^{x+6} &= 343\\ \log_7(7^{x+6}) &= \log_7 343\\ x+6 &= 3\\ x &= -3.\end{aligned}$$

(2)

$$\begin{aligned}6(4^{2x-5}) + 7 &= 391\\ 6(4^{2x-5}) &= 384\\ 4^{2x-5} &= 64\\ \log_4(4^{2x-5}) &= \log_4 64\\ 2x-5 &= 3\\ 2x &= 8\\ x &= 4.\end{aligned}$$

(3) To solve $(-5)^x = -3125$, we cannot take the logarithm of base -5 to both sides since -5 cannot be the base of a logarithm. However, we can use mental math. Since $(-5)^5 = -3125$, we have $x = 5$.

2. (1)

$$\begin{aligned}5\log_4(3x-2) &= 15\\ \log_4(3x-2) &= 3\\ 4^{\log_4(3x-2)} &= 4^3\\ 3x-2 &= 64\\ 3x &= 66\\ x &= 22.\end{aligned}$$

(2)

$$\begin{aligned}6\log_3(6x+3) + 5 &= 23\\ 6\log_3(6x+3) &= 18\\ \log_3(6x+3) &= 3\\ 3^{\log_3(6x+3)} &= 3^3\\ 6x+3 &= 27\\ 6x &= 24\\ x &= 4.\end{aligned}$$

29. Variations, Rational Functions, and Rational Equations 比例函数、有理函数与有理方程

29.1 Variations 比例函数(变分)

Direct Variation 正比例函数/正变分

See direct variation in Section 7.3.5.3.

Inverse Variation 反比例函数/逆变分

See inverse variation in Section 7.4.1.

Joint Variation 联变分

n. [dʒɔɪnt ˌveəri'eɪʃən]

Definition: When a variable z is jointly proportional to a set of variables, z is directly proportional to each variable in the set. We say that z and these variables are in a joint variation.

若 z 与一组变量中的每个变量均成正比，则 z 与这些变量为联变分。

For example, when z varies jointly with x and y, we write $z=kxy$, from which k is a constant.

Examples: Suppose x is the unit price of an item, y is the number of items, and z is the total price, and k is the constant modification (due to price increase or discount), then z varies jointly with x and y: z=10.

Questions: z varies jointly with x and y but inversely with w. When $z=5, x=3$, and $y=10, w=60$. What is the value of w when $z=4, x=3$, and $y=12$?

Answers: By the definition of inverse variation and joint variation, we have $z=k\left(\frac{xy}{w}\right)$. We want to find the value of k.

By substitution, we have $5=k\left(\frac{3(10)}{60}\right)$, from which $5=k\left(\frac{30}{60}\right)$ and $k=10$.

Now, we have $4=10\left(\frac{3(12)}{w}\right)=\frac{360}{w}$, from which $w=90$.

29.2 Rational Functions 有理函数

29.2.1 Introduction 介绍

Rational Function 有理函数

n. ['ræʃənəl 'fʌŋkʃən]

Definition: A function f of the form $f(x)=\frac{g(x)}{h(x)}$, from which $g(x)$ and $h(x)$ are polynomial functions and the degree of h is at least 1.

有理函数 f 是形式为$f(x)=\frac{g(x)}{h(x)}$的函数，其中 $g(x)$和 $h(x)$均为多项式函数，且 h 的次数至少为 1。

Notes: For $f(x)=\frac{g(x)}{h(x)}$, if $h(x)$ is a constant function (degree 0), then $f(x)$ is a polynomial function, which is not a rational function.

Examples: (1) $f(x)=\frac{4x+7}{x^2-6}$ is a rational function because both its numerator and denominator are polynomial functions. The degree of the denominator is 2.

(2) $g(x)=\frac{x^2-3x+6}{2}$ is not a rational function because this is $\frac{1}{2}x^2-\frac{3}{2}x+3$, which is a quadratic function. Polynomial functions can never be rational functions, and vice versa.

29.2.2 Parts of Rational Functions　有理函数的部分

Intercepts　截距

Definition: See x-intercept and y-intercept in Section 7.2.3.

Questions: What is the intercepts of the rational function $f(x)=\frac{x^2-6x+8}{x^3+x^2+x-14}$?

Answers: The x-intercepts of a function are all x-values a for which $f(a)=0$. To get the x-intercepts, set the output to 0:

$$0=\frac{x^2-6x+8}{x^3+x^2+x-14}.$$

We have

$$0=x^2-6x+8$$
$$0=(x-2)(x-4)$$
$$x=2,4.$$

CAUTION!!! We have to check whether $f(2)=0$ and $f(4)=0$ are indeed true. We know that the numerator is 0, but if the denominator is also 0, the fraction is undefined. By substitution, we note that $f(2)=\frac{0}{0}$, which is undefined, and $f(4)=0$. Therefore, the only x-intercept is 4.

The y-intercept of a function is the y-value b for which $f(0)=b$. Note that a function can have at most one y-intercept.

Therefore, the y-intercept of f is $f(0)=\frac{0^2-6(0)+8}{0^3+0^2+0-14}=\frac{8}{-14}=-\frac{4}{7}$.

Hole　断点

n. [hoʊl]

Definition: A point from which the graph of the function approaches it, but not actually defined on the precise x-value.　断点是指函数图像接近，但是不在此 x 值定义的点。

A hole for the blue curve is shown in Figure 29-1.

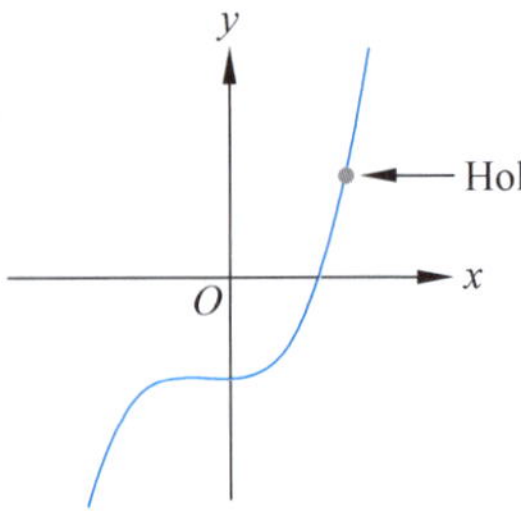

Figure 29-1

Properties: 1. For a rational function $f(x)=\dfrac{g(x)}{h(x)}$, a hole may exist on $x=a$ if a is the zero of both $g(x)$ and $h(x)$.

$f(a)=\dfrac{g(a)}{h(a)}=\dfrac{0}{0}$, which is undefined.

In other words, $g(x)$ and $h(x)$ have a common factor $x-a$.

有理函数 $f(x)=\dfrac{g(x)}{h(x)}$的一个断点可能在 $x=a$ 上,其中 a 为$g(x)$与 $h(x)$的零值。

换言之,$g(x)$与 $h(x)$有公因式 $x-a$。

2. Suppose the rational function $f(x)=\dfrac{g(x)}{h(x)}$ has a hole at $x=a$.

We know that $x-a$ is a common factor of both $g(x)$ and $h(x)$.

To find the y-coordinate of this hole, divide out as many common factors $x-a$ as possible from $g(x)$ and $h(x)$.

Suppose we end up with $f(x)=\dfrac{g(x)}{h(x)}=\dfrac{m(x)}{n(x)}$, from which $x-a$ is not a common factor for $m(x)$ and $n(x)$, and $n(a)\neq 0$, the y-coordinate of the hole is $\dfrac{m(a)}{n(a)}$.

假设有理函数 $f(x)=\dfrac{g(x)}{h(x)}$在 $x=a$ 上有断点。

可知 $x-a$ 是$g(x)$和 $h(x)$的公因式。

要求出这个断点的 y 坐标,先尽可能约去 $g(x)$和 $h(x)$的公因式 $x-a$。

假设得出 $f(x)=\dfrac{g(x)}{h(x)}=\dfrac{m(x)}{n(x)}$,其中 $m(x)$和 $n(x)$没有公因式 $x-a$,且 $n(a)\neq 0$,则这个断点的 y 坐标是$\dfrac{m(a)}{n(a)}$。

Notes: From Property 2, note that when $f(x)=\dfrac{g(x)}{h(x)}=\dfrac{m(x)}{n(x)}$, for which $x\neq a$, the graph of f is simply the graph of $y=\dfrac{m(x)}{n(x)}$, except that at $x=a$, the point $\left(a,\dfrac{m(a)}{n(a)}\right)$ is undefined (a hole). See examples below.

Examples: (1) For the function $f(x)=\dfrac{x(x+5)}{x}$:

Note that $f(x)=x+5$ for all $x\neq 0$, from which $f(0)=5$ does not exist.

We graph the function as shown in Figure 29-2.

The graph is $y=x+5$, with a hole on $(0,5)$.

(2) For the function $g(x)=\dfrac{x-9}{\sqrt{x}-3}$:

Rationalizing the denominator, we have:

$$g(x)=\frac{x-9}{\sqrt{x}-3}=\frac{(x-9)(\sqrt{x}+3)}{(\sqrt{x}-3)(\sqrt{x}+3)}=\frac{(x-9)(\sqrt{x}+3)}{x-9}.$$

Note that $g(x)=\sqrt{x}+3$ for all $x\neq 9$, from which $f(9)=6$ does not exist.

We graph the function as shown in Figure 29-3.

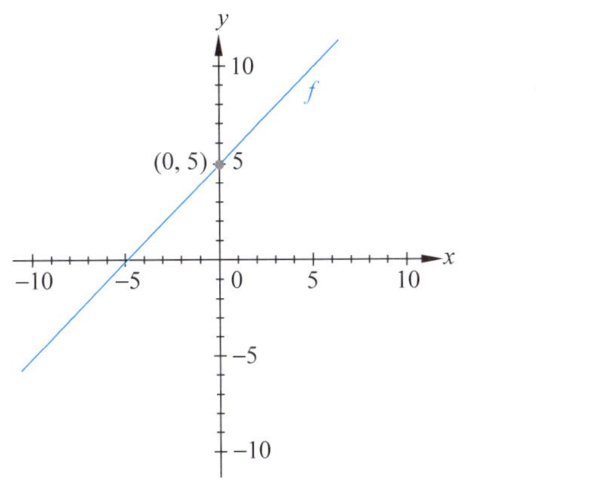

Figure 29-2

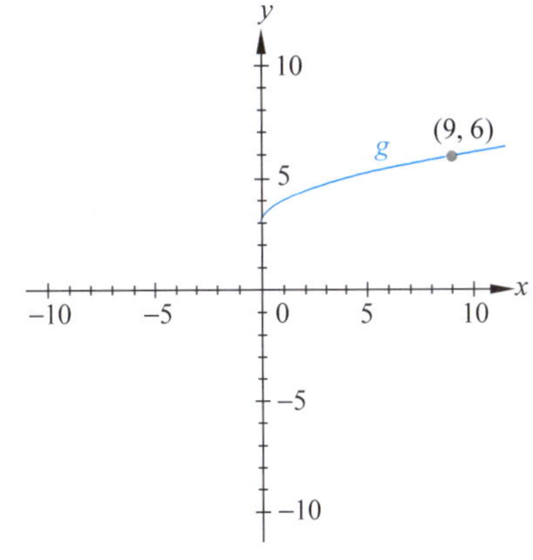

Figure 29-3

The graph is $y=\sqrt{x}+3$, with a hole on (9,6).

Asymptote　渐近线

n. ['æsəmpˌtoʊt]

Definition: When a curve gets infinitesimally close to a line as they head to positive or negative infinity, we say that the line is an asymptote for the curve.　在往正无穷或负无穷伸展时，当曲线无限接近一条直线，则说直线为曲线的渐近线。

See vertical asymptote, horizontal asymptote, and slant asymptote.

见垂直渐近线、水平渐近线、斜渐近线的词条。

Phrases: vertical～　垂直渐近线，horizontal～　水平渐近线，slant～　斜渐近线

Vertical Asymptote　垂直渐近线

n. ['vɜrtɪkəl 'æsəmpˌtoʊt]

Definition: The line $x=k$ is a vertical asymptote of the graph of the function f if $f(x)\to\pm\infty$ as $x\to k$ from either side.

在任意一边，当 $x\to k$ 时，$f(x)\to\pm\infty$，则称 $x=k$ 为 $f(x)$ 函数图像的垂直渐近线。

The vertical asymptote of the blue curve is drawn in dotted line, as shown in Figure 29-4.

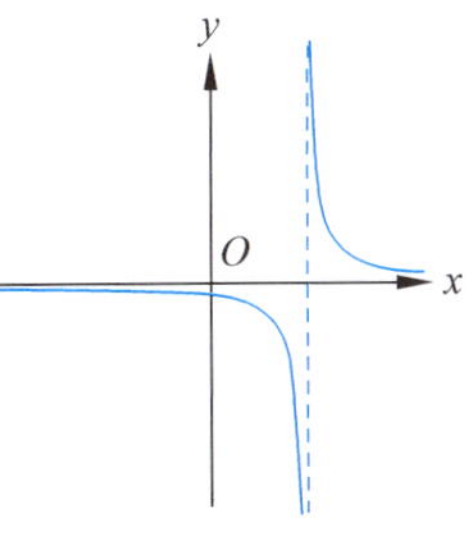

Figure 29-4

Notation: The vertical asymptote is a vertical line, in the form of $x=k$, for which k is a constant.

Properties: 1. For rational functions, a vertical asymptote occurs at $x=a$ when a is a zero of the denominator but not a zero of the numerator, after the numerator and the denominator dividing out all common factors.

Recall that if a is a zero of the original numerator and denominator, but not a zero of the denominator after they divide out all common factors, then the graph has a hole at $x=a$, not a vertical asymptote.

在有理函数中,垂直渐近线为 $x=a$,其中 a 为该函数约分后分母的零值,但不为分子的零值。

若 a 同时为该函数原分子和分母的零值,但不是函数约分后分母的零值,则图像在 $x=a$ 上有断点,而 $x=a$ 不是垂直渐近线。

2. For logarithmic functions $f(x)=\log_b(x-a)$, from which a and b are constants, the vertical asymptote occurs at $x=a$.

在对数函数 $f(x)=\log_b(x-a)$中(a 和 b 为常值),垂直渐近线为 $x=a$。

Questions: 1. What are all vertical asymptotes of the graph of the rational function $f(x)=\dfrac{x-5}{(x^2-11x+30)(x-7)}$?

2. What is the vertical asymptote of the graph of the logarithmic function $g(x)=\log_2(x+6)$?

Answers: 1. $f(x)=\dfrac{x-5}{(x^2-11x+30)(x-7)}=\dfrac{x-5}{(x-5)(x-6)(x-7)}=\dfrac{1}{(x-6)(x-7)}$, from which $x\neq 5, 6$, or 7.

The graph of f has a hole at $x=5$, namely on $\left(5,\dfrac{1}{(5-6)(5-7)}\right)=\left(5,\dfrac{1}{2}\right)$.

Excluding this hole, the graph of f is the graph of $y=\dfrac{1}{(x-6)(x-7)}$.

The vertical asymptotes of f are at $x=6$ and $x=7$.

The graph of f is shown in Figure 29-5.

Note that as x approaches 6 or 7 from either side, the absolute value of the denominator of $f(x)$ is infinitesimally close to 0, from which the absolute value of $f(x)$ is arbitrarily large. This creates the vertical asymptotes.

Review how to find holes.

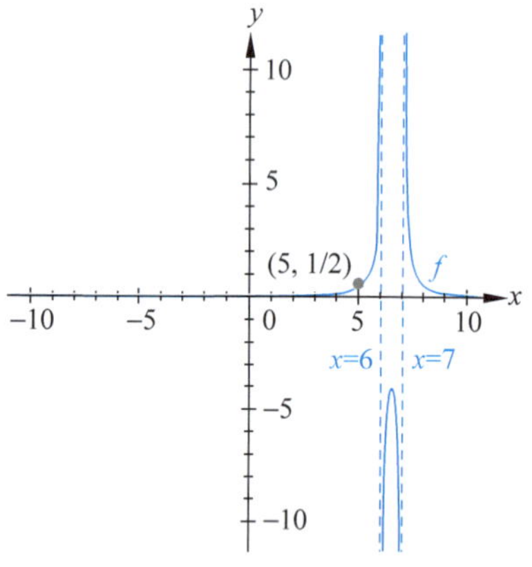

Figure 29-5

2. The graph of g has a vertical asymptote at $x=-6$. Note that as x approaches -6 from the right, $x+6$ is positive, but arbitrarily close to 0. Since $2^{\text{very negative number}}\approx 0$, $\log_2(x+6)$ is exploding to negative infinity.

The graph of g is the same as shifting the curve $y=\log_2 x$ six units to the left. Since the vertical asymptote of $y=\log_2 x$ is $x=0$, the vertical asymptote of the graph of g is $x=-6$.

Horizontal Asymptote 水平渐近线

n. [ˌhɔrəˈzɑntəl ˈæsəmpˌtoʊt]

Definition: The line $y=k$ is a horizontal asymptote of the graph of the function f if $f(x)\to k$ as

$x \to \infty$ or $x \to -\infty$. 　当 $x \to \infty$ 或 $x \to -\infty$ 时，$f(x) \to k$，则称 $y = k$ 为 $f(x)$ 函数图像的水平渐近线。

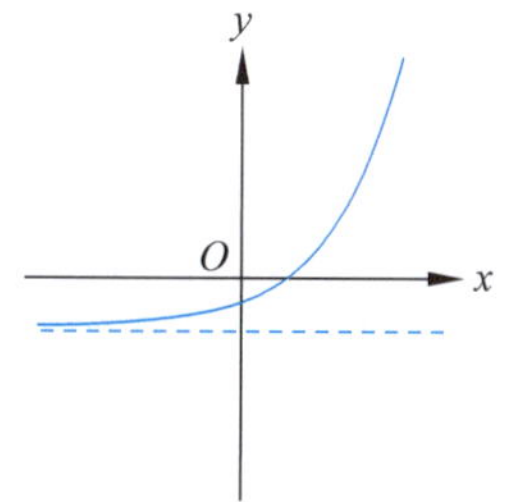

Figure 29-6

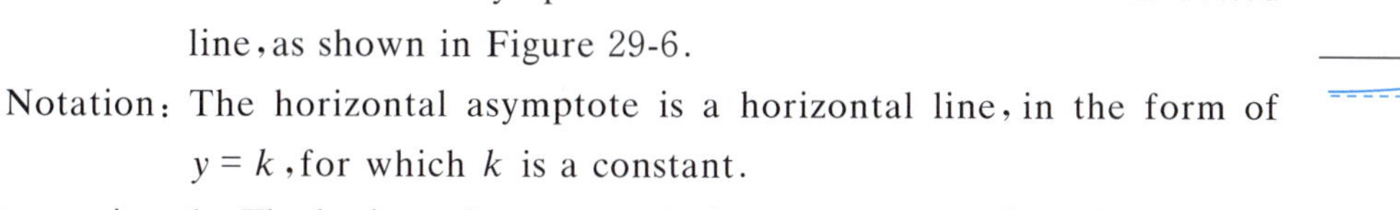

The horizontal asymptote of the blue curve is drawn in dotted line, as shown in Figure 29-6.

Notation: The horizontal asymptote is a horizontal line, in the form of $y = k$, for which k is a constant.

Properties: 1. The horizontal asymptote is the measurement of end behavior of a graph, which is based on dominance. 　水平渐近线用于测量图像的极端情况。

Recall that in a polynomial function, the dominating term is the term with the highest degree. 　在多项式函数中，主导项为最高次项。

2. To determine the horizontal asymptotes of a rational function $f(x) = \dfrac{g(x)}{h(x)}$:

- If g has a higher degree, then the graph of f does not have any horizontal asymptote. 若 g 的次数更大，则 f 没有水平渐近线。
- If h has a higher degree, then the graph of f has a horizontal asymptote $y = 0$. 若 h 的次数更大，则 f 的水平渐近线为 $y = 0$。
- If g and h have the same degree, then the graph of f has a horizontal asymptote $y = k$, for which k is the ratio of the leading coefficients of g and h. 若 g 和 h 的次数相等，则 f 的水平渐近线为 $y = k$，其中 k 为 g 和 h 的最高次项系数之比。

Notes: To show Property 2, note that in $f(x) = \dfrac{g(x)}{h(x)}$:

(1) If g has a higher degree, then the absolute value of $g(x)$ grows faster than that of $h(x)$, and $\left|\dfrac{g(x)}{h(x)}\right| \to \infty$ as $x \to \pm\infty$. The graph of f does not have a horizontal asymptote.

Figure 29-7 illustrates this case.

(2) If h has a higher degree, then the absolute value of $h(x)$ grows faster than that of $g(x)$, and $\left|\dfrac{g(x)}{h(x)}\right| \to 0$ as $x \to \pm\infty$. The graph of f has a horizontal asymptote $y = 0$.

Figure 29-8 illustrates this case.

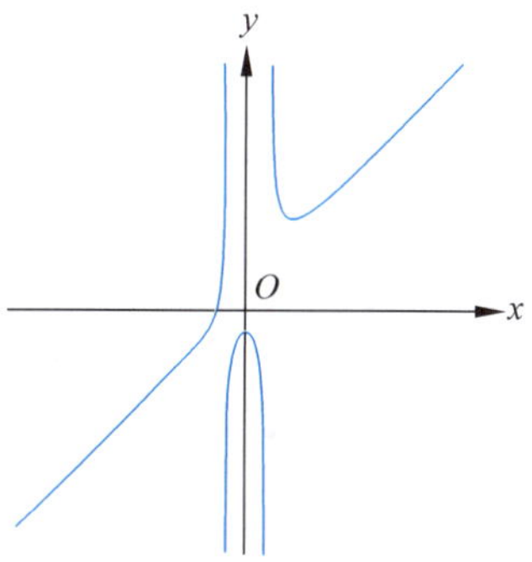

Figure 29-7

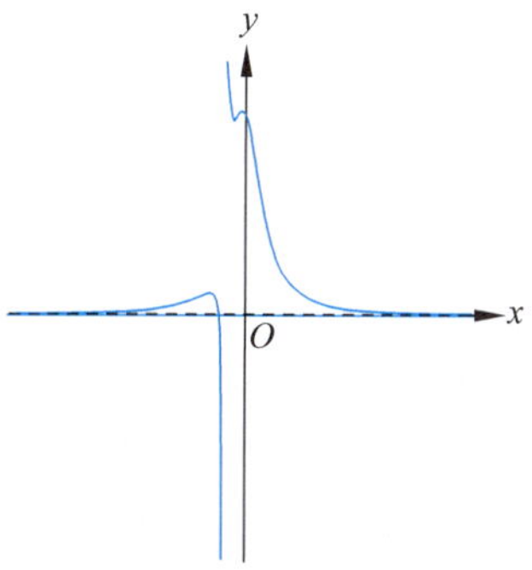

Figure 29-8

(3) If g and h have the same degree, then $\left|\dfrac{g(x)}{h(x)}\right| \to k$ as $x \to \pm\infty$, from which k is the ratio of the leading coefficients of g and h. The graph of f has a horizontal asymptote $y = k$.

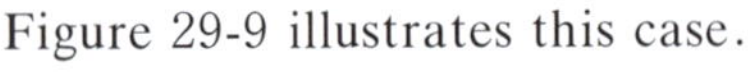

Figure 29-9 illustrates this case.

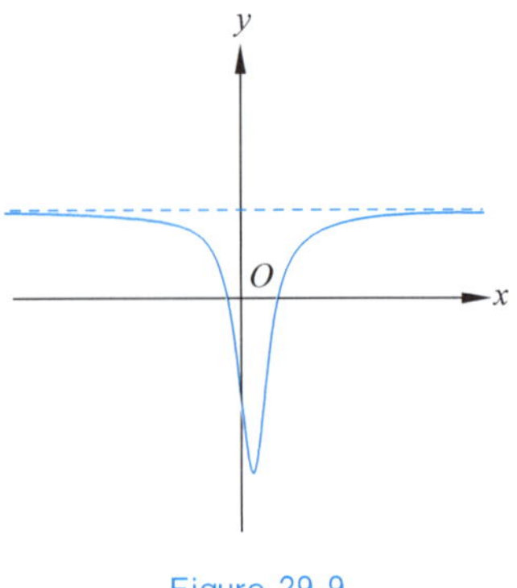

Figure 29-9

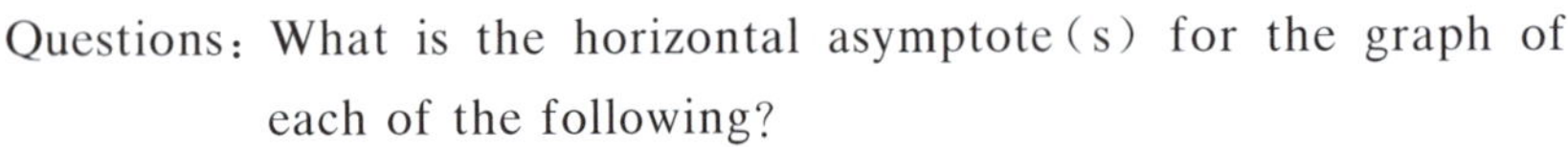

Questions: What is the horizontal asymptote(s) for the graph of each of the following?

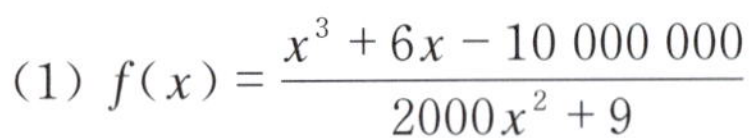

(1) $f(x) = \dfrac{x^3 + 6x - 10\,000\,000}{2000x^2 + 9}$

(2) $g(x) = \dfrac{6x^2 - 1000x + 33}{-2x^2 + 5x - 8}$

(3) $h(x) = \dfrac{-500x^2 + 300x - 5}{4x^4}$

(4) $m(x) = \dfrac{2^x + x^2}{x^4}$

Answers: (1) Since the numerator has a higher degree, the graph of f does not have a horizontal asymptote.

$|f(x)| \to \infty$ as $x \to \infty$ or $x \to -\infty$.

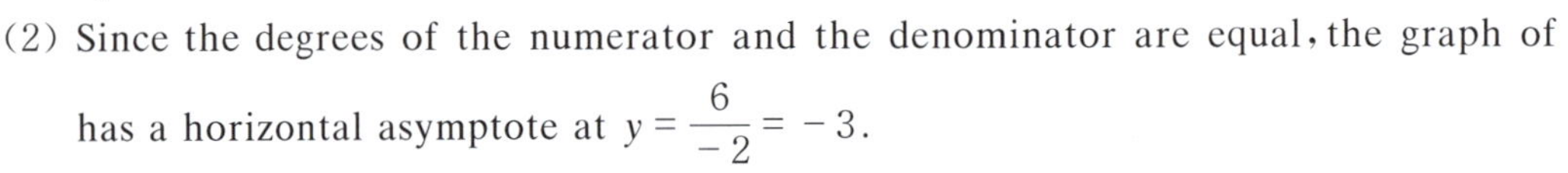

(2) Since the degrees of the numerator and the denominator are equal, the graph of f has a horizontal asymptote at $y = \dfrac{6}{-2} = -3$.

$g(x) \to -3$ as $x \to \infty$ or $x \to -\infty$.

(3) Since the denominator has a higher degree, the graph of f has a horizontal asymptote at $y = 0$.

$h(x) \to 0$ as $x \to \infty$ or $x \to -\infty$.

(4) As $x \to \infty$, the dominating term of the numerator is 2^x.

For arbitrarily large values of x, 2^x dominates x^4. Therefore, there is no horizontal asymptote for $x \to \infty$.

As $x \to -\infty$, the dominating term of the numerator is x^2.

Since x^4 has a higher degree than x^2, the graph of m has a horizontal asymptote of $y = 0$.

Slant Asymptote 斜渐近线

n. [slænt 'æsəmp,təʊt]

Definition: The asymptote that is neither vertical nor horizontal.

斜渐近线是指既非垂直也非水平的渐近线。

The slant asymptote of the blue curve is drawn in dotted line, as shown in Figure 29-10.

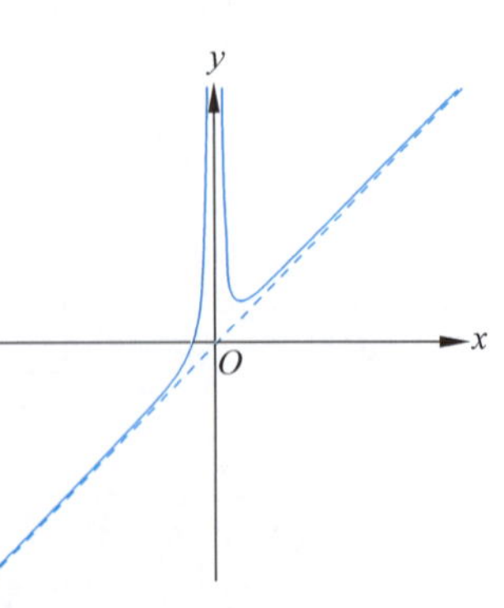

Figure 29-10

Notation: The slant asymptote is a line in the form of $y = mx + b$, for which m and b are constants, and $m \neq 0$.

Properties: For a rational function $f(x) = \dfrac{g(x)}{h(x)}$, if the degree of g is

1 higher than that of h, then there exists a slant asymptote on the graph of f.

在有理函数 $f(x)=\frac{g(x)}{h(x)}$ 中，若 g 的次数比 h 的次数大 1，则 f 的图像上存在一条斜渐近线。

Notes：For a rational function $f(x)=\frac{g(x)}{h(x)}$, if the degree of g is 1 higher than that of h, then using long division, we can rewrite $f(x)=\frac{g(x)}{h(x)}=q(x)+\frac{r(x)}{h(x)}$, from which the degree of q is 1 and the degree of r is less than that of h.

Note that $\left|\frac{r(x)}{h(x)}\right|\to 0$ as $x\to\infty$ or $x\to-\infty$.

Therefore, $q(x)+\frac{r(x)}{h(x)}\to q(x)$ as $x\to\infty$ or $x\to-\infty$.

Examples：For the function $f(x)=\frac{x^3+2x^2+3x+4}{x^2+x+1}$, we can rewrite it as $f(x)=(x+1)+\frac{x+3}{x^2+x+1}$. As $x\to\infty$ or $x\to-\infty$, the graph of f is infinitesimally close to $y=x+1$, as shown in Figure 29-11.

Branch　分支

n. [bræntʃ]

Definition：The branch of a function is one part of the function's graph separated by a vertical asymptote.　分支是指函数图像被垂直渐近线分割的一部分。

As shown in Figure 29-12, the graph has three branches, separated by two dotted vertical asymptotes.　如图 29-12 所示，图像有三个分支，被两条垂直渐近线(虚线)隔开。

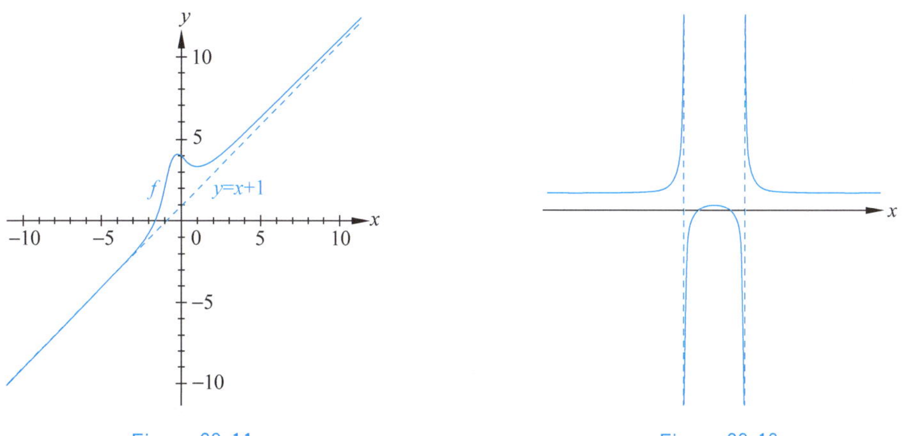

Figure 29-11　　Figure 29-12

Properties：n vertical asymptotes partition the graph of a function into $n+1$ branches.

n 条垂直渐近线把一个函数的图像分割成 $n+1$ 个分支。

Questions：How many branches does $f(x)=\frac{x^4+1}{(x+1)(x-2)(x+3)}$ have?

Answers：4.

Graphing Rational Functions 画有理函数的图像

Procedures: 1. Factor the numerator and denominator. 因式分解有理函数的分子和分母。

2. Locate all vertical asymptotes, which separate the graph into branches. 找出所有垂直渐近线,这些渐近线把图像分割成若干分支。

3. Locate all intercepts and holes. 画出所有截距和断点。

4. Locate the horizontal asymptotes and slant asymptote to draw the end behavior. 找出所有水平渐近线和斜渐近线,从而画出函数图像的极端情况。

5. For each branch, plot some points to get the general pattern of the branch, and draw the curve. 对于每个分支,先标记出已知的一些点,从而得知分支的规律,画出分支。

Questions: Graph the function $f(x)=\dfrac{(x+1)(x^2-8x+15)}{x^2-7x+10}$.

Answers: We first factor $f(x)=\dfrac{(x+1)(x^2-8x+15)}{x^2-7x+10}=\dfrac{(x+1)(x-3)(x-5)}{(x-2)(x-5)}$.

Note that $f(x)=\dfrac{(x+1)(x-3)}{x-2}$, for which $x\neq 5$.

We locate the vertical asymptote at $x=2$, as shown in Figure 29-13.

We know that the x-intercepts of f are -1 and 3, so we plot the points $(-1,0)$ and $(3,0)$.

We know that the y-intercept of f is at $f(0)=3/2$, so we plot the point $(0,3/2)$.

There is a hole at $x=5$, at $(5,4)$.

We plot the intercepts and the hole, as shown in Figure 29-14.

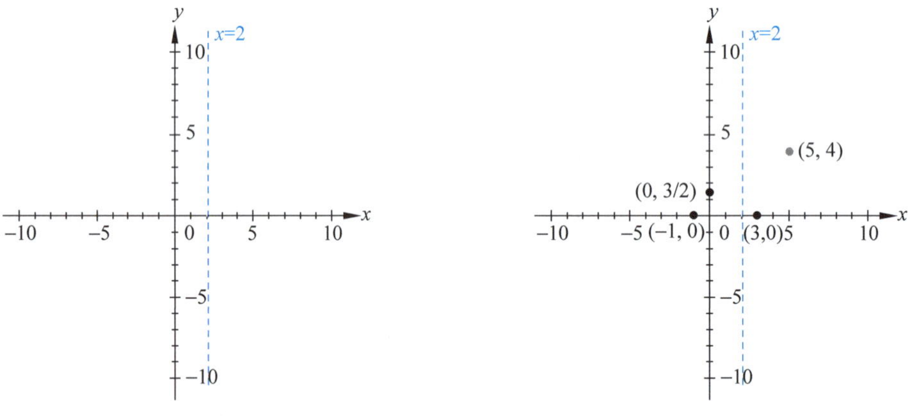

Figure 29-13　　Figure 29-14

Since the degree of the numerator is higher than that of the denominator, the graph of f does not have any horizontal asymptote. Since the degree of the numerator is 1 higher than that of the denominator, the graph of f has a slant asymptote.

We rewrite $f(x)=\dfrac{(x+1)(x-3)}{x-2}=\dfrac{x^2-2x-3}{x-2}=x-\dfrac{3}{x-2}$, from which $x\neq 2$.

The slant asymptote of the graph of f is at $y=x$, as shown in Figure 29-15.

Now, we draw each branch, as shown in Figure 29-16.

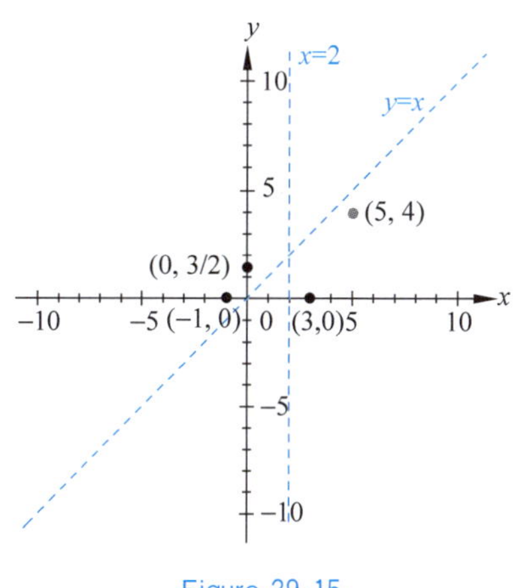

Figure 29-15

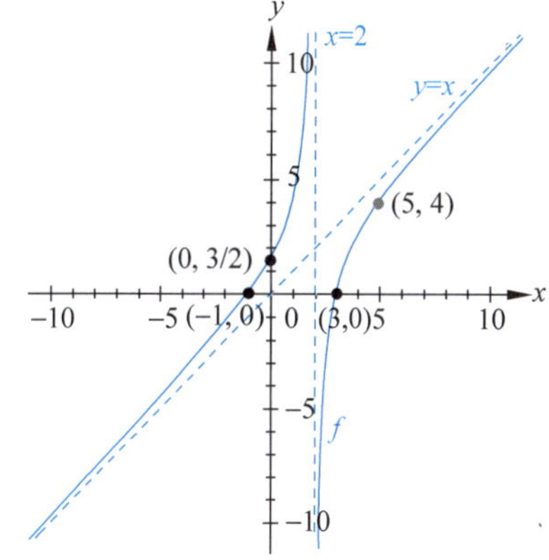

Figure 29-16

29.2.3 Indeterminate Form　不定式

Indeterminate Form　不定式

n. [ˌɪndɪˈtɜrmənət fɔrm]

Definition: When a rational function evaluates to one of $\frac{0}{0}$, $\frac{\infty}{\infty}$, $\frac{-\infty}{\infty}$, or $\frac{\infty}{-\infty}$, we say that the result is in the indeterminate form.

In this encyclopedia, we will focus on the case of $\frac{0}{0}$. The graph has either a vertical asymptote or a hole at that specific input.

不定式是指在有理函数求出的值为$\frac{0}{0}$、$\frac{\infty}{\infty}$、$\frac{-\infty}{\infty}$、$\frac{\infty}{-\infty}$中一种情况。

本书会特别集中在$\frac{0}{0}$的情况。函数的图像在那个输入上存在垂直渐近线或者断点。

Notations: One of $\frac{0}{0}$, $\frac{\infty}{\infty}$, $\frac{-\infty}{\infty}$, or $\frac{\infty}{-\infty}$, as the rational function's output.

Properties: When the rational function $f(x)$ evaluates to $\frac{0}{0}$ at $x=a$, we know that $x-a$ is a common factor of the numerator and the denominator using the Factor Theorem.

当有理函数 $f(x)$在 $x=a$ 处的值为$\frac{0}{0}$，根据因式定理，$x-a$ 是分子和分母的公因式。

In order to examine the behavior of f at $x=a$, we first divide out all common factors from the numerator and the denominator. Call g the resulting function.

要检验 f 在 $x=a$ 处的情况，先约掉分子和分母的所有公因式。设 g 为约分后的函数。

Since the numerator and the denominator of g do not have any common factors, $g(a)$ does not evaluate to $\frac{0}{0}$. Here are the cases of the indeterminate form:

因为 g 的分子和分母没有公因式，所以 $g(a)$的值不会为$\frac{0}{0}$。以下是不定式的情况：

If $g(a)=\frac{\text{nonzero}}{0}$, then the graph of f has a vertical asymptote at $x=a$.

若 $g(a)=\frac{\text{非零}}{0}$，则 f 的图像在 $x=a$ 处有垂直渐近线。

If $g(a)=\frac{\text{any number}}{\text{nonzero}}$, then the graph of f has a hole at $x=a$. The hole is $(a, g(a))$.

若 $g(a)=\frac{\text{任意数字}}{\text{非零}}$，则 f 的图像在 $x=a$ 处有断点。这个断点为$(a, g(a))$。

Examples: For the function $f(x)=\frac{(x-1)(x-2)^2}{(x-1)^2(x-2)}$, the indeterminate form 0/0 exist at $x=1$ and $x=2$.

Dividing out all the common factors from the numerator and the denominator, we get $g(x)=\frac{x-2}{x-1}$.

Note that $g(1)=\frac{-1}{0}$, which means that f has a vertical asymptote at $x=1$.

Note that $g(2)=0$, which means that f has a hole at $x=2$. The hole is $(2,0)$.

29.2.4 Summary 总结

1. Rational function $f(x)=\frac{g(x)}{h(x)}$: $g(x)$ and $h(x)$ are polynomial functions and the degree of $h(x)$ is at least 1. 有理函数 $f(x)=\frac{g(x)}{h(x)}$，其中 $g(x)$和 $h(x)$均为多项式函数，且 $h(x)$的次数至少为 1。

2. To graph a rational function $f(x)=\frac{g(x)}{h(x)}$, we perform long division $f(x)=\frac{g(x)}{h(x)}=q(x)+\frac{r(x)}{h(x)}$, from which the degree of $r(x)$ is less than that of $h(x)$. 要画出有理函数 $f(x)=\frac{g(x)}{h(x)}$的图像，先用长除法改写 $f(x)$，即 $f(x)=\frac{g(x)}{h(x)}=q(x)+\frac{r(x)}{h(x)}$，其中 $r(x)$的次数小于 $h(x)$的次数。

   Factor $r(x)$ and $h(x)$ completely. 完全分解 $r(x)$和 $h(x)$。
   x-intercept: Solve for x when $f(x)=0$. 解 $f(x)=0$ 得 x 截距。
   y-intercept: Find $f(0)$. 求 $f(0)$得 y 截距。
   To locate the holes and vertical asymptotes, review their definitions in Section 29.2.2. Also, review indeterminate form in Section 29.2.3.
   要画出断点和垂直渐近线，复习 29.2.2 节词条的定义。
   另外，复习 29.2.3 节的不定式。
   Horizontal Asymptote: The numbers that $q(x)$ approaches when $x \to \pm\infty$. If $q(x)=0$, then the horizontal asymptote is the x-axis. If $q(x)$ is other constant, then the horizontal asymptote is $y=q(x)$. If $q(x)$ has a higher degree, then the horizontal asymptote does not exist. 水平渐近线是指当 $x \to \pm\infty$时，$q(x)$无比接近的 y 值。若 $q(x)$为 0，则水平渐近线为 x 轴。若 $q(x)$为其他常值，则水平渐近线为 $y=q(x)$。若 $q(x)$的次数更高，则水平渐近线不存在。
   Slant Asymptote: $q(x)$, and it only occurs when $q(x)$ is linear.

斜渐近线：表示为 $q(x)$，只在 $q(x)$ 的次数为 1 的时候存在。

Review graphing rational functions in Section 29.2.2 for procedures and examples.

复习 29.2.2 节“画有理函数的图像”词条的步骤和例子。

29.3 Reciprocal Functions 倒数函数

Reciprocal Function 倒数函数

n. [rɪ'sɪprəkəl 'fʌŋkʃən]

Definition: Functions f and g are reciprocal functions if $f(x)=\dfrac{1}{g(x)}$.

若 $f(x)=\dfrac{1}{g(x)}$，则函数 f 和 g 互为倒数函数。

Properties: Suppose we know the graph of f (assume continuous). To graph $g(x)$:

1. Locate all points on the graph of f from which $f(x)=\pm1$. These points are on the graph of g as well, recall that the reciprocal of ±1 is itself. 在 f 的图像上画出所有使得 $f(x)=\pm1$ 的点。因为 ±1 的倒数是它们本身，所以这些点也在 g 的图像上。
2. Locate all points on the graph of f from which $f(x)=0$. There are vertical asymptotes on the graph of g at the x-values of these points, recall that the reciprocal of 0 is undefined. 在 f 的图像上画出所有使得 $f(x)=0$ 的点。因为 0 的倒数无意义，所以在这些点的 x 值上，g 的图像有垂直渐近线。
3. If $|f(x)|<1$, then $|g(x)|>1$. For every input a, $f(a)$ and $g(a)$ have the same sign, if defined.
4. $|f(x)|\to 0$ as $|g(x)|\to\infty$, and vice versa.

Questions: The graph of function f is shown in Figure 29-17. Graph its reciprocal function.

Answers: The answer is shown in Figure 29-18 in blue.

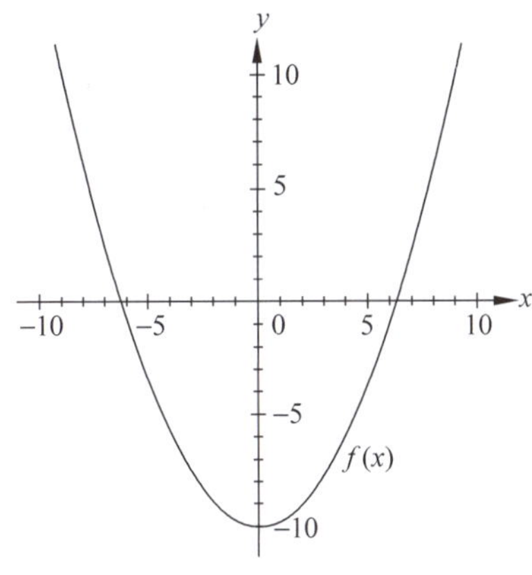

Figure 29-17

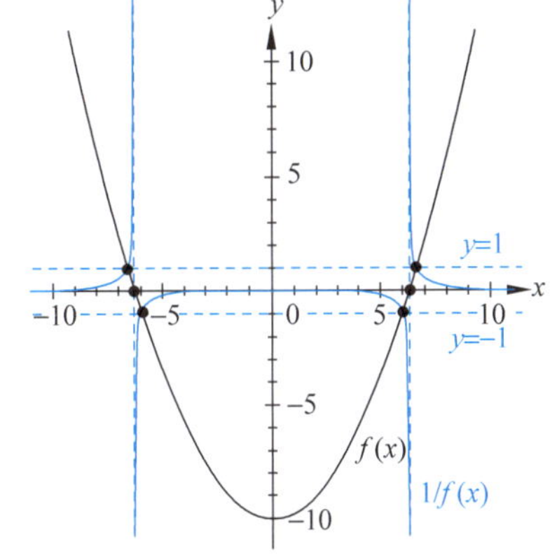

Figure 29-18

29.4 Partial Fraction Decomposition 部分分式分解

视频 42

Partial Fraction Decomposition 部分分式分解

n. ['pɑrʃəl 'frækʃən ˌdiː. kɑːm. pə'zɪʃ. ən]

Definition: The partial fraction decomposition of a rational fraction is the operation that expresses the fraction as a sum of a polynomial (possibly zero) and one or several proper fractions with denominators of lower degrees.

有理分式的部分分式分解是把这个分式改写成一个(或零个)多项式与一个(或多个)真分式的和的运算。这些分式的分母次数均比原分式的分母次数小。

Notation: The partial fraction decomposition of $\frac{P(x)}{Q(x)}$ is $g(x)+\sum\frac{a(x)}{b(x)}$, from which all every $a(x)$ has a lower degree than its corresponding $b(x)$, and every $b(x)$ has a lower degree than $Q(x)$.

每个 $a(x)$ 比它对应的 $b(x)$ 的次数小。

每个 $b(x)$ 都比 $Q(x)$ 的次数小。

Properties: From $\frac{P(x)}{Q(x)}=g(x)+\sum\frac{a(x)}{b(x)}$:

- If $P(x)$ has a lower degree than $Q(x)$, then $g(x)=0$.
- Otherwise, we perform long division of $\frac{P(x)}{Q(x)}=g(x)+\frac{r(x)}{Q(x)}$, from which $g(x)$ is the quotient and $r(x)$ is the remainder.

We want to write $\frac{r(x)}{Q(x)}$ as a sum of proper fractions with denominators of lower degrees than that of $Q(x)$. Note that the degree of $r(x)$ is less than the degree of $Q(x)$.

需要把 $\frac{r(x)}{Q(x)}$ 改写成真分式的和,其中每个真分式分母的次数均比 $Q(x)$ 的次数小。注意 $r(x)$ 的次数比 $Q(x)$ 小。

We factor $Q(x)$ completely. We write $\frac{r(x)}{Q(x)}=\sum\frac{a(x)}{b(x)}$.

- If $Q(x)$ does not have repeating factors, then we have each $b(x)$ as a factor of $Q(x)$, with lower degree. The degree of each $a(x)$ is 1 lower than that of its corresponding $b(x)$, recall that the degree of $r(x)$ is less than that of $Q(x)$.
  若 $Q(x)$ 没有重复的因式,则每个 $b(x)$ 均为 $Q(x)$ 的因式且次数均比 $Q(x)$ 小。每个 $a(x)$ 的次数比它对应的 $b(x)$ 的次数小 1,因为 $r(x)$ 的次数比 $Q(x)$ 小。
- If $Q(x)$ has repeating factors, e.g. $(x-c)^k$, then we have the following $b(x)$'s: $(x-c),(x-c)^2,(x-c)^3,\cdots,(x-c)^k$. Each corresponding $a(x)$ is a constant. We can also expand $(x-c)^k$ and decompose as if $Q(x)$ does not have repeating factors, but this is more cumbersome.
  若 $Q(x)$ 有重复的因式,如 $(x-c)^k$,则可得到以下 $b(x)$: $(x-c),(x-c)^2,(x-c)^3,\cdots,(x-c)^k$。每个对应的 $a(x)$ 为常数。也可以展开 $(x-c)^k$ 并且视为没有重复的因式,不过这种情况比较烦琐。

Examples: 1. To decompose $\frac{2x^2+13x+17}{x^2+4x+3}$, we perform long division first:

$$\frac{2x^2+13x+17}{x^2+4x+3}=2+\frac{5x+11}{x^2+4x+3}.$$

We factor the denominator: $x^2+4x+3=(x+1)(x+3)$.

We decompose the fraction as the following:

$$2+\frac{5x+11}{x^2+4x+3}=2+\frac{A}{x+1}+\frac{B}{x+3}.$$

A and B are constants.

Note that the degrees of numerators of $\frac{A}{x+1}$ and $\frac{B}{x+3}$ are less than their denominators, since $5x+11$ has a lower degree than x^2+4x+3.

To solve for A and B, we match coefficients:

$$\begin{aligned}\frac{5x+11}{x^2+4x+3}&=\frac{A}{x+1}+\frac{B}{x+3}\\&=\frac{A(x+3)}{(x+1)(x+3)}+\frac{B(x+1)}{(x+1)(x+3)}\\&=\frac{(A+B)x+3A+B}{(x+1)(x+3)}\\&=\frac{(A+B)x+3A+B}{x^2+4x+3}.\end{aligned}$$

We now have a system of equations:

$$\begin{cases}A+B=5\\3A+B=11\end{cases}.$$

We have $A=3$ and $B=2$.

Our final answer is $2+\frac{5x+11}{x^2+4x+3}=2+\frac{3}{x+1}+\frac{2}{x+3}$.

2. To decompose $\frac{x^4}{x^3-27}$, we first use long division:

$$\frac{x^4}{x^3-27}=x+\frac{27x}{x^3-27}.$$

Factor the denominator:

$$x^3-27=(x-3)(x^2+3x+9)$$

$$x+\frac{27x}{x^3-27}=x+\frac{A}{x-3}+\frac{Bx+C}{x^2+3x+9}.$$

Notice that the numerators of the partial fractions are always 1 less than the denominators.

Now, equate the coefficients and solve for A, B, and C, just like the previous example.

3. To decompose $\frac{x^3+2x^2+3}{(x+1)(x-2)^3}$, we write:

$$\frac{x^3+2x^2+3}{(x+1)(x-2)^3}=\frac{A}{x+1}+\frac{B}{x-2}+\frac{C}{(x-2)^2}+\frac{D}{(x-2)^3}.$$

Or, we can write:

$$\begin{aligned}\frac{x^3+2x^2+3}{(x+1)(x-2)^3}&=\frac{x^3+2x^2+3}{(x+1)(x^3-6x^2+12x-8)}\\&=\frac{E}{x+1}+\frac{Fx^2+Gx+H}{x^3-6x^2+12x-8}.\end{aligned}$$

The latter is much more cumbersome.

We will stick to the first approach.

29.5 Rational Equations 有理方程

Rational Expression 有理式

n. ['ræʃənəl ɪk'spreʃən]

Review rational expression in Section 26.2.1.

回顾 26.2.1 节的有理式词条。

Rational Equation 有理方程

n. ['ræʃənəl ɪ'kweɪʒən]

Definition: An equation that has at least one rational expression.

有理方程是指含有至少一个有理分式的方程。

Properties: To solve a rational equation:

解有理方程:

1. Find the least common denominator of all rational expressions. 求出所有有理分式的最小公分母。
2. Multiply both sides by this least common denominator. 方程两边乘上这个最小公分母。
3. Check for extraneous solutions **in the original equation**. They occur when the denominator is 0. **在原方程里**检验增根。增根在分母为 0 的时候出现。

Questions: Solve the equation below:

$$\frac{1}{x-6}+\frac{x}{x-2}=\frac{4}{x^2-8x+12}$$

Answers: We rewrite the equation as $\frac{1}{x-6}+\frac{x}{x-2}=\frac{4}{(x-6)(x-2)}$.

The least common denominator of the fractions is $(x-6)(x-2)$.

We multiply both sides by this LCD:

$$[(x-6)(x-2)]\left(\frac{1}{x-6}+\frac{x}{x-2}\right)=[(x-6)(x-2)]\frac{4}{(x-6)(x-2)}$$

$$(x-2)+x(x-6)=4$$

$$x-2+x^2-6x=4$$

$$x^2-5x-6=0$$

$$(x-6)(x+1)=0$$

$$x=6, -1.$$

We check our answers by substituting each $x=6$ and $x=-1$ into our original equation.

$x=6$ creates a denominator 0. We can conclude that this is an extraneous solution.

$x=1$ is the only solution for the equation.

30. Conic Sections 圆锥曲线

30.1 Introduction 介绍

Review Midpoint Formula and Distance Formula in Section 6.2. 可先复习 6.2 节的中点公式和距离公式。

Conic Section 圆锥曲线

n. [ˈkɑnɪk ˈsekʃən]

Definition: The intersection of a plane and a double-napped cone. 圆锥曲线是指一个平面和二次锥面的交线。

Properties: 1. A conic section is one of a circle, ellipse, parabola, or hyperbola. 圆锥曲线是圆、椭圆、抛物线、双曲线中的一种。

Figure 30-1 shows a circle. 圆。

Figure 30-2 shows an ellipse. 椭圆。

Figure 30-3 shows a parabola. 抛物线。

Figure 30-4 shows a hyperbola. 双曲线。

Circles and ellipses are closed. 圆和椭圆是封闭的。

Parabolas and hyperbolas are open. 抛物线和双曲线是不封闭的。

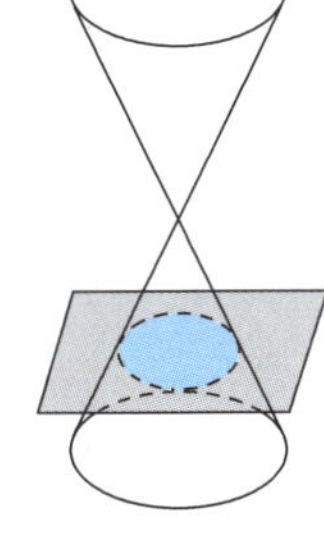

Figure 30-1

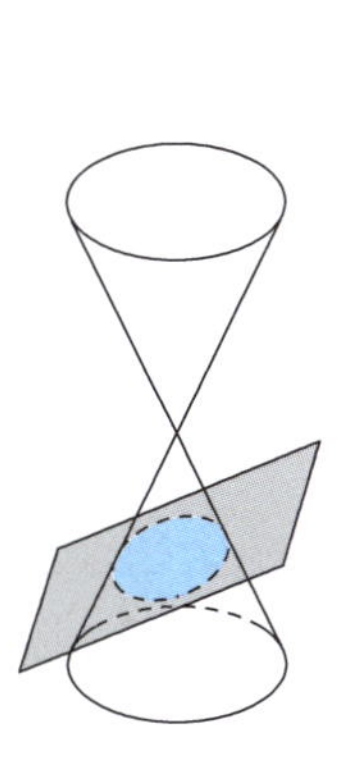

Figure 30-2

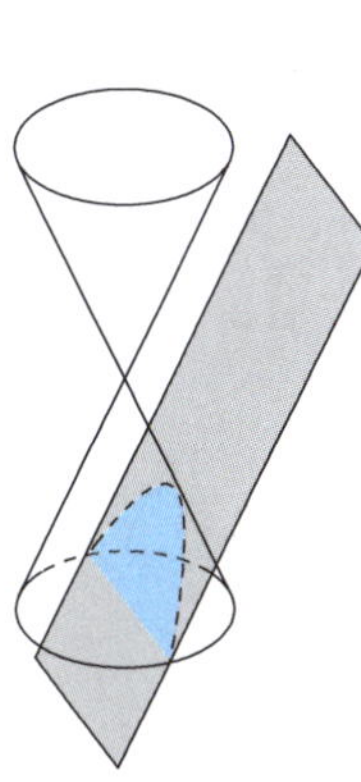

Figure 30-3

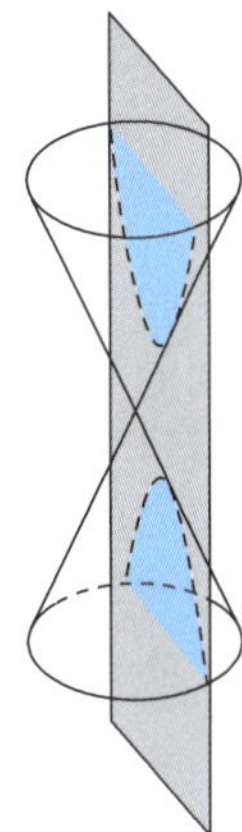

Figure 30-4

2. The degenerate cases of a conic section are: point, line, and two intersecting lines. 圆锥曲线可退化为点、线、双相交线。

Figure 30-5 shows a point. 点

Figure 30-6 shows a line.　线

Figure 30-7 shows two intersecting lines.　双相交线

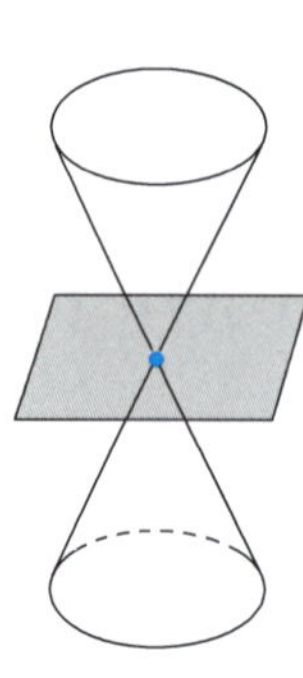

Figure 30-5

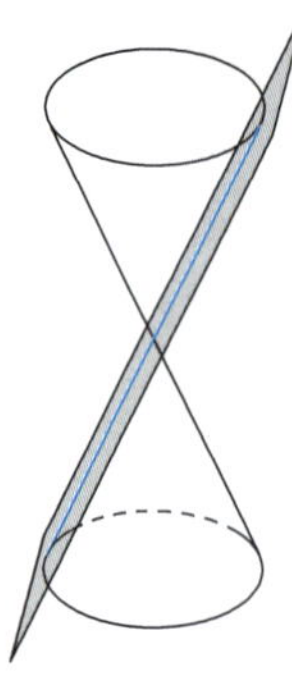

Figure 30-6

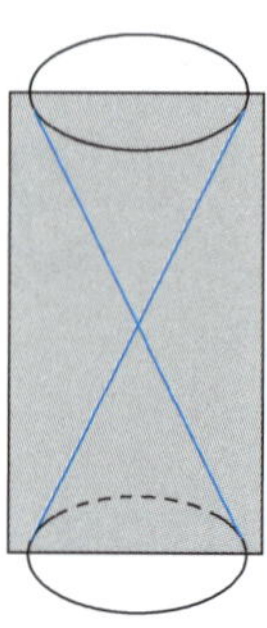

Figure 30-7

30.2　Circles　圆

Circle　圆

n. ['sɜrkəl]

Definition: A locus (set) of points for which every point is equidistant from a fixed point (h,k).

圆是一个点集,其中每个点对一个固定点(h,k)均等距。

The fixed point (h,k) is known as the center, and the line segment whose endpoints are (h,k) and one point in the set is known as the radius.

固定点(h,k)叫作圆心,端点为(h,k)和点集上一点的线段叫作半径。

Figure 30-8 shows a circle.　图 30-8 展示了一个圆。

Notation: $(x-h)^2+(y-k)^2=r^2$, from which h, k, and r are constants.

The circle has the center on (h,k) and the length of the radius is r.

圆心为(h,k),半径为 r。

If $r=0$, then this is a degenerate case of a circle—a point.

若 $r=0$,则圆退化为点。

Notes: The equation is based on the Pythagorean Theorem, as shown in Figures 30-9 and 30-10.

圆的等式可根据勾股定理得出,如图 30-9 和图 30-10 所示。

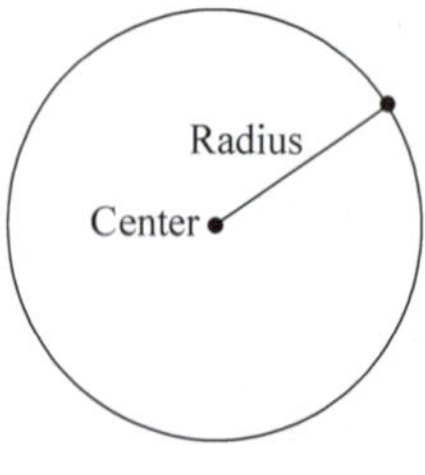

Figure 30-8

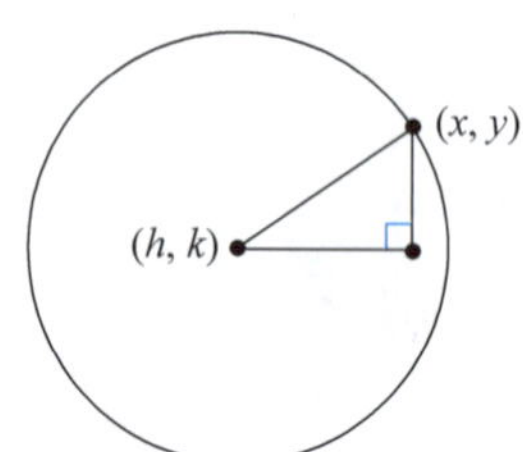

Figure 30-9

When the plane cuts the double-napped cone parallel to the bases of the cone, a circle is formed, as shown in Figure 30-11.　当平面对二次锥面以与其底平行的角度切入后,圆就

形成了，如图 30-11 所示。

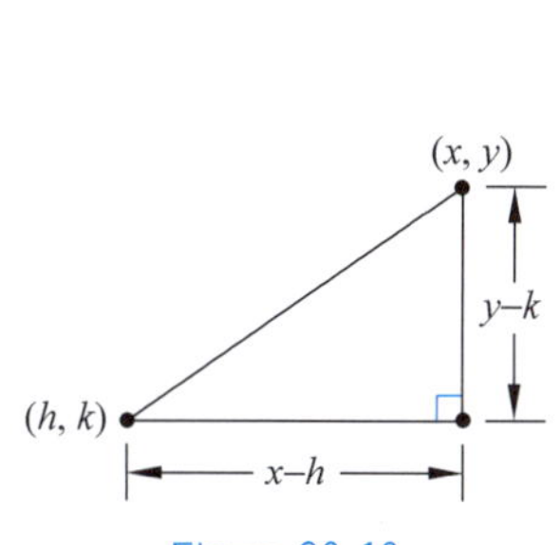

Figure 30-10

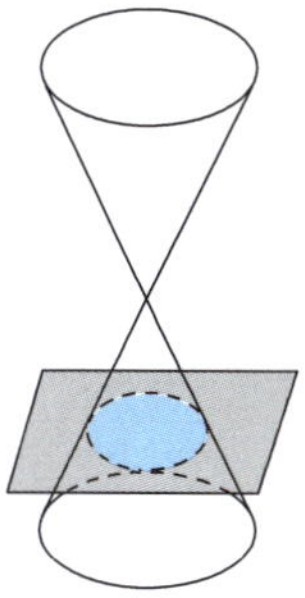

Figure 30-11

Questions：1. What is the equation of a circle whose length of radius is 5 and whose center is at (5，−3)?

2. What is the center and the length of the radius for each of the following?

(1) $(x-8)^2+(y+5)^2=49$

(2) $x^2+8x+y^2-14y=-56$

Answers：1. $(x-5)^2+(y+3)^2=25$.

2. (1) Center：(8，−5).

Radius：$\sqrt{49}=7$.

(2) We will complete the square：

$x^2+8x+y^2-14y=-56$.

$(x^2+8x+16)+(y^2-14y+49)=-56+16+49$.

$(x+4)^2+(y-7)^2=9$.

Center：(−4,7).

Radius：$\sqrt{9}=3$.

30.3　Ellipses　椭圆

Ellipse　椭圆

n. [ɪ'lɪps]

Definition：A locus (set) of points surrounding two fixed points for which every point has the same sum of distances from the fixed points.　椭圆是一个点集，其中每个点对两个固定点的距离和均相等。

Each fixed point is known as a focus，or focal point.　每个固定点称为焦点。

The line segment connecting the foci with both endpoints on the ellipse is the major axis.　通过两个焦点，并且两个端点都在椭圆上的线段叫作长轴。

The line segment passing the center and perpendicular to the major axis with both endpoints on the ellipse is the minor axis.　通过圆心并且与长轴垂直，且两个端点都在椭圆上的线段叫作短轴。

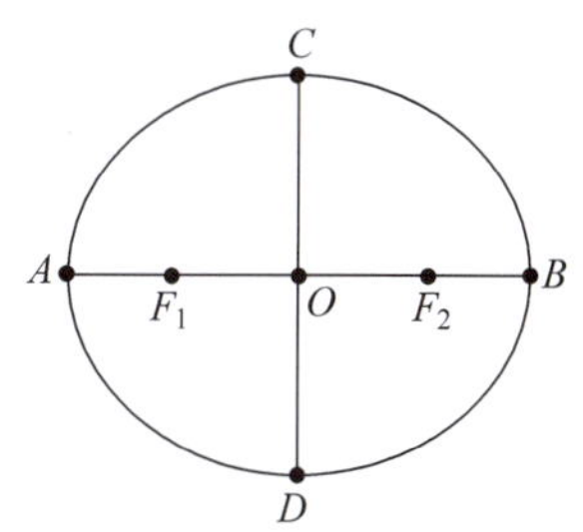

Figure 30-12

Figure 30-12 shows an ellipse. 图 30-12 展示了一个椭圆。

The center is O. O 为圆心

The foci are F_1 and F_2. F_1 和 F_2 为焦点

The major axis is $\overline{AF_1OF_2B}$. $\overline{AF_1OF_2B}$ 为长轴

The minor axis is $\overline{COD}$. $\overline{COD}$ 为短轴

For a point P on the ellipse, $\overline{PF_1}+\overline{PF_2}$ is a constant.

Notation: $\frac{(x-h)^2}{a^2}+\frac{(y-k)^2}{b^2}=1$, from which a and b are positive constants.

The ellipse has the center on (h,k).

圆心为(h,k)。

Properties: For the ellipse given by $\frac{(x-h)^2}{a^2}+\frac{(y-k)^2}{b^2}=1$:

- If $a>b$, then: 水平方向为长轴的情况

  The ellipse has a horizontal major axis.

  The endpoints of the major axis are $(h-a,k)$ and $(h+a,k)$.

  The endpoints of the minor axis are $(h,k-b)$ and $(h,k+b)$.

  Suppose c is the distance from a focus to the center. The foci are located on $(h-c,k)$ and $(h+c,k)$.

  $b^2+c^2=a^2$, for which $c>0$.

  This case is shown in Figure 30-13.

- If $a<b$, then: 垂直方向为长轴的情况

  The ellipse has a vertical major axis.

  The endpoints of the major axis are $(h,k-b)$ and $(h,k+b)$.

  The endpoints of the minor axis are $(h-a,k)$ and $(h+a,k)$.

  Suppose c is the distance from a focus to the center. The foci are located on $(h,k-c)$ and $(h,k+c)$.

  $a^2+c^2=b^2$, for which $c>0$.

  This case is shown in Figure 30-14.

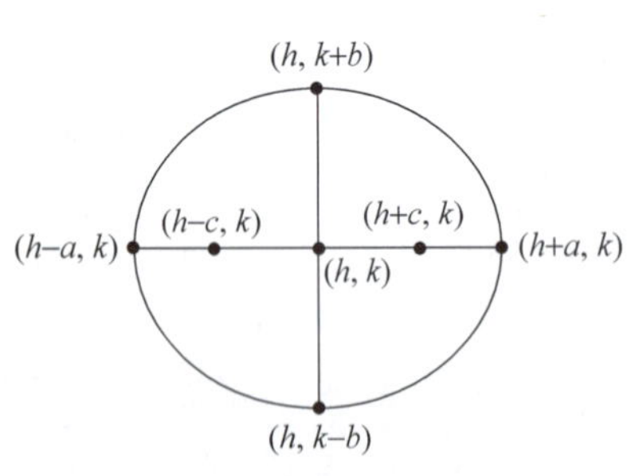

Figure 30-13

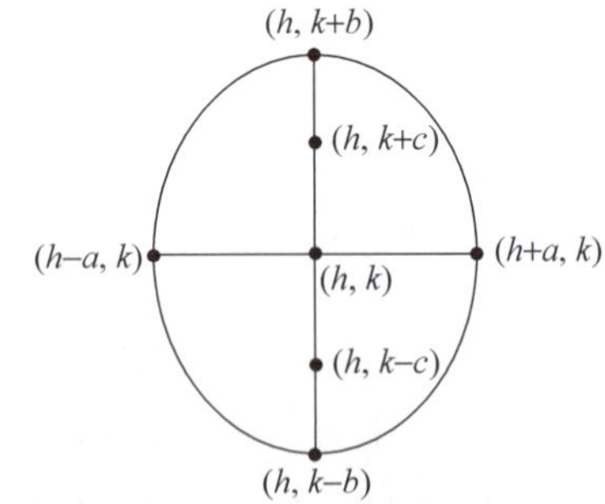

Figure 30-14

- If $a=b$, then this is a degenerative case of an ellipse—a circle.

Notes: 1. Here is how to derive the equation of an ellipse.

Without the loss of generality and for simplicity purposes, suppose the center of the ellipse is in the origin.

Without the loss of generality, suppose the ellipse has the foci on $F_1 = (-c, 0)$ and $F_2 = (c, 0)$, which means that the ellipse has a horizontal major axis. Suppose the endpoints of the major axis are $(-a, 0)$ and $(a, 0)$.

We want to find every point $P = (x, y)$ from which $PF_1 + PF_2$ is constant, by the definition of an ellipse.

Refer to Figure 30-15 for the following calculations.

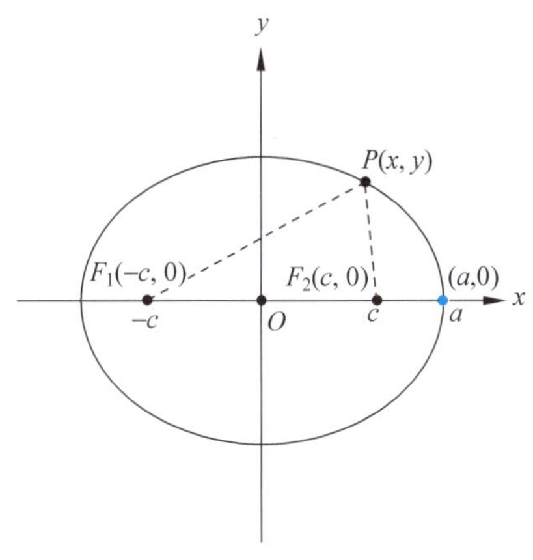

Figure 30-15

We know that $PF_1 + PF_2 = \sqrt{(x+c)^2 + y^2} + \sqrt{(x-c)^2 + y^2}$, which is a constant.

Note that for each endpoint of the major axis (one is drawn in blue), $PF_1 + PF_2 = (a-c) + (a+c) = 2a$.

We have $\sqrt{(x+c)^2 + y^2} + \sqrt{(x-c)^2 + y^2} = 2a$. Now, we manipulate the terms:

$$\sqrt{(x+c)^2 + y^2} + \sqrt{(x-c)^2 + y^2} = 2a$$

$$\sqrt{(x+c)^2 + y^2} = 2a - \sqrt{(x-c)^2 + y^2}.$$

Squaring both sides, we get

$$(x+c)^2 + y^2 = 4a^2 - 4a\sqrt{(x-c)^2 + y^2} + (x-c)^2 + y^2.$$

Expanding both $(x+c)^2$ and $(x-c)^2$, then canceling terms on both sides, we get

$$2cx = 4a^2 - 4a\sqrt{(x-c)^2 + y^2} - 2cx.$$

Moving terms, we get

$$4cx - 4a^2 = -4a\sqrt{(x-c)^2 + y^2}.$$

Dividing both sides by -4, we get

$$a^2 - cx = a\sqrt{(x-c)^2 + y^2}.$$

Squaring both sides and expanding both sides then simplify, we get

$$a^4 + c^2x^2 = a^2x^2 + a^2c^2 + a^2y^2.$$

Moving terms such that the variables are on one side and the constants are on the other side, we have

$$a^2x^2 - c^2x^2 + a^2y^2 = a^4 - a^2c^2.$$

Factoring, we get

$$x^2(a^2 - c^2) + a^2y^2 = a^2(a^2 - c^2).$$

Let $b^2 = a^2 - c^2$, we have

$$x^2b^2 + a^2y^2 = a^2b^2.$$

Dividing both sides by the RHS gives $\frac{x^2}{a^2} + \frac{y^2}{b^2} = 1$.

Deriving $\frac{(x-h)^2}{a^2} + \frac{(y-k)^2}{b^2} = 1$ would simply be the transformation of h units right and k units up.

For the case of $a < b$, similar approach follows to derive the same formula.

2. Refer to Figure 30-16 for the following calculations.

From $\frac{x^2}{a^2} + \frac{h^2}{b^2} = 1$, suppose the foci are on $F_1 = (-c, 0)$ and $F_2 = (c, 0)$. To show that $b^2 + c^2 = a^2$, note that for the point $P = (0, b)$, $PF_1 + PF_2 = \sqrt{b^2 + c^2} + \sqrt{b^2 + c^2} = 2\sqrt{b^2 + c^2}$.

For the point $P' = (a, 0)$, $P'F_1 + P'F_2 = a + c + (a - c) = 2a$.

By the definition of an ellipse, $PF_1 + PF_2$ is a constant, from which $2\sqrt{b^2 + c^2} = 2a$. From here, it is clear that $b^2 + c^2 = a^2$.

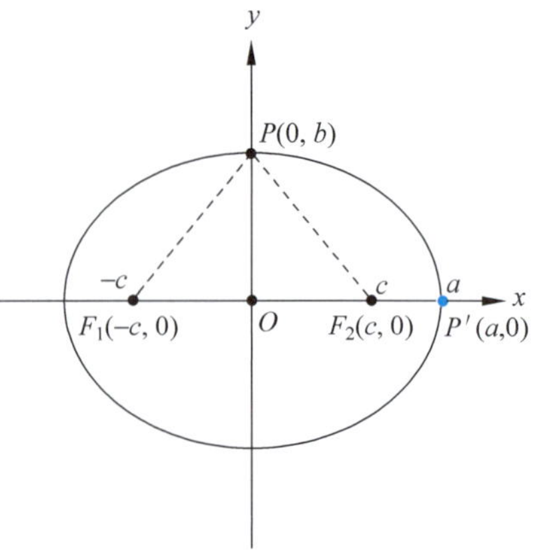

Figure 30-16

The equation $b^2 + c^2 = a^2$ is also true for $\frac{(x-h)^2}{a^2} + \frac{(y-k)^2}{b^2} = 1$, which is the transformation of $\frac{x^2}{a^2} + \frac{h^2}{b^2} = 1$ h units right and k units up.

For the case of $a < b$, similar approach follows to derive $a^2 + c^2 = b^2$.

3. When the plane cuts the double-napped cone at an angle less than an angle parallel to the side, an ellipse is formed. 当平面切入二次锥面的角度小于与边平行的角时,椭圆就形成了。

The angle parallel to the side is marked in blue in Figure 30-17.

与边平行的角在图 30-17 中用蓝色表示。

4. The eccentricity of an ellipse $\frac{(x-h)^2}{a^2} + \frac{(y-k)^2}{b^2} = 1$ is measured by $e = \frac{c}{\max(a, b)}$, from which c is the distance from the center to a focus. 椭圆的偏心率。

The eccentricity satisfies $0 < e < 1$:

- If $e \approx 0$, then the shape of the ellipse is close to a circle, as shown in Figure 30-18.
- If $e \approx 1$, then the shape of the ellipse is close to a line, as shown in Figure 30-19.

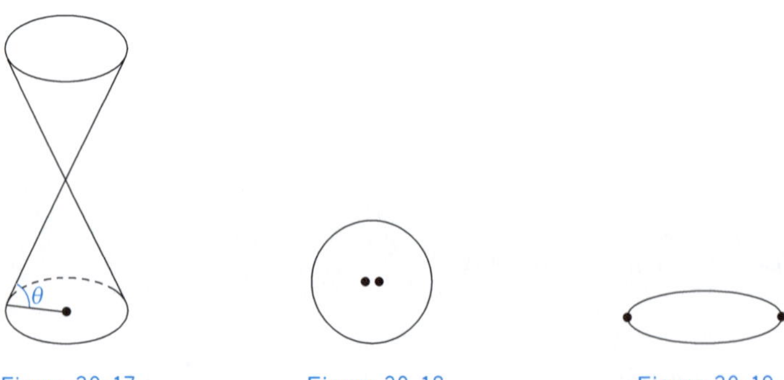

Figure 30-17　　Figure 30-18　　Figure 30-19

Questions：For the ellipse below, find the center, lengths of major and minor axes, endpoints of major and minor axes, foci, and eccentricity：

$$16x^2-96x+25y^2+100y=156.$$

Answers：$16x^2-96x+25y^2+100y=156$

$16(x^2-6x)+25(y^2+4y)=156$

We will complete the squares：

$16(x^2-6x+9)+25(y^2+4y+4)=156+144+100$

$$16(x-3)^2+25(y+2)^2=400$$

$$\frac{(x-3)^2}{25}+\frac{(y+2)^2}{16}=1$$

$$\frac{(x-3)^2}{5^2}+\frac{(y+2)^2}{4^2}=1.$$

We have $a=5$ and $b=4$. Since $a>b$, the ellipse has a horizontal major axis.

Center：$(3,-2)$

Length of major axis：$2(5)=10$

Length of minor axis：$2(4)=8$

Endpoints of major axis：$(3-5,-2)=(-2,-2)$ and $(3+5,-2)=(8,-2)$

Endpoints of minor axis：$(3,-2-4)=(3,-6)$ and $(3,-2+4)=(3,2)$

$c^2=a^2-b^2=5^2-4^2=3^2$, from which $c=3$.

Foci：$(3-3,-2)=(0,-2)$ and $(3+3,-2)=(6,-2)$

Eccentricity：3/5

30.4 Parabolas 抛物线

Parabola 抛物线

n. [pə'ræbələ]

Definition：The locus of points that are equidistant from a fixed point and a fixed line. 抛物线是一个点集，其中每个点对一个固定点和一条固定线均等距。

The fixed point is the focus. 固定点叫作焦点。

The fixed line is the directrix. 固定线叫作准线。

Figure 30-20 shows the parts of a parabola. 图 30-20 展示了抛物线的各部分。

By definition, the vertex is halfway between the focus and the directrix. 根据定义，顶点在焦点和准线中部。

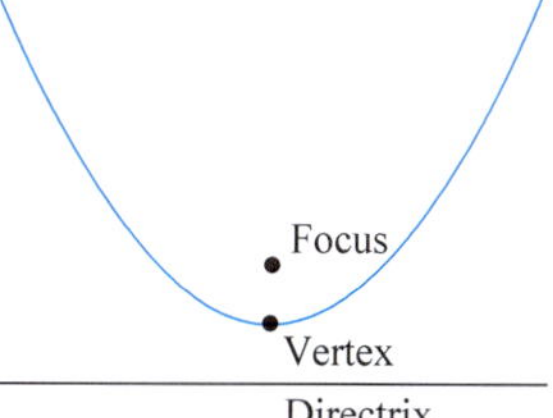

Figure 30-20

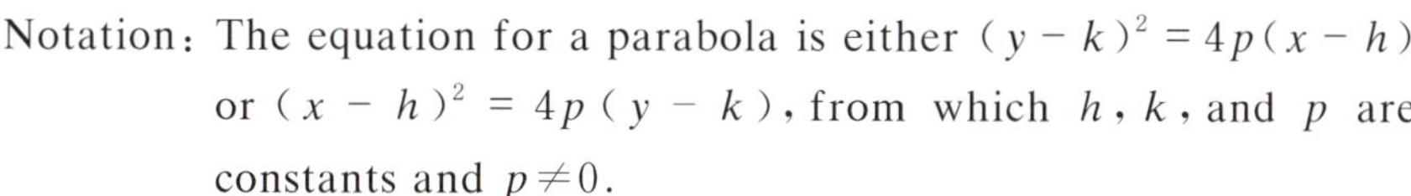

Notation：The equation for a parabola is either $(y-k)^2=4p(x-h)$ or $(x-h)^2=4p(y-k)$, from which h, k, and p are constants and $p\neq0$.

The vertex is located on (h,k). 顶点在(h,k)。

p is the signed distance from the focus to the vertex, which is the same as the signed distance from the vertex to the directrix.

p 为焦点到顶点的(带有正负符号)距离，也是顶点到准线的(带有正负符号)距离。

Properties: (1) For $(y-k)^2=4p(x-h)$:

- If $p>0$, then the graph of the parabola opens rightward.
  From left to right, we have the following order: directrix, vertex, focus, as shown in Figure 30-21.
- If $p<0$, then the graph of the parabola opens leftward.
  From left to right, we have the following order: focus, vertex, directrix, as shown in Figure 30-22.

(2) For $(x-h)^2=4p(y-k)$:

- If $p>0$, then the graph of the parabola opens upward.
  From top to bottom, we have the following order: focus, vertex, directrix, as shown in Figure 30-23.

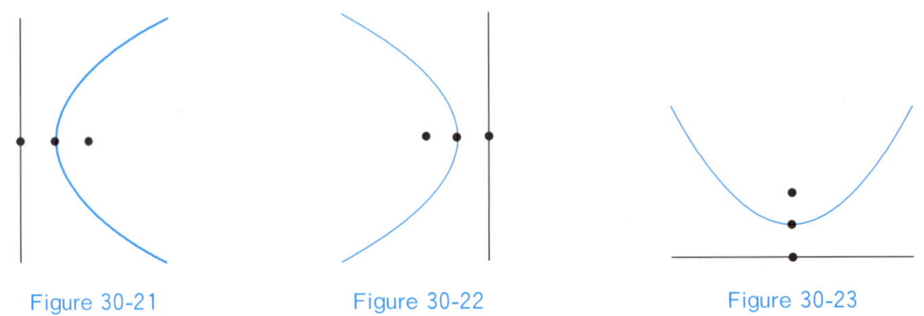

Figure 30-21　　Figure 30-22　　Figure 30-23

- If $p<0$, then the graph of the parabola opens downward.
  From top to bottom, we have the following order: directrix, vertex, focus, as shown in Figure 30-24.

Notes: 1. To derive an equation of a parabola, suppose the vertex is at (h,k), focus on $(h,k+p)$, and the directrix is $y=k-p$, so that p is the distance from the vertex to the focus, which is also the distance from the vertex to the directrix.
Figure 30-25 shows the case for a parabola that opens up. Proofs for parabolas opening in different directions are very similar.

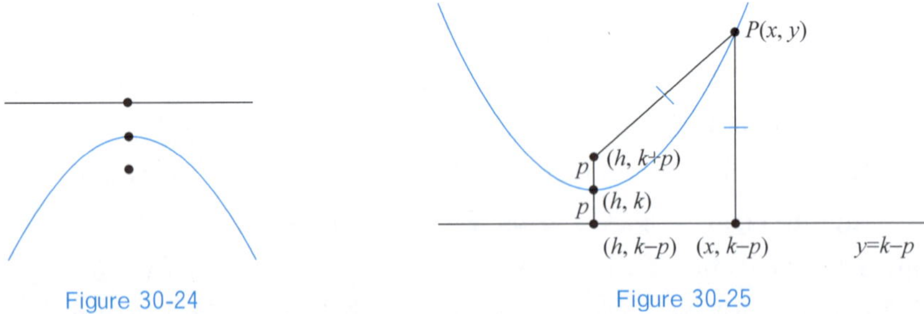

Figure 30-24　　Figure 30-25

A point $P=(x,y)$ must be equidistant from both $(h,k+p)$ and $y=k-p$.

The distance from P to $(h,k+p)$ is $\sqrt{(x-h)^2+(y-k-p)^2}$.

The distance from P to $y=k-p$ is $y-(k-p)=y-k+p$.

We have $\sqrt{(x-h)^2+(y-k-p)^2}=y-k+p$.

Squaring both sides and expanding, we have:

$(x-h)^2+y^2+k^2+p^2+2kp-2yk-2yp=y^2+k^2+p^2-2kp-2yk+2yp.$

Cancelling gives

$(x-h)^2=4py-4pk$

$(x-h)^2=4p(y-k).$

2. When the plane cuts the double-napped cone at an angle parallel to the side, a parabola is formed. 当平面切入二次锥面的角度与边平行时，抛物线就形成了。

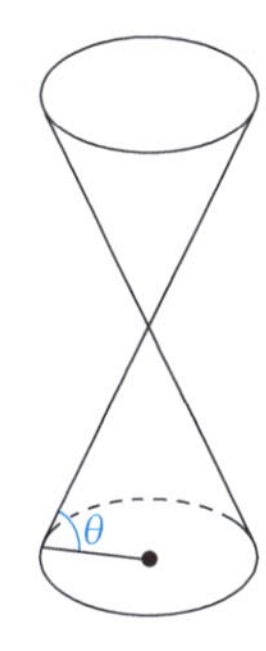

Figure 30-26

The angle parallel to the side is marked in blue in Figure 30-26.
与边平行的角在图 30-26 中用蓝色表示。

Questions: 1. Identify the vertex, focus, and directrix for the parabola below:

$$(y-4)^2=16(x+6).$$

2. For each of the following, write an equation of a parabola.

(1) directrix: $x=8$, focus: $(6,1)$

(2) vertex: $(4,7)$, directrix: $y=4$

(3) directrix: $y=5$, focus: $(7,3)$

Answers: 1. We rewrite $(y-4)^2=16(x+6)$ as $(y-4)^2=4(4)(x+6)$, from which $p=4$. This is a parabola opening rightward. The directrix is 4 units to the left of the vertex and the focus is 4 units to the right of the vertex.

Vertex: $(-6,4)$

Focus: $(-2,4)$

Directrix: $x=-10$

2. Sketching the graph of parabola helps to answer these questions.

(1) The vertex is at $(7,1)$. We have $p=-1$. The graph opens to the left.
$(y-1)^2=-4(x-7).$

(2) The vertex is at $(4,7)$. We have $p=3$. The graph opens up.
$(x-4)^2=12(y-7).$

(3) The vertex is at $(7,4)$. We have $p=-1$. The graph opens down.
$(x-7)^2=-4(y-4).$

30.5 Hyperbolas 双曲线

Hyperbola 双曲线

n. [haɪ'pɜrbələ]

Definition: A locus (set) of points surrounding two fixed points for which every point has the same absolute difference of distances to the fixed points. 双曲线是一个点集，其中每个点对两个固定点的距离差的绝对值均相等。

Each fixed point is known as a focus, or focal point. 每个固定点称为焦点。

The center is the midpoint of the line segment whose endpoints are the foci, also is the midpoint of the line segment whose endpoints are the vertices. 中心是端点为两个焦点

的线段的中点,同样是端点为两个顶点的线段的中点。

The transverse axis is a line segment whose endpoints are the vertices. The transverse axis passes through the center. 实轴是端点为两个顶点的线段。实轴穿过中心。

The conjugate axis is the axis that passes through the center and perpendicular to the transverse axis. 虚轴为穿过中心且垂直于实轴的轴。

Figure 30-27 shows a hyperbola.

The center is O. O 为中心。

The vertices are V_1 and V_2. V_1 和 V_2 为顶点。

The foci are F_1 and F_2. F_1 和 F_2 为焦点。

The transverse axis is $\overline{V_1OV_2}$. $\overline{V_1OV_2}$ 为实轴。

The conjugate axis is l. l 为虚轴。

Notation: Horizontal Hyperbolas:

$\frac{(x-h)^2}{a^2}-\frac{(y-k)^2}{b^2}=1$, from which a and b are positive constants.

Figure 30-28 shows a horizontal hyperbola.

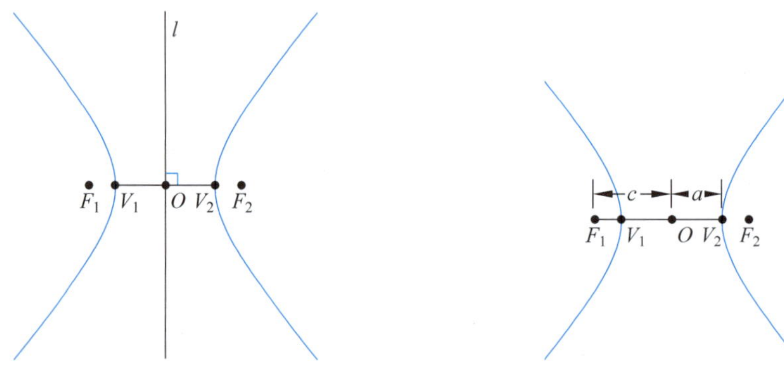

Figure 30-27　　Figure 30-28

Vertical Hyperbolas:

$\frac{(y-k)^2}{a^2}-\frac{(x-h)^2}{b^2}=1$, from which a and b are positive constants.

Figure 30-29 shows a vertical hyperbola.

For hyperbolas:

a is the distance from the center to a vertex.

c is the distance from the center to a focus.

$a^2+b^2=c^2$.

Properties: Hyperbolas have asymptotes. 双曲线有渐近线。

- For the hyperbola $\frac{(x-h)^2}{a^2}-\frac{(y-k)^2}{b^2}=1$, the asymptotes are $y=\pm\frac{b}{a}(x-h)+k$.
- For the hyperbola $\frac{(y-k)^2}{a^2}-\frac{(x-h)^2}{b^2}=1$, the asymptotes are $y=\pm\frac{a}{b}(x-h)+k$.

Notes: 1. We will derive the equation of a hyperbola.

Refer to Figure 30-30 for the following calculations.

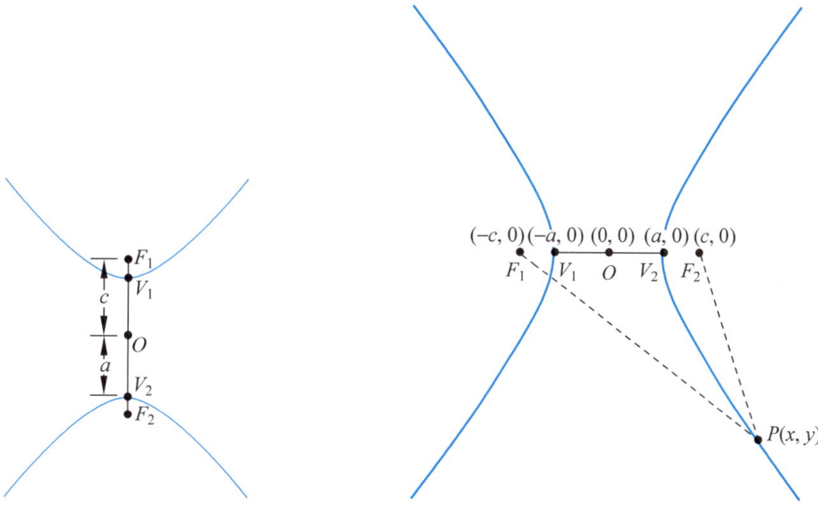

Figure 30-29

Figure 30-30

Without the loss of generality, suppose the center of the hyperbola is at the origin, the vertices are at $V_1=(-a,0)$ and $V_2=(a,0)$ and the foci are at $F_1=(-c,0)$ and $F_2=(c,0)$.

For each point P on the hyperbola, $|PF_1-PF_2|$ is a constant.

Note that $|V_2F_1-V_2F_2|=|(a+c)-(c-a)|=2a$.

For point P on the right branch, $PF_1>PF_2$, and PF_1-PF_2 is positive.

$PF_1-PF_2=\sqrt{(x-(-c))^2+(y-0)^2}-\sqrt{(x-c)^2+(y-0)^2}$

We have $\sqrt{(x-(-c))^2+(y-0)^2}-\sqrt{(x-c)^2+(y-0)^2}=2a$.

Simplifying gives $\sqrt{(x+c)^2+y^2}-\sqrt{(x-c)^2+y^2}=2a$.

Adding $\sqrt{(x-c)^2+y^2}$ to both sides, we have

$$\sqrt{(x+c)^2+y^2}=2a+\sqrt{(x-c)^2+y^2}.$$

Squaring both sides, expanding everything and simplifying, we have

$$2cx=4a^2+4a\sqrt{(x-c)^2+y^2}-2cx.$$

Isolating the radical gives

$$4cx-4a^2=4a\sqrt{(x-c)^2+y^2}.$$

Dividing both sides by 4 gives

$$cx-a^2=a\sqrt{(x-c)^2+y^2}.$$

Squaring both sides and expanding everything gives

$$c^2x^2-2a^2cx+a^4=a^2x^2-2a^2cx+a^2c^2+a^2y^2.$$

We can rearrange it as

$$c^2x^2-a^2x^2-a^2y^2=a^2c^2-a^4$$
$$x^2(c^2-a^2)-a^2y^2=a^2(c^2-a^2).$$

Let $b^2 = c^2 - a^2$, we have

$$x^2 b^2 - a^2 y^2 = a^2 b^2.$$

Dividing both sides by $a^2 b^2$ gives

$$\frac{x^2}{a^2} - \frac{y^2}{b^2} = 1.$$

Deriving $\frac{(x-h)^2}{a^2} - \frac{(y-k)^2}{b^2} = 1$ would simply be the transformation of h units right and k units up.

2. We will show why $a^2 + b^2 = c^2$.

For the hyperbola $\frac{x^2}{a^2} - \frac{y^2}{b^2} = 1$, the foci are on $(\pm c, 0)$. Note that $(\sqrt{2}a, b)$ and $(a, 0)$ are points on the hyperbola.

The positive difference of distances of $(\sqrt{2}a, b)$ to the foci is

$$\sqrt{(\sqrt{2}a + c)^2 + b^2} - \sqrt{(\sqrt{2}a - c)^2 + b^2}.$$

The positive difference of distances of $(a, 0)$ to the foci is $(a + c) - (c - a) = 2a$.

Set $\sqrt{(\sqrt{2}a + c)^2 + b^2} - \sqrt{(\sqrt{2}a - c)^2 + b^2} = 2a$. Simplifying will show that $a^2 + b^2 = c^2$.

Similarly, we can show $a^2 + b^2 = c^2$ is true for the hyperbola $\frac{(x-h)^2}{a^2} - \frac{(y-k)^2}{b^2} = 1$, which would simply be the transformation of h units right and k units up.

3. Without the loss of generality, suppose the center of the hyperbola is at the origin, the vertices are at $V_1 = (-a, 0)$ and $V_2 = (a, 0)$ and the foci are at $F_1 = (-c, 0)$ and $F_2 = (c, 0)$.

The equation for this hyperbola is $\frac{x^2}{a^2} - \frac{y^2}{b^2} = 1$.

Solving for y^2, we have $y^2 = b^2\left(\frac{x^2}{a^2} - 1\right)$.

Taking the square root of both sides, we have

$$\begin{aligned} y &= \pm b\sqrt{\frac{x^2}{a^2} - 1} \\ &= \pm b\sqrt{\frac{x^2 - a^2}{a^2}} \\ &= \pm \frac{b}{a}\sqrt{x^2 - a^2}. \end{aligned}$$

Since a is a constant, $y \to \pm \frac{b}{a}\sqrt{x^2} = \pm \frac{b}{a}x$ as $x \to \infty$ or $x \to -\infty$.

Therefore, the asymptotes for this hyperbola are $y = \pm \frac{b}{a}x$.

Similarly, the asymptotes for the hyperbola $\frac{(x-h)^2}{a^2} - \frac{(y-k)^2}{b^2} = 1$ are $y - k = \pm \frac{b}{a}(x -$

h), which is $y = \pm\frac{b}{a}(x-h)+k$. This is the transformation of h units right and k units up.

Similarly, the asymptotes for the hyperbola $\frac{(y-k)^2}{a^2}-\frac{(x-h)^2}{b^2}=1$ are $y = \pm\frac{a}{b}(x-h)+k$.

4. When the plane cuts the double-napped cone at an angle closer to the axis than the side of the cone a hyperbola is formed. 当平面切入二次锥面的角度离轴比离边近时，双曲线就形成了。

   As shown in Figure 30-31, the acute angle formed by the plane and the axis (shown in blue) must be smaller than θ, the angle formed by the side and the axis.

   如图 30-31 所示，切入平面与轴(蓝色)形成的锐角必须比 θ 小。θ 为边与轴组成的角。

5. The eccentricity of a hyperbola $\frac{(x-h)^2}{a^2}-\frac{(y-k)^2}{b^2}=1$ is measured by $e=\frac{c}{a}$,

   The eccentricity satisfies $e>1$:

   - If $e\approx 1$, then the shape of the hyperbola is two sharp curves, as shown in Figure 30-32.
   - If $e\to\infty$, then the shape of the hyperbola is straight. As shown in Figure 30-33.

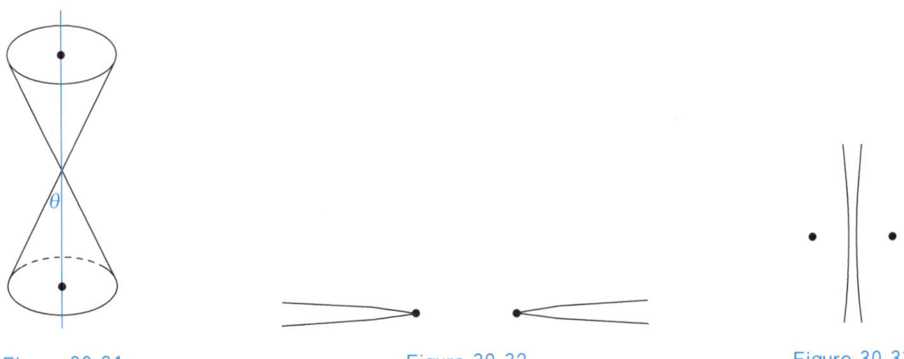

Figure 30-31 Figure 30-32 Figure 30-33

Questions: For the hyperbola below, find the center, vertices, foci, asymptotes, and eccentricity:

$$16x^2-160x-9y^2-36y+364=144.$$

Answers: We will complete the squares:

$$(16x^2-160x+400)-(9y^2+36y+36)=144$$

$$16(x^2-10x+25)-9(y^2+4y+4)=-220+400-36$$

$$16(x-5)^2-9(y+2)^2=144$$

$$\frac{(x-5)^2}{9}-\frac{(y+2)^2}{16}=1$$

$$\frac{(x-5)^2}{3^2}-\frac{(y+2)^2}{4^2}=1.$$

This is a horizontal hyperbola.

We have $a=3$ and $b=4$, we have $c=\sqrt{a^2+b^2}=\sqrt{3^2+4^2}=5$.

Center：(5，−2)

Vertices：(5±3，−2)，which are (2，−2) and (8，−2)

Foci：(5±5，−2)，which are (0，−2) and (10，−2)

Asymptotes：$y=\pm\frac{4}{3}(x-5)-2$

Eccentricity：5/3

30.6 General Formulas of Conic Sections 圆锥曲线的一般式

Degenerate Conic 退化圆锥曲线

n. [dɪ'dʒenəˌreɪt 'kɑnɪk]

Definition：The case in which the intersection of a plane and a double-napped cone is a point，line，or two intersecting lines.

退化圆锥曲线是指一个平面和二次锥面的交线为一个点、一条线或两条相交线的情况。

Figure 30-34 shows a point. 点

Figure 30-35 shows a line. 线

Figure 30-36 shows two intersecting lines. 两条双交线

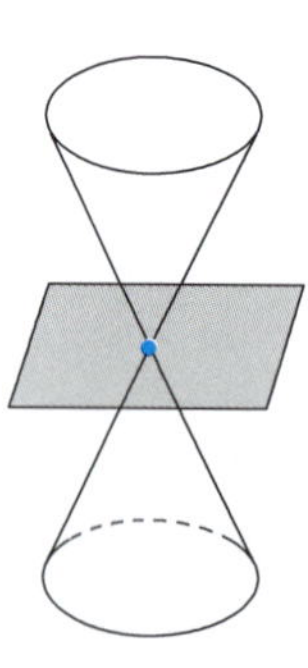

Figure 30-34

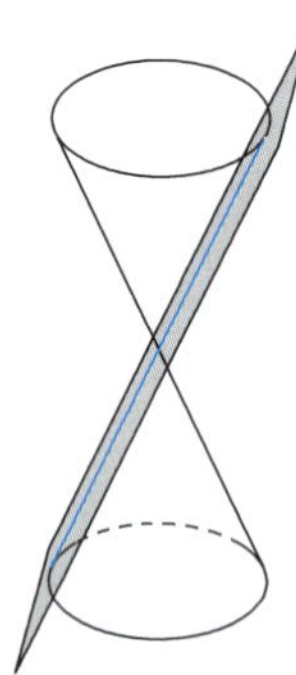

Figure 30-35

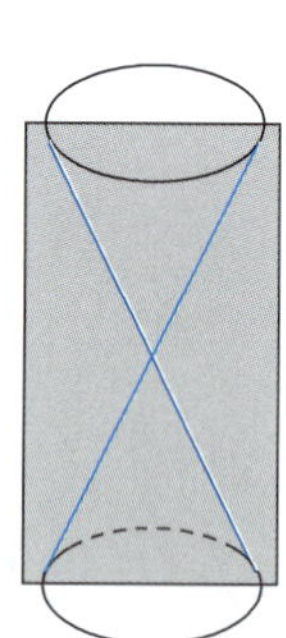

Figure 30-36

Examples：1. Equation of a point：$\frac{(x-h)^2}{a^2}+\frac{(y-k)^2}{b^2}=0$. The only solution is (h,k).

2. Equation of a line：$Dx+Ey+F=0$.

3. Equation of two intersecting lines：$\frac{(x-h)^2}{a^2}-\frac{(y-k)^2}{b^2}=0$，which is $\frac{(y-k)^2}{b^2}=\frac{(x-h)^2}{a^2}$. Solving $\frac{y-k}{b}=\pm\frac{x-h}{a}$ individually，we get $y=\frac{b}{a}x+\left(k-\frac{bh}{a}\right)$ and $y=-\frac{b}{a}x+\left(k+\frac{bh}{a}\right)$.

General Formulas of Conics 圆锥曲线的一般式

Definition：$Ax^2+Bxy+Cy^2+Dx+Ey+F=0$，from which A,B,C,D,E，and F are constants.

Properties: B^2-4AC is the discriminant, which has the property as shown in Table 30-1.
B^2-4AC 为判别式，具有以下性质，如表 30-1 所示。

Table 30-1

Discriminant　判别式	Property　属性
$B^2-4AC=0$ and ($A=0$ or $C=0$)	Parabola
$B^2-4AC<0$ and $A=C$	Circle
$B^2-4AC<0$ and $A\neq C$	Ellipse
$B^2-4AC>0$	Hyperbola

31. Sequences and Series　数列与级数

Review sequences in Chapter 8.　可先复习第 8 章的数列。

视频 43

Series　级数

n. [ˈsɪəriz]

Definition: The sum of all terms in a sequence.　级数是指数列中所有项的和。
A finite sequence has a finite series.　有限数列有有限级数。
An infinite sequence has an infinite series.　无穷数列有无穷级数。

Notation: Suppose we have a sequence

$$a_1, a_2, a_3, \cdots, a_n$$

The series is denoted by the letter S. The abbreviation of writing out the sum of all terms is the summation symbol sigma ($\sum$):

级数用 S 表示，所有项的和用求和符号 $\sum$ 简写，即

$$S=\sum_{i=1}^{n} a_i = a_1+a_2+a_3+\cdots+a_n.$$

i is the index of summation.　i 为求和指数。
1 is the lower limit of summation.　1 为求和的下限值。
n is the upper limit of summation.　n 为求和的上限值。
a_i represents the ith term of summation.　a_i 为第 i 个项的值。

Together, $\sum_{i=k}^{n} a_i$ gives the sum from the kth term to the nth term.

Properties: 1. If c is a constant, then $\sum_{i=1}^{n} c = cn$.

2. If c is a constant, then $\sum_{i=1}^{n} (ca_i) = c\sum_{i=1}^{n} a_i$.

3. For two sequences with equal length $a_1, a_2, a_3, \cdots, a_n$ and $b_1, b_2, b_3, \cdots, b_n$:

(1) $\sum_{i=1}^{n}(a_i+b_i)=\sum_{i=1}^{n}a_i+\sum_{i=1}^{n}b_i$.

(2) $\sum_{i=1}^{n}(a_i-b_i)=\sum_{i=1}^{n}a_i-\sum_{i=1}^{n}b_i$.

Proofs: 1. To prove Property 1:

$$\text{LHS}=\sum_{i=1}^{n}c=c+c+c+\cdots+c\ (\text{sum of } n\ c\text{'s})=nc=\text{RHS}.$$

2. To prove Property 2:

$$\text{LHS}=\sum_{i=1}^{n}ca_i=ca_1+ca_2+ca_3+\cdots+ca_n=c(a_1+a_2+a_3+\cdots+a_n)=c\sum_{i=1}^{n}a_i=\text{RHS}.$$

3. To prove Property 3:

(1) $\text{LHS}=\sum_{i=1}^{n}(a_i+b_i)=(a_1+b_1)+(a_2+b_2)+(a_3+b_3)+\cdots+(a_n+b_n)=(a_1+a_2+a_3+\cdots+a_n)+(b_1+b_2+b_3+\cdots+b_n)=\sum_{i=1}^{n}a_i+\sum_{i=1}^{n}b_i=\text{RHS}.$

(2) $\text{LHS}=\sum_{i=1}^{n}(a_i-b_i)=(a_1-b_1)+(a_2-b_2)+(a_3-b_3)+\cdots+(a_n-b_n)=(a_1+a_2+a_3+\cdots+a_n)-(b_1+b_2+b_3+\cdots+b_n)=\sum_{i=1}^{n}a_i-\sum_{i=1}^{n}b_i=\text{RHS}.$

Examples: 1. (1) The summation $\sum_{i=1}^{6}i^2$ represents the sum $1^2+2^2+3^2+4^2+5^2+6^2$.

(2) The summation $\sum_{i=3}^{7}[5(2)^i]$ represents the sum $5(2)^3+5(2)^4+5(2)^5+5(2)^6+5(2)^7$.

2. (1) To write the sum $4+7+10+13+16+19+22+25$ using the summation notation, note that in the sequence 4, 7, 10, 13, 16, 19, 22, 25, we have $a_i=3i+1$, from which $a_1=4$ and $a_8=25$.

Therefore, $4+7+10+13+16+19+22+25=\sum_{i=1}^{8}(3i+1)$.

Review arithmetic sequences in Section 8.2.

(2) To write the sum $1+3+9+27+81$ using the summation notation, note that in the sequence 1, 3, 9, 27, 81, we have

$a_i=3^{i-1}$, from which $a_1=1$ and $a_5=81$.

Therefore, $1+3+9+27+81=\sum_{i=1}^{5}3^{i-1}$.

Review geometric sequences in Section 8.3.

Phrases: finite～ 有限级数, infinite～ 无穷级数, arithmetic～ 等差级数, geometric～ 等比级数, finite geometric～ 有限等比级数, infinite geometric～ 无穷等比级数

Arithmetic Series 等差级数

n. [əˈrɪθməˌtɪk ˈsɪəriz]

Definition: The sum of terms of an arithmetic sequence. 等差级数为等差数列项的总和。

Properties: 1. Arithmetic series must be finite, otherwise the series will be undefined. 等差级数必

为有限，否则结果无意义。

2. Suppose we have the arithmetic sequence below：

$$1,2,3,\cdots,n.$$

We have $S=\frac{n(n+1)}{2}$.

3. Suppose we have the arithmetic sequence below，from which a is the first term and d is the common difference.

a 为首项，d 为公差。

$$a,a+d,a+2d,\cdots,a+(n-1)d.$$

We have $S=\frac{n(2a+(n-1)d)}{2}$.

Note：There are n terms in the sequence. Recall that in an arithmetic sequence，$a_n=a+(n-1)d$.

Notes：1. To prove Property 2，see proof by induction in Section 13.3.2.2 and proof by casework in Section 13.3.2.4.

2. To prove Property 3，note that

$$\begin{aligned} S&=a+(a+d)+(a+2d)+\cdots+(a+(n-1)d)\\ &=na+(1+2+3+\cdots+(n-1))d\\ &=na+\frac{(n-1)n}{2}d \qquad \text{The result from Property 2.}\\ &=\frac{2na}{2}+\frac{(n-1)nd}{2}\\ &=\frac{2na+(n-1)nd}{2}\\ &=\frac{n(2a+(n-1)d)}{2}. \end{aligned}$$

When $a=1$ and $d=1$，this becomes $S=\frac{n(n+1)}{2}$，which is the result for Property 2.

Questions：1. What is the sum of the following sequence?

$$5,7,9,11,\cdots,101$$

2. The sum of the first n positive integers is 2016. What is the value of n?

Answers：1. Since the common difference is 2 and the first term is 5，we have $d=2$ and $a=5$ in $a_n=a+(n-1)d$.

We have $a_n=5+2(n-1)$.

Since $a_1=5$ and $a_{49}=101$，this sequence has 49 terms.

Using the formula from Property 3，we have $S=\frac{n(2a+(n-1)d)}{2}=\frac{49(2(5)+2(49-1))}{2}=2597$.

2. We have $1+2+3+\cdots+n=2016$.

This is

$$\frac{n(n+1)}{2}=2016$$

$$n(n+1)=4032$$

$$n^2+n-4032=0$$

$$(n-63)(n+64)=0$$

$n=-64$ does not make sense. Therefore, $n=63$, which is our answer.

Geometric Series 等比级数

n. [ə'rɪθmə,tɪk 'sɪəriz]

Definition: The sum of terms of a geometric sequence. 等比级数是指等比数列项的总和。

Properties: 1. Geometric series can be finite or infinite. 等比级数可以为有限或无穷。

2. Suppose we have the finite geometric sequence below, from which a is the first term and r is the common ratio.
a 为首项,r 为公比。

$$a, ar, ar^2, \cdots, ar^{n-1}$$

We have $S=\dfrac{a(r^n-1)}{r-1}$.

Note: There are n terms in the sequence. Recall that in a geometric sequence, $a_n=ar^{n-1}$.

3. Suppose we have the infinite geometric sequence below, from which a is the first term and r is the common ratio.
a 为首项,r 为公比。

$$a, ar, ar^2, ar^3, \cdots$$

If $|r|>1$, then $S=a+ar+ar^2+ar^3+\cdots$ is undefined.

If $|r|<1$, then $S=\dfrac{a}{1-r}$.

4. Geometric series is a powerful tool of converting repeating decimals to common fractions.

Notes: 1. To prove Property 2, note that:

$S=a+ar+ar^2+\cdots+ar^{n-1}$

$rS=ar+ar^2+\cdots+ar^{n-1}+ar^n$

Subtracting series, we have:

$rS-S=ar^n-a$

$S(r-1)=ar^n-a$

$S(r-1)=a(r^n-1)$

$$S=\frac{a(r^n-1)}{r-1}$$

2. To prove Property 3:

- $|r|>1$

An intuitive way to think of this is that the absolute values of $a, ar, ar^2, ar^3, \cdots$ increases as the exponent increases.

If the terms are of the same sign ($r>1$), then their sum diverges.

If the terms alter in signs ($r<1$), then their sum bounces between positive and negative. As the exponent increases, the magnitude of the sum increases.

Therefore, the geometric series is undefined.

- $|r|<1$

  From Property 2, we have $S=\dfrac{a(r^n-1)}{r-1}$. Note that since $|r|<0$, r^n approaches 0 as n becomes arbitrarily large.

  Therefore, we have $S=\dfrac{a(r^n-1)}{r-1}=\dfrac{a(0-1)}{r-1}=\dfrac{-a}{r-1}=\dfrac{a}{1-r}$.

3. Below shows some examples of converting repeating decimals to common fractions. These examples can also be found in the Questions sections of Section 1.4.3, under pure recurring decimals and mixed recurring decimals.

(1) To convert $0.\overline{7}=0.777\cdots$ to a common fraction, note that

$$\begin{aligned}
0.\overline{7}&=\frac{7}{10}+\frac{7}{10^2}+\frac{7}{10^3}+\cdots\\
&=7\left(\frac{1}{10}\right)+7\left(\frac{1}{10}\right)^2+7\left(\frac{1}{10}\right)^3+\cdots\\
&=7\left[\left(\frac{1}{10}\right)+\left(\frac{1}{10}\right)^2+\left(\frac{1}{10}\right)^3+\cdots\right]\\
&=7\left[\frac{\frac{1}{10}}{1-\frac{1}{10}}\right]\\
&=7\left[\frac{1}{9}\right]\\
&=\frac{7}{9}.
\end{aligned}$$

(2) To convert $0.\overline{12}=0.121212\cdots$ to a common fraction, note that

$$\begin{aligned}
0.\overline{12}&=\frac{1}{10}+\frac{2}{10^2}+\frac{1}{10^3}+\frac{2}{10^4}+\frac{1}{10^5}+\frac{2}{10^6}+\cdots\\
&=\left(\frac{1}{10}+\frac{2}{10^2}\right)+\left(\frac{1}{10}\right)^2\left(\frac{1}{10}+\frac{2}{10^2}\right)+\left(\frac{1}{10}\right)^4\left(\frac{1}{10}+\frac{2}{10^2}\right)+\cdots\\
&=\left(\frac{12}{100}\right)+\left(\frac{1}{10}\right)^2\left(\frac{12}{100}\right)+\left(\frac{1}{10}\right)^4\left(\frac{12}{100}\right)+\cdots\\
&=\left(\frac{3}{25}\right)+\left(\frac{1}{10}\right)^2\left(\frac{3}{25}\right)+\left(\frac{1}{10}\right)^4\left(\frac{3}{25}\right)+\cdots\\
&=\frac{3}{25}\left[1+\left(\frac{1}{10}\right)^2+\left(\frac{1}{10}\right)^4+\cdots\right]\\
&=\frac{3}{25}\left[\frac{1}{1-\left(\frac{1}{10}\right)^2}\right]\\
&=\frac{3}{25}\left[\frac{100}{99}\right]\\
&=\frac{4}{33}.
\end{aligned}$$

(3) To convert $0.3\overline{2}=0.3222\cdots$ to a common fraction, note that

$$0.3\overline{2}=\frac{3}{10}+\frac{2}{10^2}+\frac{2}{10^3}+\frac{2}{10^4}\cdots$$
$$=\frac{3}{10}+2\left(\frac{1}{10}\right)^2+2\left(\frac{1}{10}\right)^3+2\left(\frac{1}{10}\right)^4+\cdots$$
$$=\frac{3}{10}+2\left[\left(\frac{1}{10}\right)^2+\left(\frac{1}{10}\right)^3+\left(\frac{1}{10}\right)^4\cdots\right]$$
$$=\frac{3}{10}+2\left[\frac{\left(\frac{1}{10}\right)^2}{1-\frac{1}{10}}\right]$$
$$=\frac{3}{10}+2\left[\frac{1}{90}\right]$$
$$=\frac{3}{10}+\frac{2}{90}$$
$$=\frac{29}{90}.$$

(4) To convert $0.3\overline{45}=0.345\,454\,5\cdots$ to a common fraction, note that

$$0.3\overline{45}=\frac{3}{10}+\frac{4}{10^2}+\frac{5}{10^3}+\frac{4}{10^4}+\frac{5}{10^5}+\frac{4}{10^6}+\frac{5}{10^7}+\cdots$$
$$=\frac{3}{10}+\left(\frac{4}{10^2}+\frac{5}{10^3}\right)+\left(\frac{1}{10}\right)^2\left(\frac{4}{10^2}+\frac{5}{10^3}\right)+\left(\frac{1}{10}\right)^4\left(\frac{4}{10^2}+\frac{5}{10^3}\right)+\cdots$$
$$=\frac{3}{10}+\left(\frac{45}{1000}\right)+\left(\frac{1}{10}\right)^2\left(\frac{45}{1000}\right)+\left(\frac{1}{10}\right)^4\left(\frac{45}{1000}\right)+\cdots$$
$$=\frac{3}{10}+\left(\frac{9}{200}\right)+\left(\frac{1}{10}\right)^2\left(\frac{9}{200}\right)+\left(\frac{1}{10}\right)^4\left(\frac{9}{200}\right)+\cdots$$
$$=\frac{3}{10}+\frac{9}{200}\left[1+\left(\frac{1}{10}\right)^2+\left(\frac{1}{10}\right)^4+\cdots\right]$$
$$=\frac{3}{10}+\frac{9}{200}\left[\frac{1}{1-\left(\frac{1}{10}\right)^2}\right]$$
$$=\frac{3}{10}+\frac{9}{200}\left[\frac{100}{99}\right]$$
$$=\frac{3}{10}+\frac{1}{22}$$
$$=\frac{19}{55}.$$

Phrases: finite～有限等比级数, infinite～无穷等比级数。

Questions: 1. What is the sum of the following sequence?

$$4, 8, 16, 32, \cdots, 4096$$

2. The sum of all terms in an infinite geometric sequence is 10. If the first term is 6, what is the common ratio?

Answers: 1. Since the common ratio is 2 and the first term is 4, we have $r=2$ and $a=4$. The formula for this geometric sequence is $a_n=4(2)^{n-1}$.

Solving $4(2)^{n-1}=4096$, we have $n=11$. There are 11 terms in this sequence.

Using the formula from Property 2, we have $S=\frac{a(r^n-1)}{r-1}=\frac{4(2^{11}-1)}{2-1}=8188$.

2. Using the formula from Property 3, we have $\frac{6}{1-r}=10$, for which $r=2/5$.

32. Combinatorics 组合学

Review counting and probability in Chapter 11. 可先复习第 11 章的计数与概率。

32.1 Permutation and Combination Review 排列与组合回顾

See permutation and combination in Section 11.1.2 for definitions. 定义见 11.1.2 节的排列与组合。

Formulas: 1. Permutation: The number of ways to arrange r out of the n objects is ${}_nP_r=\frac{n!}{(n-r)!}$. In other words, this is to pick r out of the n objects such that the order of selection matters.

排列：从 n 个物品选出 r 个物品进行排序，可表示为${}_nP_r=\frac{n!}{(n-r)!}$，须考虑选择的顺序。

2. Combination: The number of ways to pick r out of the n objects is ${}_nC_r=\binom{n}{r}=\frac{n!}{r!\,(n-r)!}$, for which the order of selection does not matter.

组合：从 n 个物品选出 r 个物品的方式，可表示为：${}_nC_r=\binom{n}{r}=\frac{n!}{r!\,(n-r)!}$，不考虑选择的顺序。

Properties: 1. $\binom{n}{r}=\binom{n}{n-r}$, for which n and r are nonnegative integers such that $n\geqslant r$.

2. $\binom{n}{0}+\binom{n}{1}+\binom{n}{2}+\cdots+\binom{n}{n}=2^n$.

Proofs: 1. To prove Property 1, interpret the expressions as followed:

$\binom{n}{r}$ represents the number of ways to choose r objects from the n objects.

$\binom{n}{n-r}$ represents the number of ways to choose $n-r$ objects from the n objects.

Of n objects, when r objects are chosen, we have $n-r$ remaining objects. Therefore, choosing r objects to take out is the same as choosing $n-r$ objects to remain, from which $\binom{n}{r}=\binom{n}{n-r}$.

在 n 个物品中，当 r 个物品被选，则余 $n-r$ 个。所以，选走 r 个物品与保留 $n-r$ 个物品是相同的，故可得出 $\binom{n}{r}=\binom{n}{n-r}$。

We can also verify this using formulas:

$$\binom{n}{r}=\frac{n!}{r!\ (n-r)!}.$$

$$\binom{n}{n-r}=\frac{n!}{(n-r)!\ r!}=\frac{n!}{r!\ (n-r)!}.$$

Therefore, we have $\binom{n}{r}=\binom{n}{n-r}$.

2. To prove Property 2, suppose we have n objects.

$\binom{n}{0}+\binom{n}{1}+\binom{n}{2}+\cdots+\binom{n}{n}$ gives the number of ways to choose a combination with any number of objects (all possible choices for which ordering does not matter).

$\binom{n}{0}+\binom{n}{1}+\binom{n}{2}+\cdots+\binom{n}{n}$ 给出了选出任意数量的物品的方法种数(顺序不重要)。

2^n also gives the number of all possible choices for which ordering does not matter—it is based on the Counting Principle, whether an object is chosen or not.

2^n 也给出了选出任意数量的物品的方法(顺序不重要)。它根据计数原理——对每个物品有两种可能性，即选与不选。

Therefore, $\binom{n}{0}+\binom{n}{1}+\binom{n}{2}+\cdots+\binom{n}{n}=2^n$.

32.2 Binomial Theorem 二项式定理

视频 44

Binomial Theorem 二项式定理

n. [baɪˈnoʊmiəl ˈθɪərəm]

Theorem: If n is a natural number, then $(x+y)^n=\sum_{i=0}^{n}\binom{n}{i}x^{n-i}y^i$.

In other words,

$$(x+y)^n=\binom{n}{0}x^ny^0+\binom{n}{1}x^{n-1}y^1+\cdots+\binom{n}{n-1}x^1y^{n-1}+\binom{n}{n}x^0y^n$$
$$=\binom{n}{0}x^n+\binom{n}{1}x^{n-1}y+\cdots+\binom{n}{n-1}xy^{n-1}+\binom{n}{n}y^n$$
$$=x^n+nx^{n-1}y+\cdots+nxy^{n-1}+y^n.$$

For example,

$(x+y)^0=1$, for which $x+y\neq 0$;

$(x+y)^1=x+y$;

$(x+y)^2=x^2+2xy+y^2$;

$(x+y)^3=x^3+3x^2y+3xy^2+y^3$.

…

Proof: We can prove the theorem by simple combinatorics. 可用组合的方法证明该定理。

Note that $(x+y)^n=(x+y)(x+y)(x+y)\cdots$ The RHS has n factors.

Each term in the expanded form of $(x+y)^n$ has the form $x^{n-i}y^i$, from which i is an integer such that $i \leqslant n$. 在$(x+y)^n$ 的一般形式里，每个项都以 $x^{n-i}y^i$ 的形式出现。i 为一个小于或等于 n 的整数。

Each term in the expanded form of $(x+y)^n$ is simply a combination of x's and y's. Each combination is of length n. The coefficient of $x^{n-i}y^i$ is the number of times this term occurs. To get this coefficient, simply count how many ways there are to choose the i number of y's from n number of factors (Each factor is either x or y.). This number is given by $\binom{n}{i}$. 在$(x+y)^n$ 的一般形式里，每个项是一个 x 和 y 的组合，且每个组合的长度为 n。

$x^{n-i}y^i$ 的系数是这个项出现的次数。要求出系数，即求出有多少种方法从 n 个因子里(每个因子均为 x 或 y)选出 i 个含有 y 的项——答案为$\binom{n}{i}$。

Therefore, $(x+y)^n = \sum\limits_{i=0}^{n} \binom{n}{i} x^{n-i}y^i$.

For example, to expand $(x+y)^3$, we have

$$\begin{aligned}(x+y)^3 &= (x+y)(x+y)(x+y)\\ &= (xx+xy+xy+yy)(x+y)\\ &= xxx+xxy+xxy+xyy+xxy+xyy+xyy+yyy.\end{aligned}$$

Each term is a combination of x's and y's. The length of each combination is the exponent, 3.

$$= \cdots x^3 + \cdots x^2y + \cdots xy^2 + \cdots y^3$$

To get the coefficient for each term, count how many ways there are to choose (y's exponent) number of y's from 3 factors.

Therefore, this becomes:

$$\begin{aligned}&= x^3+3x^2y+3xy^2+y^3\\ &= \binom{3}{0}x^3+\binom{3}{1}x^2y+\binom{3}{2}xy^2+\binom{3}{3}y^3.\end{aligned}$$

Notes: Since $\binom{n}{i} = \binom{n}{n-i}$, in the expansion $(x+y)^n$, the terms' coefficients are symmetric.

Above we write the Binomial Theorem as

$$(x+y)^n = \sum_{i=0}^{n} \binom{n}{i} x^{n-i}y^i.$$

In fact, some other valid ways to write the Binomial Theorem are:

(1) $(x+y)^n = \sum\limits_{i=0}^{n} \binom{n}{i} x^i y^{n-i}$

(2) $(x+y)^n = \sum\limits_{i=0}^{n} \binom{n}{n-i} x^{n-i} y^i$

(3) $(x+y)^n = \sum\limits_{i=0}^{n} \binom{n}{n-i} x^i y^{n-i}$.

Extension: Suppose we are doing n independent identical tasks, and the success rate (success probability) of every task is p. The probability of having exactly k successes is $\binom{n}{k}p^k(1-p)^{n-k}$. 假设有 n 个相同的独立任务。每个任务的成功率为 p，k 个任务成功的概率为 $\binom{n}{k}p^k(1-p)^{n-k}$。

$\binom{n}{k}$ represents the number of ways to choose exactly k tasks from the n tasks to succeed. $\binom{n}{k}$ 表示从 n 个任务中选出 k 个成功任务的方法种数。

$p^k \cdot (1-p)^{n-k}$ represents the probability for one such way. $p^k \cdot (1-p)^{n-k}$ 表示一种选法的概率。

Questions: 1. Expand each of the following:

(1) $(x+y)^5$.

(2) $(x+2y)^4$.

(3) $(3x-2y)^3$.

2. A factory manufactures 500 identical light bulbs. The successes/failures for the bulbs are independent of each other. For each bulb, the defection probability is 1%.

(1) What is the probability that exactly 30 bulbs are defected?

(2) What is the probability that at least 2 bulbs are defected?

Answers: 1. (1) $(x+y)^5 = \binom{5}{0}x^5y^0 + \binom{5}{1}x^4y^1 + \binom{5}{2}x^3y^2 + \binom{5}{3}x^2y^3 + \binom{5}{4}x^1y^4 + \binom{5}{5}x^0y^5$

$= x^5 + 5x^4y + 10x^3y^2 + 10x^2y^3 + 5xy^4 + y^5$.

(2) $(x+2y)^4 = \binom{4}{0}x^4(2y)^0 + \binom{4}{1}x^3(2y)^1 + \binom{4}{2}x^2(2y)^2 + \binom{4}{3}x^1(2y)^3 + \binom{4}{4}x^0(2y)^4$

$= x^4 + 4x^3(2y) + 6x^2(4y^2) + 4x^1(8y^3) + x^0(16y^4)$

$= x^4 + 8x^3y + 24x^2y^2 + 32xy^3 + 16y^4$.

(3) $(3x-2y)^3 = (3x+(-2y))^3$

$= \binom{3}{0}(3x)^3(-2y)^0 + \binom{3}{1}(3x)^2(-2y)^1 + \binom{3}{2}(3x)^1(-2y)^2 +$

$\binom{3}{3}(3x)^0(-2y)^3$

$= 27x^3 + 3(9x^2)(-2y) + 3(3x)(4y^2) + (-8y^3)$

$= 27x^3 - 54x^2y + 36xy^2 - 8y^3$.

2. (1) $\binom{500}{30}(0.01)^{30}(0.99)^{470}$.

(2) Suppose X represents the number of defected light bulbs.

$P(X=0) + P(X=1) + P(X=2) + \cdots + P(X=500) = 1$

We want to find $P(X=2) + \cdots + P(X=500)$.

Therefore, we have

$P(X=0) + P(X=1) + [P(X=2) + \cdots + P(X=500)] = 1$

$\binom{500}{0}(0.01)^0(0.99)^{500} + \binom{500}{1}(0.01)^1(0.99)^{499} + [P(X=2) + \cdots + P(X=500)] = 1$

$$0.99^{500}+500(0.01)(0.99)^{499}+[P(X=2)+\cdots+P(X=500)]=1$$
$$P(X=2)+\cdots+P(X=500)=1-0.99^{500}-500(0.01)(0.99)^{499}.$$

32.3 Challenging Questions 思考题

视频 45

Counting and Probability Template Questions 计数和概率的典型问题

Questions: 1. Reading from left to right, for how many integers greater than 9 is true that every digit after the first exceeds the digit it follows? Note: As an example, one such integer is 24 789.

从左往右读,有多少个大于 9 的整数的数位是严格递增的?注:24789 是其中一个例子。

2. If we roll a fair die 5 times and the outcomes are a, b, c, d, and e, respectively, the probability that $a \leqslant b \leqslant c \leqslant d \leqslant e$ is m/n, where m and n are relatively prime positive integers. What is $m+n$?

若掷骰子 5 次,分别得出结果 a、b、c、d、e,则 $a \leqslant b \leqslant c \leqslant d \leqslant e$ 的概率为 m/n,其中 m 和 n 为互质的正整数。求 $m+n$。

3. In how many ways can I place n indistinguishable objects into k distinguishable boxes so that each box contains at least 1 object? Write your answer in terms of n and k. Assume that $n>k$.

有多少种方法可把 n 个相同的物品放进 k 个不同的盒子里,使得每个盒子至少放一个物品?答案用 n 和 k 表示。设 $n>k$。

4. In how many ways can John select k out of the first n positive integers, disregarding the order in which these k numbers are selected, so that no two of the selected integers are consecutive? Write your answer in terms of n and k. Assume that $n>k$.

约翰从最小的 n 个正整数里选出了 k 个。不考虑这 k 个数被选择的顺序,有多少种选法不包含任意两个连续整数?答案用 n 和 k 表示。设 $n>k$。

Answers: 1. Each subset of $\{1,2,3,4,5,6,7,8,9\}$ that contains 2 or more digits has one such ordering. The total number of non-empty subsets of a 9-element set is 2^9-1. Since this total includes 9 one-digit subsets (which represent 9 one-digit numbers), we must subtract 9, making the answer $2^9-1-9=\boxed{502}$.

Alternatively, we can look at the problem this way:

How many ways can we select 2-digit numbers that fit the criterion? For any selection of 2 different digits from $\{1,2,3,4,5,6,7,8,9\}$, there is only one way to arrange them in increasing order. Thus the question becomes: How many ways are there to select 2 different digits from the 9 digits? The answer is $\binom{9}{2}$.

Similarly, there are $\binom{9}{3}$ ways to select 3-digit numbers that fit the criterion, $\binom{9}{4}$ ways to select 4-digit numbers that fit the criterion, and so on.

Our final answer is $\binom{9}{2}+\binom{9}{3}+\cdots+\binom{9}{9}=2^9-\binom{9}{0}+\binom{9}{1}=512-1-9=\boxed{502}$.

Note that $\binom{9}{0}+\binom{9}{1}+\cdots+\binom{9}{9}=2^9$.

解题思路：在集合{1,2,3,4,5,6,7,8,9}里，每个至少有两个数字的子集只有一种排列方法。举例如下：子集{3,5,8,9}对应的是四位数3589。

这道题的要求是在集合{1,2,3,4,5,6,7,8,9}里，求出有多少个至少有两个数字的子集。

2. The main difficulty is that the given inequality is not a strict one. It would be much easier to count the number of solutions if equality could be converted to inequality. Here's one way: We know that $1\leqslant a\leqslant b\leqslant c\leqslant d\leqslant e\leqslant 6$. If $A=a$, $B=b+1$, $C=c+2$, $D=d+3$, and $E=e+4$, it follows that $1\leqslant A<B<C<D<E\leqslant 10$. Every 5-tuple (A,B,C,D,E) gives rises to a solution $(A,B-1,C-2,D-3,E-4)=(a,b,c,d,e)$ whose coordinates satisfy the given inequality. For example, if we choose (A,B,C,D,E) to be $(2,4,7,8,10)$, then $(a,b,c,d,e)=(2,3,5,5,6)$ on the outcomes of the die. Any choice of 5 different integers from 1 to 10 can only be ordered one way in increasing order. The number of such ordered 5-tuples (A,B,C,D,E) is $\binom{10}{5}=252$. Finally, the total possible number of ordered 5-tuples (a,b,c,d,e) is 6^5. The desired probability is $\frac{252}{6^5}=\frac{7}{216}$; $7+216=\boxed{223}$.

解题思路：已知$1\leqslant a\leqslant b\leqslant c\leqslant d\leqslant e\leqslant 6$。根据上述解法把$(a,b,c,d,e)$转换成$(A,B,C,D,E)$，从而把所有小于或等于符号改为小于符号，题目就迎刃而解了。注意每组(A,B,C,D,E)只与一组(a,b,c,d,e)对应，亦是$(A,B-1,C-2,D-3,E-4)$。

As an extension, solve the generalized version of the problem below:

If we roll a fair die n times and the outcomes are $a_1,a_2,\cdots,a_n$ respectively, what is the probability that $a_1\leqslant a_2\leqslant\cdots\leqslant a_n$?

3. Let n_1 be the number of objects in the 1st box. Let n_2 be the number of objects in the 2nd box, and so on. Let n_k be the number of objects in the kth box. We are seeking the number of positive integer solutions $(n_1,n_2,\cdots,n_k)$ that satisfy the equation

$$n_1+n_2+\cdots+n_k=n.$$

Line up the n objects. There are $n-1$ spaces between them. Any choice of $k-1$ of those spaces divides the n objects into k portions, which is equivalent to solving the above equation. The answer is $\boxed{\binom{n-1}{k-1}}$.

解题思路：插板法。把这n个物品排成一列，从中间的$n-1$个空位中插入$k-1$块木板，分为k份(每份至少有一个物品)。第一份放进第一个盒子里，第二份放进第二个盒子里，以此类推。插板的方法种数正是$\binom{n-1}{k-1}$。

4. Method Ⅰ: Imagine that we are going to place n balls in a straight line to represent the first n positive integers. The k integers we will choose will be represented by black balls that must be stuck in among the $n-k$ we don't choose, represented by red balls. There are k black balls and $n-k+1$ spots: the $n-k-1$ that are between red balls, and the 2 that are at the ends of the straight line. Each possible positioning of the k black balls into these $n-k+1$ spots corresponds to choosing k of the first n

positive integers with no consecutive. There are $\boxed{\binom{n-k+1}{k}}$ ways of doing so.

解题思路：把 n 个球排成一行，代表最小的 n 个正整数。设被选的 k 个正整数为黑球，必须被放在 $n-k$ 个不被选的正整数中间（设为红球）。有 k 个黑球和 $n-k+1$ 个空位（$n-k-1$ 个红球中间，2 个在两端）。把这些黑球插入这些空位，有 $\binom{n-k+1}{k}$ 种插法，也是答案。

Method II: Let $a_1, a_2, \cdots, a_k$ be the k integers (no two consecutive) that are chosen at random from $S=\{1,2,\cdots,n\}$, with $a_1 < a_2 < \cdots < a_k$. Create a new set of numbers $A_1=a_1, A_2=a_2-1, A_3=a_3-2, \cdots, A_k=a_k-k+1$, so that $A_1, A_2, A_3, \cdots$, and A_k are k of the first $n-k+1$ positive integers. There is a one-to-one correspondence between each selection of k integers (no two consecutive) from set S and the corresponding set $\{A_1, A_2, A_3, \cdots, A_k\}$ from $\{1,2,3,\cdots,n-k+1\}$. Therefore, there are $\boxed{\binom{n-k+1}{k}}$ ways to choose k out of the first n positive integers such that no two are consecutive.

解题思路：与题 2 的方法相似——把数字转换，从而把“不能连续”的条件转换为“可连续”。

33. Advanced Trigonometry　进阶三角学

In Section 15.11, we studied right triangle trigonometry.　在 15.11 节学习了直角三角形的三角学。In this chapter, we will study trigonometry in general—we will mention the full definitions, relationships, and applications of trigonometric functions.　本章将学习通用的三角学，包括三角函数的完整定义、关系、应用。

33.1 Introduction　介绍

Unit Circle　单位圆

n. [ˈjuːnɪt ˈsɜrkəl]

Definition: The circle on the coordinate plane that is centered at the origin and whose radius is 1.　单位圆是指在坐标平面上圆心为坐标原点，且半径为 1 的圆。

Figure 33-1 shows the unit circle.

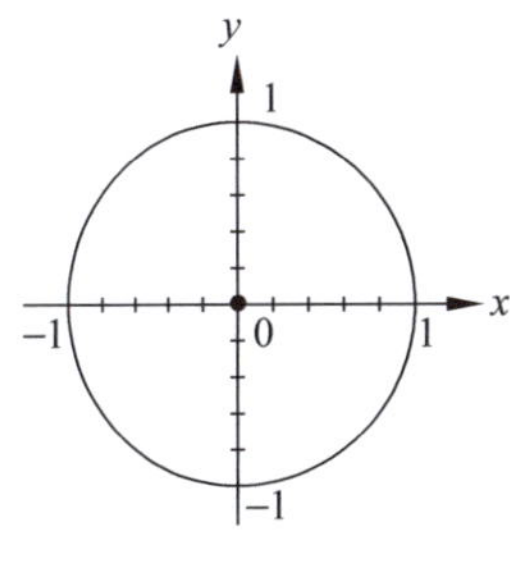

Figure 33-1

Notation: $x^2+y^2=1$. See circle in the conic sections in Section 30.2.

Questions: What are the two points on the unit circle with x-coordinate $\frac{\sqrt{3}}{2}$?

Answers: We have $\left(\frac{\sqrt{3}}{2}\right)^2 + y^2 = 1$, from which $\frac{3}{4} + y^2 = 1$. We have $y^2 = \frac{1}{4}$, from which $y = \pm\frac{1}{2}$.

The two points are $\left(\frac{\sqrt{3}}{2}, \frac{1}{2}\right)$ and $\left(\frac{\sqrt{3}}{2}, -\frac{1}{2}\right)$.

Standard Position of an Angle 角的基准位置

n. [ˈstændərd pəˈzɪʃən ʌv æn ˈæŋgəl]

Definition: The standard position of an angle satisfies all of the following:

角的基准位置满足以下所有条件:

1. The vertex of the angle is at the origin. 角的顶点在坐标原点。
2. The initial side of the angle is the positive x-axis. 角的起始边是正 x 轴。

Figure 33-2 shows the angle θ drawn in standard position. 图 33-2 展示了角 θ 的基准位置。

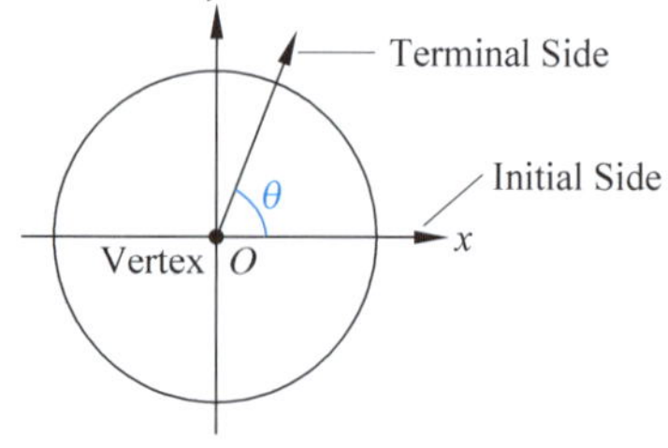

Figure 33-2

Properties: 1. An angle is formed by rotating from the initial side to the terminal side. 角是通过旋转始边至终边形成的。

If the direction of rotation is counterclockwise, then the angle is positive. 若旋转方向为逆时针方向,则角是正的。

If the direction of rotation is clockwise, then the angle is negative. 若旋转方向为顺时针方向,则角是负的。

2. An $d°$-angle is coincident with a $(d+360n)°$-angle, where n is an integer. $d°$ 的角与 $(d+360n)°$ 的角重合,其中 n 为整数。

3. We refer the quadrant an angle is in based on the position of the angle's terminal side. If the terminal side is on a coordinate axis, then it is a quadrantal angle. 一个角在哪个象限取决于它终边的位置。若终边在坐标轴上,则它是轴线角。

Let x be the measure of an angle:

(1) Quadrant Ⅰ Angles: $0° < x < 90°$ 第一象限角

(2) Quadrant Ⅱ Angles: $90° < x < 180°$ 第二象限角

(3) Quadrant Ⅲ Angles: $180° < x < 270°$ 第三象限角

(4) Quadrant Ⅳ Angles: $270° < x < 360°$ 第四象限角

(5) Quadrantal Angles: $0°, 90°, 180°, 270°$ 轴线角

Coincident angles are in the same quadrant.

重合角所在的象限相同。

Examples: Figure 33-3 shows a 225° angle.

Angles that have measures 585°, 945°, 1305°, 1665°, ⋯ have the same terminal side. They are coincident angles.

Angles that have measures −135°, −495°, −855°, −1215°, ⋯ have the same terminal side. They are also coincident angles, as shown in Figure 33-4.

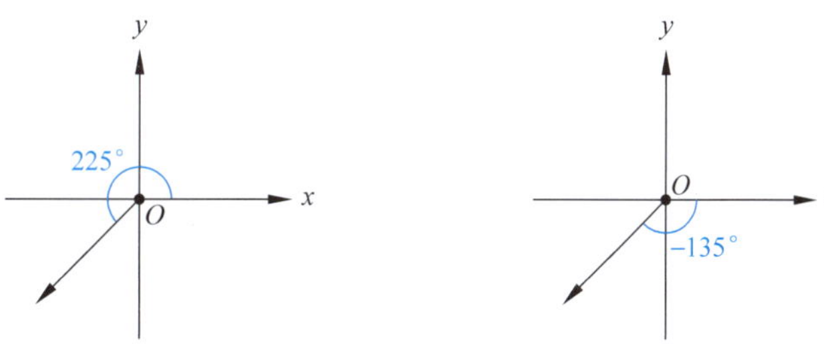

Figure 33-3　　Figure 33-4

Reference Angle　参考角

n. ['refərəns 'æŋgəl]

Definition：The acute angle formed by the terminal side and the x-axis.　参考角是指终边与 x 轴形成的锐角。

Notation：We usually name the reference angle θ.

Properties：1. Reference angles are always acute.　参考角必是锐角。

2. Reference angles always have positive measures.　参考角的度数必为正。

3. Coincident angles always have the same reference angle.　重合角总是有相同的参考角。

4. Figures 33-5 shows the reference angle for an angle in Quadrant Ⅰ.
如图 33-5 所示为第一象限的参考角。
Figures 33-6 shows the reference angle for an angle in Quadrant Ⅱ.
如图 33-6 所示为第二象限的参考角。

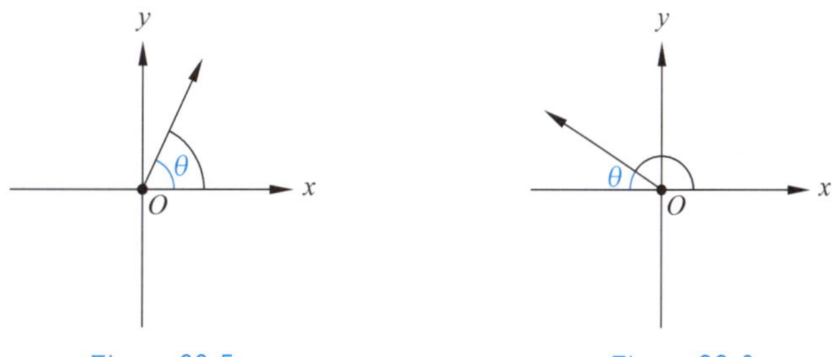

Figure 33-5　　Figure 33-6

Figures 33-7 shows the reference angle for an angle in Quadrant Ⅲ.
如图 33-7 所示为第三象限的参考角。
Figures 33-8 shows the reference angle for an angle in Quadrant Ⅳ.
如图 33-8 所示为第四象限的参考角。

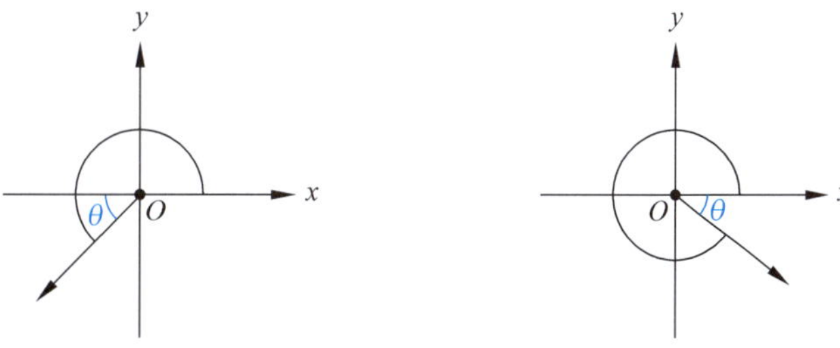

Figure 33-7　　Figure 33-8

Questions: What is the reference angle for each of the following?

(1) 40°

(2) 120°

(3) 250°

(4) 310°

(5) −30°

(6) −200°

Answers: (1) 40°

(2) 60°

(3) 70°

(4) 50°

Let $m\angle A = x°$:

If $\angle A$ is in Quadrant Ⅰ, then the reference angle of $\angle A$ is $x°$.

If $\angle A$ is in Quadrant Ⅱ, then the reference angle of $\angle A$ is $(180 - x)°$.

If $\angle A$ is in Quadrant Ⅲ, then the reference angle of $\angle A$ is $(x - 180)°$.

If $\angle A$ is in Quadrant Ⅳ, then the reference angle of $\angle A$ is $(360 - x)°$.

(5) An angle measured $-30°$ is coincident with an angle measured $-30° + 360° = 330°$. The reference angle of $-30°$ is the same as that of $330°$, which is $30°$.

(6) An angle measured $-200°$ is coincident with an angle measured $-200° + 360° = 160°$. The reference angle of $-200°$ is the same as that of $160°$, which is $20°$.

33.2 Trigonometric Functions 三角函数

视频 46

33.2.1 Definitions 定义

We use trigonometric functions on angles in standard positions. 通常对在基准位置的角应用三角函数。

In this section, suppose θ is an angle in the standard position and $P = (x, y)$ is a point (excluding the origin) on the terminal side of θ. 在本节中,设 θ 为一个在基准位置的角,且 $P = (x, y)$ 是 θ 终边上的一点(不为坐标原点)。

We draw the right triangle as shown in Figure 33-9. 画出直角三角形,如图 33-9 所示。

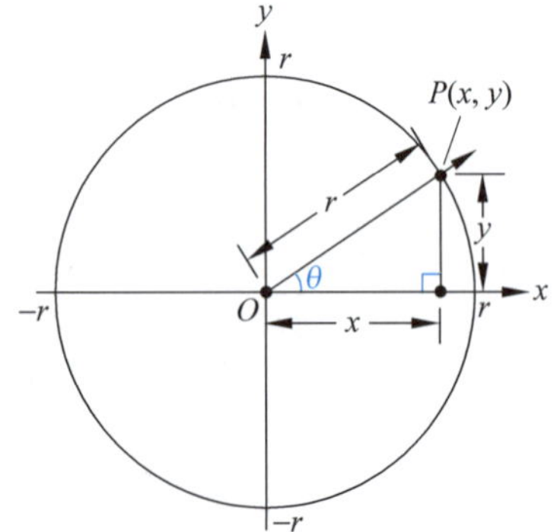

Figure 33-9

Suppose $r = \sqrt{x^2 + y^2}$ by the Pythagorean Theorem. 根据勾股定理,设 $r = \sqrt{x^2 + y^2}$。

Table 33-1 shows the definitions of the six trigonometric functions. Note that "opposite" and "adjacent" are in respect of θ. 表 33-1 展示了 6 个三角函数的定义。注意,公式中的"对边"和"邻边"是相对于 θ 而言。

Table 33-1

Trigonometric Function 三角函数	General Definition 一般定义	Special Case (when P is also a point on the unit circle, from which $r=1$) 特殊情况(当 P 在单位圆上)
$\sin\theta$	$\sin\theta=\dfrac{\text{opposite}}{\text{hypotenuse}}$ $\sin\theta=\dfrac{y}{r}=\dfrac{y}{\sqrt{x^2+y^2}}$	$\sin\theta=y$
$\cos\theta$	$\cos\theta=\dfrac{\text{adjacent}}{\text{hypotenuse}}$ $\cos\theta=\dfrac{x}{r}=\dfrac{x}{\sqrt{x^2+y^2}}$	$\cos\theta=x$
$\tan\theta$	$\tan\theta=\dfrac{\text{opposite}}{\text{adjacent}}$ $\tan\theta=\dfrac{y}{x}=\dfrac{\sin\theta}{\cos\theta}$	$\tan\theta=\dfrac{y}{x}=\dfrac{\sin\theta}{\cos\theta}$
$\csc\theta$	$\csc\theta=\dfrac{\text{hypotenuse}}{\text{opposite}}$ $\csc\theta=\dfrac{r}{y}=\dfrac{\sqrt{x^2+y^2}}{y}$	$\csc\theta=\dfrac{1}{y}$
$\sec\theta$	$\sec\theta=\dfrac{\text{hypotenuse}}{\text{opposite}}$ $\sec\theta=\dfrac{r}{x}=\dfrac{\sqrt{x^2+y^2}}{x}$	$\sec\theta=\dfrac{1}{x}$
$\cot\theta$	$\cot\theta=\dfrac{\text{adjacent}}{\text{opposite}}$ $\cot\theta=\dfrac{x}{y}=\dfrac{\cos\theta}{\sin\theta}$	$\cot\theta=\dfrac{x}{y}=\dfrac{\cos\theta}{\sin\theta}$

Notes: $\sin\theta$ and $\csc\theta$ are reciprocals.
$\cos\theta$ and $\sec\theta$ are reciprocals.
$\tan\theta$ and $\cot\theta$ are reciprocals.
Reciprocals have the same signs.
$\sin\theta$ 和 $\csc\theta$、$\cos\theta$ 和 $\sec\theta$、$\tan\theta$ 和 $\cot\theta$ 互为倒数。
倒数的正负符号相同。
$-1\leqslant\sin\theta\leqslant1$
$-1\leqslant\cos\theta\leqslant1$

Given a Point on the Terminal Side, Find the Outputs of Trigonometric Functions of an Angle 给出角的终边上的一点，求出其三角函数的输出

Questions: Each of the following gives one point P on the terminal side of angle θ drawn in the standard position. Find the values of the outputs of the six trigonometric functions.

(1) $P=\left(\dfrac{1}{2},\dfrac{\sqrt{3}}{2}\right)$

(2) $P=(8,-15)$

Answers: (1) We have $r=\sqrt{\left(\frac{1}{2}\right)^2+\left(\frac{\sqrt{3}}{2}\right)^2}=1$.

P is a point on the unit circle.

$\sin\theta=\frac{y}{r}=y=\frac{\sqrt{3}}{2}$ $\quad$ $\csc\theta=\frac{r}{y}=\frac{1}{y}=\frac{2}{\sqrt{3}}$

$\cos\theta=\frac{x}{r}=x=\frac{1}{2}$ $\quad$ $\sec\theta=\frac{r}{x}=\frac{1}{x}=2$

$\tan\theta=\frac{y}{x}=\frac{\sin\theta}{\cos\theta}=\sqrt{3}$ $\quad$ $\cot\theta=\frac{x}{y}=\frac{\cos\theta}{\sin\theta}=\frac{1}{\sqrt{3}}$

(2) We have $r=\sqrt{8^2+15^2}=17$.

$\sin\theta=\frac{y}{r}=-\frac{15}{17}$ $\quad$ $\csc\theta=\frac{r}{y}=-\frac{17}{15}$

$\cos\theta=\frac{x}{r}=\frac{8}{17}$ $\quad$ $\sec\theta=\frac{r}{x}=\frac{17}{8}$

$\tan\theta=\frac{y}{x}=-\frac{15}{8}$ $\quad$ $\cot\theta=\frac{x}{y}=-\frac{8}{15}$

Given the Measure, Find the Outputs of Trigonometric Functions of an Angle 给出角的度数,求出其三角函数的输出

Procedures: 1. Find a point on the terminal side of the angle (it would be easier if that point is also on the unit circle). 找出角的终边上的一点(如果找出的点在单位圆上,会更加容易计算)。

2. Find the outputs of trigonometric functions according to the definitions. 根据三角函数的定义求出值。

Questions: For each angle below, find the values of the outputs of the six trigonometric functions.

(1) $\theta=60°$

(2) $\theta=495°$

Answers: (1) We locate one point on the terminal side that is also on the unit circle.

We know that $r=1$.

Note that we are constructing a $30°-60°-90°$ triangle, whose side-lengths are in $1:\sqrt{3}:2$ ratio. We have $\theta=60°$, as shown in Figure 33-10.

The point we found is $\left(\frac{1}{2},\frac{\sqrt{3}}{2}\right)$.

We apply the definitions to find the outputs of trigonometric functions.

$\sin\theta=\frac{\sqrt{3}}{2}$ $\quad$ $\csc\theta=\frac{2}{\sqrt{3}}$

$\cos\theta=\frac{1}{2}$ $\quad$ $\sec\theta=2$

$\tan\theta=\sqrt{3}$ $\quad$ $\cot\theta=\frac{1}{\sqrt{3}}$

(2) $\theta = 495°$ is coincident with angle $\varphi = 135°$.

We find that the reference angle of 135° is 45°. We draw the 45° − 45° − 90° triangle as shown in Figure 33-11.

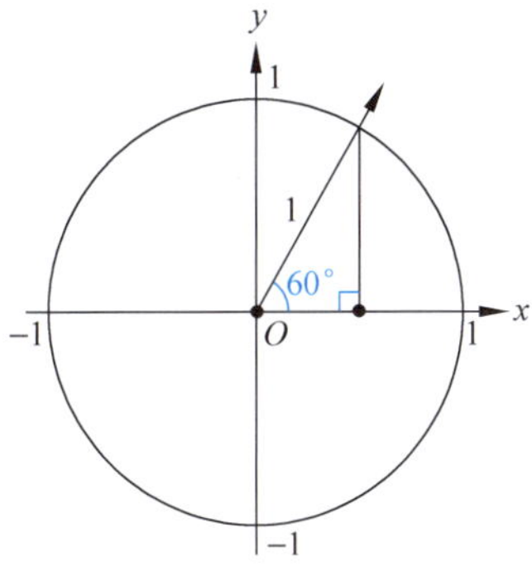

Figure 33-10

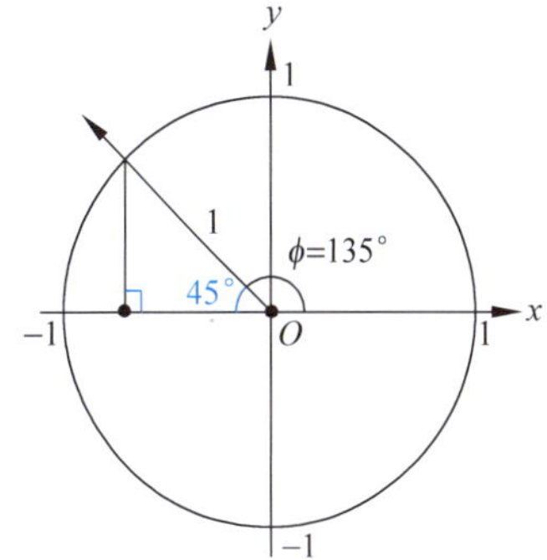

Figure 33-11

The ratio of side-lengths of this triangle is $1:1:\sqrt{2}$. So, the point on the unit circle is $\left(-\frac{1}{\sqrt{2}}, \frac{1}{\sqrt{2}}\right)$, as it is in Quadrant Ⅱ.

We apply the definitions to find the outputs of trigonometric functions.

$\sin\theta = \frac{1}{\sqrt{2}}$　　$\csc\theta = \sqrt{2}$

$\cos\theta = -\frac{1}{\sqrt{2}}$　　$\sec\theta = -\sqrt{2}$

$\tan\theta = -1$　　$\cot\theta = -1$

33.2.2 Properties 性质

ASTC 三角函数的正负规律

Definition: ASTC is the abbreviation of "All, Sine, Tangent, Cosine".

Properties: 1. Table 33-2 shows the signs for sine, cosine, and tangent functions for angles in different quadrants.

表 33-2 给出了不同象限角的正弦、余弦、正切函数的正负符号及其规律。

Table 33-2

Quadrant of θ θ 角所在象限	$\sin\theta$	$\cos\theta$	$\tan\theta$	Pattern 规律
Ⅰ	+	+	+	All Positive
Ⅱ	+	−	−	Sin Positive
Ⅲ	−	−	+	Tan Positive
Ⅳ	+	−	−	Cos Positive

We have another acronym for ASTC—"A Smart Trig Class"!

2. When we compute the outputs of trigonometric functions of an angle, we can compute the outputs of its reference angle, then adjust the signs according to ASTC.

当要算出一个角的三角函数输出时,可以先计算参考角的三角函数输出,再根据 ASTC 调整正负符号。

Questions: For each angle below, find the values of the outputs of the six trigonometric functions.

(1) 150°

(2) 225°

Answers: (1) The reference angle of 150° is 30°.

Applying the principles in "Given the Measure, Find the Outputs of Trigonometric Functions of an Angle", we find that

$\sin 30° = \frac{1}{2}$ $\csc 30° = 2$

$\cos 30° = \frac{\sqrt{3}}{2}$ $\sec 30° = \frac{2}{\sqrt{3}}$

$\tan 30° = \frac{1}{\sqrt{3}}$ $\cot 30° = \sqrt{3}$

Since 150° is in Quadrant Ⅱ, we know that only sin 150° is positive. Its reciprocal, csc 150° is also positive.

Therefore, we have

$\sin 150° = \frac{1}{2}$ $\csc 150° = 2$

$\cos 150° = -\frac{\sqrt{3}}{2}$ $\sec 150° = -\frac{2}{\sqrt{3}}$

$\tan 150° = -\frac{1}{\sqrt{3}}$ $\cot 150° = -\sqrt{3}$

(2) The reference angle of 225° is 45°.

We know that

$\sin 45° = \frac{1}{\sqrt{2}}$ $\csc 45° = \sqrt{2}$

$\cos 45° = \frac{1}{\sqrt{2}}$ $\sec 45° = \sqrt{2}$

$\tan 45° = 1$ $\cot 45° = 1$

Since 225° is in Quadrant Ⅲ, we know that only tan 150° is positive. Its reciprocal, cot 150° is also positive.

Therefore, we have

$\sin 225° = -\frac{1}{\sqrt{2}}$ $\csc 225° = -\sqrt{2}$

$\cos 225° = -\frac{1}{\sqrt{2}}$ $\sec 225° = -\sqrt{2}$

$\tan 225° = 1$ $\cot 225° = 1$

Table of Trigonometric Functions of Some Angles 一些角的三角函数表

This is the trigonometric functions evaluated at 30°, 45°, 60°, and all quadrantal angles.

30°、45°、60°，以及所有轴线角的三角函数。

Trigonometric functions for 30°,45°,and 60° angles are based on the special right triangles (45° − 45° − 90° and 30° − 60° − 90°).

Table 33-3 gives the exact values.

Table 33-3

Trigonometric Functions	0°	30°	45°	60°	90°	180°	270°
sin	0	$1/2$	$1/\sqrt{2}$	$\sqrt{3}/2$	1	0	−1
cos	1	$\sqrt{3}/2$	$1/\sqrt{2}$	$1/2$	0	−1	0
tan	0	$1/\sqrt{3}$	1	$\sqrt{3}$	—	0	—
csc	—	2	$\sqrt{2}$	$2/\sqrt{3}$	1	—	−1
sec	1	$2/\sqrt{3}$	$\sqrt{2}$	2	—	−1	—
cot	—	$\sqrt{3}$	1	$1/\sqrt{3}$	0	—	0

33.2.3 Identities 恒等式

Reciprocal Identities 倒数恒等式

n. [rɪ'sɪprəkəl ɑɪ'dentɪtis]

Identities：(1) $\sin\theta = \dfrac{1}{\csc\theta}$ and $\csc\theta = \dfrac{1}{\sin\theta}$.

(2) $\cos\theta = \dfrac{1}{\sec\theta}$ and $\sec\theta = \dfrac{1}{\cos\theta}$.

(3) $\tan\theta = \dfrac{1}{\cot\theta}$ and $\cot\theta = \dfrac{1}{\tan\theta}$.

This is based on the definition of trigonometric functions.

Pythagorean Identities 毕达哥拉斯三角恒等式

n. [pəˌθægə'riən ɑɪ'dentɪtis]

Identities：(1) $\sin^2\theta + \cos^2\theta = 1$

(2) $1 + \tan^2\theta = \sec^2\theta$

(3) $1 + \cot^2\theta = \csc^2\theta$

Proofs：(1) Proof of Identity (1) (use the definitions of trigonometric functions)：

$$\begin{aligned}\sin^2\theta + \cos^2\theta &= \left(\frac{y}{r}\right)^2 + \left(\frac{x}{r}\right)^2\\ &= \frac{y^2 + x^2}{r^2}\\ &= \frac{r^2}{r^2}\\ &= 1.\end{aligned}$$

(2) Proof of Identity (2)：

Divide both sides of Identity (1) by $\cos^2\theta$.

(3) Proof of Identity (3)：

Divide both sides of Identity (1) by $\sin^2\theta$.

Quotient Identities　商恒等式

n. [ˈkwoʊʃənt ɑɪˈdentɪtis]

Identities：(1) $\tan\theta = \dfrac{\sin\theta}{\cos\theta}$

(2) $\cot\theta = \dfrac{\cos\theta}{\sin\theta}$

This is based on the definition of trigonometric functions.

Even-Odd Identities　奇偶性恒等式

n. [ˈivən ɑd ɑɪˈdentɪtis]

Identities：(1) $\sin(-\theta) = -\sin\theta$

(2) $\cos(-\theta) = \cos\theta$

(3) $\tan(-\theta) = -\tan\theta$

(4) $\csc(-\theta) = -\csc\theta$

(5) $\sec(-\theta) = \sec\theta$

(6) $\cot(-\theta) = -\cot\theta$

This is based on the ASTC properties.

Sum-Difference Formulas　和差公式

n. [sʌm ˈdɪfrəns ˈfɔrmjələs]

Identities：(1) $\sin(u \pm v) = \sin u\cos v \pm \cos u\sin v$

(2) $\cos(u \pm v) = \cos u\cos v \mp \sin u\sin v$

(3) $\tan(u \pm v) = \dfrac{\tan(u) \pm \tan(v)}{1 \mp \tan(u)\tan(v)}$

Proofs：The proof refers to Figures 33-12 and 33-13.

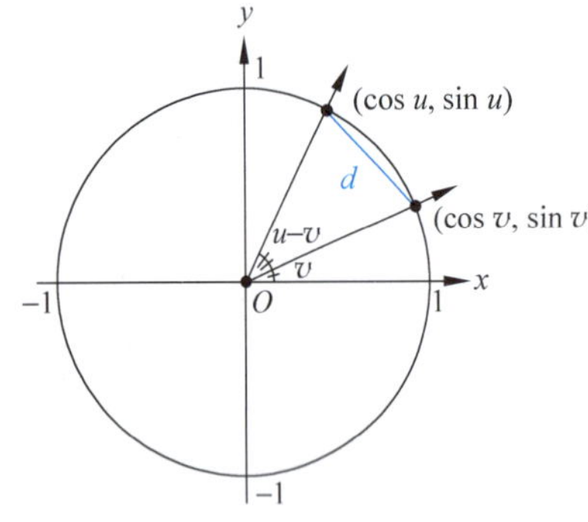

Figure 33-12

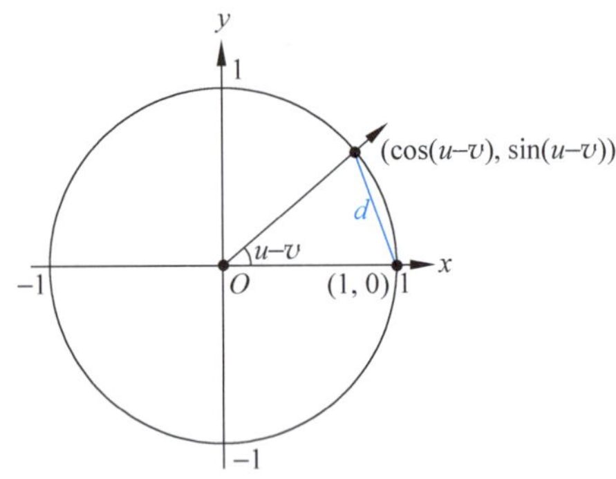

Figure 33-13

Note that the values of d in both diagrams are equal, since the central angle is $u - v$. Using the Distance Formula, from Figure 33-12 we obtain

$$d = \sqrt{(\cos u - \cos v)^2 + (\sin u - \sin v)^2}.$$

From Figure 33-13 we obtain

$$d = \sqrt{(\cos(u - v) - 1)^2 + (\sin(u - v) - 0)^2}.$$

Equating, we get

$$\sqrt{(\cos u-\cos v)^2+(\sin u-\sin v)^2}=\sqrt{(\cos(u-v)-1)^2+(\sin(u-v)-0)^2}.$$

Squaring both sides and simplify eventually gives

$\boxed{\cos(u-v)=\cos u\cos v+\sin u\sin v}$.

Using the formula above, we can show that $\cos(90°-u)=\sin u$, a co-function identity. Using this co-function identity, we can also show that $\cos(90°-(90°-u))=\sin(90°-u)=\cos u$.

Replacing v with $-v$ gives $\boxed{\cos(u+v)=\cos u\cos v-\sin u\sin v}$.

Note that sine is an odd function and cosine is an even function.

From here, replacing u by $90°-u$ gives $\cos(90°-u+v)=\cos(90°-u)\cos v-\sin(90°-u)\sin v$.

We have mentioned that $\sin(90°-u)=\cos u$ and $\cos(90°-u)=\sin u$. Therefore, this is $\cos(90°-(u-v))=\sin u\cos v-\cos u\sin v$, which is $\boxed{\sin(u-v)=\sin u\cos v-\cos u\sin v}$.

From here, replacing v with $-v$ gives $\boxed{\sin(u+v)=\sin u\cos v+\cos u\sin v}$.

We can use the sum-difference formula of sine and cosine to prove that of tangent. We use the definition of the tangent function:

$$\begin{aligned}\tan(u+v)&=\frac{\sin(u+v)}{\cos(u+v)}=\frac{\sin u\cos v+\cos u\sin v}{\cos u\cos v-\sin u\sin v}\\&=\frac{\sin u\cos v+\cos u\sin v}{\cos u\cos v-\sin u\sin v}\div\frac{\cos u\cos v}{\cos u\cos v}\\&=\frac{(\sin u\cos v+\cos u\sin v)\div(\cos u\cos v)}{(\cos u\cos v-\sin u\sin v)\div(\cos u\cos v)}\\&=\frac{\frac{\sin u}{\cos u}+\frac{\sin v}{\cos v}}{1-\frac{\sin u\sin v}{\cos u\cos v}}=\frac{\tan(u)+\tan(v)}{1-\tan(u)\tan(v)}.\end{aligned}$$

Replacing $(u+v)$ with $(u-v)$ produces

$$\tan(u-v)=\frac{\tan(u)-\tan(v)}{1+\tan(u)\tan(v)}.$$

Corollaries: 1. Co-function Identities

2. $\sin(180°-\theta)=\sin\theta$

$\cos(180°-\theta)=-\cos\theta$

$\tan(180°-\theta)=-\tan\theta$

$\csc(180°-\theta)=\csc\theta$

$\sec(180°-\theta)=-\sec\theta$

$\cot(180°-\theta)=-\cot\theta$

Co-Function Identities　余函数恒等式

n. [kəʊ'fʌŋkʃən aɪ'dentɪtis]

Identities: (1) $\sin(90°-\theta)=\cos\theta$

(2) $\cos(90°-\theta)=\sin\theta$

(3) $\tan(90°-\theta)=\cot\theta$

(4) $\csc(90° - \theta) = \sec\theta$

(5) $\sec(90° - \theta) = \csc\theta$

(6) $\cot(90° - \theta) = \tan\theta$

Proofs: Refer to right triangle trigonometric functions in Section 15.11.3

According to definitions, in a right triangle, one acute angle's sine is the other acute angle's cosine; one acute angle's tangent is the other acute angle's cotangent. 根据定义,在直角三角形里,一个锐角的正弦等于另一个锐角的余弦。一个锐角的正切等于另一个锐角的余切。

Or, we can use the sum-difference formulas to prove each of these. 或者可以用和差公式逐一证明。

Double Angle Formulas 二倍角公式

n. [ˈdʌbəl ˈæŋgəl ˈfɔrmjələs]

Identities: (1) $\sin(2u) = 2\sin u\cos u$

(2) $\cos(2u) = \cos^2 u - \sin^2 u = 2\cos^2 u - 1 = 1 - 2\sin^2 u$

(3) $\tan(2u) = \dfrac{2\tan(u)}{1 - \tan^2 u}$

Notes: We can prove these formulas using the sum-difference formulas.

Power-Reducing Formulas 降幂公式

n. [ˈpɑʊər rɪˈdusɪŋ ˈfɔrmjələs]

Identities: (1) $\sin^2 u = \dfrac{1 - \cos(2u)}{2}$

(2) $\cos^2 u = \dfrac{1 + \cos(2u)}{2}$

(3) $\tan^2 u = \dfrac{1 - \cos(2u)}{1 + \cos(2u)}$

Notes: We can prove Identities (1) and (2) by applying the Double Angle Formulas—deriving the LHS from the RHS.

Identity (3) is the quotient of Identity (1) and Identity (2).

Half-Angle Formulas 半角公式

n. [hæf ˈæŋgəl ˈfɔrmjələs]

Identities: (1) $\sin\dfrac{u}{2} = \pm\sqrt{\dfrac{1 - \cos u}{2}}$

(2) $\cos\dfrac{u}{2} = \pm\sqrt{\dfrac{1 + \cos u}{2}}$

(3) $\tan\dfrac{u}{2} = \pm\sqrt{\dfrac{1 - \cos u}{1 + \cos u}} = \dfrac{\sin u}{1 + \cos u} = \dfrac{1 - \cos u}{\sin u}$

Proofs: (1) Note that by the double-angle formula, $\cos(2u) = 1 - 2\sin^2 u$. Replacing u with $\dfrac{u}{2}$ gives $\cos(u) = 1 - 2\sin^2\dfrac{u}{2}$. Isolating the sine term, we get $\sin^2\dfrac{u}{2} = \dfrac{1 - \cos u}{2}$, from which $\sin\dfrac{u}{2} = \pm\sqrt{\dfrac{1 - \cos u}{2}}$.

The sign of $\pm\sqrt{\frac{1-\cos u}{2}}$ depends on the sign of $\sin\frac{u}{2}$. Recall that $\sin\frac{u}{2}$ is positive when $\frac{u}{2}$ is in Quadrants Ⅰ or Ⅱ, negative when $\frac{u}{2}$ is in Quadrants Ⅲ or Ⅳ.

(2) Note that by the double-angle formula, $\cos(2u)=2\cos^2 u-1$. Replacing u with $\frac{u}{2}$ gives $\cos(u)=2\cos^2\frac{u}{2}-1$. Isolating the cosine-square term, we get $\cos^2\frac{u}{2}=\frac{1+\cos u}{2}$, from which $\cos\frac{u}{2}=\pm\sqrt{\frac{1+\cos u}{2}}$.

(3) $\tan\frac{u}{2}=\dfrac{\sin\frac{u}{2}}{\cos\frac{u}{2}}=\pm\sqrt{\dfrac{1-\cos u}{1+\cos u}}$

$$\tan\frac{u}{2}=\frac{\sin\frac{u}{2}}{\cos\frac{u}{2}}=\frac{\sin\frac{u}{2}}{\cos\frac{u}{2}}\cdot\frac{2\cos\frac{u}{2}}{2\cos\frac{u}{2}}=\frac{2\sin\frac{u}{2}\cos\frac{u}{2}}{2\cos^2\frac{u}{2}}=\frac{\sin u}{2\left(\frac{1+\cos u}{2}\right)}=\frac{\sin u}{1+\cos u}$$

$$\tan\frac{u}{2}=\frac{\sin\frac{u}{2}}{\cos\frac{u}{2}}=\frac{\sin\frac{u}{2}}{\cos\frac{u}{2}}\cdot\frac{-2\sin\frac{u}{2}}{-2\sin\frac{u}{2}}=\frac{-2\sin^2\frac{u}{2}}{-2\sin\frac{u}{2}\cos\frac{u}{2}}=\frac{-2\left(\frac{1-\cos u}{2}\right)}{-\sin u}=\frac{-(1-\cos u)}{-\sin u}$$

$$=\frac{1-\cos u}{\sin u}$$

Product-to-Sum Formulas 积化和差公式

n. [ˈprɑdəkt tu sʌm ˈfɔrmjələs]

Identities: (1) $\sin u\sin v=\frac{1}{2}[\cos(u-v)-\cos(u+v)]$

(2) $\cos u\cos v=\frac{1}{2}[\cos(u-v)+\cos(u+v)]$

(3) $\sin u\cos v=\frac{1}{2}[\sin(u+v)+\sin(u-v)]$

(4) $\cos u\sin v=\frac{1}{2}[\sin(u+v)-\sin(u-v)]$

Proofs: Note that from the sum-difference formulas:

A. $\sin(u+v)=\sin u\cos v+\cos u\sin v$

B. $\sin(u-v)=\sin u\cos v-\cos u\sin v$

C. $\cos(u+v)=\cos u\cos v-\sin u\sin v$

D. $\cos(u-v)=\cos u\cos v+\sin u\sin v$

(1) Perform $D-C$ and isolate the $\sin u\sin v$ term.

(2) Perform $C+D$ and isolate the $\cos u\cos v$ term.

(3) Perform $A+B$ and isolate the $\sin u\cos v$ term.

(4) Perform $A-B$ and isolate the $\cos u\sin v$ term.

Sum-to-Product Formulas 和差化积公式

n. [sʌm tu 'prɑdəkt 'fɔrmjələs]

Identities: (1) $\sin u+\sin v=2\sin\left(\frac{u+v}{2}\right)\cos\left(\frac{u-v}{2}\right)$

(2) $\sin u-\sin v=2\sin\left(\frac{u-v}{2}\right)\cos\left(\frac{u+v}{2}\right)$

(3) $\cos u+\cos v=2\cos\left(\frac{u+v}{2}\right)\cos\left(\frac{u-v}{2}\right)$

(4) $\cos u-\cos v=-2\sin\left(\frac{u+v}{2}\right)\sin\left(\frac{u-v}{2}\right)$

Proofs: Let $A=\frac{u+v}{2}$ and $B=\frac{u-v}{2}$. We have $A+B=u$ and $A-B=v$.

From product-to-sum formulas, we know that

$$\sin A\cos B=\frac{1}{2}[\sin(A+B)+\sin(A-B)]$$

$$\sin\left(\frac{u+v}{2}\right)\cos\left(\frac{u-v}{2}\right)=\frac{1}{2}[\sin u-\sin v]$$

$$2\sin\left(\frac{u+v}{2}\right)\cos\left(\frac{u-v}{2}\right)=\sin u-\sin v$$

$$\sin u-\sin v=2\sin\left(\frac{u+v}{2}\right)\cos\left(\frac{u-v}{2}\right)\quad\text{(Formula 2)}$$

The other formulas can be proved in similar ways.

33.2.4 Graphs of Trigonometric Functions 三角函数的图像

Amplitude 振幅

n. ['æmplɪˌtud]

Definition: For sine and cosine functions, the amplitude is half the positive difference between the minimum and maximum values of the range.

对正弦和余弦函数，振幅是指值域间最小值与最大值正差的一半。

Examples: 1. In the graph of $y=\sin x$, the minimum value is -1 and the maximum value is 1. Therefore, the amplitude is $\frac{1-(-1)}{2}=1$.

2. In the graph of $y=4\cos x$, the minimum value is -4 and the maximum value is 4. Therefore, the amplitude is $\frac{4-(-4)}{2}=4$.

3. In the graph of $y=4-3\cos x$, the minimum value is 1 and the maximum value is 7. Therefore, the amplitude is $\frac{7-1}{2}=3$.

Period 周期

n. ['pɪəriəd]

Definition: The minimum distance required for the function to complete one full cycle. 周期是指函数完成一个循环的最短所需距离。

Examples: The functions $y_1=\sin x$ and $y_2=\cos x$ have periods of 360°.

Graphs of Sine, Cosine, and Tangent Functions　正弦、余弦、正切函数的图像

Properties: We plot the points using the Table for Trigonometric Functions of Some Angles and ASTC in Section 33.2.2.　根据 33.2.2 节"一些角的三角函数表"以及"ASTC 三角函数的正负规律"画点。

Figure 33-14 shows the graph $y = \sin x$.

Figure 33-15 shows the graph $y = \cos x$.

Figure 33-16 shows the graph $y = \tan x$.

As a convention, there are eight increments on the x-axis. The rightmost increment equals the period number.　根据惯例,在 x 轴上有八个增量,最右边的一个增量为周期数字。

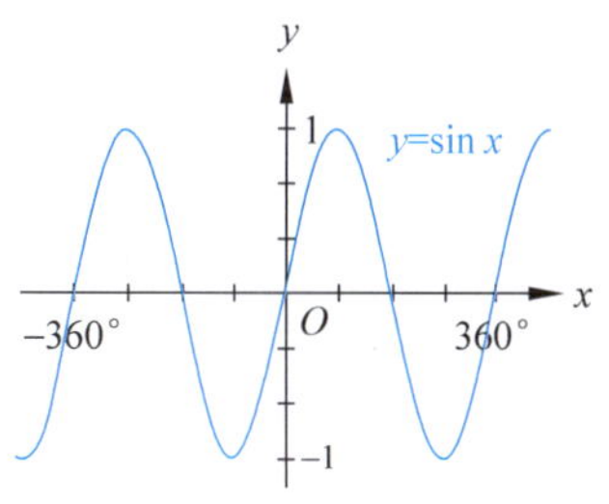

Figure 33-14

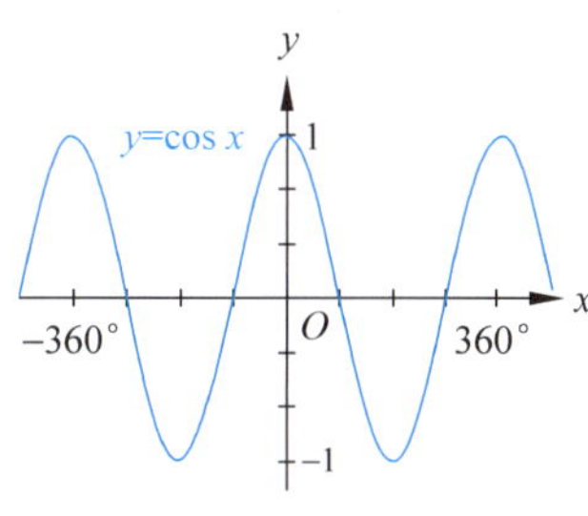

Figure 33-15

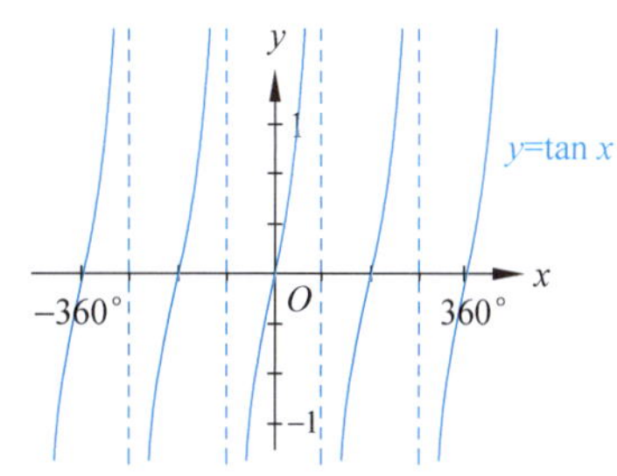

Figure 33-16

1. The graphs of sine, cosine, and tangent behave periodically. On the graphs of $\sin x$ and $\cos x$, the amplitude is 1 and the period is 360°.

   Domain: $(-\infty, \infty)$, or all real numbers

   Range: $[-1, 1]$

2. In the functions $y_1 = A\sin(Bx + C) + D$ and $y_2 = A\cos(Bx + C) + D$:

   Amplitude: $|A|$

   Period: $360°/|B|$

   Horizontal Stretch: by a factor of $1/|B|$

   Horizontal Shift: $-C/B$

   Vertical Shift: D

   Domain: $(-\infty, \infty)$, or all real numbers

   Range: $[-|A|, |A|]$

3. In the function $y = \tan x = \dfrac{\sin x}{\cos x}$, there are vertical asymptotes at $x = 90° + 180° n$, from which n is an integer.

   Period: 180°

   Vertical Asymptotes: $x = 90° + 180° n$

   Domain: all real numbers except $x = 90° + 180° n$

   Range: $(-\infty, \infty)$

4. In the function $y = A\tan(Bx + C) + D$:

   Period: $180°/|B|$

   Horizontal Stretch: by a factor of $1/|B|$

   Horizontal Shift: $-C/B$

   Vertical Stretch: by a factor of $|A|$

Vertical Shift: D

Vertical Asymptotes: $x=\frac{90°}{B}+\frac{180°\ n}{B}-\frac{C}{B}$

Domain: all real numbers except $x=\frac{90°}{B}+\frac{180°\ n}{B}-\frac{C}{B}$

Range: $(-\infty,\infty)$

Graphs of Cosecant, Secant, and Cotangent Functions 余割、正割、余切函数的图像

Properties: 1. Figure 33-17 shows the graph for $y=\csc\ x$:

Domain: all real numbers except $x=180°\ n$

Range: $(-\infty,-1]\cup[1,\infty)$

Vertical Asymptotes: $x=180°\ n$

Extrema: at $x=90°+180°\ n$

2. Figure 33-18 shows the graph for $y=\sec\ x$:

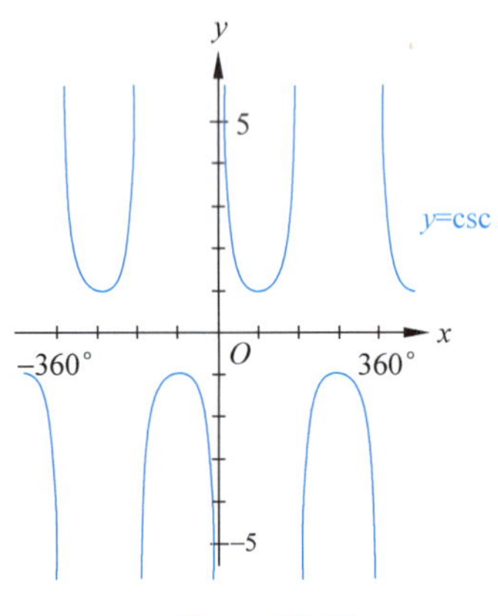

Figure 33-17

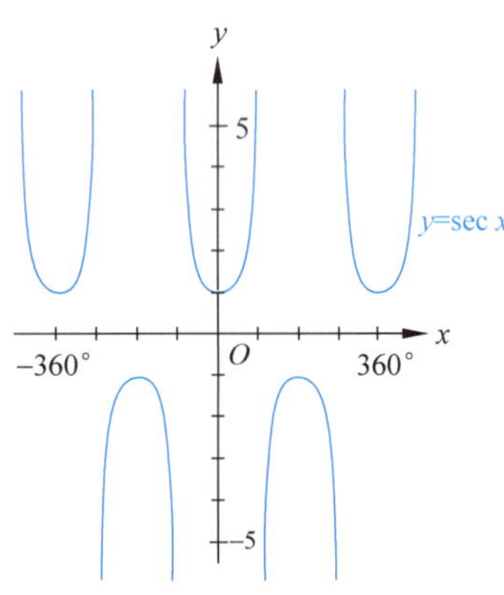

Figure 33-18

Domain: all real numbers except $x=90°+180°\ n$

Range: $(-\infty,-1]\cup[1,\infty)$

Vertical Asymptotes: $x=90°+180°\ n$

Extrema: at $x=180°\ n$

3. Figure 33-19 shows the graph for $y=\cot\ x$:

Domain: all real numbers except $x=180°\ n$

Range: $(-\infty,\infty)$

Vertical Asymptotes: $x=180°\ n$

Zeros: at $x=90°+180°\ n$

4. Refer to The Graphs of Reciprocal Functions in Section 29.3. 见29.3节倒数函数的图像。

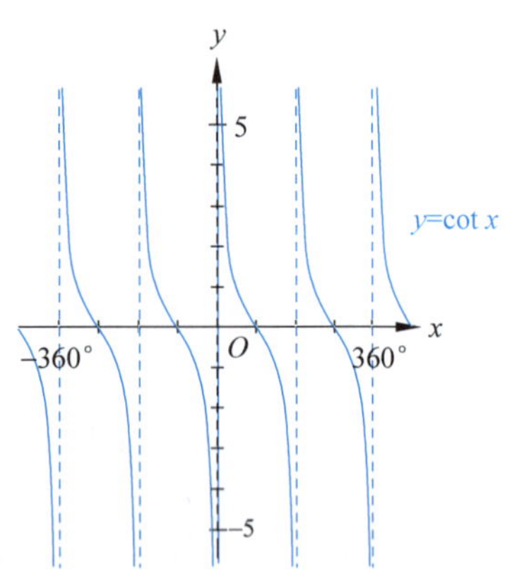

Figure 33-19

Graphs of Arcsine, Arccosine, and Arctangent Functions 反正弦、反余弦、反正切函数的图像

Notes: $f(x)=\sin x$, $g(x)=\cos x$, and $h(x)=\tan x$ are not one-to-one functions.

正弦、余弦、正切都不是一对一函数。

Their inverse functions $f^{-1}(x)=\arcsin x$, $g^{-1}(x)=\arccos x$, and $h^{-1}(x)=\arctan x$ have range restrictions.

它们的反函数：反正弦、反余弦、反正切都有值域限制。

Properties：1. For $y=\arcsin x$：

Domain：$[-1,1]$

Range：$[-90°,90°]$

The graph is shown in Figure 33-20.

2. For $y=\arccos x$：

Domain：$[-1,1]$

Range：$[0°,180°]$

The graph is shown in Figure 33-21.

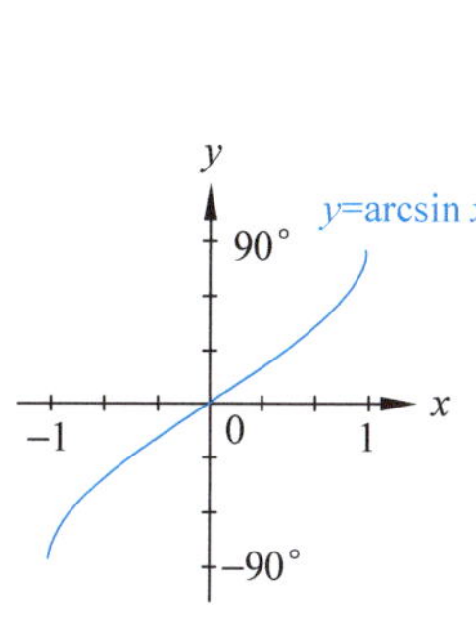

Figure 33-20

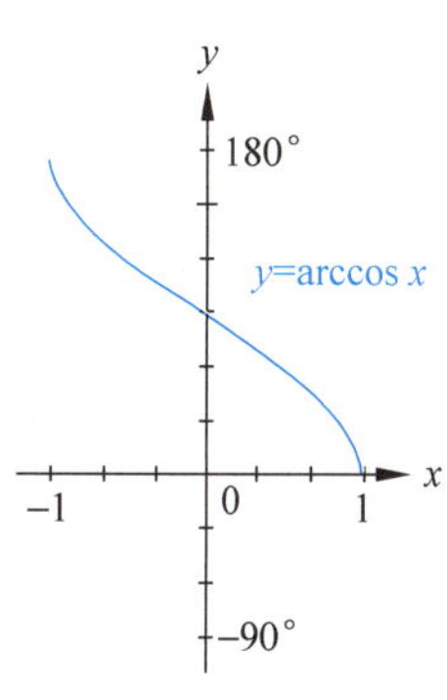

Figure 33-21

3. For $y=\arctan x$：

Domain：$(-\infty,\infty)$

Range：$(-90°,90°)$

The graph is shown in Figure 33-22.

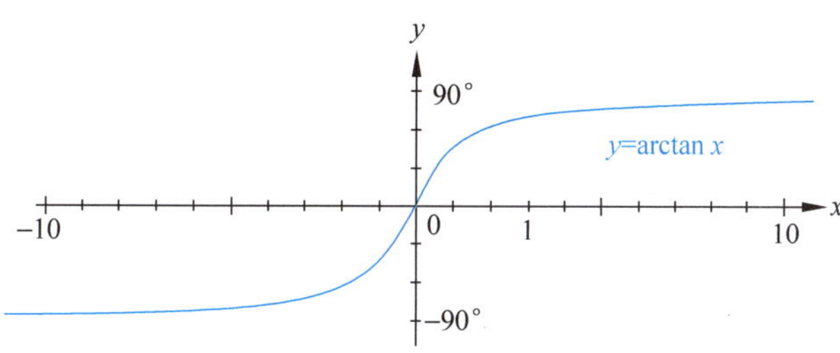

Figure 33-22

33.3 Applications of Trigonometry 三角学的应用

Law of Sines 正弦定理

n. [lɔ ʌv saɪns]

视频 47

Definition：For every $\triangle ABC$，the following is true：

$$\frac{a}{\sin A}=\frac{b}{\sin B}=\frac{c}{\sin C}.$$

Proofs：For right triangles，we can use the right triangle trigonometry，which shows the result easily.

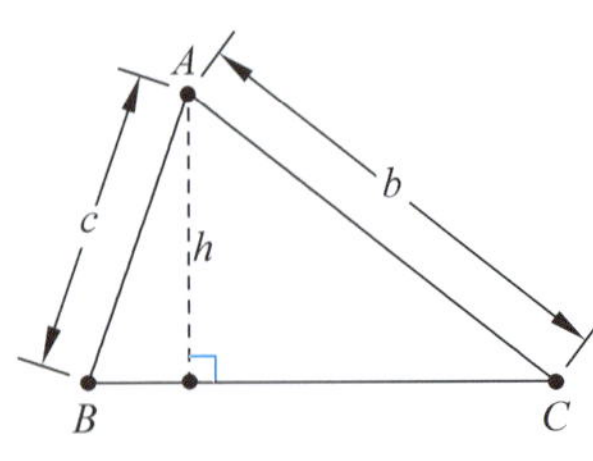

Figure 33-23

We will focus on acute triangles and obtuse triangles.

Acute Triangles:

Drop an altitude from vertex A, as shown in Figure 33-23.

We get that $\sin B=\frac{h}{c}$ and $\sin C=\frac{h}{b}$, which is $h=c\sin B$ and $h=b\sin C$.

By transitive, $c\sin B=b\sin C$.

Dividing this equation by $\sin B\sin C$ gives

$$\frac{c}{\sin C}=\frac{b}{\sin B}. \tag{1}$$

Dropping an altitude from vertex B and repeating the same process gives

$$\frac{c}{\sin C}=\frac{a}{\sin A}. \tag{2}$$

By transitive on Formula (1) and Formula (2), we have $\frac{a}{\sin A}=\frac{b}{\sin B}=\frac{c}{\sin C}$.

Obtuse Triangles:

Suppose A is the obtuse angle.

Drop an altitude from vertex A and repeating the same process gives

$$\frac{c}{\sin C}=\frac{b}{\sin B}. \tag{3}$$

Drop an altitude from vertex B, as shown in Figure 33-24. Note that $\sin(180°-A)=\sin A$. See sum-difference identities in Section 33.2.3. We have $\sin A=\frac{h}{c}$, or $h=c\sin A$.

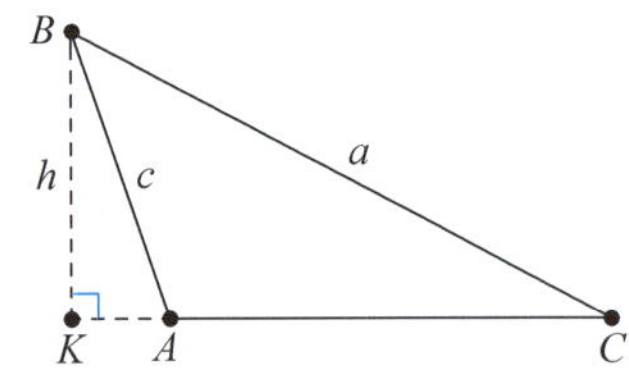

Figure 33-24

In the larger $\triangle CBK$, we have $\sin C=\frac{h}{a}$, or $h=a\sin C$.

By transitive, we have $c\sin A=a\sin C$. Dividing both sides by $\sin A\sin C$ gives

$$\frac{c}{\sin C}=\frac{a}{\sin A}. \tag{4}$$

By transitive on Formula (3) and Formula (4), we have $\frac{a}{\sin A}=\frac{b}{\sin B}=\frac{c}{\sin C}$.

Notes: We use the Law of Sines to solve a triangle when the following information are given: AAS, ASA, SSA.

For the case SSA, one of the three scenarios can occur:

1. No such triangle exists.
2. Two different triangles exist.
3. Exactly one triangle exists.

Law of Cosines 余弦定理

n. [lɔ ʌv 'koʊˌsaɪns]

Definition: For every $\triangle ABC$, the following equations are true:

$$a^2=b^2+c^2-2bc\cos A$$
$$b^2=a^2+c^2-2ac\cos B$$

$$c^2 = a^2 + b^2 - 2ab\cos C$$

Proofs: For an acute $\triangle ABC$, we drop an altitude $\overline{CD}$ for which $AD = x$ and $DB = c - x$, as shown in Figure 33-25.

From right $\triangle ADC$, we know that

$$x^2 + h^2 = b^2 \tag{5}$$

$$\cos A = \frac{x}{b} \quad \Rightarrow \quad x = b\cos A. \tag{6}$$

From right $\triangle BDC$, we know that

$$(c - x)^2 + h^2 = a^2 \quad \Rightarrow \quad c^2 - 2cx + x^2 + h^2 = a^2. \tag{7}$$

Substituting Formulas (5) and (6) into Formula (7) gives

$$c^2 - 2c(b\cos A) + b^2 = a^2.$$

This is equivalent to

$$a^2 = b^2 + c^2 - 2bc\cos A.$$

We can repeat the similar process to the altitudes dropped from the other angles.

For an obtuse $\triangle ABC$, we drop an altitude $\overline{CD}$ for which $DA = x$ and $AB = c$, as shown in Figure 33-26.

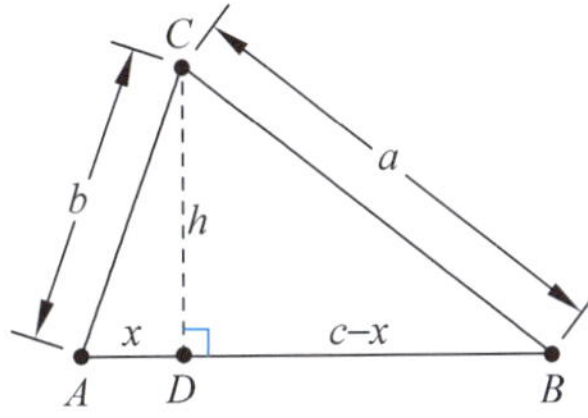

Figure 33-25

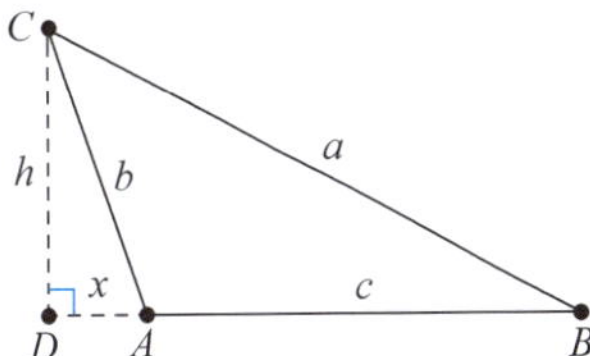

Figure 33-26

For obtuse triangles:

From right $\triangle ADC$, we know that

$$x^2 + h^2 = b^2 \tag{8}$$

$$\cos(180° - A) = -\cos A = \frac{x}{b} \quad \Rightarrow \quad x = -b\cos A. \tag{9}$$

Refer to sum-difference identities that $\cos(180° - A) = -\cos A$.

From right $\triangle BDC$, we know that

$$(c + x)^2 + h^2 = a^2 \Rightarrow c^2 + 2cx + x^2 + h^2 = a^2. \tag{10}$$

Substituting Formulas (8) and (9) into Formula (10) gives

$$c^2 - 2c(b\cos A) + b^2 = a^2.$$

This is equivalent to

$$a^2 = b^2 + c^2 - 2bc\cos A.$$

Notes: We use the Law of Cosines to solve a triangle when the following information are given: SSS and SAS.

Heron's Formula 海伦公式

n. [ˈherən ˈfɔrmjələ]

Definition: The formula that finds a triangle's area given the lengths of the three sides.

Formula: The area of a triangle T is $A=\sqrt{s(s-a)(s-b)(s-c)}$, where a, b, and c are the side-lengths of T and s is the semi-perimeter: $s=\frac{P}{2}=\frac{a+b+c}{2}$.

Proof: The proof of Heron's Formula is very cumbersome and involves with algebra heavily. We will go through the key steps here.

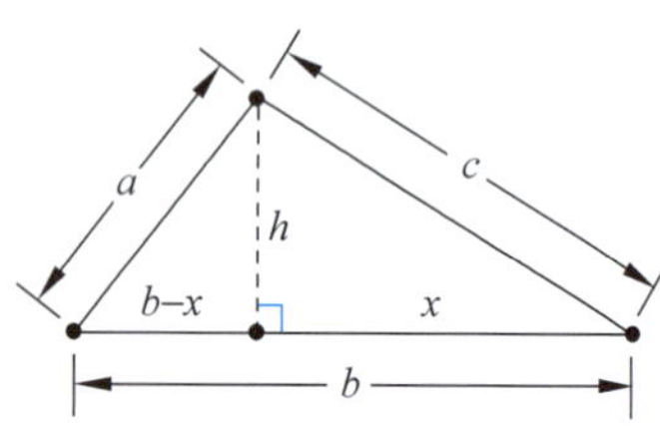

Figure 33-27

In a triangle as shown in Figure 33-27, we drop the altitude from the largest angle so that the altitude lies within the triangle.

The area is $A=\frac{1}{2}bh$. We wish to write h in terms of a, b, and c.

By the Pythagorean Theorem, we know that $x=\sqrt{c^2-h^2}$ and $(b-x)^2+h^2=a^2$.

Expanding the latter equation gives $b^2-2bx+x^2+h^2=a^2$. Now substitute the values x and x^2: $b^2-2b\sqrt{c^2-h^2}+(c^2-h^2)+h^2=a^2$.

Manipulating gives $b^2+c^2-a^2=2b\sqrt{c^2-h^2}$. Now square both sides and solve for h. The most cumbersome part is here, for which we will skip.

We will get: $h=\frac{\sqrt{P(P-2a)(P-2b)(P-2c)}}{2b}$, from which P is the perimeter: $P=a+b+c$.

Finally, $A=\frac{1}{2}bh=\frac{1}{2}b\frac{\sqrt{P(P-2a)(P-2b)(P-2c)}}{2b}=\frac{\sqrt{P(P-2a)(P-2b)(P-2c)}}{4}=$
$\sqrt{\frac{1}{16}P(P-2a)(P-2b)(P-2c)}=\sqrt{\frac{P}{2}\left(\frac{P-2a}{2}\right)\left(\frac{P-2b}{2}\right)\left(\frac{P-2c}{2}\right)}=$
$\sqrt{\frac{P}{2}\left(\frac{P}{2}-a\right)\left(\frac{P}{2}-b\right)\left(\frac{P}{2}-c\right)}=\sqrt{s(s-a)(s-b)(s-c)}$.

33.4 Radians 弧度

Radian 弧度

n. [ˈreɪdiən]

Definition: Another unit of measuring angles: In a circle, the central angle, in radians, is equal to the ratio of the length of the intercepted arc to the length of the radius. 弧度是角的另一个测量单位。圆的中心角等于截弧的长度与半径的长度之比。

As shown in Figure 33-28, we have

$$\theta=\frac{s}{r}$$

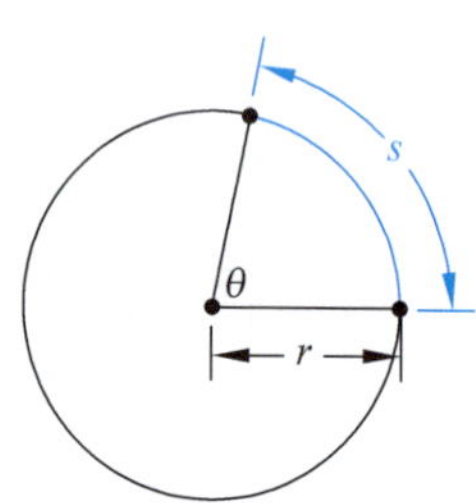

Figure 33-28

A full rotation consists of 360° ($s=2\pi r$), which is equivalent to 2π radians.

Notation: Without the degree sign "°", it is understood that an angle is expressed in radians.

Properties: 1. Radians and degrees vary directly. We know that $\frac{360°}{2\pi}=\frac{180°}{\pi}=1$.

2. (1) The conversion from degrees to radians is

$$(\text{angle in degrees}) \cdot \frac{\pi}{180^\circ}.$$

(2) The conversion from radians to degrees is

$$(\text{angle in radians}) \cdot \frac{180^\circ}{\pi}.$$

Questions: 1. Convert each of the following from degrees to radians.

(1) 45°

(2) 150°

(3) 270°

(4) 330°

2. Convert each of the following from radians to degrees.

(1) $\frac{2\pi}{3}$

(2) $\frac{3\pi}{4}$

(3) $\frac{7\pi}{6}$

(4) $\frac{11\pi}{12}$

Answers: 1. Multiply each angle by $\frac{\pi}{180^\circ}$.

(1) $\frac{\pi}{4}$

(2) $\frac{5\pi}{6}$

(3) $\frac{3\pi}{2}$

(4) $\frac{11\pi}{6}$

2. Multiply each angle by $\frac{180^\circ}{\pi}$.

(1) 120°

(2) 135°

(3) 210°

(4) 165°

33.5 Polar Coordinate System　极坐标系

Polar Coordinate System　极坐标系

n. ['poʊlər koʊ'ɔrdən,eɪts 'sɪstəm]

Definition: A two-dimensional coordinate system in which each point on a plane is determined by a distance from a reference point and an angle from a reference direction.　极坐标系是

二维坐标系,以离一个参考点的距离和一个参考方向的角度确定一个点。

The reference point is called the pole (analogue of the origin of the Cartesian Plane).

参考点又称为极,与坐标平面的坐标原点类似。

The polar coordinate system is shown in Figure 33-29.

图 33-29 展示了极坐标系。

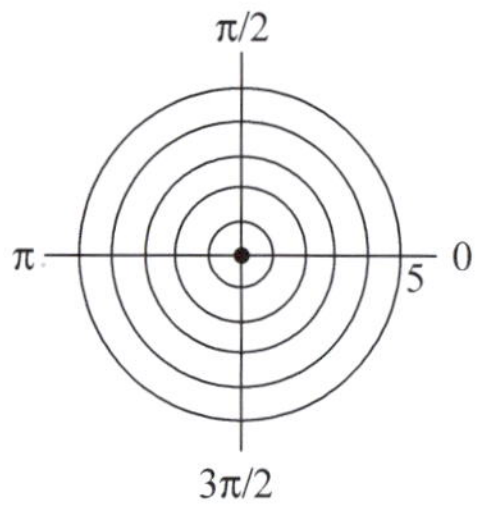

Figure 33-29

Notation: Each point in the polar coordinate system is expressed by (r, θ), from which r is the distance and θ is the angle. We presumably use radian for θ.

极坐标系的每个点以(r,θ)表示,其中 r 为距离,θ 为角度。一般用弧度表示角度。

Properties: 1. Plotting a point on the polar coordinate system is similar to drawing an angle in the standard position. To plot (r, θ):

在极坐标系找到一个点与画出角的基准位置的方法相似。要找到(r,θ),需要:

- If $r>0$ and $\theta>0$, then start in the "positive x-axis" and rotate counterclockwise.
  若 $r>0$ 且 $\theta>0$,则从"正 x 轴"开始,逆时针旋转。
- If $r>0$ and $\theta<0$, then start in the "positive x-axis" and rotate clockwise.
  若 $r>0$ 且 $\theta<0$,则从"正 x 轴"开始,顺时针旋转。
- If $r<0$ and $\theta>0$, then start in the "negative x-axis" and rotate counterclockwise.
  若 $r<0$ 且 $\theta>0$,则从"负 x 轴"开始,逆时针旋转。
- If $r<0$ and $\theta<0$, then start in the "negative x-axis" and rotate clockwise.
  若 $r<0$ 且 $\theta<0$,则从"负 x 轴"开始,顺时针旋转。

2. The polar coordinates of a point are not unique. For example, a point (r, θ) on the polar coordinate system can also be expressed as $(r, \theta+2\pi)$. 一个点的极坐标不是唯一的。

In general, for n is an integer:

- $(r,\theta)=(r,\theta+2\pi n)$
- $(r,\theta)=(-r,\theta+\pi+2\pi n)=(-r,\theta+(2n+1)\pi)$

3. To convert from polar coordinates to rectangular coordinates:

从极坐标到直角坐标的转换:

- $x=r\cos\theta$
- $y=r\sin\theta$

Every point on the rectangular coordinate system has unique coordinates (x, y).

在直角坐标系的每个点都有唯一的坐标(x,y)。

4. To convert from rectangular coordinates to polar coordinates, it is a bit less straightforward:

从直角坐标到极坐标的转换:

- $r=\sqrt{x^2+y^2}$
- $\tan\theta=y/x$

Note that the polar coordinates are not unique for a point. By convention we want to restrict $r\geqslant 0$ and $0\leqslant\theta<2\pi$.

因为一个点的极坐标不唯一，一般对 r 与 θ 有限制：$r \geqslant 0$，且 $0 \leqslant \theta < 2\pi$。

We must figure out which quadrant θ is in.

必须先求出 θ 的象限。

Figure 33-30 shows the relationship among r, θ, x, and y.

图 33-30 展示了 r、θ、x、y 的关系。

Questions: 1. For each point in the polar coordinate system as shown in Figure 33-31, (1) label the point using at least two polar coordinates (r, θ). (2) convert the polar coordinates to rectangular coordinates.

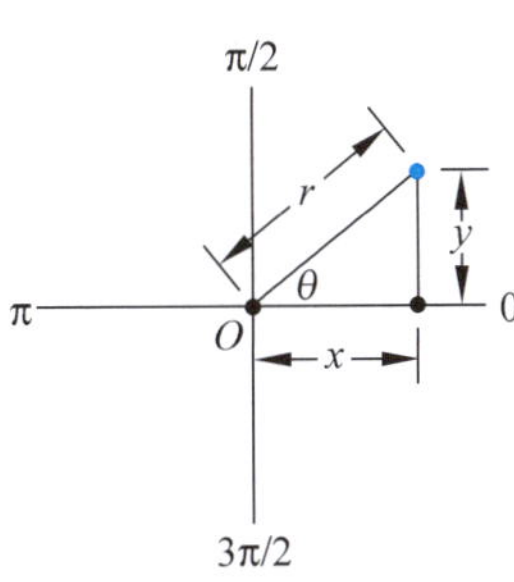

Figure 33-30

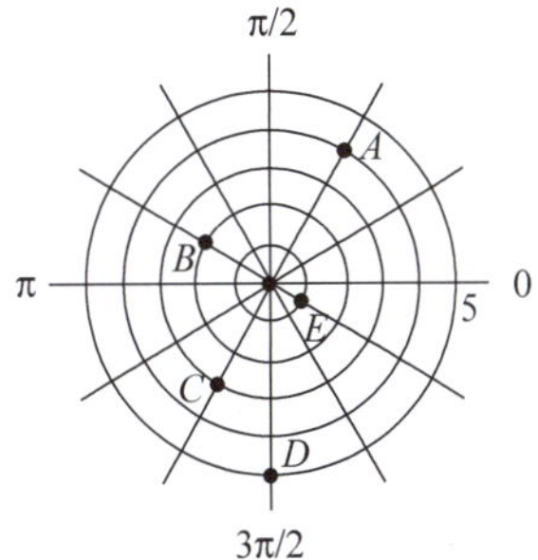

Figure 33-31

2. Convert each of the following points from rectangular coordinates to polar coordinates.

(1) $(4, 0)$

(2) $(6, 8)$

(3) $(-8, -15)$

Answers: 1. For point A:

(1) The coordinates are $(4, \pi/3)$.

Some other ways to label are $(4, 7\pi/3)$ and $(-4, 4\pi/3)$.

(2) Using the point $(4, \pi/3)$, we have

$$x = 4\cos\frac{\pi}{3} = 4\left(\frac{1}{2}\right) = 2.$$

$$y = 4\sin\frac{\pi}{3} = 4\left(\frac{\sqrt{3}}{2}\right) = 2\sqrt{3}.$$

Our answer is $(2, 2\sqrt{3})$.

For point B:

(1) The coordinates are $(2, 5\pi/6)$.

Some other ways to label are $(2, -7\pi/6)$ and $(-2, -\pi/6)$.

(2) Using the point $(2, 5\pi/6)$, we have

$$x = 2\cos\frac{5\pi}{6} = 2\left(-\frac{\sqrt{3}}{2}\right) = -\sqrt{3}.$$

$$y = 2\sin\frac{5\pi}{6} = 2\left(\frac{1}{2}\right) = 1.$$

Our answer is $(-\sqrt{3},1)$.

For point C:

(1) The coordinates are $(3,4\pi/3)$.

Some other ways to label are $(3,-2\pi/3)$ and $(-3,\pi/3)$.

(2) Using the point $(3,4\pi/3)$, we have

$$x=3\cos\frac{4\pi}{3}=3\left(-\frac{1}{2}\right)=-\frac{3}{2}.$$

$$y=3\sin\frac{4\pi}{3}=3\left(-\frac{\sqrt{3}}{2}\right)=-\frac{3\sqrt{3}}{2}.$$

Our answer is $\left(-\frac{3}{2},-\frac{3\sqrt{3}}{2}\right)$.

For point D:

(1) The coordinates are $(5,3\pi/2)$.

Some other ways to label are $(5,-\pi/2)$ and $(-5,\pi/2)$.

(2) Using the point $(5,3\pi/2)$, we have

$$x=5\cos\frac{3\pi}{2}=3(0)=0.$$

$$y=5\sin\frac{3\pi}{2}=5(-1)=-5.$$

Our answer is $(0,-5)$.

For point E:

(1) The coordinates are $(1,11\pi/6)$.

Some other ways to label are $(1,-\pi/6)$ and $(-1,5\pi/6)$.

(2) Using the point $(1,11\pi/6)$, we have

$$x=1\cos\frac{11\pi}{6}=1\left(\frac{\sqrt{3}}{2}\right)=\frac{\sqrt{3}}{2}.$$

$$y=1\sin\frac{11\pi}{6}=1\left(-\frac{1}{2}\right)=-\frac{1}{2}.$$

Our answer is $\left(\frac{\sqrt{3}}{2},-\frac{1}{2}\right)$.

2. (1) We have

$$r=\sqrt{4^2+0^2}=4.$$

$$\tan\theta=\frac{0}{4}=0,\text{so }\theta=0.$$

Our answer is $(4,0)$.

(2) We have

$$r=\sqrt{6^2+8^2}=10.$$

$$\tan\theta=\frac{8}{6}=\frac{4}{3},\text{so }\theta\approx 0.927.$$

Our answer is $(10,0.927)$.

(3) $r=\sqrt{(-8)^2+(-15)^2}=17$.

$\tan\theta=\frac{-15}{-8}=\frac{15}{8}$. Note that $\arctan\frac{15}{8}\approx 1.081$, but the point is in Quadrant Ⅲ while 1.081 an angle in Quadrant Ⅰ!

Since the period of the tangent function is π. To make θ in Quadrant III, we add π.

Our answer is $(17, 1.081+\pi)$.

34. Limits and Derivatives 极限与导数

34.1 Limits 极限

34.1.1 Introduction 介绍

Limit 极限

n. ['lɪmɪt]

Definition: If $f(x)$ becomes arbitrarily close to a unique number L as x approaches c from either side, the limit of $f(x)$ as x approaches c is L.

当 x 从两边接近 c，$f(x)$ 无比接近一个唯一值 L 时，则 x 接近 c 时 $f(x)$ 的极限为 L。

One example is shown in Figure 34-1.

Notation: To denote the definition, we write $\lim\limits_{x\to c}f(x)=L$.

Properties: 1. This definition is for the two-sided limit. Also see one-sided limit. 上面的定义是双边极限的。见本节单边极限的词条。

2. Limits can be found by direct substitution or dividing out technique, or rationalizing technique. 极限可用直接代入法、除法、有理化法求出。

3. $\lim\limits_{x\to c}f(x)=L$ does not imply that $f(c)=L$, as shown in Figure 34-2.

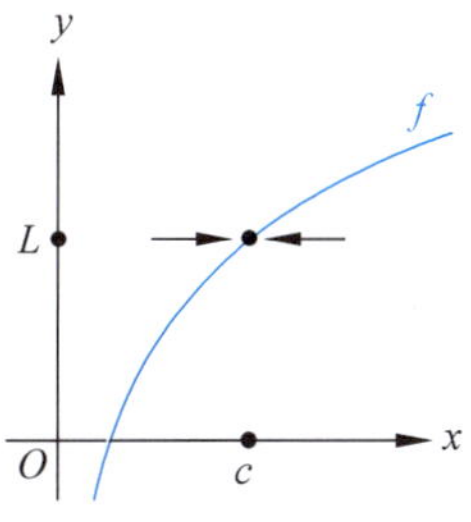

Figure 34-1

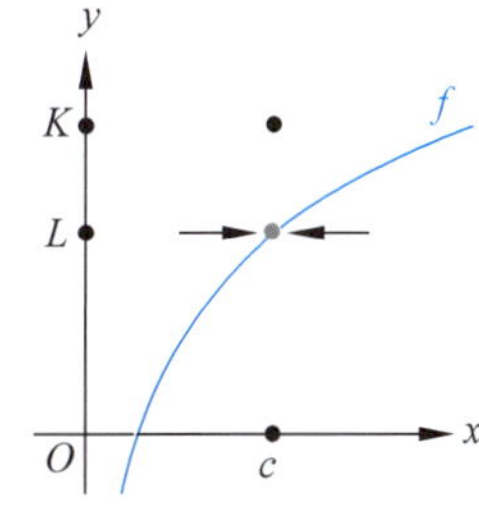

Figure 34-2

It is true that $\lim\limits_{x \to c} f(x) = L$, but $f(c) = K \neq L$.

4. $\lim\limits_{x \to c} f(x)$ does not exist if any one of the following is true:
以下任意一种情况下$\lim\limits_{x \to c} f(x)$不存在:
   (1) $f(x)$ approaches to a different number from the right side of c than from the left side of c, as shown in Figure 34-3.
   $f(x)$从 c 的两边都接近不同的数,如图 34-3 所示。
   (2) $f(x)$ increases or decreases without bound as $x \to c$, as shown in Figure 34-4.

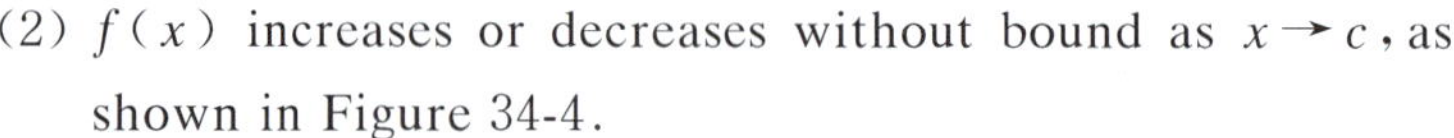

Figure 34-3

   当 $x \to c$ 时,$f(x)$没有限制地增加或者减少,如图 34-4 所示。
   (3) $f(x)$ oscillates between two fixed numbers as $x \to c$, as shown in Figure 34-5.
   当 $x \to c$ 时,$f(x)$在两个固定值间振动,如图 34-5 所示。

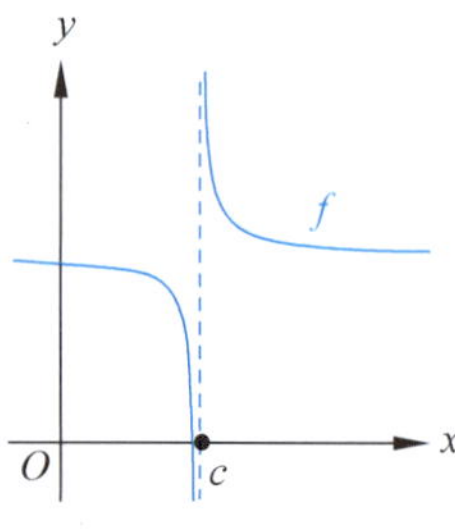

Figure 34-4

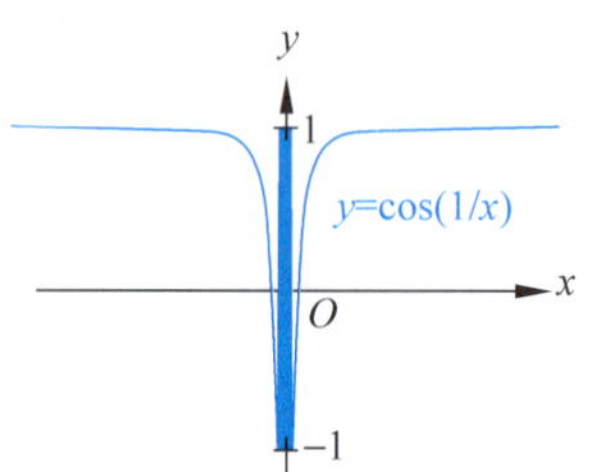

Figure 34-5

   The limit does not exist at $x = 0$.

5. Basic limits: Let b and c be real numbers and n be a positive integer. The followings are true:
   (1) $\lim\limits_{x \to c} b = b$ Limit of a constant function
   (2) $\lim\limits_{x \to c} x = c$ Limit of an identity function
   (3) $\lim\limits_{x \to c} (x^n) = c^n$ Limit of a power function
   (4) $\lim\limits_{x \to c} \sqrt[n]{x} = \sqrt[n]{c}$, if n is even, c must be nonnegative.
   Limit of a radical function

6. Properties of limits: Let b and c be real numbers, n be a positive integer, and f and g be functions such that $\lim\limits_{x \to c} f(x) = L$ and $\lim\limits_{x \to c} g(x) = K$. The followings are true:
   (1) Scalar multiple: $\lim\limits_{x \to c} [bf(x)] = bL$
   (2) Sum or difference: $\lim\limits_{x \to c} [f(x) \pm g(x)] = L \pm K$
   (3) Product: $\lim\limits_{x \to c} [f(x)g(x)] = LK$
   (4) Quotient: $\lim\limits_{x \to c} \dfrac{f(x)}{g(x)} = \dfrac{L}{K}$, from which $K \neq 0$
   (5) Power: $\lim\limits_{x \to c} [f(x)]^n = L^n$

Examples: 1. (A classic problem about limit) You have 36 inches of wire and are asked to form a rectangle with the maximum possible area. We wish to find the dimensions of the rectangle that produces such area.

We know that $2l+2w=36$, from which $l+w=18$ and $l=18-w$.

The area function is $A(w)=lw=(18-w)w=18w-w^2=-w^2+18w$, from which we want to find its maximum value.

We know that for quadratic functions, the vertex occurs at $w=-\frac{18}{2(-1)}=9$.

Plugging in, we know that $A(9)=81$.

We can verify this using a table as shown in Table 34-1:

Table 34-1

w	8.0	8.5	8.9	9.0	9.1	9.5	10.0
$A(w)$	80.00	80.75	80.99	81.00	80.99	80.75	80.00

We find out that as w is arbitrarily close to 9.0 from either side, $A(w)$ becomes arbitrarily close to a unique number, 81.00.

We say that $\lim\limits_{w\to 9}A(w)=\lim\limits_{w\to 9}(18w-w^2)=81$.

2. (1) Figure 34-6 shows the graph of the function $f(x)=3x-1$. We wish to find $\lim\limits_{x\to 3}f(x)$.

Intuitively, $\lim\limits_{x\to 3}f(x)=8$. We can construct a table to verify, as shown in Table 34-2.

Table 34-2

x	2.5	2.9	2.99	3.0	3.01	3.1	3.5
$f(x)$	6.5	7.7	7.97	8.0	8.03	8.3	9.5

(2) Figure 34-7 shows the graph of the function $g(x)=\frac{x}{\sqrt{x+4}-2}$. We wish to find $\lim\limits_{x\to 0}g(x)$.

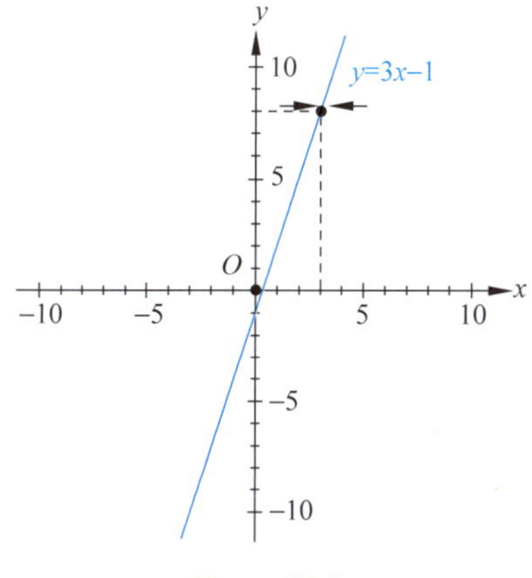

Figure 34-6

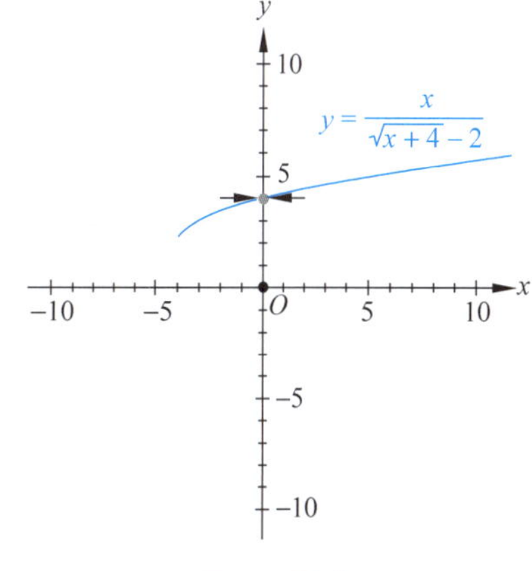

Figure 34-7

Even though $g(0)$ is undefined, our answer is $\lim_{x\to 0} g(x) = 4$ because as x is arbitrarily close to 0 from both sides, $g(x)$ approaches to 4.

We can construct a table (like that from the previous example) and verify the answer. Do it yourself.

(3) For the function $h(x) = \begin{cases} 1, & x \neq 3 \\ 0, & x = 3 \end{cases}$, let's find $\lim_{x\to 3} h(x)$.

Even though $h(3) = 0$, our answer is $\lim_{x\to 3} h(x) = 1$ because as x is arbitrarily close to 0 from both sides, $h(x)$ approaches to 1.

Graph this function and construct a table to verify the answer. Do it yourself. Use this approach whenever you need to examine the behavior of the function in the future exercises.

(4) For the function $j(x) = \sqrt{x}$, let's find $\lim_{x\to 0} j(x)$.

Note that $j(x)$ is undefined as $x \to 0$ from the left, and $j(x) \to 0$ as $x \to 0$ from the right. According to our definition of limits, $j(x)$ needs to approach the same number when x approaches 0 from both sides. Therefore, $\lim_{x\to 0} j(x)$ does not exist.

3. (1) For the function $f(x) = \dfrac{x}{|x|}$, let's find $\lim_{x\to 0} f(x)$.

Note that $f(x) \to -1$ as $x \to 0$ from the left, and $f(x) \to 1$ as $x \to 0$ from the right. This violates our definition. So, $\lim_{x\to 0} f(x)$ does not exist.

(2) For the function $g(x) = \dfrac{1}{x-2}$, let's find $\lim_{x\to 2} g(x)$.

Note that there is a vertical asymptote at $x = 2$. So, $\lim_{x\to 2} g(x)$ does not exist.

By the way, $g(x) \to -\infty$ as $x \to 2$ from the left, and $g(x) \to \infty$ as $x \to 2$ from the right.

(3) For the function $h(x) = \sin \dfrac{1}{x}$, let's find $\lim_{x\to 0} h(x)$.

Note that $h(x)$ oscillates between ± 1 as $x \to 0$ from either side. There is not a particular number that $h(x)$ approaches to as $x \to 0$ from either side. Therefore, $\lim_{x\to 0} h(x)$ does not exist.

The college-level course real analysis will rigorously prove the limit as x approaches to a number, but for now, we will find the limit either by observation or direct substitution (given continuity). Refer to the terms direct substitutions in Section 34.1.2 and continuity in Section 34.2.3.

大学里实变分析这门课会严谨地证明极限的概念,但是现在可用观察法或者直接代入法。见 34.1.2 节直接代入法和 34.2.3 节连续的词条。

One-Sided Limit 单边极限

n. [wʌn 'saɪdɪd 'lɪmɪt]

Definition: If $f(x)$ becomes arbitrarily close to a unique number L as x approaches c from the left side, the left limit of $f(x)$ as x approaches c is L.

当 x 从左边接近 c，$f(x)$ 无比接近一个唯一值 L 时，x 接近 c 时 $f(x)$ 的左极限为 L。

Same applies for the right limit.

右极限的定义也相同。

Notation: Left Limit: $\lim\limits_{x\to c^-} f(x)=L$

Right Limit: $\lim\limits_{x\to c^+} f(x)=K$

Properties: If $\lim\limits_{x\to c^-} f(x)=\lim\limits_{x\to c^+} f(x)$, then $\lim\limits_{x\to c} f(x)$ exists and $\lim\limits_{x\to c} f(x)=\lim\limits_{x\to c^-} f(x)=\lim\limits_{x\to c^+} f(x)$.

Examples: The functions in these examples are the same as those in the examples of limit in Section 34.1.1. When needed to examine the behavior of the functions, feel free to graph them or construct tables.

1. For the function $f(x)=\begin{cases}1, & x\neq 3\\ 0, & x=3\end{cases}$, we look at the limits at $x=3$. We have

   Left Limit: $\lim\limits_{x\to 3^-} f(x)=1$.

   Right Limit: $\lim\limits_{x\to 3^+} f(x)=1$.

   Since $\lim\limits_{x\to 3^-} f(x)=\lim\limits_{x\to 3^+} f(x)=1$, we know that $\lim\limits_{x\to 3} f(x)=1$.

2. For the function $g(x)=\dfrac{x}{|x|}$, we look at the limits at $x=0$. We have

   Left Limit: $\lim\limits_{x\to 0^-} g(x)=-1$.

   Right Limit: $\lim\limits_{x\to 0^+} g(x)=1$.

   Since $\lim\limits_{x\to 0^-} g(x)\neq\lim\limits_{x\to 0^+} g(x)$, we know that $\lim\limits_{x\to 0} g(x)$ does not exist.

3. For the function $h(x)=\sqrt{x}$, we look at the limits at $x=0$. We have

   Left Limit: $\lim\limits_{x\to 0^-} h(x)$ does not exist.

   Right Limit: $\lim\limits_{x\to 0^+} h(x)=0$.

   Since $\lim\limits_{x\to 0^-} h(x)\neq\lim\limits_{x\to 0^+} h(x)$, we know that $\lim\limits_{x\to 0} h(x)$ does not exist.

4. For the function $j(x)=3x-1$, we look at the limits at $x=3$. We have

   Left Limit: $\lim\limits_{x\to 3^-} j(x)=8$.

   Right Limit: $\lim\limits_{x\to 3^+} j(x)=8$.

   Since $\lim\limits_{x\to 3^-} j(x)=\lim\limits_{x\to 3^+} j(x)=8$, we know that $\lim\limits_{x\to 3} j(x)=8$.

34.1.2 Methods of Finding Limits　求出极限的方法

Direct Substitution　直接代入法

n. [daɪˈrekt ˌsʌbstɪˈtuʃən]

Definition: The method of finding limits as x approaches c by substituting c into the function. In short, $\lim\limits_{x\to c} f(x)=f(c)$.　直接代入法是指直接将 c 代入原函数求出 x 接近 c 时的极限的方法。

Properties: 1. Direct substitution works when both of the following are true:

直接代入法必须满足的条件有：

(1) $f(c)$ is defined. $f(c)$有定义。

(2) f is continuous at c. f 在 c 处是连续的。

The idea of direct substitution is shown in Figure 34-8.

It is true that $\lim\limits_{x \to c} f(x) = f(c) = L$.

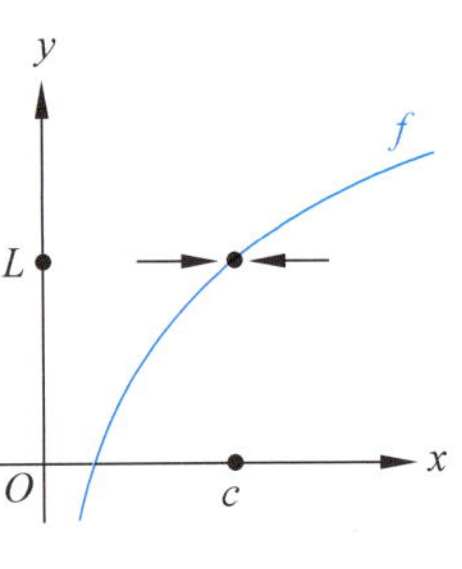

Figure 34-8

2. Refer to Properties 5 and 6 of limit in Section 34.1.1—limits of polynomial and radical functions can be found by direct substitution. Same for the limits of trigonometric and logarithmic functions. 多项式函数、根式函数、三角函数和对数函数的极限可用直接代入法求出。

These are "nice" functions because they are continuous in their domains. The exceptions are tan x, cot x, csc x, and sec x, but ignoring the vertical asymptotes, they are continuous. Their behaviors are predictable. 这些是"好"的函数,因为它们在定义域里是连续的。正切、余切、余割、正割除外,但忽略垂直渐近线,它们是连续的,其表现可被预测。

3. Limits of rational functions can also be found by direct substitution: $\lim\limits_{x \to c} \frac{p(x)}{q(x)} = \frac{p(c)}{q(c)}$, provided that $q(c) \neq 0$. 在分母不为 0 的情况下,有理函数的极限亦可以用直接代入法求出。

Questions: Evaluate each of the following:

1. $\lim\limits_{x \to 4}(3x+5)$
2. $\lim\limits_{x \to 6}\sqrt[3]{58+x}$
3. $\lim\limits_{x \to \pi}\cos x$
4. $\lim\limits_{x \to 5}\frac{x^2-7}{3x+2}$

Answers: 1. $\lim\limits_{x \to 4}(3x+5) = 3(4)+5 = 17$.

2. $\lim\limits_{x \to 6}\sqrt[3]{58+x} = \sqrt[3]{58+6} = \sqrt[3]{64} = 4$.

3. $\lim\limits_{x \to \pi}\cos x = \cos\pi = -1$.

4. $\lim\limits_{x \to 5}\frac{x^2-7}{3x+2} = \frac{5^2-7}{3(5)+2} = \frac{18}{17}$.

Dividing Out Technique 除法

n. [dɪ'vaɪdɪŋ aʊt tek'nik]

Introduction: When evaluating the limit of a rational function $\lim\limits_{x \to c}\frac{p(x)}{q(x)}$, if it turns out to be $\frac{p(c)}{q(c)} = \frac{0}{0}$, we get the **indeterminate form**, meaning that $p(x)$ and $q(x)$ have a factor in common.

有理函数的不定式情况——$p(x)$与 $q(x)$有公因式。

Review indeterminate form in Section 29.2.3. 可先复习 29.2.3 节的不定式。

Definition: The dividing out technique divides out the common factor from the numerator and denominator and finds the limit.　除法是把分子和分母都除以公因式从而找出极限的方法。

Recall that $\frac{p(c)}{q(c)}=\frac{0}{0}$ is an indeterminate form. If a hole exists at $x=c$, then the graph is not continuous at $x=c$, but the limit at $x=c$ still exists. Refer to Property 3 of limit in Section 34.1.1.

注意$\frac{p(c)}{q(c)}=\frac{0}{0}$是一个不定式。若断点在 $x=c$ 处，则图像在 $x=c$ 处不连续，但是极限仍然存在，见 34.1.1 节极限的性质 3。

The limit of as x approaches a hole is shown in Figure 34-9.

It is true that $\lim\limits_{x\to c}f(x)=L$.

Questions: Evaluate each of the following limits:

1. $\lim\limits_{x\to 6}\frac{x-6}{x-6}$

2. $\lim\limits_{x\to -3}\frac{x+3}{x^2+x-6}$

Answers: 1. Direct substitution produces $\frac{0}{0}$, which is the indeterminate form. The numerator and the denominator have a factor in common.

$\lim\limits_{x\to 6}\frac{x-6}{x-6}=\lim\limits_{x\to 6}1=1$.

The graph of the curve $y=\frac{x-6}{x-6}$ is shown in Figure 34-10.

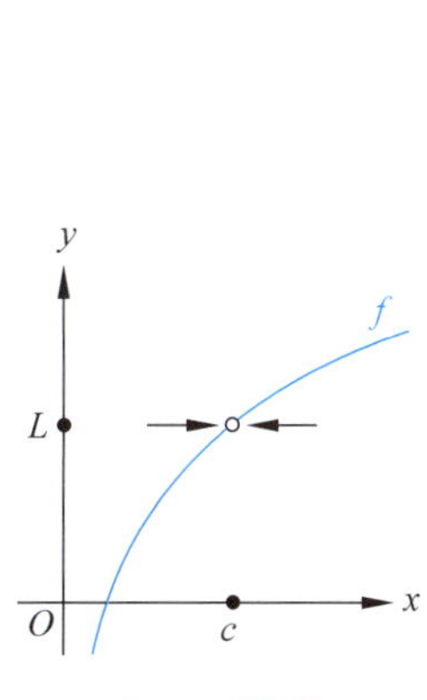

Figure 34-9

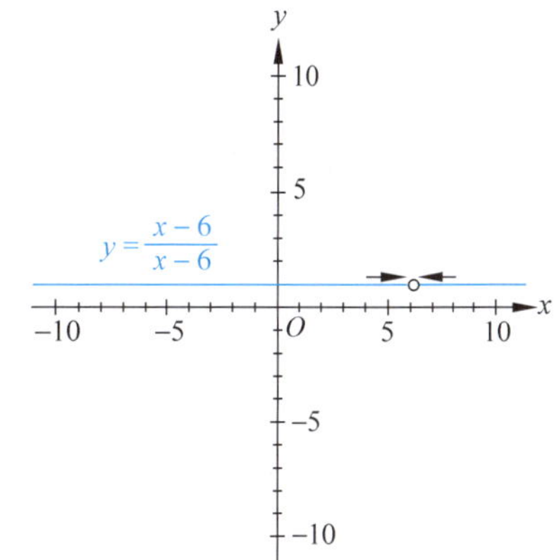

Figure 34-10

The curve is the same as $y=1$ except that there is a hole $x=6$. Table 34-3 verifies it.

Table 34-3

x	5	5.5	5.9	6	6.1	6.5	7
$\frac{x-6}{x-6}$	1	1	1	—	1	1	1

2. $\lim\limits_{x\to-3}\dfrac{x+3}{x^2+x-6}=\lim\limits_{x\to-3}\dfrac{x+3}{(x+3)(x-2)}=\lim\limits_{x\to-3}\dfrac{1}{x-2}=\dfrac{1}{-3-2}=-\dfrac{1}{5}$.

Rationalizing Technique 有理化法

n. ['ræʃənəl,aɪzɪŋ tek'nik]

Introduction: Similar to the dividing out technique, except that $p(x)$ and $q(x)$ are not necessarily polynomials—they can contain radicals as well.

与除法的介绍相似——$p(x)$和$q(x)$不一定是多项式，还可以是根式。

Definition: The rationalizing technique multiplies both the numerator and denominator by the conjugate of the denominator, then applies the dividing out technique to find the limit.

有理化法是把分子和分母都乘上分母的共轭，然后用除法求出极限的方法。

Recall that $\dfrac{p(c)}{q(c)}=\dfrac{0}{0}$ is an indeterminate form. If a hole exists at $x=c$, then the graph is not continuous at $x=c$, but the limit at $x=c$ still exists. Refer to Property 3 of limit in Section 34.1.1.

注意$\dfrac{p(c)}{q(c)}=\dfrac{0}{0}$是一个不定式。若断点在$x=c$处，则图像在$x=c$处不连续，但是极限仍然存在。见34.1.1节极限的性质3。

The limit of as x approaches a hole is shown in Figure 34-11.

It is true that $\lim\limits_{x\to c}f(x)=L$.

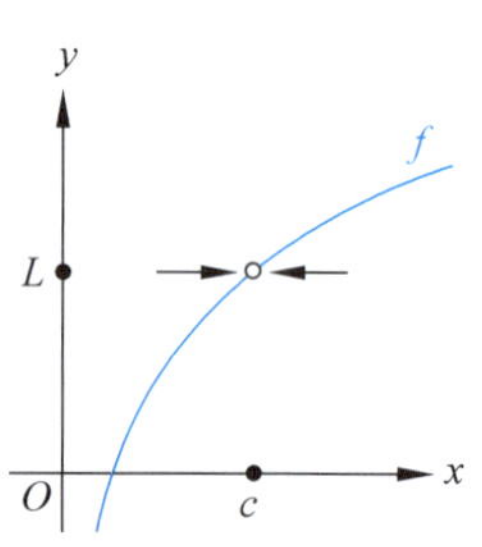

Figure 34-11

Questions: Evaluate $\lim\limits_{x\to0}\dfrac{x}{\sqrt{x+4}-2}$.

Answers: Direct substitution produces $\dfrac{0}{0}$, which is an indeterminate form. The numerator and the denominator has a factor in common.

$$\lim_{x\to0}\frac{x}{\sqrt{x+4}-2}=\lim_{x\to0}\left[\frac{x}{\sqrt{x+4}-2}\cdot\frac{\sqrt{x+4}+2}{\sqrt{x+4}+2}\right]=\lim_{x\to0}\frac{x(\sqrt{x+4}+2)}{x}=\lim_{x\to0}\frac{\sqrt{x+4}+2}{1}=$$

$$\frac{\sqrt{0+4}+2}{1}=4.$$

34.1.3 Summary 总结

1. $\lim\limits_{x\to c}f(x)=L$—two-sided limit: $f(x)\to L$ as $x\to c$ from either side.

   $\lim\limits_{x\to c^-}f(x)=L$—one-sided limit: $f(x)\to L$ as $x\to c$ from the left.

   $\lim\limits_{x\to c^+}f(x)=L$—one-sided limit: $f(x)\to L$ as $x\to c$ from the right.

   双边极限与单边极限。

2. $\lim\limits_{x\to c}f(x)=L$ does not imply that $f(c)=L$.

3. $\lim\limits_{x\to c}f(x)$ does not exist if any one of the following is true:

   以下情况$\lim\limits_{x\to c}f(x)$不存在：

   (1) $f(x)$ approaches different numbers from both sides of c.

   $f(x)$从c的两边都接近不同的数。

(2) $f(x)$ increases or decreases without bound as $x \to c$.

当 $x \to c$ 时，$f(x)$没有限制地增加或者减少。

(3) $f(x)$ oscillates between two fixed numbers as $x \to c$.

当 $x \to c$ 时，$f(x)$在两个固定值间振动。

4. If f is continuous at c and $f(c)$ is defined, then we can use direct substitution: $\lim\limits_{x \to c} f(x) = f(c)$.

若 f 在 c 处连续，并且 $f(c)$有定义，则可用直接代入法$\lim\limits_{x \to c} f(x) = f(c)$。

5. If direct substitution fails and results 0/0, then it is called the indeterminate form—it can be the case that limit does not exist or the graph has a hole. 若直接代入法得出的结果为 0/0，则为不定式情况——也许极限不存在，也许只是个图像的断点。

In case of a hole, we can use the dividing out technique or rationalizing technique to find the limit. 对于断点，可用除法或者有理化法求极限。

34.2 Derivatives 导数

视频 48

34.2.1 Introduction 介绍

Secant Line 割线

n. ['sikænt laɪn]

Definition: A secant line of the graph of a function is a line that passes through two given points of the graph. 函数图像的割线是一条穿过图像两个已知点的线。

As shown in Figure 34-12, the line l is a secant line for the graph of f.

如图 34-12 所示，l 为 f 图像的割线。

Difference Quotient 差商

n. ['dɪfrəns 'kwoʊʃənt]

Definition: The formula that computes the slope of a secant line of the graph of function f. 差商是计算函数 f 一条割线的斜率的公式。

Notation: The slope m of the graph of f between the points $(x, f(x))$ and $(x+h, f(x+h))$ is equal to the slope of its secant line and is given by $m_{\sec} = \frac{\text{rise}}{\text{run}} = \frac{f(x+h) - f(x)}{h}$, from which h is the distance between the x-coordinates of the two points (run of the slope). It is assumed that h is positive.

f 图像上的点$(x, f(x))$与$(x+h, f(x+h))$之间的斜率是割线的斜率，表示为 $m_{\sec} = \frac{\text{向上运行增量}}{\text{向右运行增量}} = \frac{f(x+h) - f(x)}{h}$。其中 h 为两点在 x 坐标的距离(斜率向右运行的增量)。可假设 h 为正数。

Figure 34-13 shows the difference quotient, which is the slope of line l. 图 34-13 展示了差商，亦是直线 l 的斜率。

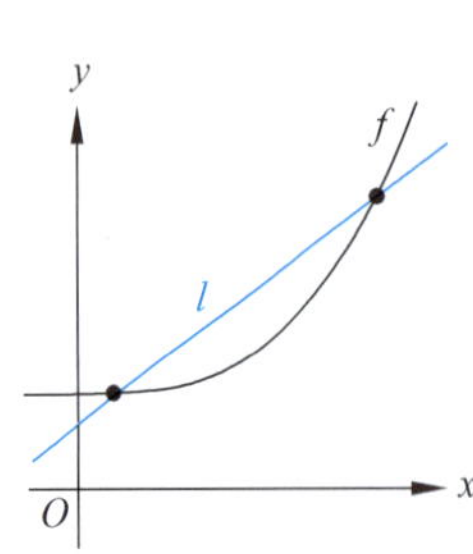

Figure 34-12

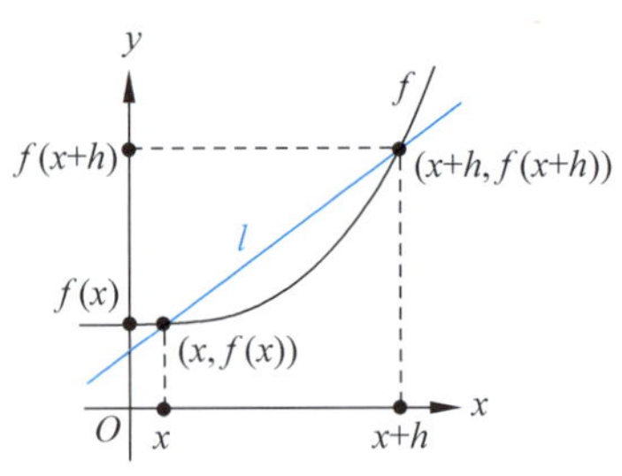

Figure 34-13

Questions: For $f(x)=x^2-5$:

(1) calculate the slope between a point at $x=2$ to a point at $x=2+h$.

(2) calculate the equation of a line that passes through these two points.

Answers: (1) $m_{\sec}=\dfrac{f(x+h)-f(x)}{h}=\dfrac{f(2+h)-f(2)}{h}=\dfrac{[(2+h)^2-5]-[2^2-5]}{h}=\dfrac{[h^2+4h-1]-[-1]}{h}=$

$\dfrac{h^2+4h}{h}=h+4.$

Once h is known, the slope is known.

(2) We are given that the slope is $h+4$ and a point is $(2,-1)$.

In point-slope form, the equation of the line is $y+1=(h+4)(x-2)$.

Tangent Line 切线

n. [ˈtændʒənt laɪn]

Definition: A tangent line of the graph of a function is a line that just "touches" the curve at one given point. 函数图像的切线是一条穿过图像一个已知点的线,且与函数图像刚刚"接触"。

The tangent line is "parallel" to the graph of the function at the point of tangency, in some sense. 切线在某些意义上是在切点处"平行"于函数的图像。

We can also understand it as a line that passes through a pair of arbitrarily close points given on the curve. 也可以将切线理解成一条经过图像两个无限接近已知点的线。

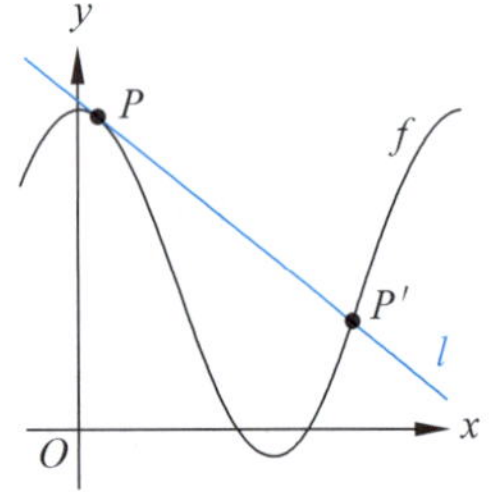

Figure 34-14

As shown in Figure 34-14, l is a tangent line for the graph of f at P, but not at P'. 如图 34-14 所示,l 是 f 图像的切线,切于 P,但不切于 P'。

Notation: The slope m of the graph of f at the specific point $(c, f(c))$ is equal to the slope of its tangent line at $(c, f(c))$ and is given by $m_{\tan}=\lim\limits_{h\to 0}m_{\sec}=\lim\limits_{h\to 0}\dfrac{f(c+h)-f(c)}{h}$, from which h is the distance between the x-coordinates of the two points (run of the slope). It is assumed that h is positive.

f 图像中某点$(c, f(c))$的斜率等于经过点$(c, f(c))$切线的斜率,为 $m_{\tan}=\lim\limits_{h\to 0}m_{\sec}=\lim\limits_{h\to 0}\dfrac{f(c+h)-f(c)}{h}$。其中 h 为两点在 x 坐标的距离(即斜率向右运行的增量)。可假设 h

为正数。

Since the two points are arbitrarily close, we take the limit of the difference quotient as h approaches to 0.

因为两点无限接近，可取当 h 接近 0 时差商的极限。

As shown in Figure 34-15, the two points on the curve have coordinates $(c, f(c))$ and $(c+h, f(c+h))$, respectively. The slope between these two points is $\dfrac{f(c+h)-f(c)}{h}$, for which h is positive and arbitrarily close to 0. More precisely the slope is $\lim\limits_{h\to 0}\dfrac{f(c+h)-f(c)}{h}$.

如图 34-15 所示，在函数曲线上的两点分别为 $(c, f(c))$ 与 $(c+h, f(c+h))$。斜率为 $\dfrac{f(c+h)-f(c)}{h}$，其中 h 为正数且无限接近 0。更准确地说，斜率为 $\lim\limits_{h\to 0}\dfrac{f(c+h)-f(c)}{h}$。

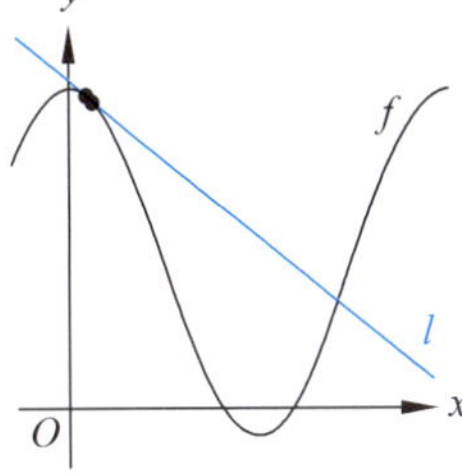

Figure 34-15

Questions: For $f(x)=x^2-5$:

(1) calculate the slope at $x=2$.

(2) calculate the equation of the tangent line that passes through $(2,-1)$ of the graph.

Answers: We have $(c, f(c))=(2, 2^2-5)=(2,-1)$.

(1) $\lim\limits_{h\to 0}\dfrac{f(c+h)-f(c)}{h}=\lim\limits_{h\to 0}\dfrac{f(2+h)-f(2)}{h}=\lim\limits_{h\to 0}\dfrac{[(2+h)^2-5]-[2^2-5]}{h}=$ $\lim\limits_{h\to 0}\dfrac{[h^2+4h-1]-[-1]}{h}=\lim\limits_{h\to 0}\dfrac{h^2+4h}{h}=\lim\limits_{h\to 0}(4+h)=4$.

(2) The slope is 4 and a point on the line is $(2,-1)$. In point-slope form, the equation is $y+1=4(x-2)$.

Derivative　导数

n. [dɪ'rɪvətɪv]

Definition: The derivative of a function $f(x)$ is a function $f'(x)$ that gives the slope of the tangent line of f at x.　函数 $f(x)$ 的导数为 $f'(x)$——给出于 x 的切线斜率的函数。

Or, the derivative of a function $f(x)$ is a function $f'(x)$ that gives the slope of f at x.

函数 $f(x)$ 的导数为 $f'(x)$——给出于 x 处 f 的斜率的函数。

Both definitions are equivalent. 上面两个定义等价。

Notation: The derivative of the function $y=f(x)$ at x has many notations, given by

$$\frac{dy}{dx}=\lim_{h\to 0}\frac{f(x+h)-f(x)}{h}$$

$$y'=\lim_{h\to 0}\frac{f(x+h)-f(x)}{h}$$

$$\frac{d}{dx}[f(x)]=\lim_{h\to 0}\frac{f(x+h)-f(x)}{h}$$

$$f'(x)=\lim_{h\to 0}\frac{f(x+h)-f(x)}{h}$$

We say $f'(x)$ is **the** derivative of $f(x)$.

$f'(x)$是 $f(x)$的导数。

$f(x)$ is **one** antiderivative of $f'(x)$.

$f(x)$是 $f'(x)$中的一个反导数。

$f'(x)$ has many antiderivatives, namely $f'(x)+C$, from which C is a constant.

$f'(x)$有很多个反导数：$f'(x)+C$,其中 C 为常数。

Properties: Let $u(x)$ and $v(x)$ be functions and a, b, and k be constants:

1. $\frac{\mathrm{d}}{\mathrm{d}x}(u \pm v)=\frac{\mathrm{d}u}{\mathrm{d}x} \pm \frac{\mathrm{d}v}{\mathrm{d}x}$
2. $\frac{\mathrm{d}}{\mathrm{d}x}(ku)=k\frac{\mathrm{d}u}{\mathrm{d}x}$
3. $\frac{\mathrm{d}}{\mathrm{d}x}(au \pm bv)=a\frac{\mathrm{d}u}{\mathrm{d}x} \pm b\frac{\mathrm{d}v}{\mathrm{d}x}$
4. $\frac{\mathrm{d}}{\mathrm{d}x}k=0$

Notes: Recall that in a tangent line, its slope is $m=\lim\limits_{h \to 0}\frac{f(c+h)-f(c)}{h}$. This is the slope at the specific point $(c, f(c))$. Note that c is a constant, and the slope is a constant. 切线在点$(c, f(c))$的斜率为 $m=\lim\limits_{h \to 0}\frac{f(c+h)-f(c)}{h}$。注意 c 为常值,所以斜率亦为常值。

The derivative $f'(x)=\lim\limits_{h \to 0}\frac{f(x+h)-f(x)}{h}$ is the slope at **any** point $(x, f(x))$. In $f'(x)$, the input is x and the output is the slope of $f(x)$. 导数 $f'(x)=\lim\limits_{h \to 0}\frac{f(x+h)-f(x)}{h}$是任意点$(x, f(x))$的斜率。$f'(x)$是一个输入为 x,输出为 $f(x)$斜率的函数。

Questions: 1. What is the derivative of $f(x)=3x^2+4x$?

2. What is the derivative of $f(x)=\sqrt{x}$? Use the derivative to find the slopes at points $(1,1)$ and $(9,3)$ respectively.

Answers: 1. $f'(x)=\lim\limits_{h \to 0}\frac{f(x+h)-f(x)}{h}=\lim\limits_{h \to 0}\frac{[3(x+h)^2+4(x+h)]-[3x^2+4x]}{h}$

$$=\lim_{h \to 0}\frac{[3(x^2+2xh+h^2)+4(x+h)]-[3x^2+4x]}{h}$$

$$=\lim_{h \to 0}\frac{[3x^2+6xh+3h^2+4x+4h]-[3x^2+4x]}{h}$$

$$=\lim_{h \to 0}\frac{6xh+3h^2+4h}{h}=\lim_{h \to 0}(6x+3h+4)=6x+4.$$

2. $f'(x)=\lim\limits_{h \to 0}\frac{f(x+h)-f(x)}{h}=\lim\limits_{h \to 0}\frac{\sqrt{x+h}-\sqrt{x}}{h}=\lim\limits_{h \to 0}\left[\left(\frac{\sqrt{x+h}-\sqrt{x}}{h}\right)\left(\frac{\sqrt{x+h}+\sqrt{x}}{\sqrt{x+h}+\sqrt{x}}\right)\right]$

$$=\lim_{h \to 0}\frac{(x+h)-x}{h(\sqrt{x+h}+\sqrt{x})}=\lim_{h \to 0}\frac{h}{h(\sqrt{x+h}+\sqrt{x})}=\lim_{h \to 0}\frac{1}{\sqrt{x+h}+\sqrt{x}}=\frac{1}{\sqrt{x}+\sqrt{x}}$$

$$=\frac{1}{2\sqrt{x}}.$$

At point (1,1), the slope is $f'(1)=\frac{1}{2\sqrt{1}}=\frac{1}{2}$.

At point (9,3), the slope is $f'(9)=\frac{1}{2\sqrt{9}}=\frac{1}{6}$.

34.2.2 Theorems and Graphs of Derivatives 导数的定理与图像

Power Rule 幂函数求导法则

n. [ˈpaʊər rul]

Definition: The rule of derivatives of a power function x^n:

$$\frac{\mathrm{d}}{\mathrm{d}x}x^n=nx^{n-1}.$$

n is a positive integer.

Proofs: Recall that $(a+b)^n=\sum_{k=0}^{n}\binom{n}{k}a^{n-k}b^k=\binom{n}{0}a^n+\binom{n}{1}a^{n-1}b+\binom{n}{2}a^{n-2}b^2+\binom{n}{3}a^{n-3}b^3+\cdots+\binom{n}{n-1}ab^{n-1}+\binom{n}{n}b^n$, from which $\binom{n}{k}=\frac{n!}{k!\,(n-k)!}$.

$$f'(x)=\lim_{h\to 0}\frac{(x+h)^n-x^n}{h}$$

$$=\lim_{h\to 0}\frac{\left[x^n+nx^{n-1}h+\frac{n(n-1)}{2}x^{n-2}h^2+\cdots+h^n\right]-x^n}{h}$$

$$=\lim_{h\to 0}\frac{nx^{n-1}h+\frac{n(n-1)}{2}x^{n-2}h^2+\cdots+h^n}{h}$$

$$=\lim_{h\to 0}\left(nx^{n-1}+\frac{n(n-1)}{2}x^{n-2}h+\cdots+h^{n-1}\right)$$

$$=\lim_{h\to 0}(nx^{n-1}+0+\cdots+0)$$

$$=nx^{n-1}$$

Notes: In fact, $\frac{\mathrm{d}}{\mathrm{d}x}x^n=nx^{n-1}$ works for every real number n. Such proof requires some differentiation techniques, which is taught in the calculus course. 公式对所有实数 n 均成立。这个证明需要用到其他求导方法，在微积分中会提及。

Questions: What is the derivative of $f(x)=x^3-2x^2+3x-4$?

Answers: $f'(x)=3x^2-4x+3$.

Graphs of Derivatives 导数的图像

Properties: To graph $f'(x)$ based on the graph of $f(x)$:

1. When $f(x)$ is decreasing (negative slope), $f'(x)<0$.
   When $f(x)$ is increasing (positive slope), $f'(x)>0$.
   When $f(x)$ is constant or has an extremum (zero slope), $f'(x)=0$.
2. When the slope of $f(x)$ is decreasing, $f'(x)$ is decreasing.
   When the slope of $f(x)$ is increasing, $f'(x)$ is increasing.

When the slope of $f(x)$ is 0, $f'(x)=0$.

Note that the sign of $f(x)$ does not reveal any information about $f'(x)$. Recall that the derivative, $f'(x)$, is a slope function of $f(x)$. Therefore, only the sign of **the slope of** $f(x)$ and the behavior of **the slope of** $f(x)$ decide the behavior of $f'(x)$.

函数 $f(x)$的正负符号与 $f'(x)$的信息无关。$f'(x)$是 $f(x)$的斜率函数。只有 $f(x)$斜率的正负值与 $f(x)$斜率的趋势决定 $f'(x)$的趋势。

Notes: To graph $f(x)$ based on the graph of $f'(x)$, same principle applies, but we need to know a point on $f(x)$ (called point of initial) before graphing!

Every function $f(x)$ has a unique derivative, $f'(x)$.

Every function $f'(x)$ has many possible antiderivatives: $f(x)+C$, from which C is a constant.

Recall that $\frac{d}{dx}[f(x)+C]=\frac{d}{dx}f(x)+\frac{d}{dx}C=f'(x)+0=f'(x)$.

要根据 $f'(x)$的图像画出 $f(x)$的图像,同理。但是画图前需要知道 $f(x)$的一个点(叫作起始点)!

每个函数 $f(x)$的导数是唯一的,叫作 $f'(x)$。

每个函数 $f'(x)$有多个反导数:$f(x)+C$,其中 C 为常值。

注意$\frac{d}{dx}[f(x)+C]=\frac{d}{dx}f(x)+\frac{d}{dx}C=f'(x)+0=f'(x)$。

Examples: 1. As shown in Figure 34-16, the function $f(x)$ is given. The derivative $f'(x)$ is unique.

2. As shown in Figure 34-17, the function $f'(x)$ is given. The function $f(x)$ is not unique, as shown in blue. In order to make $f(x)$ unique, we need a point of initial.

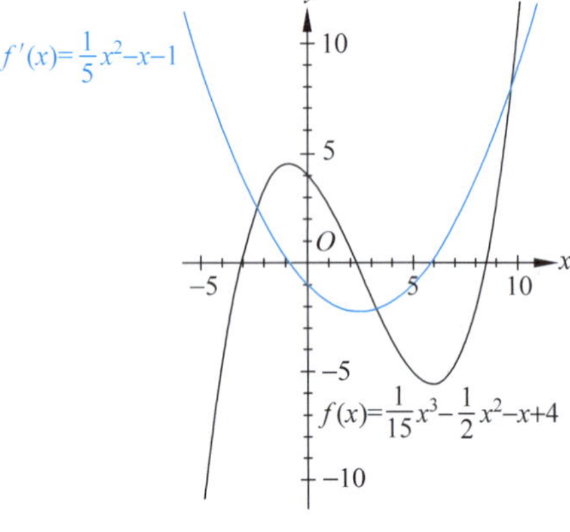

Figure 34-16

Applications of Derivatives 导数的应用

Applications: We know that $f'(x)$ is the derivative of $f(x)$, or, the slope function of $f(x)$. In real life, some relationship between $f(x)$ and $f'(x)$ are: displacement and velocity, work and power …

都知道 $f'(x)$是 $f(x)$的导数,换言之,$f'(x)$为 $f(x)$的斜率函数。

在生活中,$f(x)$ 和 $f'(x)$的关系可以是位移与速度、功与功率……

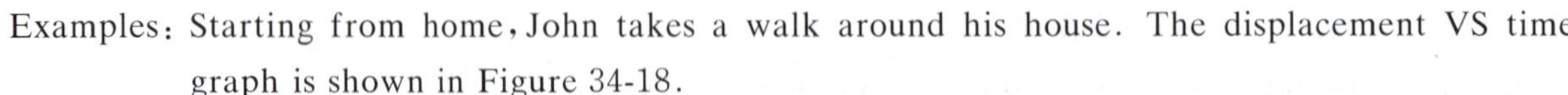

Examples: Starting from home, John takes a walk around his house. The displacement VS time graph is shown in Figure 34-18.

We plot the graph of velocity VS time as shown in Figure 34-19.

The gray dots indicate that the velocities are undefined at those times. For example, at $x=4$, the velocity has an abrupt change from 10 ft/min to 0 ft/min, so it is undefined at this x-value.

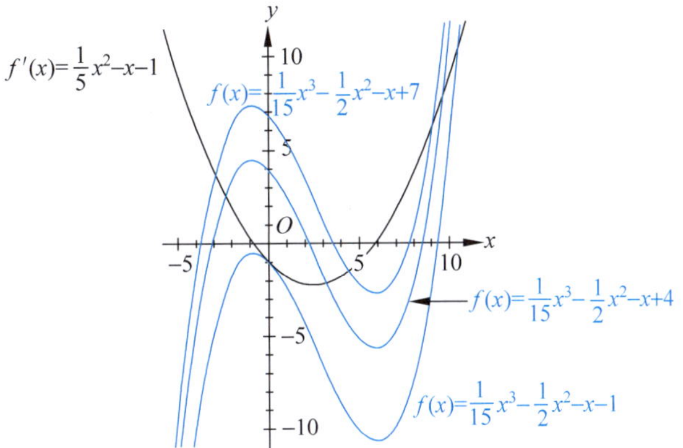

Figure 34-17

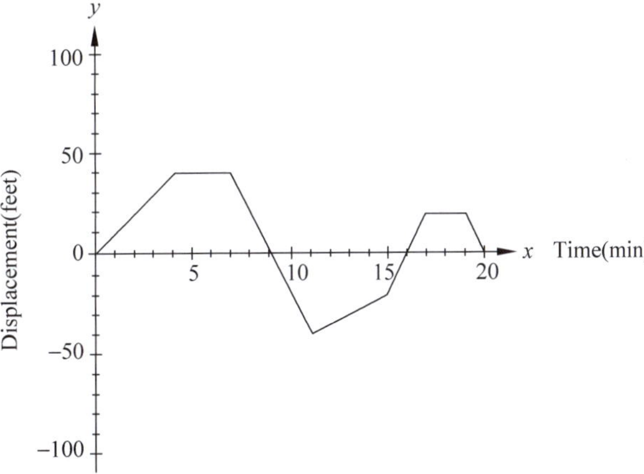

Figure 34-18

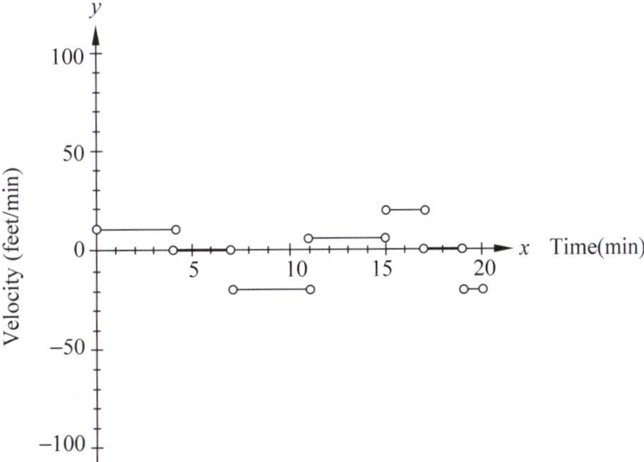

Figure 34-19

34.2.3 Important Vocabulary 重要词汇

Continuous 连续

adj. [kən'tɪnjuəs]

Definition: Roughly speaking, the graph of $f(x)$ is continuous at $x=c$ when the curve is unbroken at $x=c$ (no holes or vertical asymptote).

简单地说，当曲线不在 $x=c$ 处中断，则 $f(x)$ 的图像在 $x=c$ 处连续(没有断点或者垂直渐近线)。

More precisely, both of the following are true:

1. $\lim\limits_{x\to c}f(x)$ exists:

$$\lim_{x\to c^-}f(x)=\lim_{x\to c^+}f(x).$$

左极限 = 右极限，双边极限存在。

2. $\lim\limits_{x\to c}f(x)$ can be found by direct substitution:

$$\lim_{x\to c}f(x)=f(c).$$

双边极限可用直接代入法求出。

3. As shown in Figure 34-20, f is continuous at $x=c$.

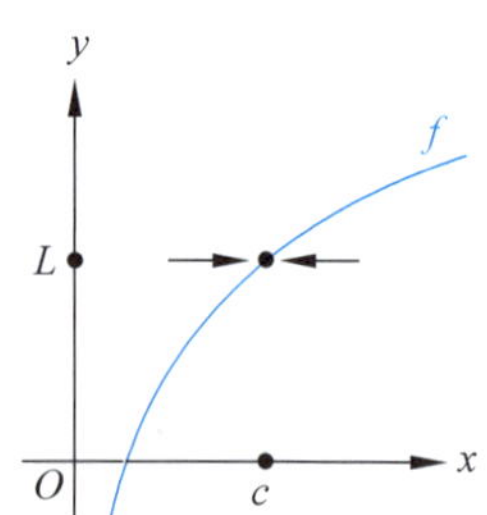

Figure 34-20

Questions: 1. Is the graph of $f(x)=\sqrt{x}$ continuous at $x=0$? Explain.

2. Is the graph of $g(x)=x^3$ continuous at $x=4$? Explain.

3. Is the graph of $h(x)=\begin{cases}x^2, & x\neq 0\\ 3, & x=0\end{cases}$ continuous at $x=0$? Explain.

Answers: 1. We have $\lim\limits_{x\to 0^+}f(x)=0$ (use the difference quotient), but $\lim\limits_{x\to 0^-}f(x)=0$ does not exist. Therefore, the graph of $f(x)$ is not continuous at $x=0$.

Summary: The left limit does not exist, so the two-sided limit does not exist.

2. We have $\lim\limits_{x\to 4^-}g(x)=\lim\limits_{x\to 4^+}g(x)=64$, and $\lim\limits_{x\to 4}g(x)=g(4)=64$. Therefore, the graph of $g(x)$ is continuous at $x=4$.

Summary: The left limit equals the right limit (two-sided limit exists), and the two-sided limit can be found by direct substitution.

3. We have $\lim\limits_{x\to 0^-}h(x)=\lim\limits_{x\to 0^+}h(x)=0$, but $h(0)=3$. Therefore, the graph of $h(x)$ is not continuous at $x=0$.

Summary: The left limit equals the right limit (two-sided limit exists), but the two-sided limit cannot be found by direct substitution.

Differentiable 可微分的

adj. [ˌdɪfə'renʃəbəl]

Definition: The graph of $f(x)$ is differentiable at $x=c$ if the slope at $x=c$ exists.

当 $x=c$ 的斜率存在，则 $f(x)$ 的图像在 $x=c$ 处可微分。

In other words, the graph of $f(x)$ is differentiable at $x=c$ if $f'(c)$ is defined.

换言之，当 $f'(c)$存在，则 $f(x)$ 的图像在 $x=c$ 处可微分。

Properties: If the graph of $f(x)$ is differentiable at $x=c$, then it is continuous at $x=c$. 若 $f(x)$的

图像在 $x = c$ 处可微分，则在 $x = c$ 处连续。

However, its converse below is not necessarily true.

If the graph of $f(x)$ is continuous at $x = c$, then it is differentiable at $x = c$.

它的逆命题不一定正确。

逆命题：若 $f(x)$的图像在 $x = c$ 处连续，则在 $x = c$ 处可微分。

Questions: What is a counterexample for the following statement:

If the graph of $f(x)$ is continuous at $x = c$, then it is differentiable at $x = c$.

Answers: One of the answers is $f(x) = |x|$ at $x = 0$. The graph is continuous at $x = 0$, but not differentiable:

$$f'(x) = \begin{cases} -1, & x < 0 \\ 1, & x > 0 \end{cases}.$$

34.2.4 Summary 总结

1. The difference quotient, $m_{\text{sec}} = \frac{\text{rise}}{\text{run}} = \frac{f(x+h) - f(x)}{h}$, calculates the slope of the secant line that intersects the graph of f at points $(x, f(x))$ and $(x+h, f(x+h))$.

   差商计算 f 图像中通过点$(x, f(x))$与点$(x+h, f(x+h))$的割线的斜率。

2. As $h \to 0$, the difference quotient $\frac{f(c+h) - f(c)}{h}$ becomes the slope of the tangent line at point $(c, f(c))$.

   当 h 接近 0 时，差商变成了在点$(c, f(c))$的切线的斜率。

   $\lim\limits_{h \to 0} \frac{f(c+h) - f(c)}{h}$ is the instantaneous rate of change at point $(c, f(c))$.

   $\lim\limits_{h \to 0} \frac{f(c+h) - f(c)}{h}$是在点$(c, f(c))$处瞬间的斜率。

3. The slope function for $f(x)$ is $f'(x) = \lim\limits_{h \to 0} \frac{f(x+h) - f(x)}{h}$. This is called the derivative of $f(x)$.

   斜率函数，也称为 $f(x)$的导数。

4. Power Rule for derivatives: $\frac{\mathrm{d}}{\mathrm{d}x} x^n = nx^{n-1}$. 幂函数求导法则。

5. Review graphs of derivatives in Section 34.2.2 复习 34.2.2 节导数的图像。

6. Differentiability implies continuity, but not the other way around. 可微分意味着连续，但连续不一定意味着可微分。

   For function f:

   To show continuity at $x = c$, show that the left-side limit is equal to the right-side limit and that limit can be found by direct substitution.

   要证出函数在 $x = c$ 处是连续的，必须证出左极限与右极限相等，并且这个极限可用直接代入法求出。

   To show differentiability at $x = c$, simply evaluate $f'(c)$.

   要证出函数在 $x = c$ 处可微分，须求出 $f'(c)$。

Appendix A　附录 A

A.1　Perimeter Formulas of Plane Shapes 平面图形的周长公式

Table A-1 shows the perimeter formulas of plane shapes.

Table A-1

Plane Shape　平面图形	Perimeter　周长
Polygons　多边形	
Triangle　三角形 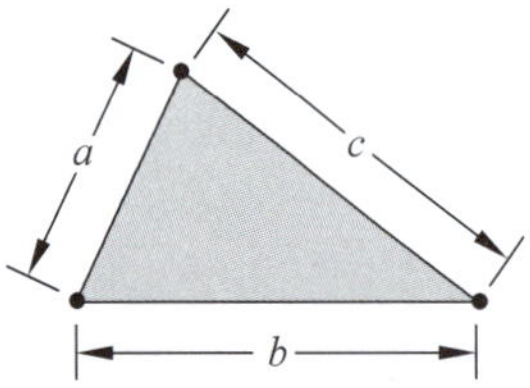	$P = a + b + c$
Parallelogram　平行四边形 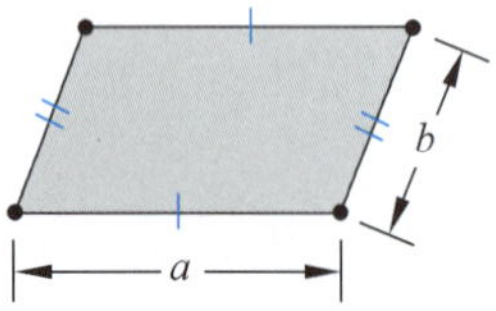	$P = a + b + a + b$ $P = 2a + 2b$ $P = 2(a + b)$
Equilateral Quadrilaterals (Rhombi and Squares) 等边四边形(菱形和正方形) 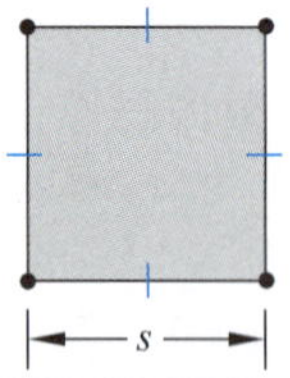	$P = 4s$
Kite　筝形 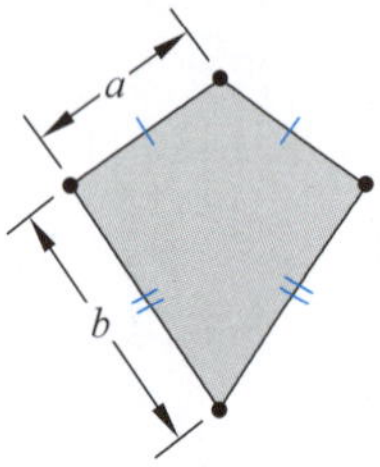	$P = a + b + a + b$ $P = 2a + 2b$ $P = 2(a + b)$

续表

<table>
<tr><th>Plane Shape 平面图形</th><th>Perimeter 周长</th></tr>
<tr><td>Quadrilaterals whose diagonals (possibly extended) are perpendicular, including kites.
对角线(可能被延长)垂直的四边形,包括筝形
Convex 凸的情况
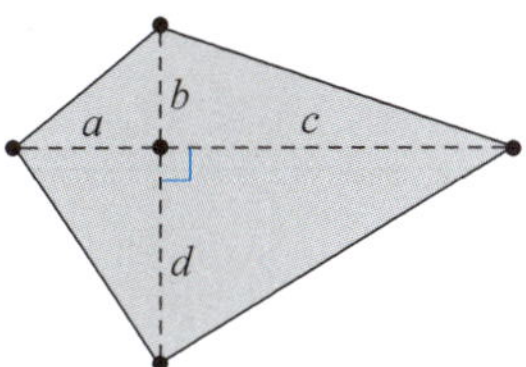
</td><td>$P=\sqrt{a^2+b^2}+\sqrt{b^2+c^2}+\sqrt{c^2+d^2}+\sqrt{d^2+a^2}$
by Pythagorean Theorem 根据勾股定理</td></tr>
<tr><td>Concave 凹的情况
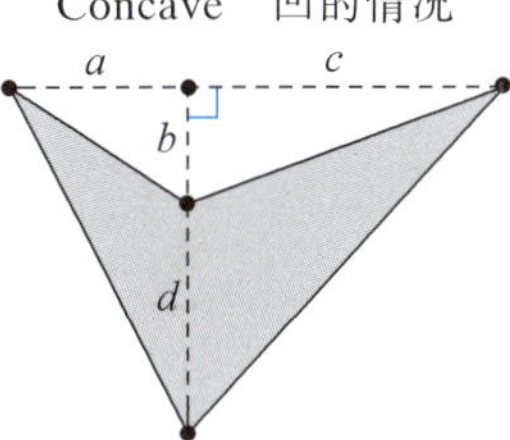
</td><td>$P=\sqrt{a^2+b^2}+\sqrt{b^2+c^2}+\sqrt{c^2+(b+d)^2}+\sqrt{(b+d)^2+a^2}$
by Pythagorean Theorem 根据勾股定理</td></tr>
<tr><td>Trapezoid and other polygons 梯形以及其他多边形</td><td>P = Sum of lengths of all sides 所有边的长度之和</td></tr>
<tr><td>Regular n-gon 正 n 边形
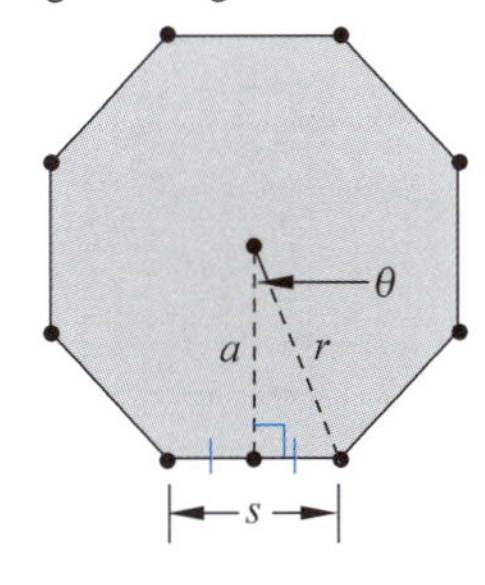
</td><td>Notes:
(1) $\theta=\left(\frac{360°}{n}\right)/2$
(2) $\left(\frac{1}{2}s\right)^2+a^2=r^2$
(3) $\cos\theta=a/r$
(4) $\sin\theta=(s/2)/r$
$P=ns$
$P=2n\sqrt{r^2-a^2}$
$P=2n\sqrt{\left(\frac{a}{\cos\theta}\right)^2-a^2}$
$P=2n\sqrt{r^2-(r\cos\theta)^2}$</td></tr>
<tr><td colspan="2">Non-Polygons 非多边形
Note: $\pi\approx3.14159265358979323846$, representing the ratio of the circumference to the diameter in a circle. This ratio can be observed by comparing the perimeter of a regular n-gon P to twice the length of a radius of P, for which n is arbitrarily large. 圆周率是圆的圆周与直径之比。这个比率可从正 n 边形 P 里观察出来:观察 P 的周长与它外接圆半径的两倍之比的值,得出圆周率。n 为任意大</td></tr>
<tr><td>Circle 圆
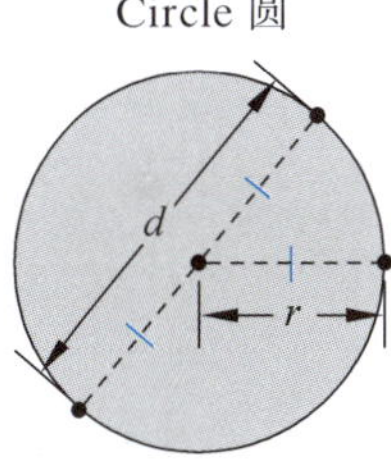
</td><td>$C=d\pi$
$C=2\pi r$
(C 为圆的周长,英文 circumference)</td></tr>
</table>

续表

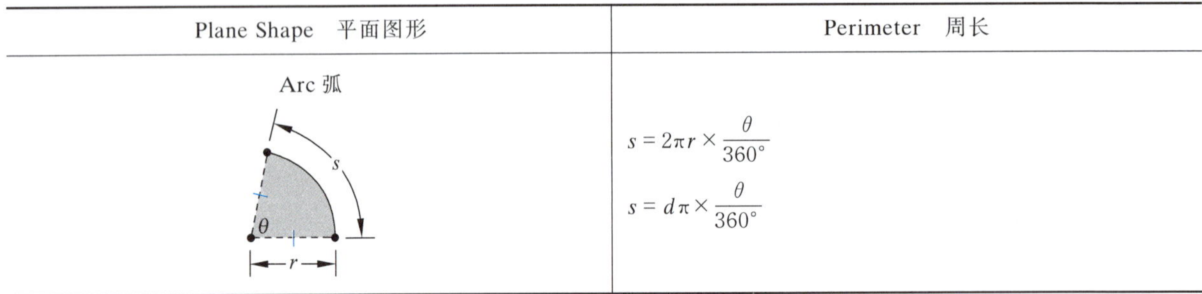

Plane Shape　平面图形	Perimeter　周长
Arc 弧	$s=2\pi r\times\dfrac{\theta}{360^\circ}$ $s=d\pi\times\dfrac{\theta}{360^\circ}$

A.2　Area Formulas of Plane Shapes 平面图形的面积公式

Table A-2 shows the area formulas of plane shapes.

Table A-2

Plane Shape　平面图形	Area (A)　面积(A)	Pictorial Proof　插图证明
	Triangle　三角形	
	$A=\dfrac{1}{2}bh$	
	$A=\dfrac{1}{2}bh$	
	$A=\dfrac{1}{2}bh$	
	Parallelogram　平行四边形	
	$A=bh$	The area formulas for rectangles and squares are self-explanatory—counting the number of unit squares that takes to cover the entire shape. 矩形和正方形的面积公式很直观——数里面单位正方形的个数
	$A=s^2$	

续表

Plane Shape 平面图形	Area (A) 面积(A)	Pictorial Proof 插图证明
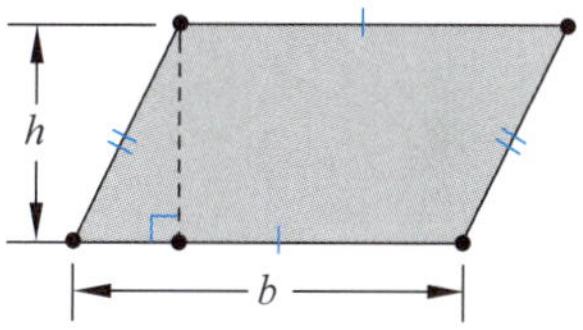 h, b	$A = bh$	For the parallelogram in the left figure, we cut along the dashed altitude and glue the pieces to a rectangle as shown in the figure below. 对于左图的平行四边形,沿着虚线高线剪开,并且拼凑成矩形,如下图所示。 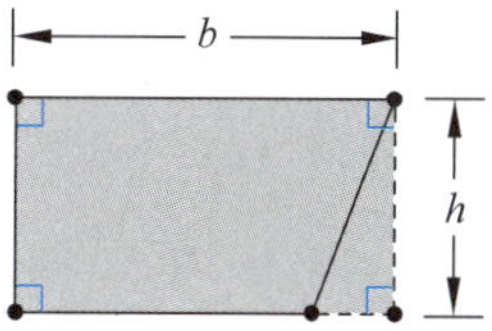 b, h
Trapezoid 梯形 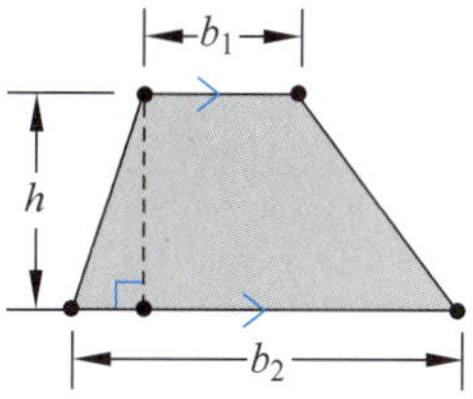 b_1, h, b_2	$A = \frac{1}{2}(b_1 + b_2)h$	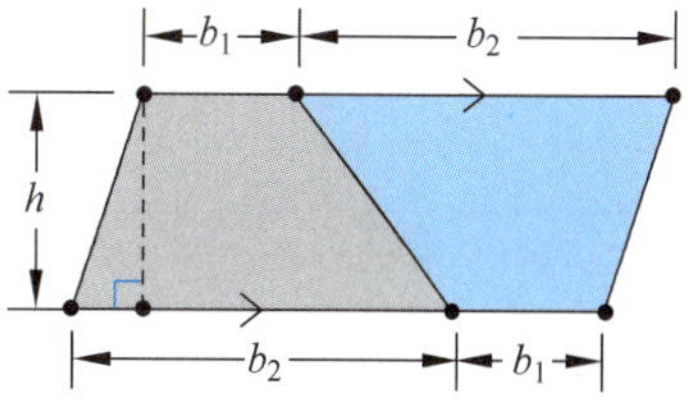 b_1, b_2, h, b_2, b_1
Quadrilaterals whose diagonals (possibly extended) are perpendicular, including kites. 对角线(可能被延长)垂直的四边形,包括筝形		
Suppose the blue dashed diagonals have lengths a and b respectively: 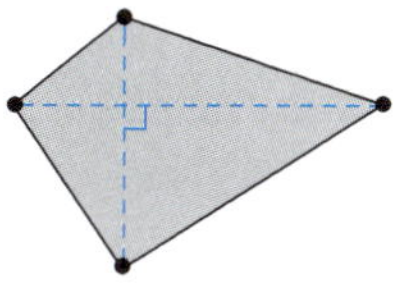	$A = \frac{1}{2}ab$	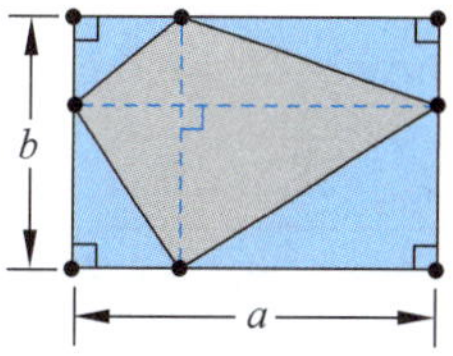 b, a
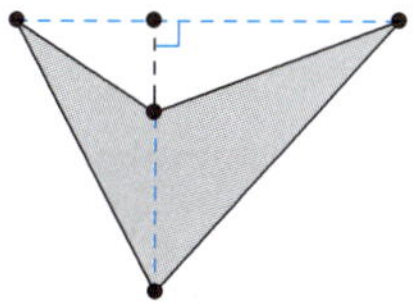	$A = \frac{1}{2}ab$	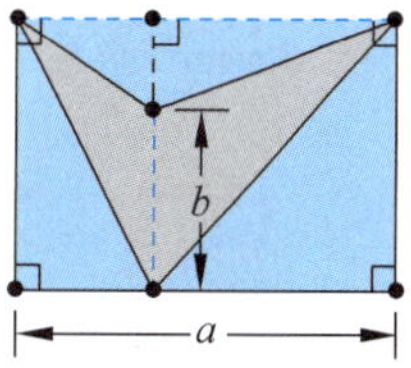 b, a
Regular n-gon 正 n 边形 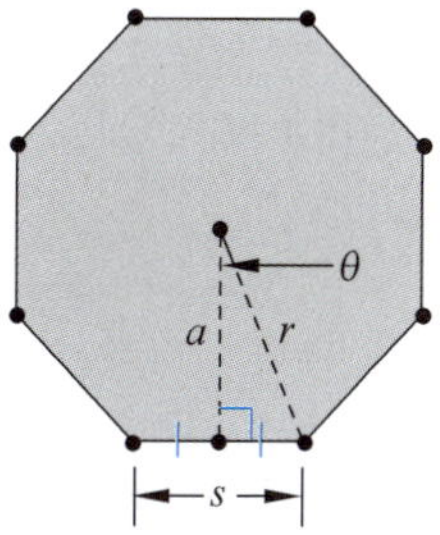 θ, a, r, s	$A = \frac{1}{2}aP$ $A = \frac{1}{2}sa \times n$ $A = \frac{1}{2}s\sqrt{r^2 - \left(\frac{1}{2}s\right)^2} \times n$ $A = (\sqrt{r^2 - a^2})a \times n$	Notes: (1) $\theta = \left(\frac{360°}{n}\right)/2$ (2) $\left(\frac{1}{2}s\right)^2 + a^2 = r^2$ (3) $\cos\theta = a/r$ (4) $\sin\theta = (s/2)/r$ (5) $P = ns$

续表

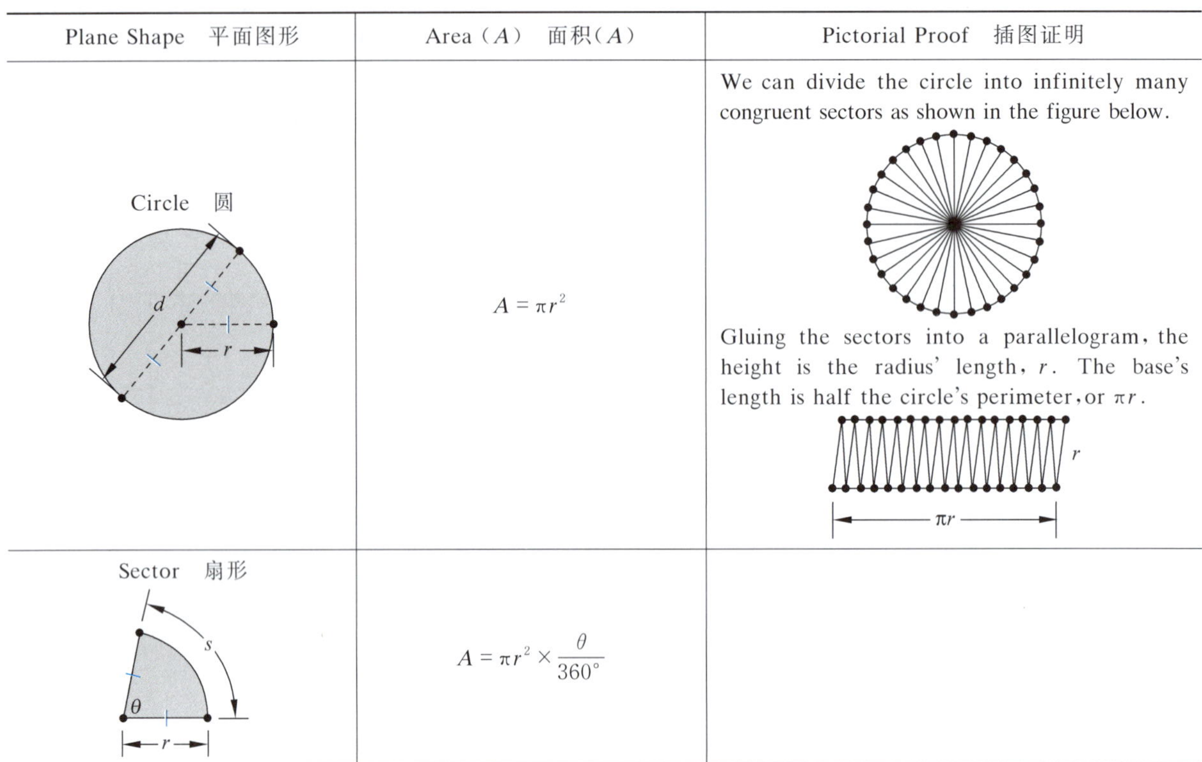

Plane Shape 平面图形	Area (A) 面积(A)	Pictorial Proof 插图证明
Circle 圆	$A=\pi r^2$	We can divide the circle into infinitely many congruent sectors as shown in the figure below. Gluing the sectors into a parallelogram, the height is the radius' length, r. The base's length is half the circle's perimeter, or πr.
Sector 扇形	$A=\pi r^2\times\dfrac{\theta}{360^\circ}$	

A.3 Surface Area Formulas of Solids 立体的表面积公式

Table A-3 shows the surface area formulas of solids.

Table A-3

Solid 立体	Surface Area 表面积(所有面的面积之和)
Polyhedrons 多面体	
Prism 棱柱	Note: SA = bases' areas + lateral faces' areas 表面积 = 底面积总和 + 侧面面积总和 $SA = 2\times$ base area + (the length of every edge in the base) $\times h$ 表面积 = $2\times$ 底面积 + 底的周长 $\times h$ Let A and P be the area and perimeter of the base, respectively, 设 A 和 P 分别为底的面积和周长, $SA=2A+Ph$
Cube 立方体	Using the formula above, we can derive that for a cube, 用于上公式推导出立方体的表面积公式, $SA=6s^2$ where s is the length of an edge of the cube. 其中 s 为立方体的棱长

续表

Solid 立体	Surface Area 表面积(所有面的面积之和)
Pyramid 棱锥	SA = sum of areas of all faces 所有面的面积总和
Regular Pyramid 正棱锥	Note: The base is a regular n-gon. 底是一个正 n 边形 The lateral faces are congruent isosceles triangles. 侧面均为全等的等腰三角形 SA = base area + the area of a lateral triangle $\times n$ 表面积 = 底面积 + 侧面三角形的面积 $\times n$ $SA = \text{base area} + \frac{1}{2} \times \text{the length of a side of the base} \times s \times n$ 表面积 $= 底面积 + \frac{1}{2} \times 底边长度 \times s \times n$ $SA = \text{base area} + \frac{1}{2} \times (\text{the length of a side of the base} \times n) \times s$ 表面积 $= 底面积 + \frac{1}{2} \times (底边长度 \times n) \times s$ Note that "(the length of a side of the base) $\times$ n" is the perimeter of the base. 注意"底边长度 $\times$ n"是底的周长。 Let A and P be the area and perimeter of the base, respectively, 让 A 和 P 分别为底的面积和周长, $$SA = A + \frac{1}{2}Ps$$
Non-Polyhedrons 非多面体	
Cylinder 圆柱 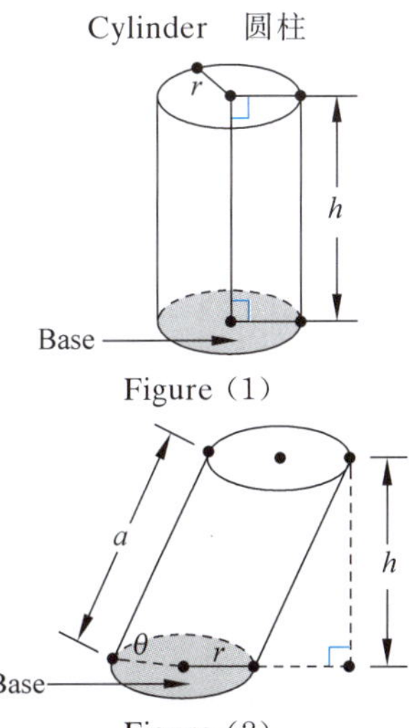 Figure (1) Figure (2)	Note: SA = bases' areas + lateral faces' area 表面积 = 底面积总和 + 侧面面积 For a right cylinder shown in Figure (1), the net is shown below. 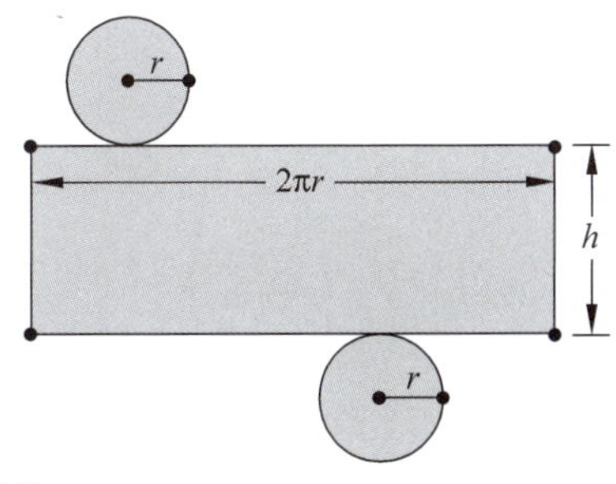 For Figure (1): $$SA = 2\pi r^2 + 2\pi rh$$ For Figure (2): $$SA = 2\pi r^2 + 2\pi ra$$ Note that $h = a \sin\theta$

续表

Solid 立体	Surface Area 表面积(所有面的面积之和)
Right Cone 直圆锥 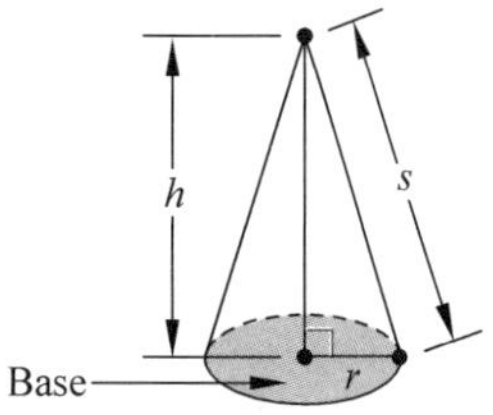 h, s, r, Base	SA = bases' area + lateral surface's area 表面积 = 底面积 + 侧面面积 Note: For a right cone, $s=\sqrt{h^2+r^2}$. The figure below shows the net for the cone in the left figure. 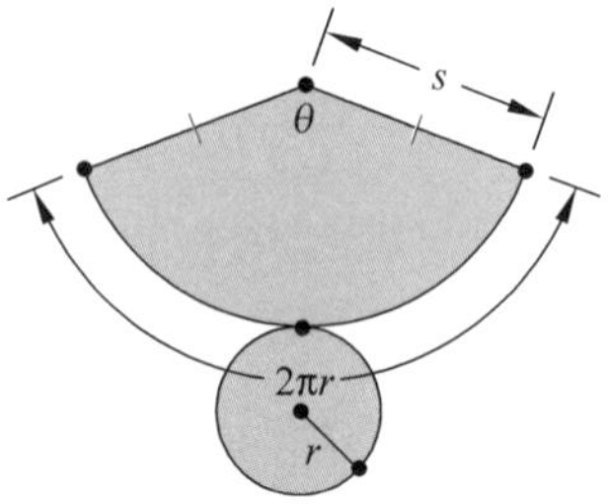 s, θ, 2πr, r $\theta=\frac{2\pi r}{2\pi s}\times 360^\circ=\frac{r}{s}\times 360^\circ$ lateral surface's area = $\frac{\pi s^2\times\left(\frac{r}{s}\times 360^\circ\right)}{360^\circ}=\pi rs$ 侧面面积 $SA=\pi r^2+\pi rs=\pi r(r+s)$ $SA=\pi r^2+\pi rs=\pi r(r+\sqrt{h^2+r^2})$
Sphere 球 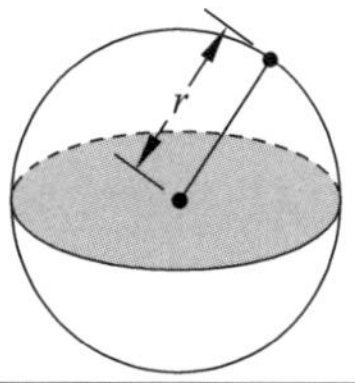 r	$SA=4\pi r^2$ The proof for this formula without using calculus is too long and is outside the scope of high school geometry course. 不用微积分的证明太长,并且不在高中几何的范围内
Frustum of a Right Cone 直圆锥的平截头体 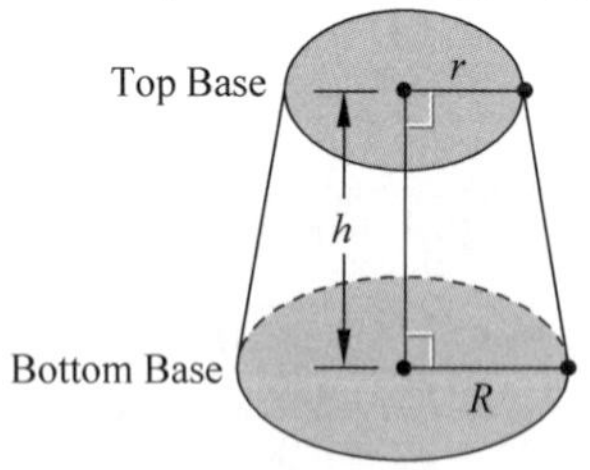 Top Base, r, h, Bottom Base, R	SA = area of top + area of bottom + areas of the lateral surfaces 表面积 = 顶面积 + 底面积 + 侧面面积总和 Note the similar right triangles. 注意相似的直角三角形 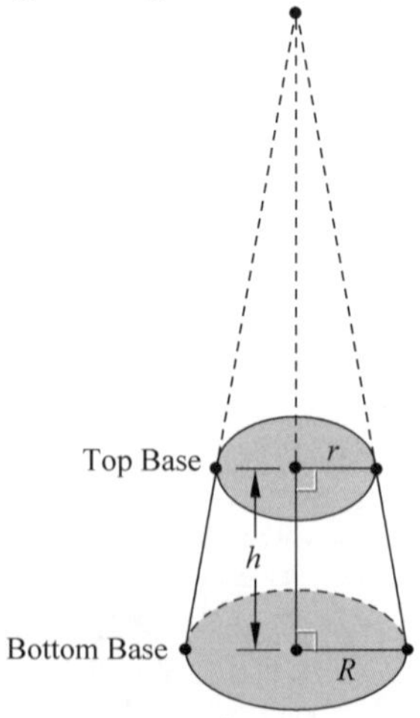 Top Base, r, h, Bottom Base, R Let LA be the abbreviation of "lateral area". LA 是侧面面积的缩写。 LA of frustum = LA of larger cone − LA of smaller cone 平截头体的侧面面积 = 大圆锥的侧面面积 − 小圆锥的侧面面积 Using algebra, 用代数, $SA=\pi[(R+r)\sqrt{(R-r)^2+h^2}+R^2+r^2]$

续表

Solid 立体	Surface Area 表面积(所有面的面积之和)
Frustum of a Regular Pyramid 正棱锥的平截头体 Top Base Bottom Base	Note: The lateral faces of a regular pyramid's frustum are congruent isosceles trapezoids. 正棱锥的平截头体侧面是全等等腰梯形。 Suppose the top and bottom are regular n-gons, 设顶和底是正 n 边形, SA = area of top + area of bottom + $\frac{1}{2}$(edge-length of top + edge-length of bottom) $\times s \times n$ 表面积 = 顶面积 + 底面积 + $\frac{1}{2}$(顶棱长 + 底棱长) $\times s \times n$ SA = area of top + area of bottom + $\frac{1}{2}$(edge-length of top $\times n$ + edge-length of bottom $\times n$) $\times s$ 表面积 = 顶面积 + 底面积 + $\frac{1}{2}$(顶棱长 $\times n$ + 底棱长 $\times n$) $\times s$ Let A and P be the area and perimeter of the bottom, respectively, and let a and p be the area and perimeter of the top, respectively: 设 A 和 P 分别是底的面积和周长, a 和 p 分别是顶的面积和周长: $SA = A + a + \frac{1}{2}(P + p)s$

A.4 Volume Formulas of Solids 立体的体积公式

Table A-4 shows the volume formulas of solids.

Table A-4

Solid 立体	Volume 体积
Prism 棱柱 Base h Base h	Note: We are computing how many unit cubes it takes to cover the entire prism. 计算有多少个单位立方体才能覆盖整个棱柱。 Let A be the base area: 设 A 为底面积: $V = Ah$
Cube 立方体 s	Using the formula above, we can derive that for a cube, 上面的公式可推导出立方体的体积公式为 $V = s^3$, where s is the length of an edge of the cube. 其中 s 为立方体的棱长

续表

<table>
<tr><th>Solid 立体</th><th>Volume 体积</th></tr>
<tr><td>Pyramid 棱锥 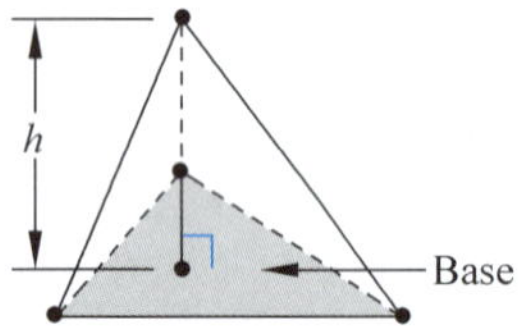

h
Base</td><td>Note: Three congruent pyramids (each with a square base, and the apex of the pyramid is vertically above one vertex of the base, and the edge-length of the base is the same as the height) can form a cube. Since the volume of a cube is s^3, the volume of one pyramid is $\frac{1}{3}s^3$. 三个(带有正方形底的，且棱锥顶尖在底的一个顶点的正上方，且底的棱长与棱锥的高相等的)全等棱锥可以拼成一个立方体。因为立方体的体积是 s^3，所以棱锥的体积是 $\frac{1}{3}s^3$。 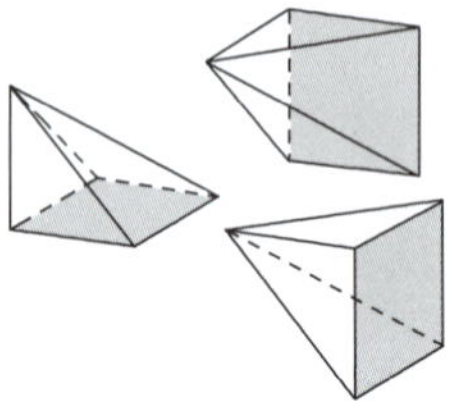

We can extend this to pyramids with different bases and apex locations using linear transformations. At the end, we will have: 对于有不同底和顶尖位置的棱锥，可以用线性变换，最终得到以下公式：
Let A and h be the base area and height (length of altitude) of a pyramid, respectively, 设 A 和 h 分别为棱锥的底面积和高，体积为
$$V=\frac{1}{3}Ah$$</td></tr>
<tr><td>Cylinder 圆柱 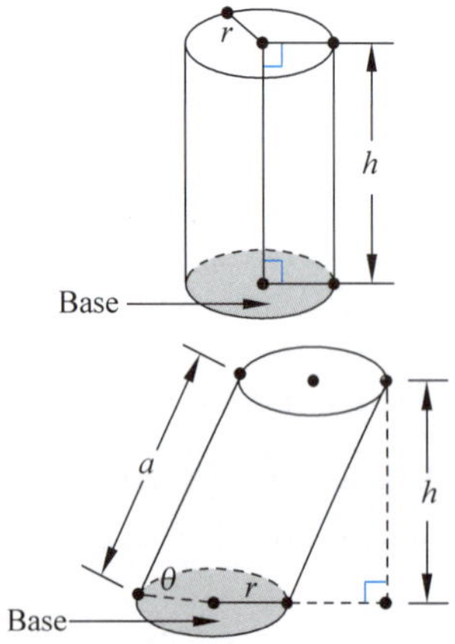

r
h
Base
a
h
θ
r
Base</td><td>Note: Similar to the bases of prisms, the bases of cylinders are parallel. Therefore, 圆柱的两底跟棱柱的两底有一个共同点——平行，所以，
Let A be the base area, 设 A 为底面积，体积为
$$V=Ah$$
$$V=\pi r^2 h$$</td></tr>
</table>

续表

Solid 立体	Volume 体积
Cone 圆锥 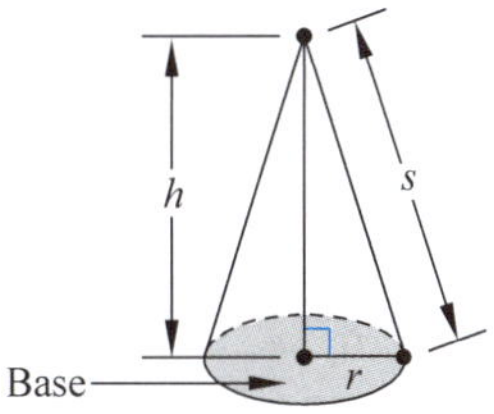 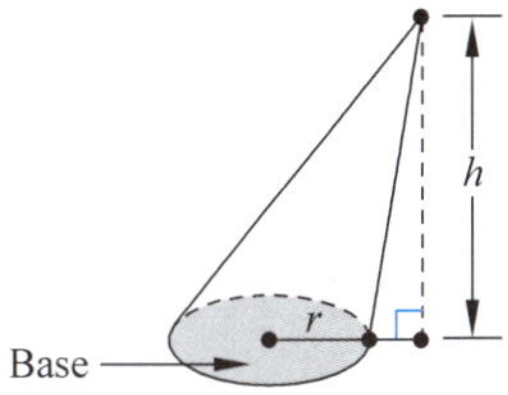	Similar to the volume of a pyramid, the volume of a cone is 与棱锥的体积相似,圆锥的体积是 $$V=\frac{1}{3}Ah,$$ where A is the base area. A 为底面积。 Note: If we were to inscribe any cone snuggedly into a pyramid whose base is a square, we have 若把圆锥的底内接在正方形上,则有 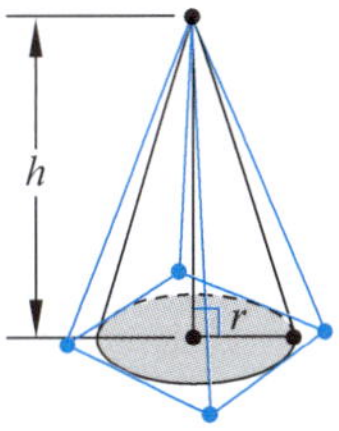 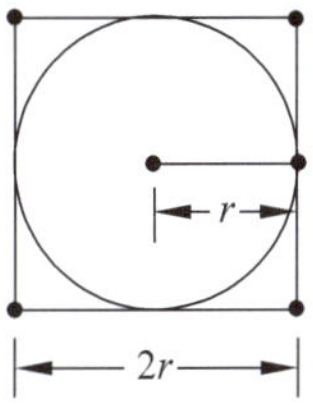 The area of the square is $4r^2$. The area of the circle is πr^2. The ratio of their areas is $4:\pi$. The volume of the pyramid is $\frac{4}{3}r^2h$. The volume of the cone is $\frac{1}{3}\pi r^2h$. $$V=\frac{1}{3}\pi r^2h$$
Sphere 球 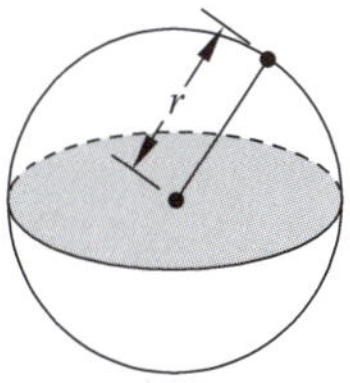	$$V=\frac{4}{3}\pi r^3$$ The proof for this formula without using calculus is too long and is outside the scope of high school geometry course. 不用微积分的证明太长了,并且不在高中几何的范围内。 Basic idea without calculus: compare the volume of a sphere with that of a cone and a cylinder such that their lengths of radii in the base are equal. 不用微积分证法的基本考虑是:对比一个球、圆锥、圆柱的体积(圆锥和圆柱底的半径与球的半径相等)

续表

Solid 立体	Volume 体积
Frustum 平截头体 Conical Frustum 圆锥的平截头体 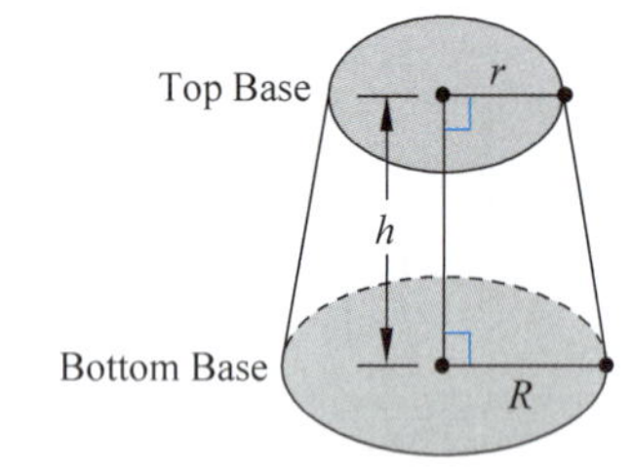 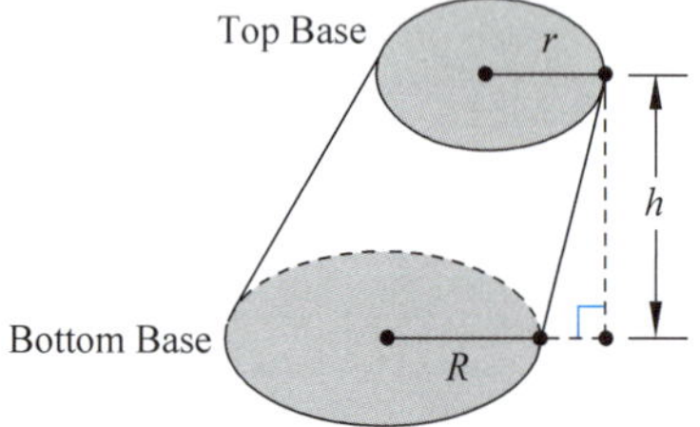 Pyramidal Frustum 棱锥的平截头体 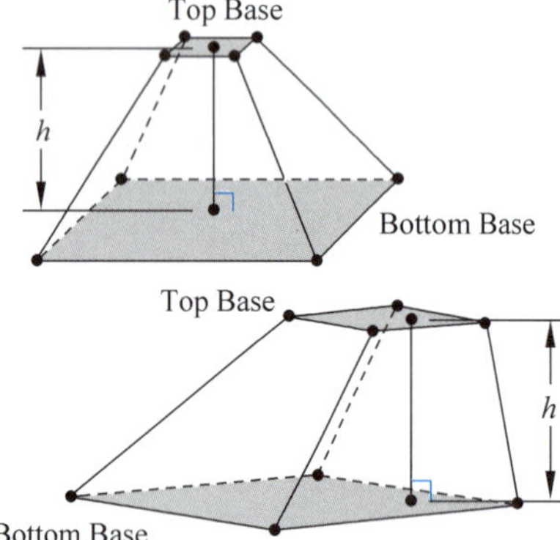	Note the similar right triangles. 注意相似的直角三角形 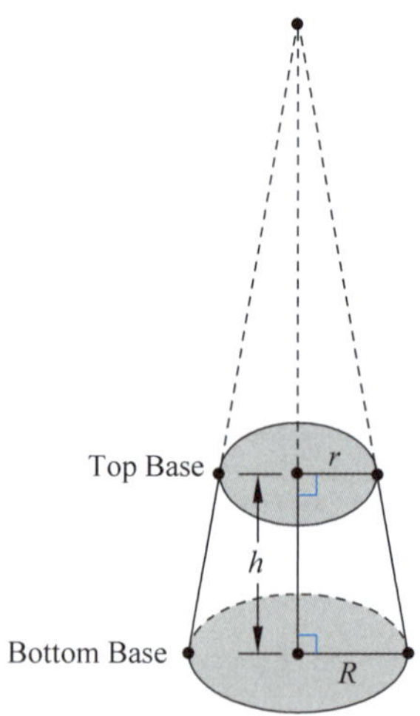 V of frustum = V of larger cone − V of smaller cone. 圆锥平截头体的体积 = 大圆锥体积 − 小圆锥体积。 Using algebra，运用代数， $$V=\frac{\pi}{3}h(R^2+r^2+R\times r)$$ Note： 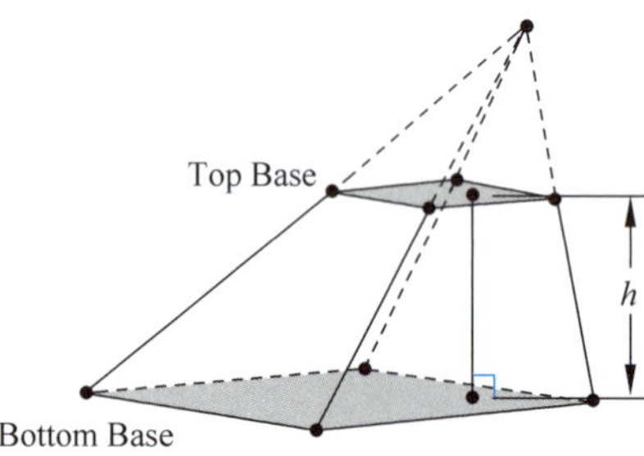 Similar to finding the volume of a conical frustum， 棱锥的平截头体与圆锥的平截头体的体积概念差不多。 V of frustum = V of larger pyramid − V of smaller pyramid. 棱锥平截头体的体积 = 大棱锥体积 − 小棱锥体积。 Let a and A be the areas of the top and bottom, respectively，设 a 和 A 分别为顶和底的面积， $$V=\frac{h}{3}\times(A+a+\sqrt{A\times a})$$

A.5 Common Geometry Symbols 常用几何符号

Table A-5 shows the common geometry symbols.

Table A-5

Symbol 符号	Name 名称	Example 例子
$\overline{AB}$ $\overline{ACB}$	line segment AB line segment ACB 线段	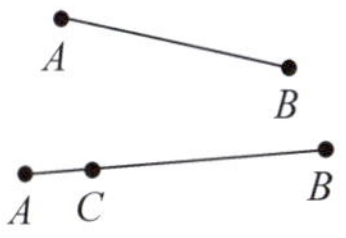 A B A C B
$\overleftrightarrow{AB}$	line AB 直线	 A B
$\overrightarrow{AB}$	ray AB 射线	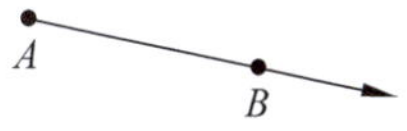 A B
$\vec{AB}$	vector AB 向量	 A B
AB	length of $\overline{AB}$ 长度	As shown in the figure below, $AB = d$ 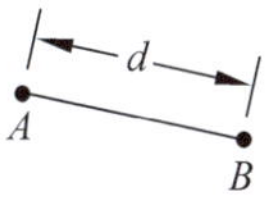 d A B
$\lvert\overrightarrow{AB}\rvert$	length of $\vec{AB}$ 长度	As shown in the figure below, $\lvert\vec{AB}\rvert = d$ 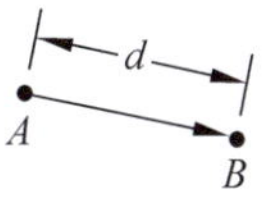 d A B
$\parallel$	parallel 平行	As shown in the figure below, $l \parallel m$ l m 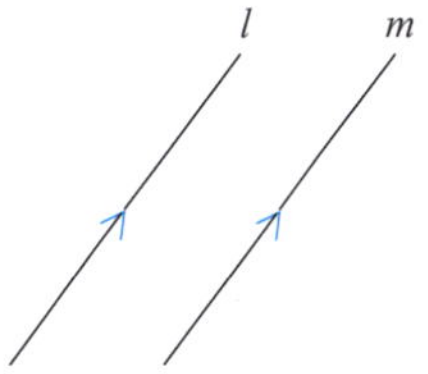
$\perp$	perpendicular 垂直	As shown in the figure below, $l \perp m$ 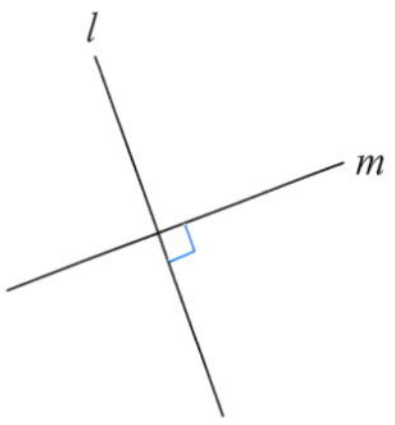 l m

续表

Symbol 符号	Name 名称	Example 例子
$\angle$	angle 角	The figure below shows $\angle ABC$, or simply $\angle B$ 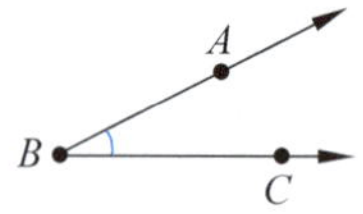 A B C
$m\angle$	measure 度数	The figure below shows $m\angle ABC = 45°$ Note: This also applies to the angular measure of arcs. 也应用到弧的度数 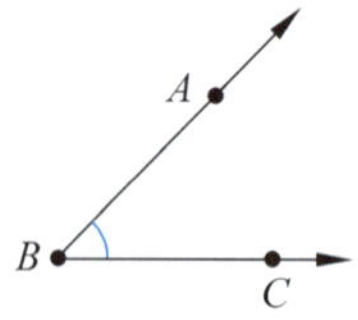 A B C
°	degree 度	The figure below shows a 45°-angle 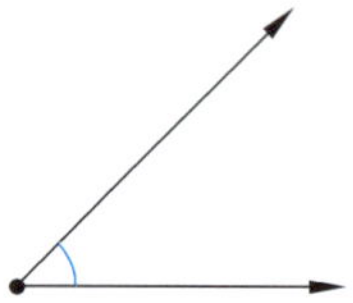
′	minute 分	$1' = \left(\frac{1}{60}\right)^{\circ}$
″	second 秒	$1'' = \left(\frac{1}{60}\right)' = \left(\frac{1}{3600}\right)^{\circ}$
$\triangle$	triangle 三角形	The figure below shows $\triangle ABC$ 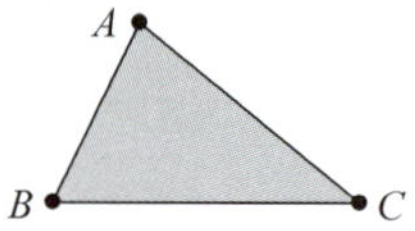 A B C
$ABC\cdots$	polygon $ABC\cdots$ 多边形	The figure below shows $ABCDE$ 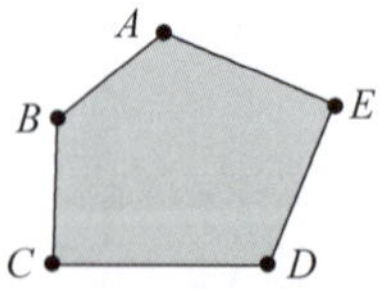 A B E C D
$[ABC\cdots]$	area of polygon $ABC\cdots$ 多边形面积	The figure below shows that $[ABCD] = 25$ 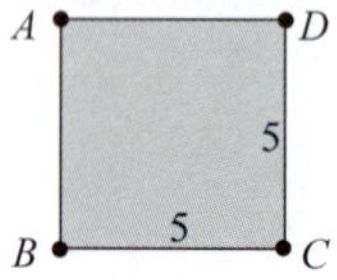 A D 5 5 B C

续表

Symbol 符号	Name 名称	Example 例子
$\odot$	circle 圆	The figure below shows $\odot$ O 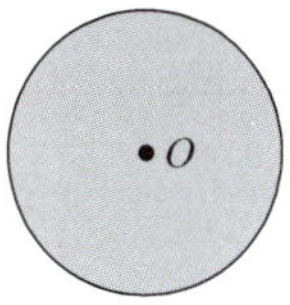
$\overset{\frown}{AB}$	minor arc 劣弧	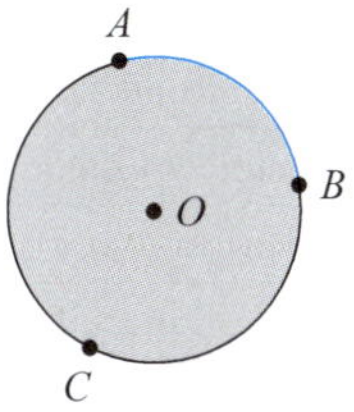
$\overset{\frown}{ACB}$	major arc 优弧	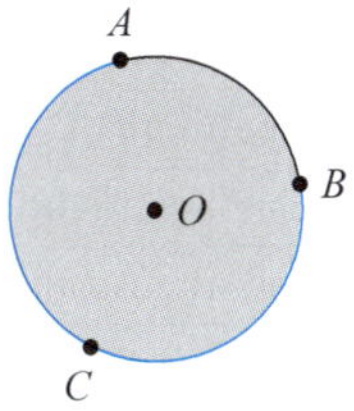
π	pi 圆周率	$\pi \approx 3.14159265358979323 85$
$\cong$	congruent 全等	The figure below shows $\overline{AB} \cong \overline{CD}$ 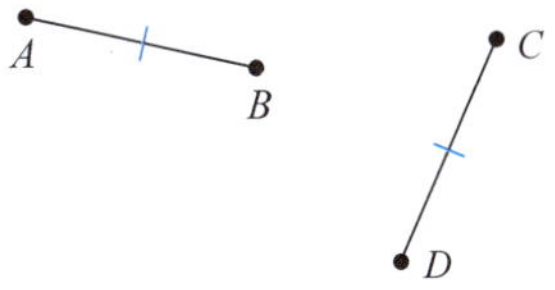 Note: The congruence symbol applies to comparing angles, vectors, arcs, curves, polygons, circles, sectors, other plane shapes and solids. 全等符号同样应用到角、向量、弧、曲线、多边形、圆、扇形、其他平面图形和立体的对比
$\sim$	similar 相似	The figure below shows $ABCDE \sim A'B'C'D'E$ 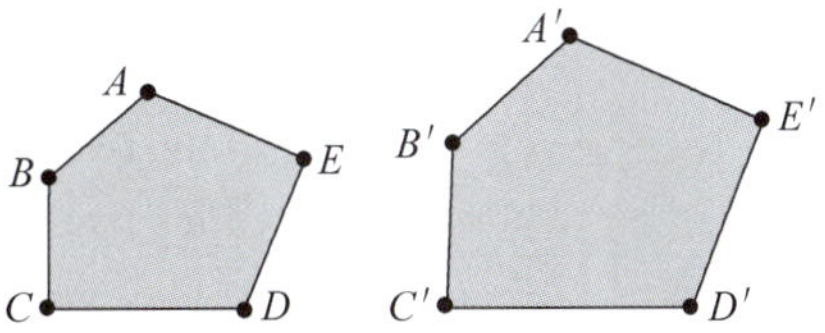 Note: The similarity symbol applies to comparing polygons, circles, sectors, other plane shapes and solids. 相似符号同样应用到多边形、圆、扇形、其他平面图形和立体的对比

A.6 Common Numbers 常用数字

Table A-6 shows the names of common numbers.

Table A-6

Number 数字	English Name 英语名称	Chinese Name 汉语名称
	Integers and Decimals 整数与小数	
1	one	一
10	ten	十
$100=10^2$	one hundred	一百
$1000=10^3$	one thousand	一千
$10000=10^4$	ten thousand	一万
$100000=10^5$	one hundred thousand	十万
$1000000=10^6$	one million	一百万
$1000000000=10^9$	one billion	十亿
$1000000000000=10^{12}$	one trillion	一万亿(一兆)
$0.1=10^{-1}$	one tenth	十分之一
$0.01=10^{-2}$	one hundredth	百分之一
$0.001=10^{-3}$	one thousandth	千分之一
$0.0001=10^{-4}$	one ten-thousandth	万分之一
$0.00001=10^{-5}$	one hundred-thousandth	十万分之一
$0.000001=10^{-6}$	one millionth	百万分之一
$0.000000001=10^{-9}$	one billionth	十亿分之一
$0.000000000001=10^{-12}$	one trillionth	万亿分之一(兆分之一)
0.26	(1) zero point two six (2) twenty-six hundredths	零点二六
6.047	(1) six point zero four seven (2) six and forty-seven thousandths	六点零四七
80.00315	(1) eighty point zero zero three one five (2) eighty and three hundred fifteen hundred-thousandths	八十点零零三一五
$0.\overline{4}$	(1) zero point four, four repeating (2) zero point four, four recurring	零点四,四循环
$2.53\overline{7}$	(1) two point five three seven, seven repeating (2) two point five three seven, seven recurring	二点五三七,七循环
$7.03\overline{82}$	(1) seven point zero three eight two, eight two repeating (2) seven point zero three eight two, eight two recurring	七点零三八二,八二循环
$9.454\overline{5}$	(1) nine point four five four five, five repeating (2) nine point four five four five, five recurring	九点四五四五,五循环
	Fractions 分数	
$\frac{1}{2}$	(1) one over two (2) one half	(1) 二分之一 (2) 一半
$\frac{1}{3}$	(1) one over three (2) one third	三分之一

续表

Number 数字	English Name 英语名称	Chinese Name 汉语名称
	Fractions 分数	
$\frac{1}{4}$	(1) one over four (2) one quarter	四分之一
$\frac{1}{5}$	(1) one over five (2) one fifth	五分之一
$\frac{1}{n}$	(1) one over n (2) one nth	n 分之一
$\frac{p}{q}$	(1) p over q (2) p qth	q 分之 p
$3\frac{1}{2}$	(1) three and one over two (2) three and one half	三又二分之一
$7\frac{3}{4}$	(1) seven and three over four (2) seven and three quarters	七又四分之三
$8\frac{4}{9}$	(1) eight and four over nine (2) eight and fourth ninths	八又九分之四

A.7 Common Variable Symbols 常用变量符号

Table A-7 shows the names of common variable symbols.

Table A-7

Symbol 符号	English Name 英语名称	Chinese Name 汉语名称
$\hat{a}$	a hat	a 帽
$\check{a}$	a check	a 勾
$\grave{a}$	a grave	a 低头符
$\acute{a}$	a acute	a 抬头符
$\bar{a}$	a bar	a 杠
$\bar{\bar{a}}$	a double bar	a 双杠
$\tilde{a}$	a tilde	a 波浪号
$\dot{a}$	a dot	a 点
$\ddot{a}$	a two dots	a 两点
$\dddot{a}$	a three dots	a 三点
a'	(1) a prime (2) a minutes	(1) a 撇 (2) a 分
a''	(1) a double prime (2) a seconds	(1) a 双撇 (2) a 秒
a'''	a triple prime	a 三撇
a_m	(1) a subscript m (2) a sub m	a 下标 m
a^m	(1) a superscript m (2) a super m	a 上标 m

A.8 Common Basic Symbols 常用基本符号

Table A-8 shows the common basic symbols.

Table A-8

Symbol 符号	Name 名称	Examples 例子
Basic Operation Symbols 基本运算符号		
$+$	plus/positive 加/正	(1) Plus 加 $3+7=10$ (2) Positive 正 $+6$ is the same as 6, which reads as "positive six", or simply "six" $+6$ 与 6 相同,读作"正六"或者直接"六"
$-$	minus/negative 减/负	(1) Minus 减 $10-3=7$ (2) Negative 负 -6 reads as "negative six" -6 读作"负六"
$\times$ or $\cdot$	multiplied by/times 乘以	$4\times5=20$ $4\cdot5=20$
$\div$ or/or $-$	divided by 除以	$20\div4=5$ $20/4=5$ $\frac{20}{4}=5$
$:$	is to/divided by 比/除以	$15:20=3:4$
$\pm$	plus or minus 加减	(1) $10\pm7=3$ or 17 (2) $(a\pm b)^2=a^2\pm2ab+b^2$ means that: • $(a+b)^2=a^2+2ab+b^2$ • $(a-b)^2=a^2-2ab+b^2$
$\mp$	minus or plus 减加	$a^3\pm b^3=(a\pm b)(a^2\mp ab+b^2)$ means that: • $a^3+b^3=(a+b)(a^2-ab+b^2)$ • $a^3-b^3=(a-b)(a^2+ab+b^2)$
$\lvert\ \rvert$	absolute value 绝对值	(1) $\lvert5\rvert=\lvert-5\rvert=5$ (2) $\lvert0\rvert=0$

续表

Symbol 符号	Name 名称	Examples 例子
	Relational Operators 关系运算符	
$\mid$	divides 能整除	$7\mid 21$ because 7 is a factor of 21 $7\mid 21$，因为 7 是 21 的因数
$\nmid$	does not divide 不能整除	$7\nmid 22$ because 7 is not a factor of 22 $7\nmid 22$，因为 7 不是 22 的因数
$\propto$	is proportional to 成正比	distance$\propto$speed 路程$\propto$速度
$=$	is equal to/equals 等于	$2-3+5=4$
$\approx$	is approximately equal to 约等于	To the nearest integer，$5.83\approx 6$ 精确到整数，$5.83\approx 6$
$\neq$	is not equal to 不等于	$1+1\neq 3$
$>$	is greater than 大于	$5>3$
$\gg$	is much greater than 远大于	$25000\gg 1$
$\geqslant$	is greater than or equal to 大于或等于	(1) $5\geqslant 3$ (2) $2\geqslant 2$
$<$	is less than 小于	$3<5$
$\ll$	is much less than 远小于	$1\ll 25000$
$\leqslant$	is less than or equal to 小于或等于	(1) $3\leqslant 5$ (2) $2\leqslant 2$
$\equiv$	is identically equal to/congruent 恒等/合同	$17\equiv 32 \pmod 5$ because 17 and 32 leave the same remainder (2) when divided by 5 $17\equiv 32 \pmod 5$，因为 17 与 32 除以 5 的余数(2)相同
$\not\equiv$	is not identically equal to/not congruent 非恒等/非合同	$17\not\equiv 41 \pmod 5$ because 17 and 41 leave different remainders when divided by 5 $17\not\equiv 41 \pmod 5$，因为 17 与 41 除以 5 的余数不同

A.9 Common Advanced Symbols 常用进阶符号

Table A-9 shows the common advanced symbols.

Table A-9

Symbol 符号	Name 名称	Examples 例子
	Set Theory Symbols 集合论符号	
$\varnothing$	empty set/null set 空集	(1) $\varnothing=\{\ \}$ (2) The set containing all positive integers less than -2 is $\varnothing$. 包含所有小于 -2 的正整数的集合为$\varnothing$

续表

Symbol 符号	Name 名称	Examples 例子
Set Theory Symbols 集合论符号		
$\mathbb{C}$	the set of complex numbers 复数集	$\{a+bi \mid a$ and b are real numbers$\}$ $\{a+bi \mid a$ 与 b 为实数$\}$
$\mathbb{R}$	the set of real numbers 实数集	$\{a \mid a$ can be located on the number line$\}$ $\{a \mid a$ 位于数轴上$\}$
$\mathbb{Q}$	the set of rational numbers 有理数集	$\left\{\frac{p}{q} \mid p\right.$ and q are integers such that $\left.q \neq 0\right\}$ $\left\{\frac{p}{q} \mid p\right.$ 与 q 为整数,其中 $\left.q \neq 0\right\}$
$\mathbb{Z}$	the set of integers 整数集	$\{\cdots,-2,-1,0,1,2,\cdots\}$
$\mathbb{N}$	the set of natural numbers 自然数集	$\{0,1,2,3,\cdots\}$ or $\{1,2,3,\cdots\}$, depending on convention $\{0,1,2,3,\cdots\}$ 或 $\{1,2,3,\cdots\}$,取决于惯例
$\mathbb{C} \setminus \mathbb{R}$	the set of imaginary numbers 虚数集	$\{a+bi \mid a$ and b are real numbers such that $b \neq 0\}$ $\{a+bi \mid a$ 与 b 为实数,其中 $b \neq 0\}$
$i\mathbb{R}$	the set of pure imaginary numbers 纯虚数集	$\{bi \mid b$ is a real number such that $b \neq 0\}$ $\{bi \mid b$ 为实数,其中 $b \neq 0\}$
		In the following examples, let $A=\{1,2,3\}$
$\subset$	proper subset 真子集	Set A has 7 proper subsets: 集合 A 有 7 个真子集: $\varnothing,\{1\},\{2\},\{3\},\{1,2\},\{1,3\},\{2,3\}$. For example, we can write: 例如,可以记作: (1) $\{3\} \subset A$ (2) $\{1,2\} \subset A$
$\subseteq$	subset 子集	Set A has 8 subsets: 集合 A 有 8 个子集: $\varnothing,\{1\},\{2\},\{3\},\{1,2\},\{1,3\},\{2,3\},\{1,2,3\}$. For example, we can write: 例如,可以记作: (1) $\{3\} \subseteq A$ (2) $\{1,2\} \subseteq A$ (3) $\{1,2,3\} \subseteq A$
$\supset$	proper superset 真超集	(1) $\{1,2,3,4\} \supset A$ (2) $\{1,2,3,7,9\} \supset A$
$\supseteq$	superset 超集	(1) $\{1,2,3,4\} \supseteq A$ (2) $\{1,2,3,7,9\} \supseteq A$ (3) $\{1,2,3\} \supseteq A$
$\in$	is an element of 是…的元素	(1) $1 \in A$ (2) $2 \in A$ (3) $3 \in A$

续表

Symbol 符号	Name 名称	Examples 例子
Set Theory Symbols 集合论符号		
$\notin$	is not an element of 不是…的元素	(1) $0 \notin A$ (2) $4 \notin A$
$\ni$	contains 包含	(1) $A \ni 1$ (2) $A \ni 2$ (3) $A \ni 3$
$\not\ni$	does not contain 不包含	(1) $A \not\ni 0$ (2) $A \not\ni 4$
$\cap$	intersection 交集	Let $A=\{2,3,5,7\}$ and $B=\{1,3,5,7,9\}$: $A \cap B=\{3,5,7\}$
$\bigcap$	intersection of all sets 所有集合的交集	Let $A_1=\{2,3,5,7\}$, $A_2=\{1,3,5,7,9\}$, $A_3=\{1,2,3,4,5\}$, and $A_4=\{3,4,5,6,7\}$: $\bigcap_{i=1}^{4} A_i=A_1 \cap A_2 \cap A_3 \cap A_4=\{3,5\}$
$\cup$	union 并集	Let $A=\{2,3,5,7\}$ and $B=\{1,3,5,7,9\}$: $A \cup B=\{1,2,3,5,7,9\}$
$\bigcup$	union of all sets 所有集合的并集	Let $A_1=\{2,3,5,7\}$, $A_2=\{1,3,5,7,9\}$, $A_3=\{1,2,3,4,5\}$, and $A_4=\{3,4,5,6,7\}$: $\bigcup_{i=1}^{4} A_i=A_1 \cup A_2 \cup A_3 \cup A_4=\{1,2,3,4,5,6,7,9\}$
$-$ or $\backslash$	minus/difference 减/差	Let $A=\{1,2,3,4,5,6,7,8,9\}$ and $B=\{2,3,5,7\}$: $A-B=A \backslash B=\{1,4,6,8,9\}$
$'$ or ^{c}	complement 补集	Let $A=\{1,2,3,4,5,6,7,8,9\}$ be the sample space, and $B=\{2,3,5,7\}$: 设 $A=\{1,2,3,4,5,6,7,8,9\}$ 为样本空间，且 $B=\{2,3,5,7\}$： $B'=B^c=A-B=A \backslash B=\{1,4,6,8,9\}$
$\times$	Cartesian Product 笛卡儿积	Let $A=\{1,2\}$ and $B=\{7,8,9\}$: (1) $A \times B=\{(1,7),(1,8),(1,9),(2,7),(2,8),(2,9)\}$ (2) $B \times A=\{(7,1),(7,2),(8,1),(8,2),(9,1),(9,2)\}$
$\lvert A \rvert$ or card(A)	number of elements/cardinal 元素个数/基数	Let $A=\{2,3,5,7\}$: $\lvert A \rvert=\text{card}(A)=4$
$\{x \mid P(x)\}$ or $\{x : P(x)\}$	the set of all x with property P 满足性质 P 的所有 x 的集合	$\{x \mid x$ is a prime number less than $20\}$ $=\{x : x$ is a prime number less than $20\}$ $=\{2,3,5,7,11,13,17,19\}$
(x,y,z) or $\langle x,y,z \rangle$	ordered set of elements x, y, and z 元素为 x、y、z 的有序集合	(1) $(1,2,3) \neq (2,1,3)$ (2) $\langle 1,2,3 \rangle \neq \langle 2,1,3 \rangle$
$\{x_1,x_2,\cdots,x_n\}$ or $\{x_i\}_{i=1}^{n}$	the set with elements $x_1,x_2,\cdots,x_n$ 元素为 $x_1,x_2,\cdots,x_n$ 的集合	Let $x_1=1$, $x_2=3$, $x_3=5$, and $x_4=7$: $\{x_i\}_{i=1}^{4}=\{x_1,x_2,x_3,x_4\}=\{1,3,5,7\}$

续表

Symbol 符号	Name 名称	Examples 例子
Set Theory Symbols 集合论符号		
$\{x_1, x_2, \cdots\}$ or $\{x_i\}_{i=1}^{\infty}$	the set with elements $x_1, x_2, \cdots$, and so on 元素为$x_1, x_2, \cdots$(如此类推)的集合	Let $x_n = 2n-1$ for all positive integers n: $\{x_i\}_{i=1}^{\infty} = \{x_1, x_2, x_3, x_4, \cdots\} = \{1,3,5,7,\cdots\}$
$\{x_i\}_{i \in I}$	the set whose elements are x_i, where $i \in I$ 元素为x_i 的集合,其中 $i \in I$	Let $x_1 = 1$, $x_2 = 3$, $x_3 = 5$, $x_4 = 7$, $x_5 = 9$, and $I = \{1,3,4\}$: $\{x_i\}_{i \in I} = \{x_1, x_3, x_4\} = \{1,5,7\}$
Complex Number Symbols 复数符号		
i	imaginary unit 虚数单位	$\mathrm{i} = \sqrt{-1}$, where $\mathrm{i}^2 = -1$
$x + y\mathrm{i}$	standard form of complex numbers 复数的标准形式	(1) $3-4\mathrm{i}$ (2) -7 (3) $5\mathrm{i}$
$\mathrm{Re}(z)$	real part of z z 的实部	(1) $\mathrm{Re}(3-4\mathrm{i}) = 3$ (2) $\mathrm{Re}(-7) = -7$ (3) $\mathrm{Re}(5\mathrm{i}) = 0$ (4) For real numbers a and b: $\mathrm{Re}(a+b\mathrm{i}) = a$
$\mathrm{Im}(z)$	imaginary part of z z 的虚部	(1) $\mathrm{Im}(3-4\mathrm{i}) = -4$ (2) $\mathrm{Im}(-7) = 0$ (3) $\mathrm{Im}(5\mathrm{i}) = 5$ (4) For real numbers a and b: $\mathrm{Im}(a+b\mathrm{i}) = b$
$r(\cos\theta + \mathrm{i}\sin\theta)$ or $r\ \mathrm{cis}\theta$ or $r\mathrm{e}^{\mathrm{i}\theta}$	polar form of complex numbers 复数的极坐标形式	$2\left(\cos\frac{\pi}{6} + \mathrm{i}\sin\frac{\pi}{6}\right) = 2\mathrm{cis}\frac{\pi}{6} = 2\mathrm{e}^{\frac{\pi}{6}\mathrm{i}} = \sqrt{3} + \mathrm{i}$
$\|z\|$	magnitude of z / norm of z z 的模/z 的范数	(1) $\|3-4\mathrm{i}\| = 5$ (2) For real numbers a and b: $\|a+b\mathrm{i}\| = \sqrt{a^2+b^2}$ (3) $\left\|2\mathrm{e}^{\frac{\pi}{6}\mathrm{i}}\right\| = 2$ (4) For real numbers r and θ: $\|r\mathrm{e}^{\mathrm{i}\theta}\| = r$
$\arg(z)$	argument of z z 的辐角	(1) $\arg\left(2\mathrm{e}^{\frac{\pi}{6}\mathrm{i}}\right) = \frac{\pi}{6}$ (2) For real numbers r and θ: $\arg(r\mathrm{e}^{\mathrm{i}\theta}) = \theta$
$\bar{z}$	conjugate of z z 的共轭	(1) $\overline{3-4\mathrm{i}} = 3+4\mathrm{i}$ (2) $\overline{-7} = -7$ (3) $\overline{5\mathrm{i}} = -5\mathrm{i}$

续表

Symbol 符号	Name 名称	Examples 例子
	Matrix Symbols 矩阵符号	
$\mathcal{M}(m,n)$ or $\mathcal{M}_{m\times n}(\mathbb{R})$	the set of all m-by-n real matrices 所有 m 行 n 列的实矩阵集	Let $A=\begin{bmatrix}1&-2&3\\-4&5&-6\end{bmatrix}$: $A\in\mathcal{M}_{2\times3}(\mathbb{R})$
a_{ij}	the entry at the ith row jth column of matrix A 矩阵 A 中第 i 行第 j 列的元素	Let $A=\begin{bmatrix}1&-2&3\\-4&5&-6\end{bmatrix}$. We have (1) $a_{11}=1$ (2) $a_{12}=-2$ (3) $a_{13}=3$ (4) $a_{21}=-4$ (5) $a_{22}=5$ (6) $a_{23}=-6$
$(a_{ij})_{m\times n}$	the entry at the ith row jth column of $m\times n$ matrix A $m\times n$ 矩阵 A 中第 i 行第 j 列的元素	From $(a_{ij})_{3\times2}=ij+1$, we have $A=\begin{bmatrix}2&3\\3&5\\4&7\end{bmatrix}$
I_n	$n\times n$ unit matrix/$n\times n$ identity matrix $n\times n$ 单位矩阵	(1) $I_1=[1]$ (2) $I_2=\begin{bmatrix}1&0\\0&1\end{bmatrix}$ (3) $I_3=\begin{bmatrix}1&0&0\\0&1&0\\0&0&1\end{bmatrix}$
A^{T} or ${}^{\mathrm{T}}A$ or A^{t} or ${}^{\mathrm{t}}A$	transpose of matrix A 矩阵 A 的转置	(1) $\begin{bmatrix}1&2&3\end{bmatrix}^{\mathrm{T}}=\begin{bmatrix}1\\2\\3\end{bmatrix}$ (2) $\begin{bmatrix}1&-2&3\\-4&5&-6\end{bmatrix}^{\mathrm{T}}=\begin{bmatrix}1&-4\\-2&5\\3&-6\end{bmatrix}$ (3) $[(a_{ij})_{m\times n}]^{\mathrm{T}}=(a_{ji})_{n\times m}$
AB	product of matrices A and B 矩阵 A 与 B 的积	(1) $\begin{bmatrix}-2&1\\-1&3\end{bmatrix}\begin{bmatrix}1&-1\\0&4\end{bmatrix}=\begin{bmatrix}-2&6\\-1&13\end{bmatrix}$ (2) $\begin{bmatrix}1&0&-1\\2&-3&2\end{bmatrix}\begin{bmatrix}-2&-1\\1&-2\\3&0\end{bmatrix}=\begin{bmatrix}-5&-1\\-1&4\end{bmatrix}$ (3) $\begin{bmatrix}-2&-1\\1&-2\\3&0\end{bmatrix}\begin{bmatrix}1&0&-1\\2&-3&2\end{bmatrix}=\begin{bmatrix}-4&3&0\\-3&6&-5\\3&0&-3\end{bmatrix}$ (4) $AB\neq BA$, as Examples 2 and 3 indicate. (5) The product $\begin{bmatrix}-2&1\\-1&3\end{bmatrix}\begin{bmatrix}-2&-1\\1&-2\\3&0\end{bmatrix}$ is undefined due to incompatible sizes

续表

Symbol 符号	Name 名称	Examples 例子
Matrix Symbols 矩阵符号		
det($\boldsymbol{A}$) or $\lvert \boldsymbol{A} \rvert$	determinant of square matrix $\boldsymbol{A}$ 方块矩阵 $\boldsymbol{A}$ 的行列式	(1) $\det([a])=a$ (2) $\det\left(\begin{bmatrix} a & b \\ c & d \end{bmatrix}\right)=a\det([d])-b\det([c])$ $=ad-bc$ (3) $\det\left(\begin{bmatrix} a & b & c \\ d & e & f \\ g & h & i \end{bmatrix}\right)$ $=a\det\left(\begin{bmatrix} e & f \\ h & i \end{bmatrix}\right)-b\det\left(\begin{bmatrix} d & f \\ g & i \end{bmatrix}\right)+c\det\left(\begin{bmatrix} d & e \\ g & h \end{bmatrix}\right)$ $=a(ei-fh)-b(di-fg)+c(dh-eg)$
$\boldsymbol{A}^{-1}$	inverse of square matrix $\boldsymbol{A}$ 方块矩阵 $\boldsymbol{A}$ 的逆矩阵	Let $\boldsymbol{A}=\begin{bmatrix} 1 & 2 \\ 3 & 4 \end{bmatrix}$ and $\boldsymbol{B}=\begin{bmatrix} -2 & 1 \\ \frac{3}{2} & -\frac{1}{2} \end{bmatrix}$. We have $\boldsymbol{AB}=\boldsymbol{BA}=\boldsymbol{I}_2$. Therefore, we conclude that $\boldsymbol{B}=\boldsymbol{A}^{-1}$
tr($\boldsymbol{A}$)	trace of square matrix $\boldsymbol{A}$ 方块矩阵 $\boldsymbol{A}$ 的迹数	$\operatorname{tr}\left(\begin{bmatrix} a_{11} & a_{12} & \cdots & a_{1n} \\ a_{21} & a_{22} & \cdots & a_{2n} \\ \vdots & \vdots & \ddots & \vdots \\ a_{n1} & a_{n2} & \cdots & a_{nn} \end{bmatrix}\right)=a_{11}+a_{22}+\cdots+a_{nn}$
Probability and Statistics Symbols 概率与统计符号		
S	sample space 样本空间	Rolling a standard six-sided die and recording its outcome, the sample space is $S=\{1,2,3,4,5,6\}$. 掷一颗标准六面骰子并记录结果，样本空间为 $S=\{1,2,3,4,5,6\}$。
X,Y	random variables 随机变量	Rolling two standard six-sided dice, let X be the sum of the outcomes, and let Y be the absolute difference of the outcomes. The sample space of X is $\{2,3,4,5,6,7,8,9,10,11,12\}$. The sample space of Y is $\{0,1,2,3,4,5\}$. 掷两颗标准六面骰子，设 X 为结果的和，Y 为结果的绝对差。 X 的样本空间为$\{2,3,4,5,6,7,8,9,10,11,12\}$。 Y 的样本空间为$\{0,1,2,3,4,5\}$。
$P(A)$	probability of event A 事件 A 的概率	Rolling a standard six-sided die, let X be the outcome. We have 掷一颗标准六面骰子，设 X 为结果。得 (1) $P(X=4)=\frac{1}{6}$ (2) $P(X\geqslant 4)=\frac{1}{2}$ (3) $P(X \text{ is prime})=\frac{1}{2}$ (4) $P(X \text{ is composite})=\frac{1}{3}$

续表

<table>
<tr><th>Symbol 符号</th><th>Name 名称</th><th>Examples 例子</th></tr>
<tr><td colspan="3">Probability and Statistics Symbols 概率与统计符号</td></tr>
<tr><td>$P(A\mid B)$</td><td>conditional probability (probability of event A given that event B occurs)
条件概率(事件 A 的概率,已知事件 B 发生)</td><td>Rolling a standard six-sided die, let X be the outcome. We have
掷一颗标准六面骰子,设 X 为结果。得
(1) $P(X=4\mid X>1)=\frac{1}{5}$
(2) $P(X\geqslant 4\mid X>2)=\frac{3}{4}$
(3) $P(X\geqslant 3\mid X \text{ is prime})=\frac{2}{3}$
(4) $P(X\geqslant 3\mid X \text{ is composite})=1$</td></tr>
<tr><td>$E(X)$</td><td>expected value of X
X 的期望值</td><td>Flip an unfair coin with $P(\text{heads})=\frac{3}{4}$ and $P(\text{tails})=\frac{1}{4}$. Suppose that you earn \$10 every time you get heads, and lose \$15 every time you get tails, the expected earning of a single flip is
$$E(X)=\frac{3}{4}\times 10+\frac{1}{4}\times(-15)=\boxed{\$3.75}.$$
抛一枚 $P(\text{正面})=\frac{3}{4}$ 且 $P(\text{反面})=\frac{1}{4}$ 的不公平的硬币。若正面朝上,你赚 \$10。若反面朝上,你亏 \$15。抛一次硬币的期望收益是
$$E(X)=\frac{3}{4}\times 10+\frac{1}{4}\times(-15)=\boxed{\$3.75}。$$</td></tr>
<tr><td>$E(X\mid Y)$</td><td>expected value of X given Y
X 的期望值,已知 Y 的值</td><td>Suppose that
$$Y=\begin{cases}1 \text{ with probability } \frac{3}{8}\\ 2 \text{ with probability } \frac{5}{8}\end{cases}$$
and
$$X\mid Y=\begin{cases}3Y \text{ with probability } \frac{3}{4}\\ 4Y \text{ with probability } \frac{1}{4}\end{cases}.$$
We have
(1) $E(X\mid Y=1)=3\times\frac{3}{4}+4\times\frac{1}{4}=\frac{13}{4}$
(2) $E(X\mid Y=2)=6\times\frac{3}{4}+8\times\frac{1}{4}=\frac{13}{2}$
(3) $E(X\mid Y=y)=\begin{cases}\frac{13}{4} & \text{if } y=1\\ \frac{13}{2} & \text{if } y=2\end{cases}$</td></tr>
</table>

续表

Symbol 符号	Name 名称	Examples 例子
Probability and Statistics Symbols 概率与统计符号		
		In the following examples: Helen wants to know the height distribution of the 1500 students in her school. She measures the heights of the 30 students in her class, and wishes to use this data to infer about the population. She finds that in her class, the average height is 172cm with a standard deviation of 5cm. On the other hand, the average height of all students in her school is 168cm with a standard deviation of 11cm. 在以下例子中： 海伦希望知道她学校 1500 名学生的身高分布。她测量了她班里 30 名同学的身高，并希望用此数据推断总体。她发现在她班里，平均身高为 172 厘米，标准差为 5 厘米。另一方面，她学校的所有学生平均身高为 168 厘米，标准差为 11 厘米
μ	population mean 总体均值	$\mu = 168$ cm
σ^2	population variance 总体方差	$\sigma^2 = 11^2 = 121$ cm^2
σ	population standard deviation 总体标准差	$\sigma = 11$ cm
$\bar{x}$	sample mean 样本均值	$\bar{x} = 172$ cm
s^2	sample variance 样本方差	$s^2 = 5^2 = 25$ cm^2
s	sample standard deviation 样本标准差	$s = 5$ cm
M	median 中位数	(1) In the data set 1,3,6,6,6,7,11,11,13,15,20, the median is 7, so $M = 7$. (2) In the data set 1,3,6,6,6,7,11,11,13,15,20,51, the median is $\frac{7+11}{2} = 9$, so $M = 9$.
ρ	population correlation coefficient 总体相关系数	(1) $\rho = 0.98$ indicates a strong positive association: $\rho^2 = 96.04\%$ of the population data can be explained by the least-square regression line. $\rho = 0.98$ 展现了强的正关联性：总体数据的 $\rho^2 = 96.04\%$可以被最小二乘法回归线所解释。 (2) $\rho = -0.15$ indicates a weak negative association: $\rho^2 = 2.25\%$ of the population data can be explained by the least-square regression line. $\rho = -0.15$ 展现了弱的负关联性：总体数据的 $\rho^2 = 2.25\%$可以被最小二乘法回归线所解释

续表

<table>
<tr><th>Symbol 符号</th><th>Name 名称</th><th>Examples 例子</th></tr>
<tr><td colspan="3">Probability and Statistics Symbols 概率与统计符号</td></tr>
<tr><td>r</td><td>sample correlation coefficient
样本相关系数</td><td>(1) $r=0.1$ indicates a weak positive association: $r^2=1\%$ of the sample data can be explained by the least-square regression line.
$r=0.1$ 展现了弱的正关联性：样本数据的 $r^2=1\%$ 可以被最小二乘法回归线所解释。
(2) $r=-0.9$ indicates a strong negative association: $r^2=81\%$ of the sample data can be explained by the least-square regression line.
$r=-0.9$ 展现了强的负关联性：样本数据的 $r^2=81\%$ 可以被最小二乘法回归线所解释</td></tr>
<tr><td colspan="3">Mathematical Logic Symbols 数理逻辑符号</td></tr>
<tr><td>¬</td><td>not
逻辑非</td><td><table><tr><th>P</th><th>¬P</th></tr><tr><td>True</td><td>False</td></tr><tr><td>False</td><td>True</td></tr></table></td></tr>
<tr><td>∧</td><td>and
逻辑与</td><td><table><tr><th>P</th><th>Q</th><th>P∧Q</th></tr><tr><td>True</td><td>True</td><td>True</td></tr><tr><td>True</td><td>False</td><td>False</td></tr><tr><td>False</td><td>True</td><td>False</td></tr><tr><td>False</td><td>False</td><td>False</td></tr></table></td></tr>
<tr><td>∨</td><td>or
逻辑或</td><td><table><tr><th>P</th><th>Q</th><th>P∨Q</th></tr><tr><td>True</td><td>True</td><td>True</td></tr><tr><td>True</td><td>False</td><td>True</td></tr><tr><td>False</td><td>True</td><td>True</td></tr><tr><td>False</td><td>False</td><td>False</td></tr></table></td></tr>
<tr><td>⊕</td><td>xor
逻辑异或</td><td><table><tr><th>P</th><th>Q</th><th>P⊕Q</th></tr><tr><td>True</td><td>True</td><td>False</td></tr><tr><td>True</td><td>False</td><td>True</td></tr><tr><td>False</td><td>True</td><td>True</td></tr><tr><td>False</td><td>False</td><td>False</td></tr></table>$P\oplus Q=(P\vee Q)\wedge\neg(P\wedge Q)$</td></tr>
<tr><td>⇒</td><td>conditional/implies
条件/蕴含</td><td><table><tr><th>P</th><th>Q</th><th>P⇒Q</th></tr><tr><td>True</td><td>True</td><td>True</td></tr><tr><td>True</td><td>False</td><td>False</td></tr><tr><td>False</td><td>True</td><td>True</td></tr><tr><td>False</td><td>False</td><td>True</td></tr></table>$P\Rightarrow Q=\neg P\vee Q$</td></tr>
</table>

续表

<table>
<tr><th>Symbol 符号</th><th>Name 名称</th><th>Examples 例子</th></tr>
<tr><td colspan="3">Mathematical Logic Symbols 数理逻辑符号</td></tr>
<tr><td>$\Leftrightarrow$</td><td>biconditional/if and only if
双条件/当且仅当</td><td>

P	Q	$P \Leftrightarrow Q$
True	True	True
True	False	False
False	True	False
False	False	True

$P \Leftrightarrow Q = (P \Rightarrow Q) \wedge (Q \Rightarrow P) = (\neg P \vee Q) \wedge (P \vee \neg Q)$</td></tr>
</table>

A.10 Common Functions 常用函数

Table A-10 shows the common functions.

Table A-10

Type of Functions 函数种类	Standard Form 标准形式	Examples 例子
Power Function 幂函数	x^a, where a is rational x^a,其中 a 为有理数	(1) x^3 (2) x^{-1} (3) x^π
Polynomial Function 多项式函数	$a_n x^n + a_{n-1} x^{n-1} + \cdots + a_1 x + a_0$, where $a_n, a_{n-1}, \cdots, a_1, a_0$ are constants with $a_n \neq 0$, and all exponents are nonnegative 其中 $a_n, a_{n-1}, \cdots, a_1, a_0$ 为常数,其中 $a_n \neq 0$,且所有指数均为非负数	(1) Linear 线性 $4x+7$ (2) Quadratic 二次 $-3x^2-5x+10$ (3) Cubic 三次 $7x^3-8x+1$ (4) Quartic 四次 x^4+2x^3-3x+4 (5) Quintic 五次 $-x^5+8x^4-3x^2+1$
Radical Function 根式函数	$x^{\frac{1}{n}}$ or $\sqrt[n]{x}$, where n an integer$\geqslant 2$ 其中 n 为$\geqslant 2$ 的整数	(1) Square Root 平方根 $x^{\frac{1}{2}}$ or $\sqrt{x}$ (2) Cube Root 立方根 $x^{\frac{1}{3}}$ or $\sqrt[3]{x}$ (3) Fourth Root 四次方根 $x^{\frac{1}{4}}$ or $\sqrt[4]{x}$ (4) nth Root n 次方根 $x^{\frac{1}{n}}$ or $\sqrt[n]{x}$

续表

Type of Functions 函数种类	Standard Form 标准形式	Examples 例子
Rational Function 有理函数	$\frac{P(x)}{Q(x)}$, where $P(x)$ and $Q(x)$ are polynomial functions 其中 $P(x)$ 与 $Q(x)$ 均为多项式函数	$\frac{-3x^2-5x+10}{x^4+2x^3-3x^2+4x-5}$
Exponential Function 指数函数	b^x, where b is a positive real number not equal to 1 其中 b 为不等于 1 的正实数	(1) 2^x (2) 0.4^x (3) π^x (4) e^x
Logarithmic Function 对数函数	$\log_b x$, where b is a positive real number not equal to 1 其中 b 为不等于 1 的正实数	(1) $\log_2 x$ (2) $\log x$ or $\lg x$ (when $b=10$) (3) $\ln x$ (when $b=e$)
Trigonometric Functions 三角函数	(1) $\sin x$ 正弦函数 (2) $\cos x$ 余弦函数 (3) $\tan x$ or $\mathrm{tg}\ x$ 正切函数 (4) $\csc x$ or $\mathrm{cosec}\ x$ 余割函数 (5) $\sec x$ 正割函数 (6) $\cot x$ or $\mathrm{ctg}\ x$ 余切函数	
Inverse Trigonometric Functions 反三角函数	(1) $\sin^{-1} x$ or $\arcsin x$ 反正弦函数 (2) $\cos^{-1} x$ or $\arccos x$ 反余弦函数 (3) $\tan^{-1} x$ or $\arctan x$ 反正切函数 (4) $\csc^{-1} x$ or $\mathrm{arccsc}\ x$ 反余割函数 (5) $\sec^{-1} x$ or $\mathrm{arcsec}\ x$ 反正割函数 (6) $\cot^{-1} x$ or $\mathrm{arccot}\ x$ 反余切函数	

A.11 Special Functions 特殊函数

Table A-11 shows the special functions.

Table A-11

Name of Function 函数名称	Notation and Definition 表示法与定义	Examples 例子
Factorial 阶乘	$n!=n\cdot(n-1)\cdot(n-2)\cdot\cdots\cdot 1$, where n is a nonnegative integer; by definition, $0!=1$ 其中 n 为非负整数；根据定义，$0!=1$	$5!=5\times4\times3\times2\times1=120$
Permutation 排列	$A_n^k=n(n-1)(n-2)\cdots(n-k+1)$ $=\dfrac{n!}{(n-k)!}$ When $n<k$, we have $A_n^k=0$ by definition. Some alternate notations are: $A(n,k)$, ${}_nP_k$, $P(n,k)$	(1) $A_{10}^3=\dfrac{10!}{(10-3)!}=720$ (2) $A_{10}^{10}=\dfrac{10!}{(10-10)!}=10!$ (3) $A_{10}^0=\dfrac{10!}{(10-0)!}=1$ (4) $A_{10}^{11}=0$
Combination 组合	$C_n^k=\dfrac{A_n^k}{k!}=\dfrac{n!}{k!(n-k)!}$ When $n<k$, we have $C_n^k=0$ by definition. Some alternate notations are: $C(n,k)$, ${}_nC_k$, $\binom{n}{k}$	(1) $\binom{10}{3}=\dfrac{10!}{3!(10-3)!}=120$ (2) $\binom{10}{7}=\dfrac{10!}{7!(10-7)!}=120$ (3) $\binom{10}{0}=\dfrac{10!}{0!(10-0)!}=1$ (4) $\binom{10}{10}=\dfrac{10!}{10!(10-10)!}=1$ (5) $\binom{10}{11}=0$
Floor Function / Greatest Integer Function 向下取整函数	$\lfloor x\rfloor$ is the greatest integer less than or equal to x $\lfloor x\rfloor$是小于或等于 x 的最大整数	(1) $\lfloor 4\rfloor=4$ (2) $\lfloor \pi\rfloor=3$ (3) $\lfloor 6.79\rfloor=6$ (4) $\lfloor -2\rfloor=-2$ (5) $\lfloor -\pi\rfloor=-4$ (6) $\left\lfloor -\dfrac{35}{9}\right\rfloor=-4$
Ceiling Function / Least Integer Function 向上取整函数	$\lceil x\rceil$ is the least integer greater than or equal to x $\lceil x\rceil$是大于或等于 x 的最小整数	(1) $\lceil 4\rceil=4$ (2) $\lceil \pi\rceil=4$ (3) $\lceil 6.79\rceil=7$ (4) $\lceil -2\rceil=-2$ (5) $\lceil -\pi\rceil=-3$ (6) $\left\lceil -\dfrac{35}{9}\right\rceil=-3$

续表

Name of Function 函数名称	Notation and Definition 表示法与定义	Examples 例子
Round Function 四舍五入函数	round(x,n) rounds the number x to the nth digit after the decimal point round(x,n)把数字x四舍五入至小数点后的第n位	Note that $\pi=3.141592\cdots$： (1) round$(100\pi,3)=314.159$ (2) round$(100\pi,2)=314.16$ (3) round$(100\pi,1)=314.2$ (4) round$(100\pi,0)=314$(nearest integer) (5) round$(100\pi,-1)=310$(nearest ten) (6) round$(100\pi,-2)=300$(nearest hundred)
Sign Function / Signum Function 符号函数	$\operatorname{sgn}(x)=\begin{cases}0 & \text{if } x=0\\ \dfrac{x}{\lvert x\rvert} & \text{if } x\neq 0\end{cases}$	(1) $\operatorname{sgn}(12.34)=1$ (2) $\operatorname{sgn}(0)=0$ (3) $\operatorname{sgn}(-\pi)=-1$
Indicator Function / Characteristic Function 示性函数	$\chi_A(x)=\begin{cases}1 & \text{if } x\in A\\ 0 & \text{if } x\notin A\end{cases}$	(1) $\chi_{\mathbf{Z}}(-8)=1$ (2) $\chi_{\mathbf{Z}}\left(\frac{4}{5}\right)=0$ (3) $\chi_{\mathbf{Z}^+}(10)=1$ (4) $\chi_{\mathbf{Z}^-}(5)=0$
Unit Step Function 阶跃函数	$\varepsilon(x)=\begin{cases}0 & \text{if } x<0\\ \dfrac{1}{2} & \text{if } x=0\\ 1 & \text{if } x>0\end{cases}$	(1) $\varepsilon(-5.67)=0$ (2) $\varepsilon(0)=\frac{1}{2}$ (3) $\varepsilon(8)=1$
Hyperbolic Functions 双曲函数	(1) $\sinh x=\dfrac{e^x-e^{-x}}{2}$ 双曲正弦 (2) $\cosh x=\dfrac{e^x+e^{-x}}{2}$ 双曲余弦 (3) $\tanh x=\dfrac{\sinh x}{\cosh x}=\dfrac{e^x-e^{-x}}{e^x+e^{-x}}$ 双曲正切 (4) $\operatorname{csch} x=\dfrac{1}{\sinh x}=\dfrac{2}{e^x-e^{-x}}$ 双曲余割 (5) $\operatorname{sech} x=\dfrac{1}{\cosh x}=\dfrac{2}{e^x+e^{-x}}$ 双曲正割 (6) $\coth x=\dfrac{1}{\tanh x}=\dfrac{e^x+e^{-x}}{e^x-e^{-x}}$ 双曲余切	

A.12 Common Function Symbols 常用函数符号

Table A-12 shows the common function symbols.

Table A-12

Symbol 符号	Name 名称	Examples 例子
$f(x)$	the value of function f at x/f of x 函数 f 在 x 处的值	For $f(x)=2x+5$, we have (1) $f(3)=11$ (2) $f(0)=5$ (3) $f(-6)=-7$ (4) $f(-10.5)=-16$
$f:A\to B$	function (mapping/transformation) from A to B (映射/变换)从 A 到 B 的函数	(1) $3x: \mathbb{R}\to\mathbb{R}$ (2) $\lvert x\rvert: \mathbb{R}\to\{0\}\cup\mathbb{R}^+$ (3) $\sqrt{x}: \{0\}\cup\mathbb{R}^+\to\{0\}\cup\mathbb{R}^+$ (4) $x!: \{0\}\cup\mathbb{Z}^+\to\mathbb{Z}^+$
$f(A)$	image of set A by f f 于集合 A 的像	(1) For $f(x)=x^2$: $f(\mathbb{R})=\{0\}\cup\mathbb{R}^+$ (2) For $g(x)=2^x$: $g(\mathbb{R})=\mathbb{R}^+$ (3) For $h(x)=\sqrt{x}$: $h([36,100])=[6,10]$
$f\mid A$	restriction of function f's domain to set A 限制函数 f 的定义域至集合 A	For $f(x)=2x+1$, the graph of $f\mid[-5,10]$ is a line segment with endpoints $(-5,-9)$ and $(10,21)$ 对于 $f(x)=2x+1$, $f\mid[-5,10]$ 的图像是端点为 $(-5,-9)$ 与 $(10,21)$ 的线段
$f(x)\mid_a^b$ or $[f(x)]_a^b$	$f(b)-f(a)$	$x^2+3x\mid_{-1}^{5}=42$
(a,b)	interval from a to b, exclusive 从 a 到 b 的区间,不含两端	(1) $x\in(2,7)$ means that $2<x<7$ (2) $x\in(-\infty,5)$ means that $x<5$ (3) $x\in(3,\infty)$ means that $x>3$ (4) $x\in(-\infty,\infty)$ means that x is any real number
$(a,b]$	interval from a to b, excluding a but including b 从 a 到 b 的区间,不含 a 但含 b	(1) $x\in(2,7]$ means that $2<x\leqslant 7$ (2) $x\in(-\infty,5]$ means that $x\leqslant 5$
$[a,b)$	interval from a to b, including a but excluding b 从 a 到 b 的区间,含 a 但不含 b	(1) $x\in[2,7)$ means that $2\leqslant x<7$ (2) $x\in[3,\infty)$ means that $x\geqslant 3$
$[a,b]$	interval from a to b, inclusive 从 a 到 b 的区间,含两端	$x\in[2,7]$ means that $2\leqslant x\leqslant 7$
$g\circ f$, $g(f(x))$	function composition 复合函数	Let $f(x)=x^2$ and $g(x)=3x+1$: (1) $g\circ f=g(f(x))=3x^2+1$ (2) $f\circ g=f(g(x))=(3x+1)^2$ Note that $g(f(x))\neq f(g(x))$
$f^{-1}(x)$	inverse of function f f 的反函数	Let $f(x)=3x+2$. We have $f^{-1}(x)=\dfrac{x-2}{3}$. Note that $f(f^{-1}(x))=f^{-1}(f(x))=x$.

续表

Symbol 符号	Name 名称	Examples 例子
$f^{-1}(A)$	preimage of set A by f f 于集合 A 的原像	Let $f(x)=3x+2$. We have $f^{-1}(x)=\dfrac{x-2}{3}$. Note that (1) $f^{-1}([-7,17])=[-3,5]$ (2) $f^{-1}([3,10])=\left[\dfrac{1}{3},\dfrac{8}{3}\right]$
$\max\{f,g\}$	maximum of functions f and g 函数 f 与 g 的最大值	$\max\{x,x^2\}=\begin{cases}x^2 & \text{if } x<0\\ x & \text{if } 0\leqslant x\leqslant 1\\ x^2 & \text{if } x>1\end{cases}$
$\max\limits_{x\in A}f(x)$	maximum of $f(x)$ when x is in set A 当 x 在集合 A 时 $f(x)$ 的最大值	(1) $\max\limits_{x\in[-5,2]}3x+4=10$ (2) $\max\limits_{x\in[-3,8]}-x^2=0$
$\min\{f,g\}$	minimum of functions f and g 函数 f 与 g 的最小值	$\min\{x,x^2\}=\begin{cases}x & \text{if } x<0\\ x^2 & \text{if } 0\leqslant x\leqslant 1\\ x & \text{if } x>1\end{cases}$
$\min\limits_{x\in A}f(x)$	minimum of $f(x)$ when x is in set A 当 x 在集合 A 时 $f(x)$ 的最小值	(1) $\min\limits_{x\in[-5,2]}3x+4=-11$ (2) $\min\limits_{x\in[-3,8]}-x^2=-64$
$\dfrac{f(x+h)-f(x)}{h}$	difference quotient 差商	Assuming that $x\neq 0$: (1) Let $f(x)=3x+7$. We have $\dfrac{f(x+h)-f(x)}{h}=3$ (2) Let $g(x)=x^2$. We have $\dfrac{g(x+h)-g(x)}{h}=2x+h$ (3) Let $p(x)=\sqrt{x}$. We have $\dfrac{p(x+h)-p(x)}{h}=\dfrac{1}{\sqrt{x+h}+\sqrt{x}}$
$\lim\limits_{x\to a}f(x)$	limit of $f(x)$ as x approaches a 当 x 趋近 a 时 $f(x)$ 的极限	(1) $\lim\limits_{x\to 3}5x-1=14$ (2) $\lim\limits_{x\to 25}\sqrt{x}=5$ (3) $\lim\limits_{x\to 0}\sqrt{x}$ does not exist (4) $\lim\limits_{x\to 0}\dfrac{x}{\lvert x\rvert}$ does not exist
$\lim\limits_{x\to a^-}f(x)$	limit of $f(x)$ as x approaches a from the left 当 x 从左侧趋近于 a 时 $f(x)$ 的极限	(1) $\lim\limits_{x\to 3^-}5x-1=14$ (2) $\lim\limits_{x\to 25^-}\sqrt{x}=5$ (3) $\lim\limits_{x\to 0^-}\sqrt{x}$ does not exist (4) $\lim\limits_{x\to 0^-}\dfrac{x}{\lvert x\rvert}=-1$

续表

Symbol 符号	Name 名称	Examples 例子
$\lim_{x\to a^+} f(x)$	limit of $f(x)$ as x approaches a from the right 当 x 从右侧趋近 a 时 $f(x)$ 的极限	(1) $\lim_{x\to 3^+} 5x-1=14$ (2) $\lim_{x\to 25^+} \sqrt{x}=5$ (3) $\lim_{x\to 0^+} \sqrt{x}=0$ (4) $\lim_{x\to 0^+} \frac{x}{\lvert x\rvert}=1$
Δx, δx	delta x/increment of x 德尔塔 x/x 的增量	In the interval $[0,10]$: 在区间 $[0,10]$ 里: (1) $\Delta x=2$ divides this interval into 5 subintervals $\Delta x=2$ 把此区间划分为 5 个子区间 (2) $\Delta x=2.5$ divides this interval into 4 subintervals $\Delta x=2.5$ 把此区间划分为 4 个子区间 (3) $\Delta x=0.01$ divides this interval into 1000 subintervals $\Delta x=0.01$ 把此区间划分为 1000 个子区间
		In the following examples, let (1) $f(x)=4x-5$ (2) $g(x)=3x^2-5x$ (3) $h(x)=\sqrt{x}$
$\mathrm{d}f$	total differential of function f 函数 f 的全微分	
$\frac{\mathrm{d}f}{\mathrm{d}x}$, $f'(x)$, $(Df)(x)$	derivative of function f with respect to x 函数 f 关于 x 的导数	(1) $f'(x)=4$ (2) $g'(x)=6x-5$ (3) $h'(x)=\frac{1}{2\sqrt{x}}$
$\left.\frac{\mathrm{d}f}{\mathrm{d}x}\right\rvert_{x=a}$, $f'(a)$, $(Df)(a)$	value at a of the derivative of function f with respect to x 函数 f 关于 x 的导数在 a 处的值	(1) $f'(7)=4$ (2) $g'(7)=37$ (3) $h'(9)=\frac{1}{6}$
$\frac{\mathrm{d}^2 f}{\mathrm{d}x^2}$, $f''(x)$, $(D^2 f)(x)$	the second derivative of function f with respect to x 函数 f 关于 x 的二阶导数	(1) $f''(x)=0$ (2) $g''(x)=6$ (3) $h''(x)=-\frac{1}{4(\sqrt{x})^3}$
$\frac{\mathrm{d}^n f}{\mathrm{d}x^n}$, $f^{(n)}(x)$, $(D^n f)(x)$	the nth derivative of function f with respect to x 函数 f 关于 x 的 n 阶导数	Let $f(x)=4x^4-3x^3+2x^2-x+1$. We have (1) $f'(x)=16x^3-9x^2+4x-1$ (2) $f''(x)=48x^2-18x+4$ (3) $f'''(x)=96x-18$ (4) $f^4(x)=96$ (5) $f^5(x)=0$
$\frac{\partial f}{\partial x}$, $\partial_x f$	partial derivative of function f with respect to x 函数 f 关于 x 的偏导数	Let $f(x,y)=x^3y^4$. We have (1) $\frac{\partial f}{\partial x}=3x^2y^4$ (2) $\frac{\partial f}{\partial y}=4x^3y^3$

续表

Symbol 符号	Name 名称	Examples 例子
		In the following examples, let (1) $f(x)=5$ (2) $g(x)=6x-7$ (3) $h(x)=\sqrt{x}$
$\int f(x)\,\mathrm{d}x$	indefinite integral of function f 函数 f 的不定积分	(1) $\int f(x)\,\mathrm{d}x = 5x + C$ for some constant C (2) $\int g(x)\,\mathrm{d}x = 3x^2 - 7x + C$ for some constant C (3) $\int h(x)\,\mathrm{d}x = \frac{2}{3}x^{\frac{3}{2}} + C$ for some constant C
$\int_a^b f(x)\,\mathrm{d}x$	definite integral of function f from a to b 函数 f 从 a 到 b 的定积分	(1) $\int_3^7 f(x)\,\mathrm{d}x = 20$ (2) $\int_2^7 g(x)\,\mathrm{d}x = 100$
$\iint_A f(x,y)\,\mathrm{d}A$	double integral of function $f(x,y)$ over set A 函数 $f(x,y)$在集合 A 的二重积分	(1) $\int_0^1 \left(\int_0^2 xy^2\,\mathrm{d}x\right)\mathrm{d}y = \int_0^1 2y^2\,\mathrm{d}y = \frac{2}{3}$ (2) $\int_0^2 \left(\int_0^{\frac{x}{2}} xy^2\,\mathrm{d}y\right)\mathrm{d}x = \int_0^2 \frac{x^4}{24}\mathrm{d}x = \frac{4}{15}$

A.13 Greek Alphabet 希腊文字母表

Table A-13 shows the Greek alphabet.

Table A-13

Ordinal 序号	Uppercase 大写	Lowercase 小写	English 英语	Chinese 汉语	Usage 用法
1	Α	α	Alpha	阿尔法	(1) angle 角度 (2) coefficient 系数 (3) first 第一
2	Β	β	Beta	贝塔	(1) angle 角度 (2) coefficient 系数
3	Γ	γ	Gamma	伽马	angle 角度
4	Δ	δ	Delta	德尔塔	(1) variation 变化量 (2) quadratic discriminant 一元二次方程的判别式
5	Ε	ε	Epsilon	艾普西隆	(1) base of logarithms 对数的底数 (2) limit 极限
6	Ζ	ζ	Zeta	泽塔	(1) coefficient 系数 (2) azimuth angle 方位角
7	Η	η	Eta	伊塔	
8	Θ	θ	Theta	西塔	(1) temperature 温度 (2) angle 角度
9	Ι	ι	Iota	约塔	smallness 微小
10	Κ	κ	Kappa	卡帕	
11	Λ	λ	Lambda	拉姆达	volume 体积

续表

Ordinal 序号	Uppercase 大写	Lowercase 小写	English 英语	Chinese 汉语	Usage 用法
12	M	μ	Mu	谬	population mean 总体均值
13	N	ν	Nu	纽	
14	Ξ	ξ	Xi	克西	random variable 随机变量
15	O	o	Omicron	奥米克戎	infinitesimal of higher order 高阶无穷小函数
16	Π	π	Pi	派	(1) pi 圆周率 (2) $\pi(n)$: number of primes not exceeding n 不大于 n 的质数个数 (3) product 求积符号
17	P	ρ	Rho	柔	radius of sphere 球的半径
18	Σ	σ	Sigma	西格马	(1) population standard deviation 总体标准差 (2) summation 求和符号
19	T	τ	Tau	陶	(1) time constant 时间常数 (2) 2π 两倍圆周率
20	Y	υ	Upsilon	宇普西隆	displacement 位移
21	Φ	ϕ	Phi	斐	(1) angle 角 (2) diameter 直径
22	X	χ	Chi	希	chi-square distribution 卡方分布
23	Ψ	ψ	Psi	普西	speed of angle 角速
24	Ω	ω	Omega	奥米伽	angular frequency/circular frequency 角频率/圆频率

A.14 Formulas/Postulates/Lemmas/Theorems/Corollaries 公式/公理/引理/定理/推论

Algebra 代数

- Fundamental Theorem of Arithmetic 算术基本定理 (Prime Factorization Property 1) 16
- Closure Property 封闭性 62
- Commutative Property 交换律 61
- Associative Property 结合律 61
- Distributive Property 分配律 61
- Identity Property 恒等性 62
- Inverse Property 逆等性 63
- Square of a Sum Formula 和的平方公式 75
- Square of a Difference Formula 差的平方公式 75
- Difference of Squares Formula 平方差公式 75
- Quadratic Formula 一元二次方程求根公式 114
- 1.5 IQR Rule 1.5 四分位距法则 208
- Counting Principle 计数原理 254

Geometry 几何

- Midpoint Formula 中点公式 139
- Distance Formula 距离公式 140
- Segment Addition Postulate 线段相加公理 280
- Angle Addition Postulate 角相加公理 283
- Vertical Angles Theorem 对顶角定理 313
- Corresponding Angles Postulate 同位角公理 317
- Converse of Corresponding Angles Postulate 同位角公理的逆公理 317
- Alternate Interior Angles Theorem 内错角定理 317
- Converse of Alternate Interior Angles Theorem 内错角定理的逆定理 317
- Alternate Exterior Angles Theorem 外错角定理 318
- Converse of Alternate Exterior Angles Theorem 外错角定理的逆定理 318
- Consecutive Interior Angles Theorem 同旁内角定理 319
- Converse of Consecutive Interior Angles Theorem 同旁内角定理的逆定理 319
- Parallel Postulate 平行公理 319
- Perpendicular Postulate 垂线公理 320
- Transitivity of Parallel Lines 平行线的传递性 320
- Two Perpendicular Lines Theorem 双垂线定理 320
- Exterior Angle Theorem 外角定理 327
- Triangle Inequality Theorem 三角不等式定理 329
- SSS Congruence Postulate 边边边全等公理 331
- SAS Congruence Postulate 边角边全等公理 331
- ASA Congruence Postulate 角边角全等公理 331
- AAS Congruence Theorem 角角边全等定理 331
- HL Congruence Theorem 斜边直角边全等定理 332
- Base Angle Theorem 等边对等角定理 333
- Corollary of Base Angle Theorem 等边对等角定理的推论 333
- Converse of Base Angle Theorem 等边对等角定理的逆定理(等角对等边定理) 333
- Converse of Corollary of Base Angle Theorem 等角对等边定理的推论 333
- Midsegment Theorem 中位线定理 336
- Perpendicular Bisector Theorem 垂直平分线定理 337
- Converse of Perpendicular Bisector Theorem 垂直平分线定理的逆定理 337
- Concurrency of Perpendicular Bisectors Theorem 垂直平分线共点定理 337
- Angle Bisector Property 角平分线性质 341
- Converse of Angle Bisector Property 角平分线性质的逆性质 342
- Angle Bisector Theorem 角平分线定理 364
- Concurrency of Angle Bisectors Theorem 角平分线共点定理 342
- Concurrency of Medians Theorem 中线共点定理 345
- Concurrency of Altitudes Theorem 高线共点定理 347
- Stability of Triangles 三角形的稳定性 351
- Hinge Theorem 大角对大边定理 351

- Converse of the Hinge Theorem 大角对大边的逆定理——大边对大角定理 352
- Corollary of the Hinge Theorem 大角对大边定理的推论 353
- Coincidence of Four Line Segments 四线合一 354
- Pythagorean Theorem 勾股定理 355
- Converse of the Pythagorean Theorem 勾股定理的逆定理 358
- AA Similarity Postulate 角角相似公理 361
- SAS Similarity Theorem 边角边相似定理 361
- SSS Similarity Theorem 边边边相似定理 362
- Inscribed Angle Theorem 圆周角定理 407
- Perpendicular Chord Theorem 垂径定理 412
- Corollary of the Perpendicular Chord Theorem 垂径定理的推论 413
- Tangent Theorem 切线定理 416
- Converse of Tangent Theorem 切线定理的逆定理 416
- Intersecting Chords Theorem 相交弦定理 419
- Chord Distance to Center Theorem 弦心距定理 419
- Parallel Lines Intercepted Arcs Theorem 平行线截弧定理 419
- Tangent-Chord Angle Theorem 弦切角定理 420
- Euler's Formula 欧拉公式 430

Pre-Calculus 微积分初步

- Midpoint Formula 中点公式 453
- Distance Formula 距离公式 454
- Cramer's Rule 克拉默法则 467
- Factor Theorem 因式定理 480
- Remainder Theorem 余式定理 481
- Rational Root Theorem 有理根定理 481
- Fundamental Theorem of Algebra 代数基本定理 485
- Linear Factorization Theorem 线性分解定理 485
- Binomial Theorem 二项式定理 536
- Law of Sines 正弦定理 557
- Law of Cosines 余弦定理 558
- Heron's Formula 海伦公式 559
- Power Rule 幂函数求导法则 577

A.15 Trigonometry Reference 三角学参考

Ⅰ. Trigonometric Functions (Right Triangle Definitions) 三角函数(直角三角形定义)

Let $0° < \theta < 90°$, as shown below:

$\sin\theta = \dfrac{\text{opp}}{\text{hyp}}$ $\csc\theta = \dfrac{\text{hyp}}{\text{opp}}$

$\cos\theta = \dfrac{\text{adj}}{\text{hyp}}$ $\sec\theta = \dfrac{\text{hyp}}{\text{adj}}$

$\tan\theta = \dfrac{\text{opp}}{\text{adj}}$ $\cot\theta = \dfrac{\text{adj}}{\text{opp}}$

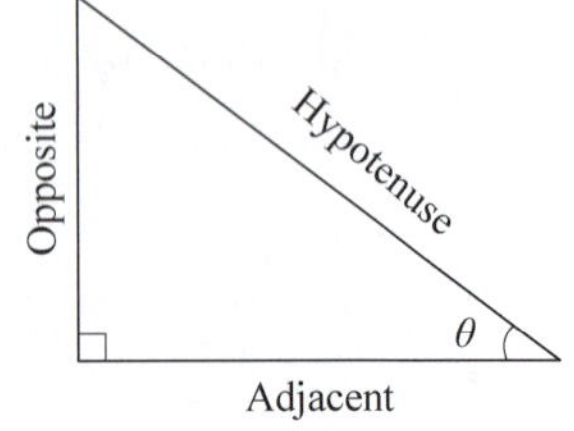

Ⅱ. Trigonometric Functions (Circular Functions) 三角函数(圆函数)

Let $r=\sqrt{x^2+y^2}$ and θ be any angle, as shown below:

$\sin\theta=\frac{y}{r}$ $\qquad$ $\csc\theta=\frac{r}{y}$

$\cos\theta=\frac{x}{r}$ $\qquad$ $\sec\theta=\frac{r}{x}$

$\tan\theta=\frac{y}{x}$ $\qquad$ $\cot\theta=\frac{x}{y}$

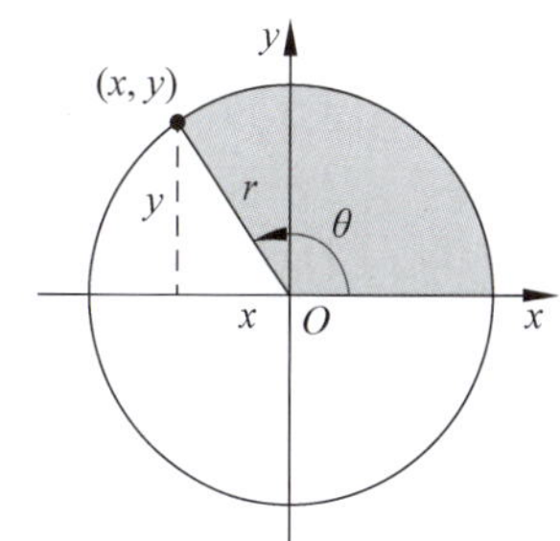

Ⅲ. Unit Circle Reference 单位圆参考

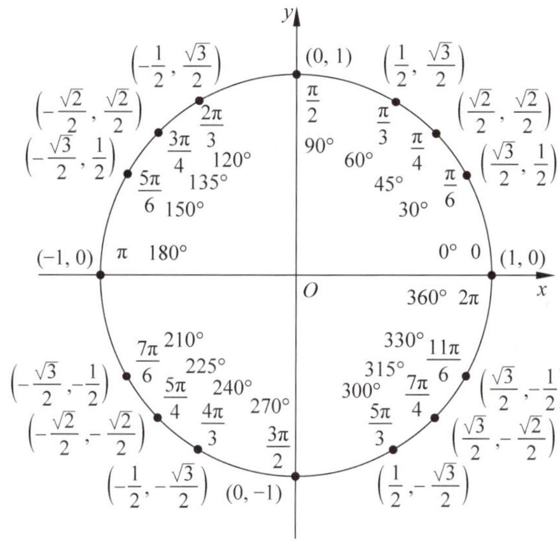

Table A-14 shows the exact values.

Table A-14

Trigonometric Functions	0°	30°	45°	60°	90°	180°	270°
sin	0	1/2	$\sqrt{2}/2$	$\sqrt{3}/2$	1	0	−1
cos	1	$\sqrt{3}/2$	$\sqrt{2}/2$	1/2	0	−1	0
tan	0	$\sqrt{3}/3$	1	$\sqrt{3}$	—	0	—
csc	—	2	$\sqrt{2}$	$2\sqrt{3}/3$	1	—	−1
sec	1	$2\sqrt{3}/3$	$\sqrt{2}$	2	—	−1	—
cot	—	$\sqrt{3}$	1	$\sqrt{3}/3$	0	—	0

Ⅳ. Trigonometric Identities/Formulas 三角恒等式/公式

1. Reciprocal Identities 倒数恒等式

$\sin\theta=\frac{1}{\csc\theta}$ $\qquad$ $\cos\theta=\frac{1}{\sec\theta}$ $\qquad$ $\tan\theta=\frac{1}{\cot\theta}$

$\csc\theta=\frac{1}{\sin\theta}$ $\qquad$ $\sec\theta=\frac{1}{\cos\theta}$ $\qquad$ $\cot\theta=\frac{1}{\tan\theta}$

2. Quotient Identities 商数恒等式

$\tan\theta=\frac{\sin\theta}{\cos\theta}$ $\qquad$ $\cot\theta=\frac{\cos\theta}{\sin\theta}$

3. Pythagorean Identities 毕达哥拉斯三角恒等式

$\sin^2\theta + \cos^2\theta = 1$

$1 + \tan^2\theta = \sec^2\theta$

$1 + \cot^2\theta = \csc^2\theta$

4. Cofunction Identities 余函数恒等式

$\sin\left(\frac{\pi}{2} - \theta\right) = \cos\theta$ $\quad \cos\left(\frac{\pi}{2} - \theta\right) = \sin\theta$ $\quad \tan\left(\frac{\pi}{2} - \theta\right) = \cot\theta$

$\csc\left(\frac{\pi}{2} - \theta\right) = \sec\theta$ $\quad \sec\left(\frac{\pi}{2} - \theta\right) = \csc\theta$ $\quad \cot\left(\frac{\pi}{2} - \theta\right) = \tan\theta$

5. Even/Odd Identities 奇偶性恒等式

$\sin(-\theta) = -\sin\theta$ $\quad \cos(-\theta) = \cos\theta$ $\quad \tan(-\theta) = -\tan\theta$

$\csc(-\theta) = -\csc\theta$ $\quad \sec(-\theta) = \sec\theta$ $\quad \cot(-\theta) = -\cot\theta$

6. Sum and Difference Formulas 和差公式

$\sin(u \pm v) = \sin u\cos v \pm \cos u\sin v$

$\cos(u \pm v) = \cos u\cos v \mp \sin u\sin v$

$$\tan(u \pm v) = \frac{\tan u \pm \tan v}{1 \mp \tan u\tan v}$$

7. Double-Angle Formulas 二倍角公式

$\sin(2u) = 2\sin u\cos v$

$\cos(2u) = \cos^2 u - \sin^2 u = 2\cos^2 u - 1 = 1 - 2\sin^2 u$

$$\tan(2u) = \frac{2\tan u}{1 - \tan^2 u}$$

8. Power-Reducing Formulas 降幂公式

$$\sin^2 u = \frac{1 - \cos(2u)}{2}$$

$$\cos^2 u = \frac{1 + \cos(2u)}{2}$$

$$\tan^2 u = \frac{1 - \cos(2u)}{1 + \cos(2u)}$$

9. Half-Angle Formulas 半角公式

$$\sin\frac{u}{2} = \pm\sqrt{\frac{1 - \cos u}{2}}$$

$$\cos\frac{u}{2} = \pm\sqrt{\frac{1 + \cos u}{2}}$$

$$\tan\frac{u}{2} = \pm\sqrt{\frac{1 - \cos u}{1 + \cos u}} = \frac{\sin u}{1 + \cos u} = \frac{1 - \cos u}{\sin u}$$

10. Sum-to-Product Formulas 和差化积公式

$$\sin u + \sin v = 2\sin\left(\frac{u + v}{2}\right)\cos\left(\frac{u - v}{2}\right)$$

$$\sin u - \sin v = 2\sin\left(\frac{u - v}{2}\right)\cos\left(\frac{u + v}{2}\right)$$

$$\cos u + \cos v = 2\cos\left(\frac{u+v}{2}\right)\cos\left(\frac{u-v}{2}\right)$$

$$\cos u - \cos v = -2\sin\left(\frac{u+v}{2}\right)\sin\left(\frac{u-v}{2}\right)$$

11. Product-to-Sum Formulas 积化和差公式

$$\sin u \sin v = \frac{1}{2}[\cos(u-v) - \cos(u+v)]$$

$$\cos u \cos v = \frac{1}{2}[\cos(u-v) + \cos(u+v)]$$

$$\sin u \cos v = \frac{1}{2}[\sin(u+v) + \sin(u-v)]$$

$$\cos u \sin v = \frac{1}{2}[\sin(u+v) - \sin(u-v)]$$

A.16 Derivatives and Integrals 导数与积分

Ⅰ. Basic Differentiation Rules 基础微分法则

Table A-15 shows the basic differentiation rules.

Table A-15

1. $\frac{d}{dx}[c]=0$	13. $\frac{d}{dx}[\sin u]=(\cos u)u'$	25. $\frac{d}{dx}[\sinh u]=(\cosh u)u'$
2. $\frac{d}{dx}[x]=1$	14. $\frac{d}{dx}[\cos u]=-(\sin u)u'$	26. $\frac{d}{dx}[\cosh u]=(\sinh u)u'$
3. $\frac{d}{dx}[cu]=cu'$	15. $\frac{d}{dx}[\tan u]=(\sec^2 u)u'$	27. $\frac{d}{dx}[\tanh u]=(\text{sech}^2 u)u'$
4. $\frac{d}{dx}[u\pm v]=u'+v'$	16. $\frac{d}{dx}[\csc u]=-(\csc u\cot u)u'$	28. $\frac{d}{dx}[\text{csch}\,u]=-(\text{csch}\,u\coth u)u'$
5. $\frac{d}{dx}[uv]=uv'+vu'$	17. $\frac{d}{dx}[\sec u]=(\sec u\tan u)u'$	29. $\frac{d}{dx}[\text{sech}\,u]=-(\text{sech}\,u\tanh u)u'$
6. $\frac{d}{dx}\left[\frac{u}{v}\right]=\frac{vu'-uv'}{v^2}$	18. $\frac{d}{dx}[\cot u]=-(\csc^2 u)u'$	30. $\frac{d}{dx}[\coth u]=-(\text{csch}^2 u)u'$
7. $\frac{d}{dx}[u^n]=nu^{n-1}u'$	19. $\frac{d}{dx}[\arcsin u]=\frac{u'}{\sqrt{1-u^2}}$	31. $\frac{d}{dx}[\sinh^{-1}u]=\frac{u'}{\sqrt{u^2+1}}$
8. $\frac{d}{dx}[\lvert u\rvert]=\frac{u}{\lvert u\rvert}(u'),u\neq 0$	20. $\frac{d}{dx}[\arccos u]=\frac{-u'}{\sqrt{1-u^2}}$	32. $\frac{d}{dx}[\cosh^{-1}u]=\frac{u'}{\sqrt{u^2-1}}$
9. $\frac{d}{dx}[\ln u]=\frac{u'}{u}$	21. $\frac{d}{dx}[\arctan u]=\frac{u'}{1+u^2}$	33. $\frac{d}{dx}[\tanh^{-1}u]=\frac{u'}{1-u^2}$
10. $\frac{d}{dx}[e^u]=e^u u'$	22. $\frac{d}{dx}[\text{arccsc}\,u]=\frac{-u'}{\lvert u\rvert\sqrt{u^2-1}}$	34. $\frac{d}{dx}[\text{csch}^{-1}u]=\frac{-u'}{\lvert u\rvert\sqrt{1+u^2}}$
11. $\frac{d}{dx}[\log_a u]=\frac{u'}{(\ln a)u}$	23. $\frac{d}{dx}[\text{arcsec}\,u]=\frac{u'}{\lvert u\rvert\sqrt{u^2-1}}$	35. $\frac{d}{dx}[\text{sech}^{-1}u]=\frac{-u'}{u\sqrt{1-u^2}}$
12. $\frac{d}{dx}[a^u]=(\ln a)a^u u'$	24. $\frac{d}{dx}[\text{arccot}\,u]=\frac{-u'}{1+u'}$	36. $\frac{d}{dx}[\coth^{-1}u]=\frac{u'}{1-u^2}$

Ⅱ. Basic Integration Formulas 基础积分公式

Table A-16 shows the basic integration formulas.

Table A-16

1. $\int du = u + C$	11. $\int \csc u\ du = -\ln\lvert \csc u + \cot u \rvert + C$
2. $\int kf(u)\,du = k\int f(u)\,du$	12. $\int \sec u\ du = \ln\lvert \sec u + \tan u \rvert + C$
3. $\int [f(u) \pm g(u)]\,du = \int f(u)\,du \pm \int g(u)\,du$	13. $\int \cot u\ du = \ln\lvert \sin u \rvert + C$
4. $\int u^n\,du = \frac{u^{n+1}}{n+1} + C, n \neq 1$	14. $\int \csc^2 u\ du = -\cot u + C$
5. $\int \frac{du}{u} = \ln\lvert u \rvert + C$	15. $\int \sec^2 u\ du = \tan u + C$
6. $\int e^u\,du = e^u + C$	16. $\int \csc u \cot u\ du = -\csc u + C$
7. $\int a^u\,du = \left(\frac{1}{\ln a}\right)a^u + C$	17. $\int \sec u \tan u\ du = \sec u + C$
8. $\int \sin u\ du = -\cos u + C$	18. $\int \frac{du}{\sqrt{a^2 - u^2}} = \arcsin \frac{u}{a} + C$
9. $\int \cos u\ du = \sin u + C$	19. $\int \frac{du}{u\sqrt{u^2 - a^2}} = \frac{1}{a}\operatorname{arcsec} \frac{\lvert u \rvert}{a} + C$
10. $\int \tan u\ du = -\ln\lvert \cos u \rvert + C$	20. $\int \frac{du}{a^2 + u^2} = \frac{1}{a}\arctan \frac{u}{a} + C$

A.17 Unit Conversions 单位转换

Table A-17 shows the unit conversions.

Table A-17

Common Imperial Unit 常用英制单位	Abbreviation 缩写	Relationship 关系
	Length 长度	
inch 英寸	in	1 in≈2.54 cm
foot 英尺	ft	(1) 1 ft=12 in (2) 1 ft≈30.48 cm
mile 英里	mi	(1) 1 mi=5280 ft (2) 1 mi≈1.609 km
league 里格	le	(1) 1 le=3 mi (2) 1 le≈4.828 km
	Area 面积	
acre 英亩		(1) 1 acre=43560 sqft (2) 1 acre≈4046.86 sq m

续表

Common Imperial Unit 常用英制单位	Abbreviation 缩写	Relationship 关系
Fluid Volume 液体容量		
teaspoon 茶匙	tspn.	1 tspn. ≈4.929 mL
tablespoon 餐匙	tbsp.	(1) 1 tbsp. = 3 tspn. (2) 1 tbsp. ≈14.787 mL
fluid ounce 液量盎司	fl oz	(1) 1 fl oz = 2 tbsp. (2) 1 fl oz≈29.574 mL
pint 品脱	pt	(1) 1 pt = 16 fl oz (2) 1 pt≈473 mL
quart 夸脱	qt	(1) 1 qt = 2 pt (2) 1 qt≈946.353 mL
gallon 加仑	gal	(1) 1 gal = 4 qt (2) 1 gal≈3.785 L
Weight 重量		
ounce 盎司	oz	1 oz≈28.350 g
pound 磅	lb	(1) 1 lb = 16 oz (2) 1 lb≈453.592 g
short ton 短吨	tn	(1) 1 tn = 2000 lb (2) 1 tn≈907.185 kg
Temperature 温度		
Fahrenheit 华氏度	℉	(1) x ℉ to Celsius: $\frac{5}{9}(x-32)$ (2) x ℃ to Fahrenheit: $\frac{9}{5}x+32$

A.18 Recommended Books 推荐书籍

为了确保中学生使用的数学书籍包含详细的出版信息，适合于学术性或专业的引用，这里将提供一些书籍的具体出版详情。这些信息可以帮助学生或教师在需要时引用这些资料：

1. ***Math Girls*** by Hiroshi Yuki
   - Publisher：Bento Books，Inc.
   - ISBN：9780983951308

   《数学女孩》by［日］结城浩
   - 出版社：人民邮电出版社
   - ISBN：9787115410351
2. 《程序员的数学》系列 by［日］结城浩
   - 出版社：人民邮电出版社
   - 《程序员的数学》ISBN：9787115293688
   - 《程序员的数学》(第 2 版)ISBN：9787115504906
3. 《数学与文化》by 齐民友
   - 出版社：大连理工大学出版社

- ISBN：9787568540827

4. ***The Beauty of Mathematics in Computer Science*** by Jun Wu
   - Publisher：Chapman and Hall/CRC
   - ISBN：9781138049604

   《数学之美》by 吴军
   - 出版社：人民邮电出版社
   - ISBN：9787115537973

5. 《吴军数学通识讲义》by 吴军
   - 出版社：新星出版社
   - ISBN：9787513344302

6. ***Fermat's Last Theorem*** by Simon Singh
   - Publisher：Fourth Estate
   - ISBN：9781857025217

   《费马大定理——一个困惑了世间智者358年的谜》by [英]西蒙·辛格
   - 出版社：广西师范大学出版社
   - ISBN：9787559828811

7. 《统计学习方法》by 李航
   - 出版社：清华大学出版社
   - ISBN：9787302517276

8. ***Head First Statistics：A Brain-Friendly Guide*** by Dawn Griffiths
   - Publisher：O'Reilly Media
   - ISBN：9780596527587

   《深入浅出统计学》by [美]道恩·格里菲斯
   - 出版社：电子工业出版社
   - ISBN：9787121338908

9. ***Automate This：How Algorithms Came to Rule Our World*** by Christopher Steiner
   - Publisher：Portfolio Hardcover
   - ISBN：9781591844921

   《算法帝国》by [美]克里斯托弗·斯坦纳
   - 出版社：人民邮电出版社
   - ISBN：9787115349002

10. ***Algorithms to Live By：The Computer Science of Human Decisions*** by Brian Christian and Tom Griffiths
    - Publisher：Henry Holt and Co.
    - ISBN：9781627790369

    《算法之美》by [美]布莱恩·克里斯汀、[美]汤姆·格里菲思
    - 出版社：中信出版集团
    - ISBN：9787508686882

11. ***One Two Three ... Infinity：Facts and Speculations of Science*** by George Gamow
    - Publisher：Dover Publications

- ISBN：9780486256641

《从一到无穷大》by［美］乔治·伽莫夫

- 出版社：科学出版社
- ISBN：9787030107596

12. ***Infinite Powers：How Calculus Reveals the Secrets of the Universe*** by Steven Strogatz

- Publisher：Mariner Books
- ISBN：9780358299288

《微积分之力量》by［美］史蒂夫·斯托加茨

- 出版社：中信出版集团
- ISBN：9787521723298

13. ***How to Solve It：A New Aspect of Mathematical Method*** by George Polya

- Publisher：Princeton University Press
- ISBN：9780691164076

《怎样解题：数学思维的新方法》by［美］乔治·波利亚

- 出版社：上海科技教育出版社
- ISBN：9787542852311

14. ***A Mathematician's Lament：How School Cheats Us Out of Our Most Fascinating and Imaginative Art Form*** by Paul Lockhart

- Publisher：Bellevue Literary Press
- ISBN：9781934137178

《一个数学家的叹息》by［美］保罗·洛克哈特

- 出版社：上海社会科学院出版社
- ISBN：9787552028218

15. ***Woo's Wonderful World of Maths*** by Eddie Woo

- Publisher：Macmillan Australia
- ISBN：9781760554217

《吴老师的趣味数学课》by［澳］埃迪·吴

- 出版社：天津科学技术出版社
- ISBN：9787557699765

16. ***Math Through the Ages：A Gentle History for Teachers and Others*** by William P. Berlinghoff and Fernando Q. Gouvea

- Publisher：Dover Publications
- ISBN：9780486832845

《这才是好读的数学史》by［美］比尔·伯林霍夫、［美］费尔南多·辜维亚

- 出版社：北京时代华文书局
- ISBN：9787569929713

17. ***Math Makers：The Lives and Works of*** 50 ***Famous Mathematicians*** by Alfred S. Posamentier and Christian Spreitzer

- Publisher：Prometheus
- ISBN：9781633885202

《他们创造了数学：50 位著名数学家的故事》by [美]阿尔弗雷德·S.波萨门蒂尔、[奥地利] 克里斯蒂安·施普赖策

- 出版社：人民邮电出版社
- ISBN：9787115595249

18. ***Math with Bad Drawings：Illuminating the Ideas That Shape Our Reality*** by Ben Orlin
   - Publisher：Black Dog & Leventhal
   - ISBN：9780316509046

   《欢乐数学：一本充满"烂插画"的快乐数学启蒙书》by [美]本·奥尔林
   - 出版社：天津科学技术出版社
   - ISBN：9787557690243
19. 《尖叫的数学：令人惊叹的数学之美》by [意]翁贝托·博塔兹尼
   - 出版社：湖南科学技术出版社
   - ISBN：9787571011703
20. ***The Mathematics Lover's Companion：Masterpieces for Everyone*** by Edward R.Scheinerman
   - Publisher：Yale University Press
   - ISBN：9780274759347

   《美丽的数学》by [美]爱德华·沙伊纳曼
   - 出版社：湖南科学技术出版社
   - ISBN：9787571000882
21. 《数学符号史》by 徐品方、张红
   - 出版社：科学出版社
   - ISBN：9787030170170
22. ***Burn Math Class：And Reinvent Mathematics for Yourself*** by Jason Wilkes
   - Publisher：Basic Books
   - ISBN：9780465053735

   《烧掉数学书：重新发明数学》by [美]杰森·威尔克斯
   - 出版社：湖南科学技术出版社
   - ISBN：9787571004071
23. ***Journey through Genius：The Great Theorems of Mathematics*** by William Dunham
   - Publisher：Penguin Books
   - ISBN：9780140147391

   《天才引导的历程：数学中的伟大定理》by [美]威廉·邓纳姆
   - 出版社：机械工业出版社
   - ISBN：9787111403296

Index 索引

A

AAS Congruence Theorem　角角边全等定理　331
AA Similarity Postulate　角角相似公理(两个角相等的两个三角形相似)　361
Absolute Value　绝对值　47
Absolute Value Equation　绝对值方程　98
Absolute Value Equation with Variable on One Side　在方程一边有未知数的绝对值方程　98
Absolute Value Equation with Variable on Both Sides　在方程两边都有未知数的绝对值方程　99
Absolute Value Inequality　绝对值不等式　128
Acute Angle　锐角　285
Acute Triangle　锐角三角形　322
Addition　加法　48
Add/Remove Parentheses　添/去括号　72
Adjacent Angles　邻角　287
Adjacent Side of an Acute Angle　锐角的邻边　369
Algebra　代数　3
Alternate Exterior Angles　外错角　318
Alternate Interior Angles　内错角　317
Altitude　高线　347
Altitude of a Cone　圆锥的高线　438
Altitude of a Cylinder　圆柱的高线　436
Altitude of a Frustum　平截头体的高线　440
Altitude of a Prism　棱柱的高线　432
Altitude of a Pyramid　棱锥的高线　434
Amplitude　振幅　554
And　且　133
And-Inequality　同时成立的不等式组　122
Angle　角　282
Angle Bisector　角平分线　341
Applications of Congruence Theorems　全等三角形的应用　334
Applications of Derivatives　导数的应用　578
Arc　弧　405
Area　面积　443
Arithmetic Mean　算术平均数　367
Arithmetic Sequence　等差数列　196
Arithmetic Series　等差级数　530
ASA Congruence Postulate　角边角全等公理　331
Ascending Order of Powers　升幂排列　69
Association　关联性　227
Associative Property　结合律　61
ASTC　三角函数的正负规律　547
Asymptote　渐近线　503

B

Bar Graph　条形图　214
Base　底数　55
　底　430
Basic Operations　(函数)基本运算　478
Bias　偏差　252
Biconditional Statement　双条件句　298
Binomial Theorem　二项式定理　536
Blinded Study　盲法研究　242
Box-and-Whisker Diagram/Boxplot　箱形图　223
Branch　分支　507

C

Cardinality　基数　131
Cartesian Plane　坐标平面　135
Categorical Variable　分类变量　200
Causation　因果关系　225
Census　普查　251
Center　中心　290
Central Angle　圆心角　407
Centroid　重心　347
Certain Event　必然事件　262
Check　验根　101
Chord　弦　404
Circle　圆　403，516
Circumcenter　外心　339
Circumcircle　外接圆　341

Circumradius 外接圆半径 341
Classify Functions as Even, Odd, or Neither 对函数进行奇偶分类 189
Closure Property 封闭性 62
Cluster Sampling 等群抽样 250
Coefficient 系数 67
Co-Function Identities 余函数恒等式 551
Coincidence of Four Line Segments 四线合一 354
Coincident 重合 171
Combination 组合 257, 535
Combine Like Terms 合并同类项 73
Combos That Cannot Conclude Two Triangles Are Congruent 不能断定两个三角形全等的组合条件 332
Common Application Vocabulary 常见应用词汇 116
Common Difference 公差 195
Common External Tangent 外公切线 425
Common Factor/Common Divisor 公约数 11
Common Factor 公因式 77
Common Internal Tangent 内公切线 425
Common Multiple 公倍数 12
Common Percent Problems 百分数常见问题 31
Common Ratio 公比 197
Common Response 共同作用关系 226
Common Tangent 公切线 424
Common Theorems of Similarity 常见的相似定理 363
Commutative Property 交换律 61
Compass 圆规 294
Complement 补集 259
Complementary Angles 余角 287
Complementary Events 对立事件 263
Complete the Square Review 配方法回顾 476
Complex Number 复数 472
Complex Plane 复平面 473
Composite/Composite Number 合数 10
Compound Inequality 复合不等式 122
Concave Polygon 凹多边形 387
Concentric Circles 同心圆 403
Concurrent 共点 336
Concurrent Circles 同圆 404
Conditional Equation 条件等式 92
Conditional Statement 条件命题 297
Cone 锥体 437
Conic Section 圆锥曲线 515
Confounding 混杂关系 226
Congruent 全等 311
Congruent Angles 等角 287
Congruent Arcs 等弧 406
Congruent Circles 等圆 404
Congruent Solids 全等立体 441
Congruent Triangles 全等三角形 330
Conjugate Complex Numbers 共轭复数 475
Consecutive Interior Angles 同旁内角 318
Constant 常数/常值/常量 68
常值 188
Constant Function 常值函数 153
Continuous 连续 580
Contrapositive 对换句/逆否命题 301
Control Group 控制组 241
Convenience Sampling 便利抽样 251
Converse 逆命题 300
Converse of the Hinge Theorem 大角对大边的逆定理——大边对大角定理 352
Converse of the Pythagorean Theorem 勾股定理的逆定理 358
Convex Polygon 凸多边形 387
Coordinate 坐标 137, 451
Coordinate Axis 坐标轴 136, 451
Coordinate Plane 坐标平面 135
Coprime 互质 13
Corollaries of Chord, Arc, Central Angles 弦、弧、圆心角的推论 412
Corollaries of Circles 圆的推论 409
Corollary 推论 304
Corollary of the Converse of the Hinge Theorem 大边对大角定理的推论 353
Corollary of the Hinge Theorem 大角对大边定理的推论 353
Corollary of the Perpendicular Chord Theorem 垂径定理的推论 413
Correlation Coefficient 相关系数 229
Corresponding Angles 同位角 317
Corresponding Parts of Congruent Triangles are Congruent (CPCTC) 全等三角形的对应部分全等 333
Cosine 余弦 369
Counterexample 反例 296
Counting and Probability Template Questions 计数和概率的典型问题 539
Counting Principle 计数原理 254
Cramer's Rule 克拉默法则 467
Cross Section 横截面 429
Cube 立方体 432
Cumulative Frequency 累积频数 212

Cumulative Relative Frequency 累积相对频数 212
Curve 曲线 153
Cyclic Polygon 圆内接多边形 409
Cylinder 柱体 436

D

Data 数据 199
Data Set 数据集 199
Decimal/Decimal Number 小数 38
Decreasing 递减 188
Degenerate Conic 退化圆锥曲线 528
Degree 次数 68
度 283
Degree-Minute-Second（DMS） 度分秒表示法 284
Denominator 分母 26
Dependent Events 相关事件 269
Dependent Variable 因变量 147
Derivative 导数 575
Descending Order of Powers 降幂排列 69
Descriptions of Graphs of One-Variable Statistics 单变量统计图表描述 215
Descriptions of Graphs of Two-Variable Statistics 双变量统计图表描述 227
Descriptive Statistics 描述统计 201
Determinant 行列式 465
Determining a Parallelogram 平行四边形的判定 376
Determining a Rectangle 矩形的判定 378
Determining a Rhombus 菱形的判定 380
Diagonal 对角线 374，388
Diameter 直径 404
Difference Quotient 差商 573
Differentiable 可微分的 580
Digit 数字/数位/位数 40
Dilation 位似变换 401
Dimension 维度 275
Direction 方向 281
Direct Proof 直接证明 305
Direct Substitution 直接代入法 569
Direct Variation 正比例函数/正变分 165，500
Disjoint Events 互斥事件 262
Distance 距离 46，314
Distance Formula 距离公式 140，454
Distributive Property 乘法分配律 61
Dividing Out Technique 除法 570
Division 除法 52
Divisor 乘数、因子（因数、约数） 10
Domain 定义域 147
Domain Restriction 定义域限制 488
Dominance of Functions 主导函数 487
Dot Plot 点图 219
Dot Product 点积 463
Double Angle Formulas 二倍角公式 552
Double-Blinded Study 双盲研究 243
Double Inequality 双重不等式 125

E

e 495
Edge 棱 428
Element 元素 130
Ellipse 椭圆 517
Empty Set 空集 131
Endpoint 端点 278
Enumeration Tree 枚举树 253
Equation 等式/方程（式） 91
Equation of a Plane 平面的方程式 455
Equiangular 等角的 388
Equiangular Triangle 等角三角形 324
Equilateral 等边的 388
Equilateral Triangle 等边三角形 325
Equivalent Statements for $f(x)=0$ $f(x)=0$的等价命题 188
Euclidean Algorithm 欧几里得算法 16
Euler's Formula 欧拉公式 430
Evaluate 求值 70
Even 偶（数） 7
偶（函数） 189
Even-Odd Identities 奇偶性恒等式 550
Event 事件 259
Exercises of Properties of Operations 运算性质的练习 63
Existential Quantifier 存在量词 296
Experiment 实验 238
Experimental Group 实验组 240
Experimental Probability 实验概率 261
Experimental Subject 实验对象 240
Explanatory Variable 解释变量 224
Exponent 指数 55
Exponential Decay 指数衰减 183
Exponential Function 指数函数 181
Exponential Growth 指数增长 182
Expression 表达式 65
Exterior Angle 外角 327，390
Extrapolation 外推 236

Extraneous Solution 增根 101

F

Face 面 428
Factor 乘数、因子(因数、约数) 10
因式 77
因式分解 78
Factor by Grouping 分组分解 79
Factor by Pulling Out the Highest Common Factor 提取最大公因式 78
Factor Completely 完全分解 79
Factored Form 交点式 177
Factorial 阶乘 255
Factorization 分解 14
Factor Theorem 因式定理 480
Factor Tree 因子树 15
False Statement 假命题 295
Favorable Outcome 有利的结果 258
Five-Number Summary 五数概括法 207
Flowchart Proof 流程图证明 307
FOIL 四项分配法 74
Formula 公式 147
Formulas for Special Products 乘法公式 75
Fraction 分数 23
Frequency 频数 211
Frequency Distribution 频数分布 213
Frustum 平截头体 439
Function 单值函数 142
Functional Inequality 函数不等式 190
Function Composition 复合函数 478
Fundamental Theorem of Algebra 代数基本定理 485

G

General Form 一般式 174
General Formulas of Conics 圆锥曲线的一般式 528
Geometric Interpretation of Systems of Linear Equations with Three Unknowns 三元一次方程组的几何分析 458
Geometric Mean 几何平均数 367
Geometric Sequence 等比数列 197
Geometric Series 等比级数 532
Geometry 几何学 273
Given a Point on the Terminal Side, Find the Outputs of Trigonometric Functions of an Angle 给出角的终边上的一点,求出其三角函数的输出 545
Given the Measure, Find the Outputs of Trigonometric Functions of an Angle 给出角的度数,求出其三角函数的输出 546
Glide Reflection 滑移反射 400
Graphing Inequalities 图解不等式法 121
Graphing Rational Functions 画有理函数的图像 508
Graphs of Arcsine, Arccosine, and Arctangent Functions 反正弦、反余弦、反正切函数的图像 556
Graphs of Cosecant, Secant, and Cotangent Functions 余割、正割、余切函数的图像 556
Graphs of Derivatives 导数的图像 577
Graphs of Inverse Functions 反函数的图像 492
Graphs of Sine, Cosine, and Tangent Functions 正弦、余弦、正切函数的图像 555
Greatest Common Factor (GCF)/Greatest Common Divisor (GCD) 最大公约数 11
Group 分组 240

H

Half-Angle Formulas 半角公式 552
Heron's Formula 海伦公式 559
Hinge Theorem 大角对大边定理 351
Histogram 直方图 217
HL Congruence Theorem 斜边直角边全等定理 332
Hole 断点 501
Horizontal Asymptote 水平渐近线 184, 504
Horizontal Line 水平线 154
Horizontal Line Test 水平线测试 493
Hyperbola 双曲线 523

I

Identity 恒等式 92
Identity Matrix 单位矩阵 468
Identity Property 单位元性质 62
Imaginary Unit 虚数单位 471
Impossible Event 不可能事件 261
Incenter 内心 343
Incircle 内切圆 344
Increasing 递增 188
Independent Events 独立事件 268
Independent Measure Design 独立测量设计 244
Independent Variable 自变量 147
Indeterminate Form 不定式 509
Inequality 不等式 117
Inequality Notation 不等式符号表示法 120
Influential Point 强影响点 230
Input 输入 146

Inradius 内切圆半径 344
Inscribed Angle 圆周角 407
Inscribed Polygon 圆内接多边形 409
Integer 整数 6
Integral Expression 整式 66
Intercept 截距 148，453，501
Interior Angle 内角 326，389
Interpolation 内推 236
Interquartile Range(IQR) 四分位距 207
Intersecting 相交 168
Intersection 交集 134
Interval Notation 区间表示法 121
Inverse 否命题 302
Inverse Function 反函数 489
Inverse Matrix 逆矩阵 468
Inverse Operation 逆运算 94
Inverse Property 逆元性质 63
Inverse Trigonometric Functions 反三角函数 370
Inverse Variation 反比例函数/逆变分 179，500
Irrational Conjugate 共轭无理数 88
Irrational Number 无理数 37
Isolate 解出/隔离 95
Isosceles Right Triangle 等腰直角三角形 325，356
Isosceles Triangle 等腰三角形 324

J

Joint Variation 联变分 500

K

Kite 筝形 384

L

Lateral Edge 侧棱 431
Lateral Face 侧面 431
Law of Sines 正弦定理 557
Law of Cosines 余弦定理 558
Leading Coefficient 最高次项系数 68
Least Common Denominator(LCD) 最小公分母 26
Least Common Multiple(LCM) 最小公倍数 13
Least-Square Regression Line(LSRL) 最小二乘法回归线 233
Lemma 引理 304
Length 长度 279
Like Terms 同类项 67
Limit 极限 565
Line 直线 153，274，278
Linear Equation with One Unknown 一元一次方程 94
Linear Equation with Two Unknowns 二元一次方程 102
Linear Factorization Theorem 线性分解定理 485
Linear Function 一次函数/线性函数 155
Linear Inequality with One Unknown 一元一次不等式 118
Linear Pair 邻补角 288
Linear Programming 线性规划 449
Line Segment 线段 279
Line Segment Bisector 线段平分线 280
Line Symmetry 轴对称 291
Logarithm 对数 495
Logarithmic Function 对数函数 498
Long Division 长除法 479

M

Major Arc 优弧 406
Matched Pair Design 配对设计 245
Mathematical Induction 数学归纳法 307
Matrix 矩阵 460
Matrix Addition and Subtraction 矩阵加法与减法 462
Matrix Multiplication 矩阵乘法 464
Maximum 最大值 203
Mean 平均数 201
Measure 测量/度数 284
Median 中位数 204
中线 345
Methods of Expressing the Solutions of Inequalities 不等式解集表示法 119
Multiple (正)倍数 12
Multiplication 乘法 51
Midpoint 中点 280
Midpoint Formula 中点公式 139，453
Midsegment 中位线 336
Minimum 最小值 203
Minor Arc 劣弧 406
Mixed Number 带分数 27
Mixed Recurring Decimal 混循环小数 43
Mode 众数 201
Monomial 单项式 68
More Properties of Parallel and Perpendicular Lines 更多平行线和垂直线的性质 319
Mutually Exclusive Events 互斥事件 262

N

Negation 否定 299
Negative 负(数) 4

Negative Mixed Number 负带分数 28
Net 展开图 429
Number Line 数轴 45
Numerator 分子 26
Nonresponse Bias 无回应偏差 253
Non-Rigid Transformation 非合同变换 401
Non-Terminating Decimal 无限小数 42
Norm 范数 474

O

Oblique Cone 斜圆锥 437
Oblique Cylinder 斜圆柱 436
Oblique Prism 斜棱柱 432
Observational Study 观察性研究 237
Obtuse Angle 钝角 285
Obtuse Triangle 钝角三角形 323
Odd 奇(数) 8
奇(函数) 189
Odds Against 不利赔率 264
Odds in Favor 有利赔率 263
One 一 5
One-Sided Limit 单边极限 568
One-To-One Function 一对一函数 492
Operation 运算 48
Opposite 相反数 46
Opposite Side of an Acute Angle 锐角的对边 369
Optical Illusion 视觉幻象 289
Or 或 133
Ordered Pair 有序对 137
Ordered Triple 有序三元组 452
Ordering 顺序 255
Order of Rigid Transformations 合同变换的顺序 399
Origin 坐标原点 138, 452
Or-Inequality 逻辑或不等式 126
Orthocenter 垂心 348
Outcome 结果 258
Outlier(of One-Variable Statistics) 离群值 207
Outlier(of Two-Variable Statistics) 离群值 231
Output 输出 146

P

Parabola 抛物线 521
Paradox Equation 矛盾等式 93
Paragraph Proof 自然段证明 306
Parallel 平行 167
Parallelogram 平行四边形 374
Parentheses 括号 60
Parts of a Regular Polygon 正多边形的部分 426
Partial Fraction Decomposition 部分分式分解 511
PEMDAS 运算顺序 60
Percent 百分数 29
Percent Change 百分数变化 33
Percent Change Application 百分数变化的应用 34
Perfect Cube 完全立方 22
Perfect nth Power 完全 n 次幂(方) 22
Perfect Square 完全平方 21
Perimeter 周长 442
Period 周期 554
Permutation 排列 256, 535
Perpendicular 垂直 170, 314
Perpendicular Bisector 垂直平分线 337
Perpendicular Chord Theorem 垂径定理 412
Piecewise-Defined Function 分段函数 185
Pie Chart 饼图(圆形分格统计图表) 214
Placebo 安慰剂 241
Placebo Effect 安慰剂效应 242
Plane 平面 274
Plane Geometry 平面几何 273
Plane Shape 平面图形 276
Point 点 138, 273, 452
Point of Maximum 最大值点 173
Point of Minimum 最小值点 173
Point-Slope Form 点斜式 160
Polar Coordinate System 极坐标系 561
Polygon 多边形 386
Polyhedron 多面体 430
Polynomial 多项式 69
Polynomial Function 多项式函数 152
Population 总体 246
Positional Relationship between a Line and a Circle 直线与圆的位置关系 418
Positional Relationship between a Point and a Circle 点与圆的位置关系 414
Positive 正(数) 3
Positive Mixed Number 正带分数 27
Possible Event 可能事件 262
Postulate 公理 303
Power 幂 55
Power Rule 幂函数求导法则 577
Power-Reducing Formulas 降幂公式 552
Prime/Prime Number 质数/素数 8
Prime Factorization 分解质因数(定义) 15

分解质因数(方法) 19
Prism 棱柱 431
Probability 概率 260
Product of a Matrix with a Constant 矩阵与常数的积 463
Product-to-Sum Formulas 积化和差公式 553
Proof 证明 305
Proof by Casework 分情况讨论法 309
Proof by Contradiction 反证法 308
Proof by Contrapositive 逆否命题证明 310
Properties of Experimental Designs 实验设计性质 243
Proportion 比例 82
Protractor 量角器 294
Pure Recurring Decimal 纯循环小数 43
Pyramid 棱锥 433
Pythagorean Identities 毕达哥拉斯三角恒等式 549
Pythagorean Theorem 勾股定理 355
Pythagorean Tree 勾股树 358

Q

Quadrant 象限 138
Quadratic Equation 一元二次方程 109
Quadratic Formula 一元二次方程求根公式 114, 477
Quadratic Function 二次函数 171
Quadrilateral 四边形 373
Quantifier 量词 295
Quantitative Variable 定量变量 200
Quartile 四分位数 205
Questions of Writing Expressions from Words 文字题 57
Quotient Identities 商恒等式 550

R

Radian 弧度 560
Radical 根式 86
Random Sampling 随机抽样 248
Range 值域 147
极差 203
Rate of Change 变动率 157
Ratio 比 81
Rationalize the Denominator 分母有理化 89
Rationalizing Technique 有理化法 572
Rational Equation 有理方程 514
Rational Expression 有理式 479, 514
Rational Function 有理函数 500
Rational Number 有理数 6
Rational Root Theorem 有理根定理 481
Ratio of Similitude 相似比 360
Ray 射线 278
Real Number 实数 3
Reciprocal 倒数 47
Reciprocal Function 倒数函数 511
Reciprocal Identities 倒数恒等式 549
Rectangle 矩形 377
Reduction of Fractions to a Common Denominator 通分 85
Reference Angle 参考角 543
Reflection 反射 396
Reflex Angle 优角 286
Regular 正的 388
Relation 多值函数 141
Relationship between Explanatory and Response Variables 解释变量与响应变量之间的关系 225
Relationship between Two Lines in a Plane 同一平面内两线的关系 167, 312
Relative Frequency 相对频数 211
Relatively Prime 互质 13
Remainder 余数 54
Remainder Theorem 余式定理 481
Repeated Measure Design 重复测量设计 245
Repetend 循环节 42
Residual 残差 234
Residual Plot 残差图 235
Response Variable 响应变量 224
Rhombus 菱形 379
Right Angle 直角 285
Right Cone 直圆锥 437
Right Cylinder 直圆柱 436
Right Prism 直棱柱 432
Right Triangle 直角三角形 323
Rigid Transformation 合同变换 395
Root 根 93
Rotation 旋转 398
Rotational Symmetry 旋转对称 292
Round 四舍五入 70
Round Angle 周角 286

S

Sample 样本 247
Sample Space 样本空间 258
SAS Congruence Postulate 边角边全等公理 331
SAS Similarity Theorem 边角边相似定理(两条边对应

成比例且夹角相等的两个三角形相似) 361
SSS Similarity Theorem 边边边相似定理(三条边对应成比例的两个三角形相似) 362
Scalene Triangle 不等边三角形 324
Scatterplot 散点图 232
Scientific Notation 科学记数法 58
Secant 割线 415，573
Semicircle 半圆/半圆形 406，408
Sequence 数列 193
Series 级数 529
Set 集合 130
Set Notation 集合表示法 121
Seven Characteristics of a Polynomial Function 多项式函数的7个特征 486
Shape 图形 276
Short Division 短除法 18
Side 边 373
Sieve of Eratosthenes 埃拉托色尼筛选法 9
Similar Polygons 相似多边形 359
Similar Solids 相似立体 441
Similar Triangles 相似三角形 360
Simple Random Sample(SRS) 简单随机抽样 247
Simplest Form of a Fraction 最简分数 85
Simplest Radical Form 最简根式 88
Simplify Fractions 约分 84
Simplify Polynomials 简化多项式 80
Sine 正弦 369
Skew Lines 异面直线 321
Slant Asymptote 斜渐近线 506
Slant Height of a Regular Pyramidal Frustum 正棱锥平截头体的斜高 440
Slant Height of a Regular Pyramid 正棱锥的斜高 434
Slant Height of a Right Cone 直圆锥的斜高 438
Slant Height of a Right Conical Frustum 正圆锥平截头体的斜高 440
Slope 斜率 155
Slope-Intercept Form 斜截式 158
Solid 立体图形 276
Solid Geometry 立体几何 273
Solution 解 93
Solving Exponential and Logarithmic Equations 解指数与对数方程 498
Solving Linear Equations with One Unknown 解一元一次方程 96
Solving Polynomial Equations 解多项式方程 484
Solving Quadratic Equations by Factoring 因式分解法 110
Solving Quadratic Equations by Completing the Square 配方法 111
Solving Right Triangles 解直角三角形 371
Solving Systems of Linear Equations by Elimination 加减消元法 106，456
Solving Systems of Linear Equations by Graphing 画图法 166
Solving Systems of Linear Equations by Substitution 代入消元法 105
Solving Systems of Linear Equations Using Matrices 用矩阵解线性方程组 469
Special Right Triangles 特殊的直角三角形 356
Sphere 球 438
Square 正方形 381
SSS Congruence Postulate 边边边全等公理 331
Stability of Triangles 三角形的稳定性 351
Standard Deviation 标准差 210
Standard Form 一般式 163
Standard Position of an Angle 角的基准位置 542
Statement 命题 294
Statistics 统计学 198
Stem-and-Leaf Plot/Stemplot 茎叶图 220
Straight Angle 平角 286
Straightedge 直尺
Stratified Random Sampling 分层随机抽样 249
Strength 强度 228
Subset 子集 131
Substitute 代入 72
Subtraction 减法 50
Sum-Difference Formulas 和差公式 550
Sum-to-Product Formulas 和差化积公式 554
Superset 超集 131
Supplementary Angles 补角 288
Surface Area 表面积 444
Survey 调查 251
Symmetric Property 对称性质 411
Symmetry 对称 290
Synthetic Division 综合除法 483
Systematic Sampling 等距抽样 248
System of Functional Inequalities 函数不等式组 191
System of Linear Equations 一次方程组 103
System of Linear Equations with Two Unknowns 二元一次方程组 104

T

Table of Trigonometric Functions of Some Angles 一些

角的三角函数表 548
Tangent 正切 370
切线、相切 415
Tangent Line 切线 574
Term 加数/项 67
数列的项 195
Terminating Decimal 有限小数 41
Theorem 定理 304
Theorem of Chord, Arc, Central Angles 弦、弧、圆心角定理 411
Theorems of Circles 圆的定理 419
Theorems and Proofs of Triangles 三角形的定理与证明 333
Theoretical Probability 理论概率 261
Three-Dimensional Space 三维空间 450
Transformation 变换 392
Translation 平移 395
Transversal 截线 316
Trapezoid 梯形 382
Treatment 实验手段 239
Triangle 三角形 321
Triangle Inequality Theorem 三角不等式定理 329
Triangle Set 三角板 294
Trigonometry 三角学 368
True Statement 真命题 294
Two-Column Proof 两列式证明 306
Type of Studies 研究类型 237

U

Undercoverage Bias 低覆盖面偏差 253
Unfavorable Outcome 不利的结果 258
Union 并集 134
Unit Circle 单位圆 541
Universal Quantifier 全称量词 295

V

Value 值 70
Variability 变异 252
Variable 变量/未知数 66
统计学的变量 200
Variance 方差 208
Vector 向量 393
Venn Diagram 文氏图 265
Verbal Model 文字模型/文字表达式 115
Vertex 顶点 173，429
Vertex Form 顶点式 175
Vertical Angles 对顶角 313
Vertical Asymptote 垂直渐近线 503
Vertical Line 垂直线 154
Vertical Line Test 垂线测试 144
Volume 体积 444
Voluntary Response Bias 自愿回应偏差 252

X

x-Axis x 轴 136
x-Coordinate x 坐标 137
x-Intercept x 截距 148

Y

y-Axis y 轴 136
y-Coordinate y 坐标 137
y-Intercept y 截距 150

Z

Zero 零 4

#

1.5 IQR Rule 1.5 四分位距法则 208
2D-Shape 平面图形 276
3D-Shape 立体图形 276
$30°-60°-90°$ Triangle $30°-60°-90°$ 三角形 357
$45°-45°-90°$ Triangle $45°-45°-90°$ 三角形 356